Lecture Notes in Computer Science 3459

Commenced Publication in 1973
Founding and Former Series Editors:
Gerhard Goos, Juris Hartmanis, and Jan van Leeuwen

Ron Kimmel Nir Sochen
Joachim Weickert (Eds.)

Scale Space and PDE Methods in Computer Vision

5th International Conference, Scale-Space 2005
Hofgeismar, Germany, April 7-9, 2005
Proceedings

Springer

Volume Editors

Ron Kimmel
Technion, Computer Science Department
518 Taub Bldg., Haifa 32000, Israel
E-mail: ron@cs.technion.ac.il

Nir Sochen
Tel Aviv University, Department of Applied Mathematics
201 Schreiber Bldg., Ramat-Aviv, Tel-Aviv 69978, Israel
E-mail: sochen@math.tau.ac.il

Joachim Weickert
Saarland University, Faculty of Mathematics and Computer Science
Building 27.1, 66041 Saarbrücken, Germany
E-mail: weickert@mia.uni-saarland.de

Library of Congress Control Number: Applied for

CR Subject Classification (1998): I.4, I.3.5, I.5, I.2.10, G.1.2

ISSN 0302-9743
ISBN-10 3-540-25547-8 Springer Berlin Heidelberg New York
ISBN-13 978-3-540-25547-5 Springer Berlin Heidelberg New York

Springer is a part of Springer Science+Business Media

springeronline.com

Printed in Germany

Typesetting: Camera-ready by author, data conversion by Scientific Publishing Services, Chennai, India
Printed on acid-free paper SPIN: 11408031 06/3142 5 4 3 2 1 0

Preface

Welcome to the proceedings of the 5th International Conference on Scale-Space and PDE Methods in Computer Vision.

The scale-space concept was introduced by Iijima more than 40 years ago and became popular later on through the works of Witkin and Koenderink. It is at the junction of three major schools of thought in image processing and computer vision: the design of filters, axiomatic approaches based on partial differential equations (PDEs), and variational methods for image regularization. Scale-space ideas belong to the mathematically best-understood approaches in image analysis. They have entered numerous successful applications in medical imaging and a number of other fields where they often give results of very high quality.

This conference followed biennial meetings held in Utrecht, Corfu, Vancouver and Skye. It took place in a little castle (Schlösschen Schönburg) near the small town of Hofgeismar, Germany. Inspired by the very successful previous meeting at Skye, we kept the style of gathering people in a slightly remote and scenic place in order to encourage many fruitful discussions during the day and in the evening.

We received 79 full paper submissions of a high standard that is characteristic for the scale-space conferences. Each paper was reviewed by three experts from the Program Committee, sometimes helped by additional reviewers. Based on the results of these reviews, 53 papers were accepted. We selected 24 manuscripts for oral presentation and 29 for poster presentation.

It is a tradition at scale-space conferences to invite keynote speakers who can provide valuable additional inspirations beyond the mainstream topics in scale-space analysis. Also this time it was our pleasure to thank three leading experts for accepting our invitation for a keynote lecture: Prof. Achi Brandt of The Weizmann Institute of Science (Rehovot, Israel), Prof. Michael Unser of the Swiss Federal Institute of Technology (Lausanne, Switzerland), and Prof. Carl-Fredrik Westin of the Harvard Medical School (Boston, USA).

We thank all authors for their excellent contributions, and the referees for their time and valuable comments. Regarding local arrangements, we are indebted to the staff at Schlösschen Schönburg, as well as to Bernhard Burgeth, Martin Welk, and Uta Merkle of Saarland University. We also thank Micha Feigin, Julia Getslev and Lori Sochen for their help with the website and Yana Katz for her help with the proceedings. Finally we are grateful to the German Pattern Recognition Society (DAGM) for sponsorship.

We wish you an exciting journey through the latest results on scale-space ideas in image analysis.

April 2005 Ron Kimmel, Nir Sochen, and Joachim Weickert

Organization

Organizing Committee

Ron Kimmel (Technion, Israel Institute of Technology)
Nir Sochen (Tel Aviv University, Israel)
Joachim Weickert (Saarland University, Germany)

Program Committee

Luis Alvarez
Alfred Bruckstein
Vicent Caselles
Antonin Chambolle
Tony Chan
Yunmei Chen
Laurent Cohen
Daniel Cremers
Rachid Deriche
Olivier Faugeras
Michael Felsberg
Luc Florack
Lewis Griffin
Atsushi Imiya
Peter Johansen
Michael Kerckhove
Renaud Keriven
Georges Koepfler
Pierre Kornprobst
Arjan Kuijper
Petros Maragos
Riccardo March
Pavel Mrázek
Karol Mikula
Mila Nikolova
Mads Nilsen
Stanley Osher
Ole Fogh Olsen
Nikos Paragios
Steve Pizer
Bart ter Haar Romeny
Martin Rumpf
Otmar Scherzer
Christoph Schnörr
Fiorella Sgallari
Jayant Shah
Kaleem Siddiqi
Stefano Soatto
Jon Sporring
Hugues Talbot
Baba C. Vemuri
Luminita Vese

Referees

Leah Bar
Alexander G. Belyaev
Rami Ben-Ari
Andres Bruhn
Bernhard Burgeth
Lorina Dascal
Stephan Didas
Florian Fruehauf
Guy Gilboa
Harald Grossauer
Nahum Kiryati
Peter Savadjiev
Sigal Trattner
Martin Welk

Invited Speakers

Achi Brandt (Weizmann Institute of Science, Israel)
Michael Unser (Swiss Federal Institute of Technology, Lausanne, Switzerland)
Carl-Fredrik Westin (Harvard Medical School, Boston, USA)

Sponsoring Institution

German Pattern Recognition Society (DAGM)

Table of Contents

Oral Presentations

Novel Linear Scale-Spaces

Image Features

Deep Structure

Image Processing

Medical Applications

Contours

Tensors

Non-linear Filters

Motion

Poster Presentations

Relativistic Scale-Spaces

Bernhard Burgeth, Stephan Didas, and Joachim Weickert

Mathematical Image Analysis Group,
Faculty of Mathematics and Computer Science, Bldg. 27,
Saarland University, 66041 Saarbrücken, Germany
{burgeth, didas, weickert}@mia.uni-saarland.de
http://www.mia.uni-saarland.de

Abstract. In this paper we extend the notion of Poisson scale-space. We propose a generalisation inspired by the linear parabolic pseudodifferential operator $\sqrt{-\Delta + m^2} - m$, $0 \leq m$, connected with models of relativistic kinetic energy from quantum mechanics. This leads to a new family of operators $\{Q_t^m \mid 0 \leq m, t\}$ which we call relativistic scale-spaces. They provide us with a continuous transition from the Poisson scale-space $\{P_t \mid t \geq 0\}$ (for $m = 0$) to the identity operator I (for $m \longrightarrow +\infty$). For any fixed $t_0 > 0$ the family $\{Q_{t_0}^m \mid m \geq 0\}$ constitutes a scale-space connecting I and P_{t_0}. In contrast to the α-scale-spaces the integral kernels for Q_t^m can be given in explicit form for any $m, t \geq 0$ enabling us to make precise statements about smoothness and boundary behaviour of the solutions. Numerical experiments on 1D and 2D data demonstrate the potential of the new scale-space setting.

Keywords: Kinetic energy, Poisson scale-space, semigroup, pseudodifferential operator.

1 Introduction

The pioneering work of Taizo Iijima [16] in the late fifties, though unrecognised in the western scientific world for decades, marks the actual beginning of modern scale-space theory. Since then the vivid research on scale-space methodologies has brought forward many valuable techniques in image processing and computer vision, as it is documented in numerous articles and books, see [24, 11, 31, 21, 28, 33] and the literature cited there. The Gaussian scale-space is the prototype of a linear scale-space. Its connection to linear diffusion processes was first pointed out by Iijima [17]. However, the field of non-linear diffusion, instigated by the influential work of Perona and Malik [25] also exhibits scale-space properties. These non-linear theories encompass anisotropic diffusion processes [33, 26], morphological operations [32, 6, 18] as well as the evolution of level curves [2, 23, 27, 19]. Non-linear differential equations are the mathematical language to describe these theories [31, 33, 14, 3, 12, 7].

Nevertheless, the exploration of the axiomatic principles of the various scale-space approaches [4, 33, 11, 22, 24, 34] usually emanates from the assumption of

R. Kimmel, N. Sochen, J. Weickert (Eds.): Scale-Space 2005, LNCS 3459, pp. 1–12, 2005.

linearity, that is to say, the validity of the superposition principle. In this linear setting the Gaussian scale-space basically had played the leading role in a one man show until the Poisson scale-space from potential theory has been made popular in image processing by Felsberg and Sommer [10].

Soon after the so-called α-scale-spaces with $\alpha \in [\frac{1}{2}, 1]$ have been advocated to bridge the gap between those two prominent representatives since they are ruled by the pseudodifferential equations $\partial_t u = (-\Delta)^\alpha u$ with initial condition $u(x,0) = f(x)$, (for more details and a histortic overview consult the very comprehensive article [8] by Duits et. al. and the literature cited therein). In this setting $\alpha = 0$ produces the family of identity operators I, $\alpha = \frac{1}{2}$ corresponds to the Poissonian, while $\alpha = 1$ delivers the Gaussian version of a linear scale-space. For the later two cases explicit integral representation formulas are known utilising the Poisson and the Gaussian kernel.

The primary tool for the investigation of the α-scale-spaces are Fourier methods since, unfortunately, no explicit integral kernel can be determined. In our paper, however, we propose a counterpart to α-scale-spaces that admits explicit kernel representations. We generalise the Poisson scale-space to a novel scale-space by exploiting the properties of a pseudodifferential operator known from Schrödinger operators in relativistic quantum mechanics [20]. The pseudodifferential operators in question read

$$\sqrt{-\Delta + m^2} - m,$$

and represent the kinetic energy operators in relativistic systems with $m > 0$ denoting mass. Therefore we will refer to these novel scale-spaces as *relativistic scale-spaces* in the sequel. Though heavily taking advantage of spectral methods during the theoretical investigation of this family of operators (indexed by m) we emphasise that the associated integral kernels can be computed explicitly. The knowledge of these kernels enables us to employ techniques from analysis to prove regularity and a maximum-minimum-principle for the solutions of the associated evolution equation.

In the sequel $\mathcal{F}(f)$ will denote the Fourier transform of a function $f \in L^2(\mathrm{I\!R}^n)$ given by

$$\mathcal{F}(f)(k) = \int_{\mathrm{I\!R}^n} e^{-2\pi i k\cdot x} f(x)\,dx\,.$$

The structure of our paper is as follows: After a very brief motivating account of some basic facts about Poisson and Gaussian scale-space we introduce and study the relativistic scale-spaces. Section 3 reports on experiments displaying the potential and limitations of the novel scale-spaces while a summary and an outlook for future research in Section 4 conclude the paper.

2 Relativistic Scale-Spaces

We recall [9, 20] that the action of the Laplace operator $\Delta = \sum_{i=1}^n \frac{\partial^2}{\partial x_i^2}$ on functions in the Fourier domain is multiplication by $-4\pi^2|k|^2$, i.e.

$$\mathcal{F}(\Delta f) = -4\pi^2|k|^2\mathcal{F}(f)$$

while the convolution with the heat or Gaussian kernel $G(x,t,y)$ means multiplication with $e^{-t\;4\pi^2|k|^2}$, $\mathcal{F}(G*f) = e^{-t\;4\pi^2|k|^2}\mathcal{F}(f)$ providing solutions of the heat equation $\partial_t u = \Delta u$.

Furthermore, the action of the pseudodifferential operator $\sqrt{-\Delta}$ is multiplication by $-2\pi|k|$, while convolution with the Poisson kernel $P(\cdot,t)$ means multiplication with $e^{-t\;2\pi|k|}$ in the Fourier domain. The Poisson kernel appears as the inverse Fourier transform $\mathcal{F}^{-1}$ of $e^{-t\;2\pi|k|}$:

$$P(x-y,t) = \mathcal{F}^{-1}\left(e^{-t\;2\pi|\cdot|}\right) = \int_{\mathbb{R}^n} e^{-t\,2\pi|k|+2\pi ik\cdot(x-y)}\,dk\,.$$

This integral can be evaluated in every dimension n yielding the well-known explicit formula for the Poisson kernel [29]

$$P(x-y,t) = \Gamma\left(\frac{n+1}{2}\right)\frac{1}{\pi^{\frac{n+1}{2}}}\frac{t}{(t^2+|x-y|^2)^{\frac{n+1}{2}}}\,. \tag{1}$$

The kernel itself and all convolutions $P(\cdot,t)*f$ with suitable functions f solve in a certain sense the pseudodifferential equation $\partial_t u = \sqrt{-\Delta}u$. The heat and the Poisson kernel generate the Gaussian, resp., the Poisson scale-space.

This can be generalised as follows: In quantum mechanics the pseudodifferential operator $L := \sqrt{-\Delta+m^2} - m$ describes the relativistic kinetic energy of a particle with mass $m \geq 0$ [20] seemingly extending the Poisson operator. In Fourier space this operator acts on function by multiplication with $\sqrt{|2\pi k|^2+m^2} - m$ as a straightforward computation shows. According to standard spectral methods the corresponding integral operator in Fourier space reads

$$e^{-t\;(\sqrt{|2\pi k|^2+m^2}-m)}\,.$$

The inverse Fourier transform of this exponential

$$T_m(x-y,t) := \mathcal{F}^{-1}\left(e^{-t\;(\sqrt{4\pi^2|\cdot|^2+m^2}-m)}\right)(x,y)$$

can be calculated explicitly yielding the expression

$$T_m(x-y,t) := 2\left(\frac{m}{2\pi}\right)^{\frac{n+1}{2}} e^{tm}\frac{t}{\left(t^2+|x-y|^2\right)^{\frac{n+1}{4}}}K_{\frac{n+1}{2}}(m\sqrt{t^2+|x-y|^2}) \tag{2}$$

for $(x-y,t) \in R^n\times]0,+\infty[$. Here K_ν stands for the modified Bessel function of the third kind [1, 13]. We briefly sketch the computational steps by pointing out the formulas

$$\int_{S^{n-1}} e^{i\langle\omega,x\rangle}\,d\omega = (2\pi)^{\frac{n}{2}}|x|^{1-\frac{n}{2}}J_{\frac{n}{2}-1}(|x|)$$

and

$$\int_{[0,+\infty[} x^{\nu+1}J_\nu(xs)\,e^{-\alpha\sqrt{x^2+\beta^2}}dx = \sqrt{\frac{2}{\pi}}\alpha\beta^{\nu+\frac{3}{2}}(s^2+\alpha^2)^{-\frac{\alpha}{2}-\frac{3}{4}}s^\nu K_{\nu+\frac{3}{2}}(\beta\sqrt{s^2+\alpha^2})\,,$$

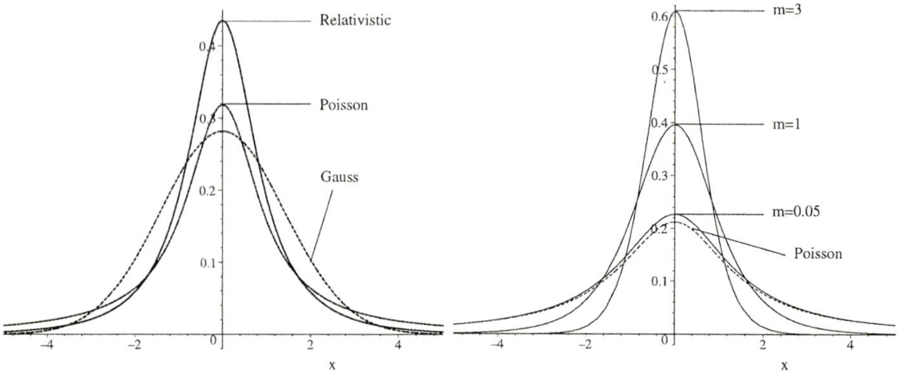

Fig. 1. *Left:* Comparison between different kernels including Poisson, eqn. (1) and relativistic kernel, eqn. (2) in 1D for $y = 0$ and $t = 1$. *Right:* Examples of the relativistic kernel (2) with $m = 3,\ 1,\ ,0.05$ in comparison with the Poisson kernel (1) for $y = 0$ and $t = 1.5$

where J_ν denotes the ν-th order Bessel function. For later use we define the operator Q_t^m on $L^2(\mathbb{R}^n)$ via the convolution

$$Q_t^m f(x) := T_m(\cdot, t) * f\ (x) = \int_{\mathbb{R}^n} T_m(x-y,t) f(y)\, dy\,. \tag{3}$$

2.1 Comparison with the Poisson Kernel

For $m \downarrow 0$ we regain the Poisson kernel which follows from

$$\mathcal{F}(Q_t^m)(k) = e^{-t\ (\sqrt{|2\pi k|^2+m^2}-m)} \longrightarrow e^{-t\ 2\pi|k|} \quad \text{if } m \downarrow 0 \tag{4}$$

for any complex number k together with the continuity of the (inverse) Fourier transform (according to a theorem of P. Levy) [5]. Furthermore, since

$$\mathcal{F}(Q_t^m)(k) = e^{-t\ (\sqrt{|2\pi k|^2+m^2}-m)} \longrightarrow 1 \quad \text{if } m \to +\infty\,,$$

a similar reasoning proves that Q_t^m approximates therefor the identity operator I if m is large. Remarkably, despite the approximation property (4), we learn from the theory of Bessel functions [1, 13] that $K_\nu(x)$ for any $\nu \geq 0$, and hence T_m as a function of x (or of y) decreases exponentially to 0 for x tending to infinity, $|x| \to +\infty$. Figure 1 displays the relativistic kernel for various values of m and also its comparison with a Poisson and a Gaussian kernel.

The relation between Poisson scale-space and the relativistic scale-spaces is sketched in the diagram to the right. $\{Q_t^m \mid 0 \leq t, m\}$ is positioned between The Poisson scale-space $\{P_t \mid 0 \leq t\}$ and $\{I \mid 0 \leq t\}$ including them as limiting cases.

$$\begin{array}{cccc}
+\infty & I_0 & \longrightarrow & I_0 \\
\uparrow & \uparrow & & \uparrow \\
m & Q_0^m = I_0 & \longrightarrow & Q_t^m \\
\uparrow & \uparrow & & \uparrow \\
0 & P_0 = I_0 & \longrightarrow & P_t \\
 & & & \\
 & 0 & \longrightarrow & t
\end{array}$$

2.2 Further Properties of the Relativistic Scale-Spaces

From the theory of contraction semigroups [15] we learn that the operator Q_t^m determines a contraction semigroup on $L^2(\mathbb{R}^n)$. Indeed, in view of Plancherel's theorem, it is enough to verify that the Fourier transforms $\mathcal{F}(Q_t^m) = e^{-t(\sqrt{|2\pi k|^2+m^2}-m)}$ of the family $\{Q_t^m\}$ satisfy the conditions

1. $\mathcal{F}(Q_{s+t}^m)\mathcal{F}(f) = \mathcal{F}(Q)_s^m\mathcal{F}(Q_t^m)\mathcal{F}(f) = \mathcal{F}(Q_t^m)\mathcal{F}(Q_s^m)\mathcal{F}(f)$ for all $s, t \geq 0$.
2. $\|\mathcal{F}(Q_t^m)\mathcal{F}(f) - \mathcal{F}(Q_s^m)\mathcal{F}(f)\|_2 \longrightarrow 0$ for $t \longrightarrow s$.
3. $\mathcal{F}(Q_0^m) = 1$, expressing the fact that $Q_0^m = I$, the identity.
4. $\|\mathcal{F}(Q_t^m)\mathcal{F}(f)\|_2 \leq \|\mathcal{F}(f)\|_2$, the contraction property.

Due to the properties of the exponentials e^{-ct} with $c > 0$ it is not difficult to check that the operator Q_t^m indeed meets these conditions. The associated *generator* is the pseudodifferential operator $L = \sqrt{-\Delta + m^2} - m$ with the Sobolev space $H^1(\mathbb{R}^n)$ as its domain $D(L)$. Here we followed [30] in the definition of the Sobolev spaces

$$H^s(\mathbb{R}^n) := \left\{u \in L^2(\mathbb{R}^n) \mid (1 + |k|^2)^{\frac{s}{2}} \mathcal{F}(u) \in L^2(\mathbb{R}^n)\right\} \tag{5}$$

of all functions in $L^2(\mathbb{R}^n)$ and $s \in \mathbb{R}$.

Next we are going to study in some detail the properties of the function $F_m(x,t)$ defined for $f \in L^2(\mathbb{R}^n)$ by

$$F_m(x,t) := Q_t^m f(x) = \int_{\mathbb{R}^n} T_m(x - y, t) f(y)\, dy$$

with $x \in \mathbb{R}^n$ and $t > 0$. Since the Bessel functions $K_\nu(x)$ are analytic for $0 < x$, the following result is not surprising.

Proposition 2.1. *F_m is analytic in $\mathbb{R}^n \times]0, \infty[$ for any function $f \in L^2(\mathbb{R}^n)$.*

Proof: Thanks to the analyticity of K_ν the function T can be expanded locally in a multivariate power series to the effect that the exchange of integration and summation yields a corresponding expansion for F_m.

Having the explicit integral kernel at our disposal will enable us to study the boundary behaviour of $F_m(x,t)$ as $t \downarrow 0$. To this end we need the next lemma.

Lemma 2.2. *For any $z \geq 0$ and $\nu \geq -\frac{1}{2}$ the following estimate holds:*

$$K_\nu(z) \leq \frac{\Gamma(\nu)}{2} \left(\frac{2}{z}\right)^\nu . \tag{6}$$

Proof: Taking advantage of an integral representation in [13], page 958, and using the well-known definition of the Γ-function we obtain

$$\begin{aligned} K_\nu(z) &= \sqrt{\pi}\left(\frac{z}{2}\right)^\nu \frac{1}{\Gamma(\nu+\frac{1}{2})}\int_1^\infty e^{-tz}(t^2-1)^{\nu-\frac{1}{2}}\,dt \\ &\le \sqrt{\pi}\left(\frac{z}{2}\right)^\nu \frac{1}{\Gamma(\nu+\frac{1}{2})}\int_0^\infty e^{-tz}t^{2\nu-1}\,dt \\ &= \sqrt{\pi}\left(\frac{z}{2}\right)^\nu \frac{1}{\Gamma(\nu+\frac{1}{2})}\cdot\frac{2}{\sqrt{\pi}}\left(\frac{2}{z}\right)^{2\nu}\Gamma(\nu)\Gamma\left(\nu+\frac{1}{2}\right) = \frac{\Gamma(\nu)}{2}\left(\frac{2}{z}\right)^\nu . \end{aligned}$$

This inequality is asymptotically ($\sim$) sharp since for $z \longrightarrow 0$,[1], page 375: $K_\nu(z) \sim \frac{\Gamma(\nu)}{2}\left(\frac{2}{z}\right)^\nu$. With this at our disposal we can proceed to the result stating that $F_m(\cdot,t) = Q_t^m\, f$ has exactly the same boundary behaviour as the corresponding functions stemming from the Gaussian or Poisson scale-space.

Theorem 2.3. *Suppose that f is a continuous and bounded on $\mathbb{R}^n$, $f \in \mathcal{C}(\mathbb{R}^n)\cap L^\infty(\mathbb{R}^n)$, then the function $F_m(x,t) = Q_t^m f(x)$ satisfies the pseudodifferential equation*

$$\partial_t F_m = (\sqrt{-\Delta + m^2} - m)F_m \tag{7}$$

for any $t > 0$ with the initial condition $\lim_{t\downarrow 0} F_m(\cdot,t) = f$.

Proof: That $F_m(x,t)$ satisfies (7) follows from the analysis above remembering that the Fourier transform of $\sqrt{-\Delta+m^2} - m$ is given by $\sqrt{4\pi^2|k|^2+m^2} - m$. Also, as stated above, the corresponding solution operator is given by Q_t^m. In order to prove the claimed boundary behaviour we observe that

$$\int_{\mathbb{R}^n} T_m(x-y,t)\,dy = e^{-t(\sqrt{0+m^2}-m)} = 1 \tag{8}$$

for all $x \in \mathbb{R}^n$ and $t > 0$, since the integral at the left side can be considered as the Fourier transform $\mathcal{F}(T_m(\cdot,t))$ of $T_m(\cdot,t)$ evaluated at $k = 0$.

Next we fix a $x_0 \in \mathbb{R}^n, \epsilon > 0$, and choose $\delta > 0$ so small that if

$$|y - x_0| < \delta \text{ for } y \in \mathbb{R}^n \text{ then } |f(y) - f(x_0)| < \epsilon . \tag{9}$$

For $(x,t) \in \mathbb{R}^n \times]0,+\infty[$ with $|(x,t)-(x_0,0)| < \frac{\delta}{2}$ we obtain the estimate

$$\begin{aligned} |F_t(x,t) - f(x_0)| &= \left|\int_{\mathbb{R}^n} T_m(x-y,t)f(y)\,dy - f(x_0)\cdot\int_{\mathbb{R}^n} T_m(x-y,t)\,dy\right| \\ &\le \int_{B(x_0,\delta)} T_m(x-y,t)|f(y)-f(x_0)|\,dy \\ &\quad + \int_{\mathbb{R}^n\setminus B(x_0,\delta)} T_m(x-y,t)|f(y)-f(x_0)|\,dy \\ &=: I_1 + I_2 . \end{aligned}$$

The equality (8) and the restriction on y in (9) yield

$$I_1 \leq \int_{\mathbb{R}^n} T_m(x-y,t)\varepsilon \, dy = \varepsilon \, .$$

If additionally $|y - x_0| \geq \delta$ we find

$$\begin{aligned} |y-x_0| &\leq |(y,0)-(x,t)| + |(x,t)-(x_0,0)| \\ &\leq |(y,0)-(x,t)| + \frac{\delta}{2} \leq |(y,0)-(x,t)| + \frac{1}{2}|y-x_0| \end{aligned}$$

which yields $|(y,0)-(x,t)| \geq \frac{1}{2}|y-x_0|$. This gives way to the estimates

$$\begin{aligned} I_2 &\leq 2\|f\|_\infty \int\limits_{\mathbb{R}^n \setminus B(x_0,\delta)} T_m(x-y,t)\, dy \\ &= 4\|f\|_\infty \left(\frac{m}{2\pi}\right)^{\frac{n+1}{2}} e^{-tm} \int\limits_{\mathbb{R}^n \setminus B(x_0,\delta)} \frac{t}{|(x,t)-(y,0)|^{\frac{n+1}{2}}} K_{\frac{n+1}{2}}(m|(x,t)-(y,0)|)\, dy \\ &\leq 4\|f\|_\infty \left(\frac{m}{2\pi}\right)^{\frac{n+1}{2}} e^{-tm} \int\limits_{\mathbb{R}^n \setminus B(x_0,\delta)} \frac{t}{\left(\frac{1}{2}|x_0-y|\right)^{\frac{n+1}{2}}} K_{\frac{n+1}{2}}\left(\frac{m}{2}|x_0-y|\right) dy \\ &\leq 2\|f\|_\infty \left(\frac{m}{\pi}\right)^{\frac{n+1}{2}} \Gamma(\nu)\, e^{-tm} \cdot t \int\limits_{\mathbb{R}^n \setminus B(x_0,\delta)} \frac{1}{|x_0-y|^{n+1}}\, dy \longrightarrow 0, \text{ as } t \downarrow 0 \, . \end{aligned}$$

The second inequality follows from the fact that $\frac{1}{|\cdot|}^{\frac{n+1}{2}}$ and K_ν are decreasing functions on $]0,\infty[$ while the last inequality is due to estimate (6) in lemma (2.2). Hence, we deduce $|F_m(x,t) - f(x_0)| \leq I_1 + I_2 \leq 2\varepsilon$ as soon as $|(x,t)-(x_0,0)|$ is sufficiently small proving the continuity of F_m on the closed set $\mathbb{R}^n \times [0,+\infty[$. Summarising the analysis above we state

Proposition 2.4. *1. The families of operators $\{Q_t^m \mid t \geq 0\}$ form for any fixed* $m \geq 0$ additive semigroups.

2. *For every $t \geq 0$ the average grey-value is preserved under the action of Q_t^m.*
3. *The operators Q_t^m are translational invariant.*

For large values of m the relativistic scale-spaces apparently approximate the trivial scale-space $\{I_t \mid t \geq 0\}$ with $I_t = I_0$ for all $t > 0$, while for small m they are very close to the Poisson scale-space.

However, with a fixed t_0 the family $\{Q_{t_0}^m \mid m \geq 0\}$ is also a scale-space, but it has no longer an additive semigroup property: $Q_t^{m_1} Q_t^{m_2} \neq Q_t^{m_1+m_2}$.

We mention briefly that $\{Q_t^{f(t)} \mid t \geq 0\}$ with an arbitrary decreasing function $f : [0,+\infty[\longrightarrow [0,+\infty[$ also describes a scale-space relying on a non-additive semigroup.

3 Numerical Experiments with Relativistic Scale-Spaces

In this section we present some numerical experiments to visualise the properties of relativistic scale-spaces. We have implemented the methods in the Fourier domain using the Discrete Fourier Transform (DFT) or Fast Fourier Transform (FFT) for suitable data dimensions. The filtering operation then can be performed as a multiplication of the Fourier coefficients with $\mathcal{F}(Q_t^m)$. Figures 2 and 3 show the simplifying effect of the relativistic scale-space in 1D and 2D for fixed stopping time t but varying parameter m. Vice versa, Fig. 4 shows a time evolution for fixed value of m and increasing time t. For $m = 0$ this we would obtain the Poisson scale-space.

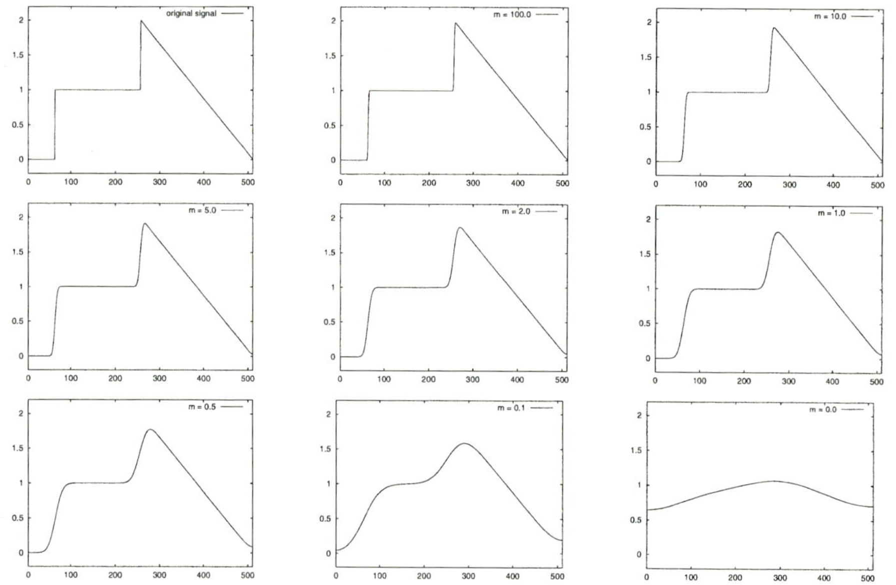

Fig. 2. Relativistic scale-space in 1D. The stopping time $t = 100$ is fixed. *Top left:* Initial signal. The mass m decreases from the top middle to the bottom right

4 Conclusion

The goal of this paper is to propose the novel two-parameter family of relativistic scale-spaces as a generalisation of the well-known Poisson scale-space, and as a counterpart to the α-scale-spaces. As such the relativistic scale-spaces are generated by pseudodifferential operators and they provide a continuous interpolation between the identity operator and the Poisson scale-space. Unlike the α-scale-spaces these new scale-spaces admit integral representations with explicit convolution kernels involving Bessel functions. This paves the way to

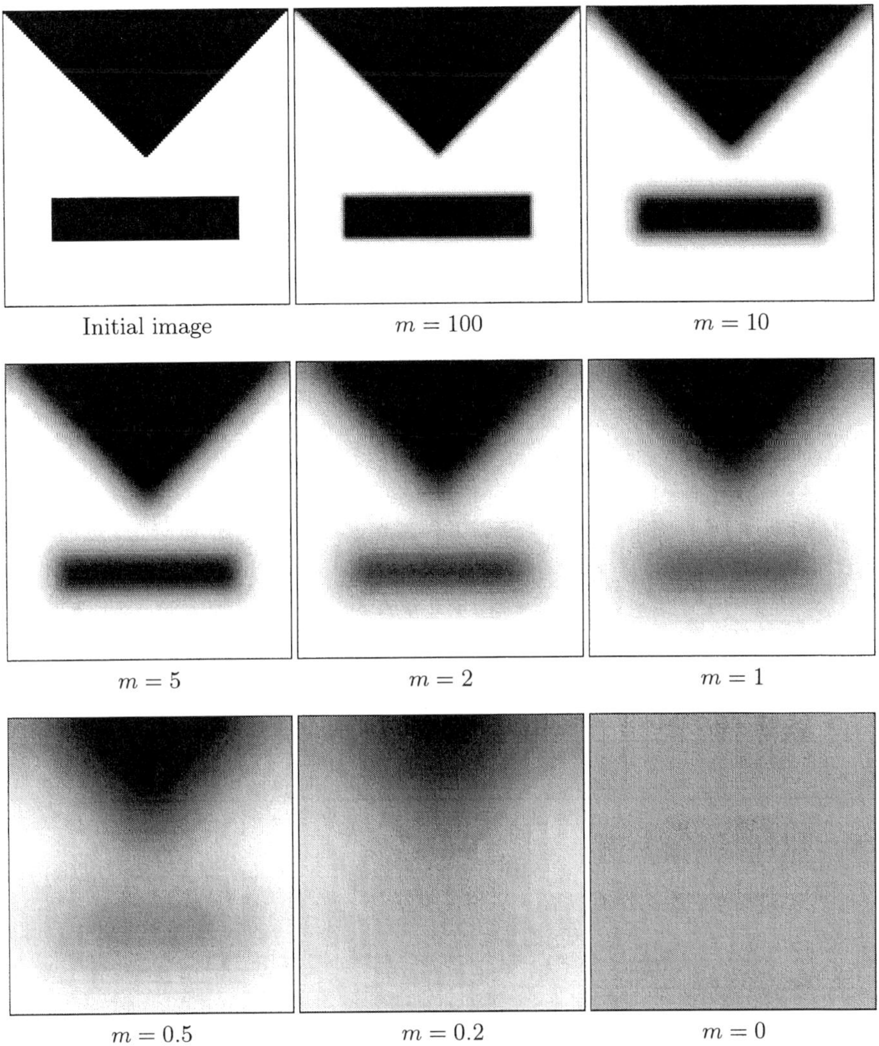

Fig. 3. Relativistic scale-space in 2D. The stopping time $t = 100$ is fixed. *Top left:* Initial image. The mass m decreases from top middle to bottom right

prove analyticity and continuous extendability of the solutions of the relativistic pseudodifferential equations.

This work evidences once more that spectral methods for pseudodifferential operators are very useful for the study and extension of scale-space concepts. Further generalisations of the relativistic scale-spaces in the framework of pseudodifferential operators are close at hand. For instance, the "α-variant" generated by $(-\Delta + m^{\frac{1}{\alpha}})^{\alpha} - m$ is the subject of ongoing research. Future research will also encompass the search for variational formulations hoping to discover

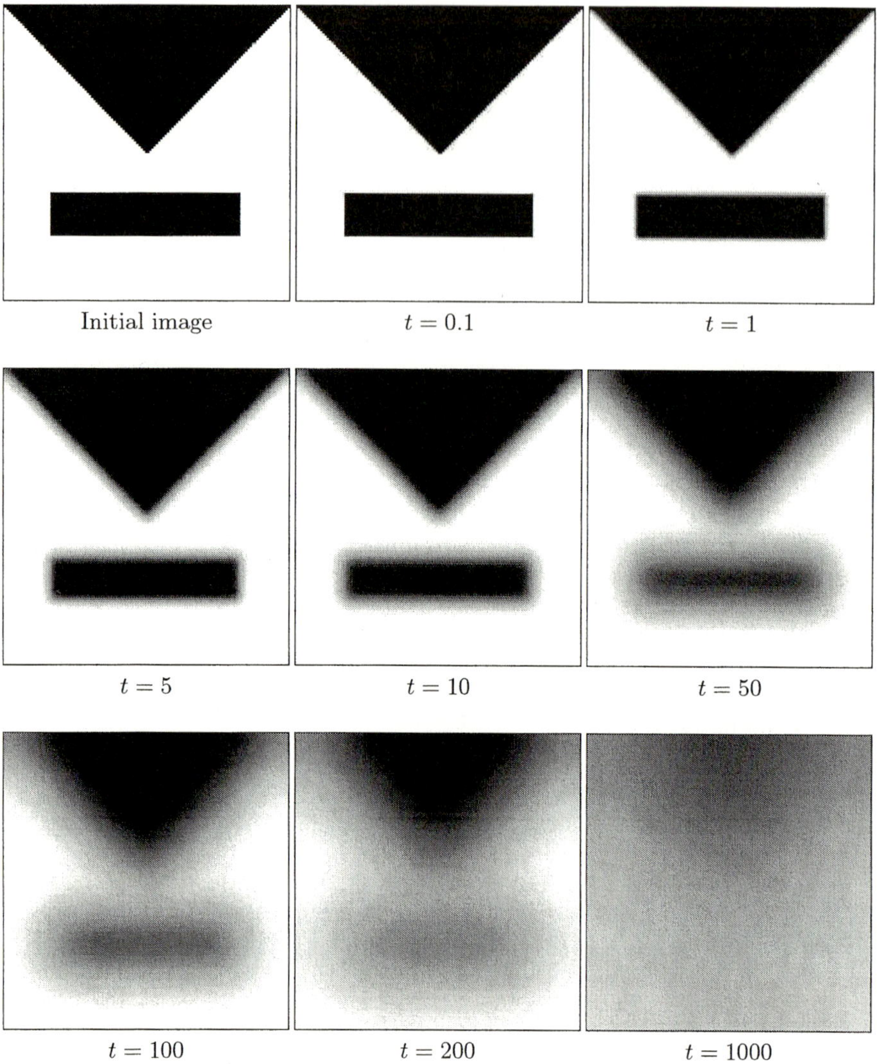

Fig. 4. Relativistic scale-space in 2D. The parameter $m = 1.0$ is fixed, and the time increases from top left to bottom right

new valuable tools for image filtering, and to enhance insight into the structure of scale-spaces.

Acknowledgements. We gratefully acknowledge partly funding by the *Deutsche Forschungsgemeinschaft (DFG)*, project WE 2602/2-2. We also would like to thank Andrés Bruhn for valuable advice concerning implementational issues.

References

1. M. Abramowitz and I.A. Stegun, editors. *Handbook of Mathematical Functions.* Dover Publications,Inc., New York, 9 edition, 1972.
2. L. Alvarez, F. Guichard, P.-L. Lions, and J.-M. Morel. Axioms and fundamental equations in image processing. *Archive for Rational Mechanics and Analysis*, 123:199–257, 1993.
3. G. Aubert and P. Kornprobst. *Mathematical Problems in Image Processing: Partial Differential Equations and the Calculus of Variations*, volume 147 of *Applied Mathematical Sciences.* Springer, New York, 2002.
4. J. Babaud, A. P. Witkin, M. Baudin, and R. O. Duda. Uniqueness of the Gaussian kernel for scale space filtering. *IEEE Transactions on Pattern Analysis and Machine Intelligence*, 8:26–33, 1986.
5. H. Bauer. *Wahrscheinlichkeitstheorie.* Walter de Gruyter, Berlin, 1991.
6. R. W. Brockett and P. Maragos. Evolution equations for continuous-scale morphological filtering. *IEEE Transactions on Signal Processing*, 42:3377–3386, 1994.
7. F. Cao. *Geometric Curve Evolution and Image Processing*, volume 1805 of *Lecture Notes in Mathematics.* Springer, Berlin, 2003.
8. R. Duits, L. Florack, J. de Graaf, and B. Ter Haar Romeny. On the axioms of scale space theory. *Journal of Mathematical Imaging and Vision*, 20:267–298, 2004.
9. L. C. Evans. *Partial Differential Equations*, volume 19 of *Graduate Studies in Mathematics.* American Mathematical Society, Providence, 1998.
10. M. Felsberg and G. Sommer. Scale-adaptive filtering derived from the Laplace equation. In B. Radig and S. Florczyk, editors, *Pattern Recognition*, volume 2032 of *Lecture Notes in Computer Science*, pages 95–106. Springer, Berlin, 2001.
11. L. Florack. *Image Structure*, volume 10 of *Computational Imaging and Vision.* Kluwer, Dordrecht, 1997.
12. G. Gilboa, N. A. Sochen, and Y. Y. Zeevi. Forward-and-backward diffusion processes for adaptive image enhancement and denoising. *IEEE Transactions on Image Processing*, 11(7):689–703, 2002.
13. I. S. Gradshteyn and I. M. Ryzhik. *Tables of Integrals, Series, and Products.* Academic Press, New York, 4 edition, 1980.
14. H. J. A. M. Heijmans. Scale-spaces, PDEs and scale-invariance. In M. Kerckhove, editor, *Scale-Space and Morphology in Computer Vision*, volume 2106 of *Lecture Notes in Computer Science*, pages 215–226. Springer, Berlin, 2001.
15. E. Hille and R. S. Philips. *Functional Analysis and Semi-Groups.* American Mathematical Society, Providence, 1957.
16. T. Iijima. Basic theory of pattern observation. In *Papers of Technical Group on Automata and Automatic Control.* IECE, Japan, December 1959. In Japanese.
17. T. Iijima. Basic theory on the construction of figure space. *Systems, Computers, Controls*, 2(5):51–57, 1971. In English.
18. P. T. Jackway and M. Deriche. Scale-space properties of the multiscale morphological dilation–erosion. *IEEE Transactions on Pattern Analysis and Machine Intelligence*, 18:38–51, 1996.
19. R. Kimmel. *Numerical Geometry of Images: Theory, Algorithms, and Applications.* Springer, New York, 2003.
20. E. H. Lieb and M. Loss. *Analysis*, volume 19. American Mathematical Society, Providence, Rhode Island, 2 edition, 1992.
21. T. Lindeberg. *Scale-Space Theory in Computer Vision.* Kluwer, Boston, 1994.

22. T. Lindeberg. On the axiomatic formulations of linear scale-space. In J. Sporring, M. Nielsen, L. Florack, and P. Johansen, editors, *Gaussian Scale-Space Theory*, volume 8 of *Computational Imaging and Vision*, pages 75–97. Kluwer, Dordrecht, 1997.
23. P. J. Olver, G. Sapiro, and A. Tannenbaum. Classification and uniqueness of invariant geometric flows. *Comptes Rendus de l'Académie des Sciences de Paris, Série I*, 319:339–344, 1994.
24. E. J. Pauwels, L. J. Van Gool, P. Fiddelaers, and T. Moons. An extended class of scale-invariant and recursive scale space filters. *IEEE Transactions on Pattern Analysis and Machine Intelligence*, 17:691–701, 1995.
25. P. Perona and J. Malik. Scale space and edge detection using anisotropic diffusion. *IEEE Transactions on Pattern Analysis and Machine Intelligence*, 12:629–639, 1990.
26. A. H. Salden. *Dynamic Scale-Space Paradigms*. PhD thesis, Faculty of Medicine, Utrecht University, The Netherlands, November 1996.
27. G. Sapiro and A. Tannenbaum. Affine invariant scale-space. *International Journal of Computer Vision*, 11:25–44, 1993.
28. J. Sporring, M. Nielsen, L. Florack, and P. Johansen, editors. *Gaussian Scale-Space Theory*, volume 8 of *Computational Imaging and Vision*. Kluwer, Dordrecht, 1997.
29. E. M. Stein and G. Weiss. *Introduction to Fourier Analysis on Euclidean Spaces*. Princeton University Press, Princeton, New Jersey, 1971.
30. M. E. Taylor. *Partial Differential Equations I – Basic Theory*. Springer, New York, 1996.
31. B. M. ter Haar Romeny, editor. *Geometry-Driven Diffusion in Computer Vision*, volume 1 of *Computational Imaging and Vision*. Kluwer, Dordrecht, 1994.
32. R. van den Boomgaard. The morphological equivalent of the Gauss convolution. *Nieuw Archief Voor Wiskunde*, 10(3):219–236, November 1992.
33. J. Weickert. *Anisotropic Diffusion in Image Processing*. Teubner, Stuttgart, 1998.
34. J. Weickert, S. Ishikawa, and A. Imiya. Linear scale-space has first been proposed in Japan. *Journal of Mathematical Imaging and Vision*, 10(3):237–252, May 1999.

Regularity and Scale-Space Properties of Fractional High Order Linear Filtering

Stephan Didas[1], Bernhard Burgeth[1], Atsushi Imiya[2], and Joachim Weickert[1]

[1] Mathematical Image Analysis Group,
Faculty of Mathematics and Computer Science, Building 27,
Saarland University, 66041 Saarbrücken, Germany
{didas, burgeth, weickert}@mia.uni-saarland.de
http://www.mia.uni-saarland.de

[2] IMIT Chiba University, Yayoi-cho 1-33, Inage-ku 262-8522, Chiba, Japan
imiya@media.imit.chiba-u.ac.jp

Abstract. We investigate the use of fractional powers of the Laplacian for signal and image simplification. We focus both on their corresponding variational techniques and parabolic pseudodifferential equations. We perform a detailed study of the regularisation properties of energy functionals, where the smoothness term consists of various linear combinations of fractional derivatives. The associated parabolic pseudodifferential equations with constant coefficients are providing the link to linear scale-space theory. These encompass the well-known α-scale-spaces, even those with parameter values $\alpha > 1$ known to violate common maximum-minimum principles. Nevertheless, we show that it is possible to construct positivity-preserving combinations of high and low-order filters. Numerical experiments in this direction indicate that non-integral orders play an essential role in this construction. The paper reveals the close relation between continuous and semi-discrete filters, and by that helps to facilitate efficient implementations. In additional numerical experiments we compare the variance decay rates for white noise and edge signals through the action of different filter classes.

1 Introduction

Regularisation and diffusion filtering belong to the most frequently used and best studied methods in image processing. In addition to the well-known Gaussian scale-space [1, 2, 3, 4, 5], other linear scale-spaces enjoy a growing popularity. Already in the 1960's Iijima [6, 7] gave an axiomatic foundation of α-scale-spaces with integer order using four axioms: linearity, translational invariance, scale invariance, and semigroup property. Later on a whole class of linear scale-spaces depending on a fractional order $\alpha > 0$ was axiomatically deduced (Pauwels et al. [8]). Duits et al. [9] further investigated the α-scale-spaces where $\alpha \in (0, 1]$ can be interpreted as fractional power the Laplacian in a pseudodifferential equation creating the scale-space. The restriction on α comes from the demand of a

R. Kimmel, N. Sochen, J. Weickert (Eds.): Scale-Space 2005, LNCS 3459, pp. 13–25, 2005.

maximum-minimum principle for the resulting filters. The most prominent representative of linear scale spaces with fractional order is the Poisson scale-space by Felsberg and Sommer [10].

In our work we use fractional powers of the Laplacian not only in partial differential equations, but also in regularisation methods. Besides the scale-space properties we are especially interested in well-posedness and regularity properties. We see that variational methods allow it to prescribe a certain fractional regularity order for a given image where diffusion methods always yield arbitrary smooth solutions. In our experiments we propose a way to construct filters with maximum-minimum property which involve both high and low fractional derivative orders.

The paper is organised as follows. In Section 2 we introduce the basic notions related to fractional powers of the Laplacian. Section 3 presents fractional order regularisation as a first application of these notions. The corresponding diffusion equations are investigated in Section 4. Section 5 reformulates both approaches in a space-discrete framework directly leading to efficient implementations. Our numerical experiments in Section 6 especially are dedicated to the question of maximum-minimum property and variance decay. Section 7 concludes the paper.

2 Fractional Powers of the Laplacian

In order to present an elegant concept for fractional powers of the Laplacian, we have to introduce some basic notions first. First we consider the Fourier transform of a function $f \in L^1(\mathrm{I\!R})$ pointwise defined by

$$\hat{f}(\xi) := \frac{1}{\sqrt{2\pi}} \int_{\mathrm{I\!R}} f(x) \exp(-ix\xi) dx$$

Let $\mathcal{F} : L^2(\mathrm{I\!R}) \longrightarrow L^2(\mathrm{I\!R})$ denote the Fourier-Plancherel transform, i. e. the extension of the mapping $L^1(\mathrm{I\!R}) \ni f \longrightarrow \hat{f}$ onto $L^2(\mathrm{I\!R})$. It is well-known that $\mathcal{F}$ is isometric with respect to the norm in $L^2(\mathrm{I\!R})$ (see [11] for details). Later on we will especially make use of the property

$$i^k \xi^k \mathcal{F} f = \mathcal{F}\left(\frac{d^k}{dx^k} f\right) \tag{1}$$

which builds the link between differentiation in the spatial domain and multiplication in the Fourier domain. For $f \in L^\infty(\mathrm{I\!R})$ let $\mathcal{M}_f : L^2(\mathrm{I\!R}) \longrightarrow L^2(\mathrm{I\!R})$ denote the multiplication operator defined by $\mathcal{M}_f g := fg$. With this notation (1) reads as $\mathcal{M}_{(i\xi)^k} \mathcal{F} f = \mathcal{F}\left(\frac{d^k}{dx^k} f\right)$.

Lemma 2.1. *For $f, g \in L^\infty(\mathrm{I\!R})$ the multiplication operator $\mathcal{M}_f$ is $L^2(\mathrm{I\!R})$-continuous with $\|\mathcal{M}_f\| \leq \|f\|_\infty$. Further, $fg \in L^\infty(\mathrm{I\!R})$ and $\mathcal{M}_f \mathcal{M}_g = \mathcal{M}_{fg}$.*

Following the notation in [12] we define the Sobolev space

$$H^s(\mathrm{I\!R}) := \left\{ u \in L^2(\mathrm{I\!R}) \mid (1 + |\xi|^2)^{\frac{s}{2}} \hat{u} \in L^2(\mathrm{I\!R}) \right\} \tag{2}$$

of all functions in $L^2(\mathbb{R})$ and $s \in \mathbb{R}$. For $s \in \mathbb{N}$ functions in $H^s(\mathbb{R})$ are weakly differentiable up to the order s. From (1) we deduce the spectral decomposition of the Laplacian $-\frac{d^2}{dx^2} = \mathcal{F}^{-1}\mathcal{M}_{|\xi|^2}\mathcal{F}$ which allows us to define fractional powers

$$\mathcal{D}^{2\alpha} := \left(-\frac{d^2}{dx^2}\right)^{\alpha} = \mathcal{F}^{-1}\mathcal{M}_{|\xi|^{2\alpha}}\mathcal{F} \qquad (\alpha > 0) \tag{3}$$

as multiplication operators in the Fourier domain (see [13, 14] for further details).

Lemma 2.2. *Applying $\mathcal{D}^{\alpha}$ to functions in a certain Sobolev space reduces the order of differentiability by α, i. e. $\mathcal{D}^{\alpha} : H^s(\mathbb{R}) \longrightarrow H^{s-\alpha}(\mathbb{R})$ for all $s \in \mathbb{R}$.*

In the next sections we are going to replace derivative operators in classical image processing approaches with operators of the type $\mathcal{D}^{\alpha}$ and investigate the properties of the resulting filter methods.

3 Regularisation with Fractional Derivative Orders

To extend linear regularisation to fractional derivative orders we consider the energy functional

$$\mathcal{E}(u) = \int_{\mathbb{R}} \left((u-f)^2 + \sum_{k=1}^{m} \beta_k \left(\mathcal{D}^{\alpha_k} u\right)^2\right) dx \tag{4}$$

with a linear combination of $m \in \mathbb{N}$ fractional derivatives of orders $\alpha_k > 0$ in the smoothness term and regularisation weights $\beta_k > 0$ for $k = 1, \ldots, m$, for short, $\alpha = (\alpha_1, \ldots, \alpha_m)$, $\beta = (\beta_1, \ldots, \beta_m) \in \mathbb{R}^+_m$. For integer derivative orders α_k, similar functionals have been considered in [15]. We assume that the signals u and f may only assume real values. With the Plancherel identity we can rewrite functional (4) in the Fourier domain as

$$\mathcal{E}(\hat{u}) = \int_{\mathbb{R}} \left(\left|\hat{u} - \hat{f}\right|^2 + \sum_{k=1}^{m} \beta_k \left|\xi^{\alpha_k} \hat{u}\right|^2\right) d\xi \tag{5}$$

depending on the complex Fourier transform $\hat{u}$. A decomposition into the real and imaginary part shows that it is necessary for a minimiser u to satisfy the Euler-Lagrange equation

$$\hat{u} - \hat{f} + \sum_{k=1}^{m} \beta_k |\xi|^{2\alpha_k} \hat{u} = 0 \quad \text{for all } \xi \in \mathbb{R} \ . \tag{6}$$

We deduce that the minimiser u of the functional $\mathcal{E}$ has the Fourier transform

$$\hat{u} = \left(1 + \sum_{k=1}^{m} \beta_k |\xi|^{2\alpha_k}\right)^{-1} \hat{f} \quad \text{for all } \xi \in \mathbb{R} \ . \tag{7}$$

To obtain a regularised version of f we transform this minimiser $\hat{u}$ in the spatial domain which motivates the following definition:

Proof. The proof of the second statement is analogous to the proof of Prop. 4.4 exchanging the operators $\mathcal{F}$ and $\mathcal{M}$ by their finite dimensional counterparts F and M. The third statement is proven as in the regularisation case. □

6 Numerical Examples

In the first numerical experiment we investigate the possibility of building linear combinations with different derivative orders such that the regularisation and diffusion filters satisfy a maximum-minimum property. Knowing from Section 4 that combinations of two orders are no longer scale-invariant we try to preserve one scale-space property at the expense of the other. To reduce the number of possible combinations we consider diffusion equations of the form

$$\frac{\partial}{\partial t}u = -\left(\sqrt{-\frac{\partial^2}{\partial x^2}} + \beta\left(-\frac{\partial^2}{\partial x^2}\right)^{\alpha}\right)u \tag{20}$$

and the corresponding regularisation. For α between 1.5 and 8, we started with $\beta = 0$ and increased it as long as nonnegative convolution kernels were obtained. The maximal values of β are shown in Fig. 3. This experiment shows the usefulness of the Poisson scale-space: Using a Gaussian scale-space instead makes it impossible to find a weight $\beta \neq 0$ that leads to a nonnegative combination. In that sense the fractional order scale-space has a clear advantage in comparison with the integer order ones.

In our second experiment we study the variance diminishing properties of different filters $\mathcal{R}$ and $\mathcal{T}$. Fig. 4 shows the variance of a white noise signal depending on regularisation weight / stopping time. We visualise the parameters needed for reducing the variance of a white noise and a step edge signal to half of its value in Fig. 5. The experiments show a similar behaviour of regularisation and corresponding diffusion techniques in terms of variance reduction. We note that higher orders lead to the same variance decay with smaller stopping times.

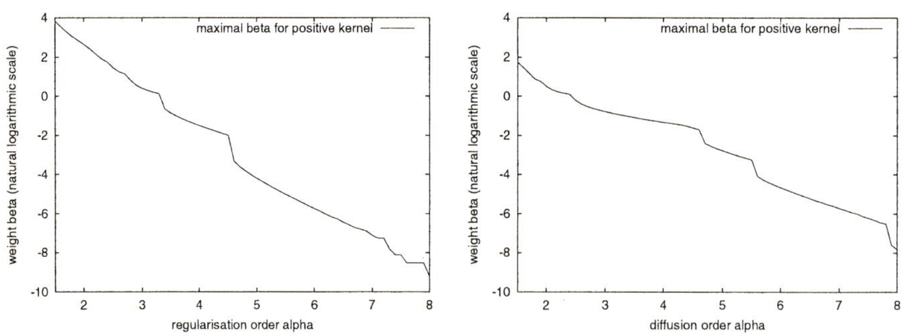

Fig. 3. Positive combinations of derivatives of order 0.5 with higher orders. Left: Regularisation. Right: Diffusion filtering

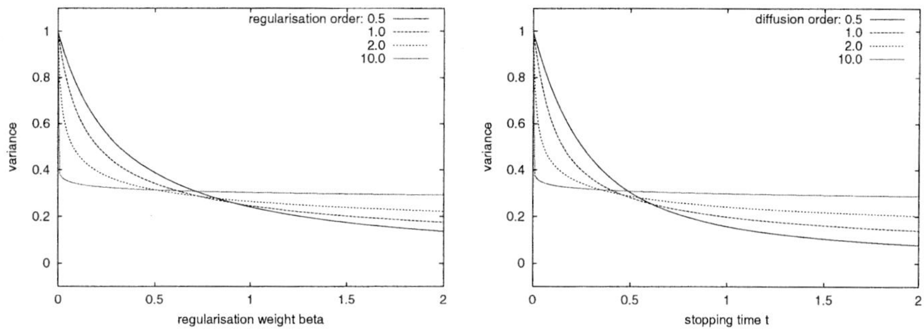

Fig. 4. Variance diminishing properties of fractional order regularisation and diffusion filtering. Variance depending on regularisaton weight/diffusion stopping time. Left: Regularisation. Right: Diffusion filtering

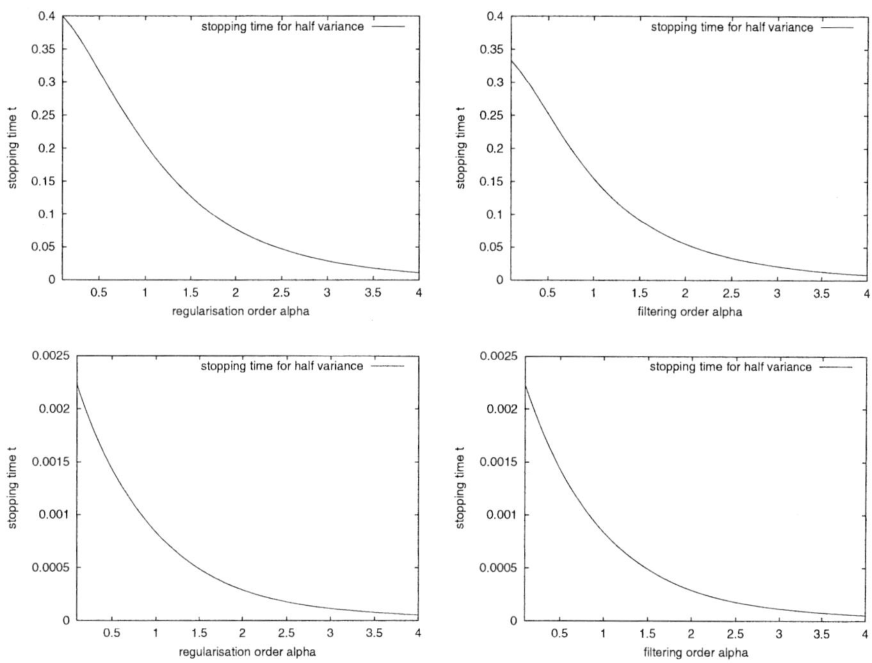

Fig. 5. Regularisation weight/diffusion stopping time for reducing the variance to half its value. Left column: Regularisation. Right column: Diffusion filtering. Top row: Experiment for white noise signal. Bottom row: Experiment for step edge signal

7 Conclusion

In this paper we have discussed regularisation techniques and diffusion methods that involve sums of fractional derivative orders. With respect to scale-space properties, fractional diffusion satisfies Iijima's axioms of linearity, translation invariance and semigroup property. If a single fractional order is used, scale invariance is satisfied as well. We have shown that both fractional diffusion and fractional regularisation are L^2-stable in the sense that the norms of the corresponding operators are bounded by 1. With respect to regularity, the regularisation approaches gain twice the minimal derivative order, while the fractional diffusion admits arbitrarily smooth solutions. For the first time in the context of α-scale-spaces, we have also presented a space-discrete theory that is in formal analogy to the continuous framework. Moreover, it gives convergence towards the average grey value, if the diffusion time / regularisation parameter tends to infinity. To our knowledge, all papers on α-scale-spaces focus their attention to the case $0 < \alpha \leq 1$, since this guarantees nonnegativity and a maximum-minimum principle. However, we have shown that it is possible to construct combinations of Poisson scale-space and diffusion scale-spaces of order $\alpha > 1$ that satisfy this principle as well. With Gaussian scale-space instead of Poisson scale-space, this is not possible. Similar statements also hold for the corresponding regularisation processes. From a practical viewpoint, we have studied the decay rates of the variance as a function of the fractional order. These studies have shown that higher orders reveal higher variance diminishing properties. In our ongoing and future work we intend to find out which of the scale-space and regularity properties of the linear methods of this paper can be generalised to nonlinear processes with higher-order derivatives.

Acknowledgements. We gratefully acknowledge partly funding by the *Deutsche Forschungsgemeinschaft (DFG)*, project WE 2602/2-2.

References

1. Witkin, A.P.: Scale-space filtering. In: Proc. Eighth International Joint Conference on Artificial Intelligence. Volume 2., Karlsruhe, West Germany (1983) 945–951
2. Koenderink, J.J.: The structure of images. Biological Cybernetics **50** (1984) 363–370
3. Lindeberg, T.: Scale-Space Theory in Computer Vision. Kluwer, Boston (1994)
4. Florack, L.: Image Structure. Volume 10 of Computational Imaging and Vision. Kluwer, Dordrecht (1997)
5. Sporring, J., Nielsen, M., Florack, L., Johansen, P., eds.: Gaussian Scale-Space Theory. Volume 8 of Computational Imaging and Vision. Kluwer, Dordrecht (1997)
6. Iijima, T.: Basic theory on normalization of pattern (in case of typical one-dimensional pattern). Bulletin of the Electrotechnical Laboratory **26** (1962) 368–388 (in Japanese).
7. Iijima, T.: Theory of pattern recognition. Electronics and Communications in Japan (1963) 123–124 (in English).
8. Pauwels, E.J., Van Gool, L.J., Fiddelaers, P., Moons, T.: An extended class of scale-invariant and recursive scale space filters. IEEE Transactions on Pattern Analysis and Machine Intelligence **17** (1995) 691–701
9. Duits, R., Florack, L., de Graaf, J., Ter Haar Romeny, B.: On the axioms of scale space theory. Journal of Mathematical Imaging and Vision **20** (2004) 267–298

10. Felsberg, M., Sommer, G.: Scale-adaptive filtering derived from the laplace equation. In Radig, B., Florczyk, S., eds.: Pattern Recognition. Volume 2032 of Lecture Notes on Computer Science., Springer (2001) 95–106
11. Rudin, W.: Real and Complex Analysis. third edn. McGraw-Hill (1986)
12. Taylor, M.E.: Partial Differential Equations I – Basic Theory. Springer, New York (1996)
13. Taylor, M.E.: Partial Differential Equations II – Qualitative Studies of Linear Equations. Springer, New York (1996)
14. Taylor, M.E.: Pseudodifferential Operators. Princeton University Press, Princeton, New Jersey (1981)
15. Nielsen, M., Florack, L., Deriche, R.: Regularization, scale-space and edge detection filters. Journal of Mathematical Imaging and Vision **7** (1997) 291–307
16. Scherzer, O., Weickert, J.: Relations between regularization and diffusion filtering. Journal of Mathematical Imaging and Vision **12** (2000) 43–63

Image Features and the 1-D, 2nd Order Gaussian Derivative Jet

Lewis D Griffin[1] and Martin Lillholm[2]

[1] Computer Science, University College London, UK
l.griffin@cs.ucl.ac.uk
2 IT University of Copenhagen, Denmark
grumse@it-c.dk

We review a previously presented proposal – Geometric Texton Theory (GTT) - that feature categories naturally arise through consideration of the maximum likelihood explanations for image measurements by gaussian derivative filters. We present results relevant to this proposal for the case of 1-D measurement by filters of 0th, 1st and 2nd order. The results are consistent with GTT.

1 Introduction

1.1 Feature-Based Vision

The grail of feature-based vision is a bottom-up process that computes representations of image structure in terms of *qualitative* descriptors (e.g. 'edge' or 'corner'). What motivates the search is the hope that feature-based description could be so successful at discarding the unimportant structure of an image and emphasizing the important that the process of vision would simply be, as David Marr wrote, that "you looked at the image, detected features on it, and used the features to classify and hence recognize what you were looking at." [1].

The Marr quote above is more than twenty years old, and he traces the feature approach back a further thirty years [2]. However, despite this relative venerability there has certainly **not** been a constant and gradual improvement over time in methods to compute features, rather the problem has been increasingly neglected. Of course, such a history does not show the approach definitely to be misconceived, but it does lay a burden on those pursuing the feature approach to explain:

(i) how progress has been possible without a theory of features, and
(ii) why development of the feature approach has been difficult.

To the first question we would answer that a fundamental problem in vision arises from having to deal with the cross-product of two spaces of very high cardinality: (I) the space of all possible images, and (C) the space of possible image contents. Feature-based image description reduces the cardinality of I. In the absence of an effective feature-based approach, research has focused on producing working solutions in constrained domains where the cardinality of C instead can be restricted. To

R. Kimmel, N. Sochen, J. Weickert (Eds.): Scale-Space 2005, LNCS 3459, pp. 26–37, 2005.

the second question we would say that a lack of ambition about how large a vocabulary of features can be stably computed has led to a misplaced pessimism about their power. For example, to quote Marr again, "think of a 5 gradually changing into a 6 – a corner disappears, a gap narrows. Almost no single feature is necessary for any numeral." [1].

1.2 Gaussian Derivative Filters

Marr's pessimism seems to have been due to an incomplete picture of visual neuron properties leading him to consider only feature vocabularies of limited expressiveness. At the time he wrote *Vision*, the linear simple cells of V1 seemed to consist of 1st and 2nd order (in a derivative sense) filters only. With such filters all that it seemed possible to detect were 'edges' and 'bars' and it is clear that much work is indeed required to discriminate a '5' from a '6' with such a limited feature vocabulary. However, it is now known that V1 simple cells frequently have more positive and negative sub-fields and are more varied than Marr knew. So, (it is hypothesized [3-6]) a set of co-localized visual neurons together have the power to characterize a substantial vocabulary of features without having to resort to the difficult process of multi-local feature combination that was Marr's answer to the expressivity problem.

The most popular model of the range of V1 simple cells is as Gabor functions [7, 8] which do indeed model the electrophysiological data very well. However gaussian derivatives (DtGs) up to 4th or 5th order provide an equally well-fitting set of models with much to recommend them [9-17] (we will refer to the measurements given by DtGs up to some order as a Jet). Although we appreciate the limited persuasiveness to some of such non-empirical considerations, to the authors the most compelling points in favour of DtGs over Gabors are the interpretations of what they measure. In particular we note two. First, that measurement of the jet up to some order is equivalent to measurement of the initial terms of the Hermite Transform, which is a local analogue of the Fourier Transform [6, 18-20]. Second that the jet is also interpretable as the initial terms of the Taylor series of the image blurred to the same degree as the scale of the DtG filters [21]. Both interpretations are very appropriate for what we consider V1 to be i.e. a fully general-purpose system for local measurement.

1.3 From Filters to Features

Even though the jet up to 4th or 5th order seems to capture local image structure richly enough to be the basis of spatially complex features it is far from obvious how to define these features from the filter responses. The majority of the relevant literature is to be found under the keyword 'textons' rather than features [22-25]. The most common position taken in this literature (implicitly or explicitly) is that textons/features correspond to clusters in the jet space.

We agree with two parts of this position – that features correspond to regions of jet space and that natural image statistics somehow determine these regions, but we disagree that these regions are revealed as simple clusters. The plain fact is that if one forms a distribution of responses in jet space for natural images there is no clustering structure (see section 3.1). The following comment is admittedly polemical, but it seems to us that the results of k-mean clustering are too often taken as intrinsic features of the data examined even though the algorithm always returns results.

An alternative approach to the 'features from filters' problem has been pioneered by Koenderink [3-6] who has stressed the relevance of metamerism: that the jet does not uniquely determine the measured image even locally. In particular he has suggested that maybe (i) it is possible non-arbitrarily to associate with each point of jet space an iconic image from the metamery class of possible images, and (ii) the equivalence relation of 'qualitative identity of icons' gives rise to a partitioning of jet space into features. Griffin et al. [26] have proposed Geometric Texton Theory (GTT) a refinement of Koenderink's proposal with the added detail that the icons should be defined as the maximum likelihood (relative to natural images) elements of metamery classes. So our GTT extension ties features to natural image statistics in the way that the (incorrect) cluster idea of the texton approaches also does.

We have previously studied [26] GTT for the 1-D, 1st order jet. In this simplest of cases, the two degrees of freedom (0th and 1st order measurements) can be eliminated by affine scaling of profiles so that they all measure to the same values. Once this was done, we found that the maximum likelihood (ML) form for profiles was a step edge. We claimed that this result, although very far from decisive, was supportive for GTT in that the ML profile had a simple qualitative structure.

1.4 Paper Notation

In the remainder of the paper we continue our study of GTT but in this case for the 1-D, 2nd order jet. This jet is measured using the 0th, 1st and 2nd order DtG filters shown in figure 1. These filters are given by:

$$G_s(x) = \tfrac{1}{\sqrt{4\pi s}} e^{-\frac{x^2}{4s}}, \qquad G_s'(x) = -\tfrac{x}{2s} G_s(x) \qquad G_s''(x) = \tfrac{1}{2s}\left(\tfrac{x^2}{2s} - 1\right) G_s(x)$$

We denote the 2nd order jet of the image I measured at the origin, at a scale s as $J_s^2(I) = \langle m_0, m_1, m_2 \rangle = \langle G_s \cdot I, -G_s' \cdot I, G_s'' \cdot I \rangle$.

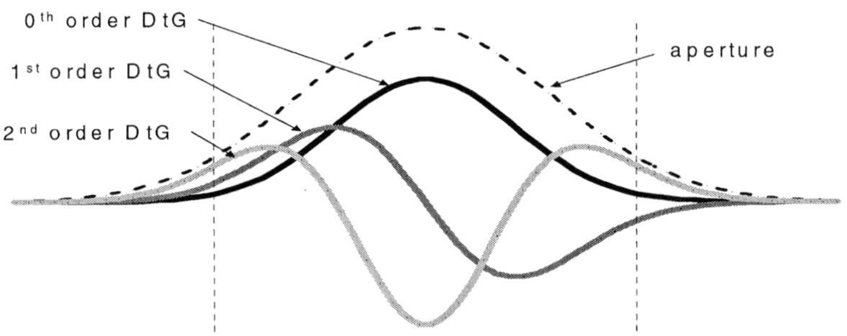

Fig. 1. Shows the three filters measurement with which is considered in this paper. The filters shown are of scale $s = 24$, and are plotted over the range $x \in [-32, 32]$. The 'aperture' function is a 0th order Gaussian of scale $2s$, it will be relevant in section 3. The vertical dashed lines mark the outermost inflexions of the 2nd order filter

2 Norm-Minimizing Metameres

Measurement of local structure with a finite number of DtG filters does not fully determine the measured image even locally. The set of profiles that measure to a given point in jet space constitute the metamery class for that point. Most of the elements of a metamery class have wild oscillations at frequencies which are sufficiently high that the DtG filters are 'blind' to them, however there are also metameres which are more conservative in their excursions. This line of thought leads to the proposal that perhaps a suitable definition of 'most conservative/simple profile in a metamery class' would supply the icons that Koenderink has suggested (section 1.3) should 'stand for' the metamery class or, equivalently, point in jet space. This is an approach that we have previously pursued [27, 28]. Although we now consider it less a well-motivated approach to icon selection than the ML approach (see again section 1.3), we feel that it is still worth studying as it illuminates what DtGs *do* measure.

The class of definitions of 'simplest' that we consider are the norm-minimizers of metamery classes. The definitions we use here of norms are slightly different from the standard mathematical definitions. For an image I, we define its luminance norms as $L^r(I)=\min_{\mu_r\in\mathbb{R}}\left(\int_{\vec{x}\in\mathbb{R}}\left|I(\vec{x})-\mu_r\right|^r\right)^{\frac{1}{r}}$, and its gradient norms as $D^r(I)=\left(\int_{\vec{x}\in\mathbb{R}}\left|\nabla I(\vec{x})\right|^r\right)^{\frac{1}{r}}$. In either case, infinity norms are defined by taking the limit $r\rightarrow\infty$.

For both types of norm, addition of a constant value to the image leaves the norm unchanged, whereas multiplication of the image by a constant similarly multiplies the norm i.e. $L^r(\alpha+\beta.I)=\beta.L^r(I)$ and $D^r(\alpha+\beta.I)=\beta.D^r(I)$. Therefore to study the norm-minimizers with respect to the 2nd order jet we do not need to consider the full range of possibilities $\langle m_0, m_1, m_2\rangle$ but only a suitable 1-D subspace. We choose $\left\langle 0, \cos\theta, (2s)^{-\frac{1}{2}}\sin\theta\right\rangle$, indexed by the phase variable θ.

The norms we have previously identified as of interest are L^2, L^∞, D^1 and D^2. The minimizers of these norms show a degree of resemblance to structures found in natural images. In contrast, the L^1 norm minimizers, for example, consist of very 'unnatural' collections of delta functions. The norm-minimizers we have determined are shown in figure 2. In the following four sub-sections we provide additional explanation.

L^2: The variance minimizers

As we have previously noted [26] the form of variance minimizing metameres can be determined by the method of Lagrangian multipliers to be a weighted sum of a constant function and the DtG filters (other than the 0th order) that measure the jet. As can be seen from figure 2 (top, right) the pure cosine-phase variance-minimizing metamere is just the 2nd order DtG, the pure sine phase the 1st order DtG, and intermediate phases are a mixture of the two.

L^∞: The range minimizers

As we have proved elsewhere [27] the range minimizers are binary-valued functions with one or two discontinuities. Only in the case of the pure sine-phase range-

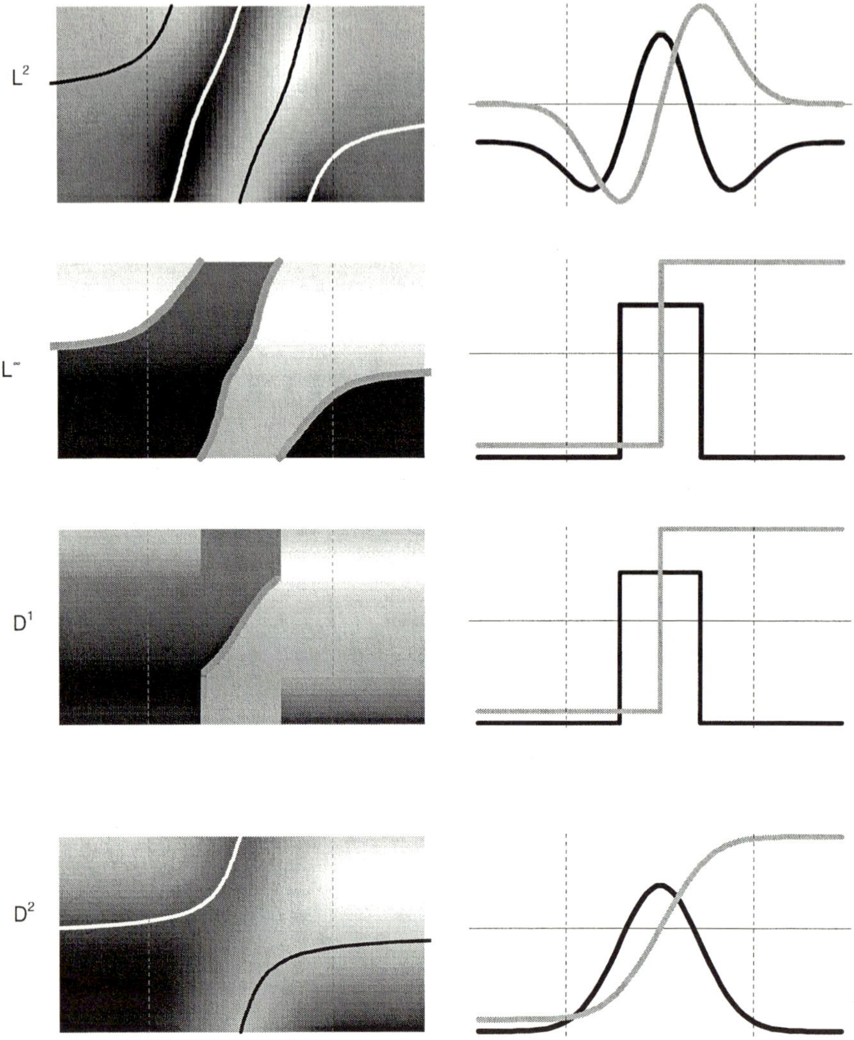

Fig. 2. Shows the norm minimizing profiles for different norms and phases. The norm for the row is indicated at left. The middle column shows the minimizers as density plots. Each row of the density plot corresponds to a minimizing profile; phases varies vertically from $-\pi/2$ (2nd order filter response is –ve, 1st order response is zero) at the bottom, through 0 (2nd order zero, 1st order +ve) in the middle, to $\pi/2$ (2nd order +ve, 1st order zero) at top. To assist visualization, lines have been added to the density plots to show the location of extrema and discontinuities. The right hand column shows the $-\pi/2$ phase (black) and the 0 phase minimizers as regular plots. For the middle and right columns the vertical dotted lines are the same as in fig 1

minimizer is there a single discontinuity (at the origin). For other phases, the two discontinuities are positioned at $x = \alpha, \beta$ such that $\alpha\beta = -2s$. For the pure cosine-phase range-minimizer the discontinuities are at $\pm\sqrt{2s}$.

The general strategy for proving the form of these minimizers comes originally from an argument as to the form of optimal spectral reflectance functions in colour science [29]. The strategy, which is to show by *reductio ad absurdum* that the form of the solution is restricted, is as follows. Suppose the jet specifying the metamery class is $\vec{m}$ and hypothesis that the minimizer s has a certain type of form. If one can show that s can always be perturbed to make s^* which has (i) the same norm value as s, and (ii) has jet $(1+\varepsilon)\vec{m}$, where $\varepsilon > 0$, then $(1+\varepsilon)^{-1} s^*$ will have jet $\vec{m}$ but will have a lower norm that s. This contradicts the hypothesis that s was a minimizer, therefore the minimizer cannot have the hypothesized form. This strategy can be used to show that (i) the range-minimizer must be binary-valued, and then that (ii) the number of discontinuities is restricted. That the appropriate types of perturbation are always possible for the proof to work follows from the full rank of the Gaussian derivatives [5].

D^1: The total variation (TV) minimizers
The proof strategy outlined in the previous sub-section applies also to TV minimization. First the strategy is used to show that the TV-minimizers must be piece-wise constant. Next the strategy is used to show that the TV-minimizers have at most two discontinuities. Then, for each possible value of the jet, one identifies the piecewise constant profiles with at most two discontinuities that measure to the correct values, and one computes which of these possibilities has the lowest TV. The resulting profiles are shown in figure 2 (third row). For phases in the range $\theta \in \left[-\frac{\pi}{4}, \frac{\pi}{4}\right]$ (i.e. close to sine-phase) the TV-minimizer has a single discontinuity somewhere in the range $x \in \left[-\sqrt{2s}, \sqrt{2s}\right]$. For phases nearer cosine-phase ($|\theta| > \frac{\pi}{4}$), the TV-minimizing form has two discontinuities at $x = \pm\sqrt{2s}$. For pure sine- and cosine-phase, and for no other phases, the TV-minimizing profiles are the same as that for range-minimization.

D^2: The roughness minimizers
In the 1-D case here studied, the problem of roughness-minimization with respect to measurements by G' and G'' can be shown to be related by integration to variance-minimization with respect to measurement by G and G'. So one can show that the roughness minimizers must be a weighted sum of a constant function, a 0th order DtG and an error function of the same scale as the filters defining the jet. These minimizers are shown in the bottom row of figure 2. Similarly to the L^2 case, the pure cosine-phase minimizers are gaussians, the pure sine-phase minimizer is an error function and the minimizers for intermediate phases are a mixture of the two.

3 Maximum Likelihood (ML) Metameres

We have investigated the ML (relative to natural images) metameres for the 1-D, 2nd order jet. That is to say we have determined the answer to questions of the form: if a randomly selected profile from a natural image has a jet that measures to $\vec{m}$ what is the most likely form of the profile? Much but not all of the relevant method is the same as that used in our earlier study [26] of the 1-D, 1st order jet.

3.1 Method

As in our previous study, we have chosen to factor out an affine component of image structure that we believe obscures the structure that we are truly interested in. We do this by transforming ($P \to \alpha + \beta.P$) each profile that we examine so that its jet has the canonical form $\left\langle 0, \cos\theta, (2s)^{-\frac{1}{2}} \sin\theta \right\rangle$. The exact value of the factor $(2s)^{-\frac{1}{2}}$ is not critical to our results but figure 3 shows that it is a natural choice, in that it causes the histogram of natural image profiles as a function of phase to be nearly flat. Note that the figure also shows that there is no hint of density clusters that could form the basis of feature categories (as is frequently assumed to be the case in texton approaches).

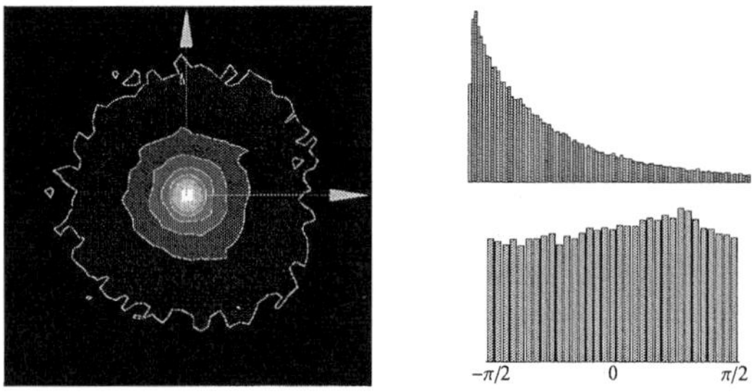

Fig. 3. At left is the density function across jet space for natural image profiles. The horizontal axis is the 1st order filter response, the vertical axis is $\sqrt{2s}$ times the 2nd order filter response. The density has been square root transformed to improve visibility. On the right are plotted the polar marginal histograms; at top for the radial variable and at bottom for the angular

3.1.1 Extracting and Preparing Profiles

As a source of natural image data we used 1220 images from the 4000 image van Hateren natural image database of linear (.iml), 1536×768, images of woods, open landscapes and urban areas [30]. In a change from our previous method, for each image we created a 5-layer quad tree by blurring and sub-sampling. The degree of blurring was chosen so that the ratio between the width of the p.s.f. and the pixel sampling was the same at all levels of the quad tree. At all levels of the quad tree we took measures to reduce quantization effects that we have previously described [26].

We extracted 2700 1-D profiles from each image, for a total of 3.3×10^6. Each profile was extracted from a random level of the quad tree (with probability proportional to the number of pixels in the level), at a random location and orientation (all real-valued) using nearest-neighbour interpolation. Each profile was 64 samples in length.

After extraction, the jet of each profile was measured ($\langle m_0, m_1, m_2 \rangle$) and the profile underwent a two-step normalization process. First, any profile that had $m_1 > 0$ had its 64 samples reversed in order. Second, each profile was individually affinely scaled according to ($I \to \left(m_1^2 + 2s.m_2^2\right)^{-\frac{1}{2}} (I - m_0)$) so that it had a canonical jet.

For comparison with natural image profiles we also constructed two sets of 3.3×10^6 synthetic profiles. One set – the Gaussian set - has each of its 64 values drawn independently from a normal distribution. The other set – Brownian profiles [31] – were generated by, for each profile, setting the value at one end with a normally distributed random variable ($\mu=0$, $\sigma=100$) then setting in turn each of the remaining 63 samples to be equal to the previous sample plus an independently generated normally distributed offset ($\mu=0$, $\sigma=1$). Both the Gaussian and Brownian sets are scaled in the same manner as the natural profiles to bring them into a canonical metamery class.

3.1.2 ML Computation

In our previous study we compute ML profiles using the method of mode estimation 'pessimistic scale space tracking' that we presented at Scale Space 2003 [32]. The method used in this study was also a scale space tracking method but with some details that make for faster more and more accurate computation. Space limitations prevent us from detailing these changes here, though in section 3.2.1 we present results that validate the algorithm on our noise profiles.

An important difference from our previous study is that the normalization step mapped profiles into a 1-D family of canonical metamery classes rather than a single metamery class. Ideally we would ascertain the ML profile for each metamery class in this 1-D family, but in practice we need a large number of profiles to determine the ML with sufficient accuracy so instead we quantize the phase parameter that indexes the 1-D family into 33 equally-sized bins and perform a separate mode estimation for each phase bin. Thus each mode estimation is based on roughly 10^5 profiles. All computations were repeated three times with fresh sets of profiles.

3.2 Results

As with our previous study we have also performed control computations to assess the stability of our results with respect to scale, image dataset, log transformation. Space prevents presentation of these results which, in summary, confirmed stability.

3.2.1 ML Noise Profiles

As shown in figure 4, the ML profiles for the gaussian noise data are an excellent match to the variance minimizing profiles, and for the Brownian data to the roughness minimizing profiles. This agrees with prediction [26] and so validates the performance of our mode estimation algorithm for this quantity and dimensionality of data.

3.2.2 ML Natural Image Profiles

The computed ML profiles for natural images are shown in figure 5. The figure also shows that we used a sufficient number of profiles to achieve low scatter across our three repeat computations.

Figure 6 shows a simple model that we have found that well captures the structure of the ML natural image profile data. The model is based on two components profiles that are shifted and added in different proportions. One component is a symmetric

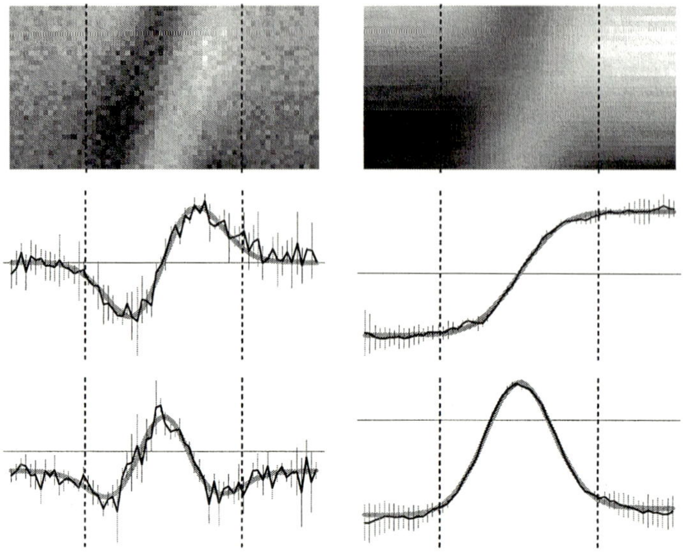

Fig. 4. Shows the ML profiles for the noise data. The left column is for gaussian noise, the right for Brownian noise. The upper panels are laid out the same as in fig. 2. The bottom two rows show the ML profiles for particular phases: the middle row is sine-phase corresponding to the central row of the top panel, the bottom row is cosine-phase corresponding to the bottom row of the top panel. The ML profiles are in black with error bars showing one sd of the scatter across the three repeat mode estimations. The thick grey curves are the L^2 and D^2 minimizers of the appropriate phase

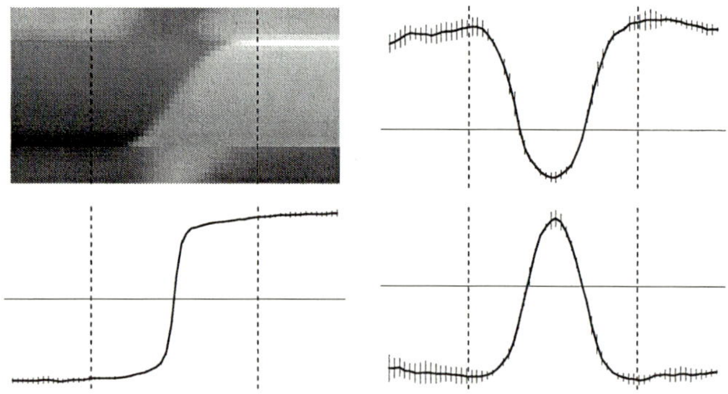

Fig. 5. The ML profiles for natural images are shown as a density plot in top-right, using the same convention as figures 2 and 4. The other panels show the sine- and cosine-phase profiles corresponding to the top, middle and bottom rows of the density plot. The error bars show 1 sd of scatter across the three repeat computations

profile having the form of a bar (like the cosine-phase forms for the range- and total variation minimizers in fig. 2) that is blurred. The other component is an anti-symmetric blurred step edge. Note the bar component is considerable more blurred than the step edge component. For near cosine phases, a shifted version of the blurred bar fits the data. For near sine phase, a shifted version of the blurred

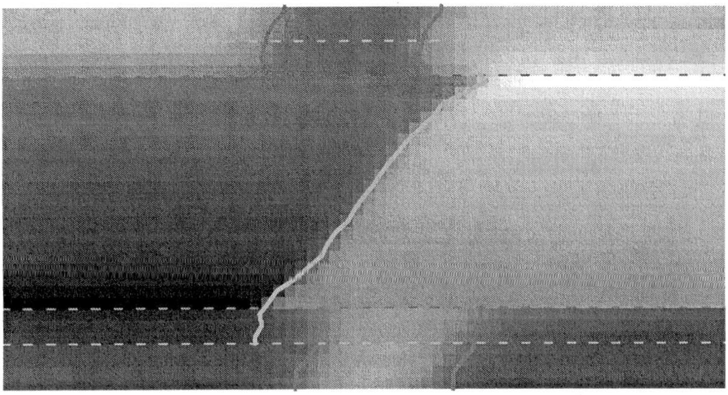

Fig. 6. Illustrates a simple model that well describes the ML natural image profile data. The background density plot is the same as the top-left panel of fig. 5. The solid red lines show the twin edges of the symmetrical blurred bar/pass part of the model, the solid green lines show the edge of the anti-symmetric blurred step edge part of the model. The dotted lines show the phase extent of the different parts of the model. Both parts of the model overlap in two bands of phases, in these bands the ML profiles are well described as a weighted mixture of the symmetric and anti-symmetric components. Outside of the bands only one component is needed

4 Discussion

Our previous result of finding that a step edge was the ML profile for the 1-D, 1st order jet was compatible with the ML profile for natural images being identical to the range- or the TV- minimizing profiles. Neither of these possibilities are borne out for the 2nd order results shown in fig.5. Initially these results seemed negative with respect to our GTT proposal for features. However on closer examination we found a simple model that describes the results. This model is compatible with our GTT proposal, which, to reiterate, is that the equivalence relation of qualitative equality between ML metameres induces feature classes on the jet space. With this proposal in mind, one can see that figure 6 shows that we have induced a categorization of the canonical 2nd order jet space into three classes separated by two fuzzy intermediate bands. Results for 2-D patches are needed to further test the GTT proposal.

References

1. Marr, D., *Vision.* 1982, New York: W H Freeman & co.
2. Barlow, H.B., *Summation and inhibition in the frog's retina.* Journal of Physiology (London), 1953. **119**: p. 69-88.

3. Koenderink, J.J. and A.J. van Doorn, *Receptive Field Assembly Specificity.* Journal of Visual Communication and Image Representation, 1992. **3**(1): p. 1-12.
4. Koenderink, J.J., *What is a feature?* J of Intelligent Systems, 1993. **3**(1): p. 49-82.
5. Koenderink, J.J. and A.J. van Doorn, *Metamerism in complete sets of image operators*, in *Adv. Image Understan. '96*. 1996. p. 113-129.
6. Koenderink, J.J. and A.J. van Doorn, *Local Image Operators and Iconic Structure*, in *Algebraic Frames for the Perception-Action Cycle*, G. Sommer and J.J. Koenderink, Editors. 1997, Springer. p. 66-93.
7. Jones, J.P. and L.A. Palmer, *The Two-Dimensional Spatial Structure of Simple Receptive Fields in Cat Striate Cortex.* J Neurophysiology, 1987. **58**(6): p. 1187-1211.
8. Daugman, J.G., *Uncertainty Relation for Resolution in Space, Spatial- Frequency, and Orientation Optimized by Two-Dimensional Visual Cortical Filters.* Journal of the Optical Society of America a-Optics Image Science and Vision, 1985. **2**(7): p. 1160-1169.
9. Georgeson, M.A. and T.C.A. Freeman, *Perceived location of bars and edges in one-dimensional images: Computational models and human vision.* Vision Research, 1997. **37**(1): p. 127-142.
10. Young, R.A., *The Gaussian derivative model for spatial vision: I. Retinal mechanisms.* Spatial Vision, 1987. **2**: p. 273-293.
11. Young, R.A., R.M. Lesperance, and W.W. Meyer, *The Gaussian Derivative model for spatial-temporal vision: I. Cortical model.* Spatial Vision, 2001. **14**(3-4): p. 261-319.
12. Young, R.A. and R.M. Lesperance, *The Gaussian Derivative model for spatial-temporal vision: II. Cortical data.* Spatial Vision, 2001. **14**(3-4): p. 321-389.
13. Koenderink, J.J. and A.J. van Doorn, *Representation of Local Geometry in the Visual-System.* Biological Cybernetics, 1987. **55**(6): p. 367-375.
14. Koenderink, J.J., *Operational Significance of Receptive-Field Assemblies.* Biological Cybernetics, 1988. **58**(3): p. 163-171.
15. Koenderink, J.J. and A.J. van Doorn, *Receptive-Field Families.* Biological Cybernetics, 1990. **63**(4): p. 291-297.
16. Koenderink, J.J. and A.J. van Doorn, *Generic Neighborhood Operators.* Ieee Transactions on Pattern Analysis and Machine Intelligence, 1992. **14**(6): p. 597-605.
17. Florack, L.M.J., et al., *Families of Tuned Scale-Space Kernels*, in *Computer Vision - Eccv 92*. 1992. p. 19-23.
18. Debnath, L., *Integral Transforms and their Applications*. 1995: CRC Press.
19. Martens, J.B., *Local orientation analysis in images by means of the Hermite transform.* Ieee Transactions on Image Processing, 1997. **6**(8): p. 1103-1116.
20. Makram-Ebeid, S. and B. Mory, *Scale-space image analysis based on hermite polynomials theory*, in *Proc. Conf. on Scale Space Methods in Computer Vision*, L.D. Griffin and M. Lillholm, Editors. 2003, Springer. p. 57-71.
21. Koenderink, J.J., *The Structure of Images.* Biological Cybernetics, 1984. **50**(5): p. 363-370.
22. Leung, T. and J. Malik, *Representing and recognizing the visual appearance of materials using three-dimensional textons.* International Journal of Computer Vision, 2001. **43**(1): p. 29-44.
23. Liu, X.W. and D.L. Wang, *A spectral histogram model for texton modeling and texture discrimination.* Vision Research, 2002. **42**(23): p. 2617-2634.
24. Zhu, S.C., et al., *What are textons?*, in *Computer Vision - Eccv 2002, Pt Iv*. 2002. p. 793-807.
25. Julesz, B., *A brief outline of the texton theory of human vision.* Trends in Neuroscience, 1984. **7**(2): p. 41-45.

26. Griffin, L.D., M. Lillholm, and M. Nielsen, *Natural image profiles are most likely to be step edges.* Vision Research, 2004. **44**(4): p. 407-421.
27. Tagliati, E. and L.D. Griffin, *Features in Scale Space: Progress on the 2D 2nd Order Jet*, in *LNCS*, M. Kerckhove, Editor. 2001, Springer. p. 51-62.
28. Griffin, L.D., *Local image structure, metamerism, norms, and natural image statistics.* Perception, 2002. **31**(3): p. 377-377.
29. Schrödinger, E., *Theorie der pigmente von grosster leuchtkraft.* Annalen den Physik, 1920. **62**: p. 603-622.
30. van Hateren, J.H. and D.L. Ruderman, *Independent component analysis of natural image sequences yields spatio-temporal filters similar to simple cells in primary visual cortex.* Proceedings of the Royal Society of London Series B-Biological Sciences, 1998. **265**(1412): p. 2315-2320.
31. Mandelbrot, B. and J. van Ness, *Fractional brownian motions, fractional noises and applications.* SIAM Review, 1968. **10**(4): p. 422-437.
32. Griffin, L.D. and M. Lillholm. *Mode Estimation by Pessimistic Scale Space Tracking*. in *Scale Space '03*. 2003. Isle of Skye, UK: Springer.

A New Technique for Local Symmetry Estimation

Matthew Mellor* and Michael Brady

Medical Vision Laboratory, University of Oxford, UK
{matt, jmb}@robots.ox.ac.uk

Abstract. This paper introduces a new approach to symmetry estimation directly from image grey values. The method was inspired by the local phase based method, proposed by Kovesi. This method is examined in the light of a strict definition of local symmetry, and found to be wanting in two respects: that it is invariant to some apparently non-trivial symmetries and that its scale is ill defined. To avoid these difficulties we propose a non-linear analog of the local phase. This leads to a family of local symmetry measures, able to provide a rich, local characterisation of shape, invariant to rotations and affine intensity transformations, and robust to affine coordinate transformations.

1 Introduction

Local symmetry analysis is a promising tool for shape analysis and image understanding. However, one of the most significant drawbacks to the approach has been that local symmetry lacks a precise definition and that, once defined, local symmetry can be difficult to estimate in practice. The original application of symmetry was shape analysis of pre-segmented image regions [1, 2, 3]. These methods are based on finding points at equal distances from shape boundaries [1], the so-called medial definition of local symmetry. This paper is concerned with the estimation of symmetry directly from grey values, rather than pre-segmented shapes. This might enable, for example, symmetry based shape representations to be used as priors in the detection and segmentation of shapes. It may also, as is suggested later, enable the symmetry approach to be applied directly to image understanding tasks, bypassing the need for segmentation altogether. Various strategies have been proposed for the estimation of local symmetry directly from an image. The most commonly used of these are 'cores', in which linear filter responses suggest feature amplitudes and are used to compose medial pairs of responses which suggest symmetry at a certain point [4]. The energy of all possible medial pairs are summed to produce a map of 'medialness', the local maxima of which give the local symmetry axes.

A somewhat different method was proposed by Kovesi [5]. In this framework, symmetry is estimated at a range of scales by convolving the image with a set of

* Matthew Mellor is funded by the EPRSC and MRC as part of the MIAS-IRC.

R. Kimmel, N. Sochen, J. Weickert (Eds.): Scale-Space 2005, LNCS 3459, pp. 38–49, 2005.

symmetric and anti-symmetric filters. Symmetric points are those at which the response of the symmetric filter dominates the response of the antisymmetric filter. In this paper, the phase technique for symmetry analysis is investigated in terms of a strict definition of the symmetry of grey-values. It is argued that the phase method has some significant practical drawbacks. Instead, we propose a non-linear method. This method maintains some of the properties of local phase, in particular the property of quadrature, whilst also supporting a well defined scale-space. Furthermore, the method leads to a family of symmetry measures. We demonstrate that two of these are adequate to find the complete set of smoothed local symmetries proposed by Brady and Asada [2].

2 Local Phase

Local phase can be estimated at each point in the image from the responses of (at least) one odd and one even filter. In two or more dimensions, the definition of local phase is rather ambiguous, but in one dimension the odd and even filter pairs may be defined through the Hilbert transform. The Hilbert transform of an even (entirely symmetric) filter can be found by taking its Fourier transform and multiplying all negative frequencies by minus one. The result is an imaginary and purely antisymmetric filter. Since multiplication in the Fourier domain is equivalent to convolution in the space domain, it is also possible to write the odd filter as the convolution of the even filter with an odd kernel.

$$\begin{aligned} f_o(x) &= f_e(x) * h(x) \\ h(x) &= \frac{i}{\pi x} \end{aligned} \tag{1}$$

The convolution of the image with the complex filter formed by summing the even and odd filters, results in a complex signal, known as the analytic signal. The magnitude of this complex signal is known as the local energy, while the angle is the local phase. Thus, the local phase ϕ of a signal s is defined as:

$$\phi = \arctan\left(\frac{s * f_e}{-i * s * f_o}\right) \tag{2}$$

Completely symmetric parts of the signal will have purely real response to the complex filter, and hence the local phase at these points is $-pi/2$ or $\pi/2$ (say); completely anti-symmetric parts of the signal will have purely imaginary responses and hence phase of 0.

When extending the concept of local phase to higher dimensions the Hilbert transform is undefined, so it is necessary to make some design choices. One way of interpreting the difficulty of defining phase in two dimensions is that it reflects the richness of symmetry in higher dimensions when compared to one dimension; simple reflection symmetry is replaced by many different forms of reflection symmetry as well as rotation symmetry. Previous applications of phase to symmetry estimation have used steerable filters [6]; this may be regarded as finding the axis about which maximal reflection symmetry (or anti-symmetry exists). A more recent method is the monogenic signal [7] which replaces the Hilbert transform with the Riesz transform. Space does not permit a detailed discussion of the monogenic signal. However, it is interesting to note that the monogenic phase may be regarded as a measure of point-reflection symmetry, as opposed to line-reflection symmetry. Arguably, it is by adopting this definition of symmetry that the monogenic signal overcomes the difficulties encountered by previous attempts to generalise the Hilbert transform to multiple dimensions.

3 Phase as a Symmetry Measure

Rather than exploring the use of the monogenic signal to estimate symmetry directly, we first consider the implications of using one dimensional (Hilbert) phase as a symmetry measure.

Unlike in higher dimensions, symmetry is uniquely defined in one dimension. Specifically, a signal s is symmetric about the point a if:

$$s(x-a) = s(-x-a) \tag{3}$$

In fact, any signal may be decomposed into the sum of a symmetric and an anti-symmetric component about any point a:

$$s(x-a) = s_e(x-a) + s_o(x-a) \tag{4}$$

where,

$$\begin{aligned} s_e(x-a) &= \frac{s(x-a)+s(-x-a)}{2} \\ s_o(x-a) &= \frac{s(x-a)-s(-x-a)}{2} \end{aligned} \tag{5}$$

In order to analyse Hilbert phase in terms of this definition of symmetry, we propose the notion of a generalised phase, defined as:

$$\phi(a) = \arctan\left(\frac{M(\{s_e(x-a)|x \in \mathcal{N}_a\})}{M(\{s_o(x-a)|x \in \mathcal{N}_a\})}\right) \tag{6}$$

where M is some measurement of the 'size' of a signal and $\mathcal{N}_a$ is some region of interest around the point a. In the case of the Hilbert transform the measurement function M_h seems somewhat unsatisfactory. By observation:

$$M_h(\{s(x-a)|x \in \mathcal{N}_a\}) = \int_{-\infty}^{\infty} s(x-a)\left(f_e(x) + f_e(x) * \frac{h(x)}{i}\right) dx. \tag{7}$$

There are two possible objections that might be made to this measurement function. Firstly, the measurement function treats the odd and even components rather differently: when measuring the even component, the right hand function

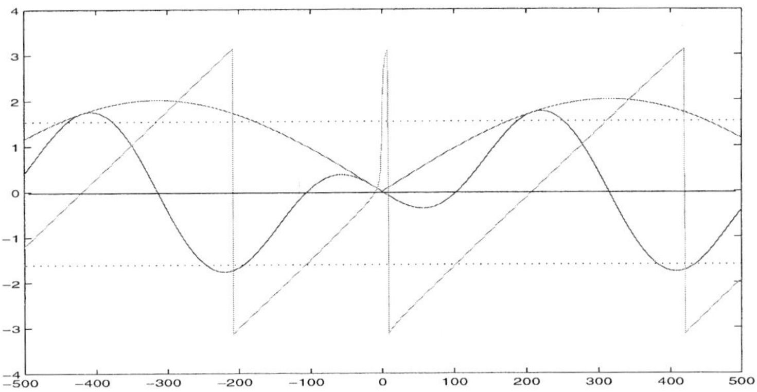

Fig. 1. Estimating the symmetry of $s(x) = \sin(x) - \sin(2x)$. The plot shows $s(x)$ in blue, its local phase in green and its local energy in red. Note that at $x = 0$ the energy is zero and hence the phase is undefined

in the bracketed term of equation 7 plays no part, while the opposite is true of the odd component. However, because of the convolution operation, the right hand function, $f_e(x) * h(x)$, will be wider, in some sense, than the left hand function $f_e(x)$. In the case of the monogenic signal, the measurement functions are slightly more involved, but the overall structure is the same: the even component is measured with one function, the odd with another, larger function. The outcome of this is that Hilbert phase cannot rightly be said to measure symmetry 'at a scale' since the scale at which the odd and even components are measured is subtly different. In practice, this manifests itself as a predilection for 'oddness' when the filter is not aligned with an obvious image structure (i.e. when the filter responses are determined by the shape of their tails). The second problem is one of more practical importance. Consider the problem of estimating the local phase of the function $s = \sin(x) - \sin(2x)$ at the point $x = 0$. Figure 1 shows $s(x)$ (blue) and its Hilbert energy (red). At $x = 0$, there is no energy and hence the phase is undefined (since it is the angle of a phasor with no length). In fact, the local phase, shown in green in fig. 1, behaves in an unexpected way near the origin where, its gradient suddenly increases; the phase rises from 0 to π extremely rapidly, suggesting a point of perfect anti-symmetry surrounded, very closely, by two points of perfect symmetry. This description of the signal seems to make little sense.

Of course, this particular problem could be solved by applying a range of different even filters, such that at least one of the filters had significant energy at each point in the signal. However, it is not obvious how to choose a family of even filters that guarantee this for real signals. What seems likely is that rather a lot of filters would be needed, and that the method would become quite unwieldy, particularly in higher dimensions. Local phase, in the conventional sense, therefore seems a poor choice of local symmetry measure.

4 A Non-linear Symmetry Measure

In order to avoid the pitfalls of Hilbert phase when measuring symmetry, it is necessary to redesign the the measurement functions $M(s)$ in equation 6. The problem of 'symmetry blindess' illustrated in fig. 1 arises because the Hilbert measurement functions are linear sums so there is always a chance that a non-zero function will measure zero. This can be avoided simply by adopting the following measurement function:

$$M(s(x-a)) = \int_{-\infty}^{\infty} (g(x-a)s(x-a))^2 dx \tag{8}$$

where $g(x)$ is a Gaussian window function. The resulting phase is given by inserting the definition of M given in equation 8 into the definition of generalised phase. The new measurement function also corrects the problem of incompatible spatial extent of the measurement function, since it operates identically on symmetric and anti-symmetric components.

There is one disadvantage to the proposed method, which is that the local mean and low frequency component contributes to only the even part of the symmetry. To avoid this problem the signals are pre-filtered to remove low frequencies before analysis; since 'low' must be taken relative to the window size, we propose that the filter consist of subtraction of the local mean estimated with the window function, i.e. the pre-filtered signal $s_f(x)$ is given by $s_f(x) = s(x) * (\delta(x) - g'(x))$.

The new measurement functions can also be used to define a non-linear analog of the analytic signal:

$$s_A(a) = \frac{1}{4}\int_{-\infty}^{\infty} g'(x)[(s(x-a)+s(-x-a))^2 + i(s(x-a)-s(-x-a))^2]dx \tag{9}$$

where $g'(x) = g(x)^2$ and a is, as before, the point about which the symmetry is measured. The phase and energy of $\sin(x) - \sin(2x)$ are shown in figure 2. The Gaussian window, $g'(x)$, used in this example has width $\sigma = 24$. The energy is non-zero throughout, the phase varies approximately periodically and the undesirable behavior near the origin is absent. The range of values of phase is now limited to the range $0 - \pi/2$, since ridges and troughs appear identically, as do edges facing in opposite directions.

So far, the discussion has been limited to one dimension. However, the method described above can be extended to any dimension, provided that point reflection is adopted as the definition of local symmetry. A two dimensional example is shown in fig. 3. Locally symmetric and anti-symmetric points appear white and black respectively. Obvious features, such as the edges of walls or the drainpipes are labelled as would be expected. More interestingly, the symmetries of some large, approximately symmetric shapes are picked up. For example the roof, the shadow of the eaves and the point on the chimney half way between the roof and the chimney top are all local symmetry maxima.

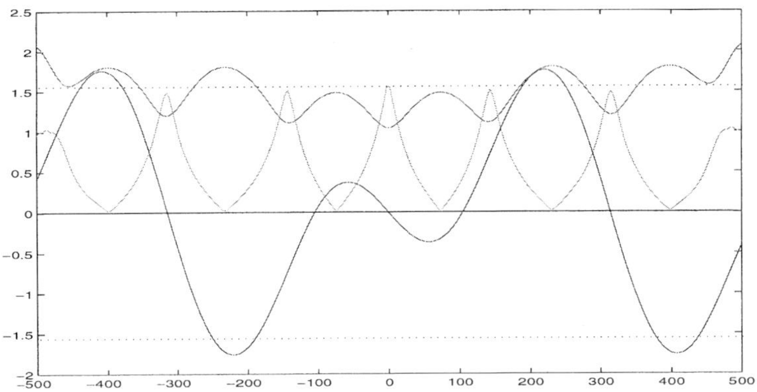

Fig. 2. Estimating the symmetry of $s(x) = \sin(x) - \sin(2x)$ using the non-linear method. $s(x)$ in blue, its local phase in green and its local energy in red. The energy no longer reaches zero at any point, and the phase varies more or less periodically, without the sudden activity near the origin

Fig. 3. Point reflection symmetry of the house image (left) estimated with window width $\sigma = 6$. White indicates symmetry, black indicates antisymmetry

4.1 Other Symmetries

The insight of the monogenic signal, applied to our new method above, is that by replacing line reflection with point reflection, simple and efficient methods for symmetry analysis may be devised for multi-dimensional signals. In the particular case of two dimensional images, however, the rotational symmetries are also uniquely defined; in fact, point reflection symmetry is equivalent to 180° degree rotation symmetry. The definition of symmetry may be extended to arbitrary rotational symmetry:

$$
\begin{aligned}
s_{e\alpha}(\bar{x}) &= \frac{s(\bar{x}) + s(R_\alpha(\bar{x}))}{2} \\
s_{o\alpha}(\bar{x}) &= \frac{s(\bar{x}) - s(R_\alpha(\bar{x}))}{2}
\end{aligned} \tag{10}
$$

where $\bar{x}$ is the coordinate vector and R_α is a matrix which rotates the coordinate vector through an arbitrary angle α. The definition of the generalised phase, eqn. 6, is easily altered to detect these symmetries: The new definitions of symmetric and anti-symmetric component may be inserted into the generalised phase, and the non-linear measurement functions applied as before. This observation opens up an infinite range of possible symmetries; for the remainder of this paper, we consider only point reflection and 90° rotation symmetry, which has some interesting properties. While the definition of 90° rotation symmetry is very similar to the point reflection case, the observed symmetries are quite different. Figure 4 shows a test image composed of simple shapes and the symmetries estimated from it. The point reflection symmetry responds well to parallel lines, but responds weakly to diverging lines and not at all to corners. According to the

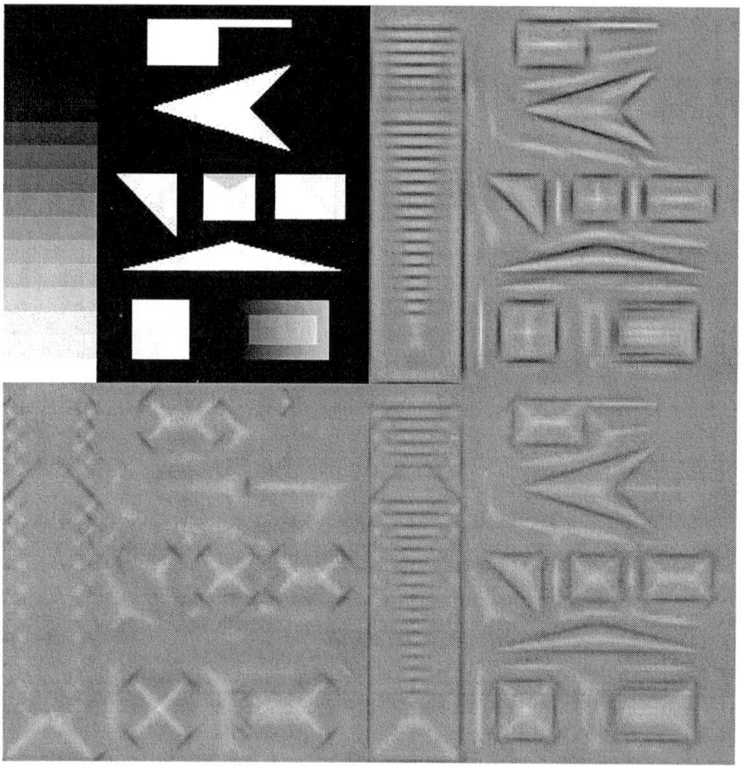

Fig. 4. Detected symmetries of some simple shapes. Top left: test image. Top right: point reflection symmetries at $\sigma = 8$. Bottom left: 90° rotation symmetry, $\sigma = 8$. Bottom right: sum of two symmetries

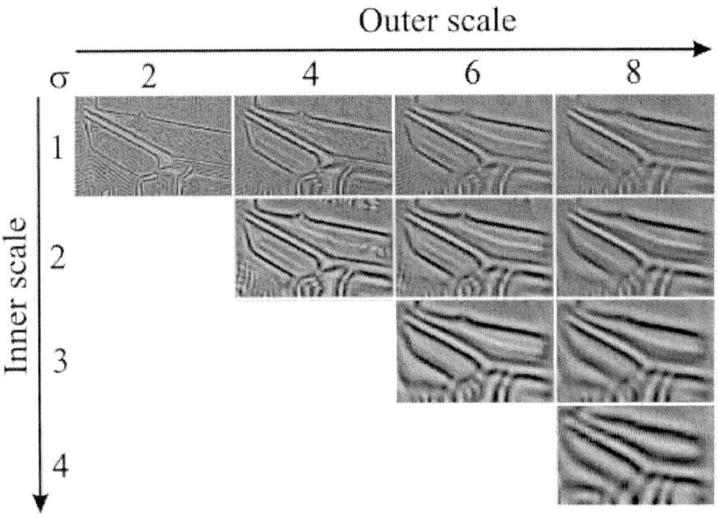

Fig. 5. Point reflection symmetry measured at a variety of a range of inner and outer scales in part of the house image

medial definition of local symmetry, corners should be intersected by a symmetry locus, so the point reflection symmetry as it stands is insufficient for detecting the full symmetry locus. The 90° rotation symmetry, however, appears to have the opposite behaviour: it responds to strongly to right angles, but not at all to parallel edges. Intuitively, it seems sensible to add these two quantities, to give the quantity shown in the bottom right of fig. 4. Between the two symmetry types, many of the symmetries of most of the objects are detected.

For shapes such as a simple rectangle, it appears that the combined symmetry detects more than just the medial locus: it actually responds to the full symmetry set, described by Brady and Asada [2]. The full symmetry set guarantees that when new image structures are introduced to an image the symmetry set is merely added to, not changed to something unrecognisable. This is an important property when superposition or occlusion of objects can occur, which is likely in real images.

4.2 Scale

The Gaussian window applied by the proposed symmetry measure controls the localization of the observed symmetry. However, the symmetry is still computed from the (weighted) pixel values, i.e. it is a measure of symmetry at the inner scale of the image, local to a region defined by the window (outer scale). One could also think of reducing the inner scale of the image by convolving with a Gaussian prior to measuring symmetry. This step, combined with the low frequency suppression, amounts to the application of a difference of Gaussians filter. Note, however, that the difference in the scales of the two Gaussians may be very large, in contrast to many of the reported linear applications. In practice, both the inner and outer scale are important, particularly when analyzing cluttered images. Figure 5 shows symmetries measured from part of the house image. Of particular interest is the symmetry of the roof. At the smallest scale (top left) the symmetry of the roof is not visible. As the outer scale increases (top right), the symmetry becomes visible. However, the symmetry axis contains a break, caused by the small extrusion in the skyline. The potentially drastic changes caused by very small changes in shape are one of the major bugbears of symmetry approaches to shape analysis [8]. In the present method, this problem can be avoided by increasing the inner scale, until the small extrusion becomes insignificant. The symmetry axis of the roof is then complete (bottom right).

5 Examples

So far the local symmetry properties have been discussed in terms of conventional symmetry analysis of (typically) extracted shapes. However, it retains many similarities with local quadrature analysis, a technique predominantly applied to feature detection and classification. In this section, we consider both applications of the method described in the previous sections. Indeed, it is hoped that since the new method combines aspects of both approaches, it may be possible to design symmetry inspired methods for detecting, labelling, grouping and interpreting features. To this end, we have applied the local symmetry estimation method to simultaneous feature detection and symmetry axis detection.

5.1 Symmetry Axes

The combined (point reflection plus 90° rotation) was used to find symmetry axes. The algorithm proceeds as follows. First, the two symmetry maps are calculated at a chosen scale. The axes are then extracted from the symmetry maps by thresholding and morphological thinning (using the Matlab function bwmorph). Although ridge detection would be more accurate in some cases, the symmetries of many objects lack obvious ridges, in which case the thresholding and thinning procedure is far more robust.

The left image in fig. 6 shows local symmetry axes found in an image of some white kidney beans, with outer scale $\sigma = 6$ and inner scale equal to the inner scale of the image. Symmetry was thresholded at $(\pi/4)$. This is a particularly challenging image for several reasons: the beans are short with respect to their width and are curved; few of the beans have complete boundaries; the beans are not truly symmetric, because of shading effects; the image contains structured noise(JPEG artifacts). The symmetry axis of every completely visible bean is at least partially extracted and in most cases completely extracted. The symmetry axes appear plausible: they are near the center of the bean and, in many cases, have curvature that is comparable to the curvature of the corresponding bean.

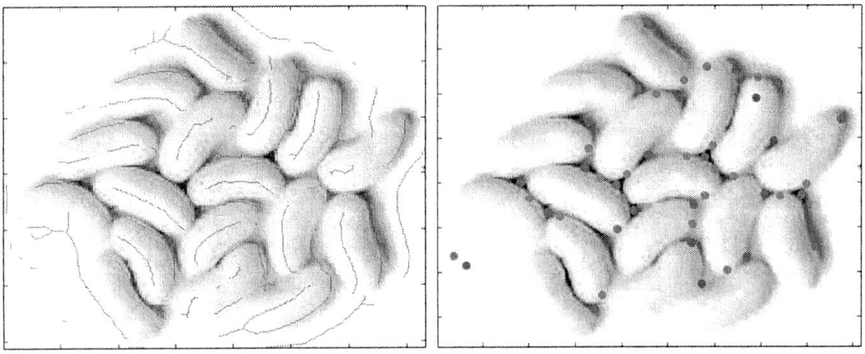

Fig. 6. Left:Approximate symmetry axes extracted from a noisy image. Right: Interested points detected by extracting minima (blue) and maxima (red) of 90° rotation symmetry

5.2 Feature Extraction

One of the interesting properties of the 90° rotation symmetry is that, by its nature, it does not respond strongly to linear structures. This, combined with the fact that it responds to an interesting range of structures, such as blobs, saddles and corners, makes it a good candidate for an interest point detector. Such a detector can be made by applying the symmetry extraction process described above to the 90° rotation symmetry alone. This detects points with strong symmetry, such as blobs or crosses. A second set of points can be found by extracting local minima of 90° rotation symmetry. These correspond to points of anti-symmetry, such as corners and saddles. The right image in fig. 6 shows the result of applying this algorithm, with appropriate thresholds (minima less than $-\pi/4$, maxima greater than $\pi/4$). The red dots represent points of high symmetry, while the blue dots represent points of high anti-symmetry. As expected, most of the blue dots appear where beans touch, which are approximate saddle points, while the read points appear in the gaps between beans, which are approximately blob like. Interestingly, many of the detected points lie on or near symmetry axes; this is unsurprising, since blobs and saddles both have high point-reflection symmetry. In fact, even the features which don't lie on detected symmetry axes do lie on local symmetry maxima that were suppressed by the threshold. Similarly, one would expect that other feature types, such as corners would appear at points of point reflection anti-symmetry.

Interest points extracted using 90° degree rotation symmetry also seem to be fairly robust to affine transformations. The features are clearly not affine invariant: the definition of point reflection symmetry is affine invariant, though the suggested scale space is clearly not, while 90° symmetry is affine dependent by definition. However, the effect of the transform is limited (since affine transformations cannot turn, say, corners into blobs), and the main effect of affine

Fig. 7. Interest points extracted from the house image, before and after affine transformation. Left: original image with extracted points. Right: transformed image with extracted points (circles) and transformed points from original image(stars)

transformations will be to alter the strength of the observed symmetry rather than its type. Figure 7 illustrates the robustness of the interest points to affine transformation. The left image shows the house image with extracted interest points, the right shows the affine transformed image, new interest points (circles) and the affine transformed points from the original image. The majority of the features are detected in both images. Interestingly, the method picks out a lot of features in the brickwork; this is to be expected since the method is contrast invariant, and the corners of bricks therefore appear as significant as the corners of houses. These features are more vulnerable to noise corruption, however, this is not a serious problem, since the susceptibility to noise can be judged from the local energy. Figure 8 shows the performance of the features on a real world data set. The two images are of the same structure, taken from significantly different viewpoints. The solid circles are interest points which were

Fig. 8. Interest points extracted from two images taken from different view points. Solid circles: points judged to represent the same features, judged by a human observer. Hollow circles: unmatched points

judged by a human observer to represent the same features. The hollow circles are unmatched points. The proportion of correctly matched features is 71%;

6 Discussion

We have explored the relationship between symmetry and local phase, by selecting a particular definition of symmetry. The resulting analysis reveals potential pitfalls in the use of phase in particular and linear filter based methods in general. Instead, we have proposed two simple non-linear local symmetry descriptors. Between them these two methods enable the complete local symmetry set (in the Brady and Asada sense) to be extracted for many shapes. By considering both inner and outer scale the method is able to avoid the sensitivity to small changes suffered by some medial methods [8]. Furthermore, the method also provides a new approach for interest point detection. The extracted feature points have the property that they lie on either the symmetry locus (blobs and saddles) or on the anti-symmetry locus (corners). The two approaches combined therefore produce both a rich set of simple interest points and an implied connectivity. Since this connectivity is determined by the local symmetry, which is a powerful description of local shape, it is hoped that this approach may provide a powerful and economical basis for image processing.

References

1. H. Blum. A transformation for extracting new descriptors of shape. *in Models for the Perception of Speech and Visual Form, W. Wathen-Dunn (ed), MIT press*, pages 363–380, 1967.
2. M. Brady; H. Asada. Smoothed local symmetries and their implementation. *International Journal of Robotics Research*, 3(3):36–61, 1984.
3. R. L; O. Kübler Ogniewicz. Hierearchic voronoi skeletons. *Pattern Recognition*, 28(3):343–359, 1993.
4. J.D. Fritsch; S.M. Pizer; B. Morse; D.H. Eberley; A. Liu. The multiscale medial axis and its applications in image registration. *Patt. Rec. Letters*, 15:445–452, 1994.
5. P. Kovesi. Symmetry and asymetry from local phase. *Tenth Australian Joint Conference on Artificial Intelligence*, 1997.
6. W. T. Freeman and E. H. Adelson. The design and use of steerable filters. *IEEE PAMI*, 13(9):891–906, 1991.
7. M. Felsberg; G. Sommer. The monogenic signal. *IEEE Trans. Sig. Proc.*, 49, 2001.
8. R.A. Katz; S.M. Pizer. Untangling the blum medial axis transform. *IJCV*, 55(2/3), 2003.

Geometry of Isophote Curves

André Diatta and Peter Giblin

University of Liverpool, Liverpool L69 3BX, England
{adiatta, pjgiblin}@liv.ac.uk

Abstract. We consider the intensity surface of a 2D image, we study the evolution of the symmetry sets (and medial axes) of 1-parameter families of iso-intensity curves. This extends the investigation done on 1-parameter families of smooth plane curves (Bruce and Giblin, Giblin and Kimia, etc.) to the general case when the family of curves includes a singular member, as will happen if the curves are obtained by taking plane sections of a smooth surface, at the moment when the plane becomes tangent to the surface.

Keywords and Phrases: Isophote curve, symmetry set, medial axis, skeleton, vertex, inflexion, shape analysis.

1 Introduction

Image data is often thought of as a collection of pixel values $I : Z^2 \mapsto Z_+$. The physical information is better captured by embedding the pixel values in the real plane, as the pixelation and quantization are artifacts of the camera, hence $I : \mathbf{R}^2 \mapsto \mathbf{R}_+$. The geometrical information of an image is even better captured looking at the level sets $I(x) = I_0$, for all $I_0 \in \mathbf{R}_+$, that is, looking at the isophote curves of the image.

Shape analysis using point-based representations or medial representations (such as skeletons) has been widely applied on an object level demanding object segmentation from the image data. We propose to combine the object representation using a skeleton or symmetry set representation and the appearance modelling by representing image information as a collection of medial representations for the level-sets of an image. As the level I_0 changes, the curves change like sections of a smooth surface by parallel planes.

The qualitive changes in the medial representation of families of isophotes fall into two types: (1) those for which the isophotes remain nonsingular (see for example [3, 8]) and (2) those for which one isophote at least is singular. The symmetry set (SS) of a plane curve is the closure of the set of centres of circles which are tangent to the curve at two or more different places. The medial axis (MA) is the subset of the SS consisting of the closure of the locus of centres of circles which are maximal, (maximal means that the minimum distance from the centre to the curve equals the radius). Our aim is to extend the investigation to the case (2) when the family includes singular curves, as is the case when one of the plane sections is tangent to the surface so that this section is a singular

R. Kimmel, N. Sochen, J. Weickert (Eds.): Scale-Space 2005, LNCS 3459, pp. 50–61, 2005.

curve. The final goal is to represent image smooth surfaces by the collection of all medial reprentations of isophotes, forming a singular surface in scale space.

In this article, which is theoretical in nature, we work with the full SS, and consider the transitions which occur in the SS of a family of plane sections of a generic smooth surface in 3-space, as the plane moves through a position where it is tangent to the surface. We investigate the local geometry of these families of curves and track the evolution of some crucial features of the SS and MA. In particular, we will trace and classify the patterns of some special points, on the sections of a surface as the section passes through a tangential point, such as *vertices* (maxima and minima of curvature), *inflexions*, *triples of points where a circle is tritangent* and the pattern of the centre of such a circle, *paires of points where a circle is bitangent with a higher order contact at one of them*, etc. The vertices are crucial to the understanding of the SS since it has branches which end at the centres of curvature at vertices. From the way in which vertices behave we can deduce a good deal about the evolution of the SS and its local number of branches. The *inflexions* correspond to where the evolute of the curve, recedes to infinity. We also classify all possible scenarii of how vertices and inflexions are distributed along the level curves.

Last, we produce examples of SS and MA illustrating the cases.

We are concerned with the local behaviour of symmetry sets (SS) and medial axis (MA) of plane sections of generic[1] smooth surfaces so we may assume that our surface M is given by an equation $z = f(x, y)$ for a smooth function f, which will often be assumed to be a polynomial of sufficiently high degree. We shall take M in Monge form, that is f, f_x and f_y all vanish at (0,0).

2 Intrinsic Geometry of Generic Isophote Curves

This section describes the geometry of isophote curves evolving on a fixed smooth surface M, under a 1-parameter family of parallel plane sections. Namely, we shall examine closely the different configurations of vertices and inflexions on the sections on our surface. We will in particular concentrate on the evolution through a plane section which is tangent to M at a point $\mathbf{p}$, so that this section is singular. For a generic surface, three situations arise, according to the contact between the tangent plane and M at $\mathbf{p}$, as measured by the singularity type of the height function in the normal direction at $\mathbf{p}$. See for example [12] for the geometry of these situations, and [4, 11] for an extensive discussion of the singularity theory.

- The contact at $\mathbf{p}$ is ordinary ('A_1 contact'), in which case the point is (i) elliptic or (ii) hyperbolic. The intersection of M with its tangent plane at $\mathbf{p}$ is locally an isolated point or a pair of transverse arcs.
- The contact is of type A_2, which means that $\mathbf{p}$ is parabolic. The intersection of M with its tangent plane at $\mathbf{p}$ is locally a cusped curve.

[1] The genericity conditions will vary from case to case. See [6].

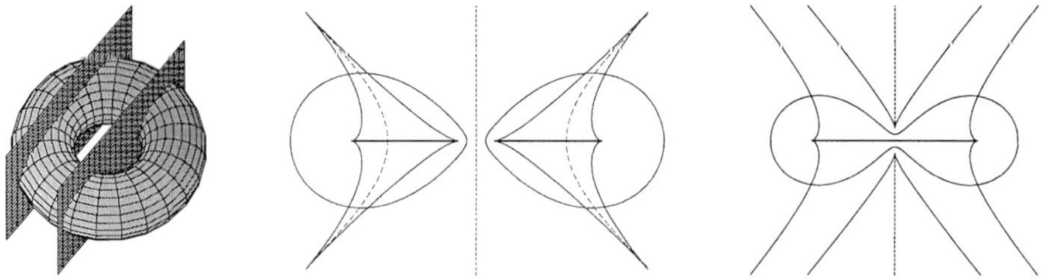

Fig. 1. Two plane sections of a torus close to a singular section, together with their evolutes. The thick lines are the MA and the dashed lines are the additional parts of the SS. As the two ovals merge, two cusps on the evolute recede to infinity, taking the branches of the SS with them. (In the right-hand figure, in fact the SS goes twice to infinity and in between these excursions it covers the whole vertical line; this part, caused by the global structure of the curve, has been omitted for clarity. In this paper we are concerned with the *local* behaviour of SS near the singular section.)

- The contact is of type A_3, which means that $\mathbf{p}$ is a cusp of Gauss, in which case it can be (i) an elliptic cusp, or (ii) a hyperbolic cusp. The intersection with the tangent plane is locally an isolated point or a pair of tangential arcs.

Elliptic and hyperbolic points occupy *regions* of M, separated by parabolic *curves* which are generically nonsingular; on the parabolic curves are isolated *points* which are cusps of Gauss.

The following gives a complete description of the behaviour of vertices and inflexions on isophotes curves near a singular point.

Theorem 1. *Let $f = k$ be a section of a generic surface M by a plane close to the tangent plane at $\mathbf{p}$, $k = 0$ corresponding with the tangent plane itself. Then for every sufficiently small open neighbourhood U of $\mathbf{p}$ in M, there exists $\varepsilon > 0$ such that $f = k$ has exactly $v(\mathbf{p})$ vertices and $i(\mathbf{p})$ inflexions lying in U, for every $0 < |k| \leq \varepsilon$, where $v(\mathbf{p})$ and $i(\mathbf{p})$ satisfy the following equalities.*

(E) *If $\mathbf{p}$ is an elliptic point, then for one sign of k the section is locally empty; in the non-umbilic case, for the sign of k yielding a locally nonempty intersection we have $v(\mathbf{p}) = 4$, $I(\mathbf{p}) = 0$. Likewise if $\mathbf{p}$ is an umbilic point, then $v(\mathbf{p}) = 6$, $I(\mathbf{p}) = 0$.*

(H) *If $\mathbf{p}$ is a hyperbolic point, $v(\mathbf{p})$ satisfies one of the following. We use $\leftrightarrow$ to indicate the transition in either direction, $m + n$ indicating the numbers of vertices on the two branches of $f = k$ for one sign of k before the $\leftrightarrow$ and for the other sign of k after it. In the most generic case (open regions of our surface) we have $2{+}2 \leftrightarrow 2{+}2$ or $1{+}1 \leftrightarrow 3{+}3$. See Figure 2. In other cases, occurring along curves or at isolated points of our surface, we can have in addition $3{+}2 \leftrightarrow 2{+}1$ or $3{+}1 \leftrightarrow 2{+}2$. Also using the same notation, $i(\mathbf{p})$ satisfies: $1 + 1 \leftrightarrow 0 + 2$ or $1 + 2 \leftrightarrow 0 + 1$. There are 8 cases in all, and the full list is given in [6].*

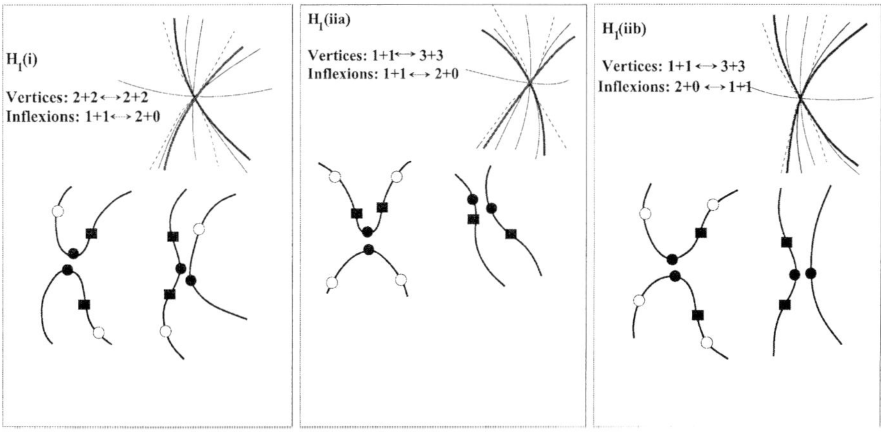

Fig. 2. Arrangements of vertices and inflexions on the level sets of f, in the most generic hyperbolic case (called $\mathbf{H}_1$ in [6]). See Theorem 1. In each case, we show, above, the vertex and inflexion curves—that is, the loci of vertices and inflexions on the level sets of f—and, below, a sketch of the level curves for $f < 0$, $f > 0$, showing the positions of these vertices and inflexions. Thick lines: $f = 0$ or $f = k$; thin solid lines: vertex curves; dashed lines: inflexion curves. Open circles: minima of curvature; solid circles: maxima of curvature; squares: inflexions

(P) *If* $\mathbf{p}$ *is a parabolic point but not a cusp of Gauss,* $v(\mathbf{p}) = 3$, $I(\mathbf{p}) = 2$.

(ECG) *If* $\mathbf{p}$ *is an elliptic cusp of Gauss,* $v(\mathbf{p}) = 4$, $I(\mathbf{p}) = 2$ *for one sign of* k, *and* $v(\mathbf{p}) = I(\mathbf{p}) = 0$ *for the other.*

(HCG) *If* $\mathbf{p}$ *is a hyperbolic cusp of Gauss,* $v(\mathbf{p})$ *satisfies* $1 + 3 \leftrightarrow 4 + 4$ *or* $2 + 2 \leftrightarrow 4 + 4$, *whereas* $I(\mathbf{p})$ *satisfies* $2 + 2 \leftrightarrow 0 + 2$ *or* $1 + 1 \leftrightarrow 0 + 0$.

For the proof and more details see [5], [6].

3 Symmetry Sets (SS) and Medial Axes (MA) of Isophote Curves

The SS of a smooth simple closed curve in $\mathbb{R}^2$ is made of piecewise smooth curves (locus of A_1^2's), triple crossings (A_1^3), cusps (A_1A_2), endpoints (A_3) and the points at 'infinity' (they correspond to bitangent *lines* to the curve). See Fig 3.

- A_1^2: The centres of bitangent circles with ordinary tangency at both points.
- A_1^3: The centres of tritangent circles with ordinary tangency at all points. They are the triple crossings on the symmetry set.
- A_1A_2: They are the centres of bitangent circles which are osculating circles at one point of the curve and have an ordinary tangency at the other point. They lie on the evolute and are cusps on the symmetry set.
- A_3: They are the centres of circles of curvature at extrema of curvature on the curve, the endpoints of the symmetry set and the cusps on the evolute.

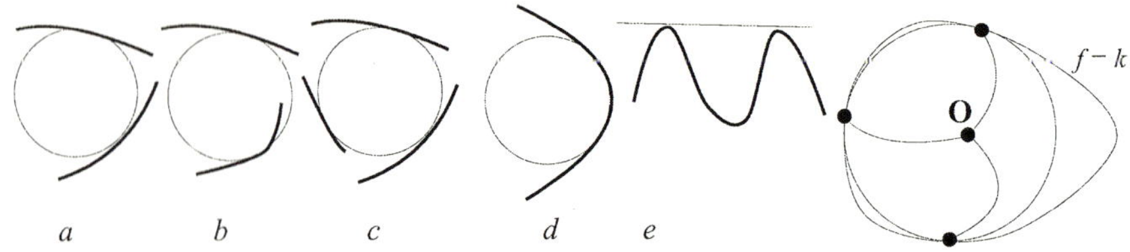

Fig. 3. (a)-(e): Illustration of the circles whose centres contribute to the symmetry set. (a) is an A_1^2, (b) an A_1A_2, (c) an A_1^3, (d) an A_3 and (e) a centre at ∞ (bitangent line). In the last case the circle has become a straight line and the centre is at infinity. Right: a schematic drawing of a tritangent circle and a level set $f = k$ for an umbilic point at the origin **O**. As $k \to 0$ the points of tangency trace out three curves which we call the 'A_1^3 curves'. Calculation of these curves is given in §3.1. Once these curves are known we can calculate the locus of centres of the tritangent circles

- Bitangent lines: the circle now has its centre at infinity so the SS goes to infinity.

At inflexions the evolute goes to infinity and the sign of curvature changes. Thus a *positive maximum* of curvature will be followed by a *negative minimum*, which in terms of the absolute value of curvature is again a maximum.

Our approach to the study of SS of families of curves which include a singular curve is to trace the A_3 points, the inflexions, the A_1A_2 points and the A_1^3 points on the curves as they approach the point at which the singularity develops. In this way we obtain significant information about the SS themselves. The patterns of vertices and inflexions have been studied in detail and for all the relevant cases in [5] and in [6], as recalled in Section 2. Subsection 3.1 and 3.2 are devoted to the study of the locus of A_1^3 and A_1A_2 points, respectively. In Subsection 3.3 we derive information on the changes on the SS of families of isophotes curves.

3.1 A_1^3 Points

The A_1^3 points are the centres of circles which are tangent (ordinary tangency) to $f = k$ (for any choice of f, such as hyperbolic or umbilic) at three distinct points. They occur at triple crossings on the SS. Instead of looking directly for the centres of those tritangent circles, we rather first look for the points where those circles are tangent to the curve $f = k$ (see Fig. 3, right, for a schematic picture of the umbilic case). Thus we expect to have three curves, the 'A_1^3 curves', having the origin as their limit point, along which the three contact points move. First, we want to find the limiting directions of these curves, ie the lines they are tangent to as $k \to 0$. After finding the limiting directions, we can then determine enough of a series expansion (possibly a Puiseux series) to decide how the A_1^3 curves lie with respect to the vertex curves, etc. which we have determined before. We will give an example of such a parametrization below.

The equations which determine the A_1^3 curves are of course highly non-linear. They are in fact 8 equations in 9 unknowns, thereby determining an algebraic variety in $\mathbb{R}^9$ which, when projected onto suitable pairs of coordinates, gives each A_1^3 curve in turn. There are two important features of these equations:

- Naturally they are symmetric in that the contact points can be permuted;
- The equations inevitably admit solutions obtained by making two of the tangency points coincide ('diagonal' solutions). This causes the algebraic variety in $\mathbb{R}^9$ to have components of dimension greater than 1 which we want in some way to discard.

We now set up the equations. Any circle has the form $C(x, y) = 0$ where

$$C = x^2 + y^2 + 2ax + 2by + c,$$

so that the centre is $(-a, -b)$ and the radius is r where $r^2 = a^2 + b^2 - c$. However we prefer the parametrization by (a, b, c) rather than (a, b, r) since it results in equations which are linear in the parameters.

Let this circle be tangent to $f = k$ at the three points $\mathbf{p}_i = (x_i, y_i)$, $i = 1, 2, 3$. There are 8 equations $F_j = 0$, $j = 1, \ldots, 8$ which connect the 9 unknowns x_i, y_i, a, b, c.

$F_1 := f(x_1, y_1) - f(x_2, y_2)$,
$F_2 := f(x_1, y_1) - f(x_3, y_3)$,
$F_{i+2} := x_i^2 + y_i^2 + 2ax_i + 2by_i + c, \quad i = 1, 2, 3,$
$F_{i+5} := a\frac{\partial f}{\partial y}(x_i, y_i) - b\frac{\partial f}{\partial x}(x_i, y_i) + x_i\frac{\partial f}{\partial y}(x_i, y_i) - y_i\frac{\partial f}{\partial x}(x_i, y_i), \quad i = 1, 2, 3.$

The meaning of the 8 equations is as follows.

eq_1: $F_1 = 0$ $\mathbf{p}_1$ and $\mathbf{p}_2$ in the same level curve of f;
eq_2 : $F_2 = 0$ $\mathbf{p}_1$ and $\mathbf{p}_3$ in the same level curve of f;
eq_{i+2} : $F_{i+2} = 0$ $\mathbf{p}_i$ lies on the circle C, $i = 1, 2, 3$;
eq_{i+5}: $F_{i+5} = 0$ C and the level set of f through $\mathbf{p}_i$ are tangent at $\mathbf{p}_i$.

First from the three equations eq_i, $i = 3, 4, 5$, we can get a, b and c as functions of x_i, y_i. Of course this is merely finding the circle through three given points, which need to be non-collinear, and in particular distinct, for a unique solution. More details about this will appear elsewhere.

Remark. In the umbilic case, we can always rotate the coordinates to make $b_0 = b_2$ in the expression of $f(x, y)$, as shown in [7]. Thus, from now on we assume $b_0 = b_2$ for an umbilic point. Once having assumed $b_0 = b_2$, we now make the genericity assumption that $b_1 \neq b_3$. We shall also look for solutions for these equations for which *the limiting directions* (limiting angles to the positive x-axis) *are distinct*. This relates to the point made earlier, that our equations inevitably admit 'diagonal' solutions which we want to suppress. Thus we are assuming here that the limiting directions of the three A_1 contact points of our tritangent circle are distinct as the oval $f(x, y) = k$ shrinks to a point with $k \to 0$.

Proposition 1. *Generically, there are no triple crossings, nor cusps on the local branches of the symmetry set of isophotes curves near a hyberbolic point.*

The limiting directions of the A_1^3 curves at an umbilic, making the assumptions in the above Remark, are at angles t_1, t_2, t_3 equal, in some order, to $90°$,

$-30°$, $-150°$ to the positive x-axis, or the 'opposites' of these, namely $-90°$, $150°$, $30°$. This suggests strongly that there are always two triples of A_1^3 contact points tending to the origin as $k \to 0$.

Proposition 1 implies, as confirmed by experimental evidence (see Fig. 6), that there are in fact two triple crossings (A_1^3) in the symmetry set in the umbilic case. The proof of the Proposition is an explicit computation[2] of the tangent cone of the algebraic variety defined by the above equations $F_i = 0$, $i = 1 \ldots 8$. The branches (x_i, y_i) corresponding to $(t_1, t_2, t_3) = (90°, -30°, -150°)$ have the form
$((-2b_1b_0 - 6b_0b_3 + 3c_3 + c_1)t^2/6(b_3 - b_1) + \ldots, t)$,
$(\frac{1}{2}\sqrt{3}t + \ldots, -\frac{1}{2}t + \ldots)$ and $(-\frac{1}{2}\sqrt{3}t + \ldots, -\frac{1}{2}t + \ldots)$

The actual locus of A_1^3 points (triple intersections) on the symmetry set close to an umbilic point where $b_0 = b_2$ as above and $b_1 \neq b_3$, is $(-a(t), -b(t))$ where

$$a(t) = \tfrac{b_0}{2}t^2 + \tfrac{1}{16}(7b_0b_1 + 9b_0b_3 - 3c_1 - c_3)t^3 + \text{h.o.t.}$$
$$b(t) = \tfrac{1}{8}(b_1 + 3b_3)t^2 + \tfrac{1}{16}(b_1^2 + 3b_1b_3 + 4b_0^2 + 5c_4 - c_2 - 3c_0)t^3 + \text{h.o.t.}$$

Generically this curve has an *ordinary cusp* at the origin.

3.2 A_1A_2 Points

The A_1A_2 points are the centres of bitangent circles which are osculating at one point and have an ordinary tangency at the other one; they produce cusps on the symmetry set. As in the case of A_1^3 points (§3.1), we look in the first instance for the points where those circles are tangent to the level sets of f.

We find these curves by taking the circle C to have equation $x^2 + y^2 + ax + by + c = 0$ as in §3.1. This time after elimination of a, b, c we obtain 3 equations in 4 unknowns instead of 5 equations in 6 unknowns. Let the circle C be tangent to the same level set $f = k$ at the two points $\mathbf{p}_i = (x_i, y_i)$, $i = 1, 2$. We proceed to write down the corresponding conditions, defining functions F_i as follows.

$$F_1 := f(x_1, y_1) - f(x_2, y_2),$$
$$F_2 := 2a(x_1 - x_2) + 2b(y_1 - y_2) + x_1^2 + y_1^2 - x_2^2 - y_2^2,$$
$$F_3 := af_y(x_1, y_1) - bf_x(x_1, y_1) + x_1f_y(x_1, y_1) - y_1f_x(x_1, y_1),$$
$$F_4 := af_y(x_2, y_2) - bf_x(x_2, y_2) + x_2f_y(x_2, y_2) - y_2f_x(x_2, y_2),$$
$$F_5 := (a + x_2)(f_{xx}f_y^2 - 2f_{xy}f_xf_y + f_{yy}f_x^2) - f_x(f_x^2 + f_y^2) \text{ (derivatives at } (x_2, y_2)\text{)}.$$

We have the corresponding equations and their interpretations:

$eq_1 : F_1 = 0$ $\mathbf{p}_1$ and $\mathbf{p}_2$ are in the same level set of f;
$eq_2 : F_2 = 0$ a circle with centre $(-a, -b)$ passes through $\mathbf{p}_1$ and $\mathbf{p}_2$;
$eq_3 : F_3 = 0$ this circle is tangent to the level set of f through $\mathbf{p}_1$;
$eq_4 : F_4 = 0$ this circle is tangent to the level set of f through $\mathbf{p}_2$;
$eq_5 : F_5 = 0$ this circle is the circle of curvature of the level set through $\mathbf{p}_2$.

We solve eq_2, eq_3 for a and b and substitute in eq_4 and eq_5. We summarize the results as follows. See Figure 4. We assume as before that the limiting angles at which the A_1 and A_2 points approach the origin are distinct.

[2] This computation, like all those underlying this article, was performed in Maple.

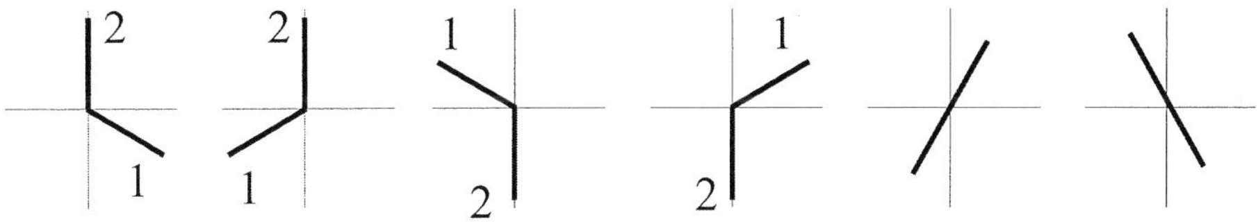

Fig. 4. The possible limiting directions of A_1 and A_2 contact points of A_1A_2 circles, in the umbilic case with axes rotated so that $b_0 = b_2$, assuming the limiting directions are unequal. Those labelled 1 are only A_1 directions and similarly for A_2. The unlabelled directions can be either, with A_1 and A_2 at 180° to one another

Proposition 2. *Generically, there are no cusps on the local branches of the symmetry set of isophotes curves near a hyperbolic point. The limiting angles in the umbilic case must be one of the following.*
$A_1 : -30°,\ A_2 : 90°$; *or* $A_1 : 150°,\ A_2 : -90°$;
$A_1 : -150°,\ A_2 : 90°$; *or* $A_1 : 30°,\ A_2 : -90°$;
$A_1 : 60°,\ A_2 : -120°$ *or vice versa;*
$A_1 : -60°,\ A_2 : 120°$ *or vice versa.*

This means that there are six cusps (A_1A_2) on the SS in this umbilic case. In that case, we expect each cusp (which requires an A_1 and an A_2 contact) to use one of the above six solutions, for a definite choice of A_1 and A_2 in the last two cases.

3.3 Symmetry Sets (SS) and Medial Axes (MA)

As suggested by Theorem 1, Propositions 2 and 2, the local structure of the SS and MA of individual isophote curves and its transitions are as follows:

- parabolic points: the local structure of SS is just 3 separate branches correponding to the 3 vertices separated by inflexions (Theorem 1), see Fig. 5.
- nonumbilic elliptic points: the SS is made of just 2 transverse arcs, one joining two centres of curvature at maxima of curvature and the other one two minimum of curvature. The SS will look like itself and disapear as the curve shrinks to a point.
- hyperbolic point: the SS and MA are made of smooth branches, which do not connect together to form cusps or crossings. This implies in particular that generically, the SS (and MA) is just given by the geometry of vertices and inflexions as well as how they are distributed along the isophote curves, as discribed in Section 1. The branches of the SS will start at endpoints which are the centres of curvature of the isophote curves at vertices and they point towards the corresponding vertex if the isophote curve has a local minimum of curvature, and away from the vertex where the curve has a maximum of curvature.

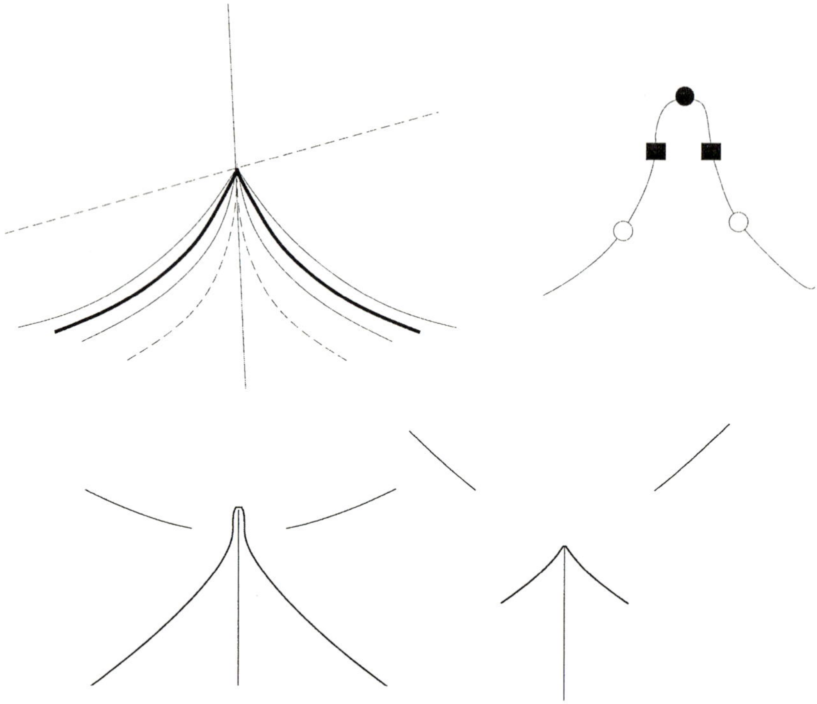

Fig. 5. Top: A schematic picture of the patterns of the vertices (vertex set $V_p = 0$: thin solid line) and inflexions (inflexion set $I_p = 0$: dashed line) of the level curves $f = k$ evolving through a parabolic point, together with the zero level set $f = 0$ (thick line), and a sketch of one level curve of f. The vertex set has two cuspidal branches and one smooth branch. The inflexion set has one cuspidal branch which is always below all cupidal branches and one smooth branch. The zero level set $f_p = 0$ has one cuspidal branch which is always between the two cuspidal vertex branches. The level set $f_p = k$ then evolves so that the number of vertices remains as 3 and the number of inflexions as 2 for both signs of k, with k small. Bottom: Symmetry sets (thin lines) of curves (thick lines) which are sections of a surface close to the tangent plane at a parabolic point. One sees 3 vertices separated by two inflexions both before and after the transition. At the transitional moment itself, the branches reach right to the curve, which then has an ordinary cusp. Figure produced with LSMP[13]

- near umbilics: the SS have generically two triple crossings and six cusps. Hence generically, the SS has one structure, as in Fig. 6.

For the drawing of the SS and MA, we will need the pre-symmetry set (preSS) which is the subset of the cartesian product $I \times I$ of the parameter space I, defined by the pairs (s,t) corresponding to points $p = \gamma(s)$ and $q = \gamma(t)$ which contribute to the SS. That is, there is a circle tangent to γ at the points $\gamma(s)$ and $\gamma(t)$. See Fig. 6

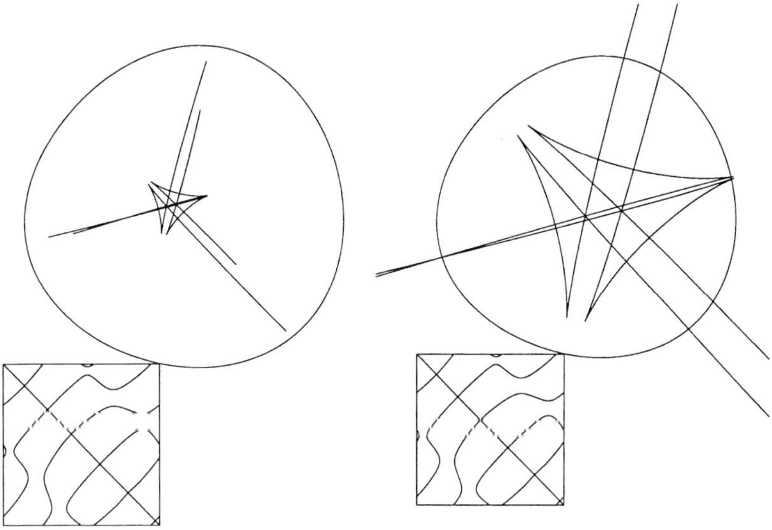

Fig. 6. Symmetry set and pre-symmetry set of $f = k$, in the umbilic case $f(x,y) = x^2 + y^2 + x^3 - xy^2 + 2y^3$ and $k = 0.09$. The figure to the right is the same as the left hand side one, but the symmetry set has been enlarged so that the two A_1^3 points (triple crossings) and the six A_1A_2 (cusps) are more visible. One can also see the six endpoints of the symmetry set, corresponding to the six vertices on the curve. Varying k then the SS will still look like itself and disappear as $k \to 0$. This figure illustrates the results of Proposition 1 and Proposition 2

4 Evolution of Symmetry Sets of Isophote Curves in 1-Parameter Families of Surfaces

As explained in Section 1, given a generic surface M, elliptic and hyperbolic points occupy *regions* of M, separated by parabolic *curves* with isolated *points* on them which are cusps of Gauss. We can then consider moving from a hyperbolic point to a parabolic point of M. We can also realise this by evolving the surface in a 1-parameter family, of the form $z = x^2 - \alpha^2 y + b_0x^3 + b_1x^2y + b_2xy^2 + b_3y^3 + \ldots$, where $\alpha \to 0$ and $b_3 \neq 0$. It turns out that, generically, the only hyperbolic points which exist sufficiently near a parabolic point are the ones corresponding to vertex transition $1 + 1 \leftrightarrow 3 + 3$ in Theorem 1. The Figure 7 shows how the vertices behave on a 2-parameter family of plane sections near the tangent plane at a hyperbolic point, evolving to a family of sections near a parabolic point.

5 Conclusion

This paper represents a step towards understanding the evolution of SS and MA of families of isophote curves, or more generally of families of plane sections of s

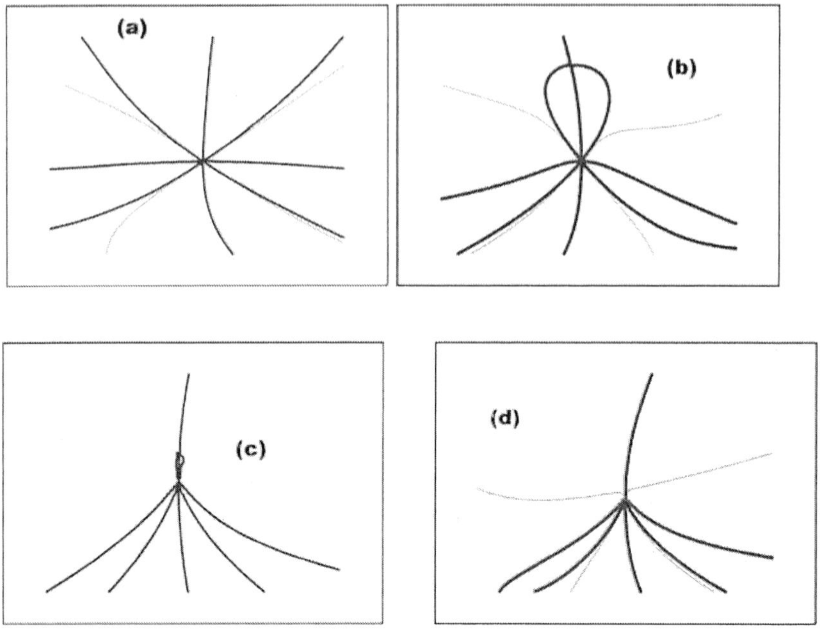

Fig. 7. Transition of the patterns of vertices (thick curve) and inflexions (thin curve) on a 2-parameter family of plane sections $f_\alpha = k$ near the tangent plane at a hyperbolic point, evolving to a family of sections $f_0 = k$ near a parabolic point. $f(x, y) = x^2 - \alpha^2 y^2 + x^3 + 2x^2 y - xy^2 + y^3$. (a) $\alpha = 1$ (hyperbolic); the vertex set has 4 branches and the inflexion 2. (b) $\alpha = 0.3$, the top part (above x-axis) of two vertex branches join together to form a loop which is shrinking to the origin as $\alpha \to 0$. The vertex branch tangent to $x = 0$ stays smooth. The other vertex branch bends to become a cusp. (c) $\alpha = 0.05$: the vanishing loop. (d) As $\alpha \to 0$, the inflexion set exchanges branches: the top part (above x-axis) join together to make a smooth branch, whereas the bottom part forms a cusp below the cuspidal vertex branches. Compare Figures 2 and 5

generic surface in 3-space. The evolution of the MA depends, in an essential way, upon the underlying evolution of the SS [10], which is why we have concentrated on the SS in this paper. An interesting follow up of this work, would be to combine into a more global represention of an image by the collection of those individual representations, as a singular surface in scale space, whose sections are the individual SS and MA.

Acknowledgements. This work is a part of the DSSCV project supported by the IST Programme of the European Union (IST-2001-35443). The authors are also grateful to Prof. Mads Nielsen and Dr. Aleksander Pukhlikov for useful discussions.

References

1. T.F.Banchoff, T.Gaffney and C.McCrory, *Cusps of Gauss Mappings*, Pitman Research Notes in Mathematics, 55, 1982.
2. J.W. Bruce and P.J. Giblin, *Curves and Singularities* Cambridge University Press, 2nd ed. (1992).
3. J.W. Bruce and P.J. Giblin, 'Growth, motion and one-parameter families of symmetry sets', *Proc. Royal Soc.Edinburgh* 104A (1986), 179–204.
4. J.W. Bruce, P.J. Giblin and F. Tari, 'Parabolic curves of evolving surfaces,' *Int. J. Computer Vision.* 17 (1996), 291–306.
5. A. Diatta and P.J. Giblin, Technical Reports 4, 11 and 23 of the project "Deep Structure, Singularities, and Computer Vision". IST Programme of the European Union (IST-2001-35443).
6. A. Diatta and P.J. Giblin, Vertices and inflexions of plane sections of smooth surfaces in $\mathbb{R}^3$. Submitted to VIII Sao Carlos International Workshop on Real and Complex Singularities. CIRM Luminy (Marseille, France), July 27-31, 2004.
7. A Diatta, P.J.Giblin, B.Guilfoyle and W.Klingenberg, 'Plane sections of surfaces and applications to symmetry sets', preprint, University of Durham/University of Liverpool, to be submitted to *Mathematics of Surfaces 2005.*
8. P.J.Giblin and B.B.Kimia, 'On the local form and Transitions of Symmetry Sets, Medial Axes, and Shocks', *Int. J. Computer Vision* 54 (2003), 143–157.
9. P.J. Giblin and B.B. Kimia, 'On the Intrinsic Reconstruction of Shape from its Symmetries', *IEEE Trans. Pattern Analysis and Machine Intelligence*, 25 (2003), 895–911.
10. P.J. Giblin, Symmetry sets and medial axes in two and three dimensions. The mathematics of surfaces, IX (Cambridge, 2000), 306-321, Springer, London, 2000.
11. P.L. Hallinan, G.G. Gordon, A.L. Yuille, P. Giblin and D. Mumford, *Two and three dimensional patterns of the face*, Natick, Massachusetts: A.K.Peters 1999.
12. J.J. Koenderink, *Solid Shape*, M.I.T. Press (1990)
13. R.J.Morris, *Liverpool Surface Modelling Package*, http://www.amsta.leeds.ac.uk/~rjm/lsmp/ See also R.J.Morris, 'The use of computer graphics for solving problems in singularity theory', in *Visualization in Mathematics*, H.-C.Hege & K.Polthier, Heidelburg: Springer-Verlag (1997), 173–187, and http://www.scs.leeds.ac.uk/pfaf/lsmp/SingSurf.html

Fig. 1. Variances of top-point displacements for all top-points projected on the xy-plane

and $\langle\eta^2\rangle$ have an extremum under the same rotation of the axes. The extrema of $\langle\xi^2\rangle$ (and $\langle\eta^2\rangle$) can be reached by rotation, when

$$\chi = \tan\varphi = \frac{A}{B} + \sqrt{1 + \left(\frac{A}{B}\right)^2} \tag{18}$$

The extremal variances are

$$\begin{aligned} X &= \langle\xi^2\rangle = \tfrac{\chi B + 2C}{2D}, \\ Y &= \langle\eta^2\rangle = \tfrac{-B + 2\chi C}{2\chi D}. \end{aligned} \tag{19}$$

X and Y are obviously invariant under rotation and translation.

By rotating the coordinate system we find directions in which the variance is maximal, respectively minimal (these two directions are orthogonal) and we construct an ellipse[1] with principal directions and axes that reflect these extremal noise variances (Fig 1).

Note, that top-points, in the neighborhood of which there is a lot of structure, have ellipses with very small radiuses (stable), and top-points in rather flat

[1] Note, that (15) does not parameterize an ellips. An elliptical "gauge figure" however is merely used for simplicity.

locations tend to have large ellipses (unstable) (Fig. 1). Another invariant is the variance of τ (scale instability), the expression of which is given in Appendix B.

5 Experiments

In order to validate the theoretical results numerical experiments have been conducted. Adding noise to the image results in changing top-points coordinates. Some of them hardly move and others move quite a lot. It is practically impossible by comparing two top-point clouds to tell which top-point of the fiducial image corresponds to which top-point of the actual image, therefore it is impossible to investigate the stability in a pure experimental way. Instead, we choose a somewhat different approach, which combines theory and experiments.

For each noise realization N^i, where $i = 1 \ldots K$ labels the experiments, we use (3) as a refining algorithm in order to estimate the coordinates of the actual top-point $(x_0 + \xi_i, y_0 + \eta_i, t_0 + \tau_i)$, taking the coordinates of the original top-point (x_0, y_0, t_0) as an initial guess. The experiment consists of $K = 500$ noise realizations. Therefore, for original top-point (x_0, y_0, t_0) we compute an array $\{(\xi_i, \eta_i, \tau_i)\}_{1 \leq i \leq K}$ of 500 displacements.

The principal directions and maximum and minimum variances for the set of points, obtained by noise perturbation, have been calculated. In order to find principal directions, the extremum problem should be solved for the averages

$$\begin{aligned} \langle \xi^2 \rangle(\chi) &= \tfrac{1}{1+\chi^2} \textstyle\sum_{i=1}^{K} (\xi_i + \chi \eta_i)^2 / K, \\ \langle \eta^2 \rangle(\chi) &= \tfrac{1}{1+\chi^2} \textstyle\sum_{i=1}^{K} (-\chi \xi_i + \eta_i)^2 / K, \end{aligned} \tag{20}$$

where T is a tangent of the angle of rotation. The extreem for both variances are reached under identical rotations, since the sum $\langle \xi^2 \rangle(T) + \langle \eta^2 \rangle(T)$ does not depend on χ.

The extremum corresponds to the angle given by

$$\tilde{\chi} = -\frac{\sum_i (\xi_i^2 - \eta_i^2)}{2 \sum_i \xi_i \eta_i} + \sqrt{\left(\frac{\sum_i (\xi_i^2 - \eta_i^2)}{2 \sum_i \xi_i \eta_i} \right)^2 + 1}. \tag{21}$$

The variance in this direction is $\tilde{X} = \langle \eta^2 \rangle(\tilde{\chi})$

$$\tilde{X} = \frac{1}{1 + \tilde{\chi}^2} \sum_{i=1}^{K} (\xi_i + \tilde{\chi} \eta_i)^2 / K \tag{22}$$

and in the orthogonal direction

$$\tilde{Y} = \frac{1}{1 + \tilde{\chi}^2} \sum_{i=1}^{K} (-\tilde{\chi} \xi_i + \eta_i)^2 / K \tag{23}$$

The comparison of theory and the experiments is depicted in Fig. 2. Since both the theory and the experiments take into account derivatives up to fourth

Fig. 2. Examples of top-point movements projected on the xy-plane under noise realizations (crosses) and theoretical predictions (ellipses). Right column shows zooming in the neighborhood of the top-point

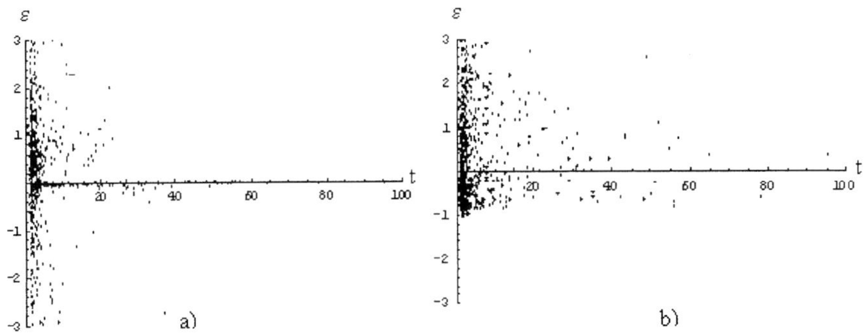

Fig. 3. Comparison of experimental and theoretical results. The value of ε denotes the ratio between theoretical and experimental variances, a) - for spatial displacement ε_X and b) - for scale-displacement ε_T

order, the scale of the top-point should be large enough to obtain reliable results. The value of ε denotes the relative difference between theoretical and experimental variances in space and scale

$$\varepsilon_X = \frac{X - \tilde{X}}{\tilde{X}} \tag{24}$$

$$\varepsilon_T = \frac{T - \tilde{T}}{\tilde{T}} \tag{25}$$

Figure 3 reveals that the relative difference between theoretical and experimental results is acceptably small for large scales and large for small scales due to computational errors in derivatives, as expected.

6 Results

In this paper we have described an algorithm for computing stability measures for top-points. The algorithm is based on a perturbation approach and uses properties of noise propagation in Gaussian scale-space.

Variances of top-point displacements can be computed on the basis of noise variance and fourth order differential structure at the top-point.

The advantage of this approach is that variances of displacements can be predicted theoretically on the basis of the local differential structure.

The experiments have shown correspondence between the analytical predictions and practice in cases where the scale of top-point is not too small for reliably computing fourth order derivatives.

Analytically computed variances can be used for several applications, such as stability measures and weight measures for top-point based image retrieval algorithm [1].

Applying the algorithm to problems listed above will be the next step in our research.

References

1. F.M.W. Kanters, B. Platel, L.M.J. Florack and B.M. ter Haar Romeny: Content based image retieval using multiscale top points. In: Lewis Griffin and Martin Lillholm, editors, Scale Space Methods in Computer Vision, 4th International Conference, Scale Space 2003, Isle of Skye, UK. Springer (June 2003) 33-43
2. J. Damon: Local Morse theory for solutions to the heat equation and Gaussian blurring. Journal of Differential Equations, vol. 12, no. 1 (January 1995) 368-401
3. L. Florack and A. Kuijper: The topological structure of scale-space images. Journal of Mathematical Imaging and Vision, vol. 12(1) (February 2000) 65-79.
4. B.M. ter Haar Romeny: Front-End Vision and Multi-Scale Image Analysis. Kluwer Academic Publisher (2003)
5. L. Florack: Image Structure. The series Computational Imaging and Vision,vol. 10. Kluwer Academic Publishers (1997)
6. R. Duits, L. Florack, J. de Graaf, B. ter Haar Romeny: On the Axioms of Scale Space Theory. Journal of Mathematical Imaging and Vision, vol. 20, iss. 3 (May 2004) 267-298(32)
7. L. G. Parratt: Probability and experimental errors in science. John Wiley and SONS, inc. (1961)
8. Matthew Brown and David G. Lowe: Invariant features from interest point groups. British Machine Vision Conference, BMVC 2002, Cardiff, Wales (September 2002) 656-665
9. R. Veltkamp P. Giannopolous: A pseudo-metric for weighted point sets. In ECCV 2002, LNCS 2352, Springer (2002) 715-730.
10. J. Blom, B. M. ter Haar Romeny, A. Bel and J. J. Koenderink: Spatial Derivatives and the Propagation of Noise in Gaussian Scale Space. Journal of Visual Communication and Image Representation, vol. 4, Issue 1 (March 1993) 1-13
11. D. Eberly: A differential geometric approach to anisotropic diffusion. In: Geometry-Driven Diffusion in computer Vision, B. M, ter Haar Romeny, Ed., vol.1 of Computational Imaging and Vision Series. Kluwer Academic Publishers, Dordrecht (1994) 371-392
12. S. Belongie, J. Malik, J. Puzicha: Shape matching and object recognition using shape contexts. IEEE Transactions on Pattern Analysis and Machine Intelligence, vol. 24, No. 24 (April 2002)

Appendix A: Displacements Under Noise Perturbation

In this appendix we give expressions for displacements in spatial and scale directions. The refining equations (3) in terms of image derivatives are given by

$$\begin{aligned}
\begin{bmatrix} \xi \\ \eta \end{bmatrix} = \Bigg((v_{xy}^2 - v_{xx}v_{yy}) \begin{bmatrix} v_{xy}(v_{xxy}+v_{yyy}) - v_{yy}(v_{xxx}+v_{xyy}) \\ v_{xy}(v_{xxx}+v_{xyy}) - v_{xx}(v_{xxy}+v_{yyy}) \end{bmatrix} - \\
(v_y(v_{xxx}+v_{xyy}) - v_x(v_{xxy}+v_{yyy})) \begin{bmatrix} -2v_{xy}v_{xyy} + v_{yy}v_{xxy} + v_{xx}v_{yyy} \\ 2v_{xy}v_{xxy} - v_{yy}v_{xxx} - v_{xx}v_{xyy} \end{bmatrix} + \\
(v_{yy}(v_{xxxx}+v_{xxyy}) + v_{xx}(v_{xxyy}+v_{yyyy}) - 2v_{xy}(v_{xxxy}+v_{xyyy})) \\
\begin{bmatrix} v_y v_{xy} - v_x v_{yy} v_x v_{xy} - v_y v_{xx} \end{bmatrix} \Bigg) / \\
\det M
\end{aligned} \tag{26}$$

The scale displacement equals

$$\begin{aligned}\tau = (&-(v_{xy}^2 - v_{xx}v_{yy})^2 + v_y(2v_{xy}^2 v_{xxy} + u_{xx}(v_{yy}v_{xxy} + v_{xx}v_{yyy}) - \\ &v_{xy}(3v_{xx}v_{xyy} + v_{yy}v_{xxx})) + v_x(2v_{xy}^2 v_{xyy} + v_{yy}(v_{xx}v_{xyy} + v_{yy}v_{xxx}) - \\ &v_{xy}(3v_{yy}v_{xxy} + v_{xx}v_{yyy})))/\mathrm{det}M\end{aligned} \tag{27}$$

In both formulas we have a denominator

$$\begin{aligned}\mathrm{det}M = &(v_{yy}v_{xxy} + v_{xx}v_{yyy} - 2v_{xy}v_{xyy})(v_{xy}(v_{xxx} + v_{xyy}) - v_{xx}(v_{xxy} + v_{yyy})) \\ &+(v_{yy}v_{xxx} + v_{xx}v_{xyy} - 2v_{xy}v_{xxy})(v_{xy}(v_{xxy} + v_{yyy}) - v_{yy}(v_{xxx} + v_{xyy})) + \\ &(v_{xx}v_{yy} - v_{xy}^2)(v_{xx}(v_{xxyy} + v_{yyyy}) + v_{yy}(v_{xxxx} + v_{xxyy})) - 2v_{xy}(v_{xxxy} + v_{xyyy}))\end{aligned} \tag{28}$$

Appendix B: Parameters for the Invariant Expressions

$$\begin{aligned}A = &3(u_{xx} - u_{yy})(u_{xx} + u_{yy})^2(u_{xx}(u_{xxy} + u_{yyy})^2 + (u_{xxx} \\ &+u_{xyy})((u_{xxx} + u_{xyy})u_{yy} - 2u_{xy}(u_{xxy} + u_{yyy}))) + 4t_0(((-2u_{xxy}u_{xy} \\ &+u_{xx}u_{xyy} + u_{xxx}u_{yy})(u_{xxy} + u_{yyy}) + (u_{xxx} + u_{xyy})(-2u_{xy}u_{xyy} + u_{xxy}u_{yy} \\ &+u_{xx}u_{yyy}) + 2u_{xy}(2u_{xy}(u_{xxxy} + u_{xyyy}) - (u_{xxxx} + u_{xxyy})u_{yy} \\ &-u_{xx}(u_{xxyy} + u_{yyyy})))^2 + 2(-2u_{xy}(u_{xxxy} + u_{xyyy})u_{yy} + (u_{xxxx} + u_{xxyy})u_{yy}^2 \\ &-(u_{xxy} + u_{yyy})(-2u_{xy}u_{xyy} + u_{xxy}u_{yy} + u_{xx}u_{yyy}) + u_{xy}^2(u_{xxyy} + u_{yyyy})) \\ &\times(-(u_{xxx} + u_{xyy})(u_{xx}u_{xyy} + u_{xxx}u_{yy}) + (u_{xxy} + u_{yyy})(u_{xxy}u_{yy} + u_{xx}u_{yyy}) \\ &+2u_{xy}(u_{xxx}u_{xxy} - u_{xyy}u_{yyy}) + (-u_{xx} + u_{yy})(2u_{xy}(u_{xxxy} + u_{xyyy}) \\ &-(u_{xxxx} + u_{xxyy})u_{yy} - u_{xx}(u_{xxyy} + u_{yyyy}))) + (-(u_{xxx} + u_{xyy})(u_{xx}u_{xyy} \\ &+u_{xxx}u_{yy}) + (u_{xxy} + u_{yyy})(u_{xxy}u_{yy} + u_{xx}u_{yyy}) + 2u_{xy}(u_{xxx}u_{xxy} - u_{xyy}u_{yyy}) \\ &+(-u_{xx} + u_{yy})(2u_{xy}(u_{xxxy} + u_{xyyy}) - (u_{xxxx} + u_{xxyy})u_{yy} - u_{xx}(u_{xxyy} \\ &+u_{yyyy})))^2 - 2((-2u_{xxy}u_{xy} + u_{xx}u_{xyy} + u_{xxx}u_{yy})(u_{xxy} + u_{yyy}) \\ &+(u_{xxx} + u_{xyy})(-2u_{xy}u_{xyy} + u_{xxy}u_{yy} + u_{xx}u_{yyy}) + 2u_{xy}(2u_{xy}(u_{xxxy} \\ &+u_{xyyy}) - (u_{xxxx} + u_{xxyy})u_{yy} - u_{xx}(u_{xxyy} + u_{yyyy})))((u_{xxx} + u_{xyy}) \\ &\times(-2u_{xy}u_{xyy} + u_{xxy}u_{yy} + u_{xx}u_{yyy}) - u_{xy}(-2u_{xy}(u_{xxxy} + u_{xyyy}) \\ &+(u_{xxxx} + u_{xxyy})u_{yy} + u_{xx}(u_{xxyy} + u_{yyyy}))))\end{aligned} \tag{29}$$

$$\begin{aligned}B = &-6u_{xy}(u_{xx} + u_{yy})^2(u_{xx}(u_{xxy} + u_{yyy})^2 + (u_{xxx} + u_{xyy})((u_{xxx} + \\ &u_{xyy})u_{yy} - 2u_{xy}(u_{xxy} + u_{yyy}))) + 4t_0(2(-2u_{xy}(u_{xxxy} + u_{xyyy})u_{yy} + \\ &(u_{xxxx} + u_{xxyy})u_{yy}^2 - (u_{xxy} + u_{yyy})(-2u_{xy}u_{xyy} + u_{xxy}u_{yy} + u_{xx}u_{yyy}) + \\ &u_{xy}^2(u_{xxyy} + u_{yyyy}))((-2u_{xxy}u_{xy} + u_{xx}u_{xyy} + u_{xxx}u_{yy})(u_{xxy} + u_{yyy}) + \\ &(u_{xxx} + u_{xyy})(-2u_{xy}u_{xyy} + u_{xxy}u_{yy} + u_{xx}u_{yyy}) + 2u_{xy}(2u_{xy}(u_{xxxy} + \\ &u_{xyyy}) - (u_{xxxx} + u_{xxyy})u_{yy} - u_{xx}(u_{xxyy} + u_{yyyy}))) + 2(-(u_{xxx} +\end{aligned}$$

$$u_{xyy}))(u_{xx}u_{xyy} + u_{xxx}u_{yy}) + (u_{xxy} + u_{yyy})(u_{xxy}u_{yy} + u_{xx}u_{yyy}) + 2u_{xy}(u_{xxx}u_{xxy} - u_{xyy}u_{yyy}) + (-u_{xx} + u_{yy})(2u_{xy}(u_{xxxy} + u_{xyyy}) - (u_{xxxx} + u_{xxyy})u_{yy} - u_{xx}(u_{xxyy} + u_{yyyy})))((u_{xxx} + u_{xyy})(-2u_{xy}u_{xyy} + u_{xxy}u_{yy} + u_{xx}u_{yyy}) - u_{xy}(-2u_{xy}(u_{xxxy} + u_{xyyy}) + (u_{xxxx} + u_{xxyy})u_{yy} + u_{xx}(u_{xxyy} + u_{yyyy}))) \tag{30}$$

$$C = 3(u_{xx} + u_{yy})^2(-(u_{xxx} + u_{xyy})u_{yy} + u_{xy}(u_{xxy} + u_{yyy}))^2 + 4t_0((-2u_{xy}(u_{xxxy} + u_{xyyy})u_{yy} + (u_{xxxx} + u_{xxyy})u_{yy}^2 + (u_{xxy} + u_{yyy}) \times (2u_{xy}u_{xyy} - u_{xxy}u_{yy} - u_{xx}u_{yyy}) + u_{xy}^2(u_{xxyy} + u_{yyyy}))^2 + ((u_{xxx} + u_{xyy}) \times (-2u_{xy}u_{xyy} + u_{xxy}u_{yy} + u_{xx}u_{yyy}) - u_{xy}(-2u_{xy}(u_{xxxy} + u_{xyyy}) + (u_{xxxx} + u_{xxyy})u_{yy} + u_{xx}(u_{xxyy} + u_{yyyy})))^2) \tag{31}$$

$$D = 8\sqrt{t_0^3}(v_{yy}v_{xxy} + v_{xx}v_{yyy} - 2v_{xy}v_{xyy})(v_{xy}(v_{xxx} + v_{xyy}) - v_{xx}(v_{xxy}+ v_{yyy})) + (v_{yy}v_{xxx} + v_{xx}v_{xyy} - 2v_{xy}v_{xxy})(v_{xy}(v_{xxy} + v_{yyy}) - v_{yy}(v_{xxx} + v_{xyy}))) \tag{32}$$

$$\langle\tau^2\rangle = 4t_0((u_{xx} + u_{yy})(u_{yy}(3u_{xx}u_{xyy} + u_{xxx}u_{yy})^2 - 2u_{xy}(3u_{xx}u_{xyy} + u_{xxx}u_{yy})\times (3u_{xxy}u_{yy} + u_{xx}u_{yyy}) + u_{xx}(3u_{xxy}u_{yy} + u_{xx}u_{yyy})^2))/D \tag{33}$$

Discrete Representation of Top Points via Scale Space Tessellation

B. Platel[1], M. Fatih Demirci[2], A. Shokoufandeh[2], L.M.J. Florack[1], F.M.W. Kanters[1], B.M. ter Haar Romeny[1], and S.J. Dickinson[3]

[1] Eindhoven University of Technology,
P.O. Box 513, 5600 MB Eindhoven, The Netherlands
[2] Drexel University, 3141 Chestnut Street,
Philadelphia, PA 19104, United States of America
[3] University of Toronto, 6 King's College Rd.,
Toronto, Ontario, Canada M5S 3G4
b.platel@tue.nl, {mdemirci, ashokouf}@cs.drexel.edu
{l.m.j.florack, f.m.w.kanters, b.m.terhaarromeny}@tue.nl
sven@cs.toronto.edu

Abstract. In previous work, singular points (or top points) in the scale space representation of generic images have proven valuable for image matching. In this paper, we propose a construction that encodes the scale space description of top points in the form of a directed acyclic graph. This representation allows us to utilize graph matching algorithms for comparing images represented in terms of top point configurations instead of using solely the top points and their features in a point matching algorithm, as was done previously. The nodes of the graph represent the critical paths together with their top points. The edge set will capture the neighborhood distribution of vertices in scale space, and is constructed through a Delaunay triangulation scheme. We also will present a many-to-many matching algorithm for comparing such graph-based representations. This algorithm is based on a metric-tree representation of labelled graphs and their low-distortion embeddings into normed vector spaces via spherical encoding. This is a two-step transformation that reduces the matching problem to that of computing a distribution-based distance measure between two such embeddings. To evaluate the quality of our representation, two sets of experiments are considered. First, the stability of this representation under Gaussian noise of increasing magnitude is examined. In the second set of experiments, a series of recognition experiments is run on a small face database.

1 Introduction

Previous research has shown that top points (singular points in the scale space representation of generic images) have proven to be valuable sparse image descriptors that can be used for image reconstruction [6, 12] and image matching [7, 14]. In our previous work, images were compared using a point matching

R. Kimmel, N. Sochen, J. Weickert (Eds.): Scale-Space 2005, LNCS 3459, pp. 73–84, 2005.

scheme which took into account the positions, scales, and differential properties of corresponding top points [7, 6]. The underlying matching framework was based on the Earth Mover's Distance, a powerful, many-to-many point matching framework. However, treating the points as an unstructured collection ignores the salient group structure that may exist within a given scale or across scales. Grouping certain top points together explicitly encodes the neighborhood structure of a point, effectively enriching the information encoded at a point – information that can be exploited during both indexing [16] and matching [17].

In this paper, we take an unstructured set of top points and impose a neighborhood structure on them. Inspired by the work of Lifshitz and Pizer [10], we will encode the scale space structure of a set of top points in a *directed acyclic graph* (DAG). Specifically, we combine the position-based grouping of the top points provided by a Delaunay triangulation with the scale space ordering of the top points to yield a directed acyclic graph. This new representation allows us to utilize powerful graph matching algorithms to compare images represented in terms of top point configurations, rather than using point matching algorithms to compare sets of isolated top points. Specifically, we draw on our recent work in many-to-many graph matching [9, 2, 3], which reduces the matching problem to that of computing a distribution-based distance measure between embeddings of labelled graphs.

We describe our new construction by first elaborating on those basics of catastrophe theory required to introduce the concept of a top point. Next, we formally define a top point, and introduce a measure for its stability that will be used to prune unstable top points. Section 4 describes the construction of the DAG through a Delaunay triangulation scheme. Section 5 reviews our many-to-many DAG matching algorithm, which will be used to evaluate the construction. In the first experiment, we examine the stability of the construction under Gaussian noise of increasing magnitude applied to the original images. In the second experiment, we examine the invariance of the graph structure to within-class image deformation, which may include minor displacements of points both within and across scales.

2 Catastrophe Theory

Critical points are points at any fixed scale in which the gradient vanishes, i.e., $\nabla u = 0$. The study of how these critical points change as certain control parameters change is called *catastrophe theory*. A Morse critical point will move along a *critical path* when a control parameter is continuously varied. In principle, the single control parameter in the models of this article can be identified as the scale of the blurring filter. The only generic morsifications in Gaussian scale space are *creations* and *annihilations* of pairs of Morse hypersaddles of opposite Hessian signature[1] [1, 4]. An example of this is given in Fig. 1.

[1] The Hessian signature is the sign of the determinant evaluated at the location of the critical point.

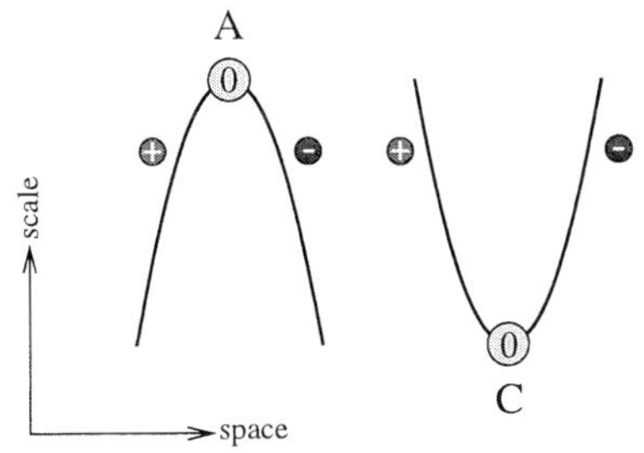

Fig. 1. The generic catastrophes in isotropic scale space. Left: an annihilation event. Right: a creation event. A positive charge ⊕ denotes an extremum, a negative charge ⊖ denotes a saddle, ⊚ indicates the singular point

The movement of critical points through scale, together with their annihilations and creations, forms *critical paths* in scale space. In this article, we will restrict ourselves to generic (non-symmetrical) 2D images, but the theory is easily adapted to higher dimensions. In the 2D case, the only generic morsification is an annihilation or creation where a saddle point and an extremum point meet. Critical paths in 2D therefore consist of an *extremum branch*, that describes the movement of an extremum through scale, and a *saddle branch*, that describes the movement of the saddle with which the extremum annihilates. Note that there is always one extremum branch continuing up to infinite scale [11]. In Fig. 2, the critical paths and their top points are shown for a picture of a face.

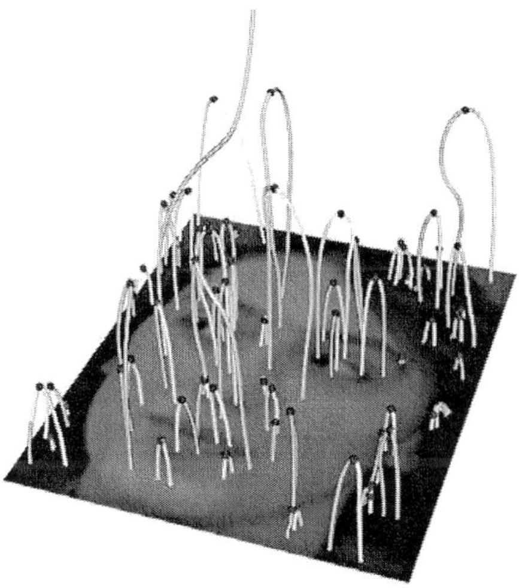

Fig. 2. Critical paths and top points of a face

3 Top Points

The points at which creation and annihilation events take place are often referred to as *top points*[2]. A top point is a critical point at which the determinant of the Hessian degenerates:

$$\begin{cases} \nabla u = 0 \\ \det(H) = 0. \end{cases} \tag{1}$$

An easy way to find these top points is by means of zero-crossings in scale space. This involves derivatives up to second order and yields sub-pixel results. Other, more elaborate methods, can be used to find or refine the top point positions. For details, the reader is referred to [4].

It is obvious that the positions of extrema at very fine scales are sensitive to noise. This, in most cases, is not a problem. Most of these extrema are blurred away at fine scales and won't affect our matching scheme at slightly coarser scales. However, problems do arise in areas in the image that consist of almost constant intensity. One can imagine that the positions of the extrema (and thus the critical paths and top points) are very sensitive to small perturbations in these areas. These unstable critical paths and top points can continue up to very high scales since there is no structure in the vicinity to interact with. To account for these instable top points, we need to have a measure of stability, so that we can either give unstable points a low weight in our matching scheme, or disregard them completely.

A top point is more stable in an area with a lot of structure. The amount of structure contained in a *spatial* area around a top point can be quantified by the *total (quadratic) variation* (TV) norm over that area:

$$TV(\Omega) \stackrel{\text{def}}{=} \frac{\sigma^2 \int_\Omega \|\nabla u(x)\|^2 dV}{\int_\Omega dV} \tag{2}$$

We calculate the TV norm in a circular area with radius $\lambda\sigma$ around a top point at position (x_c, t_c). Note that the size of the circle depends on the scale σ. The integration area of the TV norm Ω is defined by:

$$\Omega : \|x - x_c\|^2 \leq \lambda^2\sigma^2. \tag{3}$$

By using a spatial Taylor series around the considered top point, and taking into account that the first order spatial derivatives in this point are zero, we can simplify the TV-norm Eqn. (2) to what we refer to as the *differential TV-norm* by the following limiting procedure[14]:

$$tv \stackrel{\text{def}}{=} \lim_{\lambda \to 0} \frac{4}{\pi} \frac{1}{\lambda^4} TV(\lambda) = \sigma^4 \text{Tr}(H^2) \tag{4}$$

The proportionality factor $\frac{4}{\pi}$ is irrelevant for our purposes. The normalization factor $\frac{1}{\lambda^4}$ is needed prior to evaluation of the limit since $TV(\lambda) = \mathcal{O}(\lambda^4)$. Eqn. (4)

---

[2] The terminology is reminiscent of the 1D case, in which only annihilations occur generically.

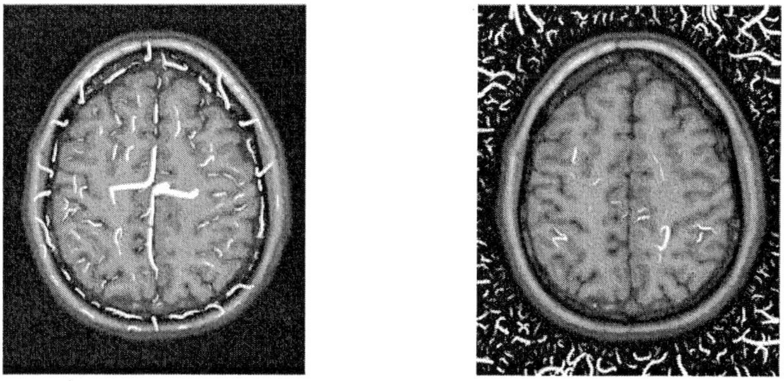

(a) Stable paths (b) Unstable paths

Fig. 3. Spatial projection of critical paths of a MR brain scan image. The paths are filtered by thresholding the stability norm of their top points. Most instabilities occur in flat regions, as expected

has been referred to by Koenderink as *deviation from flatness*, which can indeed be seen to be the differential counterpart of Eqn. (2). It enables us to calculate a stability measure for a top point *locally* by using only its second order derivatives. This stability norm can be used to weigh the importance of top points in our matching scheme, or to remove any unstable top points by thresholding them on their stability value. The latter is demonstrated in Fig. 3.

4 Construction of the Graph

The goal of our construction is two-fold. First, we want to encode the neighborhood structure of a set of points, explicitly relating nearby points to each other in a way that is invariant to minor perturbations in point location. Moreover, when local neighborhood structure does indeed change, it is essential that such changes will not affect the encoded structure elsewhere in the graph (image). The Delaunay triangulation imposes a position-based neighborhood structure with exactly these properties [15]. It represents a triangulation of the points which is equivalent to the nerve of the cells in a Voronoi tessellation, i.e., that triangulation of the convex hull of the points in the diagram in which every circumcircle of a triangle is an empty circle [13]. The edge set of our resulting graph will be based on the edges of the triangulation. Our second goal is to capture the scale space ordering of the points to yield a directed acyclic graph, with coarser scale top points directed to nearby finer scale top points.

The first step in constructing our graph G is the detection of top points and critical paths using ScaleSpaceViz [5]. The root of G, denoted as v_1, will correspond to the single critical path that continues up to infinity; note that there is no top point associated with this critical path, but simply its position at the coarsest scale. All other nodes in G, denoted as $v_2, ..., v_n$, correspond to the detected top points and their corresponding critical paths. $v_2, ..., v_n$ are ordered in decreasing order of the scale at which they are detected, e.g., v_2 is detected at a coarser scale than v_n. As we build the Delaunay triangulation of the points, we will simultaneously construct the DAG. Beginning with the root, v_1, we have a singleton point in our Delaunay triangulation, and a corresponding single node in G. Next, at the scale corresponding to v_2, we project v_1's position down to v_2's level, and recompute the triangulation. In this case, the triangulation

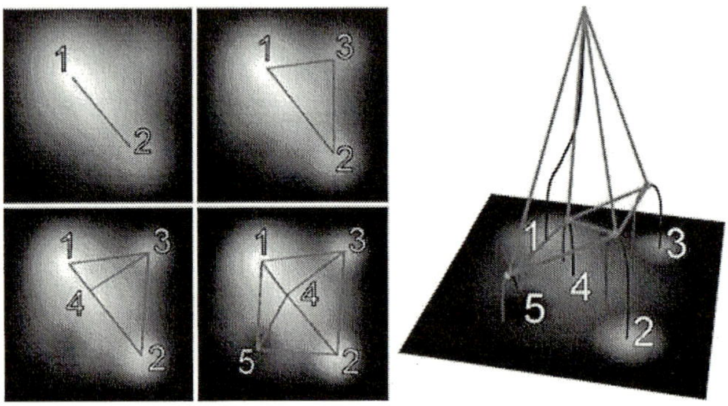

Fig. 4. Visualization of the DAG construction algorithm. Left: the Delaunay triangulations at the scales of the nodes. Right: the resulting DAG (edge directions not shown)

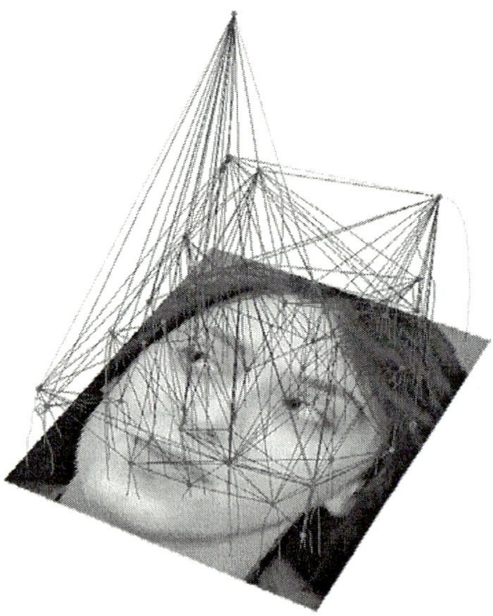

Fig. 5. The DAG obtained from applying Algorithm 1 to the critical paths and top points of the face in Fig. 2

yields an edge between v_1 and v_2. Each new edge in the triangulation yields a new edge in G, directed from coarser top points to finer top points; in this case, we add a directed edge in G from v_1 to v_2. We continue this process with each new top point, first projecting all previous top points to the new point's level, recomputing the triangulation, and using the triangulation to define new directed edges in G. A summary of this procedure is presented in Algorithm 1.

The construction is illustrated for a simple image in Fig. 4. In the top two frames in the left figure, we show the transition in the triangulation from v_2 (point 2) to v_3 (point 3); the root is shown as point 1. In the upper right frame, the triangulation consists of three edges; correspondingly, G has three edges: $(1,2),(1,3),(2,3)$, where (x,y) denotes an edge directed from node x to node y. In the lower left figure, point 4 is added to the triangulation, and the triangulation recomputed; correspondingly, we add edges $(1,4),(2,4),(3,4)$ to G (note that $(1,2)$ is no longer in the triangulation, but remains in G). Finally, in the lower right frame, point 5 is added, and the triangulation recomputed. The new edges in the triangulation yield new edges in G: (2,5),(4,5),(1,5). The right side of Fig. 4 illustrates the resulting graph (note that the directions of the edges are not shown). Fig. 5 is the result of applying this construction to the face of Fig. 2.

Algorithm 1 Top point graph construction procedure

```
Detect the critical paths.
Extract the top points from the critical paths.
Label the extremum path continuing up to infinity as v_1.
Label the rest of the nodes (critical paths, together with their top points) according
  to the scale of their top points from high scale to low as v_2, ..., v_n.
For i = 2 to n evaluate node v_i:
    Project the previous extrema into the scale of the considered node v_i.
    Calculate the 2D Delaunay triangulation of all the extrema at that scale.
    All connections to v_i in the Delaunay triangulation are stored as directed edges
  in G.
```

5 Experiments

To evaluate our construction, we explore the invariance of the construction to two types of perturbations. The first is the sensitivity of the construction to noise in the image, while the second is within-class deformation resulting in displacements of top points both within and across scales. We conduct our experiments using a subset of the Olivetti Research Laboratory face database. The database consists of faces of 20 people with 10 faces per person, for a total of 200 images; each image in the database is 112×92 pixels. The face images are in frontal view and differ by various factors such as gender, facial expression, hair style, and presence or absence of glasses. A representative view of each face is shown in Fig. 6. Invariance of a graph to noise or within-class deformation requires a measure of graph distance, so that the *distance* between the original and perturbed graphs

Fig. 6. Sample faces from 20 people

can be computed. For the experiments reported in this paper, we compute this distance using our many-to-many graph matching algorithm, which we briefly describe in the next subsection. Note that we have developed a general algorithm that is in no way specifically designed for face recognition. Therefore we have not compared our method to state-of-the-art face recognition algorithms. We present this experiment only as a proof of concept.

5.1 Overview of Matching Algorithm

The matching algorithm is based on the metric-tree representation of labelled graphs and their low-distortion embeddings into normed vector spaces via spherical coding [3, 9]. The advantage of this embedding technique is that it prescribes a single vector space into which both graphs are embedded. This two-step transformation reduces the many-to-many matching problem to that of computing a distribution-based distance measure between two such embeddings. To compute the distance between two sets of weighted vectors, we use a variation of Earth Mover's Distance under transformation sets. For two given graphs, the algorithm provides an overall similarity (distance) measure.

Fig. 7 presents an overview of the approach. For a given face, we first create its DAG according to Section 4 (Transition 1), and embed each vertex of the DAG into a vector space of prescribed dimensionality using a deterministic spherical coding (Transition 2). Finally (Transition 3), we compute the distance between the two distributions by the modified Earth Mover's Distance under transformation. The dimension of the target space in Transition 1 has a direct effect on the quality of the embedding. Specifically, as the dimensionality of the target space increases, the quality of the embedding will improve. Still, there exists an asymptotic bound beyond which increasing the dimensionality will no longer improve the quality of the embedding. Details on the many-to-many matching algorithm can be found in [3].

5.2 Graph Stability Under Additive Noise

To test the robustness of our graph construction, we first examine the stability of our graphs under additive Gaussian noise at different signal levels applied to the original face images. For this experiment, the database consists of the original 200 unperturbed images, while the query set consists of noise-perturbed

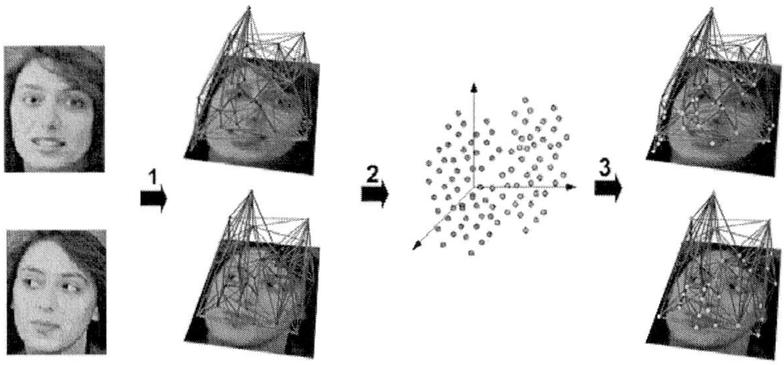

Fig. 7. Computing similarity between two given faces. (Matched point clusters are shaded with the same color.) See text

versions of the database images. Specifically, for each of the 200 images in the database, we create a set of query image by adding 1%, 2%, 4%, 8%, and 16% of Gaussian noise. Next, we compute the similarity between each query (perturbed database image) and each image in the database, and score the trial as correct if its distance to the face from which it was perturbed is minimal across all database images. This amounts to 40,000 similarity measurements for each noise level, for a total of 200,000 similarity measurements. Our results show that the recognition rate decreases down to 96.5%, 93%, 87%, 83.5%, and 74% for 1%, 2%, 4%, 8%, and 16% of Gaussian noise, respectively. These results indicate a graceful degradation of graph structure with increasing noise.

5.3 Graph Stability Under Within-Class Variation

To test the stability of the graph construction to within-class variation (e.g., different views of the same face), we first group the faces in the database by individual; these will represent our categories. Next, we remove the first image (face) from each group and compare it (the query) to all remaining database images. The image is then put back in the database, and the procedure is repeated with the second image from each group, etc., until all 10 face images of each of the 20 individuals have been used as a query. If the graph representation is invariant to within-class deformation, resulting from different viewpoints, illumination conditions, presence/absence of glasses, etc., then a query from one individual should match closest to another image from the same individual, rather than an image from another individual. The results are summarized in Table 1, Fig. 8.

The magnitudes of the distances are denoted by shades of gray, with black and white representing the smallest and largest distances, respectively. Due to symmetry, only the lower half of distance matrix is presented. Intra-object distances, shown along the main diagonal, are very close to zero.

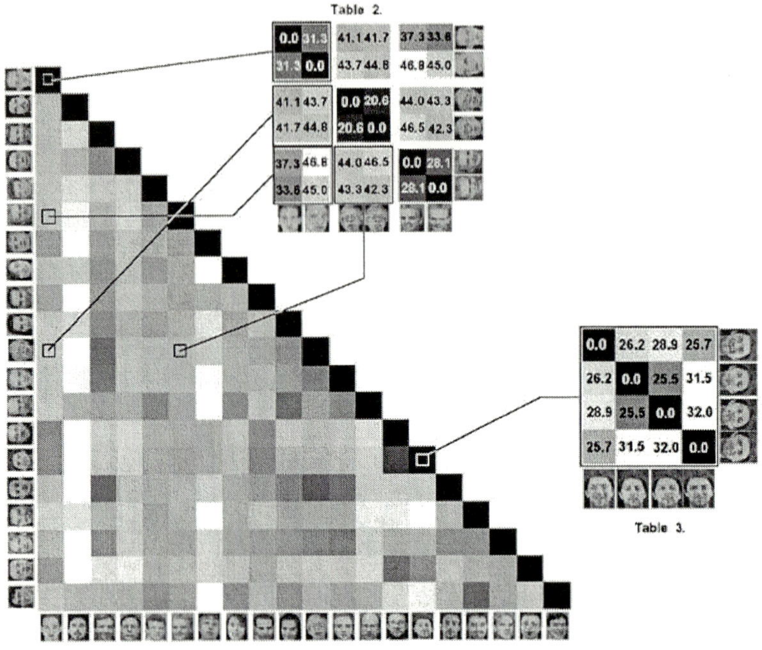

Fig. 8. Table 1: Matching results of 20 people. The rows represent the queries and the columns represent the database faces (query and database sets are non-intersecting). Each row represents the matching results for the set of 10 query faces corresponding to a single individual matched against the entire database. The intensity of the table entries indicates matching results, with black representing maximum similarity between two faces and white representing minimum similarity. Table 2: Subset of the matching results with the pairwise distances shown. Table 3: Effect of presence or absence of glasses in the matching for the same person. The results clearly indicate that the graph perturbation due to within-class deformation, including facial expression changes, illumination change, and the presence/absence of glasses is small compared to the graph distance between different classes

To better understand the differences in the recognition rates for different people, we randomly selected a subset of the matching results among three people in the database, as shown in Table 2, Fig. 8. Here, the (i, j)-th entry shows the actual distance between face i and face j. It is important to note that the distance between two faces of the same person is smaller than that of different people, as is the case for all query faces. In our experiments, one of our objectives was to see how various factors, such as the presence or absence of glasses, affects the matching results for a single person. Accordingly, we took a set of images from the database of one person, half with the same factor, and computed the distances between each image pair. Our results show that images with the same factors are more similar to each other than to others. Table 3 of Fig. 8 presents a subset of our results. As can be seen from the table, images of the same person with glasses are more similar than those of the same person with and without glasses. Still, in terms of categorical matching, the closest face always belongs to the same person. Although these results are encouraging, further evaluation on a larger database needs to be investigated to be more conclusive.

6 Conclusions

Imposing neighborhood structure on a set of points yields a graph, for which powerful indexing and matching algorithms exist. In this paper, we present a method for imposing neighborhood structure on a set of scale space top points. Drawing on the Delaunay triangulation of a set of points, we generate a graph whose edges are directed from top points at coarser scales to nearby top points at finer scales. The resulting construction is stable to noise, and within-class variability, as reflected in a set of directed acyclic graph matching experiments.

Acknowledgments

This work is part of the DSSCV project supported by the IST Program of the European Union (IST-2001-35443). Ali Shokoufandeh gratefully acknowledges the partial support provided by grants from National Science Foundation (NSF/EIA 02-05178), and the Office of Naval Research (ONR-N000140410363). Sven Dickinson gratefully acknowledges the support of NSERC, IRIS, PREA, and CITO.

References

1. J. Damon. Local Morse theory for solutions to the heat equation and Gaussian blurring. *Journal of Differential Equations*, 115(2):368–401, January 1995.
2. M. Fatih Demirci, A. Shokoufandeh, S. Dickinson, Y. Keselman, and L. Bretzner. Many-to-many matching of scale-space feature hierarchies using metric embedding. In *Proceedings, Scale Space Methods in Computer Vision, 4th International Conference*, pages 17–32, June 2003.
3. M. Fatih Demirci, A. Shokoufandeh, S. Dickinson, Y. Keselman, and L. Bretzner. Many-to-many feature matching using spherical coding of directed graphs. In *Proceedings, 8th European Conference on Computer Vision*, pages 332–335, May 2004.
4. L. Florack and A. Kuijper. The topological structure of scale-space images. *J. Math. Imaging Vis.*, 12(1):65–79, February 2000.
5. F.M.W. Kanters. Scalespaceviz. World Wide Web, `http://www.bmi2.bmt.tue.nl/image-analysis/people/FKanters/Software/ScaleSpaceViz.html`, 2004.
6. F.M.W. Kanters, L.M.J. Florack, B. Platel, and B.M. ter Haar Romeny. Image reconstruction from multiscale critical points. In *Proceedings of the 4th international conference on Scale Space Methods in Computer Vision (Isle of Skye, UK, June 2003)*, pages 464–478.

7. F.M.W. Kanters, B. Platel, L.M.J. Florack, and B.M. ter Haar Romeny. Content based image retrieval using multiscale top points. In *Proceedings of the 4th international conference on Scale Space Methods in Computer Vision (Isle of Skye, UK, June 2003)*, pages 33–43.
8. M. Kerckhove, editor. *Scale-Space and Morphology in Computer Vision: Proceedings of the Third International Conference, Scale-Space 2001, Vancouver, Canada*, volume 2106 of *Lecture Notes in Computer Science*. Springer-Verlag, Berlin, July 2001.
9. Y. Keselman, A. Shokoufandeh, M. F. Demirci, and S. Dickinson. Many-to-many graph matching via low-distortion embedding. In *Proceedings, Computer Vision and Pattern Recognition*, pages 850–857, 2003.
10. L. M. Lifshitz and S. M. Pizer. A multiresolution hierarchical approach to image segmentation based on intensity extrema. *IEEE Transactions on Pattern Analysis and Machine Intelligence*, 12(6):529–541, 1990.
11. M. Loog, J. J. Duistermaat, and L. M. J. Florack. On the behavior of spatial critical points under Gaussian blurring. a folklore theorem and scale-space constraints. In Kerckhove [8], pages 183–192.
12. M. Nielsen and M. Lillholm. What do features tell about images? In Kerckhove [8], pages 39–50.
13. A. Okabe and B. Boots. *Spatial Tesselations: Concepts and Applications of Voronoi Diagrams*. New York, 1992.
14. B. Platel, F.M.W. Kanters, L.M.J. Florack, and E.G. Balmachnova. Using multiscale top points in image matching. In *Proceedings of the 11th International Conference on Image Processing (Singapore, October 2004)*.
15. F. Preparata and M. Shamos. *Computational Geometry*. Springer-Verlag, New York, NY, 1985.
16. A. Shokoufandeh, S. Dickinson, K. Siddiqi, and S. Zucker. Indexing using a spectral encoding of topological structure. In *IEEE Conference on Computer Vision and Pattern Recognition*, pages 491–497, Fort Collins, CO, June 1999.
17. K. Siddiqi, A. Shokoufandeh, S. Dickinson, and S. Zucker. Shock graphs and shape matching. *International Journal of Computer Vision*, 30:1–24, 1999.

A Linear Image Reconstruction Framework Based on Sobolev Type Inner Products

Bart Janssen, Frans Kanters, Remco Duits, Luc Florack, and Bart ter Haar Romeny

Eindhoven University of Technology,
Den Dolech 2, Postbus 513, 5600 MB Eindhoven, The Netherlands
B.J.Janssen@student.tue.nl
{F.M.W.Kanters, R.Duits, L.M.J.Florack, B.M.terHaarRomeny}@tue.nl

Abstract. Exploration of information content of features that are present in images has led to the development of several reconstruction algorithms. These algorithms aim for a reconstruction from the features that is visually close to the image from which the features are extracted. Degrees of freedom that are not fixed by the constraints are disambiguated with the help of a so-called prior (i.e. a user defined model). We propose a linear reconstruction framework that generalises a previously proposed scheme. As an example we propose a specific prior and apply it to the reconstruction from singular points. The reconstruction is visually more attractive and has a smaller $\mathbb{L}_2$-error than the previously proposed linear methods.

1 Introduction

We describe a general method for reconstruction from scale space interest points and their differential attributes. Using the reconstruction the information content of these points can be investigated (Nielsen and Lillholm [11]).

Lillholm, Nielsen and Griffin [10, 11] have put emphasis on a "sparse" constraint set and the role of different priors. In general their priors are not given in terms of an inner product. The disadvantage of their approach is that the reconstruction algorithm is not linear and therefore slow and somewhat cumbersome to implement. Kanters et al. [9] use the assumption of a "sufficiently rich" set of constraints. The role of the prior is less significant so they chose for a standard $\mathbb{L}_2$-norm. We shall refer to this as the *standard linear reconstruction scheme*. Advantages of his approach are that the reconstruction algorithm is linear and analytical results for the generalised correlation matrix can be found. The disadvantage is that if the set of constraints is not sufficiently rich then this method is qualitatively outperformed by nonlinear reconstruction [10, 11].

We propose a general reconstruction framework which can be applied to a large set of priors. Any prior that can be described by a norm formed by an inner product can be mapped to this framework. Our method overcomes the disadvantages of the standard linear reconstruction scheme [9] while retaining

R. Kimmel, N. Sochen, J. Weickert (Eds.): Scale-Space 2005, LNCS 3459, pp. 85–96, 2005.

The filter shape in the spatial domain is somewhat harder to obtain. For two dimensions ($n = 2$) the convolution filter that represents the linear operator $(I - \gamma^2 \Delta)^{-1}$ equals

$$\phi_\gamma(x, y) = \frac{1}{2\pi\gamma^2} K_0[\frac{\sqrt{x^2 + y^2}}{\gamma}] \tag{22}$$

with K_0 representing the zeroth order modified Bessel function of the second kind. This was also noted by Florack, Duits and Bierkens [3] who worked on Tikhonov regularization and its relation to Gaussian scale space.

The calculation of the Gramm matrix G_{ij} is the computationally hardest part of the reconstruction algorithm. An analytic expression for this matrix is not available (unless $\gamma = 0$). Therefore the inner products $(\kappa_i, \kappa_j)_A$ have to be found by numerical integration. By the Parseval theorem we have (recall equations (21) and (22))

$$(\kappa_i, \kappa_j)_A = \left(\frac{1}{1 + \gamma^2 ||\omega||^2} \hat{\psi}_i, \hat{\psi}_j \right)_{\mathbb{L}_2} = \left(\frac{1}{1 + \gamma^2 ||\omega||^2}, \hat{\psi}_i^* \hat{\psi}_j \right)_{\mathbb{L}_2} = (\phi_\gamma, \psi_i^* * \psi_j)_{\mathbb{L}_2} \ . \tag{23}$$

At this point we have not yet specified the ψ_i filters. Since we are interested in the properties of singular points in Gaussian scale space we define the filters as follows.

Definition 4 (ψ_i). *A filter ψ_i is a localised derivative of the Gaussian kernel, recall equation (17), at a certain scale. Given $x, y, \xi, \eta \in \mathbb{R}$ and $m, n \in \mathbb{N}_0$*

$$\psi_i(x) \stackrel{def}{=} \varphi^{m,n}_{\sigma,\xi,\eta}(x, y) = \frac{1}{2\pi\sigma^2} \frac{\partial^{|m+n|} \varphi_s(x, y)}{\partial x^m \partial y^n} |_{x \longmapsto \frac{x-\xi}{\sigma}, y \longmapsto \frac{y-\eta}{\sigma}} \tag{24}$$

with $i \stackrel{def}{=} (m, n, \xi, \eta, \sigma) \in \mathbb{N}_0^2 \times \mathbb{R}^2 \times \mathbb{R}_+$.

Applying Definition 4 to equation (23) reveals that the inner products in the Gramm matrix can be expressed as a Gaussian derivative of the spatial representation of ϕ_γ. Note that this can be exploited for any operator that one chooses to use as a regulariser.

The singularity of $\phi_\gamma(x)$ in the origin gives rise to numerical problems. The Fourier representation $\hat{\phi}_\gamma(x)$ does not have a singularity, therefore the Fourier representation of the operator is sampled and after that a discrete inverse Fourier transform is applied to it.

At this point the Gramm matrix can be constructed. Inversion of this matrix is done by means of Singular Value Decomposition. The projection onto the filters can be done in either the frequency or the spatial domain. The image in the Fourier domain can be obtained by projecting onto the Fourier representations of the filters,

$$g = G^{ij} c_j \kappa_i = \mathcal{F}^{-1}[G^{ij} c_j \hat{\kappa}_i] \ . \tag{25}$$

This avoids problems with the singularity of the Bessel function. An Inverse Discrete Fourier Transform of the sampled reconstruction function results in the desired image.

3.4 Richer Features

Obtaining a visually appealing reconstruction from singular points can be achieved by selecting an "optimal" space for projection. This approach is discussed above. Another way to enhance the quality of the reconstruction is by using more information about the points that are used for reconstruction. In the standard case only up to second order differential structure was used. In our experiments also higher order differential properties of the singular points were taken into account. This has the side effect that the Gramm matrix will be harder to invert when more possibly dependent properties are used.

4 Evaluation

To evaluate the suggested prior and the proposed reconstruction scheme reconstructions from singular points of different images are performed. The singular points are obtained using ScaleSpaceViz [8], which is based on a zero-crossings method. After the singular points are found the unstable ones are filtered out by applying a threshold on the amount of structure that is present around a singular point. The amount of structure can be found by calculating the "differential total variation norm" or "deviation from flatness"

$$\mathrm{tv} = \sigma^4 \mathrm{Tr}\left(\mathbf{H}^2\right) \tag{26}$$

that was proposed by Platel et al. [12]. $\mathbf{H}$ represents the Hessian matrix and σ represents the scale at which the singular point appears. The reconstruction algorithm is implemented in Mathematica.

The images that are chosen to evaluate the performance of the reconstruction algorithm are those used by Kanters et al. and Lillholm et al. for the evaluation of their reconstruction algorithms [9, 10], *Lena's eye* and *MR brain*. The size of the former image is 64×64 pixels and the size of the latter image is 128×128 pixels. The pixel values of these images are integer valued ranging from 0 to 255.

4.1 Qualitative Evaluation

First we study reconstruction from singular points taking into account up to second order derivatives of the image at the locations of the singular points. Figure 2 shows the reconstruction from 31 singular points of *Lena's eye* . These points are selected using a tv-norm of 32. Note that the tv-norm scales with the square of the image range. The first image in the upper row displays the image from which the singular points were obtained. Successive images are reconstructions from these points with an increasing γ. The second image in the first row shows a reconstruction with $\gamma = 0$, which equals the reconstruction by Kanters et al. [9], and the first picture in the second row depicts the reconstruction with a minimal relative $\mathbb{L}_2$-error. The same convention is used in the reconstruction from 55 singular points of *MR brain* that is displayed in Figure 3.

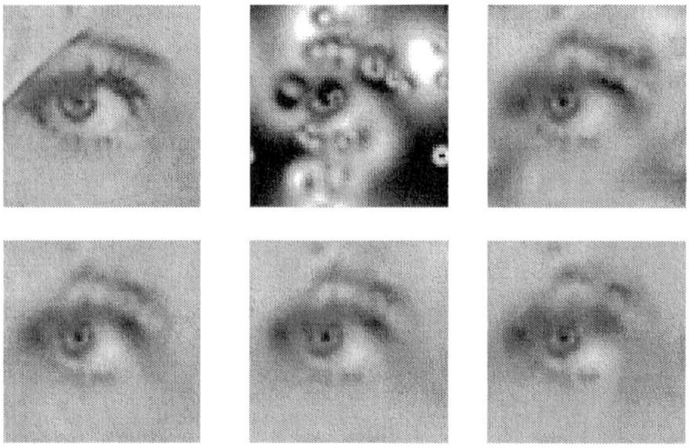

Fig. 2. Reconstruction from 31 singular points of *Lena's eye* with up to second order features. The upper row shows the original image and reconstructions with $\gamma = 0$ and $\gamma = 5$. The second row shows reconstructions with $\gamma = 22$, $\gamma = 50$ and $\gamma = 250$. The first image in the second row shows the reconstruction with the lowest relative $\mathbb{L}_2$-error

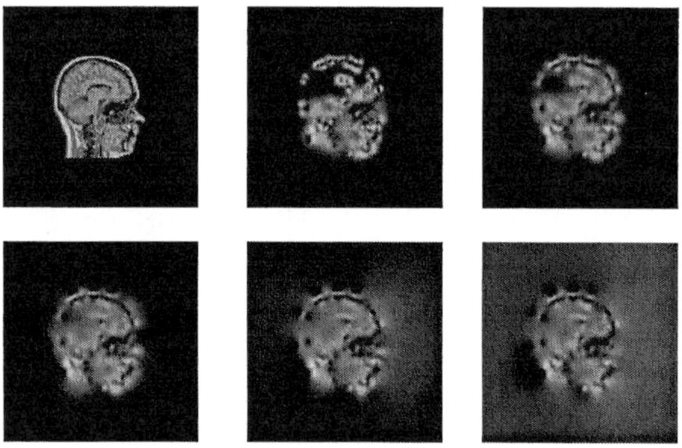

Fig. 3. Reconstruction from 55 singular points of *MR brain* with up to second order features. The upper row shows the original image and reconstructions with $\gamma = 0$ and $\gamma = 3$. The second row shows reconstructions with $\gamma = 7$, $\gamma = 50$ and $\gamma = 250$. The first image in the second row shows the reconstruction with the lowest relative $\mathbb{L}_2$-error

The singular points of this image were acquired using a tv-norm of 128. Figure 2 shows the "fill-in effect" of the smoothing prior. The reconstruction with the smallest relative $\mathbb{L}_2$-error is visually more appealing than the images with a smaller γ. A reconstruction with $\gamma = 250$ lacks details that were visible in the other reconstructions. This happens because the Gramm matrix is harder to invert when dependent basis functions are used. With an increasing γ the kernels become wider and thus more dependent on one another. The reconstructions of *MR brain* show "leaking" edges. Because the prior smoothes the image the very sharp edges of this image are not preserved and consequently the leaking effect appears.

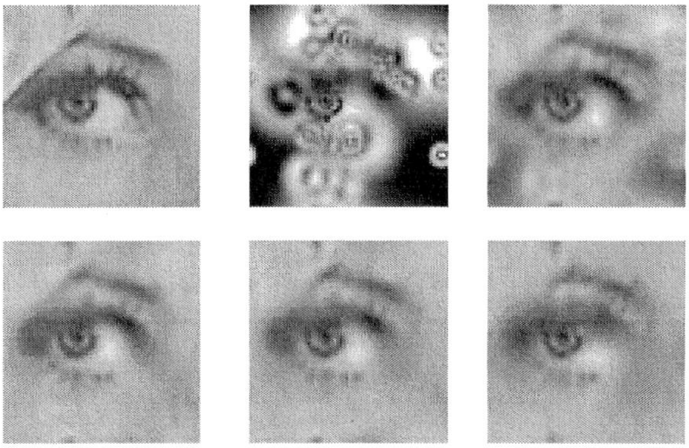

Fig. 4. Reconstruction from 31 singular points of *Lena's eye* with up to fourth order features. The upper row shows the original image and reconstructions with $\gamma = 0$ and $\gamma = 4$. The second row shows reconstructions with $\gamma = 19$, $\gamma = 50$ and $\gamma = 250$. The first image in the second row shows the reconstruction with the lowest relative $\mathbb{L}_2$-error

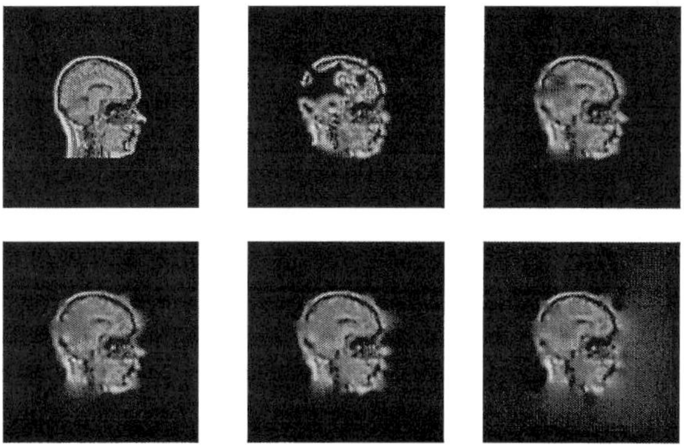

Fig. 5. Reconstruction from 55 singular points of *MR brain* with up to fourth order features. The upper row shows the original image and reconstructions with $\gamma = 0$ and $\gamma = 4$. The second row shows reconstructions with $\gamma = 8$, $\gamma = 50$ and $\gamma = 250$. The first image in the second row shows the reconstruction with the lowest relative $\mathbb{L}_2$-error

To investigate the influence of enrichment of the features the same experiments are repeated but up to fourth order derivatives are taken into account in the features. The results for the reconstruction from the singular points of *Lena's eye* can be found in Figure 4 and the results for the reconstruction from the singular points of *MR brain* are depicted in Figure 5. In both cases the images show more detail and are visually more appealing than their second order counter parts. The reconstruction of the *MR brain* image still shows leaking but this effect is reduced when compared to second order reconstruction.

4.2 Quantitative Evaluation

In order to verify the quality of the reconstructions of both images under a varying γ the relative $\mathbb{L}_2$-error,

$$\mathbb{L}_2\text{-error} = \frac{||f - g||_{\mathbb{L}_2}}{||f||_{\mathbb{L}_2}}\,, \tag{27}$$

of the reconstructed images is calculated. Figure 6 shows four graphs depicting this error for both second order and fourth order reconstruction of *Lena's eye* and *MR brain*. All graphs show that an optimal value exists for the γ parameter. This can be explained by the fact that the Gramm matrix is harder to invert with increasing γ due to increasing correlation among the filter cf. equation (23). Because of that dependent equations will be removed during the SVD. This leads to a reconstruction with less detail and thus a larger $\mathbb{L}_2$-error. The reconstructions of the *MR brain* image show an increasing $\mathbb{L}_2$- error with an increasing γ. This error becomes even larger than the $\mathbb{L}_2$-error of the reconstruction with $\gamma = 0$. This can be attributed to the sharp edges of the head that are smoothed and thus show leaking into the black surroundings of the head. The background clearly dominates the contribution to the $\mathbb{L}_2$-error. The reconstruction of *Lena's eye* does not suffer from this problem because of its smoothness.

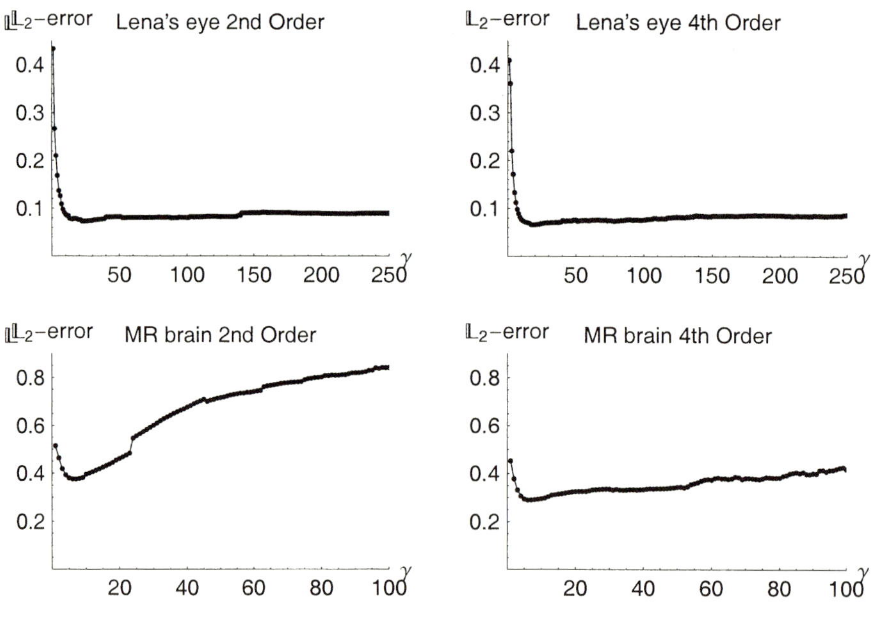

Fig. 6. The relative $\mathbb{L}_2$-error of the reconstructions from 31 singular points of *Lena's eye* (upper row) and 55 singular points of *MR brain* (lower row). The first column shows the $\mathbb{L}_2$-error for varying γ when second order reconstruction is used, i.e. up to second order derivatives are taken into account in the features. The second column displays fourth order reconstruction

5 Conclusions and Recommendations

We proposed a linear reconstruction method that leaves room for selection of arbitrary priors as long as the prior is a norm of Sobolev type. This greatly reduces the complexity of the reconstruction algorithm compared to non-linear methods.

We select one possible prior characterised by a free parameter γ that aims for a smooth reconstruction. This provides a control parameter for selecting different metameric reconstructions, i.e. reconstructions all consistent with the prescribed constraints. Comparisons with standard linear reconstruction as done by Kanters et al. [9] show it is possible to improve the reconstruction quality while retaining linearity. Reconstruction from a selection of singular points of the *MR brain* image proofs to be more difficult than reconstruction of smoother images like *Lena's eye*. The problem, that shows up as "leaking" edges, is reduced by taking higher order differential structure into account in the reconstruction algorithm. When the γ parameter is increased basis functions get more dependent on each other. This leads to a harder to invert Gramm matrix and consequently to a reduction of detail in the reconstruction.

Both, taking a $\gamma > 0$ and taking higher order features into account, lead to visually more appealing images and a smaller $\mathbb{L}_2$-error when comparing with standard linear reconstruction. It remains an open question how to select an optimal γ.

Future work will include the use of anisotropic basis functions that depend on the local image orientation and investigation of an adaptive γ parameter. Additionally other priors that fit in the proposed framework will be investigated.

A Simple Alternative Approach to Theorem 1

Recall that V is the span of the filters κ_i. Then

$$V^{\perp} = \{f \in \mathbb{L}_2(\mathbb{R}^2) \mid (\kappa_i, f)_A = 0 \; \forall \, i = 1, \dots, N\} \tag{28}$$

On the space of images $\mathbb{L}_2(\mathbb{R}^2)$ we define the following equivalence relation:

$$f \sim g \Leftrightarrow f - g \in V^{\perp} \; , \tag{29}$$

Notice that the set of equivalence/metameric classes is given by

$$\mathbb{L}_2(\mathbb{R}^2)/_{\sim} \stackrel{\text{def}}{=} \{[f] \mid f \in \mathbb{L}_2(\mathbb{R}^2)\} = \{f + V^{\perp} \mid f \in \mathbb{L}_2(\mathbb{R}^2)\} \tag{30}$$

and that an equivalence class $[f] = \{g \in \mathbb{L}_2(\mathbb{R}^2) \mid f \sim g\}$ of representant f is exactly given by those images that have the same features as image f. Notice to this end that

$$(\kappa_i, f)_A = (\kappa_i, g)_A \text{ for all } i = 1, \dots, N \Leftrightarrow f - g \in V^{\perp} \; . \tag{31}$$

Next we show that the unique element g within $[f]$ that minimizes the energy $E[g] = \|g\|_A^2$ is given by the A-orthogonal projection of f on V, $\mathcal{P}_V f$:

$$\min_{g \in [f]} \|g\|_A^2 = \min_{g \in [f]} \|g - \mathcal{P}_V f + \mathcal{P}_V f\|_A^2 = \min_{g \in [f]} \|g - \mathcal{P}_V f\|_A^2 + \|\mathcal{P}_V f\|_A^2 \tag{32}$$

and this equals $\|\mathcal{P}_V f\|_A^2$ only in the case $g = \mathcal{P}_V f$. Notice with respect to the last equality (equation (32)) is Pythogoras theorem, which can be applied since $(g - \mathcal{P}_V f) = (g - \mathcal{P}_V g) \in V^{\perp}$ and $\mathcal{P}_V f \in V$.

References

1. J. Damon. Local Morse theory for solutions to the heat equation and Gaussian blurring. *Journal of Differential Equations*, 115(2):368–401, January 1995.
2. R. Duits, L. Florack, B. ter Haar Romeny, and J. de Graaf. On the axioms of scale space theory. *Journal of Mathematical Imaging and Vision*, 20:267–298, 2004.
3. L. Florack, R. Duits, and J. Bierkens. Tikhonov regularization versus scale space: A new result. In *Proceedings of the 11th International Conference on Image Processing (Singapore, October 24–27, 2004)*, pages 271–274.
4. L. Florack and A. Kuijper. The topological structure of scale-space images. *Journal of Mathematical Imaging and Vision*, 12(1):65–79, February 2000.
5. R. Gilmore. *Catastrophe Theory for Scientists and Engineers.* Dover Publications, Inc., New York, 1993. Originally published by John Wiley & Sons, New York, 1981.
6. P. Johansen, M. Nielsen, and O. F. Olsen. Branch points in one-dimensional gaussian scale space. *Journal of Mathematical Imaging and Vision*, 13(3):193–203, 2000.
7. P. Johansen, S. Skelboe, K. Grue, and J. D. Andersen. Representing signals by their top points in scale-space. In *Proceedings of the 8th International Conference on Pattern Recognition (Paris, France, October 1986)*, pages 215–217. IEEE Computer Society Press, 1986.
8. F.M.W. Kanters. Scalespaceviz. `http://www.bmi2.bmt.tue.nl/image-analysis/people/FKanters/Software/ScaleSpaceViz.html`, 2004.
9. F.M.W. Kanters, B. Platel, L.M.J. Florack and B.M. ter Haar Romeny. Image reconstruction from multiscale critical points. In Lewis Griffin and Martin Lillholm, editors, *Scale Space Methods in Computer Vision, 4th International Conference, Scale Space 2003*, pages 464–478, Isle of Skye, UK, June 2003. Springer.
10. M. Lillholm, M. Nielsen, and L. D. Griffin. Feature-based image analysis. *International Journal of Computer Vision*, 52(2/3):73–95, 2003.
11. M. Nielsen and M. Lillholm. What do features tell about images? In M. Kerckhove, editor, *Scale-Space and Morphology in Computer Vision: Proceedings of the Third International Conference, Scale-Space 2001, Vancouver, Canada*, volume 2106 of *Lecture Notes in Computer Science*, pages 39–50. Springer-Verlag, Berlin, July 2001.
12. B. Platel, F.M.W. Kanters, L.M.J. Florack, and E.G. Balmachnova. Using multiscale top points in image matching. In *Proceedings of the 11th international conference on Image Processing (Singapore, October 2004).*
13. A. Tikhonov and V. Y. Arseninn. *Solution of Ill-Posed Problems.* John Wiley & Sons, New York, 1977.

Multi-scale Singularity Trees: Soft-Linked Scale-Space Hierarchies

Kerawit Somchaipeng[1,2], Jon Sporring[2], Sven Kreiborg[1], and Peter Johansen[2]

[1] 3D-Lab, School of Dentistry, Dept. of Pediatric Dentistry,
University of Copenhagen, Nørre Alle 20, DK-2200, Copenhagen N, Denmark
[2] Dept. of Computer Science, University of Copenhagen,
Universitetsparken 1, DK-2100, Copenhagen N, Denmark

Abstract. We consider images as manifolds embedded in a hybrid of a high dimensional space of coordinates and features. Using the proposed energy functional and mathematical landmarks, images are partitioned into segments. The nesting of image segments occurring at catastrophe points in the scale-space is used to construct image hierarchies called Multi-Scale Singularity Trees (MSSTs). We propose two kinds of mathematical landmarks: extrema and saddles. Unlike all other similar methods proposed hitherto, our method produces soft-linked image hierarchies in the sense that all possible connections are suggested along with their energies. The information added makes possible for directly estimating the stability of the connection and hence the costs of transitions. Aimed applications of MSSTs include multi-scale pre-segmentation, image matching, sub-object extraction, and hierarchical image retrieval.

1 Introduction

We are interested in the development of a multi-scale image representation which is expected to be invariant under certain image transformations and small image perturbations. Objects of any complexity when are observed at a distance, eventually reduce to an indistinguishable blob. Fine structures in the image are merged at small scales, larger structure at higher scales. To represent an image at all scales and all those discrete jumps of image complexity, tree structures instantly pop up as a natural choice. Image matching problems then can be thought of as tree matching problems. Using tree structures to represent images reduces computer vision problems to tree manipulation problems. In other words, we transform the relatively new and unfamiliar computer vision problems to the well-understood and profoundly-investigated mathematical problems. In practice, using tree structures also allows preferable possibility to compromise between speed and accuracy.

Possible applications of Multi-Scale Singularity Trees (MSSTs) include multi-scale pre-segmentation, image matching with MSSTs, sub-object extraction, hierarchical image retrieval in large image databases, etc. Together with our partners, we are currently developing an image matching algorithm based on our

R. Kimmel, N. Sochen, J. Weickert (Eds.): Scale-Space 2005, LNCS 3459, pp. 97–106, 2005.

proposed MSSTs and the Tree Edit Distance (TED) algorithm, where the distance between two images is found as the minimum cost of a series of edit operations that transforms the MSST of one image into another.

2 Gaussian Scale-Space

The $N+1$ dimensional Gaussian scale-space, $L : \mathbb{R}^{N+1} \to \mathbb{R}$, of an N dimensional image, $I : \mathbb{R}^N \to \mathbb{R}$, is an ordered stack of images, where each image is a blurred version of the former [1, 2, 3]. The blurring is performed according to the diffusion equation,

$$\partial_t L = \nabla^2 L \,, \tag{1}$$

where $\partial_t L$ is the first partial-derivative of the image in the scale direction t, and ∇^2 is the Laplacian operator, which in three dimensions reads $\partial_x^2 + \partial_y^2 + \partial_z^2$. The Gaussian kernel is the Green's function of the heat diffusion equation, i.e.

$$L(\cdot;t) = I(\cdot) \otimes g(\cdot;t) \,, \tag{2}$$

$$g(x;t) = \frac{1}{(4\pi t)^{N/2}} e^{-x^T x/(4t)} \,, \tag{3}$$

where $L(\cdot,t)$ is the image at scale t, $I(\cdot)$ is the original image, $\otimes$ is the convolution operator, $g(\cdot;t)$ is the Gaussian kernel at scale t, N is the image dimensionality, and $t = \sigma^2/2$, using σ as the standard deviation of the Gaussian kernel. The *Gaussian scale-space* is henceforth called the scale-space in this article.

The information in the scale-space is logarithmically degraded, the scale parameter is therefore often sampled exponentially using $\sigma(m) = \sigma_0 b^m$ for some base b. Since differentiation commutes with convolution and the Gaussian kernel is infinitely differentiable, differentiation of images in scale-spaces is conveniently computed as,

$$\partial_{x^n} L(\cdot;t) = \partial_{x^n} \left(I(\cdot) \otimes g(\cdot;t) \right) = I(\cdot) \otimes \partial_{x^n} g(\cdot;t) \,. \tag{4}$$

Alternative implementations of the scale-space are multiplication in the Fourier Domain, finite differencing schemes for solving the heat diffusion equation, additive operator splitting [4], and recursive implementation [5, 6].

Each method has different advantages and disadvantages. We prefer the spatial convolution, since it guarantees not to introduce spurious extrema in homogeneous regions at low scales. Typical border conditions are Dirichlet, Cyclic repetition, and Neumann boundaries. We use Dirichlet boundaries, where the image is extended with zero values in all directions according to the size of the convolution kernel.

Although the dimensionality of the constructed scale-space is one higher than the dimensionality of the original image, *critical points*, in the image at each scale are always points. The critical points treated in this article are extrema, $\partial_x L = \partial_y L = 0$, and the critical points are classified by the eigenvalues of the

Hessian matrix, the matrix of all second derivatives, computed at the critical point. Critical points with all positive eigenvalues are minima, critical points with all negative eigenvalues are maxima, and critical points with a mixture of both negative and positive eigenvalues are saddles.

As we increase the scale parameter, the critical points move smoothly forming *critical paths.* Along scale, critical points meet and annihilate or are created. Such events are called *catastrophic events*, and the points where they occur in the scale-space are called *catastrophe points.* The collection of catastrophic events at all scales is called the *deep structure* of the image.

The notion of genericity is used to disregard events that are not likely to occur for typical images, i.e. generic events are stable under slight perturbations of the image. There are only two types of generic catastrophic events in scale-space namely pairwise *creation events* and pairwise *annihilation events* [7]. It has further been shown that generic catastrophic events only involve pairs of critical points where one and only one eigenvalue of the Hessian matrix changes its sign, e.g. the annihilation of a minimum $(+,+)$ and a saddle $(+,-)$. A detailed discussion of a method for detecting critical paths and catastrophe points in the scale-space can be found in [8].

3 Building Scale-Space Hierarchies

There are already a few scale-space methods that construct image hierarchies of two-dimensional images proposed in the literatures so far [9, 10, 11]. To the best of our knowledge, an attempt to construct image hierarchies from the deep structure of three-dimensional images is first proposed in [12] followed by [8, 13, 14]. The latter scheme, which will be extended here, produces rooted ordered binary trees called *Multi-Scale Singularity Trees* (MSSTs) with catastrophe points as nodes.

In order to guarantee that the produced structures are always trees, our method only considers the linking of mathematical landmarks that exist at the original image. Only annihilation events are recorded and creation events are ignored. These creation events are generic however not frequently occur. Creation events could actually be included in the structure in the same manner but the method would inevitably produce graphs rather than trees. Theoretically, the method can be used to construct scale-space hierarchies of images of any dimensionality, assuming that critical paths and catastrophe points can be correctly detected. Our current implementation is capable of constructing MSSTs of two- and three-dimensional images.

3.1 Energy Functional and Energy Partitions

Given an image and a set of landmarks, we would like to partition the image into segments so that each segment contains exactly one landmark. Let $\Omega \subset \mathbb{R}^N$ be a compact connected domain and define $I : \Omega \to \mathbb{R}^+$ to be an image, $\boldsymbol{e} \in \Omega$ as a landmark, and $\boldsymbol{x} \in \Omega$ as a point in the domain. Consider a set of continuous functions $\gamma : [0, P] \to \Omega$ for which $\gamma(0) = \boldsymbol{e}$ and $\gamma(P) = \boldsymbol{x}$, $\gamma \in \Gamma_{\boldsymbol{ex}}$, where $\Gamma_{\boldsymbol{ex}}$ is the set of all possible paths in the domain from the landmark $\boldsymbol{e}$ to the point $\boldsymbol{x}$, and γ is parameterized using Euclidean arc-length. We define the energy $E_{\boldsymbol{e}}(\boldsymbol{x})$ with respect to an extremum $\boldsymbol{e}$ evaluated at $\boldsymbol{x}$ as,

$$E_{\boldsymbol{e}}(x) = \inf_{\gamma \in \Gamma_{\boldsymbol{ex}}} \int_0^P \sqrt{(\alpha - 1) \mid \frac{\partial \gamma(p)}{\partial p} \mid^2 + \alpha \mid \frac{\partial I(\gamma(p))}{\partial p} \mid^2 dp}\,, \tag{5}$$

for some $0 \leq \alpha \leq 1$Consider images as manifolds embedded in a high dimensional space, an N dimensional intensity image becomes an N dimensional manifold embedded in the hybrid $N + 1$ dimensional space of coordinates and features, the "space-feature" [15]. In this case the only feature is the intensity or the zeroth jet space. The energy functional can be defined for higher order jet space images, color images, or locally orderless images with scale-space histograms to handle texture [16], if a metric in the feature space is given. For two-dimensional images, image may be considered a height plot, and the energy at any point in the image with respect to a landmark can be found as the minimum weighted distance travelling up and down from the landmark to that point. The parameter α can be set to alter the emphasism of the energy functional between space and intensity. When $\alpha = 1$, the energy functional becomes the path variation, which is a generalization of the total variation [17]. The path variation depends solely on the image intensity and is invariant to affine transformation of the underlying space. Moreover, the energy is co-variant with scaling of the image intensity. When $\alpha \to 0$, the energy functional will increasingly depend on the spatial distance, and therefore become increasingly localized in space.

Let $\mathcal{E} \subset \Omega$ be the set of all landmarks in the image: An image segment or an *energy partition* S_i associated with a landmark $\boldsymbol{e}_i \in \mathcal{E}$ is defined as the set of all points in the images, where the energy $E_{\boldsymbol{e}_i}(\boldsymbol{x})$ is minimal,

$$S_i = \{\boldsymbol{x} \in \Omega | E_{\boldsymbol{e}_i}(\boldsymbol{x}) < E_{\boldsymbol{e}_i}(\boldsymbol{x}), \forall \boldsymbol{e}_j \in \mathcal{E}, i \neq j\}. \tag{6}$$

An approximation of the energy map $E_{\boldsymbol{e}_i} : \Omega \to \mathrm{I\!R}^+$, which gives the energy at every point in the image with respect to the landmark $\boldsymbol{e}_i$, can be efficiently calculated using the *Fast Marching Methods* [18].

The tessellation of the image segments obtained depends on the selection of the landmarks and the energy functional. Mathematical landmarks like the extrema seem to be the natural selection, since they are directly linked to the image content, i.e. significant features in the image usually contain at least one such points. Moreover, they can be easily and automatically detected in the scale-space, and the behavior of these critical points in the scale-space is well studied. The selection of the extrema as landmarks leads to Extrema-Based MSSTs. Another candidate for landmarks are the saddles, which leads to Saddle-Based MSSTs, which both will be discuss in the following.

3.2 Multi-scale Singularity Trees

MSSTs are constructed by connecting annihilating catastrophe points based on the nesting of image segments in the scale-space. Because of the natural pairwise interaction between critical points in the generic scale-space and the developed tree building scheme, resulting MSSTs are always rooted ordered binary tree.

MSSTs consist of nodes and their relations. Each MSST node consists of three important components: The image segment that immediately covers the area of the image segment disappearing at the catastrophe. For algorithmically convenience we denote the image 'surviving' image segment the *leftport*, the catastrophe for the *body*, and disappearing image segment for the *rightport*. Because there is exactly one image segment associated with a landmark and we choose the landmarks such that exactly one landmark disappears at an annihilation catastrophe point, then exactly one image segment also disappears.

A node $S_{left}C_{body}S_{right}$ is generated if an image segment S_{right} disappears at the catastrophe C_{body} inside an image segment S_{left}. The inclusion is easily determined by calculating the energy map with respect to the catastrophe C_{body}: the image segment S_{right} is nested in side the image segment S_{left} if the energy evaluated at the landmark of S_{left} is minimal among all landmarks existing at that scale.

As briefly mentioned above, MSSTs are ordered trees. This implies that connecting a node to another node as the leftchild or as the rightchild are semantically different events. MSSTs are built top-down starting from the catastrophe at the coarsest scale. A new node $N_{new} : S_{new,left}C_{new,body}S_{new,right}$ is connected as the leftchild of a node $N_i : S_{i,left}C_{i,body}S_{i,right}$ in the constructing MSST, if the node N_i does not have the leftchild and $S_{new,left} = S_{i,left}$, or as the rightchild, if the node N_i does not have the rightchild and $S_{new,left} = S_{i,right}$. It can easily be seen that this process is deterministic. We will now describe the algorithms for creating the Extrema- and the Saddle-Based MSSTs.

Extrema-Based MSSTs. Assuming that critical paths and catastrophe points in the scale-space are already and correctly detected, then the tree building algorithm is as follows:

1. Set the root of the tree as $BC_\infty E_{last}$, where B denotes the border of the image, E_{last} denotes the last extremum in scale, and C_∞ denotes the virtual catastrophe at scale infinity, where the border and the last extremum virtually annihilate.
2. At the highest unprocessed catastrophe C_{next} in scale, calculate the energy map with respect to the catastrophe and create a node $E_{cover}C_{next}E_{ann}$, where E_{ann} is the extremum that disappears at C_{next}, and the energy evaluated at the extremum E_{cover} is minimal among all extrema existing at that scale.
3. Link the new created node as the leftchild of a node in the tree that does not have the leftchild and where E_{cover} equals its leftport, or as the rightchild of a node in the tree that does not have the rightchild and where E_{cover} equals its rightport.
4. Repeat 2. until all catastrophe points are processed.

An example of Extrema-Based MSSTs, together with the deep structure it represents, are shown in Fig. 1.

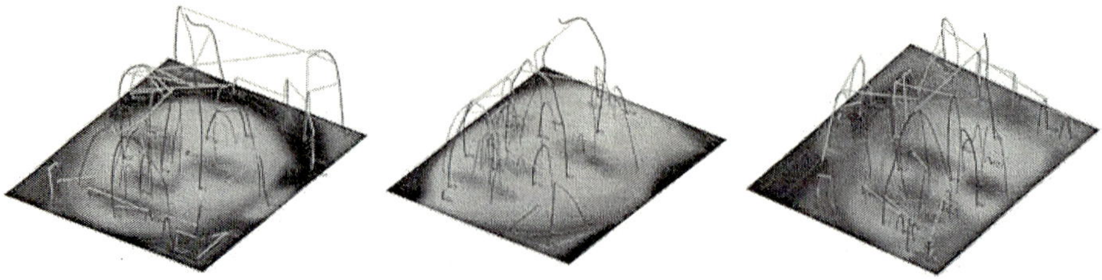

Fig. 3. Extrema-Based MSSTs of three different human faces

Notice Saddle-Based MSSTs always have one node less than that of Extrema-Based MSSTs of the same image. An example of Saddle-Based MSSTs is shown in Fig. 1, together with the deep structure it represents.

3.3 Energy Table

During the construction of MSSTs, the energies with respect to all catastrophe points in the scale-space evaluated at all present and relevant landmarks i.e. extrema for Extrema-Based MSSTs and saddles for Saddle-Based MSSTs, are stored in the *energy matrix*. It's this energy matrix that makes the tree soft-linked and makes possible the estimation of each connection stability and costs of MSST transitions.

3.4 MSST Examples of Real Images

We choose an image database of human faces [19] to test our implementation. One of the results is shown in Fig. 2. More results are shown in Fig. 3.

4 Concluding Remarks

It is important to understand the differences between two related stabilities: the stability of the positions of the catastrophe points and the stability of the relations between them (connections in the produced trees), and the fact that the stability of the constructed hierarchy depends on both. The positions of some catastrophe points are more stable than the others under small image perturbations. In general catastrophe points in an area with a lot of structures are more stable [20]. Using an appropriate stability norm, unstable catastrophe points can be eliminated before the hierarchy is constructed. In the other hand, the instability of their relations cannot and should not be avoided but have to be estimated if one wants to know the costs of the tree transitions.

We prefer our scale-space hierarchy building scheme because in contrast to hard-linked hierarchies, where only the best connection is suggested, our method produces soft-linked image hierarchies in the sense that all possible connections are suggested along with their energies. The connection with the lowest energy can be later selected in order to produce the tree that best represents the image.

The additional information, stored in the energy matrix, is useful in many ways. It makes possible for directly estimating the stability of each connection in the tree, and hence the cost of the tree transitions. For example, one might want to know how stable a particular connection is in the produced hierarchy under slight perturbations of the original images. Consider an image that is close to non-generic ones, e.g. having near-symmetric structures, only slight perturbations of the original image could change the structure of the produced hierarchy completely. For soft-linked hierarchies, the stability of each connection in the hierarchies can be directly estimated by looking at the distribution of the energies of all possible connections. If the energy of the best connection is much lower than all other connections, then the connection is not likely to switch, and hence is more stable. However, if the energies of all connections are about the same, the connection is very likely to switch and certainly is not stable. Because hard-linked hierarchy building methods naturally suggest only one best connection, there is no direct way of estimating the stability of connections of the produced hierarchies.

Acknowledgement

This work is part of the DSSCV project supported by the IST Programme of the European Union (IST-2001-35443).

References

1. Iijima, T.: Basic theory on normalization of a pattern (in case of typical one-dimensional pattern). Bulletin of Electrotechnical Laboratory **26** (1962) 368–388 (in Japanese).
2. Witkin, A.P.: Scale–space filtering. In: Proc. 8th Int. Joint Conf. on Artificial Intelligence (IJCAI '83). Volume 2., Karlsruhe, Germany (1983) 1019–1022
3. Koenderink, J.J.: The Structure of Images. Biological Cybernetics **50** (1984) 363–370
4. Weickert, J., Zuiderveld, K.J., ter Haar Romeny, B.M., Niessen, W.J.: Parallel implementations of AOS schemes: A fast way of nonlinear diffusion filtering. In: Proceedings of the 4th International Conference on Image Processing. Volume 3., Santa Barbara, CA, USA, IEEE Computer Society Press (1997) 396–399
5. Deriche, R.: Recursively Implementing the Gaussian and its Derivatives. In Srinivasan, V., Heng, O.S., Hock, A.Y., eds.: Proceedings of the 2nd Singapore International Conference on Image Processing. World Scientific, Singapore (1992) 263–267
6. van Vliet, L., Young, I., Verbeek, P.: Recursive Gaussian Derivative Filters. In: Proceedings of the 14th International Conference on Pateern Recognition ICPR'98. Volume 1., Brisbane, Australia, IEEE Computer Society Press (1998) 509–514.
7. Damon, J.: Local Morse theory for Gaussian blurred functions. In Sporring, J., Nielsen, M., Florack, L., Johansen, P., eds.: Gaussian Scale-Space Theory. Kluwer Academic Publishers, Dordrecht, The Netherlands (1997) 147–163

8. Somchaipeng, K., Sporring, J., Kreiborg, S., Johansen, P.: Software for Extracting Multi-Scale Singularity Trees. Technical report, Deliverable No.8, DSSCV, IST-2001-35443 (2003)
9. Lifshitz, L.M., Pizer, S.M.: A multiresolution hierarchical approach to image segmentation based on intensity extrema. IEEE Transaction on Pattern Analysis and Machine Intelligence **12** (1990) 529–541
10. Kuijper, A.: The deep structure of Gaussian scale space images. PhD thesis, Image Sciences Institute, Institute of Information and Computing Sciences, Faculty of Mathematics and Computer Science, Utrecht University (2002)
11. Platel, B.: Multiscale Hierarchical Segmentation. Technical Report BMT-Report no. 2002-04, Department of BioMedical Engineering, Technical Universitet of Eindhoven (2002)
12. O.F.Olsen, M.Nielsen: Multi-scale gradient magnitude watershed segmentation. In Bimbo, A.D., ed.: ICIAP '97 - 9th Int. Conf. on Image Analysis and Processing. Volume 1310 of Lectures Notes in Computer Science., Springer (1997) 6–13
13. Somchaipeng, K., Sporring, J., Kreiborg, S., Johansen, P.: Some Transitions of Extrema-Based Multi-Scale Singularity Trees. Technical report, DSAGM'04, Dept. of Computer Science, University of Copenhagen (DIKU) (2004)
14. Somchaipeng, K., Sporring, J., Kreiborg, S., Johansen, P.: Report on Assessment of MSSTs for 3D Matching Application. Technical report, Deliverable No.21, DSSCV, IST-2001-35443 (2004)
15. Kimmel, R., Sochen, N., Malladi, R.: From High Energy Physics to Low Level Vision. In: Scale-Space Theory in Computer Vision, Utrecht, The Natherlands (1997)
16. van Ginneken, B., ter Haar Romeny, B.M.: Applications of locally orderless images". In: Scale Space 1999. (1999)
17. Arbelaez, P.A., Cohen, L.D.: The Extrema Edges". In: Scale Space 2003, LNCS 2695. (2003) 180–195
18. Sethian, J.A.: Fast Marching Methods. SIAM Review **41** (1999) 199–235
19. Samaria, F., Harter, A.: Parameterisation of a Stochastic Model for Human Face Identification. In: Proceedings of 2nd IEEE Workshop on Applications of Computer Vision, Sarasota FL, December. (1994)
20. Platel, B., Florack, L.M.J., Kanters, F.M.W., Balmachnova, E.G.: Using Multiscale Top Points in Image Matching. (2004)

Image Deblurring in the Presence of Salt-and-Pepper Noise

Leah Bar[1], Nir Sochen[2], and Nahum Kiryati[1]

[1] School of Electrical Engineering
[2] Dept. of Applied Mathematics,
Tel Aviv University, Tel Aviv 69978, Israel

Abstract. The problem of image deblurring in the presence of salt and pepper noise is considered. Standard image deconvolution algorithms, that are designed for Gaussian noise, do not perform well in this case. Median type filtering is a common method for salt and pepper noise removal. Deblurring an image that has been preprocessed by median-type filtering is however difficult, due to the amplification (in the deconvolution stage) of median-induced distortion. A unified variational approach to salt and pepper noise removal and image deblurring is presented. An objective functional that represents the goals of deblurring, noise-robustness and compliance with the piecewise-smooth image model is formulated. A modified L^1 data fidelity term integrates deblurring with robustness to outliers. Elements from the Mumford-Shah functional, that favor piecewise smooth images with simple edge-sets, are used for regularization. Promising experimental results are shown for several blur models.

1 Introduction

Consider an image that has been blurred and contaminated by salt and pepper noise. Typical sources of blur are defocus and motion [3]. Salt and pepper noise is a common model for the effects of bit errors in transmission, malfunctioning pixels and faulty memory locations [5].

Significant attention has been given to image deblurring in the presence of Gaussian noise [3]. We focus on variational methods, that have an important role in modern image deblurring research, see e.g. [20, 21, 23, 14]. Most methods rely on the standard model $g = h * f + n$, that is applicable to a large variety of image degradation processes that are encountered in practice. Here h represents a known space-invariant blur kernel (point spread function), f is an ideal version of the observed image g and n is (usually Gaussian) noise. In this research, we focus on the case of salt and pepper noise.

The assumption of Gaussian noise is a fundamental element of common image deblurring algorithms. It is therefore not surprising that those algorithms produce inadequate results in the presence of salt and pepper noise. This fact is illustrated in Fig. 1. The top-left image in Fig. 1 is the 256×256 *Lena* image, blurred by a pill-box kernel of radius 3 (7×7 kernel) and contaminated

R. Kimmel, N. Sochen, J. Weickert (Eds.): Scale-Space 2005, LNCS 3459, pp. 107–118, 2005.

Fig. 1. Current image deblurring algorithms fail in the presence of salt and pepper noise. *Top-left:* Blurred image with Gaussian noise. *Top-right:* Restoration using the method of [23]. *Bottom-left:* Blurred image with salt and pepper noise. *Bottom-right:* Restoration using the method of [23]

by Gaussian noise. Successful restoration is obtained using the state of the art deblurring method of [23] (top-right). The bottom-left image in Fig. 1 is the same blurred Lena image, now contaminated by salt and pepper noise of density 0.01. In this case restoration using the method of [23] is clearly inadequate (bottom-right). Note that due to the inadequacy of the noise model, the algorithm of [23] yields poor results even at lower salt and pepper noise density. The regularization constants used to obtain Fig. 1 (top-right) and (bottom-right) are the same: 10^{-3}. Note that increasing the constant in the presence of salt and pepper noise effectively disables deblurring, while only reducing the amplitude of the noise.

Salt and pepper noise removal is considered in the literature by itself. It is commonly approached using median-type filters, see e.g [9, 13, 18]. Recently, a promising variational method for impulse denoising was proposed by [7, 16, 17].

In the absence of unified algorithms for deblurring and salt-and-pepper noise removal, the straightforward approach is to first denoise the image, then to deblur it. This two-stage method is however prone to failure, especially at high noise density. Image denoising using median-type filtering creates distortion that depends on the neighborhood size; this error can be strongly amplified by the deblurring process, even in regularized methods. Consider the example shown in Fig. 2. The top-left image is the 256×256 *Einstein* image, blurred using a

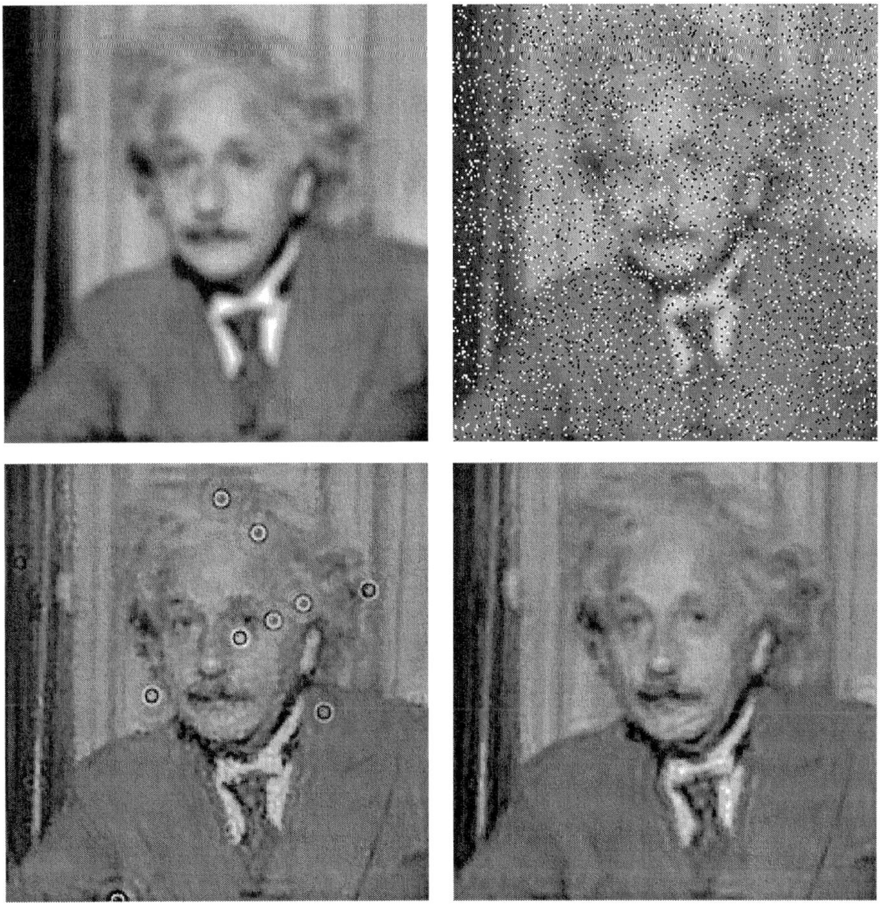

Fig. 2. The failure of the two-stage approach to salt-and-pepper noise removal and image deblurring. *Top-left:* Blurred image. *Top-right:* Blurred image contaminated by salt and pepper noise. *Bottom-left:* The outcome of 3×3 median filtering, followed by deblurring. *Bottom-right:* The outcome of 5×5 median filtering, followed by deblurring

pill-box kernel of radius 4. The blurred image with added salt and pepper noise (noise density 0.11) is shown top-right. The outcome of 3×3 median filtering followed by deblurring using the algorithm of [23] is shown bottom-left. At this noise level, the 3×3 neighborhood size of the median filter is insufficient, the noise is not entirely removed, and the residual noise is greatly amplified by the deblurring process. If the neighborhood size of the median filter is increased to 5×5, the noise is fully removed, but the distortion leads to inadequate deblurring (bottom-right).

In this paper we present a unified method for image deblurring and salt-and-pepper noise removal. Using a variational technique, we introduce a cost functional that represents the goals of deblurring, robustness to salt and pepper noise, and compliance with a piecewise-smooth image model. Experimental results exhibit effective image recovery, with various blur models and noise levels.

2 Unified Variational Framework

Image deblurring is an inverse problem, that can be formulated as a functional-minimization problem. Let Ω denote a rectangular domain in $\mathbb{R}^2$, on which the image intensity function $f : \Omega \to [0,1]$ is defined. Ideally, the recovered image $\hat{f}$ satisfies

$$\hat{f} = \arg\min_{f} \int_{\Omega} \Phi(h * f - g)dA, \tag{1}$$

where $\Phi(\cdot)$ is a metric representing data-fidelity. In the case of Gaussian noise, a quadratic data-fidelity term is used:

$$\Phi(h * f - g) = (h * f - g)^2. \tag{2}$$

The inverse problem represented by Eq. 1 is known to be ill-posed: small perturbations in the data may produce unbounded variations in the solution. To alleviate this difficulty, a regularization term, that reflects some a-priori preferences, is added. The functional to be minimized thus takes the form

$$\mathcal{F} = \int_{\Omega} \Phi(h * f - g)dA \ + \alpha \mathcal{J}(f) \tag{3}$$

where $\mathcal{J}(f)$ is the regularization operator and α is positive weighting scalar. Several regularization terms were suggested in the literature, for example Tikhonov [22] L^2 smoothness, Total variation (TV) L^1 norm [20,21], modified L^1 norm [1] and recently an integrated TV and wavelet coefficients regularization [10,11,14].

In the presence of salt and pepper noise, the quadratic data-fidelity term (2) is inadequate. In this paper, we use a robust (modified L^1 norm) data-fidelity term

$$\Phi(h * f - g) = \sqrt{(h * f - g)^2 + \eta^2}\ , \tag{4}$$

where η is a small constant. The modified L^1 norm shares the robustness to outliers of the L^1 norm, but avoids the resulting PDE from being singular at zero. Brox et al [6] have recently used the modified L^1 norm as a fidelity term for precise optical flow estimation.

The regularization terms that we use represent preference for piecewise-smooth images with simple edge sets. In the Mumford-Shah [15] functional, piecewise smooth images are favored by the term $\int_{\Omega\setminus K} |\nabla f|^2 dA$, where K is the edge set. The simplicity of the edge set is maintained in the Mumford-Shah functional by the line integral term $\int_K d\sigma$.

Ambrosio and Tortorelli [2] used the Γ-convergence framework to approximate the irregular Mumford-Shah functional by a sequence of regular functionals. The edge set K is approximated by a smooth auxiliary function v, where $v(x) \approx 0$ if $x \in K$ and $v(x) \approx 1$ otherwise. Mumford-Shah regularization, using the Γ-convergence approximation, has been recently used in electrical impedance tomography [19] and in blind image restoration [4].

The unified functional is

$$\mathcal{F}_\epsilon(f,v) = \int_\Omega \Phi(h * f - g)\, dA \;+\; \beta \int_\Omega v^2 |\nabla f|^2 dA \;+ \\ +\; \alpha \int_\Omega \left(\epsilon |\nabla v|^2 \;+\; \frac{(v-1)^2}{4\epsilon} \right) dA. \tag{5}$$

The first term in the functional is the modified L^1 data-fidelity term (4). The second term favors a piecewise smooth solution and corresponds to the term $\int_{\Omega\setminus K} |\nabla f|^2 dA$ in the Mumford-Shah functional. The third term maintains the simplicity of the edge set and corresponds to the line integral term $\int_K d\sigma$. Here ϵ is a small positive constant, and α and β are positive weights.

3 Minimization Techniques

The objective functional (5) depends on the functions f (recovered image) and v (approximated edge map). Minimization with respect to both f and v is carried out using the Euler-Lagrange (E-L) equations (6) and (8), subject to the Neumann boundary conditions $\partial v / \partial N = 0$, $\partial f / \partial N = 0$, where N denotes the normal to the boundary.

$$\frac{\delta \mathcal{F}_\epsilon}{\delta v} = 2\,\beta\, v\, |\nabla f|^2 + \alpha \left(\frac{v-1}{2\epsilon} \right) - 2\,\epsilon\, \alpha\, \nabla^2 v = 0 \tag{6}$$

$$\frac{\delta \mathcal{F}_\epsilon}{\delta f} = \Phi'(h * f - g) * h(-x,-y) - 2\beta\, \mathrm{Div}(v^2\, \nabla f) = 0 \tag{7}$$

Substituting the modified L^1 norm (4) yields

$$\frac{\delta \mathcal{F}_\epsilon}{\delta f} = \frac{(h * f - g)}{\sqrt{(h * f - g)^2 + \eta^2}} * h(-x,-y) - 2\beta\, \mathrm{Div}(v^2\, \nabla f) = 0\ . \tag{8}$$

Studying the objective functional (5), it can be seen that it is convex and lower bounded with respect to either of functions f and v if the other one is fixed. For example, given v, $\mathcal{F}_\epsilon$ is convex and lower bounded with respect to f. Therefore, following [8], the alternate minimization (AM) approach can be applied: in each step of the iterative procedure we minimize with respect to one function and keep the other one fixed.

Obviously, Eq. (6) is a linear partial differential equation with respect to v. On the contrary, (8) is a nonlinear integro-differential equation. Linearization of this equation is carried out using the fixed point iteration scheme, as in [23, 8]. We set $f = f^l$ in the denominator, and $f = f^{l+1}$ elsewhere, where l is the current iteration number. Eq. (8) can thus be rewritten as

$$\mathcal{H}(v, f^l) f^{l+1} = G(f^l), \qquad l = 0, 1, ... \tag{9}$$

where $\mathcal{H}$ is the linear integro-differential operator

$$\mathcal{H}(v, f^l) f^{l+1} = \frac{h * f^{l+1}}{\sqrt{(h * f^l - g)^2 + \eta^2}} * h(-x, -y) - 2\beta \, \mathrm{Div}(v^2 \, \nabla f^{l+1})$$

and

$$G(f^l) = \frac{g}{\sqrt{(h * f^l - g)^2 + \eta^2}} * h(-x, -y).$$

Note that (9) is now a *linear* integro-differential equation in f^{l+1}.

The two E-L equations (6) and (8) have now become two linear PDE's, that can be represented by two systems of linear equations. These systems are solved in alternation. This leads to the following iterative algorithm:

`Initialization:` $f^0 = g, \quad v^0 = 1.$

1. `Solve the Helmholtz equation for` v^{n+1}

$$\left(2\beta \, |\nabla f^n|^2 + \frac{\alpha}{2\epsilon} - 2\,\alpha\,\epsilon\,\nabla^2\right) v^{n+1} = \frac{\alpha}{2\epsilon}$$

2. `Set` $f^{n+1,0} = f^n$ `and solve for` f^{n+1} `(iterating on` l`)`

$$\mathcal{H}(v^{n+1}, f^{n+1,l}) f^{n+1,l+1} = G(f^{n+1,l}) \tag{10}$$

3. `if` $(||f^{n+1} - f^n||_{L_2} < \varepsilon_1 ||f^n||_{L_2})$ `stop.`

Here ε_1 is a small positive constant. Steps 1 and 2 both call for a solution of a system of linear equations. Step 1 was implemented using the Minimal Residual algorithm [24]. As for step 2, following Vogel and Oman [23], Eq. (10) can be expressed in a quasi-Newton like form

$$f^{n+1,l+1} = f^{n+1,l} - \left[\mathcal{H}(v^{n+1}, f^{n+1,l})\right]^{-1} R(v^{n+1}, f^{n+1,l}) \tag{11}$$

where

$$R(v, f) = \frac{(h * f - g)}{\sqrt{(h * f - g)^2 + \eta^2}} * h(-x, -y) - 2\beta \, \mathrm{Div}(v^2 \, \nabla f)$$

and $\mathcal{H}(\cdot,\cdot)$ is the approximation of the Hessian operator. It can be shown that the operator $\mathcal{H}(\cdot,\cdot)$ is self-adjoint and positive definite. Consequently $\mathcal{H}(\cdot,\cdot)^{-1}R(\cdot,\cdot)$ in (11) was computed via the Conjugate Gradients method.

Let f_{ij} denote the discretized image function. The forward and backward finite difference approximations of the derivatives $\partial f(x,y)/\partial x$ and $\partial f(x,y)/\partial y$ are respectively denoted by $\Delta^x_\pm f_{ij} = \pm(f_{i\pm1,j} - f_{ij})$ and $\Delta^y_\pm f_{ij} = \pm(f_{i,j\pm1} - f_{ij})$. Hence, the discrete form of Eq. (6) is

$$2\beta v_{ij}\left[(\Delta^x_+ f_{ij})^2 + (\Delta^y_+ f_{ij})^2\right] + \alpha\cdot\frac{v_{ij}-1}{2\epsilon} - 2\alpha\epsilon\left(\Delta^x_-\Delta^x_+ v_{ij} + \Delta^y_-\Delta^y_+ v_{ij}\right) = 0,$$

and $\mathrm{Div}(v^2\,\nabla f)$ in Eq. (8) is approximated by

$$\left(\Delta^x_+(v^2_{ij}\Delta^x_-) + \Delta^y_+(v^2_{ij}\Delta^y_-)\right) f_{ij}.$$

In the discrete case, the Neumann boundary conditions were implemented as follows. The observed image was extended by adding margins that are a few pixels wide. These margins were obtained by replicating the one-pixel thick outer frame of the image. The margins were then convolved with the blur kernel. To avoid artifacts, in the presence of salt and pepper noise, care should be taken to ensure that the outer frame of the image is noise free. This limited task can easily be achieved using a median filter.

All convolution procedures were performed in the Fourier Transform domain. The algorithm was implemented in the MATLAB environment.

4 Experimental Results

The performance of the algorithm is presented in Figs. 3, 4 and 5. Fig. 3 (left) is a blurred and noisy version of the *Einstein* image. The blur kernel was a pill-box of

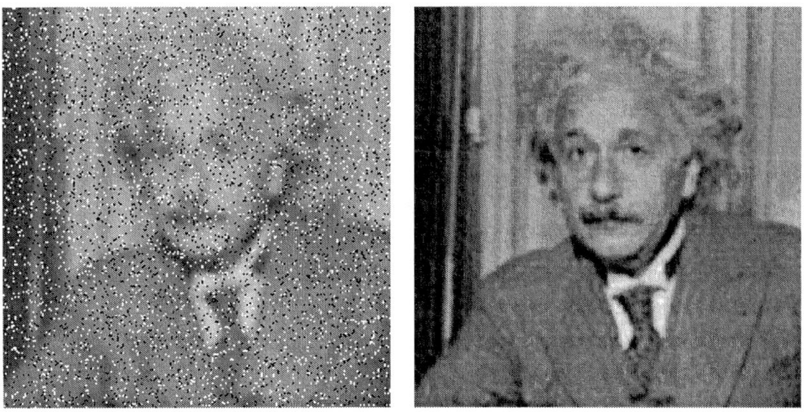

Fig. 3. Deblurring in the presence of salt and pepper noise. *Left:* Source image, blurred with a pill-box kernel of radius 4, and contaminated by noise of density 0.11. *Right:* Recovered image, using the suggested algorithm

Fig. 4. *Left column:* The *Lena* image, blurred with a pill-box kernel of radius 3, and contaminated by salt and pepper noise. The noise densities are (top to bottom) 0.01, 0.1 and 0.3. *Right column:* The corresponding recovered images

Fig. 5. The case of motion blur. *Top-left:* Blurred and noisy image. *Top-right:* Restoration using the proposed method. *Bottom-left:* The outcome of 3×3 median filtering followed by Lucy-Richardson restoration (Matlab: `deconvlucy`). *Bottom-right:* The outcome of 5×5 median filtering followed by Lucy-Richardson restoration

radius 4; the noise density was 0.11. Fig. 3 (right) is the outcome of the suggested method. The parameters were $\beta = 0.5, \alpha = 0.5, \epsilon = 0.1$. The superiority of the proposed method, with respect to the sequential one (Fig. 2), is clear.

In all the examples in this section, the convergence tolerance of $\varepsilon_1 = 1 \cdot 10^{-4}$ was reached with 3-5 external iterations (over n). The number of internal iterations (over l) was set to 5. The constant η (Eq. 4) was set to 10^{-4}.

The examples presented in Fig. 4 demonstrate the performance of the algorithm at a variety of noise levels. The images in the left column were all blurred by a pill-box kernel of radius 3. The noise densities were, from top to bottom, 0.01, 0.1 and 0.3. The corresponding recovered images are shown in the right column. Despite the large variability of the noise level, the stability of the algorithm allowed to use the same parameter set in the three cases: $\beta = 0.5, \alpha = 0.5, \epsilon = 0.1$, as in the previous example.

Recovery of motion blur in the presence of salt and pepper noise is demonstrated in Fig. 5. The 256×256 *cameraman* image was blurred by a motion blur kernel of length=8, oriented at an angle $\theta = 25^o$ with respect to the horizon. The blurred image was further contaminated by salt and pepper noise of density 0.1 (top-left). The outcome of the method suggested in this paper (with $\beta = 0.6, \alpha = 0.01, \epsilon = 0.1$) is shown top-right. The inadequacy of the sequential strategy, of median filtering followed by conventional deconvolution is demonstrated in the bottom row. The left image in that row is the outcome of 3×3 median filtering followed by the well known Lucy-Richardson restoration (Matlab: `deconvlucy`). The right image in the bottom row was obtained in a similar way, but with a 5×5 median filter.

5 Discussion

We presented a method for image deblurring in the presence of salt and pepper noise. Our unified approach to deblurring and outlier removal is novel and unique. Experimental results demonstrate the superiority of the suggested method with respect to a sequential approach, in which median-based noise removal and image deconvolution are separate steps.

The algorithm is fast, robust and stable. Computation time for 256×256 images is about 3 minutes, using interpreted MATLAB on a 2GHz PC. The robustness of the algorithm is demonstrated by the fact that similar parameters can be used in the processing of different images. For example, the same param-

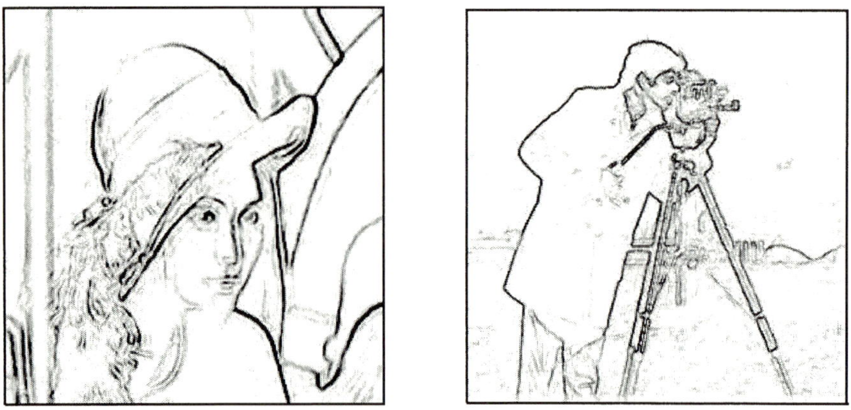

Fig. 6. Approximated edge maps obtained as a by-product of the restoration process. *Left:* The v-function that corresponds to the deblurring of the *Lena* image with a pill-box kernel and noise density 0.1. *Right:* The v-function that corresponds to the deconvolution of the *Cameraman* image with motion-blur and noise density 0.1

eter were used in Fig. 3 and in the three cases shown in Fig. 4. Furthermore, note the fast numerical convergence in our experiments.

In the variational approach, image deblurring in the presence of noise is expressed as a functional minimization problem. The functional consists of a data fidelity term and a regularization term, that stabilizes the inherent ill-posedness of the image deconvolution problem. The data fidelity term used in this study is the modified L^1 norm. It is more robust than the common L^2 norm for images contaminated by outliers, and yet it is still differentiable and convex.

Elements from the Mumford-Shah segmentation functional, in the Γ-convergence formulation, served as the regularization term. They reflect the profound piecewise-smooth image model. Unlike total variation, the alternative edge-preserving stabilizer, the selected regularization term does not induce nonlinearity beyond that of the fidelity term. An additional advantage of this method is the production of the auxiliary function v, that is an approximated edge map corresponding to the image. For example, Fig. 6 shows the v-maps obtained during the processing of the blurred and noisy *Lena* (pill-box blur, Fig. 4) and *Cameraman* (motion-blur, Fig. 5) images. Finally, Mumford-Shah regularization has profound theoretical advantages with respect to other edge preserving methods. These aspects will be discussed in the full-length version of this paper.

Acknowledgment

This research was supported by MUSCLE: Multimedia Understanding through Semantics, Computation and Learning, a European Network of Excellence funded by the EC 6th Framework IST Programme. It was supported also by the Israel Science Foundation. Leah Bar was supported by the Weinstein Center for Signal Processing Research at Tel Aviv University.

References

1. R. Acar and C.R. Vogel, "Analysis of Total Variation Penalty Methods", *Inverse Problems*, Vol.10, pp. 1217-1229, 1994.
2. L. Ambrosio and V.M. Tortorelli, "Approximation of Functionals Depending on Jumps by Elliptic Functionals via Γ-Convergence", *Communications on Pure and Applied Mathematics*, Vol. XLIII, pp. 999-1036, 1990.
3. M. Banham and A. Katsaggelos, "Digital Image Restoration", *IEEE Signal Processing Mag.*, Vol. 14, pp. 24-41, 1997.
4. L. Bar, N. Sochen and N. Kiryati, "Variational Pairing of Image Segmentation and Blind Restoration", Proc. ECCV'2004, Prague, Czech Republic, Part II: *LNCS* #3022, pp. 166-177, Springer, 2004.
5. G.R. Arce, J.L. Paredes and J. Mullan, "Nonlinear Filtering for Image Analysis and Enhancement", in A.L. Bovik (Ed.), *Handbook of Image & Video Processing*, Academic Press, 2000.
6. T. Brox, A. Bruhn, N. Papenberg and J. Weickert, "High Accuracy Optical Flow Estimation Based on a Theory for Warping", Proc. ECCV'2004, Prague, Czech Republic, Part IV: *LNCS* #3024, pp. 25-36, Springer, 2004.

7. R.H. Chan, C. Ho and M. Nikolova, "Salt-and-Pepper Noise Removal by Median-type Noise Detectors and Detail-preserving Regularization", to appear.
8. T.F. Chan and C. Wong, "Total Variation Blind Deconvolution", *IEEE Trans. Image Processing*, Vol. 7, pp. 370-375, 1998.
9. T. Chen and H.R. Wu, "Space Variant Median Filters for the Restoration of Impulse Noise Corrupted Images", *IEEE Trans. Circuits and Systems II*, Vol. 48, pp. 784-789, 2001.
10. S. Durand and J. Froment, "Reconstruction of Wavelet Coefficients Using Total Variation Minimization", *SIAM Journal of Scientific Computing*, Vol. 24, pp. 1754-1767, 2003.
11. S. Durand and M. Nikolova, "Restoration of Wavelet Coefficients by Minimizing a Specially Designed Objective Function", Proc. IEEE Workshop on Variational, Geometric and Level Set Methods in Computer Vision, pp. 145-152, 2003.
12. F. Malgouyres, "Minimizing the Total Variation Under a General Convex Constraint", *IEEE Trans. Image Processing*, Vol. 11, pp. 1450-1456, 2002.
13. H. Hwang and R. A. Haddad, "Adaptive Median Filters: New Algorithms and Results", *IEEE Trans. Image Processing*, Vol. 4, pp. 499-502, 1995.
14. J. Bect, L. Blanc-Feraud, G. Aubert and A. Chambolle, "A l^1-Unified Variational Framework for Image Restoration", Proc. ECCV'2004, Prague, Czech Republic, Part IV: *LNCS* #3024, pp. 1-13, Springer, 2004.
15. D. Mumford and J. Shah, "Optimal Approximations by Piecewise Smooth Functions and Associated Variational Problems", *Communications on Pure and Applied Mathematics*, Vol. 42, pp. 577-684, 1989.
16. M.Nikolova, "Minimizers of Cost-Functions Involving Nonsmooth Data-Fidelity Terms: Application to the Processing of Outliers", *SIAM Journal on Numerical Analysis*, Vol. 40, pp. 965-994, 2002.
17. M. Nikolova, "A Variational Approach to Remove Outliers and Impulse Noise", *Journal of Mathematical Imaging and Vision*, Vol. 20, pp. 99-120, 2004.
18. G. Pok, J.-C. Liu and A.S. Nair, "Selective Removal of Impulse Noise based on Homogeneity Level Information", *IEEE Trans. Image Processing*, Vol. 12, pp. 85-92, 2003.
19. L. Rondi and F. Santosa, "Enhanced Electrical Impedance Tomography via the Mumford-shah Functional", *ESAIM: Control, Optimization and Calculus of Variations*, Vol. 6, pp. 517-538, 2001.
20. L. Rudin, S. Osher and E. Fatemi, "Non Linear Total Variation Based Noise Removal Algorithms", *Physica D*, Vol. 60, pp. 259-268, 1992.
21. L. Rudin and S. Osher, "Total Variation Based Image Restoration with Free Local Constraints", *Proc. IEEE ICIP*, Vol. 1, pp. 31-35, Austin TX, USA, 1994.
22. A. Tikhonov and V. Arsenin, "Solutions of Ill-posed Problems", New York, 1977.
23. C. Vogel and M. Oman, "Fast, Robust Total Variation-based Reconstruction of Noisy, Blurred Images", *IEEE Trans. Image Processing*, Vol. 7, pp. 813-824, 1998.
24. E.W. Weisstein *et al*, "Minimal Residual Method", from *MathWorld–A Wolfram Web Resource*. `http://mathworld.wolfram.com/MinimalResidualMethod.html`

Phase Contrast MRI Segmentation Using Velocity and Intensity

Markus Persson[1], Jan Erik Solem[1], Karin Markenroth[2], Jonas Svensson[3], and Anders Heyden[1]

[1] Applied Mathematics Group, School of Technology and Society, Malmö University, Sweden
[2] Philips Medical Systems/MR Department, Lund University Hospital, Sweden
[3] Department of Radiation Physics, Malmö University Hospital, Sweden
markus.persson@ts.mah.se

Abstract. This paper presents a method for three-dimensional (3D) segmentation of blood vessels, i.e. determining the surface of the vessel wall, using a combination of velocity data and magnitude images obtained using phase contrast MRI. In addition to standard MRI images, phase contrast MRI gives velocity information for blood and tissue in the human body. The proposed method uses a variational formulation of the segmentation problem which combines different cues; velocity and magnitude. The segmentation is performed using the level set method. Experiments on phantom data and clinical data support the proposed method.

1 Introduction

Magnetic Resonance Imaging (MRI) provides three-dimensional images that are useful for diagnostic purposes. Phase contrast MRI provides additional velocity measurements that can be used for analysis of the blood flow and tissue motion. This paper presents a method for 3D segmentation of blood vessels. After segmentation, when the shapes of the vessels have been determined, volumetric flow data can be obtained from the velocity data. Computer aided segmentation is a well studied problem within medical image analysis and have great impact on diagnostic performance. In the case of 3D images and dynamic images it is of special importance because of the rather time-consuming task of manually segmenting huge amounts of data. Blood volume, pressure and velocity of the blood flow and the motion and shape of the vessel walls are examples of useful measurements that can aid the diagnosis and are also important for research within the medical field.

This paper deals with the problem of segmentation of moving and deforming vessels using velocity and intensity data. This covers many typical cases for medical images taken for diagnostic purposes, such as dynamic cardiac images. Many different methods have been proposed to analyze and extract shape information from cardiac images and 3D model-based methods have shown to improve the diagnostic value [1].

R. Kimmel, N. Sochen, J. Weickert (Eds.): Scale-Space 2005, LNCS 3459, pp. 119–130, 2005.

The most basic approach to segmentation is to threshold the images, cf. [2]. However, this method only works on very simple images. Later approaches makes use of moving interfaces, e.g. snakes and active contours, cf. [3] and the geodesic active contours [4]. During the latest years level set implementations have become popular since they can handle global properties, change of topology and are based on a solid mathematical framework, cf. [5, 6].

The level set method [7] is a popular technique for representing and tracking dynamic interfaces. The surface is represented implicitly as the zero level set of a function. The sign of the level set function gives a natural partitioning and is frequently used in segmentation. Our segmentation problem is solved using a variational level set framework. The optimal solution, i.e. the optimal position of the surface separating the blood from the vessel walls, corresponds to the minimum of a functional.

1.1 Relation to Previous Work

Aligning curves and surfaces to image gradient data have been analyzed in great detail in e.g. [8] in a geodesic active contour framework where curves are aligned to edges taking into account both direction and magnitude of the gradient.

In [9] blood is segmented in cardiac phase contrast MR images using the fact that the heart wall has a periodic motion and resumes its position after a completed heart cycle. The method is based on a particle trace technique for time-resolved 3D velocity vector fields, combined with magnitude image data. In [10] the myocardium is segmented from MRI intensity and phase contrast images. The segmentation is performed in 2D images using level set curve evolution. Three different constraints determine the curve motion, the intensity gradient, the velocity magnitude and the coherence of the velocity direction. For a general variational method using image intensity variations as a cue in segmentation as in the proposed method but in a statistical framework cf. [11]. In [12] segmentation of curvilinear structures in MR angiography images is performed using evolution techniques for implicit curves. In [13] blood vessels are segmented from MR angiography images using the criterion that the blood vessel boundary should be orthogonal to a vector flow field and thus minimizes the flux through the surface. Tubular structures are segmented from standard MRI in [14]. The method presented in [15] handles the additional difficulty of segmenting vessels with non-stationary walls.

1.2 Contribution of the Paper

This paper presents a method for 3D segmentation of blood vessels and determining the surface of the vessel wall by combining the velocity vector field obtained from phase contrast MRI measurements with MRI intensity gradients. The segmentation is formulated as a variational problem combining a novel functional incorporating velocity data such as velocity magnitude and discontinuities in velocity direction, using the discontinuity measure from [15], with an alignment functional introduced by [8]. The proposed method improves on the performance of the method presented in [15] especially by incorporating intensity information as well as obtaining higher numerical stability.

2 Background

2.1 Phase Contrast MRI

Phase contrast MRI is based on the property that a uniform motion of tissue in a magnetic field gradient produces a change in the MR signal phase, Φ. This change is proportional to the velocity of the tissue, $\mathbf{v}$. The MR signal from a volume element accumulates the phase [16]

$$\begin{aligned}\Phi(r,T) &= \gamma B_0 T + \gamma \mathbf{v} \int_0^T \mathbf{G}(r,t)t\,dt \\ &= \gamma B_0 T + \gamma \mathbf{v} \cdot \overline{\mathbf{G}} \ ,\end{aligned}$$

during time T, where B_0 is a static magnetic field, γ the gyro-magnetic ratio and $\mathbf{G}(r,t)$ is the magnetic field gradient. Notice that $\overline{\mathbf{G}}$ is exactly the first moment of $\mathbf{G}(r,t)$ with respect to time. If the field gradient is altered between two consecutive recordings, then by subtracting the resulting phases

$$\Phi_1 - \Phi_2 = \gamma \mathbf{v} \cdot (\overline{\mathbf{G}}_1 - \overline{\mathbf{G}}_2) \ ,$$

the velocity in the $(\overline{\mathbf{G}}_1 - \overline{\mathbf{G}}_2)$-direction is implicitly given. In this way a desired velocity component can be calculated for every volume element simultaneously. To construct the velocity vector in 3D, the natural way is to apply appropriate gradients to produce the x-, y- and z-components respectively.

2.2 Measure for Velocity Discontinuities

The velocity can be expressed in vector form $\mathbf{v} = (v_x, v_y, v_z)$, and the velocity magnitude is

$$|\mathbf{v}| = \sqrt{\mathbf{v}\cdot\mathbf{v}} = \sqrt{v_x^2 + v_y^2 + v_z^2} \ .$$

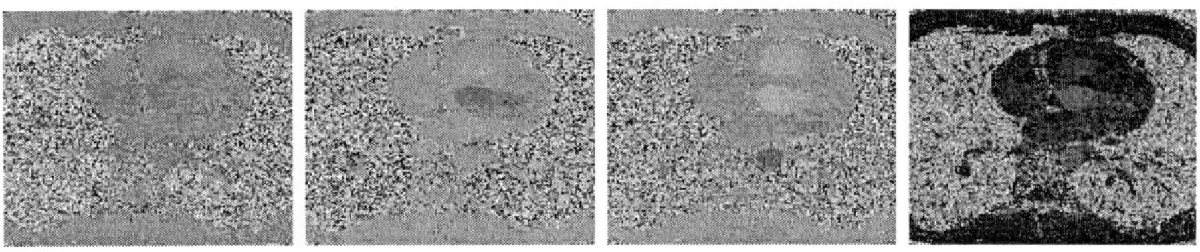

Fig. 1. Examples of the velocity vector field for one horizontal slice of the total volume. (left to right) v_x, v_y, and v_z-velocity component and velocity magnitude $|\mathbf{v}|$

Examples of the individual components are shown in Figure 1.

There are two important reasons why the velocity field $\mathbf{v}$ should be used for segmentation. First, $\mathbf{v}$ is discontinuous across the vessel boundary since the fluid inside moves parallel to the boundary surface and the walls move roughly normal to the boundary. Second, it has been noted, e.g. by [10], that the velocity magnitude, $|\mathbf{v}|$, of the fluid is large compared to the magnitude of the wall motion. These observations laid the foundation for the variational formulation introduced in [15] where the following discontinuity measure was introduced.

To encode the information in the vector field $\mathbf{v}$ a field of matrices $\mathbf{M}_\mathbf{v} = \mathbf{M}_\mathbf{v}(\mathbf{x})$, $\mathbf{x} \in \mathbf{R}^3$ was introduced, where $\mathbf{M}_\mathbf{v}$ is the positive semi-definite, symmetric matrix defined by

$$\mathbf{M}_\mathbf{v} = \mathbf{v}\mathbf{v}^T = \begin{bmatrix} v_x \\ v_y \\ v_z \end{bmatrix} \begin{bmatrix} v_x & v_y & v_z \end{bmatrix} = \begin{bmatrix} v_x^2 & v_x v_y & v_x v_z \\ v_x v_y & v_y^2 & v_y v_z \\ v_x v_z & v_y v_z & v_z^2 \end{bmatrix} .$$

$\mathbf{M}_\mathbf{v}$ has rank one with eigenvalues $\lambda_1 = |\mathbf{v}|^2$ and $\lambda_2 = \lambda_3 = 0$. Let $W \geq 0$ be a weight function (typically a Gaussian filter G_σ) and define the average matrix field to be the convolution

$$\overline{\mathbf{M}}_\mathbf{v} = W * \mathbf{M}_\mathbf{v} \ ,$$

taken componentwise. We denote this matrix field the *density matrix field*. Depending on the values of $\mathbf{v}$ in the region determined by W there can be either *i)* one dominant velocity direction, *ii)* two dominant directions or *iii)* three equally dominant directions at every point. This is reflected in the magnitude of the eigenvalues of $\overline{\mathbf{M}}_\mathbf{v}$.

To discriminate *i)* from *ii)* and *iii)*, the following real valued function is introduced, inspired by Harris [17],

$$R = \frac{4\lambda_1\lambda_2}{(\lambda_1 + \lambda_2)^2} \ ,$$

where $\lambda_1 \geq \lambda_2 \geq 0$ are the two largest eigenvalues of $\overline{\mathbf{M}}_\mathbf{v}$ and $0 \leq R \leq 1$. For case *i)* $R \approx 0$ and for *ii)* and *iii)* R will be large, i.e. $R \approx 1$. It was shown in [15] that this measure is an excellent detector for discontinuities in $\mathbf{v}$. Some examples are shown in Figure 2.

2.3 Level Set Formulation

The level set method was introduced in [7] as a tool for capturing moving interfaces. The time dependent surface $\Gamma(t)$ is implicitly represented as the zero level set of a function $\phi(\mathbf{x}, t) : \mathbf{R}^3 \times \mathbf{R}_+ \to \mathbf{R}$ as

$$\Gamma(t) = \{\mathbf{x} \ ; \ \phi(\mathbf{x}, t) = 0\} \ ,$$

where ϕ is defined such that

$$\phi(\mathbf{x}, t) \begin{cases} < 0 \text{ inside } \Gamma \\ = 0 \text{ on } \Gamma \\ > 0 \text{ outside } \Gamma \ . \end{cases}$$

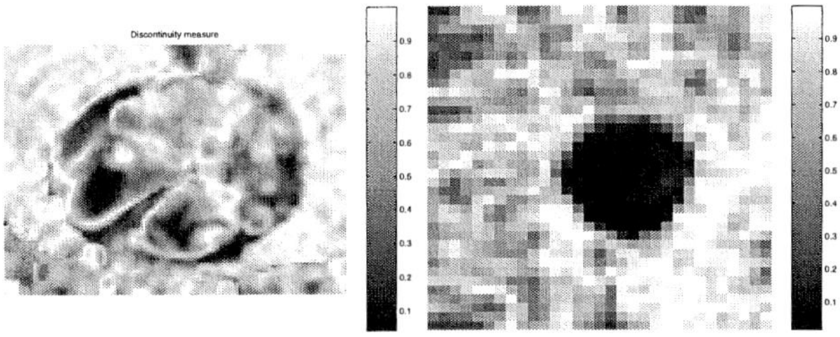

Fig. 2. Some examples of the discontinuity measure R applied to velocity data of the heart (left) and aorta (right) with white indicating high discontinuities

Using the definition above, the outward unit normal $\mathbf{n}$ and the mean curvature κ are[1]

$$\mathbf{n} = \frac{\nabla\phi}{|\nabla\phi|} \quad \text{and} \quad \kappa = \nabla \cdot \frac{\nabla\phi}{|\nabla\phi|} \ .$$

The zero set of $\phi(\mathbf{x}, t)$ represents $\Gamma(t)$ at all times t. This means that $\phi(\mathbf{x}(t), t) \equiv 0$ for a curve $\mathbf{x}(t) \in \Gamma(t)$. Differentiating with respect to t gives

$$\phi_t + \mathbf{u} \cdot \nabla\phi = 0 \ \Leftrightarrow \ \phi_t + u_n|\nabla\phi| = 0 \ , \tag{1}$$

where $\mathbf{u} = d\mathbf{x}(t)/dt$ and u_n is the normal component of the surface velocity. To move the surface according to some derived velocity, a PDE of the form (1) is solved. One of the advantages of this representation is that the topology of the surface is allowed to change as the surface evolves, thus making it easy to represent complex surfaces that can merge or split so that multiple objects are easily handled. For a more thorough treatment of level set surfaces cf. [5, 6].

3 Variational Formulation

In this section the segmentation problem is formulated as a variational problem. Given velocity data, basic dynamic characteristics of the flow and intensity images, the problem consists of finding the boundary between the blood and the vessel walls. This boundary should be a closed surface within the domain of interest i.e. within the measured volume. The segmentation is then determined by the interior of the surface representing the vessels. The interior of Γ will be denoted Ω and the exterior Ω^c. Using both velocity and intensity information an energy functional is introduced and minimized using the level set framework.

3.1 Velocity Based Terms

The first part of the energy functional is based on the velocity information given by phase contrast MRI. A first observation is that tissue moves much slower

[1] Here $\nabla\phi$ denotes the gradient of ϕ, $\nabla\phi = (\frac{\partial\phi}{\partial x}, \frac{\partial\phi}{\partial y}, \frac{\partial\phi}{\partial z})$, and $\nabla\cdot$ denotes the divergence.

than blood on average. Therefore the final surface should enclose as much flow as possible, i.e. the flow outside the surface should be minimized.

The discontinuity measure, described in Section 2.2, has shown to obtain higher values outside blood vessels than inside, not just high values on the vessel walls, as assumed in [15]. This is due to a smaller influence by image noise in high velocity areas, such as within blood vessels, than in relatively stationary areas where velocity noise to a larger extent introduce discontinuities. Based on this insight together with the fact that flow inside blood vessels is continuous the discontinuity measure should be minimized inside the surface.

To summarize the arguments above the desired surface should enclose as much of the flow in $\mathbf{v}$ as possible and the flow should be continuous inside the enclosing surface. This leads to the minimization of the following energy functional

$$E_{\mathrm{V}}(\Gamma) = \underbrace{\int_{\Omega^c} \chi(\mathbf{v}) d\mathbf{x}}_{\text{flow outside the surface}} + \underbrace{\int_{\Omega} R(\mathbf{x}) d\mathbf{x}}_{\text{discontinuities inside the surface}} , \qquad (2)$$

where Ω is the interior and Ω^c the exterior of the surface as mentioned above, $R(\mathbf{x})$ is the measure for discontinuities from Section 2.2 and $\chi(\mathbf{v})$ is a C^2 approximation to the translated Heaviside function defined by $\chi(\mathbf{v}) = \chi(|\mathbf{v}|) = H(|\mathbf{v}| - \delta) = H_\delta(|\mathbf{v}|)$, where $\delta \in \mathbf{R}_+$. Here H is defined as in [18]

$$H(x) = \begin{cases} 1 & x > \epsilon \\ 0 & x < -\epsilon \\ \frac{1}{2}[1 + \frac{x}{\epsilon} + \frac{1}{\pi}\sin(\frac{\pi x}{\epsilon})] & |x| \leq \epsilon . \end{cases}$$

This is commonly used in level set based segmentation, cf. [19]. The definition above will make $\chi(\mathbf{v})$ equal to zero for very low velocities and one otherwise. This makes it an approximate characteristic function for $\mathbf{v}$ and the measure of the first integral in (2) is then in volume units.

Representing the surface using the zero level set of a function ϕ, the energy (2) becomes

$$E_{\mathrm{V}}(\phi) = \int_{\mathbf{R}^3} \chi(\mathbf{v}) H(\phi) d\mathbf{x} + \int_{\mathbf{R}^3} R(\mathbf{x})(1 - H(\phi)) d\mathbf{x} , \qquad (3)$$

where $H(\phi)$ is again the approximation to the Heaviside function and therefore a characteristic function for Ω^c and $(1 - H(\phi))$ is analogously a characteristic function for Ω. From the Euler-Lagrange equation for (3), the motion PDE for the surface based on velocity alone, as a gradient descent, is then

$$\phi_t = (-\chi + R)|\nabla\phi| ,$$

where ϕ_t denotes derivative with respect to the evolution time.

3.2 Intensity Based Terms

The second part of the energy functional is based on the given MRI intensity images. Especially the gradients in every point of the intensity images are important and indicate tissue changes. Inside blood vessels the image intensity is relatively uniform, making the image gradient small in magnitude. At the borders of the vessels there is a jump in the intensity resulting in larger gradients. The resulting surface is supposed to be aligned to this vessel border, i.e. the surface outward normal should be parallel (or anti-parallel) to the image gradient at every point and positioned so that the magnitude of the image gradient $|\nabla I|$ is large. This results in the following energy functional

$$E_{\mathrm{I}}(\Gamma) = \underbrace{-\int_{\Gamma} |\nabla I(\mathbf{x}) \cdot \mathbf{n}| dS}_{\text{normal component of gradient at the surface}} ,$$

which has been analyzed in e.g. [8, 20]. Using Gauss' theorem this becomes

$$\begin{aligned} E_{\mathrm{I}}(\Gamma) &= -\operatorname{sign}(\nabla I(\mathbf{x}) \cdot \mathbf{n}) \int_{\Gamma} (\nabla I(\mathbf{x}) \cdot \mathbf{n}) dS \\ &= -\operatorname{sign}(\nabla I(\mathbf{x}) \cdot \mathbf{n}) \int_{\Omega} \Delta I(\mathbf{x}) d\mathbf{x} , \end{aligned} \tag{4}$$

where $(\nabla I \cdot \mathbf{n})$ is assumed not to change sign on the surface. If the surface Γ is initialized completely inside or completely outside the sought volume and the intensity images are sufficiently smooth this sign assumption is justified. Representing the surface using the zero level set of a function ϕ as in Section 3.1, the energy (4) becomes

$$E_{\mathrm{I}}(\phi) = -\operatorname{sign}\left(\nabla I(\mathbf{x}) \cdot \frac{\nabla \phi}{|\nabla \phi|}\right) \int_{\mathbf{R}^3} \Delta I(\mathbf{x})(1 - H(\phi)) d\mathbf{x} , \tag{5}$$

where $(1 - H(\phi))$ is the characteristic function of Ω. From the Euler-Lagrange equation for (5) the motion PDE for the surface based on intensity alone is

$$\phi_t = \left[-\operatorname{sign}\left(\nabla I \cdot \frac{\nabla \phi}{|\nabla \phi|}\right) \Delta I\right] |\nabla \phi| .$$

3.3 Total Energy

The combined information from Section 3.1 and 3.2 results in an energy functional containing both velocity and image intensity dependent terms

$$E_{\mathrm{Tot}}(\Gamma) = E_{\mathrm{V}} + E_{\mathrm{I}} .$$

Using the above introduced notation this becomes

$$E_{\mathrm{Tot}}(\Gamma) = \int_{\Omega^c} \chi d\mathbf{x} + \int_{\Omega} R d\mathbf{x} - \operatorname{sign}(\nabla I \cdot \mathbf{n}) \int_{\Omega} \Delta I d\mathbf{x} , \tag{6}$$

where the terms are initially scaled to the same order of magnitude. Using the level set framework on (6) results in the following functional

$$E_{\mathrm{Tot}}(\phi) = \int_{\mathbf{R}^3} \chi H(\phi) + \left[R - \operatorname{sign}\left(\nabla I \cdot \frac{\nabla\phi}{|\nabla\phi|}\right) \Delta I\right] (1 - H(\phi))\, d\mathbf{x} \ ,$$

which is a volume integral. From the Euler-Lagrange equation for the total energy the motion PDE for the surface becomes

$$\phi_t = \left[-\chi + R - \operatorname{sign}\left(\nabla I \cdot \frac{\nabla\phi}{|\nabla\phi|}\right) \Delta I\right] |\nabla\phi| \ . \tag{7}$$

This equation is numerically relatively stable as opposed to the evolution equation given in [15] which can be seen as solving a backward heat equation. Despite this some problems can occur due to the use of the Laplacian of the intensity images in (7) and hence an additional regularizing term can sometimes be useful.

4 Experiments

The segmenting performance of the proposed variational method was tested on both clinical phase contrast MRI data of the aorta as well as on a flow phantom. The constant δ used in $\chi(\mathbf{v})$ as described in Section 3.1 is set from estimates of the noise in the velocity data.

4.1 Clinical Data

The clinical data consisted of two full 3D volumes ($36 \times 31 \times 20$) of MRI intensity data and phase contrast velocity data of the aorta respectively with a resolution of $1.92 \times 1.92 \times 2.00 mm^3$. Three images based on the intensity and velocity data is shown in Figure 3. The resulting segmented vessel wall after using only the two velocity based terms of the energy functional, i.e. the two first terms in (6), is shown in 2D in Figure 4 superimposed on the corresponding magnitude,

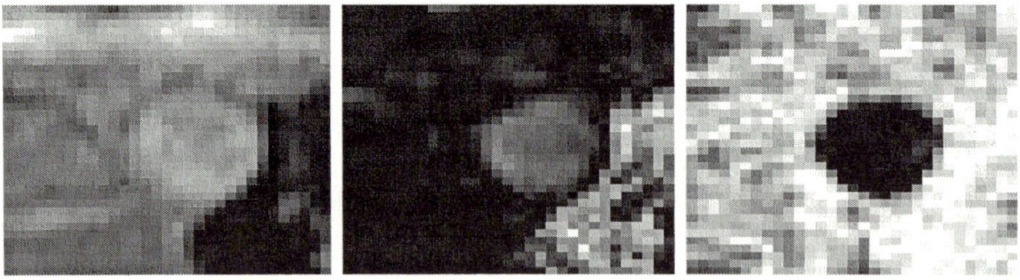

Fig. 3. (left) Intensity image of the aorta, (middle) absolute value of the velocities for the corresponding data, (right) discontinuity measure for the velocity data, white indicating higher values

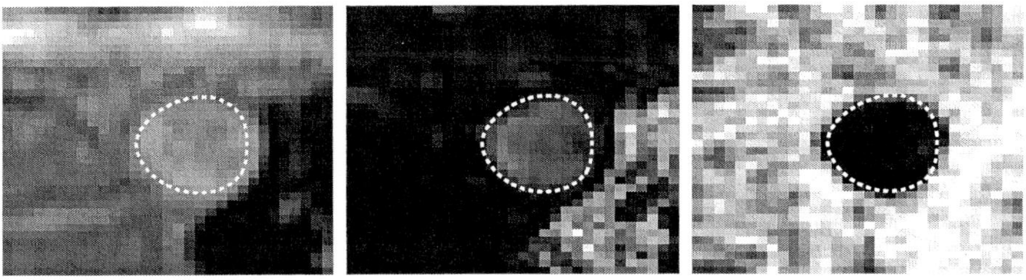

Fig. 4. Segmented aorta by using velocity information only. (left) Intensity image of the aorta with segmented vessel wall in white, (middle) absolute value of the velocities for the corresponding data, (right) discontinuity measure for the velocity data, white indicating higher values

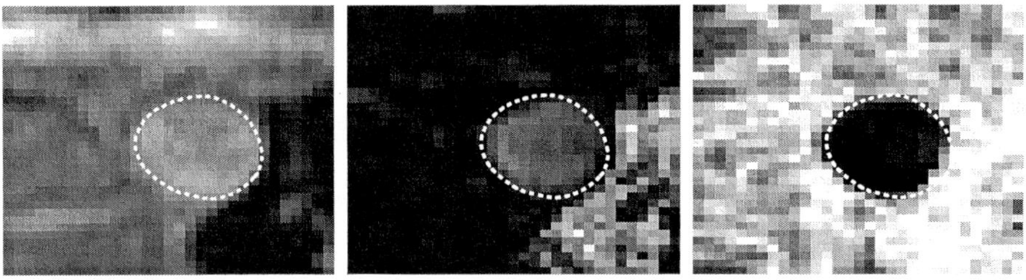

Fig. 5. Segmented aorta by using both velocity and intensity information. (left) Intensity image of the aorta with segmented vessel wall in white, (middle) absolute value of the velocities for the corresponding data, (right) discontinuity measure for the velocity data, white indicating higher values

absolute velocity and discontinuity measure images respectively. It is obvious this velocity based variety of the proposed method is not able to produce an optimal segmentation due to high velocity discontinuities to the lower right of the aorta.

In the experiments resulting in Figure 5 the same data was treated but with all three terms in (6) used. Clearly by using all three terms, i.e. utilizing intensity information as well as velocity information, the resulting segmentation was improved. The surface no longer being restrained by the high velocity discontinuities to the lower right of the aorta.

The final result of the aorta segmentation using all terms in (6) is a surface in 3D, a corresponding VRML model is shown in Figure 6.

4.2 Flow Phantom

The flow phantom consisted of a rubber hose containing flowing ion-enriched water. The hose was submerged in stationary water of the same kind. The data

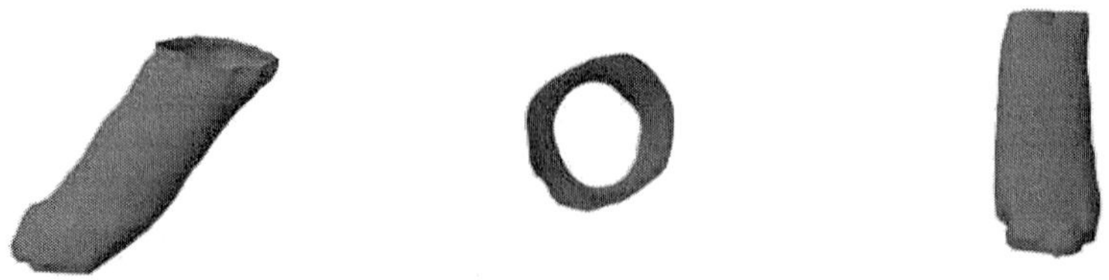

Fig. 6. Different views of an aorta segmented by the use of both velocity and intensity information. The shown segmented part of the aorta is approximately $4cm$ long

Fig. 7. Image sequence of an evolving surface, going from left to right and top to bottom, segmenting a flow phantom by using both velocity and intensity information. The surface is initialized as two spheres (top left)

sets were two full 3D volumes ($110 \times 153 \times 59$) of MRI intensity data and phase contrast velocity data of the flow phantom respectively with a resolution of $1.00 \times 1.00 \times 2.00mm^3$. The time evolving 3D segmentation of the flow phantom is shown in Figure 7 as an image sequence.

5 Conclusions

In this paper a method for 3D segmentation of blood vessels using a combination of velocity and intensity data is introduced. The segmentation problem is formulated in a variational level set setting with a functional derived from physical properties of the data. The functional incorporates velocity magnitude, velocity discontinuities as well as intensity variation. Initial tests on clinical data support the proposed method by showing higher numerical stability and an increasing tendency in segmentation performance in areas of high velocity noise as shown in the lower right of the segmented aorta in Figure 5. This experiment also proves the ability of the method to segment regions with practically equal intensities thanks to the velocity terms of the functional.

Future work will include e.g. integrating prior information using statistical shape models [21, 22], analysis of methods for initialization and automatic methods for estimating parameters from the data.

References

1. Frangi, A., Niessen, W., Viergever, M.: Three-dimensional modeling for functional analysis of cardiac images: A review. IEEE Trans. on Medical Imaging **20** (2001)
2. Otsu, N.: A threshold selection method from gray-level histograms. IEEE Trans. Systems, Man and Cybernetics **9** (1979) 62–66
3. Kass, M., Witkin, A., Terzopoulos, D.: Snakes: Active contour models. Int. J. Computer Vision **1** (1987) 321–331
4. Caselles, V., Kimmel, R., Sapiro, G.: Geodesic active contours. Int. Journal of Computer Vision (1997)
5. Osher, S.J., Fedkiw, R.P.: Level Set Methods and Dynamic Implicit Surfaces. Springer Verlag (2002)
6. Sethian, J.: Level Set Methods and Fast Marching Methods Evolving Interfaces in Computational Geometry, Fluid Mechanics, Computer Vision, and Materials Science. Cambridge University Press (1999)
7. Osher, S., Sethian, J.A.: Fronts propagating with curvature-dependent speed: Algorithms based on Hamilton-Jacobi formulations. Journal of Computational Physics **79** (1988) 12–49
8. Kimmel, R., Bruckstein, A.: Regularized laplacian zero crossings as optimal edge integrators. Int. Journal of Computer Vision **53** (2003) 225–243
9. Ebbers, T.: Cardiovascular Fluid Dynamics. PhD thesis, Departments of Biomedical Engineering & Medicine and Care, Linkoping University, Sweden (2001)
10. Wong, A., Liu, H., Shi, P.: Segmentation of myocardium using velocity field constrained front propagation. In: IEEE Applications of Computer Vision. (2002)
11. Rousson, M., Deriche, R.: A variational framework for active and adaptive segmentation of vector valued images. Technical Report 4512, INRIA (2002)
12. Lorigo, L., Faugeras, O., Grimson, W., Keriven, R., Kikinis, R., Nabavi, A., Westin, C.F.: Curves: Curve evolution for vessel segmentation. IEEE Transactions on Medical Image Analysis **5** (2001) 195–206
13. Vasilevskiy, A., Siddiqi, K.: Flux maximizing geometric flows. IEEE Trans. on PAMI **24** (2002) 1565–1578
14. Descoteaux, M., Collins, L., Siddiqi, K.: Geometric flows for segmenting vasculature in MRI: Theory and validation. In: MICCAI, France (2004)
15. Solem, J., Persson, M., Heyden, A.: Velocity based segmentation in phase contrast mri images. In: MICCAI, France (2004)
16. Pelc, N.J., Herfkens, R.J., Shimakawa, A., Enzmann, D.: Phase contrast cine magnetic resonance imaging. Magnetic Resonance Quarterly **4** (1991) 229–254
17. Harris, C., Stephens, M.: A combined corner and edge detector. In: Proc. Alvey Conf. (1988) 189–192
18. Zhao, H., Chan, T., Merriman, B., Osher, S.: A variational level set approach to multiphase motion. J. Computational Physics 127 (1996) 179–195

19. Paragios, N.: A level set approach for shape-driven segmentation and tracking of the left ventricle. IEEE Transactions on Medical Imaging **22** (2003) 773–776
20. Overgaard, N., Solem, J.: An analysis of variational alignment of curves in images. In: Scale Space 2005, Germany, Springer (2005)
21. Rousson, M., Paragios, N.: Shape priors for level set representations. In: Proc. European Conf. on Computer Vision. Volume 2351 of Lecture Notes in Computer Science., Springer (2002)
22. Cremers, D.: Statistical Shape Knowledge in Variational Image Segmentation. PhD thesis, Dept. of Mathematics and Computer Science, University of Mannheim, Germany (2002)

Active Shape Models and Segmentation of the Left Ventricle in Echocardiography

Nikos Paragios[1,*], Marie-Pierre Jolly[2], Maxime Taron[1], and Rama Ramaraj[2]

[1] Atlantis Research Group, Certis,
Ecole Nationale des Ponts et Chaussees, Champs sur Marne, France
[2] Imaging and Visualization Department,
Siemens Corporate Research, Princeton, NJ, USA

Abstract. Segmentation of the left ventricle in echocardiographic images is a task with important diagnostic power. We propose a model-based approach that aims at extracting the left ventricle for each frame of the cardiac cycle. Our approach exhibits several novel elements. Modelling consists of two separate components, one for the systolic and one for the diastolic moment. Segmentation is considered in two steps. During the first step a linear combination of the systolic and the diastolic model is to be recovered - that dictates the new model - along with a similarity transformation that projects this model to the desired image features. During the second step, a linear combination of the modes of variation for the systolic and diastolic models is recovered for precise extraction of the endocardium boundaries. The process is considered in the temporal domain where constraints are introduced to couple information across frames and to lead to a smooth solution. Promising results demonstrate the potentials of the presented framework.

1 Introduction

Cardiovascular diseases are a major health concern world-wide. The left ventricle and in particular the endocardium is a structure of particular interest since it performs the task of pumping oxygenated blood to the entire body. Echocardiographic apical views when processed can determine the ejection fraction, a critical measure of the heart cycle. While segmenting the ventricle in the systolic and diastolic frame could be sufficient to provide such a measure, continuous tracking of the endocardium could further improve diagnosis.

Portability and low acquisition cost are the most attractive elements of echocardiographic imaging [14] while the presence of low signal-to-noise (SNR) ratio is an important limitation. Model-free segmentation techniques aim at separating the intensity properties of the image entities and fail to cope with noise and speckle in echocardiography. The use of prior knowledge that encodes the geometric form of the structure of interest is a reasonable way to deal with corrupted data.

Prior art in echocardiography consists of data-driven [6] and model-based segmentation approaches [9, 1]. One can also separate the techniques that perform segmentation

* Research was carried out during the affiliation of the author with Siemens Corporate Research from November 1999 to March 2003.

R. Kimmel, N. Sochen, J. Weickert (Eds.): Scale-Space 2005, LNCS 3459, pp. 131–142, 2005.

in the polar [6] or in the raw space. Statistical/Bayesian formulations [6], active shape and appearance models [4, 1, 17], snakes and active contours [9], deformable models and templates [8] and level set techniques [2, 11] are well established techniques considered to address the segmentation of the left ventricle in echocardiographic images.

In this paper, we propose an active shape-inspired variational framework for fast, reliable and automatic segmentation of the endocardium for ultrasonic images. Our approach involves modelling, extraction of primitives, rough segmentation and border detection. We consider two separate model spaces, one for the diastolic and one for the systolic case. We recover the average shape and the modes of variations for each model through a Principal Component Analysis using a set of registered training examples.

Extraction of important primitives (ventricular walls, valve plane) that are used to initiate the segmentation process is the first step towards automatic 2D+time segmentation. Then, a linear combination of the two average models (systolic & diastolic) and the parameters of a similarity transformation between this new model and the image are incrementally recovered through a robust minimization. One should note that such a model space is dynamic. The parameters of this transformation are constrained to be smooth in the temporal domain. Precise endocardium segmentation is determined through a linear combination of the moments of variation that describe training sets, the systolic and the diastolic one. Such combination is constrained over time.

The remainder of this paper is organized as follows. In the next section, we address shape registration and modelling of the left ventricle. Global segmentation that involves a global transformation between the model-space and the image is presented in section 3 while local refinements are considered in section 4. Discussion and perspectives are addressed in section 5.

2 Modelling the Geometric Structure of the Endocardium

Building compact representations from a set of examples is a well studied problem in imaging and vision. The selection of appropriate models to represent all examples of the training set within a common pose is a critical component of such a process. Once such selection has been established, one would like to align all training examples to the same pose. Then modelling can then be performed using well known statistical techniques.

2.1 Global Registration, Mutual Information and Implicit Representations

Registration of shapes [15] is an open, interesting and challenging problem in imaging, vision and in particular in medical image analysis. Such application is not within the scope of the report, and therefore the prior art will be omitted and the adopted technique to address the problem will be briefly presented. Overviews of shape and image registration techniques are available at [10, 15]. Details on the considered approach to align the training examples can be found at [7]. Modelling requires global registration between the samples in the training set and establishment of local correspondences between them. Let us assume that a set of ground truth that consist of n components is available $[s_1, s_2, ..., s_n]$. Global alignment is equivalent with finding parametric transformations $\mathcal{A}_i$ between the training set examples and a target shape s such that

$$i \in [1, ...n] : \mathcal{A}_i(s) = s_i$$

where s is the common pose to be recovered. An emerging way to represent shapes is through the use of implicit representations. Such approaches are quite popular when the task involves tracking moving interfaces [11]. Inspired by the work proposed in [12] we represent shapes using distance transforms and implicit representations;

$$\phi_i(\omega) = \begin{cases} 0, & \omega \in s_i \\ d(\omega, s_i), & otherwise \end{cases}$$

where ω is the pixel location and $d(\omega, s_i)$ the minimum Euclidean distance between this pixel and the shape s_i.

The selected representation is translation/rotation invariant. Scale variations can be considered to be global illumination changes in the space of distance transforms. Therefore, registration under scale variations is equivalent with matching different modalities that refer to the same structure of interest. Mutual information [3, 16] is an invariant technique according to a monotonic transformation of the two input random variables. The use of such criterion to perform shape registration within the space of distance transforms was proposed in [7]. Such criterion is based on the global characteristics of the structures of interest. In order to facilitate the notation let us denote: **(i)** the source representation ϕ_i as f, and **(ii)** the target representation ϕ as g.

In the most general case, registration is equivalent with recovering the parameters $\Theta = (\theta_1, \theta_2, ..., \theta_N)$ of a parametric transformation $\mathcal{A}$ such that the mutual information between $f_\Omega = f(\Omega)$ and $g_\Omega^\mathcal{A} = g(\mathcal{A}(\Theta; \Omega))$ is maximized for a given sample domain Ω;

$$MI(X^{f_\Omega}, X^{g_\Omega^\mathcal{A}}) = \mathcal{H}\left[X^{f_\Omega}\right] + \mathcal{H}\left[X^{g_\Omega^\mathcal{A}}\right] - \mathcal{H}\left[X^{f_\Omega, g_\Omega^\mathcal{A}}\right]$$

where $\mathcal{H}$ represents the differential entropy. Such quantity represents a measure of uncertainty, variability or complexity and consists of three components: (i) the entropy of the model, (ii) the entropy of the projection of the model given the transformation, and (iii) the joint entropy between the model and the projection that encourages transformations where f explains g. One can use the above criterion and an arbitrary transformation (rigid, affine, homographic, quadratic) to perform global registration that is equivalent with minimizing:

$$E(A(\Theta)) = -MI(X^{f_\Omega}, X^{g_\Omega^\mathcal{A}}) = -\int\int_{\mathcal{R}^2} p^{f_\Omega, g_\Omega^\mathcal{A}}(l_1, l_2) log \frac{p^{f_\Omega, g_\Omega^\mathcal{A}}(l_1, l_2)}{p^{f_\Omega}(l_1) p^{g_\Omega^\mathcal{A}}(l_2)} dl_1 dl_2$$

where (i) p^{f_Ω} corresponds to the probability density in f_Ω $\left([\Phi_\mathcal{D}(\Omega)] \right)$, (ii) $p^{g_\Omega^\mathcal{A}}$ corresponds to density in $g_\Omega^\mathcal{A}$ $\left([\Phi_\mathcal{S}(\mathcal{A}(\Theta; \Omega))] \right)$, and (iii) $p^{f_\Omega, g_\Omega^\mathcal{A}}$ is the joint density. Such framework can account for various global motion models. We consider similarity registration between the training examples for the endocardium shapes.

Registration examples for the particular class of endocardium shapes are shown in [FIG. (1)]. Once training examples have been aligned, one should address the problem of recovering point(element)-wise correspondences. Such a deformation field $L(\Theta; \mathrm{x})$ can be recovered either using standard optical flow constraints or through the use of warping techniques like the free form deformations method [13], which is a popular approach in graphics, animation and rendering [5].

Such smoothness term is based on a classic error norm that has certain known limitations. One can replace this smoothness component with more elaborated norms. Within the proposed framework, an implicit smoothness constraint is also imposed by the Spline FFD. Therefore there is not need for introducing complex and computationally expensive regularization components.

The Data-driven term and the smoothness constraints term can now be integrated to recover the local deformation component of the registration and solving the correspondence problem: $E(\Theta) = E_{data}(\Theta) + \alpha E_{smoothness}(\Theta)$, where α is the constant balancing the contribution of the two terms. The calculus of variations and a gradient descent method can be used to optimize such objective function [7]. The performance of the proposed framework on the Systolic Left Ventricle dataset is demonstrated in [Fig. (2)].

2.3 Composite Model Building

Let us assume that two sets of ground truth that consist of n components are available, one for the diastolic $[\mathbf{d}_1, \mathbf{d}_2, ..., \mathbf{d}_n]$ and one for the systolic case $[\mathbf{s}_1, \mathbf{s}_2, ..., \mathbf{s}_n]$. Without loss of generality, one can assume that the elements of each set consists of m points defined on the Euclidean plane $(\mathbf{d}_i = (\mathbf{x}_1^i, \mathbf{x}_2^i, ..., \mathbf{x}_m^i))$ and are registered to a common pose.

Principle Component Analysis (PCA) can be applied to capture the statistics of the corresponding elements across the training examples. PCA refers to a linear transformation of variables that retains - for a given number o_1, o_2 of operators - the largest amount of variation within the training data, according to:

$$\mathbf{d} = \overline{\mathbf{d}} + \sum_{k=1}^{o_1} \lambda_k^d \, (\mathbf{u}_k^d, \mathbf{v}_k^d), \quad \mathbf{s} = \overline{\mathbf{s}} + \sum_{k=1}^{o_2} \lambda_k^s \, (\mathbf{u}_k^s, \mathbf{v}_k^s)$$

where $\overline{\mathbf{d}}$ (resp. $\overline{\mathbf{s}}$) is the mean diastolic shape, o_1 (resp. o_2) is the number of retained modes of variation, $(\mathbf{u}_k^d, \mathbf{v}_k^d)$ (resp. $(\mathbf{u}_k^s, \mathbf{v}_k^s)$) are these modes (eigenvectors), and λ_j^d (resp. λ_j^s) are linear factors within the allowable range defined by the eigenvalues.

Once average models for the systolic and diastolic cases are considered, one can further assume that these models are registered, therefore there is an one-to-one correspondence between the points that define these shapes. Let $(\overline{\mathbf{d}} = (\mathbf{x}_1^d, \mathbf{x}_2^d, ..., \mathbf{x}_m^d))$ be the diastolic average model and $(\overline{\mathbf{s}} = (\mathbf{x}_1^s, \mathbf{x}_2^s, ..., \mathbf{x}_m^s))$ the systolic one. Then one can define a linear space of shapes as follows:

$$\overline{\mathbf{c}}(\alpha) = \alpha \, \overline{\mathbf{s}} + (1 - \alpha) \, \overline{\mathbf{d}}, \quad 0 \leq \alpha \leq 1$$

One then can define a linear space of deformations that can account for the systolic, the diastolic frame as well as the frames in between:

$$\mathbf{c}(\alpha, \lambda_k^d, \lambda_s^d) = \overline{\mathbf{c}}(\alpha) + \sum_{k=1}^{o_1} \lambda_k^d \, (\mathbf{u}_k^d, \mathbf{v}_k^d) + \sum_{k=1}^{o_2} \lambda_k^s \, (\mathbf{u}_k^s, \mathbf{v}_k^s)$$

The most critical issue to be addressed within this process is the registration of the training examples as well as the registration of the systolic and diastolic average shapes. The approach proposed in [7] that performs registration in the implicit space of distance functions using a combination between mutual information criterion and a free-form deformation principle is used. Such an approach can provide one-to-one correspondences between shapes for any given number of sampling elements. The resulting composite model is of limited complexity, can account for the systolic and the diastolic form of the endocardium as well as for the frames between the two extrema.

2.4 Composite Active Shape Models

Active shapes assume an average model, a certain number of modes of variation and the existence of corresponding image features. Without loss of generality one can assume that for each point j on the model space $\mathbf{c}(\alpha, \lambda_k^d, \lambda_s^d)$ the corresponding image point has been recovered $\mathbf{y}_j$. Then, the objective is to recover a set of parameters that will move each point in the model space $\mathbf{c}_j$ to the corresponding location in the image space $\mathbf{y}_j$. Such a task is performed in two stages where first a global transformation $\mathcal{T}$ between the model and the image is recovered that minimizes:

$$E_{data}(\alpha, \mathcal{T}) = \sum_{j=0}^{m} \rho\left(||\mathcal{T}(\overline{\mathbf{c}}_j(\alpha)) - \mathbf{y}_j||\right)$$

according to some metric function ρ where $\mathcal{T}$ is a global transformation, similarity in our case

$$\mathcal{T}(x, y) = \begin{bmatrix} a & b \\ -b & a \end{bmatrix} \begin{bmatrix} x \\ y \end{bmatrix} + \begin{bmatrix} c \\ d \end{bmatrix}$$

that consists of a translation, a rotation and a scaling component and α defines the model space. The selection of the transformation should be consistent with the one adopted during the learning stage. It is important to point out that the model is not static since refers to a linear combination of the systolic and the diastolic model. Therefore, the process aims to recover simultaneously the combination of these two models that better accounts for the shape of the true data points and the optimal transformation between the model and the image space.

One can recover these parameters through an incremental update of the transformation. The corresponding location of the model points in the image plane could be used to improve the segmentation be seeking an incremental update on the transformation $\mathcal{T}(;)$ such that the projection of the $\overline{\mathbf{c}}_j$ moves closer to its true position $\mathbf{y}_j$ in the image.

3 Rough Segmentation of the Endocardium

The left ventricle is bounded on each side by the walls which tend to appear brighter in the ultrasound clip due to the various reflections from the tissue. In apical (both 2 chamber and 4 chamber views), the left ventricle is bounded on the bottom side by the mitral valve which connects it to the left atrium. The mitral valve is constantly moving (opening and closing) and its reflections are well recovered by the acquisition process.

We consider two parabolic equations to recover a rough approximation/detection of these walls which are the areas with the highest brightness. The parabola model the walls of the left ventricle but also outline the left atrium. The next step is to extract and track the position of the mitral valve that separates the left ventricle and the left atrium. The approach relies on the observation that if the valve is closed, the two heart chambers are clearly separated while if the valve is open, the two chambers are connected. Two ellipses are used to model the ventricle and the atrium and the plane that best separates these ellipses and is consistent over time is considered to be the valve plane.

3.1 Recovering Correspondences

The most critical part within the presented framework is solving the correspondence problem, between the actual projection of the model and the optimal position. Such task within the active shape model is solved using a normalized intensity profile in the normal direction. We consider a probabilistic formulation of the problem. One would like to recover a density $p_{border}(;)$ that can provide the probability of a given pixel ω being at the boundaries of the endocardium. Within the considered framework, one can constrain the search in the direction normal to the model projection. The ventricular area consists of blood pool and heart walls. Endocardium border detection is equivalent with finding the boundaries between these two classes.

A description on the statistical properties of the blood pool as well as cardiac wall can be recovered. Let $p_{wall}(;)$ being the probability of a given intensity being part of the endocardium walls and $p_{blood}(;)$ the density the describes the visual properties of the blood pool. Then or correspondences between the model and the image are meaningful in places where there is a transition (wall to blood pool) between the two classes. Given a local partition one can define a transition probability between these two classes. Such partition consists of two line segments $[\mathcal{L}(\mathcal{T}(\mathbf{x}_j)), \mathcal{R}(\mathcal{T}(\mathbf{x}_j))]$ that live in the normal direction $[\mathcal{T}(\mathcal{N}_j)]$ of the model curve at element $\mathcal{T}(\mathbf{x}_j)$. The origins of these line segments is the point of interest $\mathcal{T}(\mathbf{x}_j)$, the have the same slope and opposite directions. One can assume that this point is a projection of the model point $\mathbf{x}_j$:

$$p_{border}(\mathcal{T}(\mathbf{x}_j)) = p\left([wall|\omega \in \mathcal{L}(\mathcal{T}(\mathbf{x}_j))] \cap [\text{blood}|\omega \in \mathcal{R}(\mathcal{T}(\mathbf{x}_j))]\right)$$

These conditions can be considered independent, leading to the following form for the border density:

$$\begin{aligned} p_{border}(\mathcal{T}(\mathbf{x}_j)) &= p\left(wall|\omega \in \mathcal{L}(\mathcal{T}(\mathbf{x}_j))\right)\, p\left(\text{blood}|\omega \in \mathcal{R}(\mathcal{T}(\mathbf{x}_j))\right) \\ &= \prod_{\omega \in \mathcal{L}} p_{wall}(I(\omega)) \prod_{\omega \in \mathcal{R}} p_{blood}(I(\omega)) \end{aligned}$$

One can evaluate this probability under the condition that the blood pool and wall density functions are known. The use of -log function can be considered to overcome numerical constraints, that is equivalent with finding the minimum of:

$$E(\phi) = \sum_{\omega \in \mathcal{L}(\phi)} \lambda |I(\omega)| + \sum_{\omega \in \mathcal{R}(\phi)} \frac{(I(\omega) - \mu)^2}{2\sigma^2}$$

after dropping out the constant terms where blood pool is modelled using an exponential distribution (λ) and tissue/walls using a Gaussian distribution (μ, σ). Thus, the most probable correspondence is recovered through the evaluation of $E(\phi)$ where ϕ is a point in the line defined by the projected normal. The search space for ϕ is considered to be all image locations respecting two conditions; (i) live in the normal $\mathcal{T}(\mathcal{N}_j)$, and (ii) their distance from the current projection $\mathcal{T}(\overline{\mathbf{c}}_j(\alpha))$ is within a given search window. Once such correspondences were established the mechanism presented in [**Sec. 2.4**] is considered to determine the optimal solution through the estimation of the parameters of the transformation $(\alpha_t, \mathcal{T}_t)$.

3.2 Constraints on the Motion and the Position of the End-Valve Points

The motion of the valve plane is very critical to the operation of the endocardium. Such motion is consistent over time, and quite often exhibits a symmetric form. Without loss of generality, one can assume that the first $\overline{\mathbf{c}}_0(\alpha)$ and the last point $\overline{\mathbf{c}}_m(\alpha)$ of the model correspond to the valve end points. The displacement of these points from one frame to the next can be recovered in an implicit form.

Let $(\alpha_{t-1}, \mathcal{T}_{t-1})$ be the model and its transformation to the image plane towards the desired image features in the previous frame. Then, given some estimates on the current solution $(\alpha_t, \mathcal{T}_t)$ one can constrain the implicit motion of the valve points as follows:

$$E_{valve\ motion}(\alpha_t, \mathcal{T}_t) = \psi\left(|\mathcal{T}_{t-1}(\overline{\mathbf{c}}_0(\alpha_{t-1})) - \mathcal{T}_t(\overline{\mathbf{c}}_0(\alpha_t))|\right) + \psi\left(|\mathcal{T}_{t-1}(\overline{\mathbf{c}}_m(\alpha_{t-1})) - \mathcal{T}_t(\overline{\mathbf{c}}_m(\alpha_t))|\right)$$

where ψ is an error metric - the Euclidean in our case - $\mathcal{T}_{t-1}(\overline{\mathbf{c}}_m(\alpha_{t-1}))$ is the position of the valve point at frame $t-1$, $\mathcal{T}_t(\overline{\mathbf{c}}_m(\alpha_t))$ the corresponding projection at frame t and $\mathcal{T}_{t-1}(\overline{\mathbf{c}}_m(\alpha_{t-1})) - \mathcal{T}_t(\overline{\mathbf{c}}_m(\alpha_t))$ the displacement of this point from one frame to the next. Such term will constrain the motion of the valve plane to be smooth over time.

Such a term accounts for the relative motion of the valve points but not for their actual position. To this end, one can introduce constraints forcing the model projections of the valve points to be close to the valve-plane earlier recovered $(\alpha_{valve}\ x + \beta_{valve}\ y + \gamma_{valve} = 0)$. The distance between the current positions of the model valve points $(\overline{\mathbf{c}}_0(\alpha), \overline{\mathbf{c}}_m(\alpha))$ and their projections to the valve-plane $(\mathbf{p}_0(t), \mathbf{p}_m(t))$ is a term to be minimized;

$$E_{valve\ projection}(\alpha_t, \mathcal{T}_t) = \psi\left(|\mathbf{p}_0(t) - \mathcal{T}_t(\overline{\mathbf{c}}_0(\alpha_t))|\right) + \psi\left(|\mathbf{p}_m(t) - \mathcal{T}_t(\overline{\mathbf{c}}_m(\alpha_t))|\right)$$

One can consider a step further by recovering the exact position of the valve points in the image and then use these positions during the segmentation process. To this end, a model is built on the image profile for the left and the right end-valve points using an image patch cantered at the ground truth position of the valve. Then, these patches are normalized and an average model is recovered. Standard matching techniques are considered within a search area in the vicinity of the projected valve position to recover the most prominent valve points.

3.3 Smoothness Constraints on the Transformation Parameters

The motion of the ventricle also should fulfil certain constraints. It has to be periodic, exhibit a shrinking between the diastolic and the systolic frame and an expansion for the last part of the cardiac cycle. Such conditions can be imposed in various forms. Direct motion constraints (like the one earlier considered) focus on the distance of a model point in two consecutive frames. Such constraints though do not encode the continuity of the model. We consider an implicit form, where continuity is imposed on the parameters of the model ($\alpha(t)$) and the transformation ($\mathcal{T}(t)$);

$$E_{smoothness}(\alpha_t, \mathcal{T}_t) = \sum_{k=-\tau}^{\tau} \left(\omega\left(|\alpha(t) - \alpha(t+k)|\right) + w \sum_{p \in \mathcal{T}} \omega\left(|p(t) - p(t+k)|\right) \right)$$

where $p \in \mathcal{T}$ is the set of the similarity transformation parameters (a, b, c, d), ω a monotonically decreasing function and $[-\tau, \tau]$ is the interval where continuity on the rough segmentation parameters is imposed. Such a term will keep distance small between the registration parameters from the model space to the image within a couple of frames that is equivalent with constraining the motion of the endocardium from one frame to the next.

The objective function is minimized using a two-stage robust incremental estimate technique. The calculus of Euler-Lagrange equations with respect to the transformation parameters leads to a 4×4 linear system that has a closed form solution. Once such an estimate is available, the optimal model space α is recovered through an exhaustive search within the $[0, 1]$ integral according to some quantization step.

4 Refine Segmentation

Once, appropriate models and similarity transformations were recovered for all frames of the cardiac clip, the next step is precise extraction of the endocardium walls. Such a task is equivalent with finding a linear combination of the modes of variation that deforms globally the model projection towards the desired image features. The space of variations consists of the diastolic and the systolic models. Opposite to the rough segmentation case where the scale of the model is fixed, the need of a blending parameter does not exist between systolic and diastolic models of variation is not present. Under the assumption of existing correspondences $\mathbf{y}_j$ and the global transformation $(\alpha, \mathcal{T})$ for a given frame t - that is omitted from the notation -, these linear coefficients are recovered through:

$$E_{data}(\lambda_0^d, ..., \lambda_0^s, ...) = \sum_{j=0}^{m} \rho\left(||\mathcal{T}\left(\overline{\mathbf{c}}_j(\alpha)\right) + \sum_{k=1}^{o_1} \lambda_k^d \left(\mathbf{u}_k^d, \mathbf{v}_k^d\right) + \sum_{k=1}^{o_2} \lambda_k^s \left(\mathbf{u}_k^s, \mathbf{v}_k^s\right) - \mathbf{y}_j||\right)$$

Similar to the case of global transformation, one can assume now that the form of the ventricle changes gradually during the cardiac cycle. The geometry of the recovered solution is determined according to the set of coefficients $(\lambda_0^d, ..., \lambda_0^s, ...)$. Therefore, imposing constraints of smoothing deformation from one frame-to-the next is equivalent with seeking the lowest potential of

$$E_{smoothness}(\lambda_0^d, ..., \lambda_0^s, ...) =$$
$$\sum_{k=-\tau}^{\tau} \left(\sum_{l=1}^{o_1} \omega(\lambda_l^d(t) - \lambda_l^d(t+k)) + \sum_{l=1}^{o_2} \omega(\lambda_l^s(t) - \lambda_l^s(t+k)) \right)$$

Last, but not least additional constraints using the position of the valve points could be considered, that aims at moving the projections of the model valve points to the their

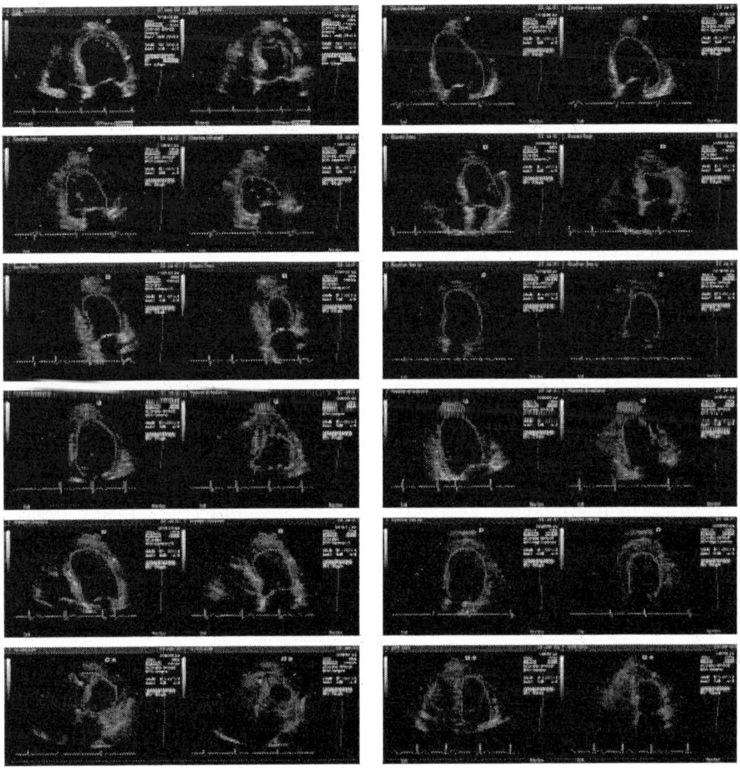

Fig. 3. Endocardium Segmentation for Apical Views for the diastolic frame and the systolic frame

true positions. The objective function is minimized using a robust incremental estimate technique. The calculus of Euler-Lagrange equations with respect to the unknown variables $(\lambda_0^d, ..., \lambda_0^s, ...)$ leads to a $[o_1 + o_2] \times [o_1 + o_2]$ linear system that has a closed form solution. Such step is repeated until convergence.

5 Conclusions

In this paper we have proposed a composite time-consistent 2D+time active shape model for the segmentation of the left ventricle in echocardiography. The approach exhibits certain novel elements, notably in the modelling and the segmentation phase.

Validation of the method was performed using a representative set of fifty patients for 2 and 4 champers views [**Fig. 3**] where the output of the proposed technique is superimposed to the ground truth. The objective was precise delineation of the ventricle, a much harder task than estimation of the ejection fraction. 50% of the time sonographers have accepted the result as it was while for the 25% of the remaining validation set, minor adjustments, notably in the valve position were sufficient to make the solution, same as the one pointed out from the clinical experts.

Future directions of our method involve epicardium segmentation and tracking. Such an objective is a natural extension that will improve results and the diagnostic power of the method since one could derive volume curves, EF radial strain, etc.

References

1. J. Bosch, S. Mitchell, B. Lelieveldt, F. Nijland, O. Kamp, M. Sonka, and J. Reiber. Automatocic Segmentation of Echocardiographic Sequences by Active Appearance Motion Models. *IEEE Transactions on Medical Imaging*, 21:1374–1383, 2002.
2. Y. Chen, H. Thiruvenkadam, H. Tagare, F. Huang, and D. Wilson. On the Incorporation of Shape Priors int Geometric Active Contours. In *IEEE Workshop in Variational and Level Set Methods*, pages 145–152, 2001.
3. A. Collignon, F. Maes, D. Vandermeulen, P. Suetens, and G. Marchal. Automated multi-modality image registration using information theory. In *Information Processing in Medical Imaging*, pages 263–274, 1995.
4. T. Cootes, A. Hill, C. Taylor, and J. Haslam. Use of Active Shape Models for Locating Structures in Medical Imaging. *Image Vision and Computing*, 12:355–366, 1994.
5. P. Faloutsos, M. van de Panne, and D. Terzopoulos. Dynamic Free-Form Deformations for Animation Synthesis. *IEEE Transactions on Visualization and Computer Graphics*, 3:201–214, 1997.
6. I. Herlin and N. Ayache. Feature extraction and analysis methods for sequences of ultrasound images. In *European Conference in Computer Vision*, pages 43–55, 1992.
7. X. Huang, N. Paragios, and D. Metaxas. Establishing Local Correspondences towards Compact Representations of Anatomical Structures. In *Medical Imaging Copmuting and Computer-Assisted Intervention*, 2003.
8. M. Kass, A. Witkin, and D. Terzopoulos. Snakes: Active Contour Models. *International Journal of Computer Vision*, 1:321–332, 1988.
9. D. Linker and V. Chalana. A Multiple Active Contour Model for Cardiac Boundary Detection on Echocardiographic Sequences. *IEEE Transactions on Medical Imaging*, 15:290–298, 1996.
10. J. Maintz and M. Viergever. A Survey for Medical Image Registration. *Medical Image Analysis*, 2:1–36, 1998.
11. S. Osher and N. Paragios. *Geometric Level Set Methods in Imaging, Vision and Graphics.* Springer Verlag, 2003.
12. N. Paragios, M. Rousson, and V. Ramesh. Non-Rigid Registration Using Distance Functions. *Computer Vision and Image Understanding*, 2003. to appear.
13. D. Rueckert, L.I. Sonda, C. Hayes, D. Hill, M. Leach, and D. Hawkes. Nonrigid registration using free-form deformations: Application to breast MR images. *IEEE Transactions on Medical Imaging*, 18:712–721, 1999.
14. C. Rumack, S. Wilson, and W. Charboneau. *Diagnostic Ultrasound.* Mosby, 1998.
15. R. Veltkamp and M. Hagedoorn. State-of-the-art in Shape Matching. Technical Report UU-CS-1999-27, Utrecht University, 1999. http://webdoc.gwdg.de/ebook/ah/2000/techrep/CS-1997/1997-27.pdf.
16. P. Viola and W. Wells. Aligment by Maximization of Mutual Information. In *IEEE International Conference in Computer Vision*, pages 16–23, 1995.
17. S. Zhou, D. Comaniciu, and A. Krishnan. Coupled-Contour Tracking through Non-orthogonal Projections and Fusion for Echocardiography. In *European Conference on Computer Vision*, volume I, pages 336–349, 2004.

A Variational Image Registration Approach Based on Curvature Scale Space

Stefan Henn and Kristian Witsch

Mathematisches Institut, Heinrich-Heine Universität Düsseldorf,
Universitätsstraße 1, D-40225 Düsseldorf, Germany
{henn, witsch}@am.uni-duesseldorf.de
http://www.am.uni-duesseldorf.de/~{henn, witsch}

Abstract. This paper presents a novel scale space approach to obtain a deformation which matches two images acquired from the same or from different medical imaging modalities. The image registration problem is known to be mathematically ill-posed due to the fact that determining the unknown components of the displacements merely from the images is an underdetermined problem. The approach presented here utilizes an auxiliary regularization term, which favors displacements with minimal curvature surface. One of the important aspects of this approach is that the kernel of the Euler-Lagrange equation is spanned by all rigid motions. Hence, the presented approach includes a rigid alignment. A minimizer is determined as the steady-state solution of the Euler-Lagrange equation namely by the biharmonic diffusion equation with higher order boundary conditions. In this framework we give a new interpretation of the underlying regularization parameter α. Finally, we present experimental results for registration problems of a Magnetic Resonance Imaging (MRI) (monomodal) registration and for a real computer tomography (CT)–magnetic resonance imaging (multimodal) registration.

1 Introduction

The aim of image registration is to find a transformation that aligns images recorded with the same or with different imaging machinery in a suitable way. Image registration problems are often multi scale problems in nature; namely, the reason for a displacement is governed by effects of different characters occurring at different scales. This phenomenon is given, e.g., in human brain mapping. Here the displacements often come from global transformations (translation and rotation) as well as from the different morphology of complex neuroanatomical shapes of the underlying brains.

There is a rich theory and also a large number of algorithms to solve image registration problems. They all ask for an "optimal" deformation which deforms one image such that there is an "optimal" correlation to another image with respect to a suitable coherence or difference measure.

The pure minimization of such difference measures typically leads to an ill-posed problem (also referred to as the "aperture problem" in computer vision,

R. Kimmel, N. Sochen, J. Weickert (Eds.): Scale-Space 2005, LNCS 3459, pp. 143–154, 2005.

see [1, 2, 3]), i.e. by the insufficient information provided solely by the available data, or by a desire to reduce noise. One effective method to overcome this problem are regularization approaches. Regularization methods for image registration (typically by adding a convex energy functional based on gradients), without additional knowledge, are an artificial way to make the problem well posed. For example, the classical Tikhonov regularization leads us to penalize large displacements, which makes no sense when displacements are the result of an elastic deformation. Therefore the underlying regularization energy should represent substantive prior information, coming, for example, from physical constraints, or laws or from information extracted from previous registration problems.

Many regularization approaches for image registration are discussed in the literature, a good survey is given in [3] and the references therein.

The novel contribution of this paper is to present a flexible image registration scheme, which treats the deformations on different spatial spaces. This is an attractive option in the situation where no a priori knowledge of the displacements is given. In order to do this, we use curvature as a fundamental description attribute for the deformation. Here, the set of decompositions defined by the curvature based regularization energy can be described as a scale space.

2 A Curvature Based Scale Space Approach

In this section, we first give a variational formulation of the image registration problem and introduce common similarity measures. In order to regularize the problem, we introduce a novel curvature based regularization term. This leads us to solve a biharmonic diffusion equation with higher order boundary conditions. In the following we describe the discretization in time and the scale space properties of the equation and discuss some computational considerations.

2.1 A Variational Image Registration Formulation

Given are two images, a reference R and a template T, of the same object obtained from the same or different imaging modalities. Usually, these images are two- or three-dimensional. Without loss of generality the problem is described in the two-dimensional case, but it is readily extendable to the three-dimensional case. We assume that in continuous variables the images can be represented by compactly supported functions $T, R : \mathbb{R}^2 \to \mathbb{R}$. This means, the map associates with each pixel (picture element) $x = (x_1, x_2)^t \in \mathbb{R}^2$ its intensities $T(x_1, x_2)$ and $R(x_1, x_2)$. We assume that T is distorted by an invertible deformation ϕ^{-1}. We search for a transformation $\phi(u)(\cdot) \; : \; \mathbb{R}^2 \to \mathbb{R}^2$, with

$$\phi(u)(x_1, x_2) \; : \; (x_1, x_2)^t \mapsto (x_1, x_2)^t - (u_1(x_1, x_2), u_2(x_1, x_2))^t$$

that depends on the unknown displacements $u(x_1, x_2) := (u_1(x_1, x_2), u_2(x_1, x_2))^t$ $: \mathbb{R}^2 \to \mathbb{R}^2$. The goal of image registration is to determine $u(x_1, x_2)$ in such a way that the transformed template $T \circ \phi(u(x_1, x_2))$ matches the reference

R. For a functional $D[R,T,u(x_1,x_2)]$, which measures the disparity between $T \circ \phi(u(x_1,x_2))$ and R in a domain $\Omega \subset \mathbb{R}^2$, the image registration problem can be identified, in that way, with a minimization problem:

$$\text{Find } u(x_1,x_2), \text{such that } D[u] = D[R,T,u(x_1,x_2)] \text{ is minimal.} \tag{1}$$

To rule out discontinuous and irregular solutions to the minimization problem (1), it is necessary to introduce a regularizing term $\alpha G[u]$ with a parameter $\alpha > 0$ and a term $G[u]$ which penalizes non-smoothness of the deformation u. This means that the solution of (1) is approximated by a solution u_α of the minimization problem:

$$\min \{ \underbrace{D[u] + \alpha G[u]}_{=J_\alpha[u]} \}. \tag{2}$$

The considered functional is nonlinear and may have many local and global minima. The parameter α allows us to balance the influence of both terms in the functional.

2.2 Distance Measures

In this section we want to collect examples of similarity measures. A lot of choices are possible depending on the application one has in mind. We describe some of the most important approaches below.

The squared differences between the images

$$D^{SQD}[u(x)] = \int_\Omega \big(T(x-u(x)) - R(x)\big)^2 dx.$$

is a common approach; cf., e.g. [1, 4, 5]. This criterion is used in situations where the intensities of the given images are comparable. This is the case, for example, if images are recorded with the same imaging machinery. This is generally referred to as the monomodal image registration. In general, if the images are recorded with different imaging machinery, the so-called multimodal registration, the D^{SQD} functional is not an appropriate measure. The main reason is that identical structures may have quite different gray values in the multimodal case.

The probability density measure (see [6]) is given by

$$D^{PLS}[u(x)] = \int_\Omega \big(d_\sigma^T(T(x-u(x))) - d_\sigma^R(R(x))\big)^2 dx,$$

where d_σ^T and d_σ^R estimate the marginal densities of T and R by

$$d_\sigma^T(x) = \big(G_\sigma \star h^T\big)(x) \quad \text{and} \quad d_\sigma^R(x) = \big(G_\sigma \star h^R\big)(x),$$

with histogram $h^{T,R} : S \to [0,1]$ and 2d-Gaussian kernel $G_\sigma(x) = \frac{1}{2\pi\sigma^2}\exp(\frac{-|x|^2}{2\sigma^2})$.

The mutual information based distance measure was introduced in the context of multimodal image-registration by Wells et al in [7]. One searches for a transformation so that the mutual information (or transinformation) is maximized; see, e.g in [8, 6, 9]. Mutual information is borrowed from information theory [10]. The mutual information based distance measure is maximal if the images are matched. Therefore the mutual information based matching energy is a measure of alignment between the images.

A morphological distance measure introduced by Marc Droske and Martin Rumpf (see [11]) is based on normal information of the level-sets. More precisely, they consider variations respectively generalizations of the energy

$$D^{MMI}[\phi] = \int_{\Omega} \left(N_T \circ \phi - N_R^{\phi}\right)^2 dx \,,$$

where N_T is the normal on the level-sets on T and N_R^{ϕ} is the transformed normal of the reference image.

2.3 A Curvature Based Regularization Energy

Fischer–Modersitzki's Curvature Approach: Bernd Fischer and Jan Modersitzki have first proposed an approach (see [12]), which explicitely penalizes the curvature of the displacement u. In their so-called curvature approach the authors propose the regularization term

$$\mathcal{S}[u,u] = \sum_{l=1}^{2} \int_{\Omega} \Delta u_l \Delta u_l \; dx_1 dx_2,$$

which involves higher order derivatives of u_l and can be seen as an approximation of the squared mean curvature

$$\left(H(u_l)\right)^2 = \left(\frac{(1+{u_l}_{x_2}^2){u_l}_{x_1x_1} - 2{u_l}_{x_1}{u_l}_{x_2}{u_l}_{x_1x_2} + (1+{u_l}_{x_1}^2){u_l}_{x_2x_2}}{(1+{u_l}_{x_1}+{u_l}_{x_2})^{3/2}} \right)^2$$

under the assumption that ∇u remains small.

Although the regularization term is neutral with respect to affine-linear displacements, the functional is not $H^2(\Omega) \times H^2(\Omega)$–coercive (see e.g. [13]), i.e. the bilinear form $\mathcal{S}$ does not satisfies the inequality

$$\gamma ||u||^2_{H^2(\Omega)\times H^2(\Omega)} \leq \mathcal{S}[u,u] \quad \forall u \in H^2(\Omega) \times H^2(\Omega)$$

and consequently there is no guaranty for existence and uniqueness of solutions of the underlying Euler-Lagrange equations. Moreover the kernel of $\mathcal{S}$ is spanned by infinitely many elements. To overcome this problem the authors have restricted their approach to the space

$$\left\{ u_l \in H^2(\Omega), \; \frac{\partial u_l}{\partial n} = \frac{\partial \Delta u_l}{\partial n} = 0, \; l = 1,2 \right\} \subset H^2(\Omega) \times H^2(\Omega)$$

of displacements. As a consequence the affine-linear displacements are penalized by the underlying function space.

Proposed Model: Consider the sum of the squared principal curvatures κ_1 and κ_2 of the displacement-field $u = (u_1, u_2)^t$:

$$\begin{aligned}\mathcal{C}(u) &= \sum_{l=1}^{2} \int_{\Omega} \left(\kappa_1^2(u_l) + \kappa_2^2(u_l)\right) dx_1 dx_2 \\ &= \sum_{l=1}^{2} \int_{\Omega} \left((\kappa_1(u_l) + \kappa_2(u_l))^2 - 2\kappa_1(u_l)\kappa_2(u_l)\right) dx_1 dx_2 \\ &= \sum_{l=1}^{2} \int_{\Omega} \left((H(u_l))^2 - 2K(u_l)\right) dx_1 dx_2,\end{aligned}$$

with mean curvature $H(u_l)$ and Gaussian curvature

$$K(u_l) = \kappa_1(u_l)\kappa_2(u_l) = \frac{u_{l x_1 x_1} u_{l x_2 x_2} - u_{l x_1 x_2}^2}{(1 + u_{l x_1} + u_{l x_2})^2}.$$

Under the assumption that $\nabla u_l \approx 0$, it follows that

$$H(u_l) \approx u_{l x_1 x_1} + u_{l x_2 x_2} = \Delta u_l \quad \text{and} \quad K(u_l) \approx u_{l x_1 x_1} u_{l x_2 x_2} - u_{l x_1 x_2}^2$$

and consequently the bilinear form

$$l(u, u) = \sum_{l=1}^{2} \langle u_l, u_l \rangle_{H^2(\Omega)},$$

with standard $H^2(\Omega)$ Sobolev inner product

$$\langle v, w \rangle_{H^2(\Omega)} = \int_{\Omega} \left(\frac{\partial^2 v}{\partial x_1^2} \frac{\partial^2 w}{\partial x_2^2} + 2 \frac{\partial^2 v}{\partial x_1 \partial x_2} \frac{\partial^2 w}{\partial x_1 \partial x_2} + \frac{\partial^2 w}{\partial x_1^2} \frac{\partial^2 v}{\partial x_2^2} \right) dx_1 dx_2$$

and corresponding semi-norm $|v|_{H^2(\Omega)}^2 = \langle v, v \rangle_{H^2(\Omega)}$, approximates the nonlinear functional $\mathcal{C}$.

This particular choice has various important aspects: Firstly, the energy l is positive semi-definite over $H^2(\Omega) \times H^2(\Omega)$ and positive definite over $V \times V$, with

$$V = \left\{ v \in H^2(\Omega), \int_{\Omega} v dx_1 dx_2 = \int_{\Omega} x_1 v dx_1 dx_2 = \int_{\Omega} x_2 v dx_1 dx_2 = 0 \right\} \subset H^2(\Omega).$$

Secondly, the kernel of the proposed energy consist only of the affine-linear displacements and consequently planar rotation and translation are not penalized by this approach.

2.4 The Biharmonic Diffusion Equation

Referring to the Riesz representation theorem, one can write the bilinear form l as

$$l(v, \varphi) = \left\langle \begin{pmatrix} L & 0 \\ 0 & L \end{pmatrix} \begin{pmatrix} v_1 \\ v_2 \end{pmatrix}, \begin{pmatrix} \varphi_1 \\ \varphi_2 \end{pmatrix} \right\rangle = \sum_{l=1}^{2} \langle L v_l, \varphi_l \rangle \tag{3}$$

for every $\varphi = (\varphi_1, \varphi_2)^t \in H^2(\Omega) \times H^2(\Omega)$. Here L is the biharmonic operator Δ^2 supplemented by the following higher order boundary conditions

$$B_1[v_l(x)] = \frac{\partial^2 v_l(x)}{\partial n^2}, \quad B_2[v_l(x)] = -\frac{\partial}{\partial n}\Delta v_l(x) - K[v_l(x)],$$

with

$$K[v_l(x)] = \frac{\partial}{\partial s}\left[\frac{\partial^2 v_l(x)}{\partial x_1 \partial x_2}(n_{x_1}^2 - n_{x_2}^2) + (\frac{\partial^2 v_l(x)}{\partial x_2^2} - \frac{\partial^2 v_l(x)}{\partial x_1^2})n_{x_1}n_{x_2}\right],$$

where $n = (n_{x_1}, n_{x_2})$ stands for the normal in outward direction, and s stands for the tangential vertical to n. Note, that this are natural boundary conditions, which are satisfied by each solution of the minimization problem (2) using the bilinear form l.

With the regularization energy (3) a minimizer $u(x) = (u_1(x), u_2(x))^t$ of (2) is characterized by the necessary condition

$$\frac{J_\alpha(u + s\varphi)}{\partial s}\bigg|_{s=0} = \alpha l(u(x), \varphi(x)) + D_u[\varphi(x)] \stackrel{!}{=} 0 \quad \forall \varphi(x) \in H^2(\Omega) \times H^2(\Omega),$$

with the Gâtaux-derivative

$$D_u[v(x)] = (\langle f_1(u(x)), v_1(x)\rangle, \langle f_2(u(x)), v_2(x)\rangle)^t = \lim_{t\to 0} \frac{\mathcal{D}[u(x) + tv(x)] - \mathcal{D}[u(x)]}{t}$$

of D and $v(x) = (v_1(x), v_2(x))^t \in L_2(\Omega) \times L_2(\Omega)$. Classical solutions fulfill

$$\alpha L u_l(x) - f_l(u(x)) = 0, \quad \text{for} \quad l = 1, 2. \tag{4}$$

A common approach to minimize $J_\alpha[u] = D[u] + \alpha l(u, u)$ is to introduce an artificial time to equation (4) and to determine the steady state solution of equation (4), i.e. to solve the biharmonic diffusion equations

$$\left.\begin{aligned} \tfrac{\partial u_l(x,t)}{\partial t} + \alpha(t)\Delta^2 u_l(x,t) &= -f_l(u(x,t)) && \text{on } \Omega \times (0,T) \\ B_1[u_l(x,t)] = B_2[u_l(x,t)] &= 0 && \text{on } \partial\Omega \times (0,T) \\ u_l(x,0) &= 0 && \text{on } \Omega \end{aligned}\right\} \tag{5}$$

for $l = 1, 2$.

2.5 Semi-implicit Time Discretization

To discretize (5), we consider the following semi-implicit time discretization scheme:

$$\left.\begin{aligned} u_l^{(0)} &= 0 && \text{for } x \in \Omega, \\ \tfrac{u_l^{(k+1)} - u_l^{(k)}}{\tau} + \alpha_k \Delta^2 u_l^{(k+1)}(x) &= f_l(u^{(k)}(x)) && \text{for } x \in \Omega, \\ B_1[u_l^{(k+1)}(x)] = B_2[u_l^{(k+1)}(x)] &= 0 && \text{for } x \in \partial\Omega, \end{aligned}\right\} \tag{6}$$

for $l = 1, 2$ and $k = 0, 1, ...,$ where $\tau > 0$ is the length of the time step. Using L, equation (6) can be written as the following linear elliptic system

$$(I + \tau\alpha_k L)u_l^{(k+1)}(x) = \underbrace{u_l^{(k)}(x) + \tau f_l(\tau, u^{(k)}(x))}_{=:g_\tau(u^{(k)}(x))}, \quad (x) \in \Omega. \tag{7}$$

In order to give a variational formulation of problem (7), we introduce the symmetric bilinear form

$$\ell(u, v) = \alpha_k\tau \sum_{l=1}^{2} l(u_l, v_l) + \sum_{l=1}^{2} \langle u_l, v_l\rangle = \sum_{l=1}^{2} \langle(\alpha_k\tau L + I)u_l, v_l\rangle \tag{8}$$

with the inner product $\langle u, v\rangle = \int_\Omega uv dx$ and associated norm $||u|| = \langle u, u\rangle^{1/2}$ for the Lebesgue space of square integrable functions on Ω.

Since α_k and τ are positive, the bilinear form $\ell(\cdot, \cdot)$ is coercive and bounded on $H^2(\Omega) \times H^2(\Omega)$. Consequently the Lax-Milgram theorem can be used to prove the existence and uniqueness of the solution of the variational equation

$$\ell(u, \varphi) = \langle g_\tau, \varphi\rangle, \quad \text{for all} \quad \varphi \in H^2(\Omega) \times H^2(\Omega) \tag{9}$$

for any bounded functional g_τ in the dual space of $H^2(\Omega) \times H^2(\Omega)$.

2.6 Spatial Discretization and Fast Numerical Solution

In order to solve equation (7) numerically we use a finite difference discretization and second order approximation of the biharmonic operator and the boundary conditions. Let L_h the discretization matrix corresponding to the operator L, then the resulting linear system

$$(\alpha_k\tau L_h + I_h)u_h = g_{\tau,h}$$

can be solved efficiently by using a multigrid method with optimal multigrid complexity $\mathcal{O}(N)$, where N is the number of picture elements, see [14].

2.7 Curvature Scale Space

Consider the diagonalization of the operator $B_h = \alpha_k\tau L_h + I_h$:

$$B_h u = \sum_{i=1}^{n} \lambda_i\langle u, \phi_i\rangle\phi_i, \tag{10}$$

where $(\phi_i)_{i=1,\dots,n}$ denotes the set of eigenvectors of the operator B_h and λ_i are the eigenvalues belonging to ϕ_i. The orthogonal system $\Phi = [\phi_1, \cdots, \phi_n]$ is given by the set of eigenvectors of B_h. For the countable set of eigenvalues $\{\lambda_i\}_{i=1}^n$ of B_h, it holds

$$1 = \lambda_1 = \lambda_2 = \lambda_3 < \lambda_4 \le \lambda_5 \le \cdots \lambda_n, \text{ with } \lambda_j = 1 + \alpha_k\tau\sigma_j, \tag{11}$$

with eigenvalues $0 = \sigma_1 = \sigma_2 = \sigma_3 < \sigma_4 \le \sigma_5 \le \dots \le \sigma_n$ of L_h. Using (10) and (11) the solution of equation (7) can be given formally by

$$u_h = \sum_{i=1}^{n} \frac{1}{1 + \alpha_k \tau \sigma_i} \langle \phi_i, g_\tau \rangle \phi_i. \tag{12}$$

The set of eigenvectors of the operator B_h can be regarded as a sequence of spatial scales, with energy $\langle B_h \phi_i, \phi_i \rangle = \lambda_i \langle \phi_i, \phi_i \rangle = \lambda_i \ge 1$.

This observation is important because it indicates that the representation can be used to access different aspects of a solution u. Here, the curvature based scale space decomposition (12) makes explicite the role of α as a scale constant. For large values of α we have the following result: The coarsest scale is given by the affine-linear functions corresponding to the eigenvalues $\lambda_1 = \lambda_2 = \lambda_3 = 1$, where no amplification of highly oscillatory functions occurs. For decreasing α, more and more terms in the sum (12) become important for the solution. They correspond to eigenvectors belonging to higher eigenvalues and consequently smaller and smaller structures of the displacements are recovered.

2.8 A Scale-Space Based Image Registration Algorithm

As a consequence of the last section, we embed the minimization of (2) into a scale space framework, which efficiently treats different scales. The minimization starts with a large initial scale parameter, i.e. $\alpha_0 \gg 0$ and an initial displacement $u^{(0)} = 0$. During the iteration the parameter is reduced by $\alpha_k := \gamma^k \alpha_0$ with some decay rate $\gamma \in (0,1)$, and the solution incorporates more and more finer scaled functions.

3 Examples

In order to demonstrate the principle and reliability of the proposed approach we present experimental results for a monomodal as well as for a multimodal image registration problem. We start with a monomodal image registration example.

3.1 Monomodal Image Registration

Figure 1 shows a magnetic resonance imaging (MRI) example. The reference image $R(x)$ is depicted in figure 1(a). The template image $T(x)$ (figure 1(b)) is given by a rotated and translated version of the reference image with preserved rows and columns flipped in the left/right direction.

For this example, we have used $\alpha_0 = 10^9$, $\tau = 2$, $\gamma = 10^{-3}$ and $u^{(0)} = 0$ (corresponds to the identity map). During the iteration the parameters α and τ are decreased down to $\alpha = 10^0$ and $\tau = 2^{-2}$. To give an idea of the effect from the decreasing scale parameter α on the deformed template, we show in figure 1(c)–1(f) the calculated results $T(x - u^{(k)}(x))$ after k =40, 143, 196 and 250 iterations. In the first iteration (with $\alpha = 10^9$) the template is transformed only

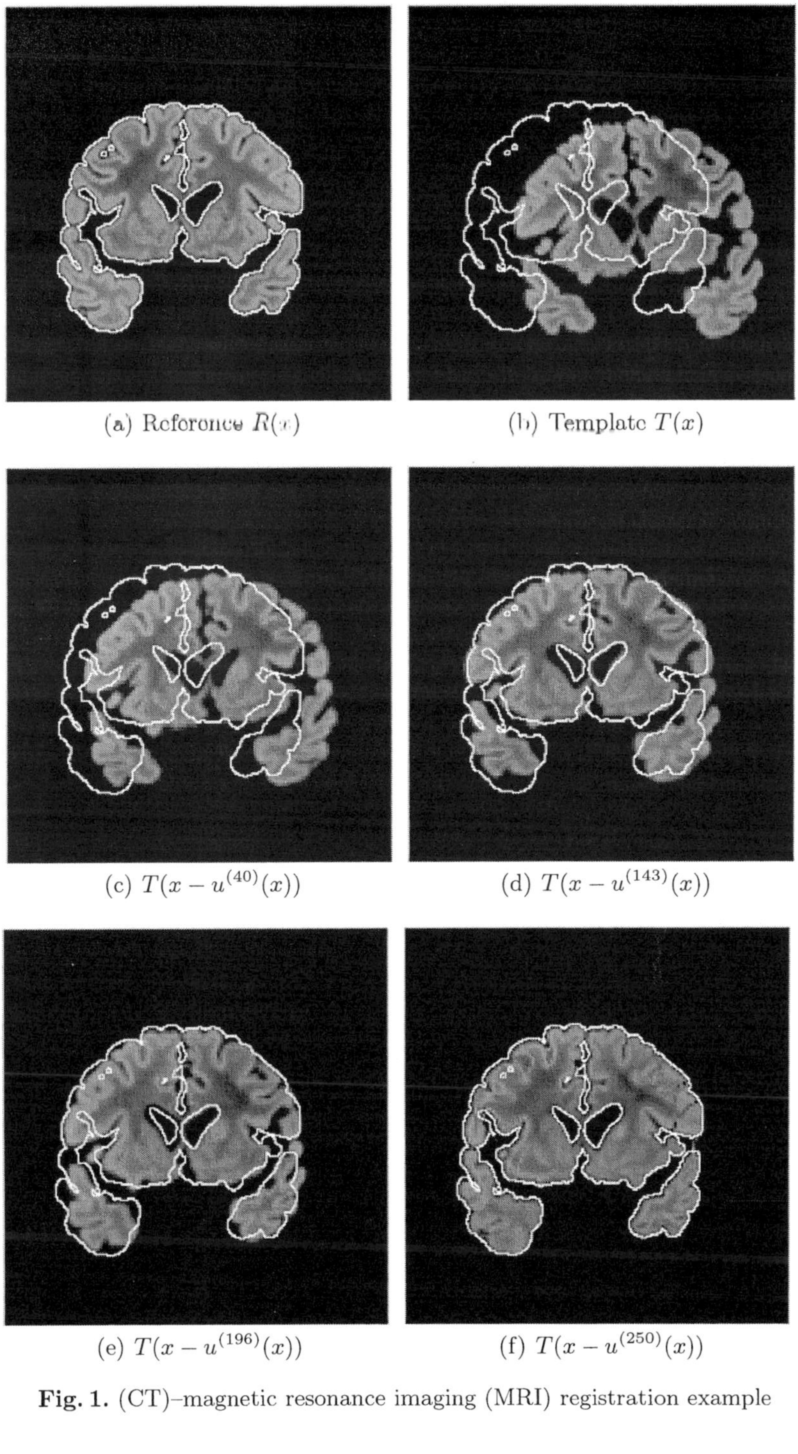

(a) Reference $R(x)$ (b) Template $T(x)$

(c) $T(x - u^{(40)}(x))$ (d) $T(x - u^{(143)}(x))$

(e) $T(x - u^{(196)}(x))$ (f) $T(x - u^{(250)}(x))$

Fig. 1. (CT)–magnetic resonance imaging (MRI) registration example

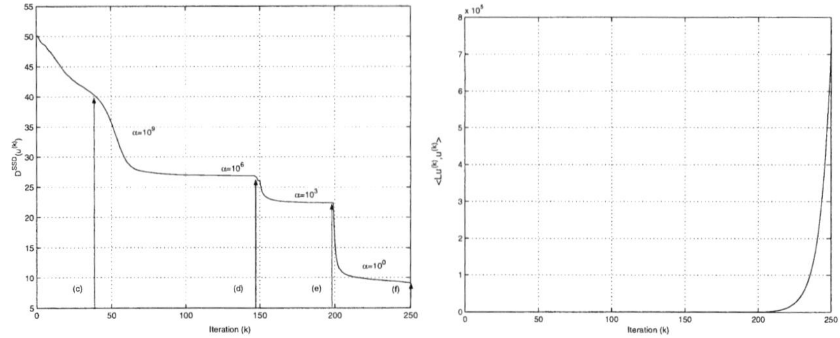

(a) Matching energy (b) Regularization energy

Fig. 2. Energy history for MRI example in figure 1

by coarse scale basis functions (rigid motion). Consequently, only coarser image structures are aligned (see figures 1(c)–1(d)). As the parameter α decreases, the solution consists of more and more fine scaled functions and hence also fine image structures are matched. These findings are stressed by figure 2. Here, we observe a strong decay of D (see figure 2(a)) and a abrupt rise of the energy $\langle Lu, u \rangle$, when the scale parameter α decreases.

3.2 Multimodal Image Registration

The CT-image (figure 3(a)) is the reference and the MRI image (figure 3(b)) is the template. Both images are displayed with superimposed contour of the reference image. For this example, we use the probability density based density measure $D^{PLS}[u]$ and the same parameter setting as in the previous example. We apply 150 iteration steps of the biharmonic diffusion iteration. As a consequence of the large regularization parameter (which focuses on coarse scales) the first iteration steps determine an affine linear transformation, see figure 3(c). Then the value of the scale parameter α is recognized as being too large, and hence decreased in the following iterations. At this point the solution incorporates more and more finer scaled functions. The result after 155 iteration steps is presented in figure 3(d). The approach matches the different structures onto the corresponding reference structures. The corresponding 4×4 checkerboard views (figure 3(e) and 3(f)) of the results show smooth transitions between the structures.

4 Summary and Conclusion

In this paper we have introduced a novel approach for digital image registration based on the biharmonic diffusion equation. Employing semi-implicit time discretization, we are facing in each time step a stationary problem given by a partial differential equation with higher order boundary conditions. From an abstract point of view we now encounter a solution which renders more precisely

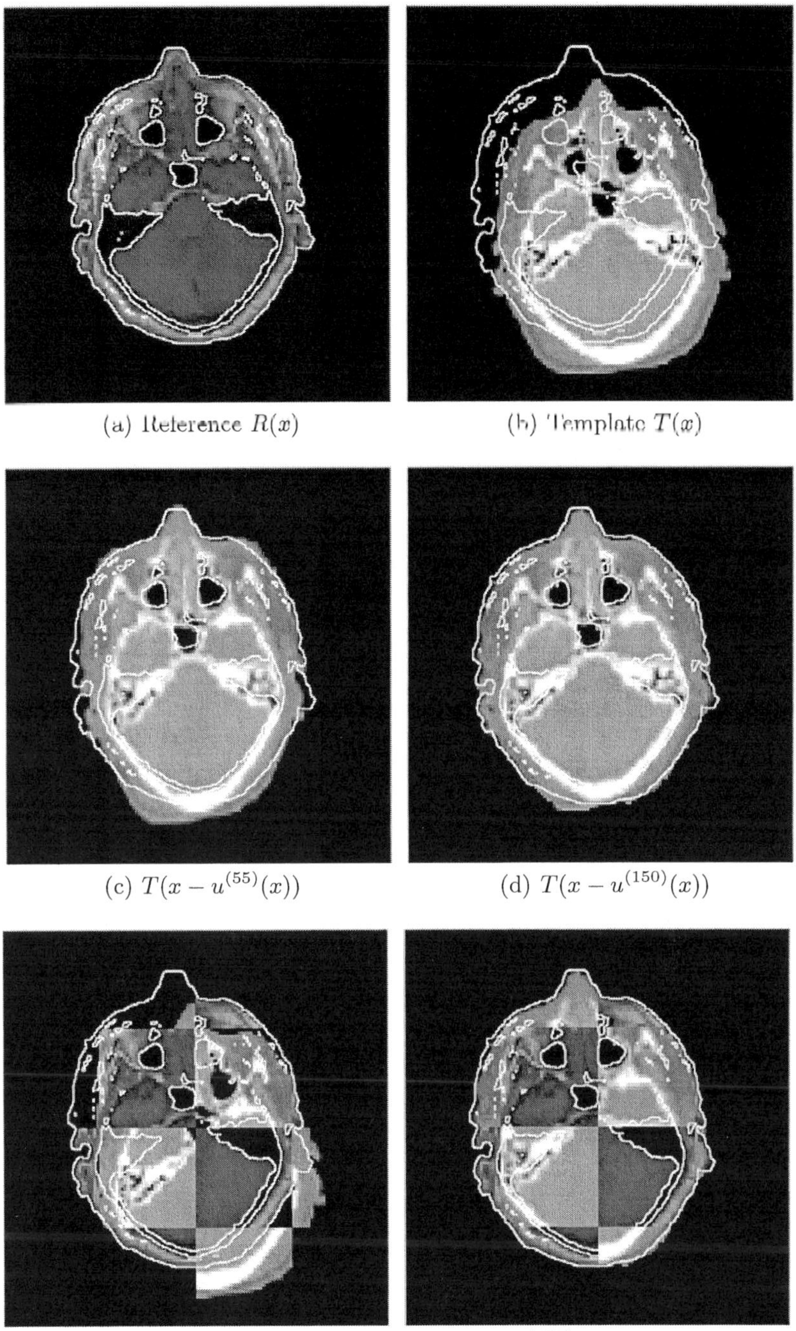

(a) Reference $R(x)$

(b) Template $T(x)$

(c) $T(x - u^{(55)}(x))$

(d) $T(x - u^{(150)}(x))$

(e) $T(x)$ and $R(x)$ mixed like a 4×4 checkerboard

(f) $T(x - u^{(150)}(x))$ and $R(x)$ mixed like a 4×4 checkerboard

Fig. 3. (CT)–magnetic resonance imaging (MRI) registration example the link between scale space and image registration. The approach provides a multiscale description of the displacement fields, but one in which the notion of scale is based on curvature scale space, rather than on conventional multigrid or Gaussian blurring minimization approaches as presented in [11, 15].

The proposed multiscale framework unifies existing image registration approaches, which are normally classified due to the underlying transformations into either affine-linear or non-linear approaches. This multiscale nature of the approach provides a description that can be made robust, because it is based on solving the registration problem successively at multiple scales. The technique has two parameters that decrease during the iteration. One of these is the underlying regularization parameter α which relates to a scale parameter, since it specifies the curvature of the resulting displacements. The other is the time-step parameter τ, which is chosen as large as possible, as long the iteration will remain stable.

References

1. Henn, S., Witsch, K.: A multigrid-approach for minimizing a nonlinear functional for digital image matching. Computing **64(4)** (1999) 339–348
2. Aubert, G., Deriche, R., Kornprobst, P.: Computing optical flow via variational techniques. SIAM J. Appl. Math. **60(1)** (1999) 156–182
3. Clarenz, U., Droske, M., Henn, S., , Rumpf, M., Witsch, K.: Computational methods for nonlinear image registration,. In: Image Registration (to appear), European Consortium for Mathematics in Industry (ECMI) (2004)
4. Amit, Y.: A nonlinear variational problem for image matching. SIAM J. Sci. Comput. **15** (1994) 207–224
5. Fischer, B., Modersitzki, J.: Fast inversion of matrices arising in image processing. Numerical Algorithms **22** (1999) 1–11
6. Henn, S., Witsch, K.: Multi-modal image registration using a variational approach. SIAM J. Sci. Comput. (SISC) **25(4)** (2004) 1429–1447
7. Wells, W., Viola, P., Atsumi, H., Nakajima, S., Kikinis, R.: Multi-modal volume registration by maximization of mutual information. Medical Image Analysis **1** (1996) 35–51
8. D'Agostino, E., Modersitzki, J., Maes, F., Vandermeulen, D., Fischer, B., Suetens, P.: Free-form registration using mutual information and curvature regularization. WBIR (2003) 11–20
9. Hermosillo, G.: Variational methods for multimodal image matching. Phd thesis, Université de Nice, France (2002)
10. Cattermole, K.: Statistische Analyse und Struktur von Information. VCH (1988)
11. Droske, M., Rumpf, M.: A variational approach to non-rigid morphological registration. SIAM Appl. Math. **64(2)** (2004) 668–687
12. Fischer, B., Modersitzki, J.: Curvature based image registration. JMIV **18** (2003) 81–85
13. Hackbusch, W.: Elliptic Differential Equations. Theory and Numerical Treatment. Springer Series in Computational Mathematics 18. Springer-Verlag, Berlin Heidelberg New York (1992)
14. Henn, S.: Multigrid method for fourth-order diffusion equation. Preprint (2004)
15. Henn, S., Witsch, K.: Iterative multigrid regularization techniques for image matching. SIAM J. Sci. Comput. (SISC) **23(4)** (2001) 1077–1093

A Scale-Space Analysis of a Contour Figure Using a Crystalline Flow

Hidekata Hontani[1], Yu Suzuki[1], Yoshikazu Giga[2],
Mi-Ho Giga[2], and Koichiro Deguchi[3]

[1] Department of Informatics, Yamagata University,
4-3-16, Jonan, Yonezawa, Yamagata, 992-8510 Japan
hontani@yz.yamagata-u.ac.jp
dsw05840@dip.yz.yamagata-u.ac.jp
[2] Graduate School of Mathematical Sciences, Tokyo University,
3-8-1, Komaba, Meguro-ku, Tokyo 153-8914 Japan
{giga, mihogiga}@math.sci.hokudai.ac.jp
[3] Department of System Information Sciences,
Graduate School of Information Sciences, Tohoku University,
Aramaki-aza Aoba01, Aoba-ku, Sendai 980-8579 Japan
deguchi@fractal.is.tohoku.ac.jp

Abstract. A method for a scale-space analysis of a contour figure based on a crystalline flow is proposed. A crystalline flow is a special family of an evolving polygons, and is a discrete version of a curvature flow. Based on a crystalline flow of a given contour, the proposed method makes a scale-space representation and extracts several sets of dominant facets from the given contour. By changing the shape of the Wulff shape that plays a role of a unit circle for computing the nonlocal curvature of each facet, the method analyses the contour shape anisotropically.

1 Introduction

Evolution based scale-space methods play an important role to characterise a contour figure[1][2][3]. Specifically in the negotiations leading to MPEG-7, such methods were intensively discussed and tested. Due to this fact, there exists an enormous amount of literature devoted to this field[4]. In this article, we propose a method for describing the shape of a simple contour figure in an image.

A contour figure in an image is often represented as a polygon. A crystalline flow[5][6] is a special family of evolving polygons, and is a discrete version of a classical curvature flow[7][8]. In the evolving process of a crystalline flow, each facet moves toward its normal direction. The velocity of a facet is determined by the nonlocal curvature, which depends on the length of the facet. The number of facets does not increase through the evolving process, and any simple polygon becomes convex at finite time. Different from a classical curvature flow, it is easy to track each facet in a given contour through the evolving process. These features of a crystalline flow help to make a scale-space representation of a given contour. In [9], the authors showed a method for computing a crystalline flow

R. Kimmel, N. Sochen, J. Weickert (Eds.): Scale-Space 2005, LNCS 3459, pp. 155–166, 2005.

from any given simple polygon, and used the method for extracting dominant facets from a given polygon[10].

In a crystalline flow, a convex polygon called Wulff shape is used to determine the nonlocal curvature of a facet. By changing the shape of the Wulff shape, we can easily obtain an anisotropic flow from a given contour. In this article, we prepare a set of Wulff shapes and extract dominant facets from a given simple contour using each Wulff shape.

2 Crystalline Flow

2.1 Admissible Crystalline Flow

We consider a motion of an evolving curve Γ_t governed by the *anisotropic curvature flow equation* of the form

$$V = \Lambda_\gamma(\boldsymbol{n}) \text{ on } \Gamma_t, \tag{1}$$

for $t > 0$. Here, $\boldsymbol{n}$ denotes the unit outward normal, and V does the velocity along $\boldsymbol{n}$. The quantity $\Lambda_\gamma(\boldsymbol{n})$ is called a weighted curvature, which is the first variation of the interfacial energy $\int_S \gamma(\boldsymbol{n})ds$ with respect to the change of the area enclosed by S. It has the form

$$\Lambda_\gamma(\boldsymbol{n}) = -\text{div } \xi(\boldsymbol{n}),$$

where $\xi = \nabla\gamma$ and γ is a given positively homogeneous function of degree one in $\boldsymbol{R}^2$; div denotes the divergence on the curve S. The quantity γ is called the interfacial energy density. If $\gamma(\boldsymbol{p}) = |\boldsymbol{p}|$, then the quantity $\Lambda_\gamma(\boldsymbol{n})$ is equal to the usual curvature κ and (1) becomes a curve shortening equation $V = \kappa$, which is widely used for a scale-space analysis of a contour figure.

There are several methods to track evolution of Γ_t; one of a typical method is the level-set method[7][8][11][12]. If γ is C^2 except the origin, global unique solvability for (1) is established by [8] (see also [13]). However, when γ has corners, conventional notion of a solution including viscosity solutions does not apply to (1).

If Frank diagram of γ :

$$\text{Frank}\gamma = \{\boldsymbol{p} \in \mathbf{R}^2 \ ; \ \gamma(\boldsymbol{p}) \leq 1\}$$

is a convex polygon, γ is called a *crystalline energy density* (see also Fig.1), and a notion of solution for (1) is proposed by [5] and [6] independently by restricting $\{\Gamma_t\}$ as a special family of evolving polygonal curves called admissible. Here and hereafter we assume that γ is a crystalline energy density, i.e., Frankγ is a convex M-polygon. Let $\boldsymbol{q}_i$ $(i = 1, 2, \cdots M)$ be the vertices of the Frankγ, and $\mathcal{N} = \{\boldsymbol{q}_i/|\boldsymbol{q}_i|\}$ denote the set of all unit vectors $\boldsymbol{q}_i/|\boldsymbol{q}_i|$ (see Fig.1(B)). We say that a simple polygonal curve S in $\boldsymbol{R}^2$ is an admissible crystal if all outward normal orientation belongs to $\mathcal{N}$ and orientations of adjacent facets point to vertices adjacent in Frankγ (see Fig.2).

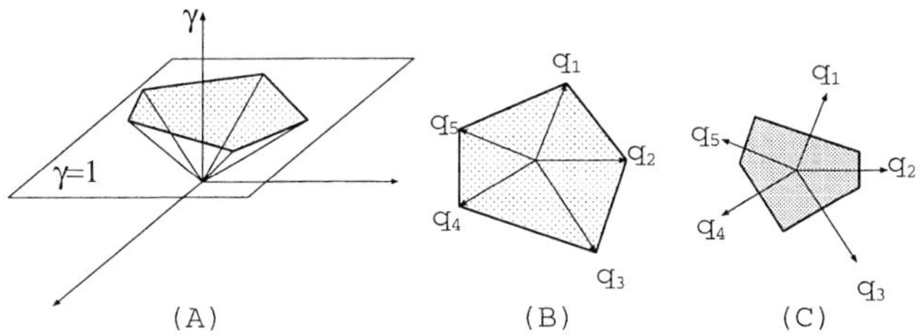

Fig. 1. The crystalline energy γ (A), the Frankγ (B), and the Wulff shape (C)

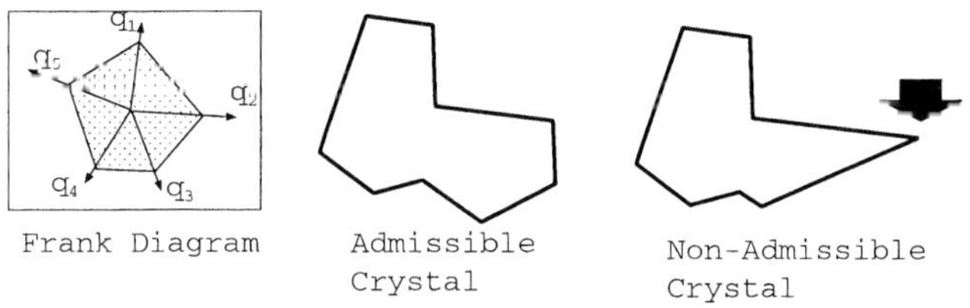

Fig. 2. The Wulff shape, an admissible crystal, and a non-admissible crystal. At the vertex indicated by the arrow, the condition is not satisfied

In [5] and [6], it was proposed to restrict the problem to an admissible crystal, and a corresponding system of ODEs (ordinary differential equations) was derived. The solution of the system is called an admissible crystalline flow. In the admissible crystalline flow, the weighted curvature is represented as follows.

$$\Lambda_\gamma(\boldsymbol{n}_j) = \chi_j \frac{\Delta(\boldsymbol{n}_j)}{L_j(t)}, \tag{2}$$

where $L_j(t)$ is the length of jth facet, and $\boldsymbol{n}_j$ is its outward unit normal. The facet number j is counted clockwise. The quantity χ_j is called a transition number, and takes +1 (resp. -1) if the jth facet is concave (resp. convex) in the direction of $\boldsymbol{n}_j$. Otherwise $\chi_j = 0$. The quantity $\Delta(\boldsymbol{n}_j)$ is the length of a facet of Wulff shape

$$W_\gamma = \{\boldsymbol{q} \in \boldsymbol{R}^2; \boldsymbol{q} \cdot \boldsymbol{p} \leq \gamma(\boldsymbol{p}) \text{ for all } \boldsymbol{p} \in R^2\},$$

of which outward normal is $\boldsymbol{n}_j$ (see Fig.1(C)). The Wulff shape is the unique minimizer of the interfacial energy among all S whose enclosed area is the same as W_γ (see e.g. [14]). The Wulff shape plays a role of a unit circle for a classical curvature flow (see Fig.3).

If the jth facet moves with the outward normal velocity V_j, then its length satisfies a geometry transport equation

$$\frac{dL_j}{dt}(t) = (\cot\phi_j + \cot\phi_{j+1})V_j - \frac{1}{\sin\phi_j}V_{j-1} - \frac{1}{\sin\phi_{j+1}}V_{j+1}, \tag{3}$$

as in [5] and [6]. Here, $\phi_j = \theta_j - \theta_{j-1}$ for $\boldsymbol{n}_j = (\cos\theta_j, \sin\theta_j)$. Combining the equations (1),(2) and (3), we obtain a system of ODEs of L_j. If a given

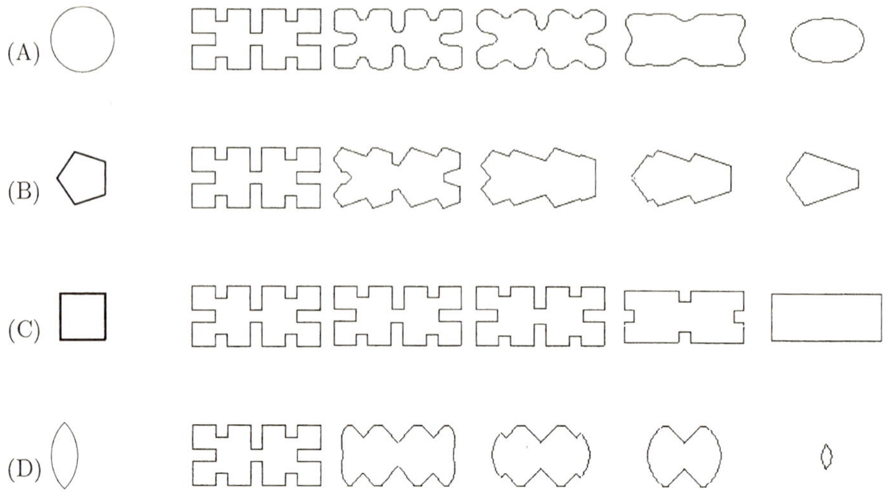

Fig. 4. Examples of the crystalline flow. The initial contour is common to all, and is shown in the second column. The Wulff shapes are shown at the left

Figure 4 shows some experimental results of crystalline flow. The initial contour is common to all, but the Wulff shape is different. As described before, the Wulff shape plays the role of a unit circle for a classical curve shortening flow. Because a crystalline flow can be obtained from a non essentially admissible crystal, any simple and convex polygon can be used for the Wulff shape.

3 Dominant Corner Facet Extraction

As mentioned, in a crystalline flow, any simple closed curve becomes convex at finite time, and it is easy to track each facet in the evolving contour. Those features of a crystalline flow are useful for a multi-scale analysis of a contour figure. In [10], we applied a crystalline flow to a multi-scale method that extracts several sets of dominant facets from a given polygon. The method is analogous to classical multi-scale methods for corner extraction[18] (see also [19]). In this article, we prepare a set of Wulff shapes, and extract sets of dominant facets from a given contour using each Wulff shape.

3.1 Scale-Space for Dominant Corner Facet Extraction

Each facet in an evolving polygon has a transition number χ, which represents the shape around the facet. It specifies whether the shape is convex, concave, or otherwise around the facet, which is a fundamental shape feature of a contour figure. If the shape is convex around some facet in an evolving polygon at some large scale, then we may interpret that the shape of the given contour is 'almost convex with small size disturbance' in a large area around the corresponding facet of the given contour. We call the facets in a given polygon whose transition numbers do not change through a long range of the scale in the crystalline flow as dominant facets.

In order to extract dominant facets, we make a scale-space representation[1] of a given polygon using a crystalline flow. The x-axis of a scale-space shows the indices j of early stage contour defined below, and the y-axis shows the time t.

As time increases, each facet of the early stage contour moves, and is contained in some (non trivial) facet in the evolved polygon $S(t)$. In the proposed method, the value of a point (j, t) in the scale-space shows the transition number of the facet in $S(t)$ which includes the facet j in the early stage contour. Referring to this scale-space representation, the proposed method extracts dominant facets whose transition numbers are inherited for a long time interval in the evolving process.

When a given polygon is not essentially admissible, then some new facets may be created instantaneously just after $t = 0$. Once new facets are created, then the polygon becomes essentially admissible and no new facet is created. Let denote an evolving contour of a crystalline flow $S(t)$. We say that $S(\tilde{t})$ is an *early stage contour* if no facet disappears and no degenerate pinching and no selfintersection occurs for all $t \in (0, \tilde{t}]$. We index all facets of an early stage contour by $j = 1, 2, \cdots, r$, clockwisely. The totality of indices denotes $\mathcal{I}$; we consider this set modulo r. We shall assign a subset $\mathcal{I}_h(t)$ of consecutive indices in $\mathcal{I}$ to each facet $F_h(t)$ of $S(t) = \cup_{h=1}^{k} F_h(t)$ and divide $\mathcal{I}$ into disjoint subsets $\{\mathcal{I}_h(t)\}_{h=1}^{k}$ in the following inductive way. We call $\mathcal{I}_h(t)$ the set of early stage indices of $F_h(t)$. Suppose that all sets of early stage indices of $S(t)$ are already known.

Suppose that $F_l(\tau)$ disappears at $t_1 > t$ and that no facet disappears at $s \in (t, t_1)$. Then, we set $\mathcal{I}_h(s) = \mathcal{I}_h(t)$ for $s \in (t, t_1)$. We shall construct the set of early stage indices at t_1 as follows. If both F_{l-1} and F_{l+1} do not disappear at t_1, then we add $\mathcal{I}_{l-1}(t)$, $\mathcal{I}_l(t)$, and $\mathcal{I}_{l+1}(t)$ to the set of early stage indices of a

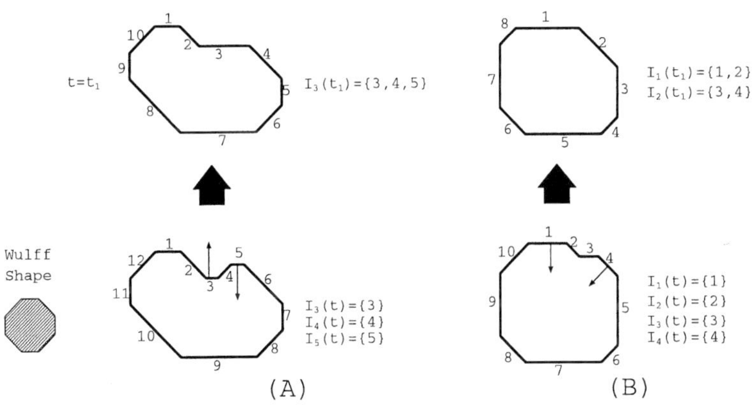

Fig. 5. Construction of indices $\mathcal{I}_l(t)$. (A): A facet $F_4(t)$ disappears at t_1. (B): Two consecutive facets $F_2(t)$ and $F_3(t)$ disappear simultaneously at $t = t_1$

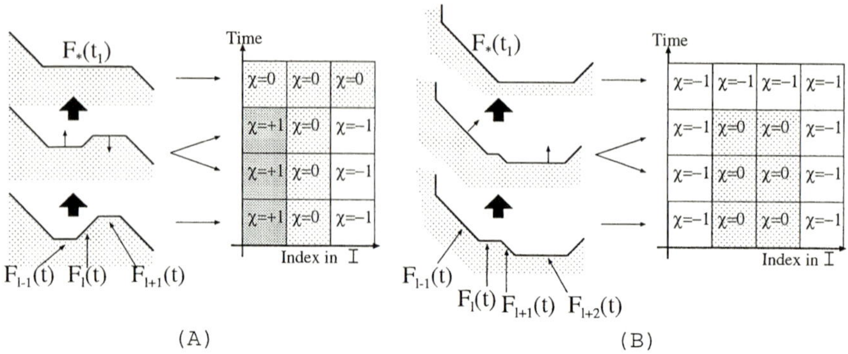

Fig. 6. A scale-space representation of the transition number. The x-axis represents the index of facet in the early stage contour, and the y axis represents the time t

(merged) facet $F_*(t_1)$ containing the limit of $F_{l-1}(s)$ and $F_{l+1}(s)$ as $s \uparrow t_1$. Figure 5(A) shows an example: At time $t = t_1$, the facet F_4 disappears. Assume that $\mathcal{I}_l(t) = \{l\}$ for $t < t_1$. Then, in this case as shown in Fig.5(A), $\mathcal{I}_3(t_1) = \{3, 4, 5\}$.

If two consecutive facets F_{l-1} (resp. F_{l+1}) and F_l disappear at t_1, then we add $\mathcal{I}_l(t)$ to the set of early stage indices of a facet $F_*(t_1)$ containing the limit $F_{l+1}(s)$ (resp. $F_{l-1}(s)$) as $s \uparrow t_1$. Figure 11(B) shows an example: At $t = t_1$, the facet F_2 and F_3 disappear simultaneously. Assume that $\mathcal{I}_l(t) = \{l\}$ for $t < t_1$. Then, in this case as shown in Fig.11(B), $\mathcal{I}_2(t_1) = \{1, 2\}$ and $\mathcal{I}_3(t_1) = \{3, 4\}$.

By this procedure, the set of early stage indices is uniquely determined for each facet of $S(t)$ as far as $S(t)$ is essentially admissible. (Note that $\mathcal{I}$ is divided into sets of early stage indices at $t > 0$.) One facet $F_h(t)$ may have several indices of $\mathcal{I}_h(t)$. Let $\chi(j, t)$ denote the transition number of the facet $F_h(t)$ such that $\mathcal{I}_h(t)$ contains j. It is proved that, if $V = \Lambda_\gamma$, all facets that disappear before the evolving contour converges to a point always have zero transition number and at

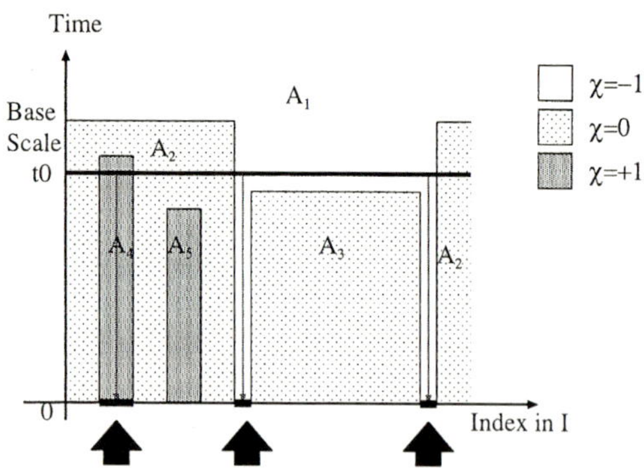

Fig. 7. Dominant facet extraction using the scale-space representation. The facets in an initial contour $\chi \neq 0$ are extracted, if they can be tracked to the base scale t_0

most two consecutive facets disappear[20]. This proof helps to track each facet through the evolution. As is shown in Fig.6, the transition number is plot at the corresponding position in the scale-space. This representation is analogous to a usual curvature scale-space.

3.2 Dominant Corner Facet Extraction

Referring to the scale-space representation of the transition number, we extract dominant facets whose transition numbers are not 0 and the values of the transition numbers are inherited for a long time interval in the evolving process. Our algorithm is as followings.

1. Make the scale-space representation of the transition number $\chi(j,t)$, where j is in $\mathcal{I}$ and t is the time.
2. Divide the scale-space into areas, so that each area has the uniform value of $\chi(j,t)$ inside, and has different value from the neighbouring areas. Let denote such the area as A_k, where $k = 1, 2, \cdots, n$ is the serial number.

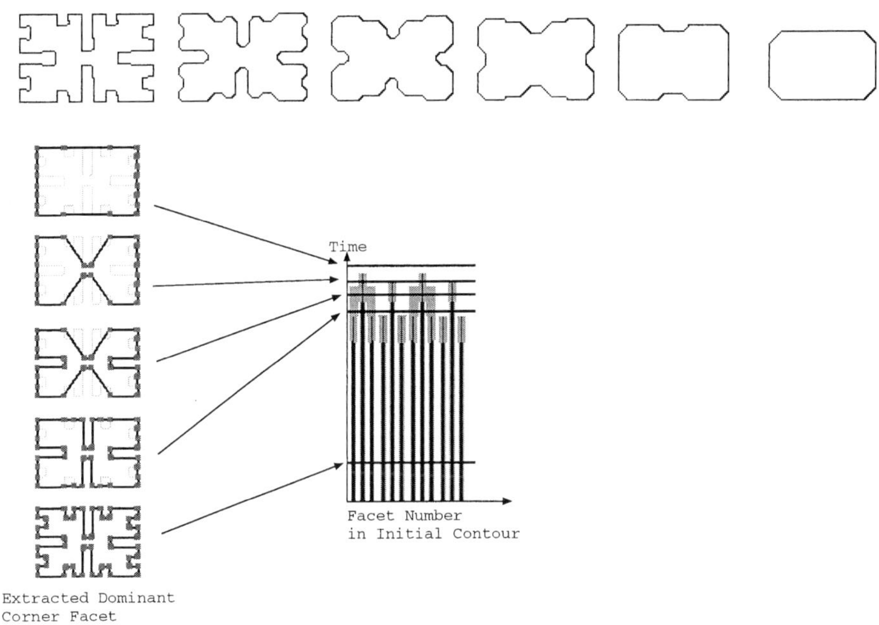

Fig. 8. An example of a crystalline flow (top) and corresponding results of dominant facet extraction (bottom). The Wulff shape is a regular octagon. The crystalline flow is shown in order of time from left to right. The left one is the initial given contour. In the scale-space, white area shows $\chi = -1$ (convex), black one does $\chi = +1$ (concave), and gray one $\chi = 0$. In the scale space, corresponding base scales are indicated by horizontal lines

3. Set the base scale t_0, and draw a line $t = t_0$ in the scale space. Then, find a set of numbers $\mathcal{U}_{t_0}$, so that the area A_k contains the line $t = t_0$ and that $\chi(j, t) \neq 0$ on A_k, if $k \in \mathcal{U}_{t_0}$.
4. Extract all indices from $\mathcal{I}$ (the set of all indices of early stage contour) that are included in the area $A_k(\tilde{t})$ for some $k \in \mathcal{U}_{t_0}$. Here, $A_k(\tilde{t})$ is the cross-section of A_k at the time $\tilde{t}$ at which $S(\tilde{t})$ is an early stage contour. We call facets of an early stage contour corresponding to such extracted indices *dominant facets* at t_0. Each of these indices corresponds to a facet of the early stage contour whose transition number is inherited to the evolving contour at t_0. In Fig.7, the indices of the dominant facets are indicated by up-arrows.
5. Increase the base scale t_0 by small amount Δt, and repeat 3, 4, and 5, if t_0 is smaller than the scale at which the evolving contour becomes convex.

We note that the set of all dominant facets may differ for different base scale t_0. If $t_0 < t_1$, then, $\mathcal{U}_{t_0} \supseteq \mathcal{U}_{t_1}$. As the result, the number of the dominant facets does not increase as t_0 increases. Figure 8 shows an example of a crystalline flow, the corresponding scale-space representation, and the extracted dominant facets.

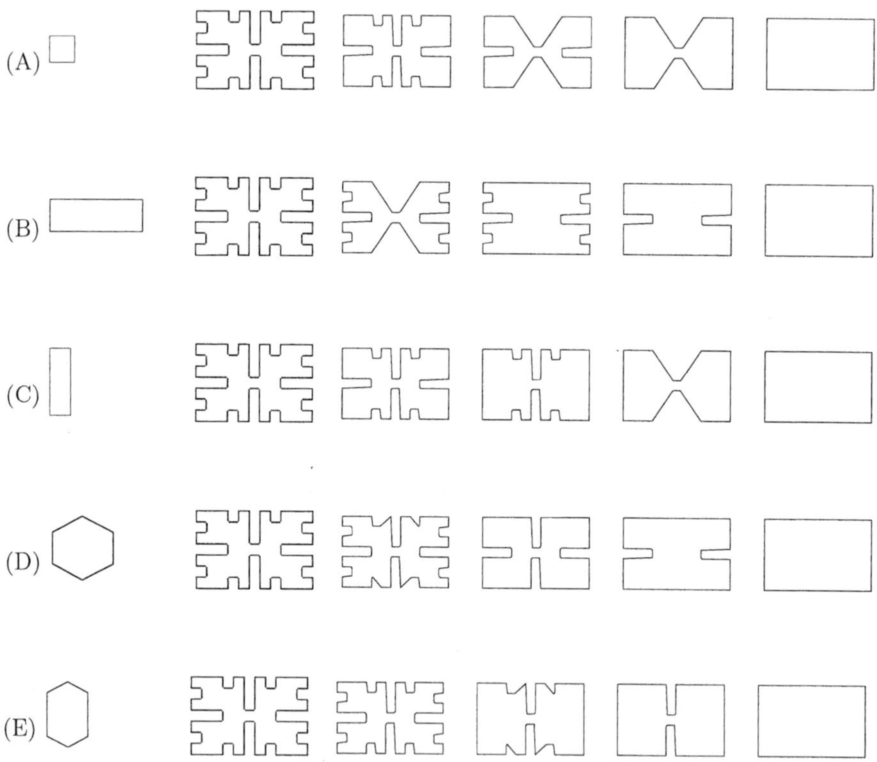

Fig. 9. Examples of extracted dominant facets. The initial contour is common to all, and is shown in the second column. The Wulff shapes are shown at the left. The extracted facets at each base scale are connected with lines

The Wulff shape is a regular octagon. The results of dominant facet extraction are indicated with polygons that are obtained by connecting extracted facets by straight lines. As shown in Fig.8, as the base scale becomes larger, fewer facets are extracted.

As mentioned, the Wulff shape plays the role of a unit circle for a classical curve shortening flow. As shown in equation (2), the nonlocal curvature Λ_γ is proportional to the length $\Delta(\boldsymbol{n})$ of the corresponding facet of the Wulff shape. If a facet of the Wulff shape is longer than other facets, then an evolving polygon tends to be squashed along the normal orientation of the longer facet.

We prepare a set of Wulff shapes, and extract sets of dominant corner facets for each Wulff shape. We believe that it enriches the ability to analyse contour shape. Figure 9 shows experimental results. In each row, the obtained sets of dominant facets are shown in order of base scale t_0. Each row corresponds to different Wulff shape. We applied five Wulff shapes. In Fig. 9, (A), (B) and (C) show the results of dominant facet extraction obtained when a rectangle is used for the Wulff shape. Figure 9 (A) corresponds to the Wulff shape of a regular rectangle, (B) corresponds to a vertically long rectangle, and (C) corresponds to a horizontally long rectangle. As shown in Fig.9, if a vertically (resp. horizontally) long shape is used as the Wulff shape, vertical (resp. horizontal))large structures tend to be extracted as dominant facets. Figure 9 (D) and (E) show other results obtained when a hexagon is used for the Wulff shape. Again, using a vertically long Wulff shape, the method extracted vertical structures as dominant facets. Only by changing the shape of the Wulff shape, we can analyse a contour shape anisotropically.

4 Summary

A crystalline flow is used for a scale-space analysis of a contour figure. A numerical method for obtaining a crystalline flow from a given polygon that is not essentially admissible is presented, and the method is applied for extracting dominant facets. In this article, we use a set of the Wulff shapes for extracting dominant facets from a given contour figure. Changing the shape of the Wulff shape, we can analyse the shape of a contour figure anisotropically.

In many cases, a contour in an image is given as a polygon. For example, a contour represented with a chain-code is a polygon that consists of short facets. Because the nonlocal curvature Λ_γ is determined by the facet length, no approximation is needed for the calculation of the curvature. In addition, because each facet moves with keeping its direction, it is not difficult to trace every facet through the evolving process. Those features of a crystalline flow are useful for multi-scale contour figure analysis.

References

1. A. P. Witkin: Scale Space Filtering: A New Approach to Multi-Scale Descriptions, in Proceedings of 8th International Joint Conference of Artificial Intelligence (1983) 1019–1022

2. Koenderink, J. J.: The Structure of Images, Biological Cybernetics, **50** (1984) 363–370
3. Alvarez, L., Guichard, F., Lions, P.-L. and Morel J.-M.: Axioms and Fundamental Equations of Image Processing, Arch. Rational Mech. Anal., **123** (1993) 199–258
4. Weickert, J.: A Review of Nonlinear Diffusion Filtering, Proceedings of 1st International Conference on Scale-Space Theories in Computer Vision (1997) 3–28
5. Angenent, S. B. and Gurtin, M. E.: Multiphase thermomechanics with interfacial structure 2. Evolution of an isothermal interface, Arch. Rational Mech. Anal., **108** (1989) 323–391
6. Taylor, J.: Constructions and conjectures in crystalline nondifferential geometry, Proceedings of the Conference on Differential Geometry, **52**, Pitman, London (1991) 321–336
7. Evans, L. C. and Spruck, J.: Motion of level-sets by mean curvature, I, J. Differential Geometry, **33** (1991) 635–681
8. Chen, Y.-G., Giga, Y. and Goto, S.: Remarks on viscosity solutions for evolution equations, J. Differential Geometry, **33** (1991) 749–786
9. Hontani, H., Giga, M.-H., Giga, Y. and Deguchi, K.: A computation of a crystalline flow starting from non-admissible polygon using expanding selfsimilar solutions, 11th International Conference DGCI 2003, LNCS 2886, Springer, Naples, (2003) 465–474
10. Hontani, H., Giga, M.-H., Giga, Y. and Deguchi, K.: Expanding selfsimilar solutions of a crystalline flow with applications to contour figure analysis, Discrete Applied Mathematics, printing
11. Osher, S. and Sethian, J. A.: Fronts propagating with curvature-dependent speed: Algorithms based on Hamilton-Jacobi formulations, J. Comput. Phys. **79** (1988) 12–49
12. Giga, Y.: A level set method for surface evolution equations, Sugaku 47 (1993) 321–340: Eng. translation, Sugaku Exposition **10** (1995), 217–241
13. Giga, Y. and Goto, S.: Motion of hypersurfaces and geometric equations, J. Mathematical Society Japan, **44** (1992) 99–111
14. Gurtin, M. E.: Thermomechanics of Evolving Phase Boundaries in the Plane, Oxford, Clarendon Press (1993)
15. Giga, M.-H. and Giga, Y.: Generalized Motion by Nonlocal Curvature in the Plane, Arch. Rational Mech. Anal., **159** (2001) 295–333
16. Giga, M. -H. and Giga, Y.: Crystalline and level-set flow – Convergence of a crystalline algorithm for a general anisotropic curvature flow in the plane, Free boundary problems: theory and applications I (ed. N. Kenmochi) Gakuto International Ser. Math. Sci. Appl., **13** (2000) 64–79
17. Giga, M.-H., Giga, Y. and Hontani, H.: Selfsimilar solutions in motion of curves by crystalline energy, Minisymposium lecture of 5th International Congress on Industrial and Applied Mathematics, Sydney, (2003) July
18. Rattarangsi, A. and Chin, R. T.: Scale-Based Detection of Corners of Planar Curves, IEEE Transactions on Pattern Analysis and Machine Intelligence, **14**, No.4 (1992) 430–449
19. Hontani, H. and DEGUCHI, K.: An Adaptive Local Smoothing for Contour Figure Approximation," 2nd International Conference, Scale-Space '99, Corfu, Greece, (1999) 483–488
20. M.-H. Giga and Y. Giga: Consistency in evolutions by crystalline curvature, Free Boundary Problems '95, Theory and Applications (eds. M. Niezgodka and P. Strzelecki), Proc. of the Zakopane Congress, Poland (1995) 186–202

Multiscale Active Contours

Xavier Bresson, Pierre Vandergheynst, and Jean-Philippe Thiran

Signal Processing Institute (ITS),
Swiss Federal Institute of Technology (EPFL),
CH-1015 Lausanne, Switzerland
{Xavier.Bresson, Pierre.Vandergheynst, JP.Thiran}@epfl.ch

Abstract. In this paper, we propose an evolution equation for the active contours in scale spaces. This evolution equation is based on the Polyakov functional that has been first introduced in physics and has been then used in image processing in [17] for image denoising. Our active contours are hypersurfaces implicitly and intrinsically represented by a level set function embedded in a scale space. The scale spaces used in our approach are defined by a family of metric tensors given by the general heat diffusion equation. The well-known scale spaces such as the linear scale space, i.e. the Gaussian scale space, the Perona-Malik scale space, the mean curvature scale space and the total variation scale space can be used in this framework. A possible application of this technique is in shape analysis. For example, our multiscale segmentation technique can be coupled with the shape recognition and the shape registration algorithms to improve their robustness and their performance.

1 Introduction and Motivation

This paper aims at introducing the *scale* parameter in the *active contour* formalism [8, 3, 9] to define an object multiscale segmentation model. One of the main motivations to develop such a technique is to deal with the shape of objects at different scales of observation/resolution. Indeed, the works of Witkin [18] and Koenderink [10] have shown that the shape of objects changes according to the scale of observation used. At large scales, the global shape of the object can be observed since smaller shape features are suppressed. And at lower scales, finer characteristics appear in the shape of the object.

As a result, it appears natural to analyze a given image not only at one scale but at several scales of observation simultaneously. This will improve the robustness of classical image analysis techniques such as the shape recognition and the shape registration methods. For instance, it could be interesting to merge our multiscale segmentation algorithm with a multiscale shape model such as the one developed by Pizer *et al.* in [14] to create a multiscale recognition method.

In [16], Schnabel and Arridge have proposed a method to extract the shape of objects at different levels of scale. They have then used the extracted multiscale shapes to localize and characterize shape changes at different levels of scale. They have applied their model to segment 3-D brain magnetic resonance images in order to quantify the structural

R. Kimmel, N. Sochen, J. Weickert (Eds.): Scale-Space 2005, LNCS 3459, pp. 167–178, 2005.

deformations for patients having epilepsy. However, they have not taken into account the *interdependance between space and scale* in their segmentation model.

Any image can be observed at different scales thanks to a multi-resolution image representation called *scale space* by Witkin in [18] (see also the pioneering work of Iijima [7]). A scale space is a hierarchical decomposition of an image according to the scale of observation. It can also be seen as a family of gradually smoother versions of the original image.

The segmentation method we use in our work is the active contour model introduced by Kass *et al.* in [8]. We want to use this segmentation model to extract objects in scale spaces. For this purpose, we need to define an evolution equation for active contours propagating in scale spaces. Two main questions arise when we try to devise such an equation. How can the active contours be introduced into scale spaces and which scale spaces can be used? An answer to the first question is given by the *Polyakov action* that we will present in the next section. For the second question, we will use the family of scale spaces proposed by Eberly in [5] which includes the linear scale space, the Perona-Malik scale space, the curvature scale spaces and the total variation scale space.

2 Polyakov Action

The Polyakov action has been introduced in image processing by Sochen *et al.* in [17]. The Polyakov action is a functional that measures the weight of a map X between the image manifold Σ and the embedding manifold M (see Figure 1). It is defined as follows:

$$P(X, \Sigma, M) = \int d^m\sigma \; g^{1/2} g^{\mu\nu} \partial_\mu X^i \partial_\nu X^j h_{ij}, \tag{1}$$

where m is the dimension of Σ, p the dimension of M, $g_{\mu\nu}$ and h_{ij} are the metric tensors of manifolds Σ and M, $g^{\mu\nu}$ is the inverse metric of $g_{\mu\nu}$, g is the determinant of $g_{\mu\nu}$, $\mu, \nu = 1, ..., m$, $i, j = 1, ..., p$, $\partial_\mu X^i = \partial X^i / \partial\sigma^\mu$. Moreover, when identical indices appear one up and one down, they are summed over according to the Einstein summation convention.

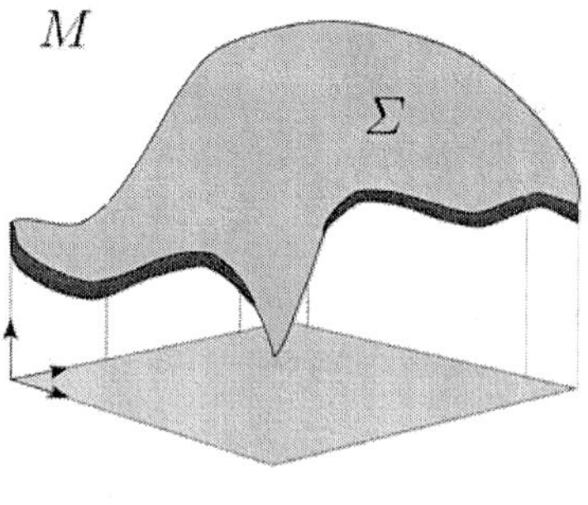

Fig. 1. The manifold Σ embedded in M, reproduced from [17]

If the Polyakov functional is minimized w.r.t. an embedding coordinate X^l, $g_{\mu\nu}$ and h_{ij} being fixed, we get the following flow acting on X^l:

$$\partial_t X^l = g^{-1/2}\partial_\mu(g^{1/2}g^{\mu\nu}\partial_\nu X^l) + \Gamma^l_{jk}\partial_\mu X^j \partial_\nu X^k g^{\mu\nu} \quad \text{for } 1 \le l \le p, \tag{2}$$

where Γ^l_{jk} is the Levi-Civita connection.

Sochen *et al.* have proved in [17] that different choices of the metric tensor $g_{\mu\nu}$ in Equation (2) give the most well-known scale spaces: the linear scale space, the scale space of Perona-Malik, the mean curvature scale space and the total variation scale space. They have also proposed a new scale space to enhance image quality. They have called it the Beltrami flow since they have used the Beltrami operator which generalizes the Laplace operator in non-flat manifolds.

3 Active Contours in Scale Spaces

3.1 Active Contours in Euclidean Spaces

Following the first model of active contours proposed by Kass *et al.* in [8], Caselles *et al.* in [3] and Kichenassamy *et al.* in [9] have proposed an energy functional invariant w.r.t. the curve parametrization:

$$F^{gac}(C) = \oint_0^{L(C)} f ds, \tag{3}$$

where ds is the Euclidean element of length, $L(C)$ is the length of the curve C and f is an edge detecting function that vanishes at object boundaries. The calculus of variations provides the Euler-Lagrange equation of the functional F^{gac} and the gradient descent method gives the flow that minimizes F^{gac}:

$$\partial_t C = (\kappa f - \langle \nabla f, \mathcal{N} \rangle)\mathcal{N}, \tag{4}$$

where κ is the curvature and $\mathcal{N}$ the normal to the curve. Osher and Sethian have introduced in [12] the implicit and intrinsic level set representation of contours to efficiently solve the contour propogation problem and to deal with topological changes. The equation (4) can be written in the level set form:

$$\partial_t \phi = \left(\kappa f + \langle \nabla f, \frac{\nabla \phi}{|\nabla \phi|} \rangle\right) |\nabla \phi|, \tag{5}$$

where ϕ is the level set function embedding the active contour C.

3.2 Scale Spaces

In the previous section, the active contours have been defined in the Euclidean space. We want to put them in scale spaces by changing the embedding Euclidean manifold into the scale spaces. The question is which scale spaces will we use. In paper [5], Eberly has studied the geometry of a large class of scale spaces and he has defined for them the general metric tensor:

$$[h_{ij}] = \operatorname{diag}\left(\frac{1}{c^2} I_n, \frac{1}{c^2 \rho^2}\right), \tag{6}$$

where n is the spatial dimension, c and ρ are two functions that physically correspond to the conductance and the density functions in the general model of heat diffusion transfert.

Eberly has also defined in [5] the natural diffusion equation in any scale space as follows: the left-hand side of the diffusion equation is given by one application of the scale derivative and the right-hand side by two applications of the spatial derivative. The natural diffusion equation in a scale space is therefore

$$\nabla^{ss}_{\sigma} u = \left(\nabla^{ss}_{x_1,\ldots,x_n}\right)^2 u, \tag{7}$$

where ∇^{ss}_{σ} is the scale derivative operator and $\nabla^{ss}_{x_1,\ldots,x_n}$ the spatial derivative operator. These operators are determined using the tensor metric (6) and the following differential geometry formulae:

$$\nabla^{ss}_{x_1,\ldots,x_n} = \begin{pmatrix} \sqrt{h^{x_1 x_1}}\partial_{x_1} \\ \vdots \\ \sqrt{h^{x_n x_n}}\partial_{x_n} \end{pmatrix} = c\nabla, \tag{8}$$

$$\nabla^{ss}_{\sigma} = \sqrt{h^{\sigma\sigma}}\partial_{\sigma} = c\rho\partial_{\sigma}. \tag{9}$$

Hence, the diffusion equation (7) is equal to

$$\partial_{\sigma} u = \frac{1}{\rho}\nabla.\left(c\nabla u\right), \tag{10}$$

which corresponds to the *general heat diffusion equation* with the conductance function c and the density function ρ. The choice of the functions c and ρ determines the scale space and the diffusion equation we use. For examples,

$$\begin{cases} \text{for } c = \sigma, \rho = 1 & u_{\sigma} = \sigma \Delta u & \textit{Linear Scale Space}, \\ \text{for } c = \exp(-\alpha|\nabla u|^2), \rho = 1 & u_{\sigma} = \nabla.(c\nabla u) & \textit{Perona-Malik Flow}, \\ \text{for } c = \rho = \frac{1}{|\nabla u|} & u_{\sigma} = \nabla.\left(\frac{\nabla u}{|\nabla u|}\right)|\nabla u| & \textit{Mean Curvature Flow}, \\ \text{for } c = \frac{1}{|\nabla u|}, \rho = 1 & u_{\sigma} = \nabla.\left(\frac{\nabla u}{|\nabla u|}\right) & \textit{Total Variation Flow}, \end{cases}$$

Eberly has proved in [5] that the linear scale space is hyperbolic and translation, rotation and scale invariant. The second scale space is given by the non-linear anisotropic diffusion equation of Perona and Malik proposed in [13]. The third one is the mean curvature flow introduced in the level set framework by Osher and Sethian in [12] and the fourth scale space is produced by the total variation flow defined by Rudin, Osher and Fatemi in [15].

3.3 General Evolution Equation for Active Contours in Scale Spaces

In the context of the Polyakov action, we look for the map X, the tensor $g_{\mu\nu}$ of the active contour manifold and the tensor h_{ij} of the embedding space that lead to an evolution equation for active contours in the scale spaces. We choose the map X as $X : (x_1, ..., x_n, \sigma) \to (x_1, ..., x_n, \sigma, \phi)$, where ϕ is the level set function representing the active contour. The metric tensor h_{ij} of the embedding space is given by the Equation (6). The last choice concerns the tensor $g_{\mu\nu}$ of the active contour manifold. We choose the *induced metric tensor* on Σ [17]:

$$g_{\mu\nu} = \partial_\mu X^i \partial_\nu X^j h_{ij}. \tag{11}$$

This choice is motivated by the classical works concerning the active contours [3, 9, 4]. The Polyakov functional corresponding to the induced metric is the Euler functional:

$$S(X) = \int d^m \sigma g^{1/2}, \tag{12}$$

which defines the (hyper-)aera of the (hyper-)surface Σ. The maps X that minimize S for any manifold Σ embedded in any manifold M are called *harmonic maps*. Harmonic maps are the generalization of geodesics and minimal surfaces (see [3, 9, 4]) to higher dimensional manifolds and for higher embedding manifolds.

The minimization of S w.r.t. the component X^l gives the generalized mean curvature flow, see [17], in any embedding manifold M defined by the metric h_{ij}:

$$\partial_t X^l = \underbrace{g^{-1/2}\partial_\mu(g^{1/2} g^{\mu\nu} \partial_\nu X^l)}_{(13.1)} + \Gamma^l_{jk} \partial_\mu X^j \partial_\nu X^k g^{\mu\nu} \equiv \mathbf{H}^l = g^{-1/2}\mathbf{K}^l \tag{13}$$

whose term (13.1) is the Laplace-Beltrami operator and $\mathbf{H}$ is the mean curvature vector generalized to any manifold (M, h_{ij}).

Proposition 1: As in [3, 9, 4], we introduce a weighting function f in the Euler fonctional:

$$S_f(X) = \int d^m \sigma f(X) g^{1/2}. \tag{14}$$

The evolution equation minimizing this functional w.r.t. the l-th component of X is [2]:

$$\partial_t X^l = f\mathbf{H}^l + \partial_k f g^{\mu\nu} \partial_\mu X^k \partial_\nu X^l - \frac{m}{2} \partial_k f h^{kl}. \tag{15}$$

Application 1: The geodesic/geometric active contour evolving in the 2-D Euclidean space proposed in [3, 9] can be recovered. Indeed, if we take $X \equiv C$ and $h_{ij} = \delta_{ij}$, the evolution equation defined in Equation (15) becomes (up to a constant) the well-known flow

$$\partial_t C = f\kappa\mathcal{N} - \frac{1}{2}\langle\nabla f, \mathcal{N}\rangle\mathcal{N}. \tag{16}$$

Application 2: The evolution equation of the level set function embedding the active contour can also be revisited. If we choose $X \equiv S : (x,y) \to (x,y,\phi(x,y))$, $h_{ij} = \delta_{ij}$, the energy is $E_{es} = \int f\sqrt{1+|\nabla\phi|^2}dxdy$ and the flow acting on the level set component ϕ is:

$$\partial_t\phi = fH_{es} + g_{es}^{-1/2}\langle\nabla f, \nabla\phi\rangle = g_{es}^{-1/2}\left(fK_{es} + \langle\nabla f, \nabla\phi\rangle\right) = g_{es}^{-1/2}F_{es} \tag{17}$$

where $g_{es} = 1 + |\nabla\phi|^2$ and K_{es} is the Euclidean mean curvature of the surface X such that $K_{es} = \frac{(1+\phi_x^2)\phi_{yy} - 2\phi_x\phi_y\phi_{xy} + (1+\phi_y^2)\phi_{xx}}{g_{es}^{3/2}}$ (see [17]). The equation of the level set function $\phi_t = g_{es}^{-1/2}F_{es}$ implies that the surface S evolves according to $S_t = F_{es}\mathcal{N}_S$ where $\mathcal{N}_S = g_{es}^{-1/2}(-\phi_x, -\phi_y, 1)$. This means that the level sets of ϕ move according to the equation:

$$C_t = P_z S_t = g_{es}^{-1/2}|\nabla\phi|\; F_{es}\,\mathcal{N}_C = r(\phi)\; F_{es}\,\mathcal{N}_C \tag{18}$$
$$= \left(fK_{es}r(\phi) - \langle\nabla f, \mathcal{N}_C\rangle r^2(\phi)\right)\mathcal{N}_C \tag{19}$$

where P_z is a projector onto the plane normal to the ϕ-axis, $\mathcal{N}_C = -\nabla\phi/|\nabla\phi|$ is the unit normal to the level sets and $r(\phi) = g_{es}^{-1/2}|\nabla\phi|$. The equation (19) is close to the evolution equation (4) up to the surface mean curvature K_{es} and the function r. Function r can be interpretated as an indicator of the height variation on the surface S (see [1]). Indeed, $g_{es}^{-1/2}$ is the ratio between the area of an infinitesimal surface in the domain (x,y) and the corresponding area on the surface S. For flat surfaces, r is equal to 0 and it is close 1 near edges. Finally the function r is constant a.e. when ϕ is a signed distance function.

We propose the following evolution equation for active contours in the scale spaces.

Proposition 2: Given the induced metric tensor, Equation (11), the harmonic map X defined by $(x_1,...,x_n,\sigma) \to (x_1,...,x_n,\sigma,\phi)$ and the weighting function $f = f(x_1,...,x_n,\sigma)$, the evolution equation of the $(n+2)$-th component of X, i.e. the level set component ϕ, is equal, according to (15), to:

$$\partial_t\phi = fH_{ss} + \langle\nabla^\star f, \nabla^\star\phi\rangle_{(g^{\mu\nu})}, \tag{20}$$

whose $H_{ss} = g^{-1/2}K_{ss} = \mathbf{H}^{n+2}$ is the $(n+2)$-th component of the mean curvature vector (13) generalized to scale spaces, $\nabla^\star = (\nabla, \partial_\sigma)$ and $\langle.,.\rangle_{(g^{\mu\nu})}$ is the inner product w.r.t. the metric $g^{\mu\nu}$ such that

$$\langle V_1, V_2\rangle_{(g^{\mu\nu})} = V_1^t(g^{\mu\nu})V_2 = V_{1\mu}g^{\mu\nu}V_{2\nu}, \tag{21}$$

where $g = 1 + |\nabla\phi|^2 + \rho^2\phi_\sigma^2$. The energy of the multiscale active contour is computed according to Equation (14):

$$E_{ss} = \int f\sqrt{1 + |\nabla\phi|^2 + \rho^2\phi_\sigma^2} \underbrace{\prod_{1\le i\le n} \frac{dx_i}{c}\frac{d\sigma}{c\rho}}_{(22.1)}, \tag{22}$$

whose term (22.1) is the infinitesimal volume in the scale spaces defined by the metric tensor (6).

The evolution equation of the level set function ϕ is:

$$\partial_t\phi = g^{-1/2} f K_{ss} + \langle\nabla^\star f, \nabla^\star\phi\rangle_{(g^{\mu\nu})} = g_{es}^{-1/2} F_{ss}, \tag{23}$$

where $g_{es} = 1 + |\nabla^\star\phi|^2$. Hence the surface S evolves according to $S_t = F_{ss}\mathcal{N}_S$ and the level sets of ϕ move according to:

$$C_t = P_z S_t = \left(f K_{ss} r_2(\phi, c, \rho) - \langle\nabla^\diamond f, \mathcal{N}_C\rangle r_2^2(\phi, c, \rho)\right)\mathcal{N}_C, \tag{24}$$

with the operator $\nabla^\diamond = \frac{1}{c^2\rho^2}(\nabla f, \rho^2\partial_\sigma)$, $\mathcal{N}_C = -\nabla^\star\phi/|\nabla^\star\phi|$ and $r_2(\phi, c, \rho) = |\nabla^\star\phi|/g^{1/2}$.

3.4 Application to the Linear Scale Space

The linear scale space is obtained when $c = \sigma$ and $\rho = 1$. In this case, the energy of the multiscale active contour is for $n = 2$:

$$E_{lss} = \int f\sqrt{1 + |\nabla\phi|^2 + \rho^2\phi_\sigma^2}\, \frac{dx dy d\sigma}{\sigma^3}, \tag{25}$$

and the flow of ϕ (embedding the active contour) is:

$$\partial_t\phi = f H_{lss} + \frac{1}{g}\langle\nabla^\star f, \nabla^\star\phi\rangle_{lss}, \tag{26}$$

where $g = \frac{1}{\sigma^6}(1 + \phi_x^2 + \phi_y^2 + \phi_\sigma^2)$, $\langle .,. \rangle_{lss}$ is the inner product in the linear scale space defined by $\langle V_1, V_2\rangle_{lss} = \frac{1}{\sigma^2}\langle V_1, V_2\rangle$ and H_{lss} is the mean curvature in the linear scale space computed using Equation (13):

$$H_{lss} = \underbrace{\frac{1}{g_{es}}\phi_{\mu\nu}g^{\mu\nu}}_{(27.1)} - \underbrace{2\frac{\phi_\mu}{\sigma}g^{\mu\sigma}}_{(27.2)}, \tag{27}$$

where $g_{es} = 1 + \phi_x^2 + \phi_y^2 + \phi_\sigma^2$ and the components of $g^{\mu\nu}$ are:

$$\begin{aligned} g^{xx} &= \tfrac{1}{\sigma^4}(1 + \phi_y^2 + \phi_\sigma^2), & g^{xy} &= -\tfrac{1}{\sigma^4}\phi_x\phi_y, \\ g^{yy} &= \tfrac{1}{\sigma^4}(1 + \phi_x^2 + \phi_\sigma^2), & g^{x\sigma} &= -\tfrac{1}{\sigma^4}\phi_x\phi_\sigma, \\ g^{\sigma\sigma} &= \tfrac{1}{\sigma^4}(1 + \phi_x^2 + \phi_y^2), & g^{y\sigma} &= -\tfrac{1}{\sigma^4}\phi_y\phi_\sigma. \end{aligned} \tag{28}$$

Roughly speaking, the term (27.1) corresponds to the Euclidean part of the mean curvature and the second term (27.2) to the Riemannian part.

4 Multiscale Image Features

4.1 Multiscale Edges

In the previous section, we have defined a multiscale segmentation model which can capture image features representing by the function f. There exists different local multiscale image features but we will use the most common one, the norm of the image gradient, which is equal to $|\nabla_{ss} f| = (c^2 f_{x_1}^2 + \ldots + c^2 f_{x_n}^2 + c^2\rho^2 f_\sigma^2)^{1/2}$ in the scale spaces. After that, we will extract the *ridges* of $|\nabla_{ss} f|$, the multiscale gradient norm.

Ridges can be defined by different ways (see [6]). In our approach, we have used the definition developed in Section 2.3 of [6]. In this case, Morse describes in [11] ridges of a 2-D feature image f as points which have local maximum in f along the direction of the greatest concavity of f. Thus, at a ridge point the direction of greatest curvature of f is the cross-ridge direction and the value of f is greater than the neighboring points on either side of it. This definition can be extended to higher dimensions. Indeed, let us consider a function f in an n-D space and let us denote $\lambda_1, \ldots, \lambda_n$ with $|\lambda_1| \leq \ldots \leq |\lambda_n|$ and $e_1, \ldots, e_n$ the eigenvalues and the corresponding eigenvectors of the $n \times n$ matrix of the second derivatives. A point in an n-D space is an m-D ridge ($m < n$) in f if for all $i < n - m$,

$$\begin{cases} \lambda_i < 0 \\ e_i \cdot \nabla f = 0 \end{cases} . \tag{29}$$

In the case of the linear scale space with $n = 2$ spatial dimensions, the Hessian matrix is different to the Euclidean one. Using the tensor metric h_{ij} defined in Equation (6), with $c = \sigma$ and $\rho = 1$, the Hessian matrix that includes the interdependence of space and scale is given in [11] by:

$$\nabla^2_{lss} = \begin{bmatrix} \sigma^2 \frac{\partial^2 f}{\partial x^2} - \sigma \frac{\partial f}{\partial \sigma} & \sigma^2 \frac{\partial^2 f}{\partial x \partial y} & \sigma \frac{\partial^2 f}{\partial x \partial \sigma} + \sigma \frac{\partial f}{\partial x} \\ \sigma^2 \frac{\partial^2 f}{\partial x \partial y} & \sigma^2 \frac{\partial^2 f}{\partial y^2} - \sigma \frac{\partial f}{\partial \sigma} & \sigma \frac{\partial^2 f}{\partial y \partial \sigma} + \sigma \frac{\partial f}{\partial y} \\ \sigma \frac{\partial^2 f}{\partial x \partial \sigma} + \sigma \frac{\partial f}{\partial x} & \sigma \frac{\partial^2 f}{\partial y \partial \sigma} + \sigma \frac{\partial f}{\partial y} & \sigma^2 \frac{\partial^2 f}{\partial \sigma^2} + \sigma \frac{\partial f}{\partial \sigma} \end{bmatrix} . \tag{30}$$

As an example, let us consider the fractal image proposed by Von Koch (see the first row on Figure 2). The magnitude of the scale space gradient is $|\nabla_{lss} I(x, y, \sigma)| = \sigma \cdot (I_x^2 + I_y^2 + I_\sigma^2)^{1/2}$ (see the second row on Figure 2). And the ridges of the multiscale gradient norm are given in the third row on Figure 2.

4.2 Gradient Vector Flow in Scale Spaces

The gradient vector flow (GVF) has been introduced by Xu and Prince in [19]. The GVF field is a non-irrotational force field, namely $\mathbf{V}$, which can capture the object boundaries far from them and can deal with concave boundary regions. It is defined in a variational

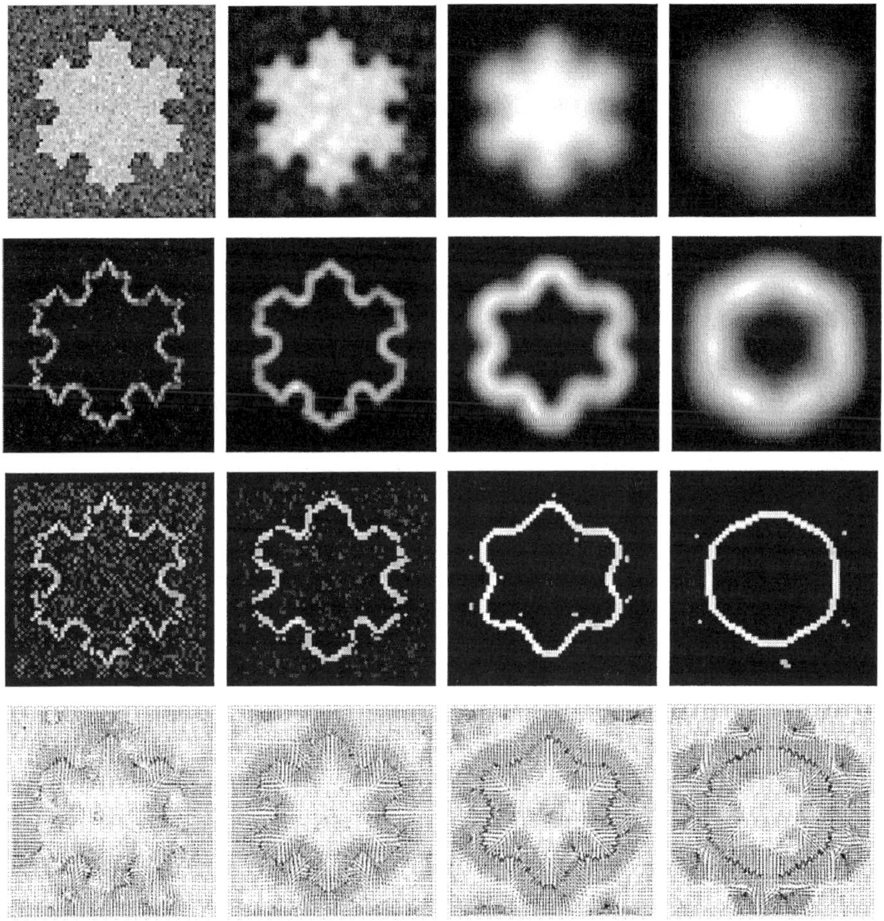

Fig. 2. First row: the Von Koch picture at four different scales of observation. The first image is the original image. Second row: the multiscale gradient of the Von Koch picture. Third row: the ridges of the multiscale gradient. Fourth row: GVFs of the ridge images

approach since the GVF field must minimize the following energy functional in the n-D Euclidean space:

$$F^{gvf}(\mathbf{V}) = \int \underbrace{\mu \sum_{i=1}^{n}(|\nabla V_i|^2)}_{(31.1)} + \underbrace{|\nabla f|^2 |\mathbf{V} - \nabla f|^2}_{(31.2)} \, d\Omega, \tag{31}$$

where V_i is the i-th component of the GVF field and μ is a constant which balances the contributions between the regularization term (31.1) and the data fidelity term (31.2). The minimization of the energy functional (31) is done using the calculus of variations and the gradient descent method which provide n flows, one per component of the GVF field.

In this section we propose to extend the GVF to the scale spaces defined by the metric tensor (6). We realize this extension by simply changing the Euclidean terms by their Riemann equivalents. Thus, we replace the Euclidean gradient ∇ by the scale space gradient ∇_{ss} and the Euclidean infinitesimal volume element $d\Omega$ by the scale space one $d\Omega_{ss}$, the energy (31) then becomes:

$$F_{ss}^{gvf}(\mathbf{V}) = \int \mu \sum_{i=1}^{n} (|\nabla_{ss} V_i|^2) + |\nabla_{ss} f|^2 |\mathbf{V} - \nabla_{ss} f|^2 d\Omega_{ss}, \tag{32}$$

The Frechet derivative of F_{ss}^{gvf} w.r.t. V_i in the ξ direction is

$$\langle \frac{\partial F_{ss}^{gvf}}{\partial V_i}, \xi \rangle = \int \xi \cdot [-\mu \left(\sum_{i=1}^{n} \partial_{x_i}(c^2 \partial_{x_i} V_i) + \partial_\sigma (c^2 \rho^2 \partial_\sigma V_i) \right) + \tag{33}$$

$$|\nabla_{ss} f|^2 |V_i - (\nabla_{ss} f)_i|^2 \,] d\Omega_{ss}. \tag{34}$$

Then, the flow minimizing F_{ss}^{gvf} w.r.t. V_i is

$$\partial_t V_i = \mu \left(\sum_{i=1}^{n} \partial_{x_i}(c^2 \partial_{x_i} V_i) + \partial_\sigma (c^2 \rho^2 \partial_\sigma V_i) \right) - |\nabla_{ss} f|^2 |V_i - (\nabla_{ss} f)_i|^2. \tag{35}$$

For the linear scale space and $n = 2$, the GVFs for $i = x, y, \sigma$ are:

$$\partial_t V_i(x, y, \sigma) = \mu(\sigma^2 \Delta V_i + 2\sigma \partial_\sigma V_i) - \sigma^2 |\nabla f|^2 (V_i - \sigma \partial_i f) \tag{36}$$

Figure 2 (fourth row) presents the GVFs of the ridges images (third row).

5 Result

We have applied the evolution equation (26) in the linear scale space to segment the Von Koch picture at different scales of observation. The Figure 3 presents our multiscale snake evolving in the linear scale space at different times and the Figure 4 shows the segmentation process at four different scales.

Fig. 3. Active contour evolving in the linear scale space

Fig. 4. Active contour evolving in the linear scale space at four different scales

6 Conclusion

In this paper, we have introduced the scale parameter in the active contour formalism by defining an evolution equation for the active contours in the scale spaces based on the general heat diffusion equation. We have supposed that the metric tensor $g_{\mu\nu}$ of the level set manifold is the induced metric tensor, i.e. the case where the active contours are harmonic maps. We could consider another choice for $g_{\mu\nu}$ such as a diagonal tensor (see [2]). Future works will be focused on integrating this multiscale segmentation technique into shape analysis methods such as the shape recognition and the shape registration methods to improve their performance.

References

1. G. Aubert and P. Kornprobst. *Mathematical Problems in Image Processing, Partial Differential Equations and the Calculus of Variations*. Springer, 2002.
2. X. Bresson, P. Vandergheynst, and J.-P-Thiran. Multiscale Active Contours - ITS Technical Report 12.04, 2004.
3. V. Caselles, R. Kimmel, and G. Sapiro. Geodesic Active Contours. *International Journal of Computer Vision*, 22(1):61–79, 1997.
4. V. Caselles, R. Kimmel, G. Sapiro, and C. Sbert. Minimal Surfaces: A Three Dimensional Segmentation Approach. *IEEE Transactions on Pattern Analysis and Machine Intelligence*, 19(4):394–398, 1997.
5. D. Eberly. A Differential Geometric Approach to Anisotropic Diffusion in Geometry-Driven Diffusion in Computer Vision. *Computational Imaging and Vision*, 1:371–392, 1994.
6. D.H. Eberly. Geometric Methods For Analysis Of Ridges In n-Dimensional Images - PhD Thesis - University of North Carolina, 1994.
7. S. Ishikawa J. Weickert and A. Imiya. Linear Scale-Space Has First Been Proposed In Japan. *Mathematical Imaging and Vision*, 10:237–252, 1999.
8. M. Kass, A. Witkin, and D. Terzopoulos. Snakes: Active Contour Models. *International Journal of Computer Vision*, pages 321–331, 1987.
9. S. Kichenassamy, A. Kumar, P. Olver, A. Tannenbaum, and A.J. Yezzi. Conformal Curvature Flows: From Phase Transitions to Active Vision. In *Archive for Rational Mechanics and Analysis*, volume 134, pages 275–301, 1996.
10. J. J. Koenderink. The Structure of Images. *Biological Cybernetics*, 50:363–370, 1984.
11. B.S. Morse. Computation Of Object Cores From Grey-Level Images - PhD Thesis - University of North Carolina, 1994.
12. S. Osher and J.A. Sethian. Fronts Propagating with Curvature-Dependent Speed: Algorithms Based on Hamilton-Jacobi Formulations. *Journal of Computational Physics*, 79(1)(12-49), 1988.
13. P. Perona and J. Malik. Scale-Space and Edge Detection Using Anisotropic Diffusion. *IEEE Transactions on Pattern Analysis and Machine Intelligence*, 1252(629-639), 1990.
14. S.M. Pizer, D. Eberly, B.S. Morse, , and D. Fritsch. Zoom-Invariant Vision of Figural Shape: The Mathematics of Cores. *Computer Vision and Image Understanding*, 69:55–71, 1998.
15. L. I. Rudin, S. Osher, and E. Fatemi. Nonlinear Total Variation Based Noise Removal Algorithms. *Physica D*, 60(1-4):259 – 268, 1992.
16. J. A. Schnabel and S.R. Arridge. Active Shape Focusing. *Image and Vision Computing*, 17(5-6):419–428, 1999.
17. N. Sochen, R. Kimmel, and R. Malladi. A General Framework For Low Level Vision. *IEEE Transactions on Image Processing*, 7(3):310 – 318, 1998.
18. A.P. Witkin. Scale-space filtering. In *Proc. 8th International Joint Conference Artificial Intelligence*, pages 1019–1022, 1983.
19. C. Xu and J. Prince. Snakes, shapes and gradient vector flow. *IEEE Transaction on Image Processing*, 7:359–369, 1998.

Riesz-Transforms Versus Derivatives: On the Relationship Between the Boundary Tensor and the Energy Tensor

Ullrich Köthe[1] and Michael Felsberg[2,*]

[1] Cognitive Systems Group, University of Hamburg
koethe@informatik.uni-hamburg.de
[2] Computer Vision Laboratory, Linköping University
mfe@isy.liu.se

Abstract. Traditionally, quadrature filters and derivatives have been considered as alternative approaches to low-level image analysis. In this paper we show that there actually exist close connections: We define the quadrature-based boundary tensor and the derivative-based gradient energy tensor which exhibit very similar behavior. We analyse the reason for this and determine how to minimize the difference. These insights lead to a simple and very efficient integrated feature detection algorithm.

1 Introduction

Image features such as edges and corners can be detected by analysing the image in the neighborhood of every (candidate) point. A compact representation of the low-order characteristics of these neighborhoods is given by the first few derivatives at every point (the n-jet [9]). Numerous feature descriptors for edges, lines, blobs, corners and so on have been defined by various combinations of low-order derivatives (see e.g. [5, 12]). However, these descriptors are usually only valid for a single feature type, and give no or wrong responses at points where the underlying feature model is violated. Improvements can be achieved by moving from scalar feature descriptors to tensor based ones. Second order tensors cannot only represent feature strength, but also allow to distinguish between intrinsically 1- and 2-dimensional features (edges vs. corners) and measure orientation.

The most common derivative-based tensor is the structure tensor [1, 6] which is obtained by spatial averaging of the outer product of the gradient. It can represent step edges and their corners/junctions but is less suitable for the detection of lines and other second order features. All these feature types are covered by the energy tensor [3], which includes higher order derivatives (details below).

A different approach to feature detection is taken by the quadrature filter method [7, 8] where derivatives are replaced with filters that are related to each

* This work has been supported by EC Grant IST-2002-002013 MATRIS and by EC Grant IST-2003-004176 COSPAL.

R. Kimmel, N. Sochen, J. Weickert (Eds.): Scale-Space 2005, LNCS 3459, pp. 179–191, 2005.

other by the Hilbert transform. These operators react, by design, uniformely to both edges and lines. This property is called phase invariance because edges and lines can be interpreted as arising from the same magnitude spectrum, but at different (namely odd and even) phase. In 2D it is common to apply a number of 1D quadrature filters at different orientations. The filter responses can then be combined into an orientation tensor [8]. However, the orientation tensor reacts in a well-defined way only to 1-dimensional features. This probem is solved by means of the boundary tensor [10], which uses truly 2-dimensional quadrature filters to also model certain 2D feature types (details also given below).

When we experimented with both the energy and the boundary tensors, we observed a striking similarity of their behavior – qualitatively, their results are almost indistinguishable. This paper is devoted to a more detailed analysis of the relationship of the two approaches. We pursue this analysis on three levels: First, we establish a formal similarity between the derivative and quadrature filter methods by basing the latter on (first and second order) Riesz transform operators [2] which closely resemble derivatives. Second, we show that the spectral properties of the two tensors are very similar when optimal scales ratios are chosen. Third, we report on experiments illustrating the similarity in practice.

2 Tensor Definitions

The structure tensor is the most common derivative based tensor. It is defined as the spatial average of the outer product of the gradient ∇f with itself [1, 6]:

$$\mathbf{S} = g \star (\nabla f)(\nabla f)^T \tag{1}$$

where g is a smoothing filter (usually Gaussian), f the original image, and derivatives are always understood to operate at a particular scale. For the purpose of our analysis, it is advantegous to approximate the gradient with a taylor series:

$$\nabla f(\mathbf{x}) \approx \nabla f\big|_{\mathbf{x}=\mathbf{x}_0} + \nabla\nabla^T f\big|_{\mathbf{x}=\mathbf{x}_0}(\mathbf{x} - \mathbf{x}_0) \tag{2}$$

where $\nabla\nabla^T f = \mathcal{H}f$ is the Hessian matrix. Inserting this into (1), we can execute the convolution analytically. If g is radially symmetric, the odd powers of $\mathbf{x}$ cancel out, whereas the even ones give a constant. We get:

$$\mathbf{S} \approx (\nabla f)(\nabla f)^T + \lambda(\mathcal{H}f)(\mathcal{H}f)^T \tag{3}$$

where the parameter λ depends on g's shape and scale. This operator is very good at detecting step edges and their corners, but often shows multi-modal or no responses at second-order features such as line edges and junctions. By adjusting λ, the behavior can be somewhat improved, but it is usually impossible to find a λ that works equally well on the entire image.

A richer signal model can be employed with the *energy tensor* [3]:

$$\mathbf{E} = (\nabla b)(\nabla b)^T - b(\mathcal{H}b) \tag{4}$$

where b is the signal to be analyzed. This is structurally similar to (3), but the square of the Hessian has been replaced with the negative product of the function b and its Hessian. We'll show later that the energy tensor achieves better feature detection results when different scales are used for different derivative orders. In a strict sense, the name "energy tensor" may not be justified because $\mathbf{E}$ is not in general positive semi-definite. But it has this property under certain conditions and degrades gracefully if these conditions are slightly violated (cf. figure 4, see [4] for an in-depth discussion of this issue). In image analysis, the energy tensor cannot be used in its pure form, because images are not DC free, so the energy would be dominated by the DC magnitude (average gray level) if b were the image f itself. Instead one computes $\mathbf{E}$ from a bandpass filtered version of the image, for example one defines $b = \nabla^T \nabla g \star f$ when the bandpass is the Laplacian of Gaussian. Since the Laplacian is a second order derivative, $\mathbf{E}$ is effectively calculated from second, third and fourth order derivatives.

Unfortunately, this means that the important first order image structure is not considered at all. Therefore, we developed a variant called *gradient energy tensor*, or GET operator [4]. Here, $b = \nabla g \star f$, so b is the Gaussian gradient vector. The gradient of b is then the Hessian of f, whereas the Hessian of b gives a third order tensor. Since the energy tensor is only a second order tensor, two indices of this third order tensor are immediately contracted, giving:

$$(\mathcal{T} f)_i = \sum_{j=1}^{N} \frac{\partial^3 f}{\partial x_i \partial x_j^2} = \nabla(\nabla^T \nabla f) \qquad \text{i.e. in 2D: } \mathcal{T} f = \begin{pmatrix} f_{xxx} + f_{xyy} \\ f_{xxy} + f_{yyy} \end{pmatrix} \tag{5}$$

(N is the space dimension). $\mathcal{T} f$ is equivalent to the gradient of the Laplacian of f, as can be seen by switching the order of differentiation and contraction. Since the outer product of two different tensors is not commutative, the gradient energy tensor must be symmetrized. This results in the following definition:

$$\mathbf{G} = \mathbf{G}^{\text{even}} + \mathbf{G}^{\text{odd}} = (\mathcal{H} f)(\mathcal{H} f)^T - \frac{1}{2}\left((\nabla f)(\mathcal{T} f)^T + (\mathcal{T} f)(\nabla f)^T\right) \tag{6}$$

The *boundary tensor* was introduced in [10] using circular harmonics. Here we base its definition on the Riesz transform [2] to emphasize the formal similarity of quadrature filters to derivatives. The Riesz transform is the N-dimensional generalization of the Hilbert transform. It is defined in the Fourier domain as:

$$\mathcal{R} \circ\!\!-\!\!\bullet \quad -\mathbb{i}\frac{\mathbf{u}}{|\mathbf{u}|} \tag{7}$$

where $\mathbf{u}$ is the frequency vector. Since the derivative is defined as $\nabla \circ\!\!-\!\!\bullet -\mathbb{i}\mathbf{u}$, the two operators differ only by a factor of $|\mathbf{u}|$ in the Fourier domain. The difference becomes clearer in polar coordinates where $\mathcal{R} \circ\!\!-\!\!\bullet -\mathbb{i}(\cos(\phi), \sin(\phi))^T$ and $\nabla \circ\!\!-\!\!\bullet -\mathbb{i}\,\rho\,(\cos(\phi), \sin(\phi))^T$. Both operators have the same angular behavior. But the derivative operator also acts as a high-pass filter, whereas the Riesz transform leaves the radial shape of the spectrum unaltered. This property is responsible for the desirable phase invariance of the boundary tensor.

The spatial domain Riesz transform operator decays only as $\mathcal{O}(|\mathbf{x}|^{-N})$, where N is the space dimension. Therefore one applies Riesz transforms to a bandpass filtered version b of the image f. The boundary tensor is then defined as

$$\mathbf{B} = \mathbf{B}^{\text{even}} + \mathbf{B}^{\text{odd}} = (\mathcal{Q}b)(\mathcal{Q}b)^T + (\mathcal{R}b)(\mathcal{R}b)^T \tag{8}$$

where $\mathcal{Q} = \mathcal{R}\mathcal{R}^T$ denotes the second order Riesz transform resulting in a matrix analogous to the Hessian (In contrast to the 1D Hilbert transform, which reproduces the negated original signal if applied twice, higher order Riesz transforms are useful because they create tensors when $N \geq 2$). Eq. (8) is formally equivalent to (3) when we set $\lambda = 1$ and replace derivatives with Riesz transforms. It should also be noted that the boundary tensor is always positive semi-definite by construction. Various bandpass filters can be used to obtain b. In [10], we used $|\mathbf{u}| \exp(-|\mathbf{u}|^2\sigma^2/2)$, but in this paper we choose the Laplacian of Gaussian $|\mathbf{u}|^2 \exp(-|\mathbf{u}|^2\sigma^2/2)$ because this allows us to establish a very strong *functional* relationship between the gradient energy tensor (6) and the boundary tensor.

3 Analysis of the Tensors

In order to analyse the behavior of the tensors, it is beneficial to express the convolution operation explicitly with integrals. For simplicity, we assume that the coordinate origin is at the center of the current window. Due to Parseval's theorem we can then express the integral in either the spatial or Fourier domains. We must only take into account that the kernels are reflected in the spatial domain expressions, which has no effect for even kernels but requires a sign-change for odd kernels. Since we are always taking products of two odd filter responses, this sign also cancels out. Using the Laplacian of Gaussian bandpass, the boundary tensor components can be expressed in the Fourier domain as

$$\begin{aligned}\mathbf{B}_{ij} = &\int -\mathbf{u}_i\mathbf{u}_k e^{-|\mathbf{u}|^2\sigma^2/2} F(\mathbf{u})\, d\mathbf{u} \int -\mathbf{u}_j\mathbf{u}_k e^{-|\mathbf{u}|^2\sigma^2/2} F(\mathbf{u})\, d\mathbf{u} \\ &+ \int -\mathrm{i}\, \mathbf{u}_i|\mathbf{u}| e^{-|\mathbf{u}|^2\sigma^2/2} F(\mathbf{u})\, d\mathbf{u} \int -\mathrm{i}\, \mathbf{u}_j|\mathbf{u}| e^{-|\mathbf{u}|^2\sigma^2/2} F(\mathbf{u})\, d\mathbf{u}\end{aligned} \tag{9}$$

where $F(\mathbf{u})$ is the image spectrum, and we use Einstein's summation convention (for index k). The components of the gradient energy tensor are

$$\begin{aligned}\mathbf{G}_{ij} = &\int -\mathbf{u}_i\mathbf{u}_k e^{-|\mathbf{u}|^2\sigma_2^2/2} F(\mathbf{u})\, d\mathbf{u} \int -\mathbf{u}_j\mathbf{u}_k e^{-|\mathbf{u}|^2\sigma_2^2/2} F(\mathbf{u})\, d\mathbf{u} \\ &- \frac{1}{2}\Big(\int -\mathrm{i}\, \mathbf{u}_i e^{-|\mathbf{u}|^2\sigma_1^2/2} F(\mathbf{u})\, d\mathbf{u} \int \mathrm{i}\, \mathbf{u}_j\mathbf{u}_k\mathbf{u}_k e^{-|\mathbf{u}|^2\sigma_3^2/2} F(\mathbf{u})\, d\mathbf{u} \\ &+ \int -\mathrm{i}\, \mathbf{u}_j e^{-|\mathbf{u}|^2\sigma_1^2/2} F(\mathbf{u})\, d\mathbf{u} \int \mathrm{i}\, \mathbf{u}_i\mathbf{u}_k\mathbf{u}_k e^{-|\mathbf{u}|^2\sigma_3^2/2} F(\mathbf{u})\, d\mathbf{u}\Big)\end{aligned} \tag{10}$$

where we allow the derivatives of different order to be applied at different scales $\sigma_1, \sigma_2, \sigma_3$. If we equate σ and σ_2, the even parts of $\mathbf{B}$ and $\mathbf{G}$ become equal, so we will require this from now on. We analyse at first how the two tensors react to intrinsically 1-dimensional images, that is when $F(\mathbf{u}) = F(t\mathbf{n})$ holds along a particular direction $\mathbf{n}$, and $F(\mathbf{u}) = 0$ otherwise. Then the $\mathbf{u}_i$ reduce to $\mathbf{n}_i t$, and the 2D integrals become 1D ones. The even part of both tensors is:

$$\mathbf{B}_{ij}^{\text{even}} = \mathbf{G}_{ij}^{\text{even}} = \mathbf{n}_i \mathbf{n}_j \left(\int t^2 e^{-t^2 \sigma_2^2 / 2} \Re(F(t))\, dt \right)^2 \tag{11}$$

and the odd parts are:

$$\mathbf{B}_{ij}^{\text{odd}} = \mathbf{n}_i \mathbf{n}_j \left(\int -\mathbb{i}\, t |t| e^{-t^2 \sigma_2^2 / 2}\, \mathbb{i} \Im(F(t))\, dt \right)^2 \tag{12}$$

$$\mathbf{G}_{ij}^{\text{odd}} = -\mathbf{n}_i \mathbf{n}_j \int -\mathbb{i}\, t\, e^{-t^2 \sigma_1^2 / 2}\, \mathbb{i} \Im(F(t))\, dt \int \mathbb{i}\, t^3 e^{-t^2 \sigma_3^2 / 2}\, \mathbb{i} \Im(F(t))\, dt \tag{13}$$

where we took advantage of the fact that the spectra of real signals have even real and odd imaginary components. It can be seen that $\mathbf{B}$ is indeed a quadrature filter: The kernels of the even and odd tensor parts are related by the Hilbert transform $-\mathbb{i}\, \text{sign}(t)$. Thus, if we shift the signal phase by $\pi/2$ (i.e. if we swap real and imaginary signal components, with the appropriate adjustment of spectrum symmetries), even and odd tensor parts are simply exchanged, but their sum remains unaltered. This is precisely the requirement of phase invariance. That requirement is not fulfilled by the GET operator: It has the same even part as the boundary tensor, but the odd parts differ. Detailed analysis of the odd parts reveals that the difference can actually be made very small. Consider at first a simple sine signal, i.e. $F(t) = \mathbb{i} \frac{a}{2} (\delta(t - \omega_a) - \delta(t + \omega_a))$. We get

$$\mathbf{B}_{ij}^{\text{odd}} = \mathbf{n}_i \mathbf{n}_j\, a^2 \omega_a^4\, e^{-\omega_a^2 \sigma_2^2}$$
$$\mathbf{G}_{ij}^{\text{odd}} = \mathbf{n}_i \mathbf{n}_j\, a^2 \omega_a^4\, e^{-\omega_a^2 (\sigma_1^2 + \sigma_3^2)/2}$$

These expressions are equal when $\sigma_2^2 = (\sigma_1^2 + \sigma_3^2)/2$ which we will require from now on. A more complicated case is the superposition of two sine waves $F(t) = \mathbb{i} \frac{a}{2} (\delta(t - \omega_a) - \delta(t + \omega_a)) + \mathbb{i} \frac{b}{2} (\delta(t - \omega_b) - \delta(t + \omega_b))$. Then we get

$$\mathbf{B}_{ij}^{\text{odd}} = \mathbf{n}_i \mathbf{n}_j \left(a\, \omega_a^2 e^{-\omega_a^2 \sigma_2^2 / 2} + b\, \omega_b^2 e^{-\omega_b^2 \sigma_2^2 / 2} \right)^2$$
$$\mathbf{G}_{ij}^{\text{odd}} = \mathbf{B}_{ij}^{\text{odd}} + \mathbf{n}_i \mathbf{n}_j\, a\, b\, \omega_a \omega_b \left(\omega_a e^{-(\omega_a^2 \sigma_3^2 + \omega_b^2 \sigma_1^2)/4} - \omega_b e^{-(\omega_a^2 \sigma_1^2 + \omega_b^2 \sigma_3^2)/4} \right)^2$$

The eigenvalue of $\mathbf{B}^{\text{odd}}$ (which we obtain by simply dropping $\mathbf{n}_i \mathbf{n}_j$) is always positive, as required for a signal energy. However, the eigenvalue of $\mathbf{G}^{\text{odd}}$ can become negative if a and b have opposite signs, i.e. if the two sines have opposite phase. This counters the intuition that the *energy* tensor $\mathbf{G}$ indeed represents signal energy. However, due to the statistical properties of natural images the situation is not so bad in practice: High energy in the derivatives typically occurs at object boundaries (edges and corners/junctions). At these points the signal components have the *same* phase over many frequencies (phase congruency, [11]).

Then the error term in **G** is positive, and the measured energy becomes too large rather than too small. Negative energy typically occurs only in flat, but noisy areas, where it is safe to simply truncate negative eigenvalues to zero.

In addition, we can try to adjust the ratio σ_3/σ_1 so that the magnitude of the error term becomes as small as possible. It is necessary to use a scale-normalized error measure, because one could otherwise make the error arbitrarily small by taking $\sigma_3 \to \infty$. The natural scale normalization for the Laplacian of Gaussian is σ_2^2 $[= (\sigma_1^2 + \sigma_3^2)/2]$ [12], so that **B** has to be multiplied with σ_2^4. To make the response of **G** comparable, we normalize it with the same factor. Then we integrate over ω_a and ω_b to calculate the average error over all frequency pairs:

$$\begin{aligned}\epsilon &= \frac{(\sigma_1^2+\sigma_3^2)^2}{4}\iint \omega_a\omega_b\left(\omega_a e^{-(\omega_a^2\sigma_3^2+\omega_b^2\sigma_1^2)/4} - \omega_b e^{-(\omega_a^2\sigma_1^2+\omega_b^2\sigma_3^2)/4}\right)^2 d\omega_a\, d\omega_b \\ &= \frac{1}{\sigma_1^2} + \frac{2}{\sigma_3^2} + \frac{\sigma_1^2}{\sigma_3^4} - \frac{2\pi}{\sigma_1^2+\sigma_3^2}\end{aligned} \tag{14}$$

(we dropped the factor $\mathbf{n}_i\mathbf{n}_j\, a\, b$ not depending on the ratio). The error is minimized for $\sigma_3/\sigma_1 = 1/\sqrt{\pi^{1/3}-1} \approx 1.47$. It is interesting to compare the optimal error with the error obtained for other ratios: If $\sigma_1 = \sigma_2 = \sigma_3$, the error becomes more than 5 times as big! If $\sigma_3/\sigma_1 = \sqrt{3}$ and $\sigma_2/\sigma_1 = \sqrt{2}$ (which means that the same first derivative filter is applied repeatedly for the higher order derivatives, resulting in a very efficient algorithm), the error is only 36% bigger.

Another possibility to find an optimal scale ratio is to start directly from (12) and (13). We transform the products of integrals in these equations into 2-dimensional integrals over the product of the integrands. Then we interpret terms not depending on the signal spectrum as quadratic filter kernels [13]:

$$\left(\int -\mathrm{i}\, t|t|\, e^{-t^2\sigma_2^2/2} F(t)\, dt\right)^2 = -\iint B(t_1,t_2)F(t_1)F(t_2)\, dt_1\, dt_2$$

$$\int -\mathrm{i}\, t\, e^{-t^2\sigma_1^2/2}F(t)\, dt \int \mathrm{i}\, t^3 e^{-t^2\sigma_3^2/2}F(t)\, dt = \iint G(t_1,t_2)F(t_1)F(t_2)\, dt_1\, dt_2$$

with (note that G is symmetric due to the symmetrization of **G**)

$$B(t_1,t_2) = t_1 t_2\, |t_1 t_2|\, e^{-(t_1^2+t_2^2)\,\sigma_2^2/2} \tag{15}$$

$$G(t_1,t_2) = \frac{1}{2}\left(t_1\, t_2^3\, e^{-(t_1^2\sigma_1^2+t_2^2\sigma_3^2)/2} + t_1^3\, t_2\, e^{-(t_1^2\sigma_3^2+t_2^2\sigma_1^2)/2}\right) \tag{16}$$

We choose the ratio σ_3/σ_1 so that the scale-normalized mean squared difference between the two kernels is minimized:

$$\epsilon^2 = \sigma_2^8 \iint (B(t_1,t_2) - G(t_1,t_2))^2\, dt_1\, dt_2 \tag{17}$$

The minimum is achieved for $\sigma_3/\sigma_1 \approx 1.55$. The choice $\sigma_1 = \sigma_2 = \sigma_3$ gives again a 5 times higher residual (see fig. 1), whereas it increases by only 23% for $\sigma_3/\sigma_1 =$

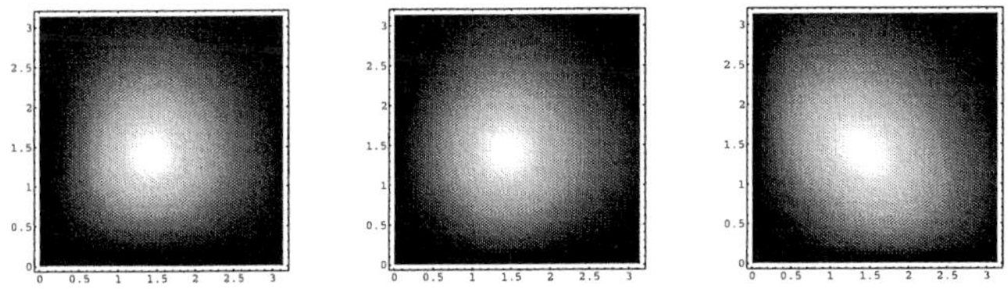

Fig. 1. Left: $B(t_1, t_2)$ (for $t_1, t_2 > 0$, $\sigma_2 = 1$); center: $G(t_1, t_2)$ with $\sigma_3/\sigma_1 = 1.55$ and $(\sigma_1^2 + \sigma_3^2)/2 = 1$; right: $G(t_1, t_2)$ for $\sigma_3/\sigma_1 = 1$: the deviation from B is much higher

$\sqrt{3}$. We also repeated the two optimizations while weighting the importance of the frequencies according to $1/\omega$, which better reflects the situation in real images. After modifying (14) and (17) accordingly, we got optimal σ_3/σ_1 ratios of $\sqrt{3}$ and 1.8 respectively, and the dependency of the residual on the ratio was reduced. Consequently, scale ratios between 1.5 and 1.8 give reasonable results, whereas it appears to be a bad idea to apply all derivatives at the same scale.

Now we analyse the response of the tensors to intrinsically 2-dimensional structures. To simplify we consider points $\mathbf{x}$ where the spectrum $F(\mathbf{u})$ computed with $\mathbf{x}$ as coordinate origin is (appproximately) polar separable within the pass band of the tensor filters. In case of the boundary tensor, the pass band is determined by the Laplacian of Gaussian, and we require $|\mathbf{u}|^2 \exp(-|\mathbf{u}|^2\sigma^2/2)F(\mathbf{u}) \approx \rho^2 \exp(-\rho^2\sigma^2/2)F_r(\rho)F_a(\phi)$. Then the integrals over $\mathbf{u}$ can be separated into products of two integrals over the radial and angular coordinates:

$$\mathbf{B}_{ij} = \int \mathbf{e}_i(\phi)F_a(\phi)d\phi \int \mathbf{e}_j(\phi)F_a(\phi)d\phi \left(\int \rho^2 e^{-\rho^2\sigma^2/2} F_r(\rho)\,\rho\,d\rho\right)^2 \\ + \int \mathbf{e}_i(\phi)\mathbf{e}_k(\phi)F_a(\phi)d\phi \int \mathbf{e}_j(\phi)\mathbf{e}_k(\phi)F_a(\phi)d\phi \left(\int \rho^2 e^{-\rho^2\sigma^2/2} F_r(\rho)\,\rho\,d\rho\right)^2 \tag{18}$$

with $\mathbf{e}(\phi) = (\cos(\phi), \sin(\phi))^T$. When we define the Fourier coefficients of $F_a(\phi)$ by $c_m = \int \cos(m\phi)F_a(\phi)\,d\phi$ and $s_m = \int \sin(m\phi)F_a(\phi)\,d\phi$, the trace of the boundary tensor becomes:

$$\mathrm{tr}(\mathbf{B}) = \frac{c_0^2 + 2c_1^2 + 2s_1^2 + c_2^2 + s_2^2}{2} \iint B(\rho_1, \rho_2)F_r(\rho_1)F_r(\rho_2)\,\rho_1\,d\rho_1\,\rho_2\,d\rho_2 \tag{19}$$

where the kernel B simplifies to $B(\rho_1, \rho_2) = \rho_1^2\rho_2^2 \exp(-(\rho_1^2+\rho_2^2)\sigma_2^2/2)$ because ρ_1 and ρ_2 are non-negative. The trace is determined by two local image properties: by the local contrast (as given by the radial integrals), and by how well the angular shape variation is captured with low-order circular harmonics (as given by the magnitude of the first five Fourier coefficients). It is interesting to compare this with the gradient at a polar separable location:

$$(\nabla f)^2 = (c_1^2 + s_1^2) \iint S(\rho_1, \rho_2)F_r(\rho_1)F_r(\rho_2)\,\rho_1\,d\rho_1\,\rho_2\,d\rho_2 \tag{20}$$

where $S(\rho_1, \rho_2) = \rho_1\rho_2 \exp(-(\rho_1^2 + \rho_2^2)\sigma_1^2/2)$. Again we obtain an expression of the form "contrast times Fourier coefficients". Since all Fourier coefficients in

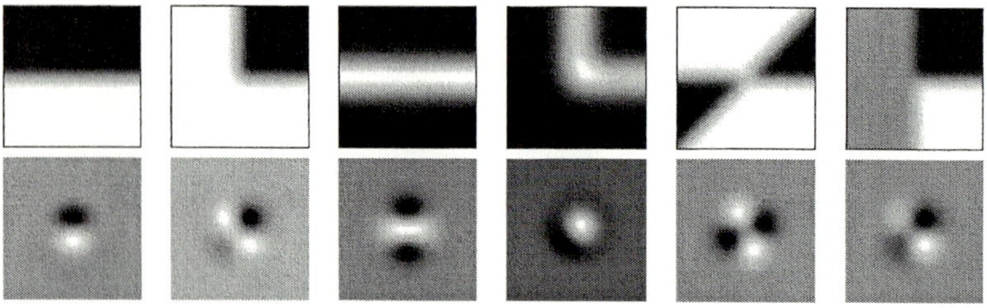

Fig. 2. Top: original images; bottom: reconstruction obtained by a weighted sum of the boundary tensor filter kernels, where the weights correspond to the normalized filter responses at the center pixel

(19) and (20) are weighted by only one radial integral, the form of this integral is not crucial (gradients can be defined with many filters, the boundary tensor originally used the kernel S above, see [10]). Thus, the key difference between the boundary tensor and the gradient squared is that the former includes *three additional Fourier coefficients*: The boundary tensor can be interpreted as a natural generalization of the gradient towards a more sophisticated local signal model. Fig. 2 illustrates this generalization by means of a local image reconstruction from the filter responses that constitute the boundary tensor. This reconstruction essentially shows how the boundary tensor "sees" certain shapes. Obviously large part of the shape information is already contained in five filter responses (only the first two patterns could be reconstructed from the gradient filters). A similar generalization to five Fourier coefficients is achieved by the structure tensor (3). At a polar separable point, its trace can be written as:

$$\begin{aligned} \mathrm{tr}(\mathbf{S}) = \lambda\frac{c_0^2+c_2^2+s_2^2}{2} &\iint B(\rho_1,\rho_2)F_r(\rho_1)F_r(\rho_2)\,\rho_1\,d\rho_1\,\rho_2\,d\rho_2 \\ +(c_1^2+s_1^2) &\iint S(\rho_1,\rho_2)F_r(\rho_1)F_r(\rho_2)\,\rho_1\,d\rho_1\,\rho_2\,d\rho_2 \end{aligned} \tag{21}$$

But here the even and odd Fourier coefficients are weighted by *different* radial integrals. One can try to optimize λ and σ_2/σ_1 in order to minimize the difference between B and S, but it turns out that good agreement can only be achieved for a few frequencies at a time. This means in practice that at many image locations the contributions of even and odd tensor parts are not well balanced, which results in multiple responses for a single boundary or boundary gaps. Fortunately, the trace of the GET operator shows much better behavior:

$$\begin{aligned} \mathrm{tr}(\mathbf{G}) = \frac{c_0^2+c_2^2+s_2^2}{2} &\iint B(\rho_1,\rho_2)F_r(\rho_1)F_r(\rho_2)\,\rho_1\,d\rho_1\,\rho_2\,d\rho_2 \\ +(c_1^2+s_1^2) &\iint G(\rho_1,\rho_2)F_r(\rho_1)F_r(\rho_2)\,\rho_1\,d\rho_1\,\rho_2\,d\rho_2 \end{aligned} \tag{22}$$

Although even and odd Fourier coefficients are still weighted differently, we have shown above (see fig. 1) that the kernels B and G can be made extremely similar, so that the GET operator **G** can be considered a very good approximation of the boundary tensor **B**. Strictly speaking this applies only at polar separable image locations, but we have found experimentally that this desirable behavior carries over to many interesting images features.

4 Experimental Comparison of Tensor Based Feature Detectors

The local shape information represented by the gradient energy and boundary tensors can be extracted in the usual way. The most important tensor characteristic in this context is the *tensor trace* which indicates the local contrast independently of feature type (edge, line, corner, or junction) and thus acts as a general boundary indicator. Intrinsically 1- and 2-dimensional parts of the boundary can be distinguished by the tensors' *eigenvalues*: The smaller eigenvalue indicates corner and junction strength, whereas the difference of the two eigenvalues represents edge and line strength. If the eigenvalues indicate a 1D structure, the *eigenvector* corresponding to the large eigenvalue points in the direction perpendicular to the edge or line. In all experiments we compare the following tensors: (i) the boundary tensor computed with the Laplacian of Gaussian at $\sigma = 0.9$, (ii) the gradient energy tensor computed from Gaussian derivatives with $\sigma_2 = 0.9$ and various ratios σ_3/σ_1 (images are shown for $\sigma_3/\sigma_1 = 1.5$), and (iii) the gradient energy tensor computed by applying Scharr's optimal 3×3 derivative filter $(3, 10, 3)^T(1, 0, -1)/32$ one, two and three times [14].

In the first experiment, we computed the tensors for simple test images. Fig. 3 shows typical results. We found that all tensor variants have very similar trace (boundary strength) and small eigenvalue (corner strength). The trace is phase invariant (to very good approximation in case of the GET operator), i.e. responds uniformly to step edges and lines. The step edge response of the

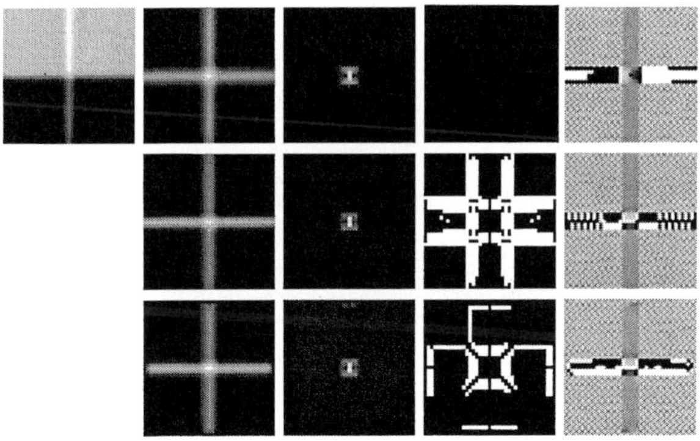

Fig. 3. Top left: original image; col. 2: tensor trace (row 1: boundary tensor, row 2: GET operator, Gaussian derivatives, row 3: GET operator, 3×3 filter); col. 3: junction strength; col. 4: locations with negative junction strength; col. 5: edge orientation (hatched: not a 1D feature, black/white: horizontal edge, gray: vertical edge)

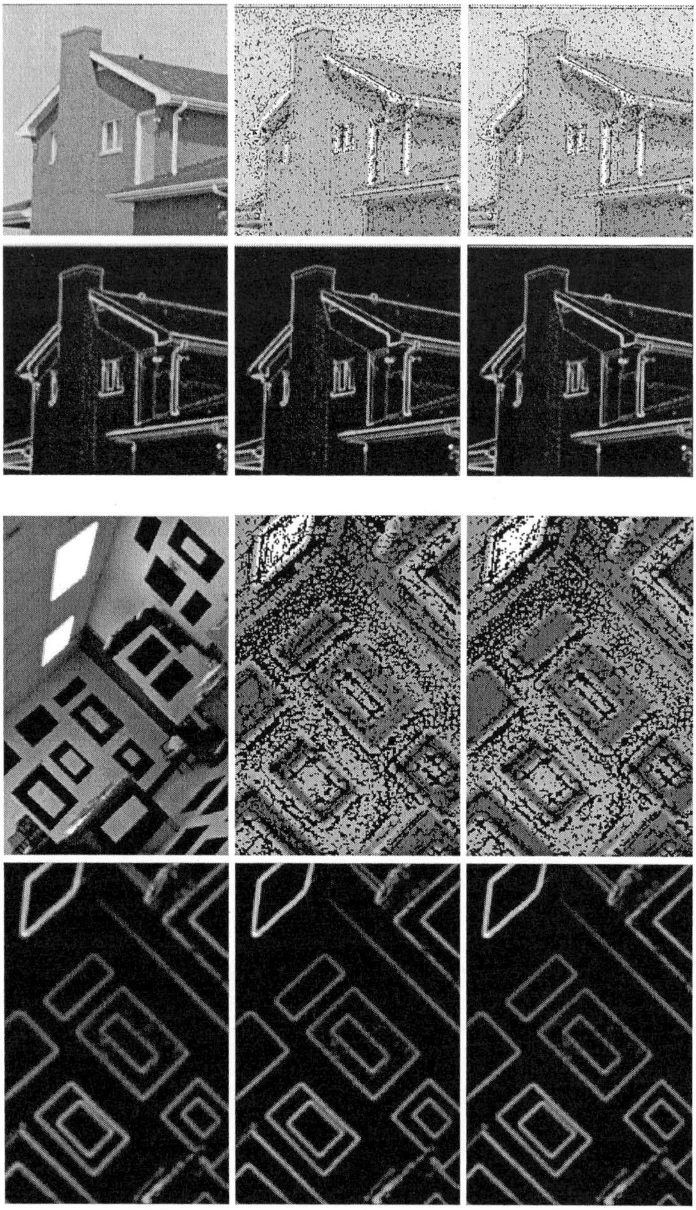

Fig. 4. Rows 1 and 3: original image, negative small eigenvalues of GET operator with $\sigma_3/\sigma_1 = 1.5$, negative small eigenvalues of GET operator with Scharr filter; rows 2 and 4: square root of tensor trace for boundary tensor and the two GET operators

GET operator is slightly narrower than that of the boundary tensor, which may be desirable in practice as it reduces the likelihood that nearby edges blend into each other. On the other hand, there are several locations where the small

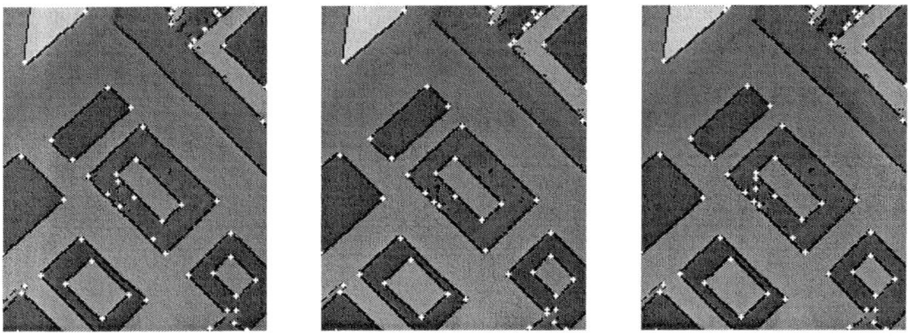

Fig. 5. Integrated boundary detection from boundary tensor, GET operator ($\sigma_3/\sigma_1 = 1.5$) and GET operator (Scharr filter)

eigenvalues of the GET operators are negative, but this only occurs away from junctions. The large eigenvalues are always positive.

The second experiment illustrates the same properties on real images (fig. 4). Again the traces are almost indistinguishable. The small eigenvalue is negative at about 10...35% of the pixels, but never at corners or junctions (we checked this against the corner locations detected with the boundary tensor). Negative values in the trace occur much less frequently (about 1...10% of the pixels, and never on edges) because the large eigenvalue was never negative in the experiments (formal proof of this fact is subject to further research). Gaussian derivatives and the Scharr filter perform similarly, with the exception of derivatives at $\sigma_3/\sigma_1 = 1$, where the number of negative pixels increases 1.5...3-fold.

In the last experiment we show that the three tensors can be used for integrated edge and junction detection as described in [10]. The tensor at each image location is decomposed into its corner/junction (small eigenvalue) and edge/line (difference of eigenvalues times main eigenvector) parts. Then local maxima above a certain threshold are detected in the corner/junction map, and oriented non-maxima supression and thresholding is performed in the edge/line map. The resulting boundaries are overlayed over the original image, see fig. 5. Again, the results are extremely similar.

5 Conclusions

Traditionally, quadrature filters and derivatives have been used by what might be considered different schools of low-level image analysis. In this paper we demonstrated a very close relationship between two typical methods from both camps: the boundary tensor and the GET operator. It turned out that these operators behave almost identically in experiments. Theoretical analysis supports this finding: We established a close formal relationship by giving a new boundary tensor definition using Riesz transforms. And we showed for typical

1- and 2-dimensional image structures that the resulting integral expressions are very similar for the two methods, if suitable operator scales are chosen.

Boundary tensor and GET operator can be interpreted as natural generalizations of the gradient, which uses filters whose angular behaviour corresponds to the first two odd circular harmonics: they add filters representing the first three even circular harmonics. It should be stressed that the feature detection capability depends mainly on this angular behavior – the radial filter shape can be altered considerably, as long as it remains approximately equal for all filters (in the Fourier domain): The boundary tensor can be defined with other band-pass filters, and slightly different radial shapes for even and odd filters can be tolerated in the GET operator. But the angular behavior has to be equal.

Some differences remain: The boundary tensor is always positive semi-definite by construction, whereas the GET operator sometimes measures negative corner strength. Since this does not occur at true corners, it is safe to truncate negative values at zero. On the other hand, the filters constituting the GET operator are simpler then the ones for the boundary tensor (in the spatial domain). The GET operator can already be computed accurately with a 3×3 filter mask, and only seven convolutions with this mask are needed. This is roughly the same effort as needed for the structure tensor, but the underlying feature model is much richer, containing not only edges but also lines, corners, and junctions. Extension to 3D and to multiple scales will likely be easier for the GET operator due to the huge existing body of established analysis for derivative filters.

References

1. J. Bigün, G. Granlund: *Optimal Orientation Detection of Linear Symmetry*, in: ICCV 87, Proc. 1st Intl. Conf. on Computer Vision, pp. 433-438, 1987
2. M. Felsberg, G. Sommer: *The Monogenic Signal*, IEEE Trans. Image Processing, 49(12):3136-3144, 2001
3. M. Felsberg, G. Granlund: *POI Detection Using Channel Clustering and the 2D Energy Tensor*, in: Rasmussen et al. (Eds.): Pattern Recognition, Proc. 26th DAGM Symposium, pp. 103-110, Springer LNCS 3175, 2004
4. M. Felsberg, U. Köthe: *GET: The Connection Between Monogenic Scale-Space and Gaussian Derivatives*, in: ScaleSpace 2005 (this volume)
5. L.M.J. Florack, B.M. ter Haar Romeny, J.J. Koenderink, and M.A. Viergever: *Cartesian differential invariants in scale-space*, J. of Mathematical Imaging and Vision, 3(4), pp. 327-348, 1993
6. W. Förstner: *A Feature Based Correspondence Algorithm for Image Matching*, Intl. Arch. of Photogrammetry and Remote Sensing, vol. 26, pp. 150-166, 1986
7. W.T. Freeman, E.H. Adelson, *The Design and Use of Steerable Filters*, IEEE Trans. on Pattern Analysis and Machine Intelligence, 13(9), pp. 891-906, 1991
8. G. Granlund, H. Knutsson: *Signal Processing for Computer Vision*, Kluwer, 1995
9. J.J. Koenderink: *The Structure of Images*, Biological Cybernetics, 50:363-370, 1984
10. U. Köthe: *Integrated Edge and Junction Detection with the Boundary Tensor*, in: ICCV 03, Proc. of 9th Intl. Conf. on Computer Vision, Nice 2003, vol. 1, pp. 424-431, Los Alamitos: IEEE Computer Society, 2003

11. P. Kovesi: *Image Features From Phase Congruency*, Videre: A Journal of Computer Vision Research. MIT Press. Volume 1, Number 3, 1999
12. T. Lindeberg: *Scale-Space Theory in Computer Vision*, Kluwer, 1994
13. G. Sicuranza: *Quadratic Filters for Signal Processing*, Proc. of the IEEE, 80(8):1263-1285, 1992
14. J. Weickert, H. Scharr: *A scheme for coherence-enhancing diffusion filtering with optimized rotation invariance*, J. Visual Communication and Image Representation, 13(1/2):103-118, 2002

GET: The Connection Between Monogenic Scale-Space and Gaussian Derivatives

Michael Felsberg[1,*] and Ullrich Köthe[2]

[1] Linköping University, Computer Vision Laboratory,
SE-58183 Linköping, Sweden
mfe@isy.liu.se
http://www.isy.liu.se/cvl/

[2] University of Hamburg, Cognitive Systems Group,
D-22527 Hamburg, Germany
koethe@informatik.uni-hamburg.se
http://kogs-www.informatik.uni-hamburg.de/

Abstract. In this paper we propose a new operator which combines advantages of monogenic scale-space and Gaussian scale-space, of the monogenic signal and the structure tensor. The gradient energy tensor (GET) defined in this paper is based on Gaussian derivatives up to third order using different scales. These filters are commonly available, separable, and have an optimal uncertainty. The response of this new operator can be used like the monogenic signal to estimate the local amplitude, the local phase, and the local orientation of an image, but it also allows to measure the coherence of image regions as in the case of the structure tensor. Both theoretically and in experiments the new approach compares favourably with existing methods.

1 Introduction

In this paper we derive a connection between features of the monogenic scale-space [1] of an image and its Gaussian scale-space [2], respectively the derivatives of the latter. Thus, it becomes possible to compute monogenic features from Gaussian derivatives. The advantages of the proposed method are:

- Many people have implementations of Gaussian derivatives available so that they can use monogenic features without implementing new basis filters.
- The Gaussian derivatives are separable and decay faster than the Poisson filter and its Riesz transform resulting in more efficient computational schemes.
- The additional feature (coherence) of the derivative-based method directly indicates the validity of the monogenic phase model which is based on the assumption of locally 1D signals.

* This work has been supported by EC Grant IST-2002-002013 MATRIS and by EC Grant IST-2003-004176 COSPAL.

R. Kimmel, N. Sochen, J. Weickert (Eds.): Scale-Space 2005, LNCS 3459, pp. 192–203, 2005.

A key assumption of this paper is of course that the local phase is useful for the processing and analysis of images. Therefore, we give a short motivation of phase-based image processing in the subsequent section. Although most of the discussions focus on images, the reflections about phase based signal processing generalize to signals of arbitrary dimension.

1.1 Phase-Based Image Processing

First of all, there is some evidence that the human visual system makes use of local phase to analyze the image contents [3]. Since the human visual system performs remarkably well in analyzing images, it is reasonable to design technical systems accordingly. However, there are also purely technical observations which motivate the use phase. In [4] the authors present several experiments which show that the Fourier phase contains the major part of the signal information. The same applies to the local phase. For the definition of local phase, we assume an image (patch) model according to

$$I(\mathbf{x}) = A(\mathbf{x})\cos(\varphi(\mathbf{x})) + \bar{I} \tag{1}$$

where $\mathbf{x} = (x, y)^T$ indicates the spatial coordinate vector, $I(\mathbf{x})$ the image (patch), $\bar{I}$ the average intensity (DC level), $A(\mathbf{x})$ the local amplitude (non-negative), and $\varphi(\mathbf{x})$ the local phase. The average intensity is irrelevant for the analysis of the image contents and is largely compensated already during the image acquisition in the human visual system. What remains is to analyze the relation of local amplitude and local phase. Although the decomposition in (1) seems to be ambiguous, this is not the case due to the non-negativity of the amplitude. Due to the latter, zero crossings in $I(\mathbf{x}) - \bar{I}$ must be covered by zeros of $\cos(\varphi(\mathbf{x}))$ and zero crossings are in direct correspondence to the full phase [5]. Therefore, the local phase becomes a uniquely defined feature.

If the image is decomposed into its amplitude and phase information, it becomes evident that the local amplitude is basically a measure for the confidence of the extracted phase, i.e., in technical terms it represents the signal-to-noise ratio (SNR), cf. Fig. 1, center. The local phase represents most of the image structure, cf. Fig. 1, left. In the areas where the amplitude is close to zero, thus meaning 'no confidence', the local phase contains mainly noise. In the regions of non-zero confidence, the cosine of the local phase results in a visual impression which comes very close to the original, bandpass-filtered image, cf. Fig. 1, right.

1.2 The Monogenic Scale-Space: A Brief Survey

The monognic scale-space is a framework to estimate the local phase, the local orientation, and the local amplitude from an image at different scales [1]. The starting point is to compute the Poisson scale-space $p(\mathbf{x}, s)$ of the image. The corresponding figure flow is obtained as the Riesz transform $\mathbf{q} = (q_1, q_2)^T$

$$\mathbf{q}(\mathbf{x}, s) = \int_{\mathrm{R}^2} \frac{\mathbf{x}'}{2\pi|\mathbf{x}'|^3} p(\mathbf{x}-\mathbf{x}', s)\, d\mathbf{x}' = \int_{\mathrm{R}^2} \frac{\mathbf{x}'}{2\pi(|\mathbf{x}'|^2+s^2)^{3/2}} p(\mathbf{x}-\mathbf{x}', 0)\, d\mathbf{x}' \tag{2}$$

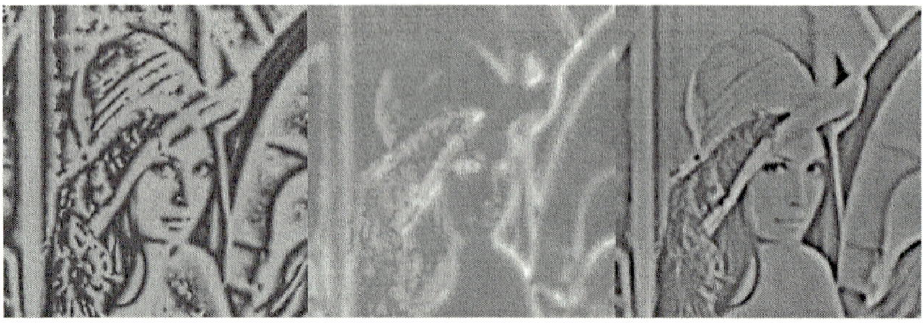

Fig. 1. Decomposing a bandpass image into its local phase and its local amplitude. From left to right: $\cos(\varphi(\mathbf{x}))$, $A(\mathbf{x})$, and $I(\mathbf{x})$, where the intensities where adjusted to obtain similar intensity ranges. Grey means zero, white means positive values, and black means negative values. $I(\mathbf{x})$ is obtained from a bandpass-filters with center frequency $\pi/6$

of the image at each scale s. Together, the blurred image and its Riesz transform form a monogenic signal [6] at the respective scale.

The monogenic signal contains of three components at each position, i.e., for a fixed scale s_0 it is a function $\mathrm{R}^2 \to \mathrm{R}^3 : \mathbf{x} \mapsto (q_1(\mathbf{x}, s_0), q_2(\mathbf{x}, s_0), p(\mathbf{x}, s_0))^T$. For convenience, we sometimes omit the arguments $\mathbf{x}$ and s in the following. The 3D co-domain is then transformed into polar coordinates, cf. Fig. 2, left, resulting in a triplet $(A, \varphi, \theta) \in \mathrm{R}^+ \times [0, 2\pi) \times [0, \pi)$ where $A = \sqrt{q_1^2 + q_2^2 + p^2}$ is

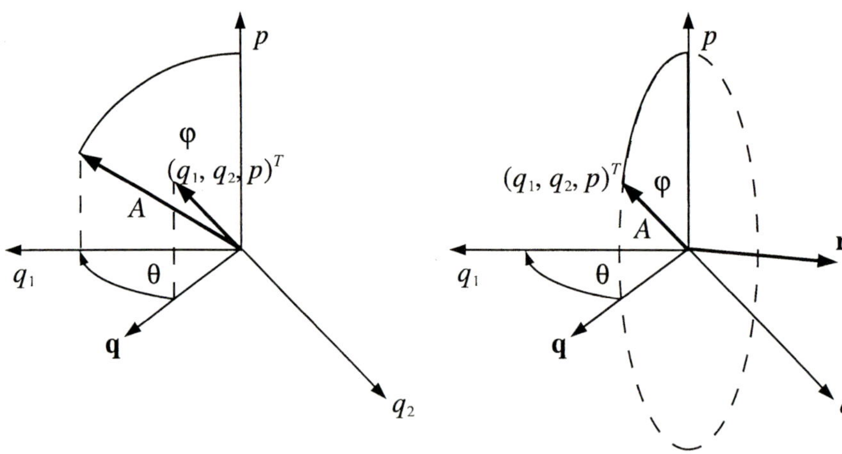

Fig. 2. Phase models used in context of the monogenic signal. Left: the 3D vector is derotated by the local orientation θ, such that it lies in the (q_1, p)-plane. The amplitude and phase are then extracted like in the 1D case as vector length and argument. Right: the 3D vector together with the p-axis define a plane in which the rotation takes place. The normal of this plane multiplied by the (directed) rotation angle φ results in the rotation vector $\mathbf{r}$

an estimate for the local amplitude, $\varphi = \arg(p + i\text{sign}(q_1)|q|)$ for the local phase, and $\theta = \tan^{-1}(q_2/q_1)$ for the local orientation of the image if the image varies locally only in the orientation θ (intrinsic dimensionality of one [7]). Since the direction of an image does not follow from its local orientation [8], an ambiguity concerning the sign of the phase is obtained. In order to obtain a continuous representation of orientation and phase, they can be combined to form a 2D phase vector $\mathbf{r} = \varphi(-\sin\theta, \cos\theta)^T = \mathbf{q}^\perp/|\mathbf{q}| \tan^{-1}(|\mathbf{q}|/p)$, cf. Fig. 2, right.

Further features can be derived from the local features of the monogenic signal respectively the monogenic scale-space, e.g., local frequency and phase congruency as the spatial and scale derivatives of the local phase. The consideration of these features is however out of the scope of this paper.

In order to estimate the local features, implementations of the monogenic signal and the monogenic scale-space are required. This can either be done by local operators, which combine a radial bandpass filter with its Riesz transform [9, 10], or by a global eigentransform solution [11]. The problem is, however, that the involved Poisson kernel decays quite slowly, resulting in either large truncation errors of the filter masks or non-locality of the output.

Even and odd filters based on, e.g., Gaussian derivatives, are preferable concerning locality, but these filters do not allow to estimate the local phase or phase invariant features in a linear framework, since their respective amplitude responses differ. To combine the locality of Gaussian derivatives with phase invariant feature extraction and phase estimation is the main topic of the present paper. The key idea is to use a quadratic operator in order to avoid using the Riesz transform. This idea is based on the concept of the 1D energy operator, which is briefly introduced in the subsequent section.

1.3 The 1D Energy Operator

This brief review of the 1D energy operator is based on [12]. The purpose of the energy operator is to compute directly the squared local amplitude of a signal without using the Hilbert transform, since the Hilbert transform based methods suffer from the same phenomena as the implementations of the monogenic scale-space. The energy operator is defined for continuous 1D signals $s(t)$ as

$$\Psi_c[s(t)] = [\dot{s}(t)]^2 - s(t)\ddot{s}(t) \ . \tag{3}$$

It is obviously not positive semi-definite, but it tracks the energy of simple harmonic oscillators. Moreover, for constants A, r, and ω_0 and for any s_1, s_2

$$\Psi_c[Ar^t \cos(\omega_0 t + \varphi_0)] = A^2 r^{2t} \omega_0^2 \tag{4}$$

$$\Psi_c[s_1(t)s_2(t)] = s_1(t)^2 \Psi_c[s_2(t)] + s_2(t)^2 \Psi_c[s_1(t)] \ . \tag{5}$$

If we instead just consider $[\dot{s}(t)]^2$, likewise the orientation tensor in higher dimensions, we obtain

$$[\frac{d}{dt} A\cos(\omega_0 t + \varphi_0)]^2 = \frac{1}{2} A^2 \omega_0^2 (1 - \cos(2\omega_0 t + 2\varphi_0)) \ ,$$

which is obviously not phase invariant and might even suffer from aliasing if ω_0 is larger than half the Nyquist frequency. Apparently, the second part $s(t)\ddot{s}(t)$ of the energy operator exactly compensates the spurious modulation components at $2\omega_0$. A possible 2D generalization of the energy operator is the energy tensor [13], which we introduce in the subsequent section.

1.4 The 2D Energy Tensor

For continuous, 2D bandpass signals $b(\mathbf{x})$, the 2D energy tensor is defined as [13]

$$\Psi_c[b(\mathbf{x})] = [\nabla b(\mathbf{x})][\nabla b(\mathbf{x})]^T - b(\mathbf{x})[\mathbf{H}b(\mathbf{x})] \ , \tag{6}$$

where $\nabla = (\partial_x, \partial_y)^T$ indicates the gradient and $\mathbf{H} = \nabla\nabla^T$ indicates the Hessian. Likewise in the 1D case, this operator is not positive semi-definite in general, but for a simple harmonic oscillator it results in a energy-frequency-weighted orientation tensor. Moreover, we obtain

$$\Psi_c[Ar^{x+y}\cos(\mathbf{u}_0^T\mathbf{x} + \varphi_0)] = A^2 r^{2x+2y}\mathbf{u}_0\mathbf{u}_0^T \tag{7}$$

$$\Psi_c[s_1(\mathbf{x})s_2(\mathbf{x})] = s_1(\mathbf{x})^2\Psi_c[s_2(\mathbf{x})] + s_2(\mathbf{x})^2\Psi_c[s_1(\mathbf{x})] \ . \tag{8}$$

If we just consider the first part of (6), i.e., the structure / orientation tensor according to [14, 15] (but without spatial averaging), we obtain

$$[\nabla A\cos(\mathbf{u}_0^T\mathbf{x} + \varphi_0)][\nabla A\cos(\mathbf{u}_0^T\mathbf{x} + \varphi_0)]^T = \frac{1}{2}A^2\mathbf{u}_0\mathbf{u}_0^T(1 - \cos(2\mathbf{u}_0^T\mathbf{x} + 2\varphi_0)) \ ,$$

which is likewise in the 1D case not phase invariant and might show aliasing artifacts.

The energy tensor is a second order symmetric tensor like the structure tensor. The latter is included in the energy tensor, but it is combined with a product of even filters, which provides phase invariance for simple harmonic oscillators and products thereof. The energy tensor can hence be classified as a phase invariant, orientation equivariant second order tensor [16]. Same as the 2D structure tensor, the energy operator can be converted into a complex double angle orientation descriptor [17]:

$$o(\mathbf{x}) = \Psi_c[b(\mathbf{x})]_{11} - \Psi_c[b(\mathbf{x})]_{22} + i2\Psi_c[b(\mathbf{x})]_{12} \tag{9}$$

which is equivalent to the 2D energy operator defined in [18]. As one can easily show, $|o(\mathbf{x})| = |\lambda_1(\mathbf{x}) - \lambda_2(\mathbf{x})|$, where $\lambda_1(\mathbf{x}), \lambda_2(\mathbf{x})$ are the eigenvalues of the energy tensor. Since the trace of the tensor is given by the sum of eigenvalues, we obtain $2\lambda_{1,2} = \mathrm{tr}(\Psi_c[b(\mathbf{x})]) \pm |o(\mathbf{x})|$, which can be subject to the same analysis in terms of coherence as suggested in [19, 8] or for the Harris detector [20].

If the signal is not a (product of) simple harmonic oscillations, the operator (6) does not result in a positive response in general. However, if the local signal region adheres to the model (1) with slowly varying amplitude and frequency, the response is positive. This is the case if we prefilter the signal with a bandpass, which avoids low frequencies (DC component and changes of local amplitude) and high frequencies. Removing high frequencies can be considered as a regularization that allows the computation of derivatives for discrete data. The prefiltering is necessary in most practical situations, since natural images $I(\mathbf{x})$ are typically no bandpass signals $b(\mathbf{x})$.

2 The GET Operator

As pointed out above, the signal needs to be bandpass filtered in order to obtain small frequency ranges, and hence, positive responses. For the high frequency regularization, we prefer Gaussian functions due to their high localization in both domains. However, Gaussian filters are not DC-free, which is a central requirement in context of the energy tensor. If we consider a difference of Gaussian filters as in [13], we implicitly lift the level of spatial differentiation by two. According to the equation of linear diffusion [2], the scale derivative of a Gaussian filter is equivalent to the Laplacian of the Gaussian, i.e., a combination of second order derivatives. Hence, applying the Hessian to the Laplacian of the Gaussian means to consider fourth order derivatives instead of second order derivatives. Due to angular aliasing however, one cannot compute fourth order derivatives on a local support [10]. Therefore, we propose an operator below which makes use of Gaussian derivatives up to order three, but avoids the zeroth order Gaussian, i.e., the DC-component is removed.

2.1 The Gradient Energy Tensor

The idea to define the gradient energy tensor (GET) follows from the previous considerations. We introduce the tensor in three steps. First, we plug the gradient of the image into (6) and use tensor notation instead of matrix notation:

$$\begin{aligned} \mathrm{GET}\,\{I(\mathbf{x})\} &= \Psi_c[\nabla I(\mathbf{x})] \\ &= [\nabla \otimes \nabla I(\mathbf{x})] \otimes [\nabla \otimes \nabla I(\mathbf{x})] \\ &\quad -\frac{1}{2}([\nabla I(\mathbf{x})] \otimes [\nabla \otimes \nabla \otimes \nabla I(\mathbf{x})] + [\nabla \otimes \nabla \otimes \nabla I(\mathbf{x})] \otimes [\nabla I(\mathbf{x})]) \end{aligned} \tag{10}$$

where we symmetrized the tensor by replacing the second term with the corresponding anticommutator term. The obtained operator has 16 coefficients, where 6 can be omitted due to symmetry and one further coefficient is a linear combination of two others. Hence, 9 independent coefficients are left. However, all components are formed from sums of even derivative products and odd derivative products. Considering these separately, it turns out that the even part results in just 3 degrees of freedom (the Hessian) and the odd part results in 6 d.o.f. .

In a second step, we contract the tensor. This becomes possible, since there is no practical gain from the coefficients that are omitted in the contraction:

$$\begin{aligned} \mathrm{GET}\,\{I(\mathbf{x})\} &= [\nabla \otimes \nabla I(\mathbf{x})] \cdot [\nabla \otimes \nabla I(\mathbf{x})] \\ &\quad -\frac{1}{2}([\nabla I(\mathbf{x})] \otimes [\nabla \cdot \nabla \otimes \nabla I(\mathbf{x})] + [\nabla \otimes \nabla \cdot \nabla I(\mathbf{x})] \otimes [\nabla I(\mathbf{x})]) \\ &= [\mathbf{H}I(\mathbf{x})][\mathbf{H}I(\mathbf{x})] - \frac{[\nabla I(\mathbf{x})][\nabla \Delta I(\mathbf{x})]^T + [\nabla \Delta I(\mathbf{x})][\nabla I(\mathbf{x})]^T}{2} \end{aligned} \tag{11}$$

In this formula $\Delta = \nabla^T \nabla$ denotes the Laplacian. Due to the non-linearity of the operator, it is difficult to show which degrees of freedom are lost in the contraction, but we can consider certain different cases. Assuming a simple harmonic oscillation $I(\mathbf{x}) = \cos(u\,x + v\,y + \phi)$, we obtain for the full tensor

$$\text{GET}\{I(\mathbf{x})\} = \begin{bmatrix} \begin{bmatrix} u^4 & u^3 v \\ u^3 v & u^2 v^2 \end{bmatrix} & \begin{bmatrix} u^3 v & u^2 v^2 \\ u^2 v^2 & u v^3 \end{bmatrix} \\ \begin{bmatrix} u^3 v & u^2 v^2 \\ u^2 v^2 & u v^3 \end{bmatrix} & \begin{bmatrix} u^2 v^2 & u v^3 \\ u v^3 & v^4 \end{bmatrix} \end{bmatrix} = \begin{bmatrix} u^2 \begin{bmatrix} u^2 & u v \\ u v & v^2 \end{bmatrix} & u v \begin{bmatrix} u^2 & u v \\ u v & v^2 \end{bmatrix} \\ u v \begin{bmatrix} u^2 & u v \\ u v & v^2 \end{bmatrix} & v^2 \begin{bmatrix} u^2 & u v \\ u v & v^2 \end{bmatrix} \end{bmatrix}$$

and for the contracted tensor

$$\text{GET}\{I(\mathbf{x})\} = \begin{bmatrix} u^2\,(u^2+v^2) & u v\,(u^2+v^2) \\ u v\,(u^2+v^2) & v^2\,(u^2+v^2) \end{bmatrix} . \tag{12}$$

Hence, no information is lost by the contraction under the assumed signal model. If we extend the model to two different frequencies in the same direction, the tensor coefficients are multiplied by a spurious modulation factor.[1] However, this modulation is the same for all coefficients, and therefore, the full tensor does not provide additional information. By repeating this procedure for more frequencies in the same direction, the result will always be the same, and hence, of locally 1D signals there is no gain from the full tensor.

Due to the non-linear behavior of the tensor it is impossible to calculate the response for a general 2D signal. However, one can analyze it in terms of null-spaces and it turns out that the contraction does not change the null-space of the operator. The GET becomes zero for

$$I(\mathbf{x}) = A\exp(ax+by) + D \ , \tag{13}$$

where A, a, b, D are complex constants. Hence, the three degrees of freedom which are lost in the contraction of the odd part of the tensor do not reduce the null-space and are therefore of minor importance. Deeper investigations of the null-spaces and the number of independent components will be subject to a future publication.

Finally, we would like to point out here that the contraction can be done in an alternative way by taking the inner product at a different grade of the odd tensor:

$$[\nabla I(\mathbf{x})] \cdot [\nabla \otimes \mathbf{H} I(\mathbf{x})] = \Psi_c[\partial_x I(\mathbf{x})] + \Psi_c[\partial_y I(\mathbf{x})] \ . \tag{14}$$

The behavior for locally 1D signals is the same, but for 2D signals we get different results.

2.2 Regularization and Gaussian Derivatives

The results from the previous section are obtained for idealized, continuous signals. In practice, however, we have to deal with non-ideal, noisy, and discrete signals. The most common thing to do is therefore to regularize the derivative operators from (11) with Gaussian kernels. A Gaussian regularization is the optimal choice if nothing is known about the signal and its noise characteristic. Therefore, we replace the derivatives in (11) with Gaussian derivatives of order one to three.

[1] The spatial modulation is undesired. The response should have constant amplitude.

The scales for the regularization are chosen such that the variance increases linearly with the order of the derivative, cf. (13) in [21]. In [22] we discuss the choice of scales and different regularizations more in detail.

2.3 Extraction of Monogenic Features

The monogenic signal provides three features: local amplitude, local phase, and local orientation [6]. In case of signals with intrinsic dimensionality one, i.e., $I(\mathbf{x}) = s(\mathbf{n}^T\mathbf{x})$ $(s : \mathrm{R} \to \mathrm{R},\ |\mathbf{n}| = 1)$, the GET is of rank one:

$$\begin{aligned}\mathrm{GET}\{I(\mathbf{x})\} &= [\mathbf{nn}^T\ddot{s}(\mathbf{n}^T\mathbf{x})][\mathbf{nn}^T\ddot{s}(\mathbf{n}^T\mathbf{x})] \\ &\quad - \frac{[\mathbf{n}\dot{s}(\mathbf{n}^T\mathbf{x})][\mathbf{n}\dddot{s}(\mathbf{n}^T\mathbf{x})]^T + [\mathbf{n}\dddot{s}(\mathbf{n}^T\mathbf{x})][\mathbf{n}\dot{s}(\mathbf{n}^T\mathbf{x})]^T}{2} \\ &= \mathbf{nn}^T[\ddot{s}(\mathbf{n}^T\mathbf{x})^2 - \dddot{s}(\mathbf{n}^T\mathbf{x})\dot{s}(\mathbf{n}^T\mathbf{x})] \ .\end{aligned}$$

The first eigenvector of this expression is $\pm\mathbf{n}$, i.e., the local orientation of the signal. The first eigenvalue (or its trace, aka the second eigenvalue is zero) of the GET is more difficult to analyze, except for the single-frequency case, where we obtain according to (12) $|\mathbf{u}|^4 A^2$ for an oscillation with amplitude A.

Much more interesting is the extraction of the local phase, which is obtained in two steps. First, we consider the two addends of the GET separately. The first one represents the symmetric (even) parts of the signal, whereas the second one represents the antisymmetric (odd) parts of the signal. However, both parts are quadratic expressions, such that we have to consider their square-roots:

$$q_{\mathrm{even}} = \pm\sqrt{\mathrm{trace}(\mathbf{T}_{\mathrm{even}})} \qquad \text{and} \qquad q_{\mathrm{odd}} = \pm\sqrt{\mathrm{trace}(\mathbf{T}_{\mathrm{odd}})}$$

where

$$\mathbf{T}_{\mathrm{even}} = [\mathbf{H}I(\mathbf{x})][\mathbf{H}I(\mathbf{x})] \qquad \text{and} \tag{15}$$

$$\mathbf{T}_{\mathrm{odd}} = -\frac{[\nabla I(\mathbf{x})][\nabla\Delta I(\mathbf{x})]^T + [\nabla\Delta I(\mathbf{x})][\nabla I(\mathbf{x})]^T}{2} \ . \tag{16}$$

In a second step, the correct signs for the even and the odd parts are selected, such that $\arg(q_{\mathrm{even}} + iq_{\mathrm{odd}})$ gives the local phase of the signal. Comparing the signs in different quadrants of a harmonic oscillation results in the following procedure.

Let $\mathbf{T} = \mathbf{T}_{\mathrm{even}} + \mathbf{T}_{\mathrm{odd}}$ denote the GET response, $o = \mathbf{T}_{11} - \mathbf{T}_{22} + i2\mathbf{T}_{12}$ its complex double angle representation [17], and $\mathbf{o} = (\mathrm{real}(\sqrt{o}), \mathrm{imag}(\sqrt{o}))^T$ the orientation vector. We then define the two signs as

$$s_{\mathrm{even}} = -\mathrm{sign}\,(\mathbf{o}^T[\mathbf{H}I(\mathbf{x})]\mathbf{o}) \qquad \text{and} \qquad s_{\mathrm{odd}} = -\mathrm{sign}\,(\mathbf{o}^T\nabla I(\mathbf{x})) \tag{17}$$

such that

$$\varphi = \arg(q_{\mathrm{even}} + iq_{\mathrm{odd}}) = \arg(s_{\mathrm{even}}\sqrt{\mathrm{trace}(\mathbf{T}_{\mathrm{even}})} + is_{\mathrm{odd}}\sqrt{\mathrm{trace}(\mathbf{T}_{\mathrm{odd}})}) \tag{18}$$

is consistent with the definition of the monogenic phase. This can easily be verified by inserting $\cos(ux + vy)$ into the previous four expressions, resulting in $\varphi = ux + vy$ if $(u, v)^T$ lies in the upper half-plane and $\varphi = -ux - vy$ otherwise. This behavior is correct since we obtain the same sign ambiguity for the monogenic phase [6].

If the underlying signal is non-simple, i.e., it has intrinsic dimensionality two, the analysis becomes more difficult. Following the strategy of the structure multivector in [10], the first eigenvector is extracted from $\mathbf{T}$. Then, the even tensor and the odd tensor are projected onto the first eigenvector and onto the orthogonal vector (aka the second eigenvector). This gives two even components and two odd components, which are then combined with appropriate signs to extract two phases for the two perpendicular orientations.

Note also that in the latter case not a single amplitude is obtained, but two eigenvalues, which correspond to the local amplitudes of the two perpendicular components. These eigenvalues can then be used for coherence analysis or corner detection likewise the eigenvalues of the structure tensor.

3 Comparisons

In this section we compare the results of the GET operator with those of the DCT-based implementation [11], the spherical quadrature filters [10], and the structure tensor (ST) by outer products of gradients (see e.g. [19]). The latter approach is not suited for phase-estimation per se, but one can easily extend it for this purpose in the following way. Assuming that the outer product of gradients of a cosine oscillation results in a trace which is $A\sin^2(\varphi)$ and assume further that local averaging can be replaced with integration over entire periods, the trace of the (averaged) tensor becomes $\frac{A}{2}$. Hence, the sine and the cosine are obtained up to a sign-ambiguity by

$$q_{\mathrm{odd}} = \pm\sqrt{t(\mathbf{x})} \qquad \text{and} \qquad q_{\mathrm{even}} = \pm\sqrt{-t(\mathbf{x}) + 2\sum t(\mathbf{x})} \ ,$$

where $t(\mathbf{x}) = \mathrm{trace}([\nabla I(\mathbf{x})][\nabla I(\mathbf{x})]^T)$. For the subsequent comparison only the second sign needs to be recovered. If we have locally vanishing DC components and single frequency, it is obtained by the sign of $I(\mathbf{x})$, otherwise we use the sign of $-\Delta I$. In order to remove some outliers, the signs are median-filtered.

3.1 Experiment: Extraction of Phase and Orientation

In this experiment, we applied all three methods to a synthetic pattern, cf. Fig. 3, top left, and a real image, cf. Fig. 3, bottom row. We added Gaussian noise to the pattern (SNR 3.0dB) and selected a mask for the feature comparison. The scales of all methods were chosen such that the local amplitude estimates were comparably similar, although the higher spatial-frequency localization of the GET and the ST leads to a narrower ridge then for the other two methods, cf. Fig. 3, second row. Instead of showing the phase estimates, we reconstructed the signal from the phase estimates (cf. Fig. 3, third row) and computed their SNRs. Furthermore, we computed the orientation error according to [23]. The results

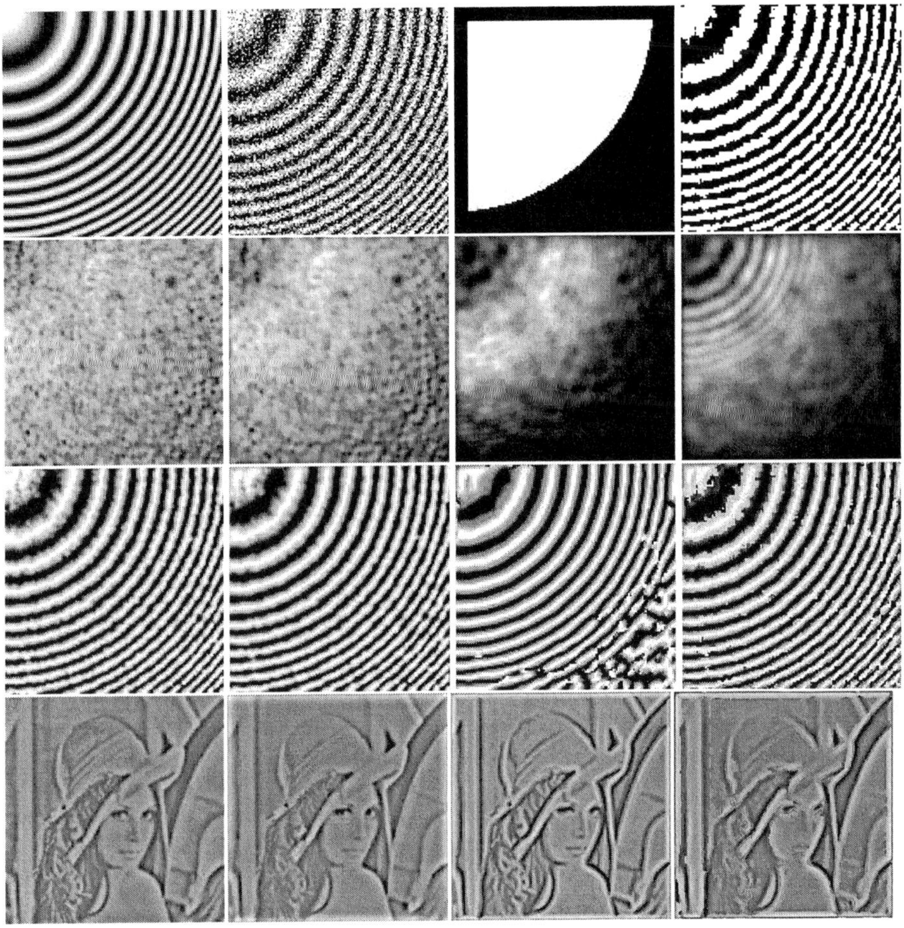

Fig. 3. Top row from left to right: test-pattern, test-pattern with noise (SNR 3.0dB), mask for error evaluation, and sign extraction for ST. Second row: amplitudes of DCT, SQF, GET, and ST (from left to right). Third row: respective reconstructions. Bottom row: respective amplitude-weighted reconstructions of a real image

Table 1. SNR of reconstruction and orientation error from estimates, cf. Fig. 3

method	DCT	SQF	GET	ST
SNR	12.8dB	13.7dB	13.5dB	11.1dB
$\Delta\theta$	19.6°	18.5°	7.0°	3.6°

are summarized in Tab. 1, which shows that if we are interested in simultaneously estimating orientation and phase, the GET gives the best results.

3.2 Complexity Analysis

The computational complexity might also be an important aspect when it comes to the selection of suitable methods. Since the extraction of phase and orientation has to be done in all cases, we only compare the complexities up to that point.

The complexity of the DCT-based method is given by ten 2D FFTs [11], since we have to compute three scales. Hence, we obtain $30N^2 \log_2 N$ floating point operations (FLOPs) for an image of size $N \times N$ if N is a power of two. For our test image we have $N = 128$, such that we applied about $3.4 \cdot 10^6$ FLOPs.

For the SQF filter set, the complexity depends on the filter size. In our experiment, we used three 23×23 filters. The filters are not separable, but we can exploit a four-fold symmetry for the even filter and an eight-fold symmetry for the odd filter pair. Hence, we end up with $1850N^2$ FLOPs, which is about $3 \cdot 10^7$ FLOPs in our test.

For the GET operator, the complexity also depends on the filter size. In our experiment, we used 2σ-truncation (for the largest scale) resulting in seven 17×17 filters. These filters are separable and each of the 1D filters can exploit a two-fold symmetry. Hence, we get $357N^2$ FLOPs, i.e., $6 \cdot 10^6$ FLOPs in our special case. For the structure tensor, the computational effort is about the same if we take into account the calculation of the sign (bandpass filter and median filter).

One problem with these complexity estimates are the missing complexities for memory accesses, which become more and more important nowadays. As an side-effect of this, the SQF filters are 1.5 times faster than the DCT based method and the GET operator is 2 times faster than the SQF filters.

4 Conclusion

In this paper we have described an alternative way of extracting the image features of the monogenic signal, i.e., local amplitude, local phase, and local orientation, by using a quadratic form. The proposed method of the gradient energy tensor is the contraction of a fourth order tensor built from image derivatives of order one to three. The new tensor is compatible to the structure tensor concerning eigensystem analysis, but it is phase-invariant without spatial averaging. Using Gaussian regularization of the derivatives leads to a connection of monogenic scale-space and Gaussian scale-space via the quadratic form.

We provided formulas to extract the local phase from the two different parts of the GET, and compared the extracted features phase and orientation to those of previous approaches. Considering both estimates at once, the GET provides the best estimates and it is also among the fastest operators. For non-simple signals, it even provides the two additional features of second eigenvalue and second phase, which makes it comparable to the much slower structure multivector.

References

1. Felsberg, M., Sommer, G.: The monogenic scale-space: A unifying approach to phase-based image processing in scale-space. JMIV **21** (2004) 5–26

2. Koenderink, J.J.: The structure of images. Biol. Cybernetics **50** (1984) 363–370
3. Mechler, F., Reich, D.S., Victor, J.D.: Detection and discrimination of relative spatial phase by V1 neurons. Journal of Neuroscience **22** (2002) 6129–6157
4. Oppenheim, A., Lim, J.: The importance of phase in signals. Proc. of the IEEE **69** (1981) 529–541
5. Hurt, N.E.: Phase Retrieval and Zero Crossings. Kluwer Academic, Dordrecht (1989)
6. Felsberg, M., Sommer, G.: The monogenic signal. IEEE Transactions on Signal Processing **49** (2001) 3136–3144
7. Krieger, G., Zetzsche, C.: Nonlinear image operators for the evaluation of local intrinsic dimensionality. IEEE Trans. on Image Processing **5** (1996) 1026–1041
8. Granlund, G.H., Knutsson, H.: Signal Processing for Computer Vision. Kluwer Academic Publishers, Dordrecht (1995)
9. Felsberg, M.: On the design of two-dimensional polar separable filters. In: 12th European Signal Processing Conference, Vienna, Austria. (2004)
10. Felsberg, M.: Low-Level Image Processing with the Structure Multivector. PhD thesis, Christian-Albrechts-University of Kiel (2002) TR no. 0203.
11. Felsberg, M., Duits, R., Florack, L.: The monogenic scale space on a rectangular domain and its features. IJCV (2004) accepted publ. in future issue.
12. Maragos, P., Bovik, A.C., Quartieri, J.F.: A multi-dimensional energy operator for image processing. In: SPIE Conference on Visual Communications and Image Processing, Boston, MA (1992) 177–186
13. Felsberg, M., Granlund, G.: POI detection using channel clustering and the 2D energy tensor. In: 26. DAGM Symposium Mustererkennung, Tübingen. (2004)
14. Förstner, W., Gülch, E.: A fast operator for detection and precise location of distinct points, corners and centres of circular features. In: ISPRS Intercommission Workshop, Interlaken. (1987) 149–155
15. Bigün, J., Granlund, G.H.: Optimal orientation detection of linear symmetry. In: Proc. IEEE First International Conference on Computer Vision (1987) 433–438
16. Nordberg, K.: Signal Representation and Processing using Operator Groups. PhD thesis, Linköping University (1995) Dissertation No 366, ISBN 91-7871-476-1.
17. Bigün, J., Granlund, G.H., Wiklund, J.: Multidimensional orientation estimation with applications to texture analysis and optical flow. PAMI **13** (1991) 775–790
18. Larkin, K.G., Oldfield, M.A., Bone, D.J.: Demodulation and phase estimation of two-dimensional patterns. Australian patent AU 200110005 A1 (2001)
19. Jähne, B.: Digitale Bildverarbeitung. Springer, Berlin (1997)
20. Harris, C.G., Stephens, M.: A combined corner and edge detector. In: 4th Alvey Vision Conference. (1988) 147–151
21. Florack, L., Duits, R.: Regularity classes for locally orderless images. In Griffin, L.D., Lillholm, M., eds.: Scale-Space Theory in Computer Vision. Volume 2695 of LNCS., Springer, Heidelberg (2003) 255–265
22. Köthe, U., Felsberg, M.: Riesz-transforms vs. derivatives: On the relationship between the boundary tensor and the energy tensor. In: Proc. Scale Space Conference (this volume). (2005)
23. Knutsson, H., Andersson, M.: Robust N-dimensional orientation estimation using quadrature filters and tensor whitening. In: Proceedings of IEEE International Conference on Acoustics, Speech, & Signal Processing, Adelaide, Australia (1994)

Matrix-Valued Filters as Convex Programs

Martin Welk[1], Florian Becker[2], Christoph Schnörr[2], and Joachim Weickert[1]

[1] Mathematical Image Analysis Group,
Faculty of Mathematics and Computer Science, Bldg. 27,
Saarland University, 66041 Saarbrücken, Germany
{welk, weickert}@mia.uni-saarland.de
http://www.mia.uni-saarland.de

[2] Computer Vision, Graphics, and Pattern Recognition Group,
Faculty of Mathematics and Computer Science,
University of Mannheim, 68131 Mannheim, Germany
flbecker@rumms.uni-mannheim.de
schnoerr@uni-mannheim.de
http://www.cvgpr.uni-mannheim.de

Abstract. Matrix-valued images gain increasing importance both as the output of new imaging techniques and as the result of image processing operations, bearing the need for robust and efficient filters for such images. Recently, a median filter for matrix-valued images has been introduced. We propose a new approach for the numerical computation of matrix-valued median filters, and closely related mid-range filters, based on sound convex programming techniques. Matrix-valued medians are uniquely computed as global optima with interior point solvers. The robust performance is validated with experimental results for matrix-valued data including texture analysis and denoising.

1 Introduction

In this paper, we are concerned with the processing of images where the value attached to each pixel or voxel is a symmetric matrix. Image data of this kind appear in a variety of different contexts in modern image acquisition and processing. For example, diffusion tensor magnetic resonance imaging (DT-MRI) is an upcoming medical image acquisition technique which measures the diffusion characteristics of water molecules in tissue, yielding valuable insights into the structure and function of tissues, particularly fibre connectivity in the brain [13]. Moreover, structure tensors arise as derived quantities in motion detection, texture analysis and segmentation and other fields of image processing [8]. Tensor data also occur in solid and fluid mechanics. The latter can have eigenvalues of either sign while diffusion tensors and structure tensors are positive semidefinite.

All of these data, be they directly measured or computed, are often degraded by noise. One of the basic tasks in matrix-valued image processing as in other fields of image processing is therefore denoising. A simple but effective denoising filter is the matrix-valued median filter introduced in [18]. Based on generalising

R. Kimmel, N. Sochen, J. Weickert (Eds.): Scale-Space 2005, LNCS 3459, pp. 204–216, 2005.

the minimisation property of the scalar-valued median, it inherits from its scalar counterpart the robustness and capability to preserve discontinuities. To compute matrix-valued medians, in [18] a gradient descent algorithm was proposed.

In this paper, we introduce a new and efficient algorithm for the computation of matrix-valued medians according to the (slightly generalised) definition from [18]. The new approach is based on convex conic programming methods and can easily be adapted to closely related problems like the computation of matrix-valued mid-range filters. We apply the new algorithm to DT-MRI data to demonstrate its use. Furthermore, we use matrix-valued medians to smooth structure tensor data from textured grey-value images as a preprocessing step for texture segmentation.

We proceed as follows. In Section 2 we describe the local matrix-valued image filters that we are concerned with. Sections 3 shows how these filters can be rewritten as convex optimisation problems which are then solved in Section 4. Experiments on DT-MRI data and local orientation estimation of grey-value images are presented and discussed in Section 5. Conclusive remarks are given in Section 6.

Related Work. Median filtering of matrix-valued data is closely related to that of vector valued data. Indeed, the definition from Welk et al. [18] has an obvious vector-valued analog. For earlier approaches to vector-valued median filtering in the image processing literature we refer to Astola et al. [1] and Caselles et al. [7]. While Caselles et al. [7] require that the median has always to be one of the given data vectors, Astola et al. [1] relax this condition somewhat while the definition given in [18] does not make such a restriction at all. Barni et al. [3] define a vector median using the Euclidean distance sum minimisation similarly as [18], but again restricted to the given data vectors. Interestingly, the exact analog to the definition from [18] for 2-D vectors has already been proposed in 1959 by Austin [2] along with a graphical algorithm which is closely related to the gradient descent procedure from [18]. The problems of this procedure and improvements have been discussed in Seymour's 1970 reply [15]. Vector-valued medians as well as vector-valued mid-range values (often called 1-centres) have also been studied in the context of facility location problems, see e.g. Megiddo [12], Fekete et al. [9] and the references therein.

The concept of the structure tensor goes back to Förstner and Gülch [8]. It is common in image analysis to smooth the rank one matrices which arise directly from the gradient vectors in single points by Gaussian convolution which leads in general to rank two matrices which integrate directional information from a neighbourhood and suffer less from noise sensitivity. The observation that the Gaussian convolution used in this process is essentially a linear diffusion of the directional information, thus introducing a blurring that is unwished at times, led to the definition of a nonlinear structure tensor by Weickert and Brox [17], [6] in which Gaussian convolution is replaced by nonlinear diffusion. For its better preservation of discontinuities, the nonlinear structure tensor is well-suited for texture segmentation [4], [14] and optical flow analysis [6]. Smoothing structure tensors with medians is also related to the robust structure tensor introduced by van den Boomgaard and van der Weijer in [16] and which for a particular choice of the penaliser function ϱ also amounts to a minimisation similar to that in the matrix-valued median.

Regarding convex programming, all concepts we use can be found in corresponding textbooks (e.g., Boyd and Vandenberghe [5]). Recently, these optimisation methods have been also successfully applied to various other image processing problems by Keuchel et al. [11].

Notation and Preliminaries. Throughout the paper, e denotes the vector $(1,\ldots,1)^\top \in \mathbb{R}^n$. By I_d we denote the $d \times d$ unit matrix. Further, $\mathcal{L}^d$ is the convex cone of vectors $\{x \in \mathbb{R}^d \mid x_d \geq \sqrt{x_1^2 + \cdots + x_{d-1}^2}\}$ while $\mathcal{S}^d$ is the linear space of symmetric $d \times d$ real matrices. The ith eigenvalue of $X \in \mathcal{S}$ in the order $\lambda_1(X) \geq \cdots \geq \lambda_d(X)$ will be denoted by $\lambda_i(X)$. Finally, by $\mathcal{S}_+^d$ we mean the convex cone of positive semidefinite symmetric matrices $\{X \in \mathcal{S}^d \mid \lambda_d(X) \geq 0\}$.

2 Problem Statement: Local Matrix Filters

Given n real numbers $a_1, a_2, \ldots, a_n$, their median is defined as the middle value in the sequence that contains all the numbers ordered by size. The median concept gives rise to a class of image filters, called *median filters*, which are known for their outstanding capability for edge-preserving denoising of images. Median filtering of a discrete grey-value image requires the specification of a pixel mask, the so-called structure element, which is used to select a neighbourhood for each pixel. The new grey-value of each pixel is taken to be the median of the old grey-values of all pixels within its neighbourhood. Median filtering can be iterated, thereby performing a progressive edge-preserving smoothing. This can be compared to the approximation of the (non-edge-preserving) Gaussian smoothing by iterated box averaging.

The matrix-valued generalisation of median filtering introduced in [18] is based on an interesting energy minimisation property of the scalar-valued median: The median of $a_1, a_2, \ldots, a_n$ is exactly the real number x for which $\sum_{i=1}^n |x - a_i|$ is minimal. The median of n matrices $A_1, \ldots, A_n \in \mathcal{S}^d$ is then defined as

$$\operatorname{med}(A_1, \ldots, A_n) := \operatorname{argmin}_{X \in \mathcal{S}^d} \sum_{i=1}^{n} d(X, A_i)$$

where d is a suitable, rotationally invariant metric on $\mathcal{S}^d$. In [18], the Frobenius norm was used,

$$d(X, A_i) = \|X - A_i\|_2 = \sqrt{\operatorname{tr}\big[(X - A_i)(X - A_i)\big]}\ ;$$

another possible choice is the spectral norm,

$$d(X, A_i) = |X - A_i| = \max_{i=1,\ldots,d} \big|\lambda_i(X - A_i)\big|\ .$$

Interestingly, the so-called mid-range value of real numbers $a_1, a_2, \ldots, a_n$ which is defined as the arithmetic mean of their maximum and minimum, can be described by an extremality property very similar to that of the median – instead of the sum of the distances $|x - a_i|$, their maximum is minimised. The transfer to matrices is therefore straightforward. We define

$$\text{midr}(A_1, \ldots, A_n) := \text{argmin}_{X \in \mathcal{S}^d} \max\left\{d(X, A_1), \ldots, d(X, A_n)\right\}$$

with the same requirements for d as in the case of the median. Midrange filtering is less attractive by itself but stands in close relation to other matrix filters.

3 Convex Optimisation

In this section, we show that each filter introduced in the previous section is defined as global optimum of a *convex* optimisation problem.

3.1 Median Filter: Frobenius Norm

We consider the optimisation problem:

$$\text{med}_F(A_1, \ldots, A_n) := \text{argmin}_{X \in \mathcal{S}^d} \sum_{i=1}^{n} \|X - A_i\|_2 \tag{1}$$

and identify the unknown matrix $X \in \mathcal{S}^d$ with a vector $X \in \mathbb{R}^{d^2}$. Introducing n additional variables $t = (t_1, \ldots, t_n)^\top$, we rewrite (1):

$$\inf_{X \in \mathcal{S}^d, t \in \mathbb{R}^n} \langle e, t \rangle \,, \quad \|X - A_i\|_2 \leq t_i \,, \quad i = 1, \ldots, n \tag{2}$$

Each constraint is convex, because $(X^\top, t_i)^\top$ varies in the convex cone $\mathcal{L}_i^{d^2+1}$ translated by $(A_i^\top, 0)^\top$. Denoting the corresponding convex constraint sets with C_i, $i = 1, \ldots, n$, problem (2) reads:

$$\inf_{X \in \mathcal{S}^d, t \in \mathbb{R}^n} \langle e, t \rangle \,, \quad \begin{pmatrix} X \\ t \end{pmatrix} \in \bigcap_{i=1}^{n} C_i \tag{3}$$

This optimisation problem is convex, since the objective function is linear, and since the intersection of convex sets is convex, too.

3.2 Median Filter: Spectral Norm

We consider the optimisation problem:

$$\text{med}_S(A_1, \ldots, A_n) := \text{argmin}_{X \in \mathcal{S}^d} \sum_{i=1}^{n} |X - A_i| \tag{4}$$

Similarly to section 3.1, we introduce auxiliary variables $t \in \mathbb{R}^n$ and corresponding constraints:

$$|X - A_i| \leq t_i \ , \quad i = 1, \ldots, n$$

These constraints are satisfied if

$$t_i I_d - (X - A_i) \in \mathcal{S}^d_+ \qquad \text{and} \qquad t_i I_d + (X - A_i) \in \mathcal{S}^d_+ \ , \qquad i = 1, \ldots, n$$

Again, the variables (X, t_i) are constrained to convex sets, defined by the intersection of affine sets (left hand sides) with the convex cone $\mathcal{S}^d_+$. Denoting the constraint sets with $C_{i,+}$, $C_{i,-}$, $i = 1, \ldots, n$, we can rewrite problem (4):

$$\min_{X \in \mathcal{S}^d, t \in \mathbb{R}^n} \langle e, t \rangle \ , \quad \begin{pmatrix} X \\ t \end{pmatrix} \in \bigcap_{i=1}^{n} (C_{i,+} \cap C_{i,-}) \tag{5}$$

This optimisation problem is convex, since the objective function is linear, and since the intersection of convex sets is convex, too.

We remark that for positive semidefinite data $A_i \in \mathcal{S}^d_+$, $i = 1, \ldots, n$, the constraints represented by the sets $C_{i,-}$ are redundant and can be dropped.

3.3 Midrange Filters

For midrange filters defined by

$$\operatorname{midr}(A_1, \ldots, A_n) := \operatorname{argmin}_{X \in \mathcal{S}^d} \max \left\{ d(X, A_1), \ldots, d(X, A_n) \right\} \ , \tag{6}$$

we introduce the *scalar* auxiliary variable $t := \max \left\{ d(X, A_1), \ldots, d(X, A_n) \right\}$. Similar to the derivation of (3) and (5), problem (6) results in two convex optimisation problems, depending on which norm we choose. We focus on the median filters in the remainder of this paper.

4 Convex Programming and Duality

We represent the optimisation problems defined in the previous section as convex programs. This allows to implement matrix-valued median filters using corresponding numerical interior-point algorithms. The corresponding dual programs reveal that solutions automatically satisfy plausible conditions whose direct computation (without convex programming) would be more involved.

4.1 Convex Conic Programs

Conic programs generalise linear programs by replacing the standard cone $\mathbb{R}^n_+$ with more general convex cones $\mathcal{K}$:

$$\inf_x \langle c, x \rangle \ , \quad Fx - g \in \mathcal{K} \tag{7}$$

The corresponding dual conic program reads:[1]

$$\sup_{y} \langle g, y \rangle \,, \quad F^{\top} y = c \,, \quad y \in \mathcal{K} \tag{8}$$

If at least one of these problems is bounded and strictly feasible, then $\{x, y\}$ is a pair of optimal solutions if and only if the duality gap is zero:

$$\langle c, x \rangle = \langle g, y \rangle \tag{9}$$

4.2 Medians as Conic Programs

We consider problem (1) and identify again matrices $X, A_i \in \mathcal{S}^d$ with vectors $X, A_i \in \mathbb{R}^{d^2}$. (2) and (3) corresponds to (7):

$$\inf_{X \in R^{d^2}, t \in \mathbb{R}^n} \langle e, t \rangle \,, \quad F \begin{pmatrix} X \\ t \end{pmatrix} - g \in \mathcal{K} \,, \tag{10}$$

where F and g are obtained by stacking the matrices resp. vectors

$$\begin{pmatrix} I_{d^2} & 0_{d^2 \times n} \\ 0^{\top} & e_i^{\top} \end{pmatrix} \quad \text{and} \quad \begin{pmatrix} A_i \\ 0 \end{pmatrix} \,, \quad i = 1, \dots, n$$

together, e_i is the i-th unit vector, and $\mathcal{K} = \mathcal{L}^{d^2+1} \times \cdots \times \mathcal{L}^{d^2+1}$.

Below, $X \in \mathcal{S}^d$ is again regarded as a matrix. Problem (4) or (5), respectively, directly lead to (7), formulated as semidefinite program:

$$\inf_{X \in \mathcal{S}^d, t \in \mathbb{R}^n} \langle e, t \rangle \,, \qquad \text{subject to } \mathcal{F} \begin{pmatrix} X \\ t \end{pmatrix} - G \in \mathcal{S}_+^{n \times d^2} \,, \tag{11}$$

with the linear mapping:

$$\mathcal{F}(X, t) = \operatorname{Diag}\{ \dots, t_i I_d - X, \dots, t_i I_d + X, \dots \} \tag{12}$$

and:

$$G = \operatorname{Diag}\{ \dots, -A_i, \dots, +A_i, \dots \} \tag{13}$$

4.3 Dual Programs and Optimality Conditions

Evaluating (8), the conic dual program to (10) reads:

$$\sup_{Y_i \in \mathbb{R}^{d^2}} \sum_{i=1}^{n} \langle Y_i, A_i \rangle \,, \quad \sum_{i=1}^{n} Y_i = 0 \,, \quad \|Y_i\|_2 \le 1 \,, \quad \forall i \tag{14}$$

---

[1] In general, conic duals are defined w.r.t. dual cones $\mathcal{K}_*$. In this paper, however, we consider only self-dual cones $\mathcal{K}_* = \mathcal{K}$.

Since $\langle\sum_{i=1}^n Y_i, X\rangle = 0$, we can rewrite the objective function as $\sum_{i=1}^n \langle Y_i, A_i - X\rangle$. Using (9), we obtain:

$$\sum_{i=1}^n \|X - A_i\|_2 = \sum_{i=1}^n \langle Y_i, A_i - X\rangle$$

The constraints $\|Y_i\|_2 \le 1$ suggest as solution to (14):

$$Y_i = \frac{A_i - X}{\|A_i - X\|_2}\,, \qquad i = 1, \ldots, n$$

Inserting this into the constraint $\sum_{i=1}^n Y_i = 0$ yields the stationarity conditions of the original problem (1):

$$\sum_{i=1}^n \frac{X - A_i}{\|X - A_i\|_2} = 0 \tag{15}$$

Using this condition for the computation of X, however, leads to a non-trivial numerical optimisation problem, the need of choosing suitable damping parameters to achieve convergence, and differentiability problems in cases where the median X coincides with some data point A_i (in this case, the corresponding term in (15) is ill-defined, whereas Y_i in (14) is not). In contrast, all these problems can be avoided by the convex programming formulation presented above.

In order to compute the dual program to (11), we first have to clarify the meaning of $F^\top$ in (8) for the mapping $\mathcal{F}$ in (12). According to (12), the mapping $\mathcal{F}z = \sum_i z_i F_i$ defines elementary matrices F_i for each single variable $z_i = X_{j,k}$ or $z_i = t_j$. $F^\top$ in (8) is then given by the adjoint mapping[2] $\mathcal{F}^* Y = (\ldots, \langle F_i, Y\rangle, \ldots)^\top$. Computing the dual program to (11) then results – analogously to (12) and (13) – in a block-diagonal matrix of the dual variables:

$$Y = \mathrm{Diag}\{Y_1^-, \ldots, Y_n^-, Y_1^+, \ldots, Y_n^+\}\,,$$

and, using the definition

$$Y_i := Y_i^+ - Y_i^-\,, \quad \forall i\,,$$

to the optimisation problem:

$$\sup_{Y_i \in \mathcal{S}^d} \sum_{i=1}^n \langle Y_i, A_i\rangle\,, \quad \sum_{i=1}^n Y_i = 0\,, \quad \mathrm{tr}[Y_i^+ + Y_i^-] = 1\,, \quad Y_i \in S_+^d\,, \quad \forall i \tag{16}$$

Note the similarity of (16) and (14). Using the same reasoning as after (14), we obtain:

$$\sum_{i=1}^n |X - A_i| = \sum_{i=1}^n \langle Y_i, A_i - X\rangle$$

---

[2] $\langle F_i, Y\rangle$ denotes the matrix inner product $\mathrm{tr}[F_i^\top Y]$.

Again, the dual matrices Y_i seem to play the role of normalised gradients of the original objective function (4). Because the spectral norm $|\cdot|$ is non-smooth, it is not obvious how to make this more explicit. More important, however, are the computational advantages of the convex programming formulation presented above, as compared to directly optimising (4).

5 Experiments and Discussion

In our first experiment (Fig. 1) we demonstrate the capability of matrix-valued median filtering to remove outliers from structure tensor data. The photograph (a) shows a texture with randomly interspersed inhomogeneities. The outer products $\nabla u \nabla u^\top$ have been computed by 3×3 derivative-of-Gaussian (DoG) filters and smoothed with a 15×15 Gaussian mask. In (b) a subsampling of the resulting matrix field is shown. Outliers are removed from this matrix field by applying a 7×7 median filter (with Frobenius norm) as can be seen in (c).

Fig. 1. Left to right: (a) Image containing oriented texture with inhomogeneities. **(b)** Structure tensors computed by smoothing the outer products $\nabla u \nabla u^\top$ with 15×15 Gaussian. The gradients themselves have been calculated by 3×3 derivative-of-Gaussian filtering. The final matrix field has been subsampled for visualisation. **(c)** Result of median filtering of (b) with 7×7 structure element and Frobenius norm (subsampled)

Fig. 2. Left to right: (a) Synthetic image with oriented textures, inspired by [16]. **(b)** Local orientations computed via DoGs and visualised as grey-values. **(c)** Orientations after median filtering of the orientation matrices with Frobenius norm and a disk-shaped structure element of diameter 7. **(d)** Same with structure element of diameter 9. **(e)** Spectral norm median filtering, diameter 9

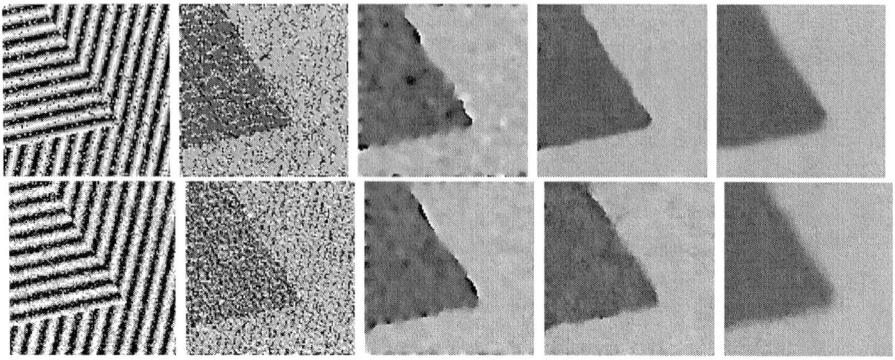

Fig. 3. Top, left to right: (a) Test image with 20 % impulse noise. **(b)** Orientation field of (a). **(c)** Structure tensor orientation obtained by Gaussian smoothing of the outer product matrices with standard deviation 19. **(d)** Same after median filtering with Frobenius norm and disk-shaped structure element of diameter 9. **(e)** Median filtering of (a) with Frobenius norm and disk-shaped structure element of diameter 19. **Bottom, left to right: (f)** Test image perturbed by Gaussian noise of standard deviation 0.2 (where grey-values vary between 0 and 1). **(g)** Orientation field of (f). **(h)** Structure tensor orientation as in (c). **(i)** Median filtering as in (d). **(k)** Median filtering as in (e)

In the following experiments we show the application of matrix-valued median filtering in the context of texture analysis. The synthetic test image in Fig. 2 (a) contains two oriented texture regions separated by a sharp edge. We compute the gradient ∇u at each pixel using a 3×3 DoG filter and the outer product matrix $\nabla u \nabla u^\top$ (of rank one) which estimates the local orientation. We visualise the orientations of the principal eigenvectors by mapping angles directly into grey-values (b). The direct transitions between black and white at image boundaries and along the texture edge are caused by the fact that black and white in fact represent orientations which are very close to each other because of the cyclic nature of angles. Median filtering of the outer product matrices yields new matrix fields. We visualise their orientation in the same way as before (c–e). Juxtaposing orientation fields obtained with Frobenius norm (d) and spectral norm (e) shows that the two distance measures yield no significantly different results. In the following we therefore restrict ourselves to the Frobenius norm.

Let us turn now to investigate orientation estimation in noisy images. Fig. 3 shows two noisy versions of the test image (Fig. 2 (a)) together with their local orientation estimates. Each orientation matrix field is then smoothed by matrix-valued median filtering. For comparison, we show also the orientation of the standard structure tensor obtained by Gaussian smoothing of the orientation matrices. While in (a–e) impulse noise is shown where the grey values at 20 % of all pixels have been replaced with random values from $[0, 1]$, images (f–k) show perturbation by Gaussian noise. While for impulse noise the median filter de-noises orientation better and also better preserves the discontinuity, the removal

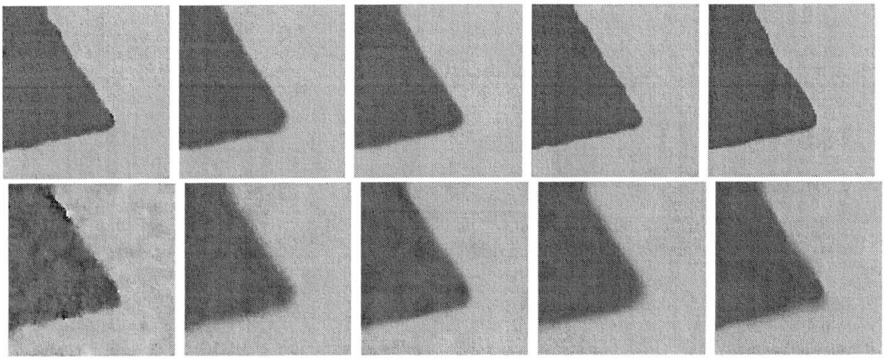

Fig. 4. Top, left to right: (a) Median filtering of local orientation derived from normalised gradients of Fig. 3 (a) with Frobenius norm and disk-shaped structure element of diameter 9. **(b)** As (a) but with structure element of diameter 19. **(c)** Median filtering without normalisation of gradients as in Fig. 3 (d), iterated four times. **(d)** Median filtering with normalisation, structure element of diameter 5, iterated five times. **(e)** Orientation estimate from Boomgaard–Weijer's robust structure tensor, parameters (see [16]) $m = 0.05$, $s = 5$. **Bottom, left to right:** Filtering of Fig. 3 (f). **(f)** Median filtering (Frobenius norm) with normalisation, structure element of diameter 9. **(g)** Same with diameter 19. **(h)** Filtering as in Fig. 3 (i), iterated five times. **(i)** Median filtering with normalisation, structure element of diameter 15, iterated four times. **(k)** Boomgaard–Weijer's robust structure tensor, $m = 0.05$, $s = 9$

for Gaussian noise is still less satisfactory. Increasing the size of the structure element reduces noise at the cost of blurring also the discontinuity and rounding corners, see Fig. 3 (e, k). We are therefore led to propose two modifications which improve the quality of the orientation estimation by median filtering in the case of noisy images.

The first modification is to normalise the gradients before computing the outer products and applying the median filter. This leads to a sharper representation of the discontinuity in the case of impulse noise as shown in Fig. 4 (a, b). With Gaussian noise, however, only a marginal improvement is achieved (f, g).

Our second modification is to iterate median filtering. While the improvement achieved for the impulse-noise image is comparable to that of the normalisation procedure, see Fig. 4 (c), it outperforms it in the case of Gaussian noise as shown in (h). Compared to a single median filtering step with the same structure element, corners are rounded slightly more but less than with a single step with larger structure element. The sharpness of the discontinuity is not reduced considerably compared to a single iteration while noise is removed more effectively.

Both presented modifications can be combined: In case of impulse noise, see Fig. 4 (d), the edges are sharpened and the corner is reconstructed more precisely. However in connection with Gaussian noise, as in Fig. 4 (d), this combination cannot improve the results.

Table 1. Average angular errors (AAE) for orientation estimation. Values in brackets are method-specific parameters: for median filtering, diameter of structure element and number of iterations; for Boomgaard–Weijer method, m and s (see [16])

Method	AAE undisturbed	AAE impulse noise	AAE Gaussian noise
gradient direction	3.387°	20.612°	31.429°
Frobenius median	1.591° (7, 1)	1.914° (9, 4)	3.207° (9, 5)
Frobenius median, norm.	1.312° (7, 1)	1.655° (5, 5)	3.434° (15, 4)
Boomgaard–Weijer	1.634° (0.1, 3)	1.489° (0.05, 5)	3.657° (0.05, 9)

The smoothing of outer product matrices by iterated median filtering can be interpreted as computation of a robust structure tensor. When computing classical structure tensors as in Fig. 3 (c, h), the outer product matrices are smoothed by Gaussian filtering, thus by linear diffusion. Nonlinear structure tensors as established by Weickert and Brox [17] use instead nonlinear diffusion to achieve a better representation of orientation discontinuities. The robust structure tensor introduced by van den Boomgaard and van der Weijer [16] smoothes the outer product matrices by minimising an energy in which a function ϱ is applied to matrix distances. In the case $\varrho(s) = s$, their robust structure tensor is similar to a single step of median filtering, with the difference that not a sharp structure element but Gaussian weights are used. In our filtering procedure, *iterated* matrix-valued median filtering takes the role of the smoothing process. This is primarily a change in theoretic perspective since it means that linear filtering is replaced by robust filtering more consequently. Orientation estimates obtained by Boomgaard and Weijer's method are shown in Fig. 4 (e, k). In Table 1, we compare the different orientation estimation methods by their average angular errors. As the experiments show, both types of robust structure tensors yield comparable results.

6 Conclusion and Further Work

In this paper, we have introduced a novel numerical algorithm for the computation of matrix-valued median filters which in their basic form have been introduced in [18], and for closely related mid-range filters. This algorithm is based on convex programming ideas. It uses interior-point techniques to compute the filtered matrices as global optima. Further, we have demonstrated the application of matrix-valued median filtering as a discontinuity-preserving denoising technique for orientation data obtained from grey-value images with oriented textures. It has become evident that median filtering of local orientation matrices is an attractive alternative to Gaussian–smoothed structure tensors. It also leads in a natural way to a concept of robust structure tensor in which matrix-valued median filtering takes the role of the smoothing process.

Future work will include the embedding of matrix-valued median filtering into texture segmentation procedures. Moreover, it will address a better understanding of the properties of the so defined type of robust structure tensor and its comparison to the already existing concepts of nonlinear and robust structure tensors.

References

1. J. Astola, P. Haavisto, and Y. Neuvo. Vector median filters. Proc. IEEE, vol. 78 no. 4, 678–689, 1990
2. T. L. Austin. An approximation to the point of minimum aggregate distance. Metron, 19, 10–21, 1959
3. M. Barni, F. Buti, F. Bartolini, and V. Cappellini. A quasi-Euclidean norm to speed up vector median filtering. IEEE Trans. Image Processing, vol. 9 no. 10, 1704–1709, 2000
4. T. Brox, M. Rousson, R. Deriche, J. Weickert. Unsupervised segmentation incorporating colour, texture, and motion. In N. Petkov, M. A. Westenberg, editors, *Computer Analysis of Images and Patterns,* vol. 2756 of *Lecture Notes in Computer Science,* pages 353–360, Berlin, 2003, Springer
5. S. Boyd, and L. Vandenberghe. *Convex Optimization.* 2004, Cambridge University Press
6. T. Brox, J. Weickert. Nonlinear matrix diffusion for optic flow estimation. In L. Van Gool, editor, *Pattern Recognition,* vol. 2449 of *Lecture Notes in Computer Science,* pages 446–453, Berlin, 2002, Springer
7. V. Caselles, G. Sapiro, and D. H. Chung. Vector median filters, inf-sup operations, and coupled PDE's: Theoretical Connections. J. Mathematical Imaging and Vision 8, 109–119, 2000
8. W. Förstner and E. Gülch. A fast operator for detection and precise location of distinct points, corners and centres of circular features. In *Proc. ISPRS Intercommission Conference on Fast Processing of Photogrammetric Data,* pages 281–305, Interlaken, Switzerland, June 1987
9. S. P. Fekete, J. S. B. Mitchell, K. Beurer. On the continuous Fermat–Weber problem. To appear in Operations Research. Preprint: http://arxiv.org/abs/cs/0310027 (last visited: Sept. 27, 2004)
10. G. H. Granlund and H. Knutsson. *Signal Processing for Computer Vision.* Kluwer, Dordrecht, 1995
11. J. Keuchel, C. Schnörr, C. Schellewald, and D. Cremers. Binary partitioning, perceptual grouping, and restoration with semidefinite programming. IEEE Trans. Pattern Analysis and Machine Intelligence, vol. 25, no. 11, pages 1364–1379, 2003
12. N. Megiddo. The weighted Euclidean 1-center problem. Mathematics of Operations Research, vol. 8, no. 4, pages 498–504, 1983
13. C. Pierpaoli, P. Jezzard, P. J. Basser, A. Barnett, and G. Di Chiro. Diffusion tensor MR imaging of the human brain. *Radiology,* 201(3) pages 637–648, Dec. 1996
14. M. Rousson, T. Brox, R. Deriche. Active unsupervised texture segmentation on a diffusion based feature space. Technical report no. 4695, Odyssée, INRIA Sophia-Antipolis, France, 2003
15. D. R. Seymour. Note on Austin's "An approximation to the point of minimum aggregate distance". Metron, 28, 412–421, 1970

16. R. van den Boomgaard, and J. van de Weijer. Least squares and robust estimation of local image structure. In L. D. Griffin and M. Lillholm, editors, *Scale Space Methods in Computer Vision,* volume 2695 of *Lecture Notes in Computer Science,*, pages 237–254, Berlin, 2003, Springer
17. J. Weickert, T. Brox. Diffusion and regularization of vector- and matrix-valued images. In M. Z. Nashed and O. Scherzer, editors, *Inverse Problems, Image Analysis, and Medical Imaging,* vol. 313 of *Contemporary Mathematics,* pages 251–268, Providence, 2002, AMS
18. M. Welk, C. Feddern, B. Burgeth, and J. Weickert. Median filtering of tensor-valued images. In B. Michaelisi and G. Krell, editors, *Pattern Recognition,* volume 2781 of *Lecture Notes in Computer Science,*, pages 17–24, Berlin, 2003, Springer

Retinex by Two Bilateral Filters

Michael Elad

The Computer Science Department, The Technion - Israel
Institute of Technology, Haifa 32000 Israel
elad@cs.technion.ac.il

Abstract. Retinex theory deals with the removal of unfavorable illumination effects from images. This ill-posed inverse problem is typically regularized by forcing spatial smoothness on the recoverable illumination. Recent work in this field suggested exploiting the knowledge that the illumination image bounds the image from above, and the fact that the reflectance is also expected to be smooth. In this paper we show how the above model can be improved to provide a non-iterative retinex algorithm that handles better edges in the illumination, and suppresses noise in dark areas. This algorithm uses two specially tailored bilateral filters – the first evaluates the illumination and the other is used for the computation of the reflectance. This result stands as a theoretic justification and refinement for the recently proposed heuristic use of the bilateral filter for retinex by Durand and Dorsey. In line with their appealing way of speeding up the bilateral filter, we show that similar speedup methods apply to our algorithm.

1 Introduction

Retinex theory deals with the removal of unfavorable illumination effects from a given image. A commonly assumed model suggests that any given image S is the pixel-wise multiplication of two images, the reflectance R and the illumination L, i.e., $S = R \cdot L$. A look-up-table log operation transfers this multiplication into an addition, resulting with $s = \log(S) = \log(L) + \log(R) = \ell + r$. Clearly, the recovery of ℓ from s is an ill-posed inverse problem. Solving it is typically done by introducing a regularization that forces a spatial smoothness on the recoverable illumination. Thus, early heuristic and successful retinex methods, such as the homomorphic filtering algorithm [1] and many others (e.g., [2, 3, 4, 5]), proposed a low-pass filter on s, or an algorithm that amounts to this effect, to obtain a rough estimate of ℓ. In this paper we refrain from reviewing this literature, and limit our approach to this topic by building upon a recent study presented in [6].

The work in [6] describes several improvements to the classical retinex models. One improvement refers to the passivity of the reflectance, assumed to satisfy $0 \leq R \leq 1$. As a direct consequence we have that $L \geq S$, implying that the illumination image should be an envelope image bounding S from above. Due to the monotonicity of the log operation we have $\ell \geq s$. Merging the above with the desire to get spatially smooth ℓ may lead to the trivial and meaningless result of

R. Kimmel, N. Sochen, J. Weickert (Eds.): Scale-Space 2005, LNCS 3459, pp. 217–229, 2005.

a constant image, $\ell = \max(s)$. The remedy, as proposed in [6], is to assume that $\|\ell - s\|_2$ should be small, implying that ℓ should upper envelope s while being close to it. Based on these modifications, the reconstruction of the illumination can be posed as the following quadratic programming (QP) problem

$$\min_{\ell \geq s} \lambda \|\ell - s\|_2^2 + \left\{ \|\mathbf{D}_x \ell\|_2^2 + \|\mathbf{D}_y \ell\|_2^2 \right\}. \tag{1}$$

The operators $\mathbf{D}_x$ and $\mathbf{D}_y$ represent horizontal and vertical discrete derivatives, forcing this way spatial smoothness.

A second ingredient introduced in [6, 7] is a smoothness penalty forced also on the reflectance image $r = s - \ell$. This added to (1) gives

$$\min_{\ell \geq s} \lambda \|\ell - s\|_2^2 + \left\{ \|\mathbf{D}_x \ell\|_2^2 + \|\mathbf{D}_y \ell\|_2^2 \right\} + \alpha \left\{ \|\mathbf{D}_x (s - \ell)\|_2^2 + \|\mathbf{D}_y (s - \ell)\|_2^2 \right\} \tag{2}$$

Note that, since $s = \ell + r$ is enforced, the new term contradicts the illumination smoothness, as r and ℓ cannot be jointly smooth. Thus, the effect is to gain some smoothness in r at the expense of losing some of it in ℓ. The justification for this is the desire to lead r to be "nice-looking", as natural images should be.

Based on the above model, an efficient multi-scale algorithm has been proposed in [6] to estimate ℓ and thus r. The work in [7] used the same model to propose a simplified estimate solvers based on known implementation constraints. More recently, [8] further simplified the computation of ℓ by introducing a spatial recursive smoothing filter.

While the above model is general enough and covers the correct forces to be used in the solution of the retinex problem, it has several flaws:

- **Hallows:** A commonly encountered artifact with retinex algorithms is the existence of hallows. This is a direct consequence of the smoothness assumption discussed above. When passing from a strongly illuminated region to a dark zone (e.g., on a border of a shadowed area), the smoothness forces the illumination to remain high in the dark region near this edge and smoothly descend to grasp the illumination within the dark region. Thus, when removed, the dark regions near such edges remain dark, resulting with these hallow effects. Such effects can be also obtained in the bright areas near such illumination edges, if the constraint $\ell \geq s$ is not practiced. Then those bright areas become further brighter.
- **Noise:** In dark regions of the image these retinex algorithms are expected to yield a contrast stretching, very much similar to the effect caused by standard Gamma correction. The stretching causes a magnification of the noise, and this becomes evident especially in low-quality images, or ones with noticeable compression. The constraint $s = \ell + r$ implies that the noise migrates as a whole to the two ingredients, rather than being suppressed.
- **Iterative Solution:** The above model formulation leads naturally to the need for an iterative solver. The work in [7] and [8] bypassed this limitation, but with a price on the final outcome's quality.

In this paper we propose an alternative model for retinex, and a numerical algorithm that builds on it. The new model is similar to the one in (2), in the sense that all the presented forces are included. However, this new model is enhanced to solve the above mentioned shortcomings.

More specifically, smoothness of the illumination and the reflectance are forced using a robust statistics method, and hallows are avoided. The smoothness terms used are very much in the spirit of the bilateral filter [9, 10], having a wide stencil effect that enables avoiding the need for an iterative or multi-scale solver. We use different smoothing formulation for the reflectance and the illumination to handle them differently, and absorb the constraint $\ell \geq s$ in a natural way. This leads to a two-stage algorithm that applies two variations of the bilateral filter, first estimating the illumination, and then the reflectance. The new model suppresses noise by allowing $\ell + r$ to deviate from s, implying that the residual should be the additive noise we want to discard of. The new model and accompanying algorithm stand as a theoretic justification and refinement for the recently proposed heuristic use of the bilateral filter for retinex as appeared in [11]. In line with their appealing way of speeding up the bilateral filter, we show similar speedup methods for our two bilateral filter variations.

This paper is organized as follows: Section 2 presents the bilateral filter that this work is building on. Section 3 then turn to describe the new model for retinex, and the algorithm that emerges from it. Speedup methods are discussed in Section 4. Section 5 presents some results, and Section 6 concludes this paper.

2 Denoising by the Bilateral Filter

In this Section we present the bilateral filter, designed for the removal of additive noise from images [9]. We also discuss its origins as described in [10, 11, 12, 13]. These will serve us as we turn later to consider the retinex problem.

Consider an image s contaminated by additive noise. Our goal is to develop an edge-preserving smoothing algorithm that effectively removes most of the noise while preserving the image details. A maximum a-posteriori probability (MAP) formulation of this problem as presented in [10] yields

$$\min_{\hat{s}} \lambda \|\hat{s} - s\|_2^2 + \sum_{m=-P}^{P} \sum_{n=-P}^{P} \left(\mathbf{C}_{m,n}\hat{s} - \hat{s}\right)^T \mathbf{W}_{[m,n]}(s) \left(\mathbf{C}_{m,n}\hat{s} - \hat{s}\right). \quad (3)$$

The operators $\mathbf{C}_{m,n}$ are shift operators, moving the image $\hat{s}$ by m pixels horizontally and n pixels vertically. The matrices $\mathbf{W}_{[m,n]}$ are diagonal matrices that down-weight large edge entries in s so as not to smooth over edges of the image. The choice $\mathbf{W}_{[m,n]}(s) = \mathbf{I} \; \forall \; m, n$ leads to the non-robust option that makes the overall problem QP as in (2). Choosing these weights to be inversely proportional to $|\mathbf{C}_{m,n}s - s|$ leads to the ability to handle edges in the image better. Note that using weighting here parallels the use of robust statistics - more on this relationship and can be found in [10, 12].

decomposition that seeks both ℓ and r as unknown, and one does not imply the other as before.

Instead of optimizing with respect to both ℓ and r in parallel (which is an option we have not explored in our work, but one that can certainly be addressed based on the model we have posed), we adopt a two stage process, first estimating ℓ, based on the first part in (11), and then given ℓ, we evaluate r.

Starting with the quest for ℓ, let us attempt to evaluate it such that it addresses only the first term in (11). Thus, we seek a solution to the problem

$$\min_{\ell:\ \ell \geq s} \quad \lambda_\ell \cdot \|\ell - s\|_2^2 + \mathbf{B}_{\mathbf{W}_\ell, P_\ell}\{\ell\}. \tag{12}$$

Clearly, without the constraint $\ell \geq s$, the above is equivalent to the problem posed in (3), and as such, the bilateral filter is an excellent solver candidate. Thus, the natural question we should pose here is how the constraint should be accommodated, in a way that preserves the convenience of the bilateral filter. We propose to introduce a special choice of weights $\mathbf{W}_\ell$ that handle the constraint implicitly. These new weights are based on Equation (7), but using a one-sided robust function ρ,

$$\rho(x) = \begin{cases} 1 - \exp(-x^2/2\sigma_r^2) & x \leq 0 \\ \infty & x > 0 \end{cases}. \tag{13}$$

This alternative choice of weights introduces a simple modification to the bilateral filter, where, among the $(2P+1)$-by-$(2P+1)$ neighbors per each pixel, we consider only those that satisfy $s[k,j] \leq s[k-m, j-n]$. This way, the local averaging is done with non-negative normalized weights, while combining only pixels that have higher gray values than the center pixel, resulting with a final outcome that must satisfy $\ell_1[k,j] \geq s[k,j]$. Thus, this new filter will necessarily achieve both a satisfaction of the constraint (by virtue of the weights), while reducing the newly defined penalty term that still considers smoothness as we desire. We refer hereafter to this filter as the envelope-bilateral filter.

In practice, the above implies that the bilateral filter as presented in section 2 is slightly changed. Parallel to (8) and (9), in the envelope-bilateral filter every output pixel $\ell_1[k,j]$ is evaluated by

$$\ell_1[k,j] = \sum_{m=-P}^{P} \sum_{n=-P}^{P} a[m,n,k,j] s[k-m, j-n], \tag{14}$$

where

$$a[m,n,k,j] = \frac{\exp\left(-\frac{m^2+n^2}{2\sigma_s^2} - \frac{(s[k,j]-s[k-m,j-n])^2}{2\sigma_r^2}\right) \cdot \mu\{s[k-m,j-n] - s[k,j]\}}{Z[k,j]}.$$

The notation $\mu\{x\}$ stands for the step-function, being 1 for non-negative x and zero elsewhere. The term $Z[k,j]$ normalizes these weights to sum to one, as before. Note that from the above description it is clear that if $s[k,j]$ is the peak of its $(2P+1)^2$ neighborhood, then its filtering amounts to $\ell_1[k,j] = s[k,j]$, since in this case all weights are zero and only $a[0,0,k,j] = 1$.

Assuming that the above stage has been completed, we have an estimate of ℓ and we now turn to evaluate r. We consider the second term in (11), solving

$$\min_r \lambda_r \cdot \|r - (s-\ell)\|_2^2 + \mathbf{B}_{\mathbf{W}_r, P_r}\{r\}. \tag{15}$$

Since the image $s - \ell$ is given, this is the very bilateral filter formulation in (2). Thus, an application of the bilateral filter on the image $s - \ell$ should lead to the desired r. However, due to the transform to the log-domain, the noise that should be discarded from the reflectance image resides mostly in the regions where s is low. Thus, we can better direct the above bilateral filter by using σ_r to be inversely proportional to s to reflect this matter. A choice of the form $\sigma_r[k,j] = (C_1 \cdot s[k,j]^p + C_2)^{-1}$ could be used to this effect[2]. Nothing in the definition of the implementation of the bilateral filter prevents having such spatially adaptive parameter. This will ensure that r is hardly smoothed in regions where s is bright, while it is being smoothed in darker regions.

4 Speeding Up the Retinex Algorithm

In their paper, Durand and Dorsey proposed a wonderful speedup algorithm for the bilateral filter, and this algorithm can be applied directly to both our two bilateral filter versions. Here we outline the basic ideas of this speedup, starting from Equation (14), although everything said applies just as well to the second bilateral filter.

Referring to $s[k,j]$ in these equations as a constant c, we can re-write these equations as

$$\begin{aligned}\ell_1[k,j] &= \frac{1}{Z[k,j]} \cdot \sum_{m=-P}^{P} \sum_{n=-P}^{P} \exp\left(-\frac{m^2+n^2}{2\sigma_s^2}\right) \cdot \\ &\cdot \left[\exp\left(-\frac{(c - s[k-m,j-n])^2}{2\sigma_r^2}\right) \cdot \mu\{s[k-m,j-n]-c\} s[k-m,j-n]\right] \\ &= \frac{1}{Z[k,j]} \cdot \sum_{m=-P}^{P} \sum_{n=-P}^{P} \exp\left(-\frac{m^2+n^2}{2\sigma_s^2}\right) \cdot g[k-m,j-n].\end{aligned} \tag{16}$$

This expression is a convolution between the image $g[k,j]$, being

$$g[k,j] = \left[\exp\left\{-\frac{(c-s[k,j])^2}{2\sigma_r^2}\right\} \cdot \mu\{s[k,j]-c\} s[k,j]\right], \tag{17}$$

and the Gaussian blur. Thus, we could apply a sequence of such convolutions, scanning the values of $s[k,j]$ in the range $[0, \log_e 255]$, and then merging the

[2] Recall that $S[k,j] \in [1, 255]$ as we shift by 1 to avoid singularities, and we have also $0 \le s[k,j] \le 5.54$.

results, choosing the proper values from each output, based on the $s[k,j]$ values. Note that the value of $Z[k,j]$ is given by a similar expression

$$Z[k,j] = \sum_{m=-P}^{P} \sum_{n=-P}^{P} \exp\left(-\frac{m^2+n^2}{2\sigma_s^2}\right) u[k-m, j-n], \tag{18}$$

where

$$u[k,j] = \exp\left\{-\frac{(c-s[k,j])^2}{2\sigma_r^2}\right\} \cdot \mu\{s[k,j]-c\}. \tag{19}$$

Thus, its computation can also be done using a sequence of similar convolutions.

Durand and Dorsey proposed two ways to further speed-up the evaluation of ℓ_1: (i) piece-wise linear approximation; and (ii) multi-scale implementation. The first idea is to scan the values of $s[k,j]$ in the range $[0, \log_e 255]$ with jumps, and interpolate in between. Practically speaking, using $30-50$ equispaced jumps in the range $[0, \log_e 255]$ are found to induce almost no change to the outcome. Since our weights include a step-function discontinuity, the interpolation should be done as a one-sided operation, always preferring to adopt the larger c to avoid a violation of the $\ell \geq s$ constraint. This causes the interpolation to loose some of of its accuracy, but our experiments show that this lose is mild and unnoticeable.

As to the multi-scale option, since images are convolved in the above expressions with wide-range Gaussian smoothers, a pre down-scale and post up-scale yield a substantial gain in run-time with almost no change in the outcome. The gain is especially noticed for wide supports ($P \gg 1$, and $\sigma_s \gg 1$). On top of these two ideas, note that the required convolutions required are all separable. Furthermore, when σ_s is large enough, the effective convolving kernel is the square step function. In such a case further speedup can be obtained using the computation of the integral image [17].

5 Results

An interesting idea reported in [6] is to return some of the illumination to the reflectance when presenting the final output image. Thus, the output image is computed as $\text{Out} = R[k,j] \cdot L[k,j]^{1/\gamma} = S[k,j] \cdot L[k,j]^{1/\gamma - 1}$. Reflectance images are typically unrealistic looking, and with a modest and reduced effect of illumination returned to it, the final image enjoys both the desired brightness and the natural appearance. We have made use of this idea in the following presented results. The illumination is returned to the original image by applying Gamma-correction on it using $\gamma = 3$, and multiplying it back by the estimated reflectance.

Figures 1-2 present two pairs of original images[3] and their retinex results. In these two cases, the use of spatially varying σ_r has very little effect because the

[3] These images and the one in Figure 3 are from the NASA retinex web-page.

Fig. 1. Example 1 - An original image (left) and its retinex result (right)

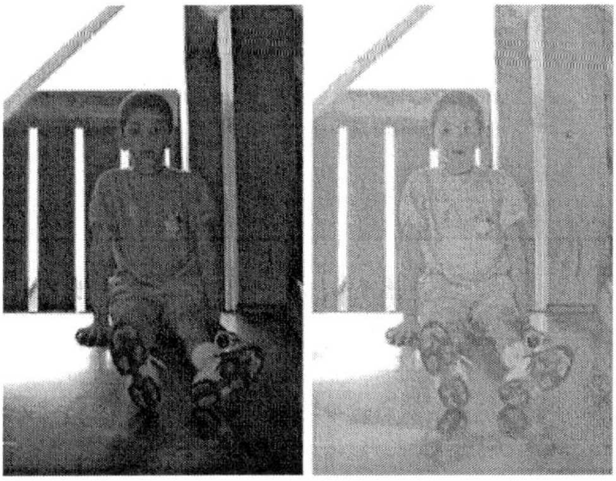

Fig. 2. Example 2 - An original image (left) and its retinex result (right)

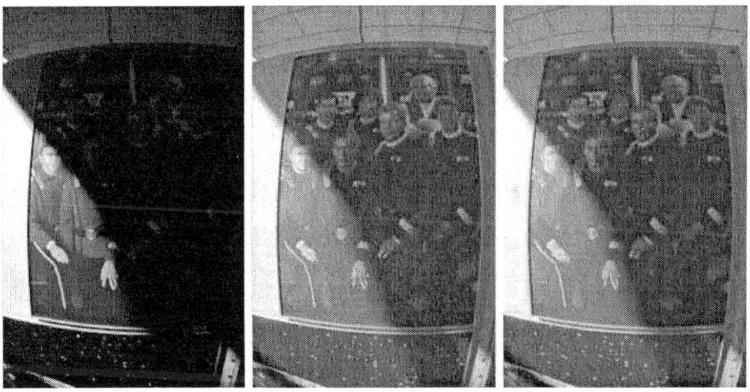

Fig. 3. Example 3 - An original image (left) and its retinex results using a regular bilateral filter for computing r (middle) and using the spatially adaptive σ_r (right)

images are of high quality, and thus we do not show it. Figure 3 presents results for a third image, where the dark region is noisy, and thus the two versions are shown side-by-side for comparison (with parameters $p = 8$, $C_1 = 5e -$

Fig. 4. The reflectance and the illumination images in Example 1

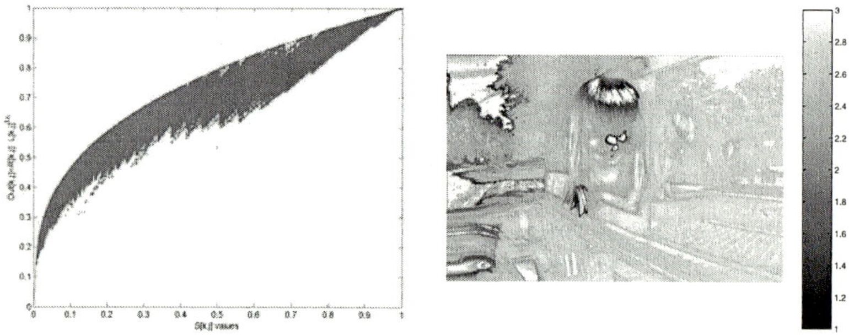

Fig. 5. Example 1 - Right: The Input-Output mapping, and overlayed on it is the Gamma-correction that corresponds to $\gamma = 3$; Left: The effective Gamma correction value per pixel

3, $C_2 = 0.3$). We should note that in processing color images we apply the retinex algorithm to the luminance (V) layer in the HSV color representation, and leave the chromatic layers unchanged. In all three cases we used the following setup parameters: the envelope bilateral filter parameters used are $P_\ell = 15$, $\sigma_r = 0.3$, and $\sigma_s = 100$. The second bilateral filter used $P_r = 4$, $\sigma_r = 0.3$ or an adaptive method as described earlier, and $\sigma_s = 100$. The speedup algorithm was used with scale down factor of 2 : 1, and grey-value steps of 0.1.

Figures 4 and 5 return to the first example, presenting several accompanying results. Figure 4 shows the obtained reflectance and illumination results (gray-value images referring to the V-layer). As can be seen, what we call 'reflectance' is far from being satisfactory to describe the image, and indeed there is room to return of illumination. This Figure indicates that our separation is not perfect and there is a leakage between r and ℓ. In fact, our ℓ are r stand for large scale intensity components and small scale corrections, respectively, both estimated with preservation of discontinuities. Still, the final outcome is satisfactory because of the illumination return.

Fig. 6. An original image (top), the new retinex algorithm (bottom left), and the one reported in[6] (bottom right)

Figure 5 describes the Out-to-In correspondence of the overall retinex algorithm, showing that while the retinex process generally resembles a Gamma-correction effect, is has a different effect of varying Gamma. This idea is further expanded, showing in Figure 5 the image $\gamma_{\text{effective}} = \log In / \log Out$ as a function of the location. This gives the effective Gamma correction that should be applied in every pixel to reproduce the obtained result.

The overall improvement in speed introduced by the speedup algorithm depends on many of the parameters that are mentioned above, and on implementation issues. We compared two efficient implementations of the bilateral filter - both implemented with Matlab. The first sweeps through the support of the filter, applying operations on complete images, and the other being the speedup algorithm mentioned above. For the parameters used here we obtained a factor of $5 - 10$ shorter run time with the speedup algorithm.

Figure 6 presents a comparison between the new algorithm and the one reported in [6] on a severely degraded image[4]. For this comparison we changed the color space to YCbCr, and choose $\gamma = 2.3$ in the illumination return, both done to match with the alternative algorithm. The results show strong hallows in the previous method, while those are fully suppressed by our algorithm.

[4] Curtesy of Eyal Gordon, The CS department - The Technion.

6 Conclusion

In this paper we have presented a new model for the retinex problem – removal of undesired illumination effects from an image. The new model enables a better handling of edges in the illumination that causes hallow effects, and it enables the suppression of noise in dark areas. An algorithm based on this model has been developed, leading to two specially tailored bilateral filters, the first evaluates the illumination and the second is used for the computation of the reflectance. Our work stands as a theoretic justification and refinement for the recently proposed heuristic use of the bilateral filter for retinex by Durand and Dorsey. We have used their way of speeding up the bilateral to propose a similar speedup methods for our filters.

References

1. Faugeras, O.D. (1979) "Digital image color processing within the framework of a human visual system", *IEEE Trans.on ASSP*, 27: 380–393.
2. Land, E.H. (1983) "Recent advances in the Retinex theory and some implications for cortical computations: Color vision and the natural image", *Proc. Nat. Acad. Sci. USA*, 80: 5163–5169.
3. Blake, A. (1985)"Boundary conditions of lightness computation in mondrian world", *Computer Vision Graphics and Image Processing*, 32: 314–327.
4. Terzopoulos, D. (1986) "Image analysis using multigrid relaxation methods", *IEEE Trans. on PAMI*, 8: 129–139.
5. Funt, B., Ciurea, F., and McCann, J. (2004) Retinex in Matlab, *Journal of the Electronic Imaging*, 13(1):48-57.
   Terzopoulos, D. (1986) "Image analysis using multigrid relaxation methods", *IEEE Trans. on PAMI*, 8: 129–139.
6. Kimmel, R., Elad, M., Shaked, D., Keshet, R., and Sobel. I. (2003) "A variational framework for retinex", *International Journal of Computer Vision* 52(1): 7–23.
7. M. Elad, M., Kimmel, R., Shaked, D., and Keshet, R. (2003) "Reduced complexity retinex algorithm via the variational approach", *Journal on Visual Communication and Image Representation*, 14(4): 369–388.
8. Shaked, D. and Keshet, R. (2002) "Robust recursive envelope operators for fast retinex", *Hewlett-Packard Research Laboratories Technical Report*, HPL-2002-74R1.
9. Tomasi, C. and Manduchi, R. (1998) "Bilateral filtering for gray and color images", *Proc. 6th Int. Conf. Computer Vision*, New Delhi,India, 839–846.
10. Elad, M. (2002) "On the bilateral filter and ways to improve it", *IEEE Trans. On Image Processing*, 11(10): 1141–1151.
11. Durand, F. and Dorsey, J. (2002) "Fast bilateral filtering for the display of high-dynamic-range images", *SIGGRAPH* 2002: 257–266.
12. Barash, D. (2001) "A fundamental relationship between bilateral filtering, adaptive smoothing, and the nonlinear diffusion equation", *IEEE Trans. Pattern Anal. Mach. Intell.*, 24(6): 844–847.
13. Mrazek, P., Weickert, J., and Bruhn, A. (2005) On robust estimation and smoothing with spatial and tonal kernels. In *Geometric Properties from Incomplete Data*. Kluwer, Dordrecht, 2005, to appear.

14. Sochen, N., Kimmel, R., and Malladi, R. (1998) "A geometrical framework for low level vision", *IEEE Trans. Image Processing*, 7: 310–318.
15. Sochen, N., Kimmel, R. and Bruckstein, A.M. (2001) "Diffusions and confusions in signal and image processing", *J. Math. Imag. Vis.*, 14(3): 195–209.
16. Chan, T.F., Osher, S., and Shen, J. (2001) "The digital TV filter and nonlinear denoising", *IEEE Trans. Image Processing*, 10: 231–241.
17. Viola, P. and Jones, M. (2001) "Rapid object detection using a boosted cascade of simple features", *Conference on Computer Vision and Pattern recognition.*

Estimation of the Optimal Variational Parameter via SNR Analysis

Guy Gilboa[1,*], Nir A. Sochen[2,**], and Yehoshua Y. Zeevi[3,***]

[1] Department of Mathematics, UCLA, Los Angeles, CA 90095, USA
[2] Department of Applied of Mathematics, Tel-Aviv Univ., Tel-Aviv 69978, Israel
[3] Department of Electrical Engineering, Technion, Haifa 32000, Israel

Abstract. We examine the problem of finding the optimal weight of the fidelity term in variational denoising. Our aim is to maximize the signal to noise ratio (SNR) of the restored image. A theoretical analysis is carried out and several bounds are established on the performance of the optimal strategy and a widely used method, wherein the variance of the residual part equals the variance of the noise. A necessary condition is set to achieve maximal SNR. We provide a practical method for estimating this condition and show that the results are sufficiently accurate for a large class of images, including piecewise smooth and textured images.

1 Introduction

Variational methods have been increasingly applied for purposes of image denoising and restoration (for some examples see [3, 6, 8, 11, 12]). The basic concept is to view the restoration process as a task of energy minimization. Classically, the restored image is a minimization of a weighted sum of two fundamental energy terms:

$$E(u) = E_{smooth}(u) + \lambda E_{fidelity}(u, f), \tag{1}$$

where u is the restored image, and f is the input (noisy) image. E_{smooth} is a smoothing term which rewards smooth signals and penalizes oscillatory ones. $E_{fidelity}$ accounts for fidelity, or closeness, to the input image f. The underlying assumption is that the original clean image is smoother than the noisy image. By minimizing both terms we seek a compromise between a smooth solution (often in the TV sense, so edges are preserved) and one which is "close enough" to the original image. Any minimization of one of the terms by itself

* Supported by grants from the NSF under contracts ITR ACI-0321917, DMS-0312222, and the NIH under contract P20 MH65166.

** Supported by MUSCLE, a European Network of Excellence funded by the EC 6th Framework IST Programme, the Israeli Ministry of Science, the Israel Science Foundation, the Tel-Aviv University fund and the Adams Center.

*** Supported by the Ollendorf Minerva Center, the Fund for the Promotion of Research at the Technion and the Israel Academy of Science.

R. Kimmel, N. Sochen, J. Weickert (Eds.): Scale-Space 2005, LNCS 3459, pp. 230–241, 2005.

leads to degenerate solutions which are not interesting (a constant or the input noisy image). The appropriate compromise then highly depends on λ, the weight parameter between these two energies. When it is too low, the restored image is over-smoothed. When it is too high, u still contains too much noise. Finding the right value of λ for the problem at hand is therefore imperative. A similar problem has been investigated in regularization theory, in the context of operator inversion by Tikhonov-type methods (e.g. [4, 9]). As we are concerned with denoising of images (therefore our operator is the identity and the regularization preserves edges), different approaches should be used. In our field of PDE-base image processing, the problem was seriously addressed by only a few researchers: by [11] for total-variation denoising and by [7] and [13] for a closely related problem of finding the right stopping time in nonlinear scale-space. We refer in this paper only to the variational setting, but our method has shown to be very effective also for selecting the proper stopping time [5].

An analysis of the optimal parameter choice from SNR perspective is presented. We examine the widely used denoising strategy of [11] where the weight of the fidelity term is set such that the variance of the residual part equals that of the noise. Lower bound on the SNR performance of this strategy is established as well as a proof of non existence of an upper bound. Examples which illustrate worst- and best-case scenarios are presented and discussed.

Next, we derive a necessary condition for optimality in the SNR sense. From a theoretical viewpoint, this facilitates the computation of upper and lower bounds of the optimal strategy. From a practical viewpoint, the condition suggests the numerical method that should be followed for the purpose of maximizing the SNR of the filtered image. An algorithm for parameter calculation is suggested based on the above condition, resulting in fairly accurate estimates.

2 SNR Bounds for the Scalar Φ Process

2.1 Denoising Model, Definitions and Assumptions

We assume that the input signal f is composed of the original signal s and additive uncorrelated noise n of variance σ^2. Our aim is to find a decomposition u, v such that u approximates the original signal s and v is the residual part of f:

$$f = s + n = u + v. \tag{2}$$

We accomplish that by finding the minimum to the following energy

$$\tilde{E}_\Phi(u) = \int_\Omega \left(\Phi(|\nabla u|) + \tilde{\lambda}(f-u)^2\right) d\Omega. \tag{3}$$

Φ is assumed to be convex in this paper. Some of the following results, though, can also apply to the more general case of monotonically increasing Φ. The standard condition $\int_\Omega f d\Omega = \int_\Omega u d\Omega$ is set, (corresponding to the Neumann boundary condition of the evolutionary equations). Then $\int_\Omega v dx dy = 0$, rescaling $\tilde{\lambda}$ by the area of the domain $|\Omega|$: $\lambda = \tilde{\lambda}|\Omega|$, we get

$$E_\Phi(u,v) = \int_\Omega \Phi(|\nabla u|)d\Omega + \lambda V(v), \quad f = u + v. \tag{4}$$

where $V(q)$ is the variance of a signal q: $V(q) \doteq \frac{1}{|\Omega|}\int_\Omega (q-\bar{q})^2 d\Omega$, and $\bar{q}$ is the mean value: $\bar{q} \doteq \frac{1}{|\Omega|}\int_\Omega q d\Omega$. The covariance of two signals is defined as: $\text{cov}(q,r) \doteq \frac{1}{|\Omega|}\int_\Omega (q-\bar{q})(r-\bar{r})d\Omega$. We remind the identity $V(q+r) = V(q) + V(r) + 2\text{cov}(q,r)$.

Let us denote u^z as the solution of (4) for $f = z$. For example, u^s is the solution where $f = s$. The decorrelation assumption is taken also between s and n with respect to the Φ process:

$$\text{cov}(u^s, n) = 0, \quad \text{cov}(u^n, s) = 0, \quad \forall \lambda \geq 0. \tag{5}$$

We further assume the Φ process applied to $f = s + n$ does not amplify or sharpen either s or n. This can be formulated in terms of covariance as follows:

$$\text{cov}(u^{s+n}, s) \leq \text{cov}(f, s), \quad \text{cov}(u^{s+n}, n) \leq \text{cov}(f, n), \quad \forall \lambda \geq 0. \tag{6}$$

Both of the above assumptions were verified numerically on a collection of natural images. We are investigating the possibility to characterize in an analytical manner the appropriate spaces of s and n such that (5) and (6) are followed. In this paper this question is left open and we resort to the following definition:

Definition 1 ((s,n) pair). *An (s,n) pair consists of two uncorrelated signals s and n which obey conditions (5) and (6).*

Theorem 1. *For any (s,n) pair and an increasing Φ ($\Phi'(q) > 0, \forall q \geq 0$) the covariance matrix of $U = (f,s,n,u,v)^T$ has only non-negative elements.*

For proof see the appendix. Theorem 1 implies that the denoising process has smoothing properties and consequently, there is no negative correlation between any two elements of U. This basic theorem will be later used to establish several bounds in our performance analysis.

We define the Signal-to-Noise Ratio (SNR) of the recovered signal u as

$$\text{SNR}(u) \doteq 10\log\frac{V(s)}{V(u-s)} = 10\log\frac{V(s)}{V(n-v)}, \tag{7}$$

where $\log \doteq \log_{10}$. The initial SNR of the input signal, denoted by SNR_0, where no processing is carried out ($u = f$, $v = 0$), is according to (7) and (2):

$$SNR_0 \doteq \text{SNR}(f) = 10\log\frac{V(s)}{V(n)} = 10\log\frac{V(s)}{\sigma^2}. \tag{8}$$

Let us define the optimal SNR of a certain Φ process applied to an input image f as:

$$SNR_{opt} \doteq \max_\lambda \text{SNR}(u_\lambda) \tag{9}$$

where $u = u_\lambda$ attains the minimal energy of (4) with weight parameter λ (for a given f, v is implied). We denote by (u_{opt}, v_{opt}) the decomposition pair (u, v) that reaches SNR_{opt}, and define $V_{opt} \doteq V(v_{opt})$.

Equivalently, the desired variance could be set as $V(v) = P$, where P is some constant, and then (4) is reformulated to a constrained convex optimization problem

$$\min_u \int_\Omega \Phi(|\nabla u|) d\Omega \text{ subject to } V(v) = P. \tag{10}$$

In this formulation λ is viewed as a Lagrange multiplier. The value λ can be computed using the Euler-Lagrange equations and the pair (u, v):

$$\lambda = \frac{1}{P} \int_\Omega \text{div} \left(\Phi' \frac{\nabla u}{|\nabla u|} \right) v d\Omega. \tag{11}$$

The problem then transforms to which value P should be imposed.

The strategy of [11] is to assume $v \approx n$ and therefore impose

$$V(v) = \sigma^2. \tag{12}$$

We define

$$SNR_{\sigma^2} \doteq \text{SNR}(u)|_{V(v)=\sigma^2}. \tag{13}$$

We denote by $(u_{\sigma^2}, v_{\sigma^2})$ the (u, v) pair that obeys (12) and minimizes (4). We will now analyze this method for selecting u in terms of SNR.

Proposition 1 (SNR lower bound). *Imposing (12), for any (s, n) pair SNR_{σ^2} is bounded from below by*

$$SNR_{\sigma^2} \geq SNR_0 - 3dB, \tag{14}$$

where we use the customary notation $3dB$ for $10 \log_{10}(2)$.

Proof. From Theorem 1 we have $\text{cov}(n, v) \geq 0$, therefore,

$$\begin{aligned} SNR_{\sigma^2} &= 10 \log \tfrac{V(s)}{V(n-v)} \\ &\geq 10 \log \tfrac{V(s)}{V(n)+V(v)} \\ &= 10 \log \tfrac{V(s)}{2\sigma^2} \\ &= SNR_0 - 3dB. \end{aligned}$$

□

The lower bound of proposition 1 is reached only in the very rare and extreme case where $\text{cov}(n, v) = 0$. This implies that only parts of the signal were filtered out and no denoising was performed.

Proposition 2 (SNR upper bound). *Imposing (12), then there does not exist an upper bound $0 < M < \infty$, where $SNR_{\sigma^2} \leq SNR_0 + M$, that is valid for any given (s, n) pair.*

Proof. To prove this we need to show only a single case where the SNR cannot be bounded. Let us assume $V(s) = h\sigma^2$, $0 < h < 1$. Then $SNR_0 = 10 \log h$. As signal and noise are not correlated we have $V(f) = V(s) + V(n) = (1+h)\sigma^2$. We can write $V(f)$ also as $V(u+v) = V(u) + V(v) + 2\text{cov}(u, v)$. From (12), $V(v) = \sigma^2$, and from Theorem 1, $\text{cov}(u, v) \geq 0$, therefore $V(u) \leq h\sigma^2$. Since $\text{cov}(u, s) \geq 0$ (Theorem 1) we get $V(u - s) \leq 2h\sigma^2$. This yields $SNR_{\sigma^2} \geq 10 \log \frac{1}{2}$ and

$$SNR_{\sigma^2} - SNR_0 \geq 10 \log \frac{1}{2h}.$$

For any M we can choose a sufficiently small h where the bound does not hold. □

Simulations that illustrate worst- and best-case scenarios are presented in Figs. 1 and 2. A signal that consists of a single very contrasted step function is shown in Fig. 1. This example illustrates a best-case scenario for an edge preserving Φ. SNR resulting from the PDE-based denoising is greatly increased (by $\sim 20dB$). Note that this case approximates an ideal decomposition $u \approx s$, $v \approx n$ which differs from the simple case used in the proof of Proposition 2. A worst-case scenario is illustrated in Fig. 2 by means of the Checkered-board example. A very oscillatory signal s is being denoised and, in the process, is heavily degraded. The reduction in SNR, compared to SNR_0, is $\sim 2.9dB$, close to the theoretical $3dB$ bound.

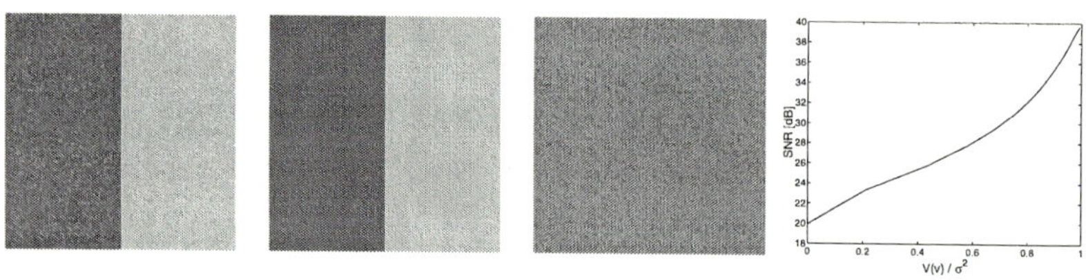

Fig. 1. Approaching best-case scenario in piece-wise constant images. In this example SNR increases by almost $20dB$ from 19.9dB to 39.6dB (variance of noise is $\approx \frac{1}{100}$ of the input noise). From left: f, u, v, SNR as a function of $V(v)/\sigma^2$

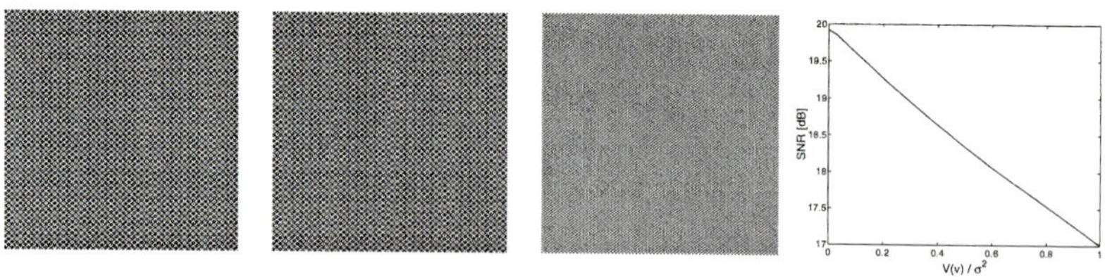

Fig. 2. Approaching worst-case scenario in a checkered-board image. SNR decreases by almost $3dB$ from 19.9dB to 17.0dB. From left: f, u, v, SNR as a function of $V(v)/\sigma^2$

2.2 Condition for Optimal SNR

We will now develop a necessary condition for the optimal SNR. As discussed, we have a single degree of freedom of choosing $V(v)$. We therefore regard SNR as a function $\text{SNR}(V(v))$ and assume that it is smooth. A necessary condition for the maximum in the range $V(v) \in (0, V(f))$ is:

$$\frac{\partial\,\text{SNR}}{\partial V(v)} = 0. \tag{15}$$

Rewriting $V(n-v)$ as $V(n) + V(v) - 2\text{cov}(n,v)$, and using (15) and (7), yields

$$\frac{\partial \text{cov}(n,v)}{\partial V(v)} = \frac{1}{2}. \tag{16}$$

The meaning of this condition may not appear at first glance to be very clear. We therefore resort to our intuition: let us think of an evolutionary process with scale parameter $V(v)$. We begin with $V^0(v) = 0$ and increment the variance of v by a small amount $dV(v)$, so that in the next step $V^1(v) = dV(v)$. The residual part of f, v, contains now both part of the noise and part of the signal. As long as in each step the noise is mostly filtered, that is $\frac{\partial \text{cov}(n,v)}{\partial V(v)} > \frac{1}{2}$, then one should keep on with the process and SNR will increase. When we reach the condition of (16), noise and signal are equally filtered and one should therefore stop. If filtering is continued, more signal than noise is filtered (in terms of variance) and SNR decreases.

There is also a possibility that the maximum is at the boundaries: If SNR is dropping from the beginning of the process we have $\frac{\partial \text{cov}(n,v)}{\partial V(v)}|_{V(v)=0} < \frac{1}{2}$ and $SNR_{opt} = SNR_0$. The other extreme case is when SNR increases monotonically and is maximized when $V(v) = V(f)$ (the trivial constant solution $u = \bar{f}$). We will see later (Proposition 3) that this can only happen when SNR_0 is negative or, equivalently, when $V(s) < \sigma^2$.

In light of these considerations, provided that one can estimate $\text{cov}(n,v)$, our basic numerical algorithm should be as follows:

1. Set $\text{cov}^0(n,v) = 0$, $V^0(v) = 0$, $i = 1$.
2. $V^i(v) \leftarrow V^{i-1}(v) + dV(v)$. Compute $\text{cov}^i(n,v)$.
3. If $\frac{\text{cov}^i(n,v) - \text{cov}^{i-1}(n,v)}{dV(v)} \leq \frac{1}{2}$ then stop.
4. $i \leftarrow i+1$. Goto step 2.

In the next section we suggest a method to approximate the covariance term.

Definition 2 (Regular SNR). *We define the function* $\text{SNR}(V(v))$ *as regular if (16) is a sufficient condition for optimality or if the optimum is at the boundaries.*

Proposition 3 (Range of optimal SNR). *If SNR is regular, then for any* (s,n) *pair* $0 \leq V_{opt} \leq 2\sigma^2$.

Proof. Let us first show the relation $\text{cov}(n,v) \leq \sigma^2$: $\text{cov}(n,f) = \text{cov}(n,n+s) = V(n)+\text{cov}(n,s) = \sigma^2$. On the other hand $\text{cov}(n,f) = \text{cov}(n,u+v) = \text{cov}(n,u)+\text{cov}(n,v)$. The relation is validated by using $\text{cov}(n,u) \geq 0$ (Theorem 1).

We reach the upper bound by the following inequalities:

$$\sigma^2 \geq \text{cov}(n,v)|_{V_{opt}} = \int_0^{V_{opt}} \frac{\partial \text{cov}(n,v)}{\partial V(v)} dV(v) \geq \int_0^{V_{opt}} \frac{1}{2} dV(v) = \frac{1}{2} V_{opt}.$$

The inequality on the right is based on that $\frac{\partial \text{cov}(n,v)}{\partial V(v)} \geq \frac{1}{2}$ for $V(v) \in (0, V_{opt})$.

The lower bound $V_{opt} = 0$ is reached whenever $\frac{\partial \text{cov}(n,v)}{\partial V(v)}|_{V(v)=0} < \frac{1}{2}$. □

Theorem 2 (Bound on optimal SNR). *If SNR is regular, then for any (s,n) pair and $V_{opt} \in \{[0,\sigma^2), (\sigma^2, 2\sigma^2]\}$,*

$$0 \leq SNR_{opt} - SNR_0 \leq \begin{cases} -10\log(1 + V_{opt}/\sigma^2 - 2\sqrt{V_{opt}/\sigma^2}), & 0 \leq V_{opt} < \sigma^2 \\ -10\log(V_{opt}/\sigma^2 - 1), & \sigma^2 < V_{opt} \leq 2\sigma^2 \end{cases} \tag{17}$$

Proof. By the SNR definition, (7), and expanding the variance expression, we have

$$SNR_{opt} - SNR_0 = 10\log(\frac{\sigma^2}{\sigma^2 + V_{opt} - 2\text{cov}(n, v_{opt})}). \tag{18}$$

For the lower bound we use the relation shown in Proposition 3: $\text{cov}(n, v_{opt}) \geq \frac{1}{2}V_{opt}$. For the upper bound we use two upper bounds on $\text{cov}(n, v_{opt})$ and take their minimum. The first one, $\text{cov}(n, v_{opt}) \leq \sigma\sqrt{V_{opt}}$, is a general upper bound on covariance. The second relation, $\text{cov}(n, v_{opt}) \leq \sigma^2$, is outlined in Proposition 3. □

A plot of the upper bound of the optimal SNR with respect to V_{opt}/σ^2 is depicted in Fig. 3, left.

In practice, the flow is not performed by directly increasing $V(v)$, but by decreasing the value of λ. Therefore, it is instructive to check how $V(v)$ varies, as well as the other energies, as λ varies. In the next proposition we show that as λ decreases the total energy strictly decreases, $E_v(v) \doteq V(v)$ increases and $E_u(u) \doteq \int_\Omega \Phi(|\nabla u|) d\Omega$ decreases.

Proposition 4 (Energy change as a function of λ). *The energy parts of Eq. (4) vary as a function of λ as follows:*

$$\frac{\partial E_\Phi}{\partial \lambda} > 0, \quad \frac{\partial E_v}{\partial \lambda} \leq 0, \quad \frac{\partial E_u}{\partial \lambda} \geq 0. \tag{19}$$

For proof see [5].

3 Estimating cov(n, v)

The term cov(n, v) is unknown, as we do not know the noise, and therefore should be estimated. We are showing here for the first time a representation of denoising by a family of curves which connects the variance of the noise, λ and cov(n, v) of pure noise. This can be regarded as some sort of nonlinear statistics of noise with respect to a specific Φ process. It appears that cov(n, v) as a function of λ is almost independent from the underlying image and can be estimated with quite a good accuracy.

First we need to compute the "statistics" by processing a patch of pure noise and measuring cov(n, v) with respect to λ. This is done a single time for each noise variance and can be regarded as a look-up-table (see Fig. 3, right). For each processed image the behavior of λ with respect to $V(v)$ is measured. Combining the information, it is possible to approximate how cov(n, v) behaves with respect to $V(v)$. In other words, this is simply the chain-rule for differentiation:

$$\begin{aligned} \frac{\partial \mathrm{cov}(n,v)}{\partial V(v)} &= \frac{\partial \mathrm{cov}(n,v)}{\partial \lambda}\frac{\partial \lambda}{\partial V(v)} \\ &\approx \frac{\partial \mathrm{cov}(n,v)}{\partial \lambda}\Big|_{f=patch}\frac{\partial \lambda}{\partial V(v)}\Big|_{f=s+n}. \end{aligned} \tag{20}$$

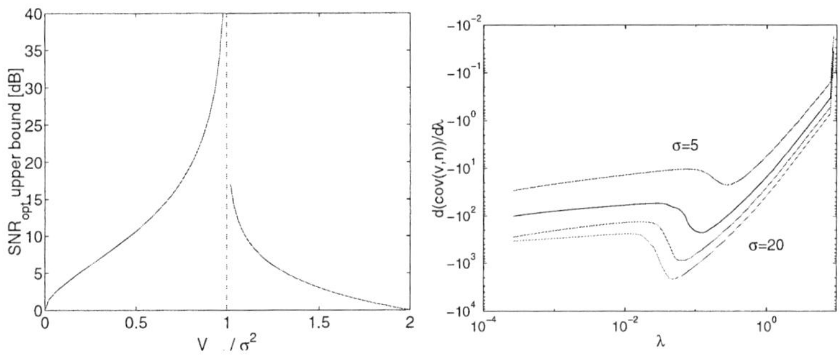

Fig. 3. Left: Visualization of Theorem 2: Upper bound of $SNR_{opt} - SNR_0$ as a function of V_{opt}/σ^2. For $V_{opt} \to \sigma^2$ the bound approaches ∞. Right: Precomputed function $\partial \mathrm{cov}(n, v)/\partial\lambda$ plotted as a function of λ (log scale). Graphs depict plots for values of σ: 5, 10, 15, 20, from upper curve to lower curve, respectively

3.1 Experimental Results

We compare our method for finding λ with the standard method of imposing (12) and with the optimal λ, which maximizes the SNR. Six classical benchmark images are processed: Cameraman, Lena, Boats, Barbara, Toys and Sailboat. The summary of the results is shown in Table 1. Our method is quite close to the optimal denoising (less than $0.1dB$ difference on average) and performs better than the method of [11].

We used $\Phi(s) = \sqrt{1 + s^2}$, which can be viewed as the Vogel-Oman [12] regularization of TV [11] with $\epsilon = 1$ or the Charbonnier [2] process. The image

Table 1. Denoising results of several images widely used in image processing. The original images were degraded by additive white Gaussian noise ($\sigma = 10$) prior to their processing

Image	SNR_0	SNR_{opt}	$SNR_{\sigma 2}$	SNR_{ours}
Cameraman	15.86	19.56	19.32	19.50
Lena	13.47	18.19	17.65	18.18
Boats	15.61	20.23	19.83	20.22
Barbara	14.73	16.86	16.21	16.64
Toys	10.00	17.69	17.29	17.65
Sailboat	10.36	15.51	15.16	15.48
Average difference from SNR_{opt}	4.67	0.00	0.43	0.06

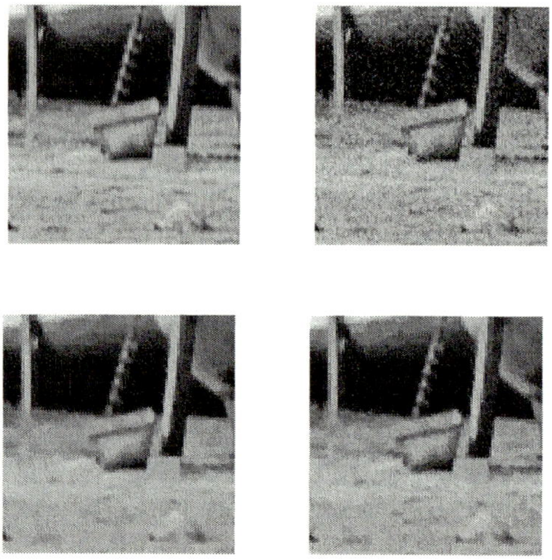

Fig. 4. Part of Boats image. Top (left to right): s, f. Bottom (left to right): u by standard method ($V(v) = \sigma^2$), u by our estimation method. More textural information is preserved by our method

grey-level range is $1 : 256$ so edges are well preserved. Other details about this experiment can be found in [5].

In Fig. 4 we show example results of processing the Boats image. The main visual difference from the standard method is that textural information is better preserved, as we approach the optimal λ. In Fig. 5 the terms SNR(u) and $\frac{\partial \text{cov}(n,v)}{\partial \lambda}$ are plotted as functions of the normalized variance $V(v)/\sigma^2$. It is apparent that the SNR is smooth and behaves regularly, in accordance with our assumptions. An interesting phenomenon is that the covariance derivative esti-

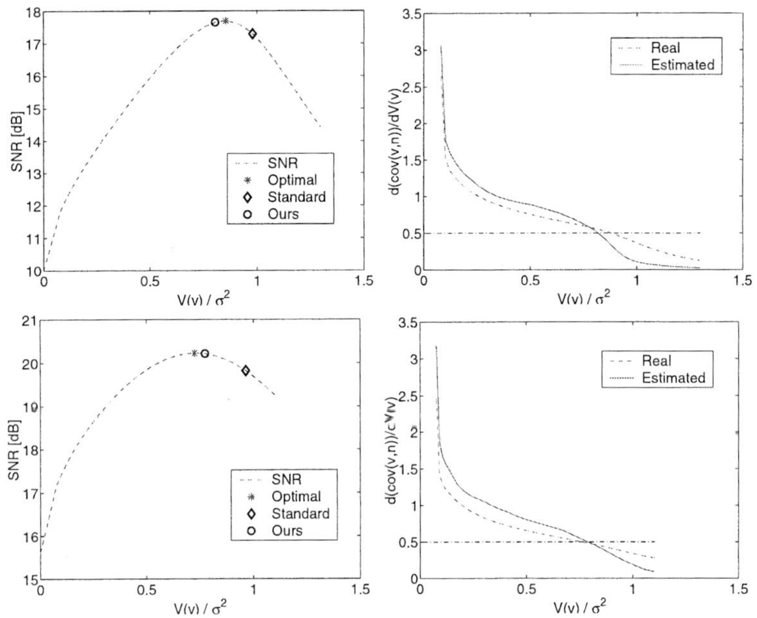

Fig. 5. SNR as a function of $V(v)/\sigma^2$ (left). $d\text{cov}(n, v)/dV(v)$ as a function of $V(v)/\sigma^2$ (right), as computed by our estimation method (solid) and the ground truth (dashed). Graphs depict processing of Toys (top) and Boats (bottom)

mation tends to be more accurate near the critical value of $\frac{1}{2}$. Naturally, this is advantageous to our algorithm. We currently have no explanation for this behavior.

4 Conclusion

Most image denoising processes are quite sensitive to the choice and fine tuning of various parameters. This is a major obstacle for fully automatic algorithms. This problem motivated us to develop a criterion for the optimal choice of the fidelity weight parameter in variational denoising. Our criterion is to maximize the SNR of the resultant image. Bounds on the SNR as well as on the optimal variance are obtained. We demonstrate our method on a series of benchmark images and show that the performance is only slightly worse than optimal (less than 0.1dB difference).

We should comment that the SNR criterion is not always in accordance with human-based quality evaluations. Other, more sophisticated criteria, may also be applied for parameter selection using the spirit of the method presented here.

The basic ingredients of the proposed method, namely the covariance condition (16) and its estimation (20), are quite universal and do not depend on the specific denoising algorithm. The method was generalized for selecting the stopping time in nonlinear diffusion [5] and for regularizations based on BV and Hilbert-space norms [1].

A Proof of Theorem 1

We present the main steps of the proof. A full version is given in [5]. Since $\mathrm{cov}(q,r) = \mathrm{cov}(r,q)$, the matrix is symmetric. The diagonal is the variance of each element, which is non negative. Therefore we have to consider all 10 possible signal pairs and show that their covariance is non-negative.

$\mathbf{cov}(s,n)$, $\mathbf{cov}(f,s)$, $\mathbf{cov}(f,n)$. Since s and n are not correlated, we have $\mathrm{cov}(s,n) = 0$, $\mathrm{cov}(f,s) = \mathrm{cov}(s+n,s) = V(s) \geq 0$, $\mathrm{cov}(f,n) = \mathrm{cov}(s+n,n) = V(n) \geq 0$.

$\mathbf{cov}(u,v)$, $\mathbf{cov}(f,u)$, $\mathbf{cov}(f,v)$. Once we prove $\mathrm{cov}(u,v) \geq 0$, then we readily have $\mathrm{cov}(f,u) = \mathrm{cov}(u+v,u) = V(u) + \mathrm{cov}(u,v) \geq 0$ and $\mathrm{cov}(f,v) = \mathrm{cov}(u+v,v) = V(v) + \mathrm{cov}(u,v) \geq 0$.

We follow the spirit of the proof of Meyer [8]. As the (u,v) decomposition minimizes the energy of Eq. (4), we can write for any function $h \in BV$ and scalar $\epsilon > 0$ the following inequality:

$$\int_\Omega \Phi(|\nabla(u-\epsilon h)|)d\Omega + \lambda V(v+\epsilon h) \geq \int_\Omega \Phi(|\nabla u|)d\Omega + \lambda V(v). \qquad (21)$$

Replacing $V(v+\epsilon h)$ by $V(v) + \epsilon^2 V(h) + 2\epsilon \mathrm{cov}(v,h)$ and then changing h to u and dividing both sides by ϵ we get

$$2\lambda \mathrm{cov}(v,u) \geq \frac{1}{\epsilon}\int_\Omega \left(\Phi(|\nabla u|) - \Phi(|\nabla(u-\epsilon u)|)\right) d\Omega - \lambda\epsilon V(u).$$

In the limit as $\epsilon \to 0$, the right term on the right-hand-side vanishes. Since Φ is increasing, the term in the integral is non-negative.

$\mathbf{cov}(s,u)$, $\mathbf{cov}(n,u)$. By writing $V(v)$ as $V(s+n-u)$, expanding the variance expression and omitting expressions that do not involve u, we can reach the following minimization problem equivalent to minimizing (4): $u = \mathrm{argmin}_u\{\hat{E}_\Phi(u)\}$ where

$$\hat{E}_\Phi(u) = \int_\Omega \Phi(|\nabla u|)d\Omega + \lambda(V(u) - 2\mathrm{cov}(s,u) - 2\mathrm{cov}(n,u)). \qquad (22)$$

Since $\mathrm{cov}(s,u) + \mathrm{cov}(n,u) = \mathrm{cov}(f,u) \geq 0$ at least one of the terms $\mathrm{cov}(s,u)$ or $\mathrm{cov}(n,u)$ must be non-negative. We will now show, by contradiction, that it is not possible that the other term be negative. Let us assume, without loss of generality, that $\mathrm{cov}(s,u^{s+n}) \geq 0$ and $\mathrm{cov}(n,u^{s+n}) < 0$. We denote the optimal (minimal) energy of (22) with $f = s+n$ as $\hat{E}^*_\Phi|_{f=s+n}$. The energy can be written as

$$\begin{aligned}\hat{E}^*_\Phi|_{f=s+n} &= \hat{E}_\Phi|_{f=s+n}(u^{s+n}) \\ &= \textstyle\int_\Omega \Phi(|\nabla u^{s+n}|)d\Omega + \lambda(V(u^{s+n}) - 2\mathrm{cov}(s,u^{s+n}) - 2\mathrm{cov}(n,u^{s+n})).\end{aligned} \qquad (23)$$

On the other hand, according to condition (5), $\mathrm{cov}(u^s,n)=0$ and we have

$$\hat{E}_\Phi|_{f=s+n}(u^s)=\int_\Omega \Phi(|\nabla u^s|)d\Omega+\lambda(V(u^s)-2\mathrm{cov}(s,u^s))$$
$$=\hat{E}^*_\Phi|_{f=s}\le \hat{E}_\Phi|_{f=s}(u^{s+n})=\int_\Omega \Phi(|\nabla u^{s+n}|)d\Omega+\lambda(V(u^{s+n})-2\mathrm{cov}(s,u^{s+n})).$$

In the above final expression, adding the term $-\lambda 2\mathrm{cov}(n,u^{s+n})$ we obtain the right hand side of expression (23). Since we assume $\mathrm{cov}(n,u^{s+n})<0$, we get the following contradiction: $\hat{E}_\Phi|_{f=s+n}(u^s)<\hat{E}^*_\Phi|_{f=s+n}$. Similarly, the opposite case $\mathrm{cov}(n,u^{s+n})\ge 0$ and $\mathrm{cov}(s,u^{s+n})<0$ is not possible.

cov(s,v), cov(n,v). This follows directly from condition (6) as $\mathrm{cov}(f,s)=\mathrm{cov}(u,s)+\mathrm{cov}(v,s)$ and $\mathrm{cov}(f,n)=\mathrm{cov}(u,n)+\mathrm{cov}(v,n)$. □

References

1. J.F. Aujol, G. Gilboa, "Implementation and parameter selection for BV-Hilbert space regularizations", UCLA CAM Report 04-66, November 2004.
2. P. Charbonnier, L. Blanc-Feraud, G. Aubert, M. Barlaud, "Two deterministic half-quadratic regularization algorithms for computed imaging", Proc. IEEE ICIP '94, Vol.2, 168-172, 1994.
3. R. Deriche, O. Faugeras, "Les EDP en traitement des images et vision par ordinateur", Traitement du Signal, 13(6), 1996.
4. H.W. Engl, H. Gfrerer, "A posteriori parameter choice for general regularization methods for solving linear ill-posed problems", Appl. Numer. Math, 4(5), 395 - 417, 1988.
5. G. Gilboa, N. Sochen, Y.Y. Zeevi, "Estimation of optimal PDE-based denoising in the SNR sense", CCIT report No. 499, Technion, August 2004.
6. G. Gilboa, N. Sochen, Y.Y. Zeevi, "Texture preserving variational denoising using an adaptive fidelity term", Proc. VLSM 2003, Nice, France, pp. 137-144, 2003.
7. P. Mrázek, M. Navara, "Selection of optimal stopping time for nonlinear diffusion filtering", IJCV, v. 52, no. 2/3, pp. 189-203, 2003.
8. Y. Meyer, *Oscillatory Patterns in Image Processing and Nonlinear Evolution Equations*, Vol. 22 University Lecture Series, AMS, 2001.
9. S. Pereverzev, E. Schock, "Morozov's discrepancy principle for Tikhonov regularization of severely ill-posed problems in finite-dimensional subspaces", Numer. Funct. Anal. Optim, 21, pp. 901-916, 2000.
10. P. Perona, J. Malik, "Scale-space and edge detection using anisotropic diffusion", PAMI 12(7), pp. 629-639, 1990.
11. L. Rudin, S. Osher, E. Fatemi, "Nonlinear Total Variation based noise removal algorithms", Physica D 60 259-268,'92.
12. R.V. Vogel and M.E. Oman, "Iterative methods for total variation denoising", SIAM J. Scientific Computing, 17(1):227–238, 1996.
13. J. Weickert, "Coherence-enhancing diffusion of colour images", IVC, Vol. 17, 201-212, 1999.

A Contrast Invariant Approach to Motion Estimation

V. Caselles, L. Garrido, and L. Igual*

Universitat Pompeu Fabra, Barcelona (Spain)
{vicent.caselles, luis.garrido, laura.igual}@upf.edu

Abstract. Motion estimation is one of the key tools in many video processing applications. Most of the existing motion estimation approaches use the brightness constancy assumption in order to model the movements of the objects present in the scene. In this paper the motion of objects is modeled from a geometrical-based point of view, leading thus to a contrast invariant formulation. The present approach is region-based and assumes affine motion model for each region.

1 Introduction

Computing the apparent motion of objects in a sequence of images is one of the key problems in computer vision known as the optical flow computation. Its numerous applications make it the object of current research (see [23] for an account of it).

Most known motion estimation methods, in one form or another, employ the *optical flow constraint* which states that the image intensity remains unchanged from frame to frame along the true motion path. The *optical flow equation* is derived from the optical flow constraint:

$$\partial_x I \; u + \partial_y I \; v + \partial_t I = 0 \tag{1}$$

where $I(t, x, y)$ denotes the image sequence and (u, v) the motion vector field. The movement of the objects present in the scene may be recovered by minimizing an error measure based on the optical flow equation [23]. Furthermore, it is known that motion estimation is an "ill-posed" problem, indeed, the solution may not be unique, and/or solutions may not depend continuously on the data [4]. Current motion estimation approaches try to solve the latter issue by imposing additional assumptions about the structure of the 2D motion field. The latter constraints are introduced into the error measure either by adding a smoothness term to it, or by restricting it to a particular motion model. The

* The first and second authors acknowledge partial support by the Departament d'Universitats, Recerca i Societat de la Informació de la Generalitat de Catalunya and by PNPGC project, reference BFM2003-02125. L. Igual acknowledges support by the French Space Agency (CNES) and the company THALES (France).

R. Kimmel, N. Sochen, J. Weickert (Eds.): Scale-Space 2005, LNCS 3459, pp. 242–253, 2005.

former strategies are called *dense motion field estimation* approaches, whereas the latter ones are usually called *parametric motion estimation* approaches.

The Horn-Shunck's method is a classical method for dense motion field estimation. It seeks for a motion field that satisfies the optical flow equation (1) with a minimum pixel-to-pixel variation between the flow vectors:

$$\min \int_{\Omega} (\partial_x I \; u + \partial_y I \; v + \partial_t I)^2 + \alpha^2 \left((\partial_x u)^2 + (\partial_y u)^2 + (\partial_x v)^2 + (\partial_y v)^2 \right)$$

where Ω is the image domain and α may be used to control the influence of the constraint. Larger values of α^2 increase the influence of the constraint.

The Lucas-Kanade method can be considered a *parametric motion estimation*, since it estimates the motion by assuming that the motion vector associated to the optical flow equation remains unchanged over a particular block of pixels. The method thus allows to estimate a translational motion vector for that block. A very interesting combination of both previous methods with an efficient implementation has been proposed in [7]. For the interested reader, a good review of current motion estimation techniques can be found in [3, 22].

The optical flow constraint assumption is generally violated in image sequences taken from the real world. Global or local changes in illumination due to, for instance, a moving camera or a change in the shade of an object may prevent the correct motion to be estimated. Alternatives to the classical brightness constancy assumption have been already proposed in the literature. A common approach to handle non constant intensity is through explicit modelling of the illumination change in the optical flow equation [18]. The approach requires complex minimization since, in addition to the motion field, illumination fields must also be estimated.

In [4] a constraint based on spatial gradient's constancy is proposed. It relaxes the classical assumption, but requires that the amount of dilation and rotation in the image be negligible, a limitation often satisfied in practice according to [22]. The technique has been demonstrated to be very robust in the presence of time-varying illumination. More recently, is has been shown that the direction of the intensity gradient is invariant to global illumination changes [10]. The work presented in [8] is based on this property.

In this paper we propose to substitute the optical flow equation, derived from the brightness constancy assumption, by the assumption that the shapes of the image move along the sequence. We identify the shapes of the image with the family of its level lines [9] and we assume that they move along the image sequence (with possible deformation). This assumption permits us to design contrast invariant estimate of the optical flow. The approach is in fact based on the invariance of the gradient direction to contrast changes. However, no restriction to the amount of dilation and rotation is imposed.

The paper is organized as follows. Section 2 describes the contrast-invariant model that has been developed, whereas Sect. 3 introduces the region-based strategy that has been implemented. Sect. 4 gives some details about the implementation. Finally, Sect. 5 presents the results obtained with the proposed method and Sect. 6 ends up with the conclusions and future research work.

2 A Contrast-Invariant Functional

Let Ω be the image domain, which we may assume to be normalized to $[0,1]^2$. Let $I : \Omega \to R$ be a given image. Mathematical morphology offers an image description in terms of its level sets, be upper $X^\lambda I = \{\mathbf{p} \in \Omega : I(\mathbf{p}) \geq \lambda\}$, or lower $X_\lambda I = \{\mathbf{p} \in \Omega : I(\mathbf{p}) \leq \lambda\}$. Level sets provide a complete image description, in particular, the image I can be reconstructed from its (upper) level sets by the formula $I(\mathbf{p}) = \sup\{\lambda : \mathbf{p} \in X^\lambda I\}$ (a similar formula exists for the lower level sets) where sup denotes the supremum operator, and $\mathbf{p} = (x,y)^T$ denotes a point in Ω. Level sets give a contrast invariant representation of the image [21].

We call level lines the boundaries of the level sets. In the discrete framework, any level set can be described in terms of its boundary. Indeed, the connected component of each level set can be described in terms of its external and the family of its internal boundaries [9, 21]. Thus, we may use the family of level lines as basic contrast invariant geometric description of the image I. As an analytical tool, we shall use the unit normals to the level lines to describe them.

Let $I(t,\mathbf{p})$ be a given image sequence, t being in the time interval $[T_0, T_1]$ and $\mathbf{p} \in \Omega$. We assume that the image sequence has been sampled at points multiple of Δt, the sampling points being $t_j = T_0 + j\Delta t$, $j = 0, \ldots, N$ ($t_N = T_1$). Let us denote by $\phi^j(\mathbf{p})$ the coordinates at time $t_j + \Delta t$ of the point whose coordinates at time t_j are $\mathbf{p}$, $j = 0, \ldots, N-1$. The map $\phi^j : \Omega \to \Omega$ is nothing else than the motion path starting from time t_j and we may think about it as a deformation. We do not assume in this section any particular motion model for ϕ^j. That is, the image objects may suffer any deformation over time. For simplicity, when no confusion arises, the arguments of the previous function will be dropped out.

Assume for a while that j is fixed and let $\phi = (\phi_1, \phi_2)$ be any of the maps ϕ^j, where ϕ_1 and ϕ_2 are the components of ϕ^j. Let $X = (x(s), y(s))^T$ be the arclength parameterization of a given level line $\mathcal{C}$ of the image $I(t_j, \mathbf{p})$, s being the arc length parameter. The curve $\mathcal{C}$ may be described by its normal vectors $Z = (-y'(s), x'(s))^T$, where $(.)'$ denotes the first derivative with respect to s. Note that Z has unit norm.

Let us describe the normal vectors to the curve $\phi(\mathcal{C})$ in terms of ϕ and the normal vectors to $\mathcal{C}$. Since the curve $\phi(\mathcal{C})$ is described by

$$\overline{X} = (\overline{x}(s), \overline{y}(s))^T = \phi(x(s), y(s)),$$

the tangent vector to the deformed curve $\phi(\mathcal{C})$ is given by

$$\overline{X}' = \begin{pmatrix} \overline{x}'(s) \\ \overline{y}'(s) \end{pmatrix} = \begin{pmatrix} \partial_x \phi_1 & \partial_y \phi_1 \\ \partial_x \phi_2 & \partial_y \phi_2 \end{pmatrix} \begin{pmatrix} x'(s) \\ y'(s) \end{pmatrix} = D\phi \; X'$$

where ∂_x and ∂_y denote the partial derivative with respect to x and y respectively. Thus, the normal vector Z of the deformed curve is

$$Z = \begin{pmatrix} -\overline{y}'(s) \\ \overline{x}'(s) \end{pmatrix} = \begin{pmatrix} \partial_y \phi_2 & -\partial_x \phi_2 \\ -\partial_y \phi_1 & \partial_x \phi_1 \end{pmatrix} \begin{pmatrix} -y'(s) \\ x'(s) \end{pmatrix}. \tag{2}$$

Observe that the matrix in the right hand side of (2) is the *cofactor matrix* associated to $D\phi$ which we shall denote by $(D\phi)^\dagger$. Thus, the normal vectors of the deformed curve are related to the original normal vectors by means of the cofactor matrix. Observe that s is not necessarily the arclength parameter of $\phi(\mathcal{C})$, hence Z is not, in general, a unit vector. We normalize it to be of unit norm by redefining

$$\overline{Z}_\phi = \frac{(D\phi)^\dagger\, Z}{\|(D\phi)^\dagger\, Z\|} \quad \text{if } (D\phi)^\dagger Z \neq 0; \quad 0 \text{ otherwise}, \tag{3}$$

where $\|.\|$ denotes the modulus of a vector in R^2 .

In the context of this work, $Z^t(\mathbf{p})$ will be the vector field of unit normals to the level lines of $I(t, \mathbf{p})$. Usually, the energy functional whose minimum gives the optical flow tries to impose the brightness constancy equation (1). Instead, our main assumption will be that shapes move with possible deformation along the sequence. We interpret it in the following way: (*) we may find the boundary of a connected component of the level set $[I(t_j, \cdot) \geq \lambda]$, $\lambda \in R$, eventually deformed by $\phi^j(\cdot)$, as a level curve of $I(t_{j+1}, \cdot)$ at some other level λ'. Observe that if two consecutive frames are related by the motion model and a global illumination change, i.e., if (**) $I(t_{j+1}, \phi^j(\mathbf{p})) = g_j(I(t_j, \mathbf{p}))$ for some contrast change g_j, then (*) holds (in this case $\lambda' = g_j(\lambda)$). Our assumption (*) is more general than (**) since the former is local: the level at which we may find the boundary of the connected component of $[I(t_j, \cdot) \geq \lambda]$ may depend on the connected component itself, besides of depending on λ. Thus, our purpose will be to align the level lines of two consecutive frames at times t_j and t_{j+1} by a map ϕ^j. Using the description of level lines in terms of unit normals, we propose to compute the optical flow ϕ^j by aligning the unit normal vector field $Z^{t_{j+1}}(\mathbf{p})$ with the transformed vector field of $Z^{t_j}(\mathbf{p})$ by the map ϕ^j (i.e., the vector field obtained by (3)). Thus, we propose to compute the motion estimation by minimizing the energy functional

$$E(\phi) = \sum_{j=0}^{N-1} \int_\Omega \left\| Z^{t_{j+1}}(\phi^j(\mathbf{p})) - \overline{Z}_{\phi^j}(\mathbf{p}) \right\|^2 \mu_j(\mathbf{p})\, dx\, dy. \tag{4}$$

where $\mu_j(\mathbf{p})$ represent weight functions that will be later discussed. The vector field $Z^{t_j}(\mathbf{p})$ is computed by

$$Z^{t_j}(\mathbf{p}) = \frac{\nabla I(t_j, \mathbf{p})}{\|\nabla I(t_j, \mathbf{p})\|} \quad \text{if } \nabla I(t_j, \mathbf{p}) \neq 0; \ 0 \text{ otherwise}, \tag{5}$$

where $\nabla := (\partial_x, \partial_y)^T$ denotes the 2D gradient. Note that (5) computes the normal vector of the level line that passes through point $\mathbf{p}$.

Since for any smooth strictly increasing function we have $\nabla g(I) = g'(I)\nabla I$, it is easy to check that if $\mu_j(\mathbf{p}) = 1$, then the energy (4) is contrast invariant. In case that we decide to give more weight to edges, we may take $\mu_j(\mathbf{p}) = \|\nabla I(t_j, \mathbf{p})\|$, $j = 0, \ldots, N-1$, in this case, if $I(t_j + 1, \phi^j(\mathbf{p})) = g_j(I(t_j, \mathbf{p})))$ for some contrast change g_j, then the estimate of ϕ^j obtained by minimizing the corresponding term in (4) does not depend on the contrast change g_j.

parameters that minimize (6) is the *conjugate-gradient* method. In this paper the conjugate update directions are computed using Polak-Ribiere method. At each iteration, Brent's line minimization is used to find the minimum along the selected update direction [20].

Coordinate System Selection. Let us discuss how the selection of the coordinate system may affect the convergence of the descent method to minimize E_R. For simplicity, assume that Z^{t+1} and $\overline{Z}^t_{\phi^t}$ are unit vectors. The differentiation of each term in (7) with respect to any motion parameter $m \in \mathbf{m}$ gives the contribution

$$<\partial_m \overline{Z}^t_{\phi^t}, Z^{t+1}> + <\overline{Z}^t_{\phi^t}, \partial_m Z^{t+1}> .$$

Let us compute $\partial_m Z^{t+1}$. For that, assume that the origin of the coordinate system is located at the upper left corner of the image, whereas samples are placed on a squared grid at a distance of one. Let us denote by $(\hat{x},\hat{y})^T$ the coordinates of $\mathbf{p}$ in this coordinate system. Let Z_i^{t+1}, $i=1,2$, be the x and y components, respectively, of Z^{t+1}. Observe that the partial derivative of Z_1^{t+1} with respect to the motion parameter a is

$$\partial_a Z_1^{t+1}(\phi(\mathbf{p})) = \partial_{\phi_1} Z_1^{t+1}\ \partial_a \phi_1 + \partial_{\phi_2} Z_1^{t+1}\ \partial_a \phi_2 = \partial_{\phi_1} Z_1^{t+1}\ \hat{x} \tag{9}$$

since $\partial_a \phi_1 = \hat{x}$ and $\partial_a \phi_2 = 0$. The derivative of Z_1^{t+1} with respect to the motion parameter e is

$$\partial_e Z_1^{t+1}(\phi(\mathbf{p})) = \partial_{\phi_1} Z_1^{t+1}\ \partial_e \phi_1 + \partial_{\phi_2} Z_1^{t+1}\ \partial_e \phi_2 = \partial_{\phi_1} Z_1^{t+1} \tag{10}$$

since $\partial_e \phi_1 = 1$ and $\partial_e \phi_2 = 0$. With similar computations we see that the partial derivatives of Z_1^{t+1} and Z_2^{t+1} with respect to a, b, c or d depend proportionally on the pixel coordinates (either $\hat{x}$ or $\hat{y}$), while its partial derivatives with respect to e or f do not exhibit such dependence. Thus the derivatives with respect to a, b, c, d have a stronger contribution in the descent than the ones with respect to e, f and affect its convergence.

In [14], the authors propose to normalize the pixel coordinates simply by the image dimensions. We shall use a different normalization for each region. The origin of the coordinate system, $(\tilde{x}^c, \tilde{y}^c)$, will be the centroid of the region, and the axis will be re-scaled by σ where $\sigma^2 = \sum_{\mathbf{p}\in R}\left((\hat{x}-x^c)^2 + (\hat{y}-y^c)^2\right)$ is the variance of the distance of the pixel coordinates to the centroid.

If $(x,y)^T$ denote the coordinates of $\mathbf{p}$ in this new coordinate system, then its relation with $(\hat{x},\hat{y})^T$ is given by

$$x = \frac{\hat{x}-x^c}{\sigma} \qquad y = \frac{\hat{y}-y^c}{\sigma}. \tag{11}$$

The partial derivatives of Z_1^{t+1} with respect to a and e have the same form as in (9), (10), with a different interpretation of the coordinates:

$$\partial_a Z_1^{t+1}(\phi(\mathbf{p})) = \partial_{\phi_1} Z_1^{t+1}\ \frac{(\hat{x}-x^c)}{\sigma} \quad \text{and} \quad \partial_e Z_1^{t+1}(\phi(\mathbf{p})) = \partial_{\phi_1} Z_1^{t+1}. \tag{12}$$

Comparing these expressions with (9) and (10), we notice that the normalized coordinate system has the effect that no particular derivative value is predominant with regard to the others. Our experiments have shown that in this case a large range of motions may be recovered since no particular type of motion is priorized.

On the other hand, if the $(\hat{x}, \hat{y})^T$ coordinate system is used the conjugate gradient algorithm selects update directions in which the first order motion parameters are predominant. As a result, the algorithm will "try to explain" the motion present in the image using only first order parameters (zoom, rotations and skew) even if only translational motion is present in the image. Motions such as translations may not be recovered in this case. Thus, the selection of the proper coordinate system affects directly the gradient values and thus the convergence of our estimator.

Multiresolution Analysis. Both to avoid local mimima and for computational efficiency, motion estimation is usually embedded in a multiresolution scheme [1, 14, 22, 23]. The basic idea is to obtain a set of coarse to fine images which are obtained by means of a low-pass filter. Starting the parameter estimation at the coarsest resolution level, the motion is estimated on each level successively using the resulting motion parameters of the current resolution level as input to initialize the gradient descent on the next level. Lower resolution levels allow to obtain an approximation of the motion parameters, whereas finer resolution levels are used to improve and fine-tune the motion parameter estimation. Multiresolution representations allow to deal with large zero and first order parameters.

In our experiments a set of three (including the original image) levels are used. At each level the image is lowpass filtered with respect to the previous level [24]. As proposed in [23], the downsampling step is skipped. Thus the pyramid contains images that are all the same size by successively more blurred as we go to the coarser resolution levels. This permits us to maintain the geometry of the region.

However, Δx and Δy are divided by two between successive levels, hence the area of the region is scaled by 4. Thus, Δx and Δy act as a scale parameter. If such area is small at a fixed level, only the zero-order parameters are estimated. This is due to that the texture information of small regions present at the coarser levels is poor, and thus the minimum may not be well defined for the first-order parameters.

Differentiation. Differentiation is an ill-posed problem [4], and regularization may be used to obtain good numerical derivatives. Such regularization may be accomplished with a low-pass filter such as the Gaussian, and is essential for motion estimation [3, 22]. More recently, [15] proposes to use a matched pair of low pass and differentiation filters as a gradient operator.

Notice that, for motion estimation applications, it may be necessary to compute the gradient at non integer points, since non integer displacements are allowed. In such cases, a simple way to proceed is a two step process: in a first step, the image is interpolated at the required points using an interpolation kernel such as [16], and in the second step the derivative is computed. Since both are linear operators they may be performed in one step: the derivative filter is interpolated at the required non-integer positions, then the derivative can be computed at integer pixel positions using the interpolated filter taps.

5 Results

In all experiments below, we assume that $\mu_j(\mathbf{p}) = 1$ (see (6)).

The original images in our first example correspond to the table tennis sequence (frames #4 and #1), and are displayed in Fig. 1a and Fig. 1b respectively. In these images, the ball moves downwards, the arm moves upwards and the background is static. The associated partition has been computed using the Mumford-Shah segmentation functional subordinated to the topographic map [2], and is shown in Fig. 1c. The motion field recovered by our estimator is shown in Fig. 1d and corresponds to our above description.

An interesting point is to compare our results with those obtained with the classical motion estimation approach based on minimization of the squared prediction error, defined as: $E_R^{\rm int}(\phi) = \sum_{\mathbf{p}\in R} \left(I(t+1, \phi(\mathbf{p})) - I(t, \mathbf{p})\right)^2$. The previous error measure is in fact based on the brightness constancy assumption. The latter approach has been implemented using the techniques described in Sect. 4. In order to simplify nomenclature, the latter approach is called *intensity*-based

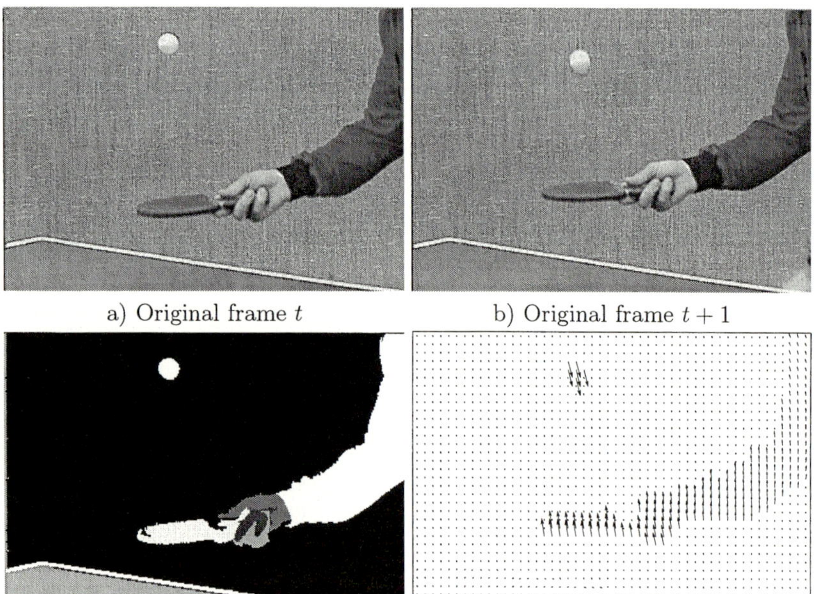

a) Original frame t b) Original frame $t+1$

c) Partition d) Motion field

Fig. 1. Region-based motion example. a) Original frame t, b) Original frame $t+1$, c) Partition of original frame t, d) Recovered motion field

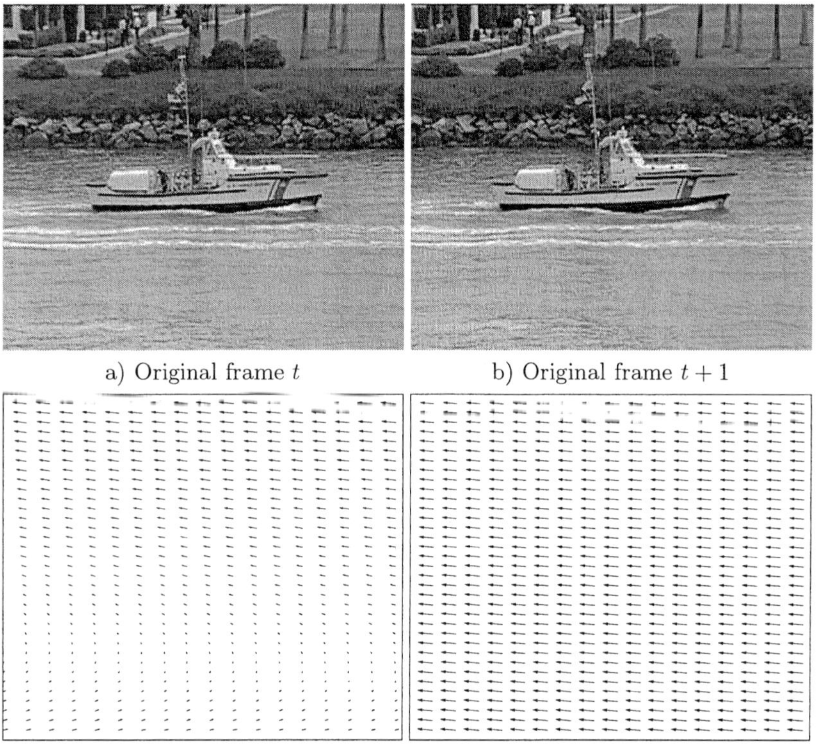

a) Original frame t b) Original frame $t+1$

c) Intensity-based d) Shape-based

Fig. 2. Global motion estimation example. The purpose is to extract camera's movement between images a) and b). Results for intensity and shape-based approach are shown in images c) and d) respectively

motion estimation, whereas the technique presented in this paper is called *shape-based* motion estimation.

The next experiments deal with global motion estimation, that is, the extraction of the camera motion. This is very useful in many scene analysis approaches, where first the camera motion is detected and then the moving objects in the scene are detected and tracked. Fig. 2 shows two frames from the *coastguard* sequence (frames #170 and #176). The frames show a moving boat and a static background. In these frames the camera follows the displacement of the boat, thus the apparent motion of the boat is zero (i.e. no motion) whereas the background has an apparent motion which corresponds to the camera's movement. We choose a partition made up of *one region* which includes the whole image support. Thus, the global motion between the two frames is estimated. The resulting motion vector fields are shown in Fig. 2c and Fig. 2d. Note that our approach has been able to properly extract the camera's motion. We believe that the intensity-based motion estimation has failed due to the strong influence of the high gradient of the boat. Since the apparent motion of the boat is

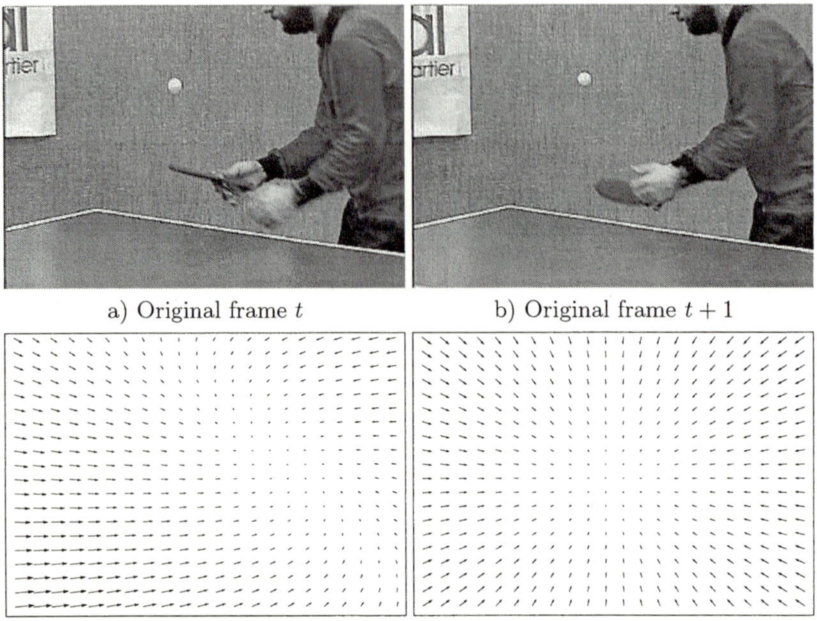

Fig. 3. Global motion estimation example. The purpose is to extract camera's movement between images a) and b). They correspond to frames #45 and #47 of the table tennis sequence. Recovered motion fields for intensity and shape-based approach are shown in images c) and d) respectively

zero, the motion estimation algorithm tries to set to zero the motion at the boat pixel locations. This is not the case of the shape-based approach, where gradient modulus has no effect. Motion in the latter case is recovered by interpreting the image as a set of moving level lines. Thus, the boat is treated as an outlier. Note that the correct motion parameters may be recovered using an intensity based energy if robust estimation techniques are used [5, 22].

Fig. 3 shows another example of global motion estimation. The camera performs a zoom out of the scene. Even though the tennis player and the ping-pong ball is moving, our approach has been able to properly recover the zoom.

6 Conclusions and Future Work

We have presented a contrast invariant model for the computation of the optical flow. We interpret the image sequence as a set of moving level lines and we propose to compute the deformation between the level lines of two consecutive frames. Several topics have to be further developed in the future: a) the selection of regions bounded by level lines where motion is estimated by an affine model, b) joint motion segmentation techniques, c) the computation of a dense motion field from the image sequence without imposing a particular motion model.

References

1. L. Alvarez, J. Weickert, and J. Sánchez. A scale-space approach to nonlocal optical flow calculations. In *Scale-Space Theories in Computer Vision*, pages 235–246. Springer Verlag, 1999.
2. C. Ballester, V. Caselles, and L. Igual. Minimal morphological shape selection for segmentation and encoding. *Preprint*, 2004.
3. J. Barron, D. Fleet, and S. Beauchemin. Performance of optic flow techniques. *IJCV*, 12:43–77, 1994.
4. M. Bertero, T.A. Poggio, and V. Torre. Ill-posed problems in early vision. *Proceedings of the IEEE*, 76(8):869–889, August 1988.
5. M.J. Black and P. Anandan. The robust estimation of multiple motions: parametric and piecewise-smooth flow fields. *CVIU*, 63(1):75–104, January 1996.
6. T. Brox, A. Bruhn, N. Papenberg, and J. Weickert. High accuracy optical flow estimation based on a theory for warping. In *European Conf. on Computer Vision*, pages 25–36. Springer Verlag, 2004.
7. A. Bruhn, J. Weickert, C. Feddern, T. Kohlberger, and C. Schnoerr. Real-time optic flow computation with variational methods. In *Int. Conf. on Computer Analysis of Images and Patterns*, pages 222–229. Springer Verlag, August 2003.
8. P.Y. Burgi. Motion estimation based on the direction of intensity gradient. *Image and Vision Computing*, 22:637–653, 2004.
9. V. Caselles, B. Coll, and J-M. Morel. Topographic maps and local contrast changes in natural images. *IJCV*, 33(1):5–27, 1999.
10. H. Chen, P. Belhumeur, and D. Jacobs. In search of illumination invariants. In *Int. Conf. on Computer Vision and Pattern Recognition*, pages 254–261, 2000.
11. D. Cremers and S. Soatto. Motion competition: a variational approach to piecewise parametric motion segmentation. *IJCV*, 2004.
12. C. Davatzikos. Spatial transformation and registration of brain images using elastically deformable models. *CVIU*, 1997.
13. M. Droske and M. Rumpf. A variational approach to non-rigid morphological image registration. *SIAM Journal Applied Mathematics*, 2004.
14. J.L. Dugelay and H. Sanson. Differential methods for the identification of 2D and 3D motion models in image sequences. *Image Communication*, 7:105–127, 1995.
15. H. Farid and E.P. Simoncelli. Differentiation of discrete multidimensional signals. *IEEE Trans. on IP*, 13(4), April 2004.
16. R.G. Keys. Cubic convolution interpolation for digital image processing. *IEEE Trans. on ASSP*, 29(6):1153–1160, December 1981.
17. J.M. Morel and S. Solimini. *Variational methods in image segmentation.* Birkhäuser Verlag: Basel, 1995.
18. J.M. Obobez and P. Bouthemy. Robust multiresolution estimation of parametric motion models applied to complex scenes. *JVCIR*, 6(4):348–365, December 1995.
19. J.P.W. Pluim, J.B.A. Maintz, and M.A. Viergever. Mutual-information-based registration of medical images. *IEEE Trans. on MI*, 22(8):986–998, August 2003.
20. W.H. Press, S.A. Teukolsky, W.T. Vetterling, and B.P. Flannery. *Numerical Recipes in C: the art of scientific computing.* Cambridge Univ. Press, 1992.
21. J. Serra. *Image Analysis and Mathematical Morphology.* Academic Press, 1982.
22. C. Stiller and J. Konrad. Estimating motion in image sequences. *IEEE Signal Processing Magazine*, pages 70–91, July 1999.
23. A.M. Tekalp. *Digital Video Processing.* Prentice-Hall, 1995.
24. M. Unser, A. Aldroubi, and M. Eden. Enlargement or reduction of digital images with minimum loss of information. *IEEE Trans. on IP*, 4(3):247–258, March 1995.

Vortex and Source Particles for Fluid Motion Estimation

Anne Cuzol and Etienne Mémin

IRISA, Université de Rennes 1,
Campus de Beaulieu, 35 042 Rennes Cedex, France
{acuzol, memin}@irisa.fr

Abstract. In this paper we propose a new motion estimator for image sequences depicting fluid flows. The proposed estimator is based on the Helmholtz decomposition of vector fields. This decomposition consists in representing the velocity field as a sum of a divergence free component and a curl free component. The objective is to provide a low-dimensional parametric representation of optical flows by depicting them as a flow generated by a small number of vortex and source particles. Both components are approximated using a discretization of the vorticity and divergence maps through regularized Dirac measures. The resulting so called irrotational and solenoidal fields consist then in linear combinations of basis functions obtained through a convolution product of the Green kernel gradient and the vorticity map or the divergence map respectively. The coefficient values and the basis function parameters are obtained by minimization of a functional relying on an integrated version of mass conservation principle of fluid mechanics. Results are provided on real world sequences.

1 Introduction

The observation, understanding and control of complex fluid flows is a major scientific issue. For instance, in environmental sciences such as oceanography, meteorology and climatology, the monitoring or the forecasting of the atmosphere or the ocean is becoming more and more crucial for our everyday life. Due to their very complex nature and also to unknown or inaccurate border conditions, we have a lack of complete physical understanding of these flows. Accurate and dense measurements can hardly be recovered by probes or by numerical evaluation of current physical models. Imaging sensors are very attractive in this context as they provide multi-modal data at high spatio-temporal resolution.

The analysis of dynamic structures and the estimation of velocities for fluid image sequences gave rise to a great attention from the computer vision community since several years [6, 7, 10, 12, 15, 16]. These works concern application domains such as experimental visualization in fluid mechanics, environmental sciences (oceanography, meteorology, ...), or medical imagery.

Recently, several dedicated approaches have been proposed for fluid flow velocity estimation [4, 9]. Unlike most of the motion estimator based on the brightness consistency assumption and a first order smoothness function, these techniques rely on a data-model derived from the *continuity equation* of fluid mechanics and second order *div-curl* regularizers. In the same way as a first order regularizer (eventually associated

R. Kimmel, N. Sochen, J. Weickert (Eds.): Scale-Space 2005, LNCS 3459, pp. 254–266, 2005.

to a robust cost function) favors piecewise translational motion fields by penalizing high gradients of the solution, second order *div-curl* penalizers encourage solutions with blobs of piecewise constant divergence and curl. These methods are conceptually much more satisfying as they comply with the brightness variations and the motions observed in fluid image sequences. Nevertheless, these method have to face much greater numerical complexity. Besides, these estimators are dense estimator and the solutions associated belong therefore to spaces of great dimension. It is desirable for some applications to provide low dimensional solutions. This is the purpose of this paper.

We propose here a technique to estimate low dimensional motion field from image sequences depicting a fluid flow. This method relies on the Helmholtz decomposition of a motion field which consists to decouple the vector into a divergence free component and a curl free component. The method we devise is based on a discrete representation of the curl (also called vorticity) and divergence map. This discretization enables to define implicitly adapted regularizers for fluid motion estimation problems.

2 Definitions and Properties of Vector Fields

In this section, we present first known analytic results on planar vector fields. We shall rely on them to develop an original method for fluid motion estimation.

A two-dimensional vector field $\mathbf{w}$ is a $\mathbb{R}^2$-valued map defined on a bounded set Ω of $\mathbb{R}^2$. We denote it $\mathbf{w}(\mathbf{x}) = (u(\mathbf{x}), v(\mathbf{x}))^T$, where $\mathbf{x} = (x, y)$ and x and y are the spatial coordinates. Each component of the vector field will be supposed twice continuously differentiable: $u, v \in C^2(\Omega, \mathbb{R})$.

Noting $\nabla = (\frac{\partial}{\partial x}, \frac{\partial}{\partial y})$ the operator whose components are the partial derivatives with respect to the coordinates x and y, we define the *divergence*: $\text{div}\,\mathbf{w} = \dfrac{\partial u}{\partial x} + \dfrac{\partial v}{\partial y} = \nabla.\mathbf{w}$ and the scalar *vorticity* of the vector field: $\text{curl}\,\mathbf{w} = \dfrac{\partial u}{\partial y} - \dfrac{\partial v}{\partial x} = \nabla.\mathbf{w}^{\perp}$, where $\mathbf{w}^{\perp} = (-v, u)$ is the orthogonal counterpart of $\mathbf{w}$.

The vorticity accounts for the presence of a rotating motion, while the divergence is related to the presence of *sinks* or *sources* in the flow. A vector field whose divergence is null at every point is called *solenoidal*. Similarly, a field with zero vorticity will be called *irrotational*. It is well known that for irrotational fields there exists a scalar function ϕ, called the *velocity potential*, such that $\boldsymbol{w} = \nabla\phi$. Similarly, for solenoidal fields there exists a scalar function ψ called the *stream function* such that $\boldsymbol{w}^{\perp} = \nabla\psi$.

Any continuous vector field that vanishes at infinity can be decomposed into a sum of an irrotational component with null vorticity and a solenoidal component with null divergence. This is called the *Helmholtz Decomposition*. When the null border condition can not be imposed, an additional component, named the *laminar* component, which is both irrotational and solenoidal, has to be included. The decomposition reads then: $\mathbf{w} = \mathbf{w}_{irr} + \mathbf{w}_{sol} + \mathbf{w}_{lam}$. This last component can be approximated using the Horn and Schunck estimator with a strong regularization coefficient [5]. In the sequel we will assumed that the laminar component has been previously computed and that its associated motion has been removed from the image sequence. We will consequently assume a null boarder condition at infinity knowing that the image sequence, $I(\mathbf{x}, t)$, is related to the original image sequence, $I_o(\mathbf{x}, t)$, by $I(\mathbf{x}, t) = I_o(\mathbf{x} + \mathbf{w}_{lam}(\mathbf{x}, t), t)$.

Substituting the two components $\mathbf{w}_{irr}$ and $\mathbf{w}_{sol}$ by their expressions in terms of potential functions and considering the divergence and the curl of the motion field enables to write the potential function as solution of two Poisson equations:

$$\Delta\phi = \text{div}\mathbf{w}_{irr} \quad \text{and} \quad \Delta\psi = -\text{curl}\mathbf{w}_{sol}, \tag{1}$$

where Δ denotes the Laplacian operator. These solutions may be expressed as convolution products:

$$\phi = \int G(\mathbf{x} - \mathbf{u})\text{div } \mathbf{w}_{irr}(\mathbf{u})d\mathbf{u} = G \otimes \text{div } \mathbf{w}_{irr}, \tag{2}$$

$$\psi = -\int G(\mathbf{x} - \mathbf{u})\text{curl } \mathbf{w}_{sol}(\mathbf{u})d\mathbf{u} = -G \otimes \text{curl } \mathbf{w}_{sol}, \tag{3}$$

where G is the Green's function associated to the two-dimensional Laplacian:

$$G(\mathbf{x}) = \frac{1}{2\pi}\ln(|\mathbf{x}|). \tag{4}$$

As the vector fields $\mathbf{w}_{irr}$ and $\mathbf{w}_{sol}$ are respectively the gradient and the orthogonal gradient of the potential functions ϕ and ψ, equation (2-3) may be rewritten as:

$$\mathbf{w}_{irr} = K \otimes \text{div } \mathbf{w}_{irr} \text{ and } \mathbf{w}_{sol} = -K^{\perp} \otimes \text{curl } \mathbf{w}_{sol}, \tag{5}$$

where K denotes the gradient of the Green kernel. The second equation of (5) is known as the *Bio-Savart* integral. These two equations state that the solenoidal and the irrotational components (and consequently the whole vector field) may be recovered through a convolution product knowing the divergence and the vorticity of the velocity field.

3 Vortex Particles

The idea of vortex particles methods [2, 11] consists in approximating the vorticity of a field $\mathbf{w}$ by a discrete sum of delta functions located at *point vortices* $\mathbf{z}_i$:

$$\text{curl } \mathbf{w}(\mathbf{x}) \approx \sum_{i=0}^{n} \gamma_i \delta(\mathbf{x} - \mathbf{z}_i), \tag{6}$$

with δ denoting the Dirac measure.

This discretization of the vorticity into a limited number of elements enables to evaluate the velocity field directly from the Bio-Savart integral (equ. 5). Due to the singularity of the Green kernel gradient, K, the induced field develops $\frac{1}{r}$-type singularities, where r is the distance to the point vortices. These singularities can be removed by smoothing the Dirac measure with a *cutt-off* or *blob* function, leading to a smoothed version of K. Let f_ϵ be such a blob function scaled by a parameter ϵ: $f_\epsilon(\mathbf{x}) = \frac{1}{\epsilon^2} f(\frac{\mathbf{x}}{\epsilon})$. The smoothed kernel is defined as $K_\epsilon = K \otimes f_\epsilon$. The amount of smoothing is determined by the value of ϵ. If $\epsilon \to 0$, f_ϵ tends to the Dirac function and $K_\epsilon \to K$.

In the same way, for the divergence map a *source* particles representation reads then:

$$\text{div } \mathbf{w}(\mathbf{x}) \approx \sum_{i=0}^{n} \gamma_i f_{\epsilon_i}(\mathbf{x} - \mathbf{z}_i), \tag{7}$$

where $\mathbf{z}_i$ denotes the center of each basis function f_{ϵ_i}, the coefficient γ_i is the strength associated to the particle i, and ϵ_i represents its influence domain. These parameters are free to vary from a function to another.

4 Fluid Motion Estimation from Image Sequences

In this section we present how a vortex and source particles representation may be used in conjunction with an appropriate cost function to devise a motion estimator for image sequences depicting fluid flows.

4.1 Motion Representation

As we saw previously, discretizing the vorticity map with vortex particles together with a Gaussian smoothing of the Dirac measure leads through Bio-Savart integral to the following representation of the solenoidal component of the motion field:

$$\mathbf{w}_{sol}(\mathbf{x}) \approx \sum_{i=0}^{n^{sol}} \gamma_i^{sol} K^{\perp} \otimes f_{\epsilon_i^{sol}}(\mathbf{z}_i^{sol} - \mathbf{x}) = \sum_{i=0}^{n^{sol}} \gamma_i^{sol} K^{\perp}_{\epsilon_i^{sol}}(\mathbf{z}_i^{sol} - \mathbf{x}), \tag{8}$$

where $K^{\perp}_{\epsilon_i}$ is a new kernel function obtained by convolving the orthogonal gradient of the Green kernel with the blob function. Obviously, a similar representation of the irrotational component can be obtained using source particles.

As a result, we exhibit an approximation of the complete motion field as weighted sums of *basis functions* defined by their center location and respective spatial influence. With a Gaussian smoothing function which allows to derive analytically the associated smoothed kernel K_ϵ, the final expressions of the motion field components are:

$$\mathbf{w}_{sol}(\mathbf{x}) = \sum_{i=0}^{n^{sol}} \gamma_i^{sol} \frac{(\mathbf{z}_i^{sol} - \mathbf{x})^{\perp}}{2\pi |\mathbf{x} - \mathbf{z}_i^{sol}|^2} \left(1 - e^{-\frac{|\mathbf{x} - \mathbf{z}_i^{sol}|^2}{\epsilon_i^{sol\,2}}}\right), \tag{9}$$

and

$$\mathbf{w}_{irr}(\mathbf{x}) = \sum_{i=0}^{n^{irr}} \gamma_i^{irr} \frac{\mathbf{x} - \mathbf{z}_i^{irr}}{2\pi |\mathbf{x} - \mathbf{z}_i^{irr}|^2} \left(1 - e^{-\frac{|\mathbf{x} - \mathbf{z}_i^{irr}|^2}{\epsilon_i^{irr\,2}}}\right). \tag{10}$$

This representation will be incorporated within a spatio-temporal variation model of the luminance function in order to devise fluid motion recovery as an estimation problem from the image sequence data.

4.2 Integrated Continuity Equation as a Brightness Variation Model

For image sequences showing evolving fluid phenomena, the usual brightness consistency assumption ($\frac{dI}{dt} = 0$) doesn't allow to model temporal distortions of luminance patterns caused by 3D matter transportation. For such kind of sequences, several works have shown that a data model build from an analogy with the mass conservation constraint of fluid mechanics (also known as *continuity equation*) constitutes a better model [1,4, 13, 15]. This data model reads:

$$\frac{dI}{dt} + I \mathrm{div} \mathbf{w} = 0. \tag{11}$$

Such a constraint relates the effect of a divergent motion to a brightness change. By this way it is possible to modelize the effect of the apparent disappearance/appearance of matter caused by 3D motions which are not in the visualization plane. For a null divergence this data model reduces exactly to the usual brightness consistency equation.

For long range displacements (i.e. fast flows or long time latency between two images as in meteorology) an integrated form of this constraint can be obtained[4]:

$$I(\mathbf{x} + \mathbf{w}(\mathbf{x}), t+1) \exp(\mathrm{div}\mathbf{w}(\mathbf{x})) - I(\mathbf{x}, t) = 0. \tag{12}$$

According to this constraint the displaced image at time $t+1$ is related to the image at time t by a scale factor which depends on the motion divergence. This constraint comes to the standard displaced frame formulation of brightness consistency for a null divergence.

Considering this constraint holds almost everywhere on the whole image plane leads to seek a motion field minimizing the following cost function:

$$\mathcal{F}(I, \mathbf{w}) = \int_{\Omega} \left[I(\mathbf{x} + \mathbf{w}(\mathbf{x}), t+1) \exp(\mathrm{div}\, \mathbf{w}(\mathbf{x})) - I(\mathbf{x}, t) \right]^2 d\mathbf{x}. \tag{13}$$

4.3 General Minimization Problem

Considering such a cost function for an unknown motion field approximated through vortex and source particles representations comes down to solve the following minimization problem:

$$\hat{\beta} = \arg\min_{\beta} \mathcal{F}(I, \mathbf{w}(\beta)), \tag{14}$$

with $\beta = (\{\mathbf{z}_i^{sol}, \gamma_i^{sol}, \epsilon_i^{sol}\}_{i=1:n^{sol}}, \{\mathbf{z}_i^{irr}, \gamma_i^{irr}, \epsilon_i^{irr}\}_{i=1:n^{irr}})$.

One seeks therefore the minimizer of the cost function $\mathcal{F}$ in terms of particles location, strength coefficients and influence domains. Due to the peculiar form of the data model this minimization problem is highly non linear. To face this difficult optimization problem we have chosen to rely on a non linear least square process embedded in a multi-resolution framework and associated to a generalized conjugated gradient optimization known as *Fletcher-Reeves* method.

We present more precisely in the next section how this difficult global optimization issue is handled.

5 Estimation

The non linear cost function we consider can be seen as a weighted displaced frame differences cost function. As most of the standard motion estimators based on such a non linear formulation we will consider an incremental minimization framework to remove the non linearity of the displaced image brightness function. This scheme consists in applying successive linearizations around previous estimates. This kind of techniques, in the same spirit as Gauss-Newton non linear least squares, is in most of the case embedded within a multi-resolution framework. We will also rely on such a data representation.

5.1 Incremental Estimation Scheme

We assume first that a previous estimate of the set of unknowns is available. All these unknowns combine together with respect to our modelization to give a motion field $\tilde{\mathrm{w}}$. Considering a linearization around $(\boldsymbol{x} + \tilde{\mathrm{w}}, t+1)$ and dropping the time indices of the intensity function for sake of clarity we end up with the following functional to be minimized according to $\mathbf{h}$, an unknown correction motion field:

$$\mathcal{F}(\mathbf{h}) = \int_{\Omega} \Big[\exp(\mathrm{div}\ \tilde{\mathrm{w}}(\mathrm{x}))\{(\tilde{I}(\mathrm{x})\nabla div\tilde{\mathrm{w}}(\mathrm{x}) + \nabla\tilde{I}(\mathrm{x}))^T\mathbf{h}(\mathrm{x}) + \tilde{I}(\mathrm{x})\} - I(\mathrm{x})\Big]^2 d\mathrm{x}.$$

In this equation we have introduced a compact notation $\tilde{I}(\mathrm{x})$ for the backward registered image $I(\boldsymbol{x} + \tilde{\mathrm{w}}, t+1)$. The correction field $\mathbf{h}$ is a combination of a solenoidal component $\mathbf{h}_{sol}$ and an irrotational component $\mathbf{h}_{irr}$ according to the Helmholtz decomposition. Like the field $\tilde{\mathrm{w}}$, this correction field is parameterized on the basis of a set of vortex and source particles. In practice, this kind of scheme is embedded into a pyramidal multiresolution data representation scheme. Such a representation is obtained through low-pass filtering and sub-sampling. At a given level, the known motion estimate $\tilde{\mathrm{w}}$ is fixed to be the projected estimate obtained at the previous level. For the first level this field is a null field.

5.2 Resulting Minimization Problem

The incremental estimation scheme transforms the original non linear optimization problem (14) into a succession of simpler minimization problems with respect to some of the unknowns. As a matter of fact, considering the derivatives with respect to the different types of unknowns gives:

$$\frac{\partial \mathcal{F}(\mathbf{h})}{\partial \gamma_i} = \int_{\Omega} \frac{r_i(\mathrm{x})}{\pi |r_i(\mathrm{x})|^2}(1 - e^{-\frac{|r_i(\mathrm{x})|^2}{\epsilon_i{}^2}})y(\mathrm{x})[y(\mathrm{x})^T\mathbf{h}(\mathrm{x},\gamma_i) + z(\mathrm{x})]d\mathrm{x}, \tag{15}$$

$$\left.\frac{\partial \mathcal{F}(\mathbf{h})}{\partial \beta_i}\right|_{\beta_i = \frac{1}{\epsilon_i}} = \int_{\Omega} \frac{2\gamma_i}{\pi\epsilon_i}\frac{r_i(\mathrm{x})}{|r_i(\mathrm{x})|^2} e^{-\frac{|r_i(\mathrm{x})|^2}{\epsilon_i{}^2}} y(\mathrm{x})[y(\mathrm{x})^T\mathbf{h}(\mathrm{x},\epsilon_i) + z(\mathrm{x})]d\mathrm{x}, \tag{16}$$

$$\nabla_{\mathbf{z}_i}\mathcal{F}(\mathbf{h}) = \begin{pmatrix} \dfrac{\partial \mathcal{F}(\mathbf{h})}{\partial x_i} \\ \dfrac{\partial \mathcal{F}(\mathbf{h})}{\partial y_i} \end{pmatrix}, \tag{17}$$

Finally, in order to restrict the displacements of the different particles to localized areas we combine these functions with an *a priori* prior on the particles location.

A Priori Probability Distribution for Particles Location. Considering $\mathbf{z}_i^k$ the random vector denoting the location of particle i at step k, we propose to fix a distribution of $\mathbf{z}_i^{k+1}$, knowing $\mathbf{z}_{1:n}^k$, where $\mathbf{z}_{1:n}^k$ represents the set of the n vectors $(\mathbf{z}_1^k, ..., \mathbf{z}_n^k)$ at step k. We assume this probability distribution is Gaussian, defined as $\mathbf{z}_i^{k+1}|\mathbf{z}_{1:n}^k \sim \mathcal{N}(\mathbf{z}_i^k, \sigma_i^k)$. The standard deviation σ_i^k is set to the half of the distance between $\mathbf{z}_i^k$ and the closest center among $\{\mathbf{z}_j^k\}_{j=1,..,n,j\neq i}$. The distribution takes into account the previous location of the particles through a Gaussian prior of mean $\mathbf{z}_i^k$ but also the dependency between $\mathbf{z}_i^{k+1}$ and all the other particles through the expression of σ_i^k.

Conditional Version of the Probability Distribution. Combining the *a priori* distribution $p_{\mathbf{z}_i^{k+1}|\mathbf{z}_{1:n}^k}$ defined above with the surface described before, denoted $\mathcal{S}_{\mathbf{z}_{1:n}^k}$ and characterized by (22) or (23), we can define a conditional probability distribution function of a particle $\mathbf{z}_i^{k+1}$ given the others:

$$p_{\mathbf{z}_i^{k+1}|\mathbf{z}_{1:n}^k,\mathcal{S}_{\mathbf{z}_{1:n}^k}}(\mathbf{x}) \propto \mathcal{S}_{\mathbf{z}_{1:n}^k}(\mathbf{x}).p_{\mathbf{z}_i^{k+1}|\mathbf{z}_{1:n}^k}(\mathbf{x}). \quad (24)$$

This pdf balances an *a priori* for the location of one given particle (whose role is to confine the particle to stay in a certain area between two iterates) and the information brought by the characteristic surface (associated to all the particles locations) in the neighborhood of this position. Once known this distribution for each particle we propose to shift $\mathbf{z}_i^k$ towards the pdf local mode in order to adjust optimally the location of the particles set.

Shifting the Particles Towards the Pdf Modes. From the sample $\{\mathcal{S}_{\mathbf{z}_{1:n}^k}(\mathbf{s})\}_{\mathbf{s}\in\mathcal{S}}$ evaluated at pixel coordinates $\mathbf{s}$, and the probability distribution $p_{\mathbf{z}_i^{k+1}|\mathbf{z}_{1:n}^k}$, a statistical non parametric estimate of the conditional probability distribution $p_{\mathbf{z}_i^{k+1}|\mathbf{z}_{1:n}^k,\mathcal{S}_{\mathbf{z}_{1:n}^k}}$, may be obtained [14] as:

$$\hat{p}_{\mathbf{z}_i^{k+1}|\mathbf{z}_{1:n}^k,\mathcal{S}_{\mathbf{z}_{1:n}^k}}(\mathbf{x}) \propto \frac{\sum_{\mathbf{s}\in\mathcal{S}} \mathcal{S}_{\mathbf{z}_{1:n}^k}(\mathbf{s}) p_{\mathbf{z}_i^{k+1}|\mathbf{z}_{1:n}^k}(\mathbf{s}) K(\frac{\mathbf{x}-\mathbf{s}}{h})}{\sum_{\mathbf{s}\in\mathcal{S}} K(\frac{\mathbf{x}-\mathbf{s}}{h})}, \quad (25)$$

where K is a kernel and h is its corresponding window size.

The continuous pdf $\hat{p}_{\mathbf{z}_i^{k+1}|\mathbf{z}_{1:n}^k,\mathcal{S}_{\mathbf{z}_{1:n}^k}}(\mathbf{x})$ is thus expressed as a linear combination of basis functions with weighted coefficients given by $w(\mathbf{s}) = \mathcal{S}_{\mathbf{z}_{1:n}^k}(\mathbf{s}) p_{\mathbf{z}_i^{k+1}|\mathbf{z}_{1:n}^k}(\mathbf{s})$.

To shift a center $\mathbf{z}_i^k$ towards the nearest mode of $\hat{p}_{\mathbf{z}_i^{k+1}|\mathbf{z}_{1:n}^k,\mathcal{S}_{\mathbf{z}_{1:n}^k}}$ we rely on the *mean shift* estimate of the gradient of a density function [3, 8]. This estimate called the *mean shift vector* reads:

$$M_{h,G}(\mathbf{x}) = \frac{\sum_{\mathbf{s}\in\mathcal{S}} w(\mathbf{s})\mathbf{s}G(\frac{\mathbf{x}-\mathbf{s}}{h})}{\sum_{\mathbf{s}\in\mathcal{S}} w(\mathbf{s})G(\frac{\mathbf{x}-\mathbf{s}}{h})} - \mathbf{x}, \quad (26)$$

where G is the kernel obtained by derivation of the kernel K. This vector gives at each point the direction of the maximum increase of the density function estimated through the weights $w(\mathbf{s})$ and the kernel K. Different choices can be done for this kernel. Usual choices are the Epanechnikov kernel or a Gaussian kernel. The gradient of the Epanechnikov kernel is a box function kernel whereas G remains Gaussian for a Gaussian kernel K.

Given this estimate of the pdf gradient, an iterative convergent [3] process called mean shift naturally arises. This process consists in moving iteratively the kernel center $\mathbf{x}$ following $M_{h,G}(\mathbf{x})$ until a stationary point (i.e., zero gradient) of the underlying density is found.

In our case, the mean shift procedure is applied to the $n^{sol} + n^{irr}$ centers of the basis functions (or particles) involved in our motion field modelization. Through this process, each particle is shifted towards the nearest mode of the conditional density $\hat{p}_{\mathbf{z}_i^{k+1}|\mathbf{z}_{1:n}^k, S_{\mathbf{z}_{1:n}^k}}$. We have chosen to use the Epanechnikov kernel. Besides, the choice of the window size is crucial. Different choices can be made. In our case we have settled adaptive window sizes. They are fixed to the distance of the nearest particles. Such a choice make sense in our case. As a matter of fact, for distant particles only a rough and smooth estimate of the pdf function is needed whereas for close particles an accurate estimate of the density is at the opposite required to approximate at best the vorticity and divergence maps.

5.5 Overall Estimation Scheme

The overall estimation scheme consists in an alternate updating of the different unknowns. It is composed by the following two steps, repeated in turn until convergence:

1. For a given set of particles at fixed locations, the strength coefficients and the influence domains attached to the particles blob function are estimated through the generalized conjugated gradient optimization described in section 5.3.
2. The vortex and source particles locations are shifted toward the nearest local mode of the corresponding pdf. This shift is realized applying the mean shift procedure described in section 5.4.

The whole process is stopped when the divergence and vorticity reach a certain stability. This criterion is expressed as:

$$\left(\frac{\|\text{div}\,\tilde{\mathbf{h}}^{k+1} - \text{div}\,\tilde{\mathbf{h}}^k\|_2}{\|\text{div}\,\tilde{\mathbf{h}}^k\|_2}\right)^2 + \left(\frac{\|\text{curl}\,\tilde{\mathbf{h}}^{k+1} - \text{curl}\,\tilde{\mathbf{h}}^k\|_2}{\|\text{curl}\,\tilde{\mathbf{h}}^k\|_2}\right)^2$$

6 Results

In this section we present some results given by our method on real sequences.

The first example corresponds to the motion of smoke behind a landing passenger air plane. A strong vortex is located in the center of the image, and a second weaker one begins to appear just below. The particles are initialized on a grid, without *a priori*. The

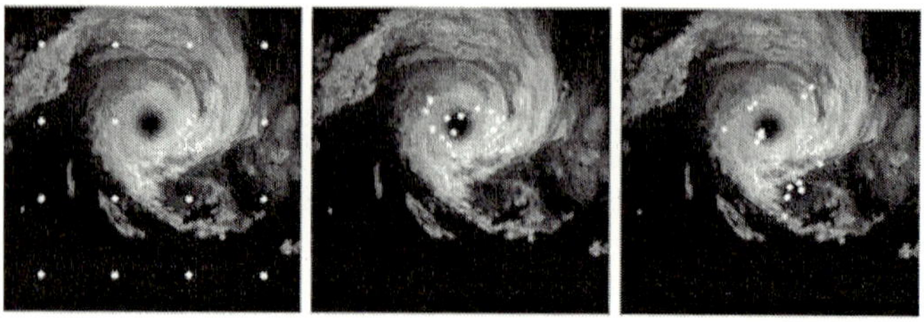

(a) (b) (c)

Fig. 1. Plane sequence. (a) Initial uniform disposition of the particles; (b) Final position of the particles at the first level of multiresolution; (c) Final position of the particles at the second level

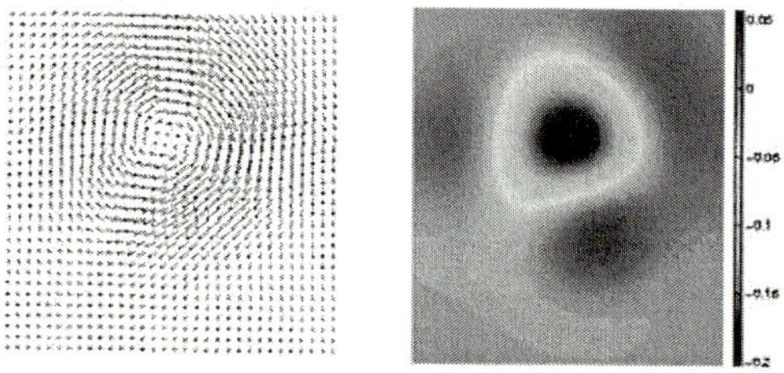

(a) (b)

Fig. 2. Plane sequence. (a) Resulting motion field; (b)Associated vorticity

estimation method allows to guide the vortex particles towards the regions of interest of the image and to estimate an accurate motion field (see the vector field and the associated vorticity map fig. 2). For this sequence we used a multi-resolution pyramid of two levels. At the first level, the particles move all towards the strong vortex (fig. 1 (b)). At the finest level, the particles cloud splits up into two parts (fig. 1 (c)). A set of particles has moved towards the weaker vortex, authorizing them to capture its motion.

The second example shows results on two consecutive images of the infra-red channel of Meteosat. The sequence represents a depression with a vortex in the left part of the image domain and presence of convective clouds in the center. In this example, we want to observe the motion in specific areas, we dispose thus the vortex and source particles

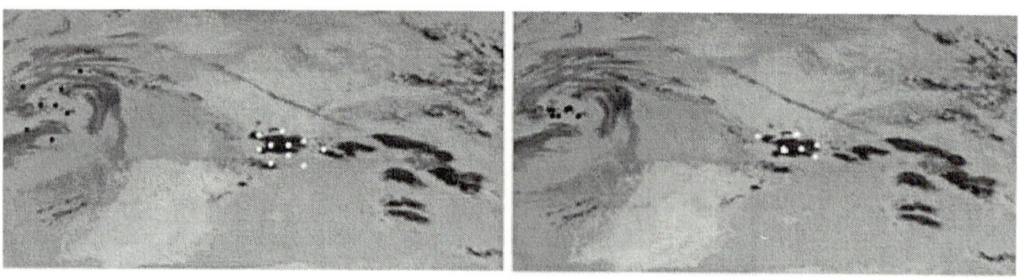

(a) (b)

Fig. 3. Depression sequence. (a) Initial manual disposition of the particles. Black points represent the vortex particles, white points the source ones; (b) Final position of the same particles

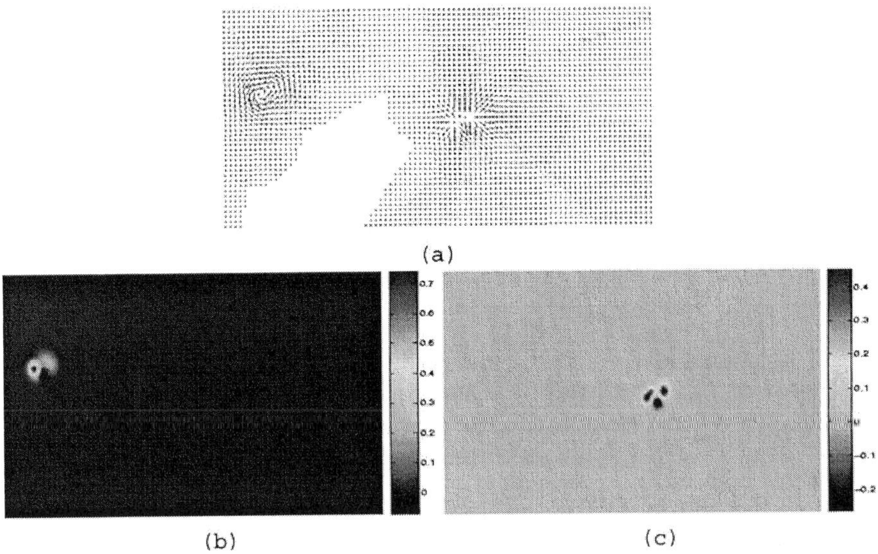

Fig. 4. Depression sequence. (a) Resulting motion field; (b) Associated vorticity; (c) Associated divergence

manually in the regions of interest (fig. 3(a)). During the estimation, the particles location fits locally automatically. At convergence, the vortex particles remains mostly concentrated in the center of the vortex, while the source particles are located on the convective cloud (fig. 3(b)).

7 Conclusion

In this paper we have presented an optical flow estimator dedicated to image sequences depicting fluid flows. The proposed estimator provides a low dimensional parametric representation of fluid motion. This parameterization has been obtained through a peculiar discretization of the divergence and the vorticity map by means of adapted basis function centered at elements named particles. To handle the associated estimation problem we have proposed an efficient strategy based on the coupling of a generalized conjugated gradient and a mean shift process.

References

1. D. Béréziat, I. Herlin, and L. Younes. A generalized optical flow constraint and its physical interpretation. In *Proc. Conf. Comp. Vision Pattern Rec.*, volume 2, pages 487–492, Hilton Head Island, South Carolina, USA, 2000.
2. A. Chorin. Numerical study of slightly viscous flow. *J. Fluid Mech.*, 57:785–796, 1973.
3. D. Comaniciu and P. Meer. Mean shift: A robust approach toward feature space analysis. *IEEE Trans. Pattern Analysis Machine Intelligence*, 24(5):603–619, 2002.
4. T. Corpetti, E. Mémin, and P. Pérez. Dense estimation of fluid flows. *IEEE Trans. Pattern Anal. Machine Intell.*, 24(3):365–380, 2002.

5. T. Corpetti, E. Mémin, and P. Pérez. Extraction of singular points from dense motion fields: an analytic approach. *J. Mathematical Imaging and Vision*, 19(3):175–198, 2003.
6. J.M. Fitzpatrick. A method for calculating velocity in time dependent images based on the continuity equation. In *Proc. Conf. Comp. Vision Pattern Rec.*, pages 78–81, San Francisco, USA, 1985.
7. R.M. Ford, R. Strickland, and B. Thomas. Image models for 2-d flow visualization and compression. *Graph. Mod. Image Proc.*, 56(1):75–93, 1994.
8. K. Fukanaga and L.D. Hostetler. The estimation of the gradient of a density function, with applications in pattern recognition. *IEEE Trans. on Info. Theory*, 21(1):32–40, 1975.
9. T. Kohlberger, E. Mémin, and C. Schnörr. Variational dense motion estimation using the Helmholtz decomposition. In *Int. conf on Scale-Space theories in Computer Vision(Scale-Space '03)*, Isle of Skye, june 2003.
10. R. Larsen, K. Conradsen, and B.K. Ersboll. Estimation of dense image flow fields in fluids. *IEEE trans. on Geoscience and Remote sensing*, 36(1):256–264, Jan. 1998.
11. A. Leonard. Vortex methods for flow simulation. *J. Comp. Phys.*, 37, 1980.
12. E. Mémin and P. Pérez. Fluid motion recovery by coupling dense and parametric motion fields. In *Int. Conf. on Computer, ICCV'99*, pages 620–625, 1999.
13. B.G. Schunk. The motion constraint equation for optical flow. In *Proc. Int. Conf. Pattern Recognition*, pages 20–22, Montreal, 1984.
14. R.A. Thisted. *Elements of statistical computing*. Chapman and Hall, 1988.
15. R. Wildes, M. Amabile, A.M. Lanzillotto, and T.S. Leu. Physically based fluid flow recovery from image sequences. In *Proc. Conf. Comp. Vision Pattern Rec.*, pages 969–975, 1997.
16. L. Zhou, C. Kambhamettu, and D. Goldgof. Fluid structure and motion analysis from multi-spectrum 2D cloud images sequences. In *Proc. Conf. Comp. Vision Pattern Rec.*, volume 2, pages 744–751, Hilton Head Island, South Carolina, USA, 2000.

Discrete Orthogonal Decomposition and Variational Fluid Flow Estimation

Jing Yuan[1], Paul Ruhnau[1], Etinne Mémin[2], and Christoph Schnörr[1]

[1] Department of Mathematics and Computer Science,
Computer Vision, Graphics, and Pattern Recognition Group,
University of Mannheim, 68131 Mannheim, Germany
{yuanjing, ruhnau, schnoerr}@uni-mannheim.de
www.cvgpr.uni-mannheim.de

[2] IRISA Rennes, Campus Universitaire de Beaulieu,
35042 Rennes Cedex, France
momin@irisa.fr
www.irisa.fr/vista

Abstract. The decomposition of motion vector fields into components of orthogonal subspaces is an important representation for both the analysis and the variational estimation of complex motions. Common finite differencing or finite element methods, however, do not preserve the basic identities of vector analysis. Therefore, we introduce in this paper the mimetic finite difference method for the estimation of fluid flows from image sequences. Using this discrete setting, we represent the motion components directly in terms of potential functions which are useful for motion pattern analysis. Additionally, we analyze well-posedness which has been lacking in previous work. Experimental results, including hard physical constraints like vanishing divergence of the flow, validate the theory.

1 Introduction

The estimation of highly non-rigid image flows is an important problem in various application areas of image analysis like remote sensing, medical imaging, and experimental fluid mechanics. Such flows, which cannot be represented by a single parametric model, are typically estimated by variational approaches. In contrast to standard approaches, however, higher-order regularization is necessary in order to accurately recover important flow structures like vortices, for example, and to incorporate physically plausible constraints, like vanishing divergence of the flow.

The basis for our paper is early work on second-order regularizers constraining the gradients of the flow components divergence and curl [1, 2, 3]. This regularization approach has been elaborated in a series of papers by Mémin and co-workers [4, 5, 6]. Moreover, the decomposition and representation of *continuous* vector fields by velocity potentials and stream functions [7] has been adopted to

R. Kimmel, N. Sochen, J. Weickert (Eds.): Scale-Space 2005, LNCS 3459, pp. 267–278, 2005.

derive piecewise parametric representations of relevant flow structures. Recently, the direct estimation of this representation has been studied in [8].

Contribution. From numerical fluid dynamics, it is well known that standard discretizations, like piecewise linear finite elements, are not approproriate. Imposing the constraint of vanishing divergence, for example, may result in a constant flow. Therefore, we introduce the mimetic finite difference method [9, 10, 11] to the field of image sequence analysis, which uses basic integral identities of vector analysis to derive discrete differential operators preserving these relationships *after* discretization. Based on this *exact* discrete representation, we study div-curl regularization, detect and remove a corresponding sensitivity of this regularizer to "boundary noise", state precise conditions for well-posedness, and present a *provably convergent* iterative implementation for directly estimating velocity potentials and stream functions by iterative subspace correction. Most importantly, our approach makes the estimation of *accurate* solenoidal (non-divergent) flows feasible. The theory is validated by numerical experiments.

2 Vector-Field Representation

2.1 Discretization and Vector Spaces

We use the mimetic finite difference method for discretization [9, 10] in order to preserve basic relationships of continuous vector analysis. This discretization will be applied in section 2.2 to accurately represent and decompose vector fields.

Figure 1 illustrates the definitions of the following finite-dimensional vector spaces:

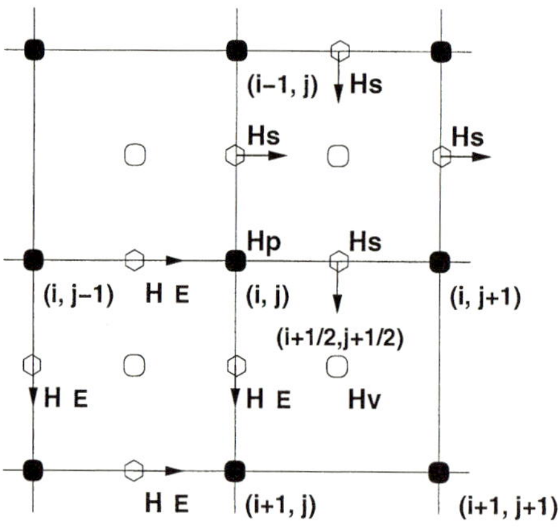

Fig. 1. Definition of finite-dimensional spaces of scalar fields and vector fields on a rectangular grid. Filled circles depict *nodes* or *vertices*, the other circles indicate *cells*. The positions of diamonds are referred to as *sides*

H_V: the space of *scalar fields* defined on cells,
H_P: the space of *scalar fields* defined on vertices,
H_E: the space of *vector fields* defined tangential to sides,
H_S: the space of *vector fields* defined normal to sides.

Furthermore, we define the following *primal* discrete first-order differential operators:

$\mathbb{G} : H_P \to H_E$ the discrete gradient operator representing ∇,
$\mathbb{G}^\perp : H_P \to H_S$ the discrete directional derivative along level curves representing $\nabla^\perp$ in the discrete case. This operator is specific to the 2D case considered here.
$\mathbb{D}iv : H_S \to H_V$ the discrete divergence operator,
$\mathbb{C}url : H_E \to H_V$ the discrete curl operator.

In order to construct the discrete second-order differential operators by combining first-order operators, *dual* discrete first-order differential operators

$$\mathbb{G}^* : H_V \to H_S\,,\ \mathbb{G}^{\perp *} : H_V \to H_E\,,\ \mathbb{D}iv^* : H_E \to H_P\,,\ \mathbb{C}url^* : H_S \to H_P$$

are defined so as to solve the incompatibilities of domains and ranges of the primal operators defined above [10]. For example, $\mathbb{G}$ and $\mathbb{D}iv$ cannot be regarded as mutually adjoint operators like in the continuous case, whereas $\mathbb{G}$, $\mathbb{D}iv^*$ and $\mathbb{G}^*$, $\mathbb{D}iv$ do.

2.2 Orthogonal Decomposition

We represent vector fields directly in terms of their irrotational and solenoidal components. These components are defined by the first-order variations of velocity potentials ψ and stream functions ϕ, respectively [11]:

Theorem 1 (Vector Field Decomposition). *For any 2D vector field $u \in H_S$, the representation of u in terms of ψ, ϕ:*

$$u = \mathbb{G}^*\psi + \mathbb{G}^\perp\phi, \quad u_{\partial\Omega} = \partial_n\psi, \tag{1}$$

where $\phi_{\partial\Omega} = 0$, is unique up to a constant of ψ.

Here, Ω denotes the image section (grid), n the corresponding outer normal vector, and $f_{\partial\Omega}$ the boundary values of f. Let

$$u = v + w\,, \quad v = \mathbb{G}^*\psi\,, \ w = \mathbb{G}^\perp\phi$$

according to (1). Since the operators defined in the previous section satisfy [11]:

$$\mathbb{D}iv\,\mathbb{G}^\perp \equiv 0\,, \qquad \mathbb{C}url^*\mathbb{G}^* \equiv 0\,,$$

we have

$$\mathbb{D}iv\, w = 0\,, \qquad \mathbb{C}url^* v = 0\,, \tag{2}$$

and:

$$\langle w, v\rangle_{H_S} = \left\langle \mathbb{G}^*\psi, \mathbb{G}^\perp\phi\right\rangle_{H_S} = \langle \mathbb{C}url^*\mathbb{G}^*\psi, \phi\rangle_{H_P} \equiv 0 \tag{3}$$

This shows:

Theorem 2 (Orthogonality). *The decomposition (1) is orthorgonal, that is:*

$$\left\langle \mathbb{G}^*\psi, \mathbb{G}^\perp\phi \right\rangle_{H_S} = 0\,, \quad \forall u \in H_S \tag{4}$$

Let S_{ir} express the subspace of all vectors which can be written as $\mathbb{G}^*\psi$ and S_{sol} the subspace of vectors which can be reprented as $\mathbb{G}^\perp\phi$. Then the previous theorem asserts that the direct sum holds:

$$H_S = S_{ir} \oplus S_{sol} \tag{5}$$

Representation (1) is motivated by analogous decompositions of continuous vector fields [7]. However, discretizing such vector fields with standard finite differences or finite elements yields *approximate* decompositions only, which may lead to numerical instabilities in applications. In contrast, theorem 1 provides an *exact* orthogonal decomposition of the *finite*-dimensional space of vector fields H_S. Furthermore, as detailed below, the decomposition allows to estimate ψ, ϕ directly, and in parallel, using variational optical flow approaches and subspace correction methods (cf. section 5.1).

Alternatively, we may first estimate u and then compute ψ and ϕ in a subsequent step by solving the Neumann and Dirichlet problems:

$$\triangle_D\psi = \mathbb{D}iv\ u\,, \quad \partial_n\psi = u_{\partial\Omega}\,, \tag{6}$$

$$\triangle_C\phi = \mathbb{C}url\ ^*u\,, \quad \phi_{\partial\Omega} = 0\,, \tag{7}$$

where the discrete Laplacians are defined by:

$$\triangle_D := \mathbb{D}iv\ \mathbb{G}^*\,, \qquad \triangle_C := \mathbb{C}url\ ^*\mathbb{G}^\perp \tag{8}$$

and the additional constraint $\sum_{cells}\psi = 0$. In the remainder of this paper, however, we show that directly estimating ψ, ϕ from image sequence data is feasible.

3 Regularization and Optimization Problems

3.1 Representation of the Data Term and Linearization

We consider pixels as cells and define accordingly $I \in H_V$ for a given image.

We use the conventional data term for optical flow estimation, along with regularizers $L(u)$ to be specified below (section 3.2):

$$\min_{u \in H_S} F(u)\,, \quad F(u) := \|I(x+u) - I(x)\|^2_{H_V} + L(u) \tag{9}$$

Note that this data term could be made robust against outliers by using some robust estimators or the L^1-norm [12]. In this paper, however, we focus on higher-order regularization in connection with the representation (1).

In order to alleviate the local minima problem, we apply the standard procedure of minimizing $F(u)$ using a sequence of linearizations of the data term:

$$F^l(u^l) := \left\|\mathbb{G}^* I_1^l \cdot u^l + \partial_t I^l\right\|_{H_V}^2 + L(u^l)\,, \tag{10}$$

where $\{I_1^l, I_2^l\}_{l=0,1,\ldots,m}$ denote linear scale-space representations of a given image pair, and $\partial_t I^l = I_1^l(x) - I_2^l(x + u^{l+1}(x))$.

3.2 Regularization

We wish to apply the following second-order regularizer (cf. the discussion of related work in section 1):

$$\int_\Omega \lambda_1 |\nabla \mathrm{div}\, u|^2 + \lambda_2 |\nabla \mathrm{curl}\, u|^2 dx \tag{11}$$

where λ_1 and λ_2 are two positive constants. This term measures the variation of the basic flow components divergence and curl, but *does not penalize* the components itself. However, both standard finite differences or finite elements discretization lead to finite-dimensional representations which do not satisfy (1), (4). As a result, penalizing one component may affect the other component too. Therefore, we adopt the framework sketched in section 2.1 which leads to the following discretization of (11):

$$L(u) := L_{div}(u) + L_{curl}(u) := \lambda_1 \left\|\mathbb{G}^* \mathbb{D}iv\; u\right\|_{H_S}^2 + \lambda_2 \left\|\mathbb{G}\mathbb{C}url\;^* u\right\|_{H_E}^2\,, \tag{12}$$

3.3 Estimation of Non-rigid Flows

Based on (12), we consider the functional:

$$\min_{u \in H_S} F(u) := \|I(x+u) - I(x)\|_{H_V}^2 + L_{div}(u) + L_{curl}(u) \tag{13}$$

Inserting the decomposition (1), we obtain the minimization problem:

$$\begin{aligned}\min_{\psi,\phi} F(\psi,\phi) = {} & \left\|I(x + \mathbb{G}^*\psi + \mathbb{G}^\perp \phi) - I(x)\right\|_{H_V}^2 \\ & + \lambda_1 \left\|\mathbb{G}^* \triangle_D \psi\right\|_{H_S}^2 + \lambda_2 \left\|\mathbb{G} \triangle_C \phi\right\|_{H_E}^2\end{aligned} \tag{14}$$

subject to the linear constraints:

$$\sum_{cells} \psi = 0\,, \quad \phi_{\partial\Omega} = 0 \tag{15}$$

Note that the first constraint fixes the free constant mentioned in theorem 1. Furthermore, the arguments of (14) are elements of orthogonal subspaces (5), and thus may be determined in parallel by subspace correction methods.

3.4 Estimation of Solenoidal Flows

An important special case, particularly in applications of experimental fluid dynamics, concerns the estimation of solenoidal (divergence-free) flows. In this case the decomposition (1) reduces to:

$$u = \mathbb{G}^* \psi_l + \mathbb{G}^\perp \phi := u_l + \mathbb{G}^\perp \phi \tag{16}$$

where the *laminar flow* u_l can be computed through the full flow u by solving:

$$\triangle_D \psi_l = 0 \,, \quad \partial_n \psi_l = u_{\partial\Omega} \tag{17}$$

and $u_l = \mathbb{G}^* \psi_l$. Since $\mathbb{C}url\,^* \mathbb{G}^* \equiv 0$, the laminar flow u_l is both divergence and curl free. In order for (17) to be solvable, we require the compatible condition $\int_{\partial\Omega} u_{\partial\Omega} dl = 0$ (cf., e.g., [13]).

Let $S_{div0} = \{u \in H_S \mid \mathbb{D}iv\ u = 0\}$ be the linear space of vector fields with vanishing divergence. Then the following direct sum holds:

$$S_{div0} = S_{lam} \oplus S_{sol} \tag{18}$$

with laminar and solenoidal flows as basic components.

In order to estimate solenoidal flows, we consider instead of (13) the functional:

$$\min_{u \in S_{div0}} F_{sol}(u) := \|I(x+u) - I(x)\|^2_{H_V} + L_{curl}(u) \tag{19}$$

which is well-defined by (18). Inserting the decomposition (16), we obtain the minimization problem:

$$\min_{\psi_l, \phi} F_{sol}(\psi_l, \phi) = \left\| I(x + \mathbb{G}^* \psi_l + \mathbb{G}^\perp \phi) - I(x) \right\|^2_{H_V} + \lambda \left\| \mathbb{G} \triangle_C \phi \right\|^2_{H_E} \tag{20}$$

subject to the constraints:

$$\triangle_D \psi_l = 0 \,, \quad \sum_{cells} \psi_l = 0 \,, \quad \phi_{\partial\Omega} = 0 \tag{21}$$

Note that the arguments of (20) are elements of orthogonal subspaces (18), and thus may be determined in parallel by subspace correction methods.

4 Well-Posedness and Stability

4.1 Well-Posedness

We state the conditions under which the functionals (13) and (19) with *linearized* data terms (cf. (10)) are strictly convex. To this end, we consider the spaces:

$$\begin{aligned}
S_d &= \{u \in H_S \mid \mathbb{D}iv\ u = C \,,\ \mathbb{C}url\,^* u = 0 \,,\ C \in \mathbb{R} \text{ arbitrary}\} \\
S_c &= \{u \in H_S \mid \mathbb{D}iv\ u = 0 \,,\ \mathbb{C}url\,^* u = C \,,\ C \in \mathbb{R} \text{ arbitrary}\} \\
S_{dc} &= \{u \in H_S \mid u = u_1 + u_2 \,,\ u_1 \in S_d \,,\ u_2 \in S_c\} \\
S_g &= \{u \in H_S \mid \mathbb{G}^* I_1 \cdot u = 0\}
\end{aligned}$$

As we work with finite-dimensional vector fields, the following two assertions are obvious: problem

$$\min_{u \in H_S} F(u) := \|\mathbb{G}^* I_1 \cdot u + \partial_t I\|^2_{H_V} + \lambda_1 \|\mathbb{G}^* \mathbb{D}iv\ u\|^2_{H_S} + \lambda_2 \|\mathbb{G}\mathbb{C}url\,^* u\|^2_{H_E} \tag{22}$$

is strictly convex iff the subspaces S_g and S_{dc} trivially intersect, that is there is no vector $0 \neq u \in S_{dc}$ which is perpendicular to $\mathbb{G}^* I_1$. Similarly, problem

$$\min_{u \in S_{div0}} F_{sol}(u) := \|\mathbb{G}^* I_1 \cdot u + \partial_t I\|^2_{H_V} + \lambda \|\mathbb{G}\mathbb{C}url\, ^* u\|^2_{H_E} \tag{23}$$

is strictly convex iff S_g and S_c trivially intersect.

4.2 Stability

It is well-known that existence of a unique solution, as established in the previous section, does not say much about *numerical* stability. Indeed, inspection of the second-order regularizer (11) reveals a particular sensivity of u with respect to the image data, and suggests using a corresponding regularizer.

To motivate this additional term, we consider the following representation of vector fields u in terms of functions ρ, ω and boundary data f:

$$\operatorname{div} u = \rho\,, \quad \operatorname{curl} u = \omega\,, \quad u_{\partial\Omega} = f$$

Provided the compatibility condition:

$$\int_\Omega \rho\, dx = \int_{\partial\Omega} f\, dl \tag{24}$$

holds, u is uniquely defined, both in the continuous case [13] and in the discrete case, using the discretization of section 2.1.

It is clear that the regularizer (11) only constrains ρ and ω, but *not* f which is weakly constrained only through the data terms of the functionals considered above. Therefore, in practice, it turned out to be useful to reduce this sensivity of u by including a regularizer which weakly constraints the boundary values:

$$\int_{\partial\Omega} (\partial_n u)^2\, dl. \tag{25}$$

By virtue of the orthogonal decomposition, this constraint can be expressed in terms of ψ.

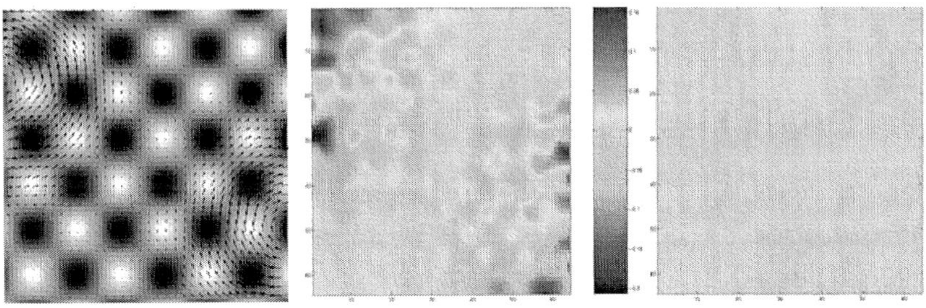

Fig. 2. Left: Synthetic image and solenoidal velocity field. **Middle:** Divergence error using Horn-Schunck regularization. **Right:** Divergence error using our approach

5 Experiments and Discussion

5.1 Implementation Details

Minimization of the functionals (14) and (20), respectively, with linearized data terms (see (10)) can be done by alternating partial minimizations with respect

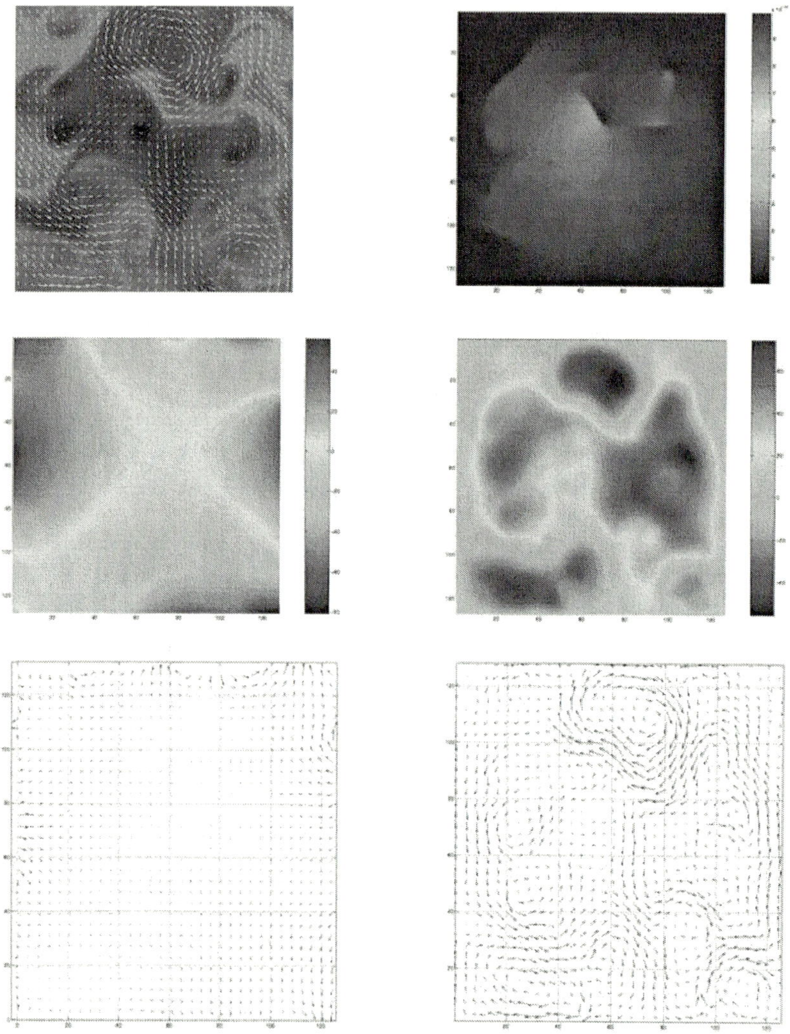

Fig. 3. Top Left The first image I_1 with the restored solenoidal flow. **Top Right** The divergence field of the flow which is less than $3 * 10^{-12}$. **Middle Left** The potential field $\psi_l(\Omega)$ related to the laminar flow. **Middle Right** The potential field $\phi(\Omega)$. **Bottom Left** The first component of flow: the laminar flow u_{lam}. **Bottom Right** The second component of flow related to potential $\phi(\Omega)$. The comparison with standard regularization is depicted in Figure 4

to ψ, ϕ and subsequent subspace corrections. The proof of convergence and further details are given in [14, 15]. In the case of solenoidal flows, the first linear constraint in (21) is incorporated by using the Augmented Lagrangian Method [16]. The remaining two constraints can be taken into account by directly modifying the two linear systems involved.

5.2 Experiment Results

Error Measures. In pactice, evaluating non-rigid flows by computing the average angular and norm error, respectively, induced by the inner product of the space $(L^2(\Omega))^2 = L^2(\Omega) \times L^2(\Omega)$ [17] appeared to us too insensitive to the im-

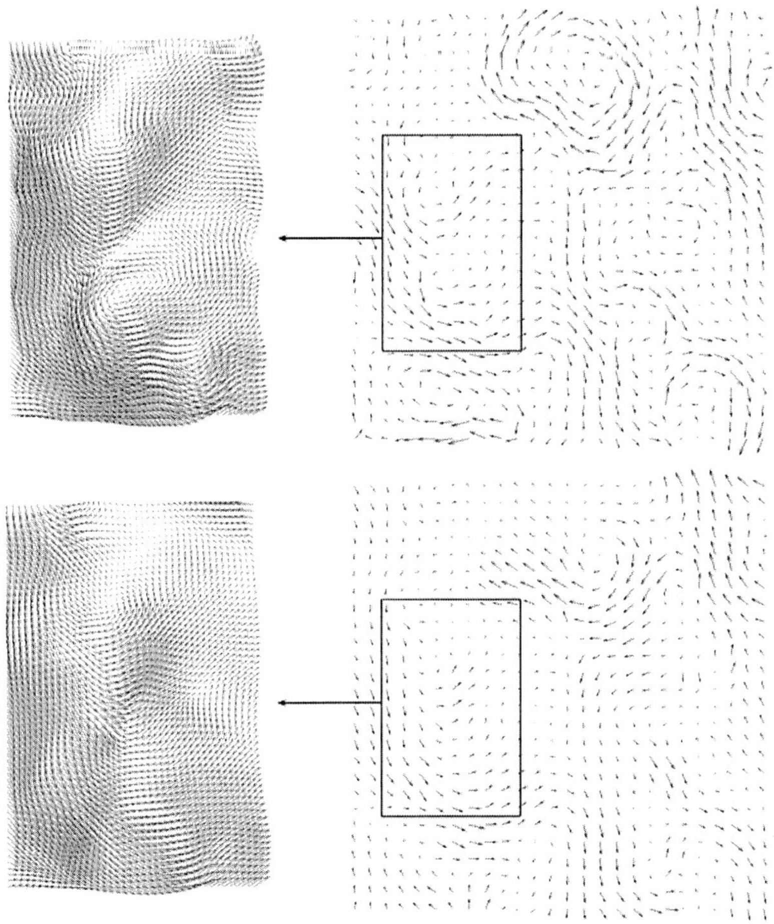

Fig. 4. Top The restored solenoidal flow $u(\Omega)$. **Bottom** The restored flow $u_{hs}(\Omega)$ using the Horn-Schunck regularization. This results clearly show that vortex structures are better recovered by our approach. Furthermore, the magnitude of the divergence is below 10^{-11} throughout the image plane

portant flow structures. Therefore, we suggest error measures that also take into account divergence and curl of flow structures:

$$e_{norm} = \frac{\langle w, w \rangle_{DC}}{N}; \quad e_{ang.} = \arccos \frac{\langle u, v \rangle_{DC} + 1}{\sqrt{\langle u, u \rangle_{DC} + 1}\sqrt{\langle v, v \rangle_{DC} + 1}}. \tag{26}$$

where we adopt the average angular and norm error measures but use the inner products of the space $H(\text{div}; \Omega) \cap H(\text{curl}; \Omega)$ (see, e.g., [7]):

$$\langle u, v \rangle_{DC} = \langle u, v \rangle_{H_S} + \langle \mathbb{D}iv\, u, \mathbb{D}iv\, v \rangle_{H_V} + \langle \mathbb{C}url\,^* u, \mathbb{C}url\,^* v \rangle_{H_P}. \tag{27}$$

Ground Truth Experiments. Figure 2 shows a real image which was warped by the indicated flow. The corresponding errors for the approach (20) $e_{norm} = 6.1 * 10^{-3}$, $e_{ang.} = 6.51°$ are smaller than the approach with Horn-Schunck regularization, for which $e_{norm} = 2.95 * 10^{-2}$, $e_{ang.} = 13.52°$. Note, that these error

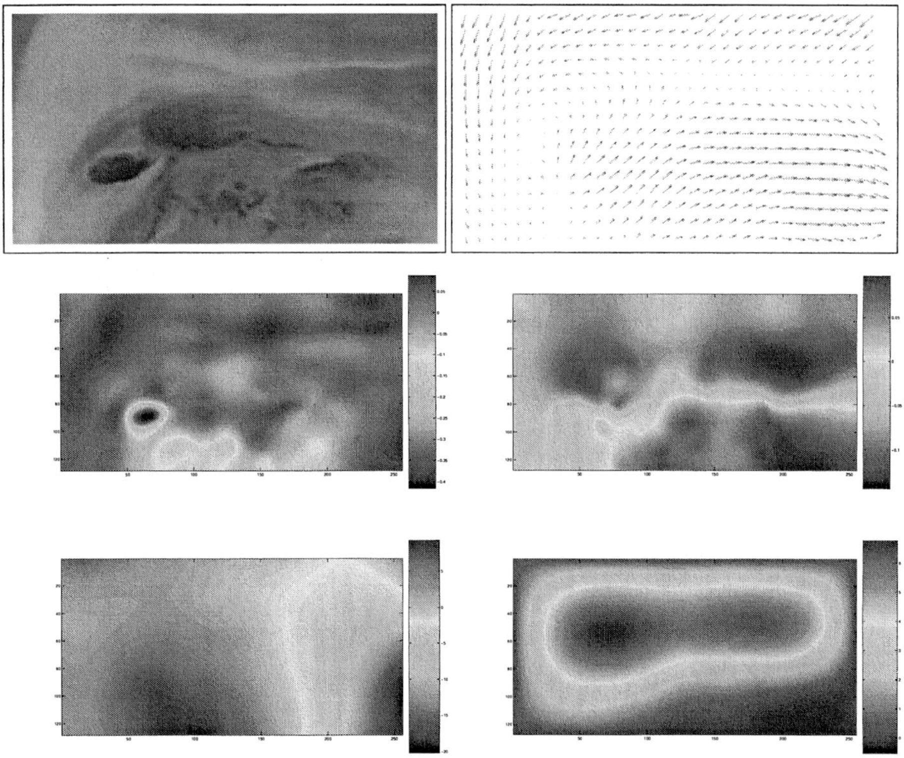

Fig. 5. Top Image I with the restored flow field u. **Middle Left** The divergence field of u. **Middle Right** The curl field of u. **Bottom Left** The potential field $\psi(\Omega)$. **Bottom Right** The potential field $\phi(\Omega)$. The divergence field, for example, which clearly detects a "source" (blue blob), illustrates the quality and usefulness of the results

measures include flow derivatives as opposed to common measures used in the literature.

Estimating Solenoidal Flows. Figure 3 shows the result of estimating the solenoidal flow for a real image sequence. The comparison with first-order regularization (Horn-Schunck approach) in Figure 4 cleary reveals the superiority of our approach regarding the reconstruction of vortex structures. Furthermore, the (in this case) physically plausible constraint of vanishing divergence is satisfied quite accurately.

Estimating General Non-rigid Flows. Figures 5 and 6 show general non-rigid flow estimated for two different real image sequences. As in the solenoidal case, the potential functions provide a useful representation of qualitative properties of the flow.

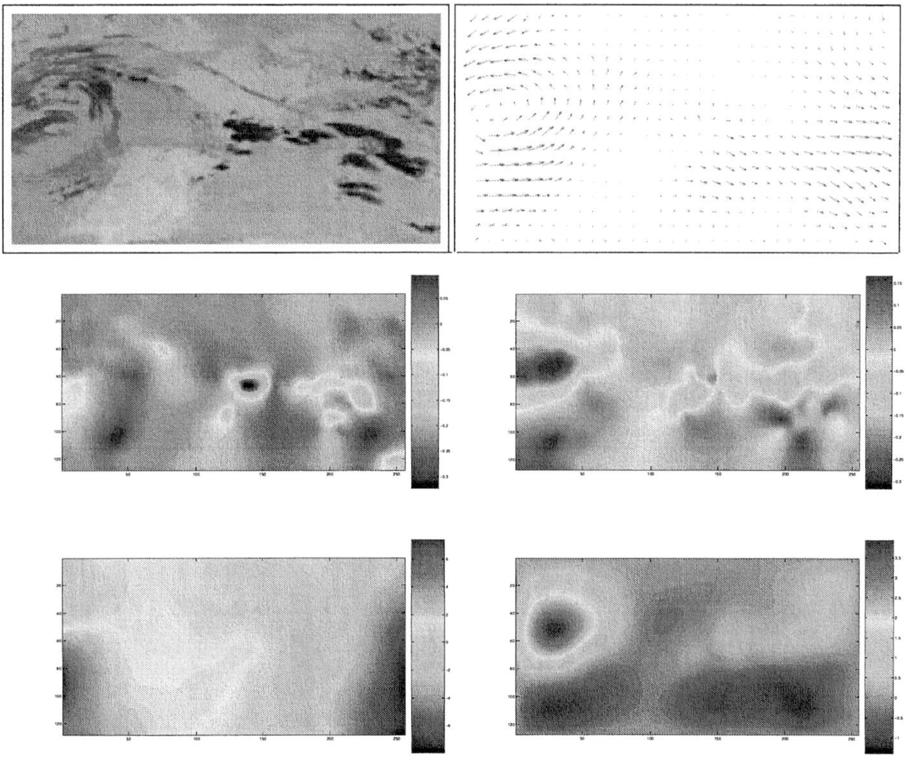

Fig. 6. Top Image I with the restored flow field u. **Middle Left** The divergence field of u. **Middle Right** The curl field of u. **Bottom Left** The potential field $\psi(\Omega)$. **Bottom Right** The potential field $\phi(\Omega)$. As in the previous figure, the potential functions provide a useful representation of qualitative properties of the flow

6 Conclusion and Future Works

We presented a high-quality discrete representation of flow estimation schemes for non-rigid flows. Our further work will focus on the extension to 3D image sequences.

References

1. L. Amodei and M. N. Benbourhim. A vector spline approximation. *J. Approx. Theory*, 67(1):51–79, 1991.
2. D. Suter. Motion estimation and vector splines. In *Proceedings of the Conference on Computer Vision and Pattern Recognition*, pages 939–942, Los Alamitos, CA, USA, June 1994. IEEE Computer Society Press.
3. S. Gupta and J. Prince. Stochastic models for div-curl optical flow methods. *Signal Proc. Letters*, 3(2):32–34, 1996.
4. T. Corpetti, É. Mémin, and P. Pérez. Dense motion analysis in fluid imagery. *Lecture Notes in Computer Science*, 2350:676–691, 2002.
5. T. Corpetti, E. Mémin, and P. Pérez. Dense estimation of fluid flows. *IEEE Trans. Pattern Anal. Machine Intell.*, 24(3):365 –380, 2002.
6. T. Corpetti, E. Mémin, and P. Pérez. Extraction of singular points from dense motion fields: an analytic approach. *J. of Math. Imag. Vision*, 19(3):175–198, 2003.
7. V. Girault and P.-A. Raviart. *Finite Element Methods for Navier-Stokes Equations.* Springer, 1986.
8. T. Kohlberger, E. Mémin, and C. Schnörr. Variational dense motion estimation using the helmholtz decomposition. In L.D. Griffin and M. Lillholm, editors, *Scale Space Methods in Computer Vision*, volume 2695 of *LNCS*, pages 432–448. Springer, 2003.
9. James M. Hyman and Mikhail J. Shashkov. Natural discretizations for the divergence, gradient, and curl on logically rectangular grids. *Comput. Math. Appl.*, 33(4):81–104, 1997.
10. James M. Hyman and Mikhail J. Shashkov. Adjoint operators for the natural discretizations of the divergence, gradient and curl on logically rectangular grids. *Appl. Numer. Math.*, 25(4):413–442, 1997.
11. James M. Hyman and Mikhail J. Shashkov. The orthogonal decomposition theorems for mimetic finite difference methods. *SIAM J. Numer. Anal.*, 36(3):788–818 (electronic), 1999.
12. M. J. Black and P. Anandan. The robust estimation of multiple motions: Parametric and piecewise-smooth flow fields. *Computer Vision and Image Understanding*, 63(1):75–104, 1996.
13. Lawrence C. Evans. *Partial differential equations*, volume 19 of *Graduate Studies in Mathematics*. American Mathematical Society, Providence, RI, 1998.
14. J. Xu. Iterative methods by space decomposition and subspace correction: A unifying approach. *SIAM Review*, 34:581–613, 1992.
15. Xue-Cheng Tai and Jinchao Xu. Global and uniform convergence of subspace correction methods for some convex optimization problems. *Math. Comp.*, 71(237):105–124 (electronic), 2002.
16. D. P. Bertsekas. *Nonlinear Programming.* Athena Scientific, Belmont, MA, 1995. 2nd edition 1999.
17. J. L. Barron, David J. Fleet, and S. S. Beauchemin. Performance of optical flow techniques. *International Journal of Computer Vision*, 12(1):43–77, 1994.

Discontinuity-Preserving Computation of Variational Optic Flow in Real-Time

Andrés Bruhn[1], Joachim Weickert[1], Timo Kohlberger[2], and Christoph Schnörr[2]

[1] Mathematical Image Analysis Group,
Faculty of Mathematics and Computer Science, Building 27.1,
Saarland University, 66041 Saarbrücken, Germany
{bruhn, weickert}@mia.uni-saarland.de

[2] Computer Vision, Graphics, and Pattern Recognition Group,
Department of Mathematics and Computer Science,
University of Mannheim, 68131 Mannheim, Germany
{kohlberger, schnoerr}@uni-mannheim.de

Abstract. Variational methods are very popular for optic flow computation: They yield dense flow fields and perform well if they are adapted such that they respect discontinuities in the image sequence or the flow field. Unfortunately, this adaptation results in high computational complexity. In our paper we show that it is possible to achieve real-time performance for these methods if bidirectional multigrid strategies are used. To this end, we study two prototypes: i) For the anisotropic image-driven technique of Nagel and Enkelmann that results in a linear system of equations we derive a regular full multigrid scheme. ii) For an isotropic flow-driven approach with total variation (TV) regularisation that requires to solve a nonlinear system of equations we develop a full multigrid strategy based on a full approximation scheme (FAS). Experiments for sequences of size 160×120 demonstrate the excellent performance of the proposed numerical schemes. With frame rates of 6 and 12 dense flow fields per second, respectively, both implementations outperform corresponding modified explicit schemes by two to three orders of magnitude. Thus, for the first time ever, real-time performance can be achieved for these high quality methods.

Keywords: computer vision, optical flow, differential techniques, variational methods, multigrid methods, partial differential equations.

1 Introduction

In computer vision, the estimation of motion information from image sequences is one of the key problems. Typically, one is thereby interested in finding the displacement field between two consecutive frames, the so-called *optic flow*. In this context, variational methods play a very important role, since they allow for both a precise and dense estimation of the results. Such techniques are based on the minimisation of a suitable energy functional that consists of two terms: A data term that imposes temporal constancy on certain image features, e.g. on the grey value, and a smoothness term that regularises the often non-unique solution of the data term by an additional smoothness assumption.

R. Kimmel, N. Sochen, J. Weickert (Eds.): Scale-Space 2005, LNCS 3459, pp. 279–290, 2005.

2.2 The TV-Based Regularisation Method

In contrast to image-driven regularisation methods, *flow-driven* techniques reduce smoothing where edges in the *flow field* occur during computation. Our prototype for this class of variational optic flow techniques is an *isotropic* method that penalises deviations from the smoothness assumption with the L_1 norm of the flow gradient magnitude. This corresponds to total variation regularisation [28] and can be related to statistically robust error norms [21]. Thereby large deviations are penalised less severely than in the frequently used quadratic setting (L_2 norm). As a consequence, large gradient features such as edges are better preserved. The energy functional for this approach is given by

$$E(u,v) = \int_{\Omega} \left((f_x u + f_y v + f_t)^2 + \alpha \sqrt{|\nabla u|^2 + |\nabla v|^2 + \epsilon^2} \right) dx dy, \tag{8}$$

where ϵ serves as small regularisation parameter. A related functional that approximates TV regularisation is proposed in [34], while variational approaches for rotationally not invariant versions of TV regularisation have been researched in [11, 12, 23]. At first glance, the corresponding Euler-Lagrange equations that are given by

$$0 = f_x^2\, u + f_x f_y\, v + f_x f_t \;-\; \frac{\alpha}{2}\, \mathcal{L}_{\mathrm{TV}}(u,v), \tag{9}$$

$$0 = f_x f_y\, u + f_y^2\, v + f_y f_t \;-\; \frac{\alpha}{2}\, \mathcal{L}_{\mathrm{TV}}(v,u) \tag{10}$$

have a very similar structure than those in (5)-(6). However,

$$\mathcal{L}_{\mathrm{TV}}(z(x,y), \tilde{z}(x,y)) = \mathrm{div}\ (D(\nabla z(x,y), \nabla \tilde{z}(x,y))\ \nabla z(x,y)) \tag{11}$$

is evidently a *nonlinear* differential operator in z and $\tilde{z}$, since

$$D(\nabla z, \nabla \tilde{z}) = \frac{1}{\sqrt{|\nabla z|^2 + |\nabla \tilde{z}|^2 + \epsilon^2}}\, I =: \begin{pmatrix} a & b \\ b & c \end{pmatrix}, \tag{12}$$

where I is the identity matrix, $b = 0$ and $c = a$. As we will see later, this nonlinearity of the differential operator $\mathcal{L}_{\mathrm{TV}}$ has serious impact on the resulting discrete system of equations and on the derived multigrid strategy.

3 Discretisation

3.1 General Discretisation Aspects

Let us now discuss a suitable discretisation for the Euler-Lagrange equations (5)-(6) and (9)-(10). To this end we consider the unknown functions $u(x,y,t)$ and $v(x,y,t)$ on a rectangular pixel grid with cell size $\mathbf{h} = (h_x, h_y)^\top$, and we denote by $u_{i,j}^{\mathbf{h}}$ the approximation to u at some pixel i,j with $i = 1,...,N_x$ and $j = 1,...,N_y$. Spatial derivatives of the image data are approximated using a fourth-order approximation with the stencil $(1, -8, 0, 8, -1)/(12h)$, while temporal derivatives are computed with a simple two-point stencil. In order to discretise the divergence expressions in the differential operators L_{NE} and L_{TV} we propose the following finite difference approximations:

Table 1. Discretisations of averaging and differential operators

One-sided averaging	$M_x^{\pm,\mathbf{h}}\,(z_{i,j}) := \frac{z_{i\pm1,j}+z_{i,j}}{2}$	(13)
	$M_y^{\pm,\mathbf{h}}\,(z_{i,j}) := \frac{z_{i,j\pm1}+z_{i,j}}{2}$	(14)
One-sided differences	$D_x^{\pm,\mathbf{h}}\,(z_{i,j}) := \pm\frac{z_{i\pm1,j}-z_{i,j}}{h_x}$	(15)
	$D_y^{\pm,\mathbf{h}}\,(z_{i,j}) := \pm\frac{z_{i,j\pm1}-z_{i,j}}{h_y}$	(16)
Central differences	$D_x^{\mathbf{h}}\,(z_{i,j}) := \frac{z_{i+1,j}-z_{i-1,j}}{2h_x}$	(17)
	$D_y^{\mathbf{h}}\,(z_{i,j}) := \frac{z_{i,j+1}-z_{i,j-1}}{2h_y}$	(18)
Squared differences	$D_x^{2,\mathbf{h}}\,(z_{i,j}) := \frac{1}{2}\left(D_x^{+,\mathbf{h}}(z_{i,j})\right)^2 + \frac{1}{2}\left(D_x^{-,\mathbf{h}}(z_{i,j})\right)^2$	(19)
	$D_y^{2,\mathbf{h}}\,(z_{i,j}) := \frac{1}{2}\left(D_y^{+,\mathbf{h}}(z_{i,j})\right)^2 + \frac{1}{2}\left(D_y^{-,\mathbf{h}}(z_{i,j})\right)^2$	(20)
Gradient magnitude	$\left\|D^{2,\mathbf{h}}\,(z_{i,j})\right\| := \sqrt{D_x^{2,\mathbf{h}}(z_{i,j}) + D_y^{2,\mathbf{h}}(z_{i,j})}$	(21)

$$\partial_x\,(\,a(x,y)\,\partial_x z(x,y)) \approx D_x^{-,\mathbf{h}}\left(M_x^{+,\mathbf{h}}(a_{i,j})\;D_x^{+,\mathbf{h}}(z_{i,j})\right), \tag{22}$$
$$\partial_x\,(\,b(x,y)\,\partial_y z(x,y)) \approx D_x^{\mathbf{h}}\left(b_{i,j}\;D_y^{\mathbf{h}}(z_{i,j})\right), \tag{23}$$
$$\partial_y\,(\,b(x,y)\,\partial_x z(x,y)) \approx D_y^{\mathbf{h}}\left(b_{i,j}\;D_x^{\mathbf{h}}(z_{i,j})\right), \tag{24}$$
$$\partial_y\,(\,c(x,y)\,\partial_y z(x,y)) \approx D_y^{-,\mathbf{h}}\left(M_y^{+,\mathbf{h}}(c_{i,j})\;D_y^{+,\mathbf{h}}(z_{i,j})\right), \tag{25}$$

where the coefficients a, b and c are entries of the matrices $D(\nabla f)$ and $D(\nabla u, \nabla v)$ as shown in (4) and (12). Details on the required averaging and differential operators within the approximations are given in Table 1.

3.2 The Method of Nagel and Enkelmann

We are now in the position to write down the discrete Euler-Lagrange equations for the method of Nagel and Enkelmann. They are given by

$$0 = f^{2,\mathbf{h}}_{x\,i,j}\; u^{\mathbf{h}}_{i,j} + f^{\mathbf{h}}_{x\,i,j} f^{\mathbf{h}}_{y\,i,j}\; v^{\mathbf{h}}_{i,j} + f^{\mathbf{h}}_{x\,i,j} f^{\mathbf{h}}_{t\,i,j} \;-\; \alpha\; L^{\mathbf{h}}_{\mathrm{NE\ i,j}}\; u^{\mathbf{h}}_{i,j}, \tag{26}$$
$$0 = f^{\mathbf{h}}_{x\,i,j} f^{\mathbf{h}}_{y\,i,j}\; u^{\mathbf{h}}_{i,j} + f^{2,\mathbf{h}}_{y\,i,j}\; v^{\mathbf{h}}_{i,j} + f^{\mathbf{h}}_{y\,i,j} f^{\mathbf{h}}_{t\,i,j} \;-\; \alpha\; L^{\mathbf{h}}_{\mathrm{NE\ i,j}}\; v^{\mathbf{h}}_{i,j}, \tag{27}$$

for $i = 1, .., N_x$ and $j = 1, .., N_y$, where $L^{\mathbf{h}}_{\mathrm{NE\ i,j}}$ denotes the discrete version of the linear operator $\mathcal{L}_{\mathrm{NE}}$ at some pixel i, j. These $2N_xN_y$ equations constitute a *linear* system for the unknowns $u^{\mathbf{h}}_{i,j}$ and $v^{\mathbf{h}}_{i,j}$. One should note that there are two different types of coupling between the equations. The pointwise coupling between $u^{\mathbf{h}}_{i,j}$ and $v^{\mathbf{h}}_{i,j}$ in the data term and the anisotropic neighbourhood coupling via the operator $L^{\mathbf{h}}_{\mathrm{NE\ i,j}}$ in the smoothness term (for $u^{\mathbf{h}}_{i,j}$ and $v^{\mathbf{h}}_{i,j}$ separately).

3.3 The TV-Based Regularisation Method

Analogously, we discretise the Euler Lagrange equations for the TV-based regularisation method. The obtained *nonlinear* system of equations then reads

$$0 = f_{x\ i,j}^{2,\mathrm{h}}\ u_{i,j}^{\mathrm{h}} + f_{x\ i,j}^{\mathrm{h}} f_{y\ i,j}^{\mathrm{h}}\ v_{i,j}^{\mathrm{h}} + f_{x\ i,j}^{\mathrm{h}} f_{t\ i,j}^{\mathrm{h}} - \frac{\alpha}{2}\ L_{\mathrm{TV\ i,j}}^{\mathrm{h}}(u_{i,j}^{\mathrm{h}}, v_{i,j}^{\mathrm{h}})\ u_{i,j}^{\mathrm{h}}, \quad (28)$$

$$0 = f_{x\ i,j}^{\mathrm{h}} f_{y\ i,j}^{\mathrm{h}}\ u_{i,j}^{\mathrm{h}} + f_{y\ i,j}^{2,\mathrm{h}}\ v_{i,j}^{\mathrm{h}} + f_{y\ i,j}^{\mathrm{h}} f_{t\ i,j}^{\mathrm{h}} - \frac{\alpha}{2}\ L_{\mathrm{TV\ i,j}}^{\mathrm{h}}(u_{i,j}^{\mathrm{h}}, v_{i,j}^{\mathrm{h}})\ v_{i,j}^{\mathrm{h}}, \quad (29)$$

for $i = 1, .., N_x$ and $j = 1, .., N_y$. Here, the finite difference approximation of $\mathcal{L}_{\mathrm{TV}}(u, v)$ and $\mathcal{L}_{\mathrm{TV}}(v, u)$ results in the product of a common nonlinear operator $L_{\mathrm{TV\ i,j}}^{\mathrm{h}}(u_{i,j}^{\mathrm{h}}, v_{i,j}^{\mathrm{h}})$ and the pixel $u_{i,j}^{\mathrm{h}}$ and $v_{i,j}^{\mathrm{h}}$, respectively. Evidently, this constitutes a third way of coupling.

4 Multigrid

4.1 Basic Concept

In general, the obtained linear and nonlinear systems of equations are solved by using non-hierarchical iterative schemes; e.g. variants of the Jacobi or the Gauß-Seidel method [27, 36]. However, such techniques are not suitable for equation systems that are only coupled via a small local neighbourhood: It may take thousands of iterations to transport local information between unknowns that are not coupled directly. A Fourier analysis of the error confirms this observation: While high frequency components (small wavelength, local impact) are attenuated efficiently, lower frequency components (large wavelength, global impact) remain almost un-dampened. In order to overcome this problem multigrid methods are based on a sophisticated strategy. They make use of correction steps that compute the error (not a coarser version of the fine grid solution) on a coarser grid. Thus, lower frequency components of the error reappear as higher ones and allow for an efficient attenuation with standard iterative methods. In the following we explain this strategy in detail for both the linear and the nonlinear case by the example of a basic bidirectional two-grid cycle.

4.2 The Linear Two-Grid Cycle

For the sake of clarity, let us reformulate the linear equation system of the method of Nagel and Enkelmann (26)-(27) as

$$A^{\mathrm{h}} x^{\mathrm{h}} = f^{\mathrm{h}}. \quad (30)$$

Here x^{h} denotes the concatenated vector $((u^{\mathrm{h}})^\top, (v^{\mathrm{h}})^\top)^\top$, A^{h} is a symmetric positive definite matrix and f^{h} stands for the right hand side.

I) Multigrid methods starts by performing several iterations with a basic iterative solver. This is the so-called presmoothing relaxation step, where high frequency components of the error are removed. If we denote the result after these iterations by $\tilde{x}^{\mathrm{h}}$, the error is given by

$$e^{\mathrm{h}} = x^{\mathrm{h}} - \tilde{x}^{\mathrm{h}}. \quad (31)$$

II) Evidently, one is interested in finding $e^{\mathbf{h}}$ in order to correct the approximated solution $\tilde{x}^{\mathbf{h}}$. Although $e^{\mathbf{h}}$ cannot be computed directly, the linearity of $A^{\mathbf{h}}$ allows its computation via

$$A^{\mathbf{h}}e^{\mathbf{h}} = A^{\mathbf{h}}(x^{\mathbf{h}} - \tilde{x}^{\mathbf{h}}) = A^{\mathbf{h}}x^{\mathbf{h}} - A^{\mathbf{h}}\tilde{x}^{\mathbf{h}} = f^{\mathbf{h}} - A^{\mathbf{h}}\tilde{x}^{\mathbf{h}} = r^{\mathbf{h}}, \tag{32}$$

where $r^{\mathbf{h}}$ is called residual. Since high frequencies of the error have already been removed, we can speed up the computation by solving this equation system at a coarser resolution with grid cell size $\mathbf{H} = (H_x, H_y)^\top$:

$$A^{\mathbf{h}}e^{\mathbf{h}} = r^{\mathbf{h}} \quad \rightarrow \quad A^{\mathbf{H}}e^{\mathbf{H}} = r^{\mathbf{H}}. \tag{33}$$

One should note that at this point, a transfer of the original equation system to a coarser grid makes no sense: Unlike the error, the solution very probably contains (desired) high frequency components. A restriction of these components would severely deteriorate the approximative solution (aliasing).

III) After we have solved the residual equation system on the coarse grid with a method of our choice, we transfer the solution back to the fine grid and correct our approximation by the computed error

$$\tilde{x}^{\mathbf{h}}_{\text{new}} = \tilde{x}^{\mathbf{h}} + e^{\mathbf{h}}. \tag{34}$$

IV) In general, the transfer of the computed correction from a coarse grid by means of interpolation introduces some new high frequency components. To this end, a so-called postsmoothing relaxation step is performed, where once again some iteration of the basic iterative solver are applied.

4.3 The Nonlinear (FAS) Two-Grid Cycle

Also in this case, let us start with a reformulation of the nonlinear equation system resulting from the TV-based regularisation method (28)-(29) as

$$A^{\mathbf{h}}(x^{\mathbf{h}}) = f^{\mathbf{h}} \tag{35}$$

where $A^{\mathbf{h}}(x^{\mathbf{h}})$ is a nonlinear operator. The FAS strategy [5] works as follows:

I) We perform a presmoothing relaxation step with a nonlinear basic solver.

II) However, since $A^{\mathbf{h}}(x^{\mathbf{h}})$ is a nonlinear operator, the way of deriving a suitable coarse grid correction is significantly different from the linear case. The (implicit) relation between the error and the residual is given by

$$A^{\mathbf{h}}(\tilde{x}^{\mathbf{h}} + e^{\mathbf{h}}) - A^{\mathbf{h}}(\tilde{x}^{\mathbf{h}}) = f^{\mathbf{h}} - A^{\mathbf{h}}(\tilde{x}^{\mathbf{h}}) = r^{\mathbf{h}}. \tag{36}$$

In order to compute the desired correction we transfer the following nonlinear equation system to the coarse grid

$$A^{\mathbf{h}}(\tilde{x}^{\mathbf{h}} + e^{\mathbf{h}}) = r^{\mathbf{h}} + A^{\mathbf{h}}(\tilde{x}^{\mathbf{h}}) \quad \rightarrow \quad A^{\mathbf{H}}(\boxed{\tilde{x}^{\mathbf{H}}} + e^{\mathbf{H}}) = r^{\mathbf{H}} + \boxed{A^{\mathbf{H}}(\tilde{x}^{\mathbf{H}})}\,. \tag{37}$$

Here, frames visualise the additional terms compared to the linear case.

III) After we have solved the nonlinear residual equation system on the coarse grid, we subtract $\tilde{x}^{\mathbf{H}}$ from the solution in order to obtain $e^{\mathbf{H}}$. Its transfer to the fine grid then allows to perform the correction step.

IV) We perform a postsmoothing relaxation step with a nonlinear basic solver.

4.4 Advanced Multigrid Strategies

In order to increase the computational efficiency, the presented two-grid cycles are generally applied in a hierarchical way. While *V–cycles* make one recursive call of a two-grid cycle per level, faster converging *W–cycles* perform two. Nevertheless, multiple of such advanced cycles are required to reach the desired accuracy. Refining the original problem step by step (unidirectional coarse-to-fine approach) and solving the resulting linear or nonlinear equation system at each level by using some bidirectional V– or W–cycles, the multigrid strategy with the best performance is obtained: *full multigrid* [6]. For both the linear and nonlinear case we have developed such full multigrid schemes. Let us now sketch some implementation details.

4.5 Implementation Details

For the method of Nagel and Enkelmann we implemented a full multigrid scheme with four W–cycles per level each one based on one pre- and one postsmoothing iteration. In order to overcome the problematic anisotropic coupling we made use of a Gauß-Seidel method with alternating line relaxation (ALR) [35] as basic solver. For our second prototype, the TV-based regularisation method, we designed a FAS full multigrid scheme with two W–cycles per level each one based on two pre- and two postsmoothing iterations. In this case we embedded a Gauß-Seidel method with coupled point relaxation (CPR) [8] and frozen coefficients [16]. In order to allow for a complete multigrid hierarchy we thereby considered the use of non-dyadic intergrid transfer operators. As proposed in [8] they were realised by constant interpolation and simple averaging. Coarser versions of the linear and nonlinear operators were created by a discretisation coarse grid approximation (DCA) [35].

5 Experiments

In our first experiment we compare the efficiency of different numerical schemes for the discussed prototypes (Nagel and Enkelmann with $\alpha = 1000$ and $\epsilon = 10^{-2}$, TV-based

Table 2. Performance benchmark on a standard desktop computer with 3.06 GHz Pentium 4 CPU. Run times refer to the computation of a single flow field from the 160×120 dancing sequence

(a) Linear : Image-driven anisotropic regularisation (Nagel-Enkelmann)

Solver	Iterations	Time $[s]$	FPS $[s^{-1}]$	Speedup
Mod. Explicit Scheme ($\tau = 0.1666$)	36558	47.053	0.021	1
Gauß-Seidel (ALR)	607	3.608	0.277	13
Full Multigrid	1	0.171	5.882	275

(b) Nonlinear : Flow-driven isotropic regularisation (TV)

Solver	Iterations	Time $[s]$	FPS $[s^{-1}]$	Speedup
Mod. Explicit Scheme ($\tau = 0.0025$)	10631	30.492	0.033	1
Gauß-Seidel (CPR)	2679	6.911	0.145	4
FAS - Full Multigrid	1	0.082	12.172	372

Fig. 1. *Left to right:* dancing sequence, waving sequence, rotating thumb sequence. *Top to bottom:* first frame, second frame, our CLG multigrid implementation from [8], our Nagel-Enkelmann multigrid implementation, our TV-based regularisation FAS multigrid implementation. *Brightness code:* The magnitude of a flow vector is encoded by its brightness. Brigther pixels stand for larger displacements. Color versions of the flow fields are available at `http://www.mia.uni-saarland.de/bruhn/scsp05/flowfields/`

19. W. Hackbusch. *Multigrid Methods and Applications*. Springer, New York, 1985.
20. B. Horn and B. Schunck. Determining optical flow. *Artificial Intelligence*, 17:185–203, 1981.
21. P. J. Huber. *Robust Statistics*. Wiley, New York, 1981.
22. A. Kimmel and I. Yavneh. An algebraic multigrid approach for image analysis. *SIAM Journal on Scientific Computing*, 24(4):1218–1231, 2003.
23. A. Kumar, A. R. Tannenbaum, and G. J. Balas. Optic flow: a curve evolution approach. *IEEE Transactions on Image Processing*, 5(4):598–610, Apr. 1996.
24. M. R. Luettgen, W. C. Karl, and A. S. Willsky. Efficient multiscale regularization with applications to the computation of optical flow. *IEEE Transactions on Image Processing*, 3(1):41–64, 1994.
25. E. Mémin and P. Pérez. A multigrid approach for hierarchical motion estimation. In *Proc. 6th International Conference on Computer Vision*, pages 933–938, Bombay, India, Jan. 1998.
26. H.-H. Nagel and W. Enkelmann. An investigation of smoothness constraints for the estimation of displacement vector fields from image sequences. *IEEE Transactions on Pattern Analysis and Machine Intelligence*, 8:565–593, 1986.
27. J. M. Ortega and W. C. Rheinboldt. *Iterative Solution of Nonlinear Equations in Several Variables*, volume 30 of *Classics in Applied Mathematics*. SIAM, Philadelphia, 2000.
28. L. I. Rudin, S. Osher, and E. Fatemi. Nonlinear total variation based noise removal algorithms. *Physica D*, 60:259–268, 1992.
29. C. Schnörr. Segmentation of visual motion by minimizing convex non-quadratic functionals. In *Proc. Twelfth International Conference on Pattern Recognition*, volume A, pages 661–663, Jerusalem, Israel, Oct. 1994. IEEE Computer Society Press.
30. D. Terzopoulos. Image analysis using multigrid relaxation. *IEEE Transactions on Pattern Analysis and Machine Intelligence*, 8(2):129–139, Mar. 1986.
31. U. Trottenberg, C. Oosterlee, and A. Schüller. *Multigrid*. Academic Press, San Diego, 2001.
32. C. R. Vogel. A multigrid method for total variation-based image denosing. *Computation and Control IV*, 20:323–331, 1995.
33. J. Weickert and C. Schnörr. A theoretical framework for convex regularizers in PDE-based computation of image motion. *International Journal of Computer Vision*, 45(3):245–264, Dec. 2001.
34. J. Weickert and C. Schnörr. Variational optic flow computation with a spatio-temporal smoothness constraint. *Journal of Mathematical Imaging and Vision*, 14(3):245–255, May 2001.
35. P. Wesseling. *An Introduction to Multigrid Methods*. Wiley, Chichester, 1992.
36. D. M. Young. *Iterative Solution of Large Linear Systems*. Academic Press, New York, 1971.
37. G. Zini, A. Sarti, and C. Lamberti. Application of continuum theory and multi-grid methods to motion evaluation from 3D echocardiography. *IEEE Transactions on Ultrasonics, Ferroelectrics, and Frequency Control*, 44(2):297–308, Mar. 1997.

The Structure of Shapes Scale Space Aspects of the (pre-) Symmetry Set*

Arjan Kuijper and Ole Fogh Olsen

Image Group, IT-University of Copenhagen,
Rued Langgaardsvej 7, DK-2300 Copenhagen, Denmark

Abstract. Shapes simplify under to the intrinsic heat equation - the Mean Curvature Motion (MCM) - forming a shape scale space. The same holds for a representation of the shape, viz. the Symmetry Set (SS), a superset of the Medial Axis. Its singularities under the MCM are known, opening possibilities to investigate its deep structure. As data structure we use so-called Gauss diagrams, structures that depend on the pre-Symmetry Set, the SS in parameter space. Its properties, as well as its evolution and singularities under MCM, are presented. The set of all possible Gauss diagrams under MCM form a directed graph with one end point, in which the shape's scale space describes a specific path. These paths can be used for shape description and comparison.

1 Introduction

Among the numerous shape representations [28], the skeleton [1], or Medial axis, takes an important role as simplifying structure [12]. It is defined as the closure of the centres of maximal circles tangent to the shape at at least two points. Modifications of the skeleton made it more stable [22]. The Shock Graph approach [25] incorporates distance information at some points. Promising results were presented on matching of these descriptions [21] using the possible changes of the Shock Graphs.

The Medial Axis is a subset of the Symmetry Set [3]. Changes of this set (transitions, singularities) [2] are directly responsible for changes of the Medial Axis [8]. The Symmetry Set can easily be computed and appears to be able to be represented as a string-like data structure that allows operations with very low computational complexity [18]. In contrast to Medial Axis related approaches, all extremal curvature points are taken into account.

All these methods start from the given shape and do not take into account the *scale of the shape* as a free parameter. In general, the radius of the circle is considered as scale, but the radius is introduced by definition of the Medial

* This work is part of the DSSCV project supported by the IST Programme of the European Union (IST-2001-35443). WWW home page: `http://www.itu.dk/Internet/sw1953.asp`.

R. Kimmel, N. Sochen, J. Weickert (Eds.): Scale-Space 2005, LNCS 3459, pp. 291–302, 2005.

Axis. Scale, or inverse resolution [5], is, however, an intrinsic problem that has to be taken into account when analyzing images and shapes.

For images, a solution to this problem lies in the use of so-called test-functions [24], that regularize the data by performing a local integration at observation points. One suitable testfunction is the Gaussian filter [5]. By keeping the variance of the filter a free parameter, one obtains a multiscale extension of the original image, as originally proposed by Koenderink [13], and followed by many others, see e.g. [5, 10, 19]. Investigation of the deep structure, i.e. the image at all scales simultaneously, led, among other things, to the discovery of a topological hierarchy within the image extended with a scale variable [15].

This idea can be directly transferred to shapes. For shapes, the intrinsic heat equation is the Mean Curvature Motion, see e.g. [4, 11]. The changes in the (pre-)Symmetry Set (the local situations) are theoretically known [2, 17], as well as transitions of the Symmetry Set under the influence of Mean Curvature Motion (MCM) [26]. More detailed descriptions are given in Section 2.

To derive a multi-scale shape hierarchy based on Symmetry Sets, two new ingredients are needed. Firstly necessary properties of the multi-scale pre-Symmetry Set are derived in section 3. Secondly, a novel representation, called Gauss diagram, is presented in section 4.

Using these results, we propose to embed the Gauss diagram representation of the pre-Symmetry Set in the MCM multiscale context. This novel approach creates a multiscale shape representation that reveals a hierarchical simplification of the pre-Symmetry Set as the scale increases. Properties of this shape scale space are given and discussed in section 5.

2 Background

We firstly provide some background theory regarding shape evolution, and Symmetry Set-based representations.

2.1 Evolution

Let $S(x(p), y(p))$ be a closed curve - a shape - given in its parameterized coordinates, and p taken on the unit circle S^1. Then the simplest way of regularizing the curve is by convolving it with the simplest smoothing kernel, a Gaussian (see Cao [4] for more details). Then the heat equation is solved for each of the coordinates:

$$\frac{\partial x}{\partial t} = \frac{\partial^2 x}{\partial p^2}, \quad \frac{\partial y}{\partial t} = \frac{\partial^2 y}{\partial p^2}.$$

Now the coordinates $x(p, t)$ and $y(p, t)$ are smooth, but the curve may become non-smooth.

The reason for this is the fact that the parameterization parameter p describes a curve that shrinks as the scale increases. The overcome aforementioned problem, the curve needs a renormalization s_σ at every scale σ, [11, 20]. Solving the heat equation between two infinitesimal close instances (and renormalizations) yields

$$\frac{\partial x}{\partial t} = \frac{\partial^2 x}{\partial s^2}, \quad \frac{\partial y}{\partial t} = \frac{\partial^2 y}{\partial s^2},$$

called the intrinsic heat equation. The parameter s is also called the arc-length. This gives the Mean Curvature Motion (MCM):

$$\frac{\partial C}{\partial t} = \frac{\partial^2 C}{\partial s^2} = \kappa N,$$

where κ is the curvature and N the unit length normal. Note that in this equation the arc-length s depends on t, so the equation is non-linear. Each point on the curve moves in the direction of the normal, proportional to the curvature. During evolution, the curve becomes convex and shrinks to a point and remains smooth (no self intersections or cusp points). This compares to the properties of Gaussian blurring for images. More mathematical details are given by Gage and Hamilton [6] and Grayson [9], while Kimia and Siddiqi [11] present theory and applications of MCM on shapes and images. Discussions on the abovementioned relation with Gaussian filtering, and the volumetric blurring proposed by Koenderink and Van Doorn [14], can be found in [4, 11].

When the curve is considered as a level line (an isophote) of an image L, κN equals $\|\nabla L\| \nabla \cdot (\nabla L / \|\nabla L\|)$, or in short gauge coordinates, L_{vv}: the second order derivative in the direction tangentional to the isophote. The image evolution is often called Euclidean Shortening Flow [10, 11]. By tracing the zero-crossings of the curvature over scale one obtains a Curvature Scale Space [20].

2.2 Symmetry Set

The Medial Axis ($\mathcal{MA}$) of a shape is defined as the closure of the set of centres of circles that are tangent to the shape at least two points and that contain no other tangent circles: the are so-called maximal circles. The Symmetry Set $\mathcal{SS}$ is defined as the closure of the set of centres of circles that are tangent to the shape at least two points [2, 3, 8, 7]. The $\mathcal{MA}$ is a subset of the $\mathcal{SS}$ [7].

To calculate these sets from above definition, the following procedure can be used [3]: Let a circle with unknown location be tangent to the shape at two points. Then its centre can be found by using the normal vectors at these points: it is located at the position of each point minus the radius of the circle times the normal vector at each point. To find these two points, the location of the centre and the radius, do the following: Given two vectors p_i and p_j (right, with $i = 1$ and $j = 2$) pointing at two locations at the shape, construct the difference vector $p_i - p_j$. Given the two unit normal vectors N_i and N_j at these locations, construct the vector $N_i + N_j$. If the two constructed vectors are non-zero and perpendicular,

$$(p_i - p_j).(N_i + N_j) = 0, \tag{1}$$

the two locations give rise to a tangent circle. The radius r and the centre of the circle are given by

$$p_i - rN_i = p_j - rN_j. \tag{2}$$

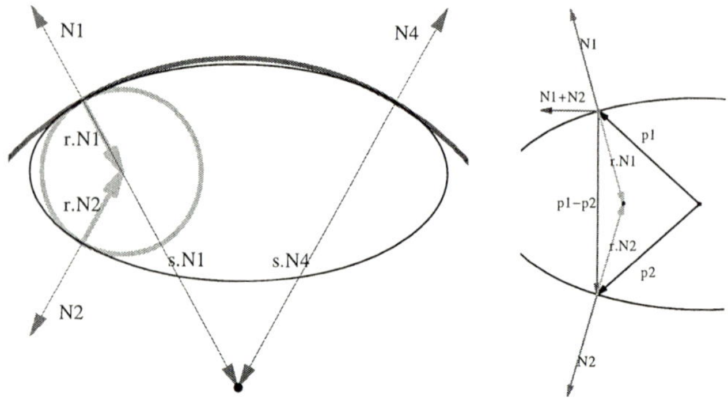

Fig. 1. a) Circles tangent to a shape b) Computing the Symmetry Set (see text)

In Figure 1a, the shape is given by the oval. Inside a circle is tangent to it at two locations, so the *unit* normals $\mathcal{N}_1$ and $\mathcal{N}_2$ are equal for the shape and the circle. The centre of the circle is found by multiplying minus the radius r with the normals. Note that this is also a $\mathcal{MA}$ point. Next, also outside a circle is tangent to the shape at two locations, where the unit normals $\mathcal{N}_1$ and $\mathcal{N}_4$ are equal for the shape and the circle. From this image it follows immediately that a point on the shape relates to at least two points on the $\mathcal{SS}$, in contrast with the $\mathcal{MA}$. Changes in the shape yield changes in the Symmetry Set and are well-known [2]. The Symmetry Set can be represented as a string structure (while the MA requires a graph), whose changes are directly inherited from the Symmetry Set [18]. The transitions of the $\mathcal{SS}$ under MCM have been described by Teixeira [26].

2.3 pre-Symmetry Set

The pre-Symmetry Set is defined as the Symmetry Set in parameter space: instead of the centres of the circles defining the Symmetry Set, the points on the shape where these circles are tangent, are taken. This yields the same (data), but in this case the representation is clearer [16, 18]. The pre-Symmetry Set representation of Figure 2 is shown in Figure 3. In a pre-Symmetry Set diagram, the two axes represent points on the shape. If two points p_i and p_j give rise to a Symmetry Set point, the corresponding points (p_i, p_j) and, due to symmetry, (p_j, p_i) are marked in the diagram. The diagram shows curves that continue along the boundaries. Each curve represents a distinct part of the Symmetry Set. Curves in the pre-$\mathcal{SS}$ represent branches of the $\mathcal{SS}$. The endpoints of the $\mathcal{SS}$ are located at the diagonal of the pre-$\mathcal{SS}$. They relate to points of extremal curvature on the shape.

On the pre-Symmetry Set, the changes of the structure can be detected as well, they can even be labelled with relevance with respect to changes in number

Fig. 2. The fish shape top left transforms into a circle under influence of Mean Curvature Motion

Fig. 3. The pre-Symmetry Set of the fish shape transforms to four parallel lines (two essential loops) under influence of Mean Curvature Motion

of elements (curves), smoothness of elements, swapping of branches of two curves, and changes in the number of special points related to the junctions of the skeleton [17].

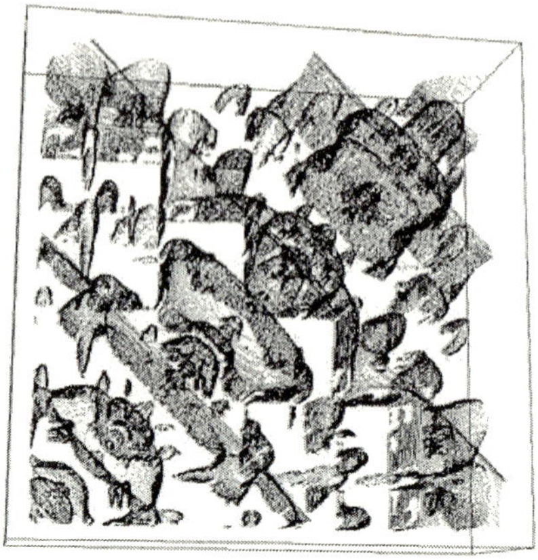

Fig. 4. The pre-Symmetry Set of the fish shape in full MCM scale space forms disappearing domes and two remaining sheets

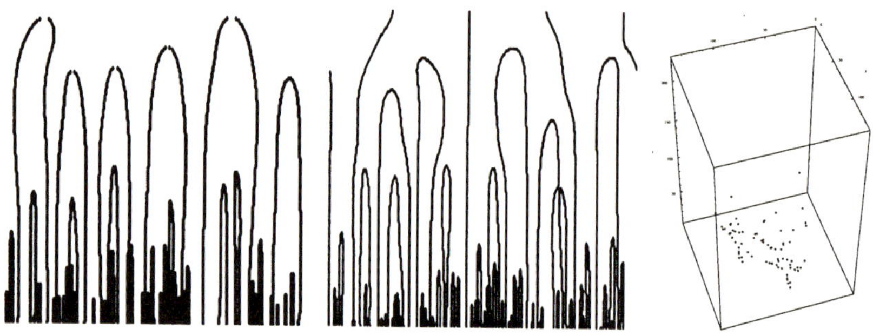

Fig. 5. Left: CSS of the fish. Middle: evolution of the extrema of the curvature of the fish. Right: Position of the curvature extrema of the fish in scale space

CSS investigates the zero crossings of the curvature over scale and is not related to Medial Axis methods. For instance, all convex shapes have the same (zero) CSS representation. Here zero crossings of derivative of curvature are considered (i.e. the extrema of κ) instead of zero crossings of the curvature itself. In Figure 5 the CSS of the fish is shown on the left. Note that all branches end at a certain scale, when the shape becomes convex. In the middle, the extrema of the curvature are shown. Now exactly 4 branches remain, resembling the 4 extremal curvatures of an oval. On the right these curves are shown as function of their spatial and scale positions. The essential difference is that these curves show the connections with respect to their annihilations, while the Symmetry Set connections are related to all their intermediate connections, which change.

5.2 Gauss Diagrams Under Mean Curvature Motion

Next, also the Gauss diagrams change under MCM. They inherit the transitions from the pre-$\mathcal{SS}$ scale space. They are: annihilations of chords, swapping of chords and changing of labels on a chord. The diagrams representing the simplifying fish sequence is shown in Figure 6.

For all possible shapes we can construct a space of Gauss diagrams. Given the simplification of structure as scale increases, we can construct a directionality into this space, as shown in Figure 7. At the top the simplest shape, an ellipse, is represented by two intersecting chords. At each subsequent level, a chord is added. At a specific level, the positions of chords can change due to swappings.

One thus obtains hierarchical metameric classes. Obviously, adding information to the chords enlarges the possible diagrams and decreases the size of the class. The swapping is not the only reason that the graph is not a tree. Also the disappearing of chords can relate to causes a child node to be possibly related to multiple parents, in contrast to trees, as shown in Figure 8a. At a certain level, two subsequent swappings can as well take place in one swap, see Figure 8b.

Given an arbitrary closed non-intersecting curve, applying MCM yields a convex shape shrinking to a point at a sufficiently large scale. So the accompanying Gauss diagrams are related to a path through the space of all possible diagrams. Each shape will have its own path, and "more of less" similar shapes will have paths that coincide at some stage. So the difference of the shapes can be expressed as the difference in paths. This opens new ways to describe and compare shapes in a well-defined topological manner.

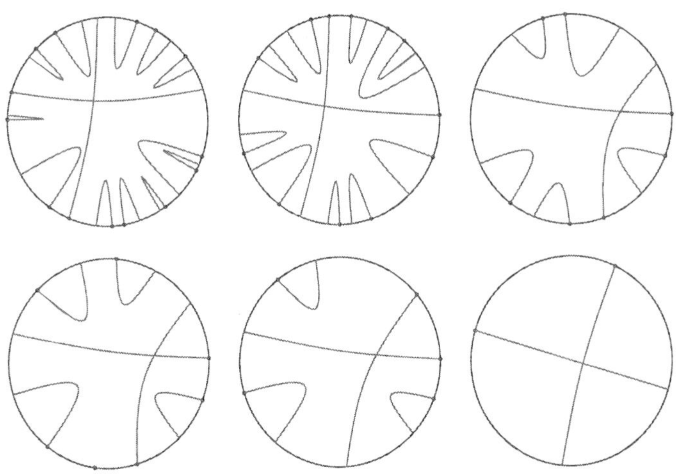

Fig. 6. The Gauss diagrams of the fish shape transforms into a circle with two intersecting chords under influence of Mean Curvature Motion

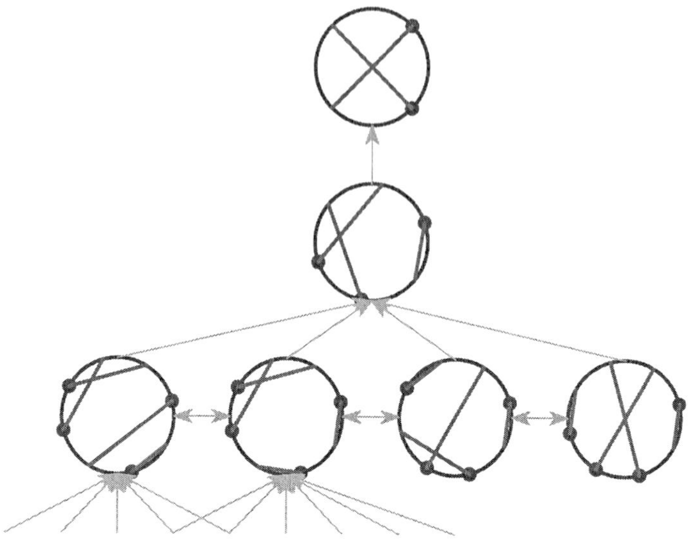

Fig. 7. The Gauss diagrams under influence of Mean Curvature Motion form a hierarchy. The horizontal level represents the number of chords (or A_3 points, or extrema of curvature). They change under A_4 transitions [2] in a simplifying manner: chords disappear. At the horizontal levels chords swap due to A_2^2nib transitions [2]. A further refinement can be achieved by adding additional special points along the chords

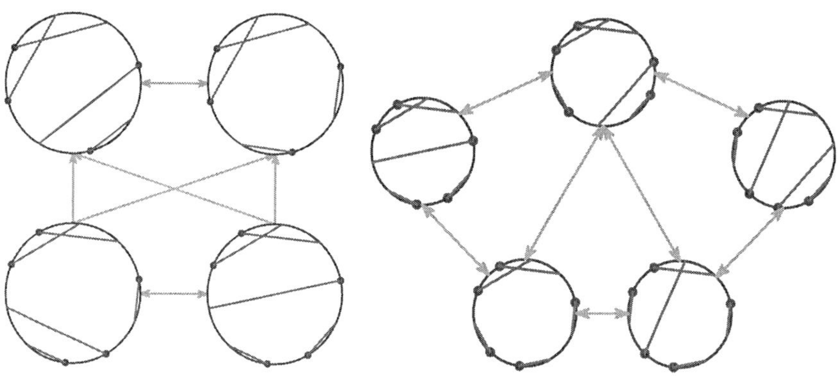

Fig. 8. a) The state space of Gauss diagrams is a directed graph. Simplification can be achieved in different ways. b) At a specific level multiple swap transitions are possible for changing the chords of the Gauss diagrams

6 Conclusions

In this paper we presented the lay-out of a truly multiscale hierarchy for shapes, based on the Symmetry Set. The multi-scale hierarchy is obtained by evolving the shape and its Symmetry Set under Mean Curvature Motion. As data structure the pre-Symmetry Set, visualized in a Gauss diagram is taken. This new approach allows one to map the data on a circle, and effectively represent the multiscale representation as a series of stacked circles with connected points at each scale. Changes as scale increases are well-defined and known for the Symmetry Set. We derived them for the pre-Symmetry Set representation. It makes this structure suitable and well-defined as a descriptive space for shapes. Since all shapes converge to the same structure, this completely novel method allows one to express differences in shapes as differences of convergence paths. Experiments need to be carried out to validate the practical use with the presented theory.

Acknowledgements. The authors thank Prof. Giblin from Liverpool University and Prof. Siersma from Utrecht University for fruitful discussions, and the reviewers for useful comments.

References

1. H. Blum. Biological shape and visual science (part I). *Journal of Theoretical Biology*, 38:205–287, 1973.
2. J. W. Bruce and P. J. Giblin. Growth, motion and 1-parameter families of symmetry sets. *Proceedings of the Royal Society of Edinburgh*, 104(A):179–204, 1986.
3. J. W. Bruce, P. J. Giblin, and C. Gibson. Symmetry sets. *Proceedings of the Royal Society of Edinburgh*, 101(A):163–186, 1985.
4. F. Cao. *Geometric Curve Evolution and Image Processing*, volume 1805 of *Lecture Notes in Mathematics*. Springer Verlag, 2003.
5. L. M. J. Florack. *Image Structure*, volume 10 of *Computational Imaging and Vision Series*. Kluwer Academic Publishers, Dordrecht, The Netherlands, 1997.
6. M. Gage and R. S. Hamilton. The heat equation shrinking convex plane curves. *J. Differential Geom.*, 23(1):69–96, 1986.
7. P. J. Giblin and B. B. Kimia. On the intrinsic reconstruction of shape from its symmetries. *IEEE Transactions on Pattern Analysis and Machine Intelligence*, 25(7):895–911, July 2003.
8. P. J. Giblin and B. B. Kimia. On the local form and transitions of symmetry sets, medial axes, and shocks. *International Journal of Computer Vision*, 54(1/2):143–156, 2003.
9. Matthew A. Grayson. The heat equation shrinks embedded plane curves to round points. *J. Differential Geom.*, 26(2):285–314, 1987.
10. B. M. ter Haar Romeny. *Front-end vision and multi-scale image analysis*, volume 27 of *Computational Imaging and Vision Series*. Kluwer Academic Publishers, Dordrecht, The Netherlands, 2003.
11. B. Kimia and K. Siddiqi. Geometric heat equation and non-linear diusion of shapes and images. *Computer Vision and Image Understanding*, 64(3):305–322, 1996.
12. B.B. Kimia. On the role of medial geometry in human vision. *Journal of Physiology - Paris*, 97(2-3):155–190, 2003.
13. J. J. Koenderink. The structure of images. *Biological Cybernetics*, 50:363–370, 1984.
14. J. J. Koenderink and A. J. van Doorn. Dynamic shape. *Biological Cybernetics*, 53:383–396, 1986.

15. A. Kuijper. *The Deep Structure of Gaussian Scale Space Images.* PhD thesis, Utrecht University, 2002.
16. A. Kuijper and O.F. Olsen. On extending symmetry sets for 2D shapes. In *Proceedings of S+SSPR 2004, LNCS 3138*, pages 512–520, 2004.
17. A. Kuijper and O.F. Olsen. Transitions of the pre-symmetry set. In *Proceedings of the 17th ICPR*, volume III, pages 190–193, 2004.
18. A. Kuijper, O.F. Olsen, P.J. Giblin, Ph. Bille, and M. Nielsen. From a 2D shape to a string structure using the symmetry set. In *Proceedings of the 8th ECCV, LNCS 3022*, pages 313–326, 2004.
19. T. Lindeberg. *Scale-Space Theory in Computer Vision.* The Kluwer International Series in Engineering and Computer Science. Kluwer Academic Publishers, 1994.
20. Farzin Mokhtarian and Alan K. Mackworth. A theory of multiscale, curvature-based shape representation for planar curves. *IEEE Transactions on Pattern Analysis and Machine Intelligence*, 14(8):789–805, 1992.
21. M. Pelillo, K. Siddiqi, and S. Zucker. Matching hierarchical structures using association graphs. *IEEE Transactions on Pattern Analysis and Machine Intelligence*, 21(11):1105–1120, 1999.
22. S.M. Pizer, K. Siddiqi, G. Székely, J.N. Damon, and S.W. Zucker. Multiscale medial loci and their properties. *International Journal of Computer Vision*, 55(2/3):155–179, 2003.
23. M. Polyak and O. Viro. Gauss diagram formulas for vassiliev invariants. *International Mathematical Research Notes*, 11:445–454, 1994.
24. L. Schwartz. *Théorie des Distributions*, volume I, II of *Actualités scientifiques et industrielles; 1091,1122.* Publications de l'Institut de Mathématique de l'Université de Strasbourg, Paris, 1950–1951.
25. K. Siddiqi, A. Shokoufandeh, S. Dickinson, and S. Zucker. Shock graphs and shape matching. *International Journal of Computer Vision*, 30:1–22, 1999.
26. R. C. Teixeira. *Curvature Motions, Medial Axes and Distance Transforms.* PhD thesis, Harvard University, Cambridge, MA, USA, June 1998.
27. V. A. Vassiliev. Homology of spaces of knots in any dimensions. *Philosophical Transactions: Mathematical, Physical and Engineering Sciences*, 359(1784):1343–1364, 2001.
28. D. Zhang and G. Lu. Review of shape representation and description techniques. *Pattern Recognition*, 37(1):1–19, 2004.

A Non-convex PDE Scale Space

Markus Grasmair, Frank Lenzen, Andreas Obereder, Otmar Scherzer, and Matthias Fuchs

Department of Computer Science, University of Innsbruck, Technikerstr. 25, A–6020 Innsbruck, Austria
Mathematics Competence Center, Mathconsult GmbH, Altenbergerstr. 69, A–4040 Linz, Austria

Abstract. For image filtering applications, it has been observed recently that both diffusion filtering and associated regularization models provide similar filtering properties. The comparison has been performed for regularization functionals with convex penalization functional. In this paper we discuss the relation between non-convex regularization functionals and associated time dependent diffusion filtering techniques (in particular the Mean Curvature Flow equation). Here, the general idea is to approximate an evolution process by a sequence of minimizers of iteratively convexified energy (regularization) functionals.

Keywords: Morphological regularization, diffusion filtering, equivalence relations.

1 Introduction

Let $A : X \to 2^X$ be a *maximal monotone* operator on a real Hilbert space X. Here, we call A maximal monotone, if for every $x, x' \in X$ the implication

$$x' \in Ax \quad \Leftrightarrow \quad \langle x' - Ay, x - y \rangle \geq 0 \text{ for every } y \in X$$

holds. Then there exists a solution of

$$\frac{du}{dt}(t) + A(u(t)) \ni 0 \quad (t \geq 0)\,, \quad u(0) = u^0\,. \tag{1}$$

For the precise mathematical formulation of this statement we refer to Zeidler [1–Theorem 32.P]. The solution of (1) is given by

$$u(t) = \lim_{\mathcal{N} \to \infty} \left(I + \frac{t}{\mathcal{N}} A \right)^{-\mathcal{N}} u^0\,.$$

See e.g. Crandall & Liggett [2]. We define

$$u_k^{\mathcal{N}} := \left(I + \frac{t}{\mathcal{N}} A \right)^{-k} u^0 \quad (k = 0, 1, \ldots, \mathcal{N}) \text{ and } u^{\mathcal{N}} := u_{\mathcal{N}}^{\mathcal{N}}\,.$$

R. Kimmel, N. Sochen, J. Weickert (Eds.): Scale-Space 2005, LNCS 3459, pp. 303–315, 2005.

From this formula it is evident that $u_k^{\mathcal{N}}$ solves

$$u + \frac{t}{\mathcal{N}} A(u) \ni u_{k-1}^{\mathcal{N}} \quad (k = 1, \ldots, \mathcal{N}) . \tag{2}$$

An important example of a maximal monotone operator is the subdifferential $A = \partial J$ of a convex functional $J : X \to \mathbb{R} \cup \{+\infty\}$ defined on a real Hilbert space X. In this case (1) is a *gradient flow equation* and $u_k^{\mathcal{N}}$ minimizes the functional

$$u \to \frac{1}{2} \| u - u_{k-1}^{\mathcal{N}} \|_{L^2(\Omega)}^2 + \frac{t}{\mathcal{N}} J(u) \quad (k = 1, 2, \ldots, \mathcal{N}) . \tag{3}$$

That is, the solution of the gradient flow equation can be approximated by iterative regularization.

In [3, 4, 5] we performed a systematic comparison of regularization, iterative regularization, and the solution of the according gradient flow equation for image filtering. The experiments show similar solutions for all three methods. Recently Mrázek, Steidl, and Weickert [6, 7] proved analytically for the one-dimensional discrete bounded variation functional $J(u)$ that both regularization and the solution of the discretized gradient flow equation are exactly the same. The similarity relation between the three methods has been validated for gradient flow equations with $A = \partial J$ maximal monotone (which follows from the convexity of J). In this paper we show that the solution of the *Mean Curvature Motion (MCM)*

$$\frac{dv}{dt}(t) = |\nabla v(t)| \nabla \cdot \left(\frac{\nabla v(t)}{|\nabla v(t)|} \right) \quad (t > 0) , \quad v(0) = v^0 , \tag{4}$$

is approximated by the $\mathcal{N}$-th minimizer of a *non-convex* iterative regularization technique, where in each iteration step a regularization parameter $\alpha = T/\mathcal{N}$ is used. Here, in contrast to (2) we determine $u_k^{\mathcal{N}}$ by solving an equation of the form

$$u + \frac{t}{\mathcal{N}} A_{\frac{t}{\mathcal{N}}}(u) \ni u_{k-1}^{\mathcal{N}} .$$

Note that the operator A now depends on $t/\mathcal{N}$. Provided that the limit $u^{\mathcal{N}} = u_{\mathcal{N}}^{\mathcal{N}}$ exists for $t/\mathcal{N} \to 0$, we expect to have a solution of

$$\frac{\partial u}{\partial t}(t) \in - \lim_{s \to 0+} A_s(u(t)) .$$

This provides a formal relation between the Mean Curvature Flow equation by mimicking nonlinear semi-group theory.

The MCM equation has been extensively studied. For instance, it is well-known that it attains a unique *viscosity solution* for given continuous and bounded initial data $v^0 : \mathbb{R}^n \to \mathbb{R}$ (see e.g. Evans [8]). Only in very special cases the solution can be calculated analytically. Invariance properties and the use of MCM for image processing applications have been studied by Alvarez & Guichard & Lions & Morel [9]. MCM is an example of a morphological filtering technique.

Therefore, we call the associated non-convex variational principle investigated in this paper *morphological regularization* method.

In [10] a variational form related to the mean curvature flow equation has been derived and a relaxation technique has been used to prove existence of a generalized minimizer. This approach is impractical for a numerical solution since the functional has to be redefined via Γ-limits first, and the relaxed functional eventually has to be minimized. The variational formulation reveals interesting properties (see [11]): it can be motivated as a regularization functional to clean noisy images with random perturbations of the level lines.

The outline of this paper is as follows: In Section 2 we recall the formal relation between the Mean Curvature Flow equation and the according variational principle. In Section 3 we prove well-posedness of iterative regularization based on the concept of convexification. Moreover, a nontrivial part is the characterization of the relaxed functional on the nonreflexive Banach space of functions of bounded variation. Previously, we computed the convex envelope for approximations on Sobolev spaces (see [13, 12, 11]). In Section 4 we discuss the numerical minimization of the nonconvex variational principle and review solving the Mean Curvature Flow equation. The results extend previous numerical experiments in [10] for the minimization of the variational principles, which have been implemented for relatively large regularization parameters. In Section 5 we compare iterative regularization and the solution of the Mean Curvature Flow equation.

2 The Link Between MCM and Iterative Regularization

In order to establish the link relation between Mean Curvature Flow and variational forms we study the following energy functional:

$$I(u) := I_{\alpha,u^0}(u) := \int f\big(x, u(x), \nabla u(x)\big)\, dx \quad (\alpha > 0)\;, \tag{5}$$

where $f : \Omega \times \mathbb{R} \times \mathbb{R}^n \to \mathbb{R} \cup \{+\infty\}$ is defined by

$$f(x, \xi, A) = \frac{\big(\xi - u^0(x)\big)^2}{2|A|} + \alpha |A|\;. \tag{6}$$

We can interpret I as a regularization functional with *fit-to-data* term $\int \frac{(u-u^0)^2}{2|\nabla u|}$ and the total variation semi-norm as *fidelity term*.

Aside from the theoretical interest in this functional we use it for solving imaging problems with discontinuous solutions. This motivates the usage of the *total variation semi-norm* for penalization, which has turned out to be quite successful for this purpose (cf. Rudin & Osher & Fatemi [14, 15]).

The following computations are purely formal and not mathematically rigorous. The steepest descent direction of the functional I is

$$\partial I(u) := \frac{u - u^0}{|\nabla u|} + \nabla \cdot \left(\left(\frac{(u - u^0)^2}{2|\nabla u|^2} - \alpha \right) \frac{\nabla u}{|\nabla u|} \right)\;. \tag{7}$$

Therefore, a minimizer of I satisfies the optimality condition

$$u + \alpha A_\alpha(u) := u + \alpha|\nabla u|\nabla \cdot \left(\left(\frac{(u-u^0)^2}{2\alpha|\nabla u|^2} - 1 \right) \frac{\nabla u}{|\nabla u|} \right) \ni u^0 . \tag{8}$$

We set $\alpha = t/\mathcal{N}$ and perform iterative regularization by minimization of the functionals $I^k_{t/\mathcal{N}}$, $(k = 1, \ldots, \mathcal{N})$, defined by $I^k_{t/\mathcal{N}} := I_{t/\mathcal{N}, u^{\mathcal{N}}_{k-1}}$ with $u^{\mathcal{N}}_0 := u^0$. The minimizer of $I^k_{t/\mathcal{N}}$ is denoted by $u^{\mathcal{N}}_k$.

With the change of notation $\Delta T := T/\mathcal{N}$, $v(T) := u^{\mathcal{N}}_{\mathcal{N}}$, $v(T-\Delta T) := u^{\mathcal{N}}_{\mathcal{N}-1}$, we find from the according optimality condition for the functional $I^{\mathcal{N}}_{T/\mathcal{N}}$ (cf. (8)) which we multiply by $|\nabla v(T)|/\Delta T$ that

$$\frac{v(T) - v(T-\Delta T)}{\Delta T} \in |\nabla v(T)|\nabla \cdot \left(A(T, \Delta T, v) \frac{\nabla v(T)}{|\nabla v(T)|} \right) , \tag{9}$$

where

$$A(T, \Delta T, v) := 1 - \frac{\Delta T}{2} \frac{(v(T) - v(T-\Delta T))^2}{(\Delta T)^2} \frac{1}{|\nabla v(T)|^2} .$$

Taking $\Delta T \to 0^+$ and considering $\frac{dv}{dt}(T) = \lim_{\Delta T \to 0^+} \frac{v(T)-v(T-\Delta T)}{\Delta T}$, we recover (4).

For the regularization functional (3), if J is convex, there exists a unique minimizer of the associated regularization functional. Here this is no longer trivial and is a first step of an analysis.

3 Minimizers of Non-convex Energy Functionals

In this section we prove existence of a minimizer of the functional

$$I(u) := \int_\Omega \frac{\left(u(x) - u^0(x)\right)^2}{2|\nabla u(x)|} \, dx + \alpha|Du|(\Omega) \quad (\alpha > 0) . \tag{10}$$

Here Ω is a bounded domain with Lipschitz boundary and $|Du|(\Omega)$ denotes the *total variation* semi-norm. By Du we denote the distributional derivative of u, which is a Radon measure on Ω. Thus we can use the Lebesgue decomposition $Du = \nabla u \, dx + D^s u$, where $\nabla u \in L^1(\Omega)$ denotes the absolutely continuous part of Du and $D^s u$ is the singular part (cf. Rudin [19])[1]. In (10) we define

$$\frac{\left(u(x) - u^0(x)\right)^2}{2|\nabla u(x)|} := 0 \text{ if } u(x) = u^0(x) .$$

[1] We follow the terminology of Ambrosio & Fusco & Pallara [17] and call $D^s u$ the singular part. Other publications denote by $D^s u$ the jump part of the distributional gradient, which belongs to discontinuities in the function u. In particular, in this paper $D^s u$ also contains the Cantor part of u

Minimization of the functional I is considered over the space $\mathrm{BV}(\Omega)$, the space of functions of bounded variation (cf. Evans & Gariepy [16] or Ambrosio & Fusco & Pallara [17]). There are two major difficulties associated with the functional:

1. For a convex function g and a measure m the functional $J(m) = \int_\Omega g\big(m(x)\big)$ is well-defined (see e.g. Temam [18]). Here, this theory is not applicable, since the functional I is non-convex with respect to the measure Du, the derivative of the function $u \in BV(\Omega)$.
2. The functional I is *not* lower semi-continuous with respect to the weak* topology on $\mathrm{BV}(\Omega)$, and compensated compactness arguments are not applicable to prove existence of a minimizer.

A standard approach to obtain a meaningful interpretation of I is via *relaxation* (cf. [20]) For a functional $J : X \to \mathbb{R} \cup \{\infty\}$ and $\emptyset \neq X \subseteq \mathrm{BV}(\Omega)$, and $u \in \mathrm{BV}(\Omega)$ its relaxation is defined by

$$\mathcal{R}(J,X)(u) := \begin{cases} +\infty \text{ if } u \notin \overline{X} \cap \mathrm{BV}(\Omega) \\ \inf\left\{\liminf_{k\to\infty} J(u^{(k)}) : \{u^{(k)}\} \subset X, \|u^{(k)} - u\|_{L^1(\Omega)} \to 0\right\} \end{cases} \tag{11}$$

Here $\overline{X}$ is the closure of X with respect to the $L^1(\Omega)$-norm. In order to simplify the notation we define $\mathcal{R}(I) := \mathcal{R}\big(I, \mathrm{BV}(\Omega)\big)$. In the following we show that $\mathcal{R}(I)$ attains a minimizer that can be considered a generalized minimizer of I.

Theorem 1. *Let $u^0 \in L^\infty(\Omega)$, then the functional $\mathcal{R}(I)$ attains a minimizer in $\mathrm{BV}(\Omega)$ that can be considered a generalized minimizer of I, i.e., if the minimum of I is attained in $u \in \mathrm{BV}(\Omega)$, then u is a minimizer of $\mathcal{R}(I)$.*

Proof. The functional $\mathcal{R}(I)$ is lower semi-continuous with respect to the L^1-topology on $\mathrm{BV}(\Omega)$, coercive, and proper (i.e., $\mathcal{R}(I) \not\equiv \infty$). Thus it attains a minimizer in $\mathrm{BV}(\Omega)$. To see that $\mathcal{R}(I)$ is proper take $u(x) = x_1$ if $x = (x_1, \ldots, x_n)$. Then $|\nabla u(x)| = 1$. Thus, $I(u) < \infty$ and consequently $\mathcal{R}(I) < \infty$ showing that $\mathcal{R}(I)$ is proper. The coercivity assertion follows from the characterization of $\mathcal{R}(u)$ given in Theorem 2. To show that each minimizer of I is a minimizer of $\mathcal{R}(I)$ we take $c := \inf\{I(u)\}$. The definition of the relaxed functional implies that $\inf\{\mathcal{R}(I)(u)\} \geq c$. Since I attains the minimum value c, we also have that $\mathcal{R}(I)(u) \leq c$ by using the constant sequence $\{u\}$ in the right hand side of (11).

We now turn to characterizing the relaxed functional.

Theorem 2. *If $u^0 \in L^\infty(\Omega)$, then*

$$\mathcal{R}(I)(u) = I_c(u) := \int_\Omega f_c\big(x, u(x), \nabla u(x)\big)\, dx + \alpha |D^s u|(\Omega) \quad (u \in \mathrm{BV}(\Omega))\,. \tag{12}$$

Here $Du = \nabla u\, dx + D^s u$ is the Lebesgue decomposition of the distributional gradient of u and

$$f_c(x,\xi,A) := \begin{cases} \dfrac{\big(\xi - u^0(x)\big)^2}{2|A|} + \alpha|A|\,, & \text{if } \sqrt{2\alpha}|A| > |\xi - u^0(x)|\,. \\ \sqrt{2\alpha}|\xi - u^0(x)|\,, & \text{if } \sqrt{2\alpha}|A| \leq |\xi - u^0(x)| \end{cases} \tag{13}$$

Before we prove this theorem, we require some properties of the function f_c, which are summarized in the following lemma:

Lemma 1. *Let $u^0 \in L^\infty(\Omega)$. For almost every $x \in \Omega$*

(a) $f_c(x,\cdot,\cdot)$ is convex,
(b) $f_c(x,\cdot,\cdot)$ is continuously differentiable in every point $(\xi, A) \neq (u^0(x), 0)$.

Proof. For $x \in \Omega$ let $U_1 := \{(\xi, A) : \sqrt{2\alpha}|A| < |\xi - u^0(x)|\}$ and $U_2 := \{(\xi, A) : \sqrt{2\alpha}|A| > |\xi - u^0(x)|\}$. For $(\xi, A) \in U_1$ we have

$$\nabla f_c(x,\xi,A) := \nabla_{\xi,A} f_c(x,\xi,A) = \left(\sqrt{2\alpha}\,\mathrm{sgn}(\xi - u^0(x)), 0\right),$$

and for $(\xi, A) \in U_2$ we have

$$\nabla f_c(x,\xi,A) = \left(\frac{\xi - u^0(x)}{|A|}, \left(\alpha - \frac{\left(\xi - u^0(x)\right)^2}{2|A|^2}\right)\frac{A}{|A|}\right).$$

For $\sqrt{2\alpha}|A| - |\xi - u^0(x)| \to 0$ both gradients coincide, and thus f_c is continuously differentiable. Obviously $f_c(x,\cdot,\cdot)$ is convex on U_1. Since the Hessian of $f(x,\cdot,\cdot)$ is positive definite, $f_c(x,\cdot,\cdot)$ is convex on U_2. From [21–Sec. 42, Thm. B] it follows that the differentiable function f_c is convex, iff ∇f_c is monotone, i.e., $(\nabla f_c(x,\xi,A) - \nabla f_c(x,\zeta,B)) \cdot ((\xi,A) - (\zeta,B)) \geq 0$ for all (ξ,A), (ζ,B). Since f_c is continuously differentiable and monotone on $\mathrm{int}(U_1)$ and $\mathrm{int}(U_2)$ it follows that ∇f_c is monotone on $\mathrm{int}(\bar{U}_1 \cup \bar{U}_2) = \mathbb{R} \times \mathbb{R}^n$, which shows the convexity of f_c.

From Lemma 1 it follows that the operator $\int_\Omega f_c(x, u(x), v(x))\,dx$ is well-defined for $u, v \in L^1(\Omega) \times \left(L^1(\Omega)\right)^n$. In particular $\int_\Omega f_c(x, u(x), \nabla u(x))\,dx$ is well-defined, if $u \in L^1(\Omega)$ and ∇u is the absolutely continuous part of Du.

Proof (of Theorem 2). Let

$$I^*(u) := \begin{cases} \int_\Omega f\left(x, u(x), \nabla u(x)\right) dx & \text{for } u \in W^{1,1}(\Omega), \\ +\infty & \text{else.} \end{cases}$$

It is immediate that $I(u) \leq I^*(u)$, and since $f_c \leq f$ we also have $I_c(u) \leq I(u)$. Consequently, it follows that

$$\mathcal{R}(I_c)(u) \leq \mathcal{R}(I)(u) \leq \mathcal{R}(I^*)(u). \tag{14}$$

Therefore, to prove the assertion of this theorem, it suffices to show that $\mathcal{R}(I^*)(u) = I_c(u)$. Since $I^*(u) = +\infty$ for $u \notin W^{1,1}(\Omega)$, we have

$$\mathcal{R}(I^*)(u) = \mathcal{R}\left(I^*, W^{1,1}(\Omega)\right)(u).$$

Every $u \in \mathrm{BV}(\Omega)$ can be approximated by a sequence $\{u^{(k)}\}_{k\in\mathbb{N}} \subset W^{1,1}(\Omega)$ satisfying $\|u^{(k)} - u\|_{L^1(\Omega)} \to 0$. Moreover, from the definition of $\mathcal{R}(I^*)$ it follows that for every $k \in \mathbb{N}$ there exists $\tilde{u}^{(k)} \in W^{1,1}(\Omega)$ satisfying $\|\tilde{u}^{(k)} - u^{(k)}\|_{L^1(\Omega)} \leq 1/k$ and

$$\mathcal{R}(I^*)(u^{(k)}) \geq I(\tilde{u}^{(k)}) - 1/k \,. \tag{15}$$

For $u \in W^{1,1}(\Omega)$ it follows from the general results in [22] that

$$\mathcal{R}(I^*)(u) = \mathcal{R}\big(I^*, W^{1,1}(\Omega)\big)(u) = I_c(u) \,. \tag{16}$$

From (15), (16), and the fact that $\|\tilde{u}^{(k)} - u\|_{L^1(\Omega)} \to 0$, it follows that

$$\begin{aligned}\mathcal{R}(I^*)(u) \leq \liminf_{k\to\infty} I^*(\tilde{u}^{(k)}) = \liminf_{k\to\infty} I(\tilde{u}^{(k)}) \leq \\ \leq \liminf_{k\to\infty} \mathcal{R}(I^*)(u^{(k)}) = \liminf_{k\to\infty} I_c(u^{(k)}) \,.\end{aligned}$$

Thus, $\mathcal{R}(I^*)(u) = \mathcal{R}\big(I_c; W^{1,1}(\Omega)\big)(u)$ for $u \in \mathrm{BV}(\Omega)$. We note that for $u \in \mathrm{BV}(\Omega) \cap L^\infty(\Omega)$ and $\varepsilon > 0$, we may choose a sequence $u^{(k)} \in W^{1,1}(\Omega)$ satisfying $I_c(u^{(k)}) \to \mathcal{R}(I^*)(u)$, which satisfies $\|u^{(k)}\|_{L^\infty} < \|u\|_{L^\infty} + \varepsilon$ for all $k \in \mathbb{N}$. In other words, setting $X^r := \{u \in \mathrm{BV}(\Omega) : \|u\|_{L^\infty} < r\}$ we have

$$\mathcal{R}(I^*)(u) = \mathcal{R}\big(I_c; X^r \cap W^{1,1}(\Omega)\big)(u) \text{ for } u \in X^r \,. \tag{17}$$

For $r > 0$ and $u \in W^{1,\infty}(\Omega)$ let

$$f^r(x,\xi,A) := \begin{cases} \dfrac{\big(\xi - u^0(x)\big)^2 \wedge r^2}{2|A|} + \alpha|A| \,, & \text{if } \sqrt{2\alpha}|A| > |\xi - u^0(x)| \wedge r \,, \\ \sqrt{2\alpha}\big(|\xi - u^0(x)| \wedge r\big) \,, & \text{if } \sqrt{2\alpha}|A| \leq |\xi - u^0(x)| \wedge r \,, \end{cases}$$

and

$$I_c^r(u) := \int_\Omega f^r\big(x, u(x), \nabla u(x)\big)\, dx \,.$$

Here $a \wedge b, a \vee b$ denote the minimum, maximum of a and b, respectively. Since $\|u^0\|_{L^\infty} =: r_0 < \infty$ it follows that for every $u \in \mathrm{BV}(\Omega)$ satisfying $\|u\|_{L^\infty} < r - r_0$ we have $I_c(u) = I_c^r(u)$. Thus, from (17) we find that for $u \in X^{r-r_0}$

$$\mathcal{R}(I^*)(u) = \mathcal{R}\big(I_c^r; X^r \cap W^{1,1}(\Omega)\big)(u) \,.$$

Using [23–Thm. 4.1.4] it follows that for $u \in X^{r-r_0}$ we have

$$\mathcal{R}(I^*)(u) = \int_\Omega f^r\big(x, u(x), \nabla u(x)\big)\, dx + |D^s u|(\Omega) = I_c(u).$$

Using [24–Prop. 2.4] it follows that for every $u \in \mathrm{BV}(\Omega)$

$$\mathcal{R}(I^*)(u) = \lim_{r\to+\infty} \mathcal{R}(I^*)\big((u \wedge r) \vee -r\big) = \lim_{r\to\infty} I_c\big((u \wedge r) \vee -r\big) \,.$$

From this and the monotone convergence theorem (see e.g. [16]) the assertion follows. □

We recall that the assumption $u^0 \in L^\infty(\Omega)$ is needed in order to satisfy the growth conditions required in [23]. The functional $\mathcal{R}(I)$ is coercive with respect to the total variation semi-norm. It can be shown by a truncation argument that there exists a minimizer of $\mathcal{R}(I)$ with L^∞-norm less than $\|u^0\|_{L^\infty(\Omega)}$. Thus the functional attains a minimizer u in BV.

We note that convexification of non-convex functionals on BV is a recent research topic. We mention the papers [23, 25, 26, 27].

4 Numerics

We describe a finite element method for minimization of the functional $\mathcal{R}(I)(u)$. We use $M := (n-1) \times (m-1)$ quadratic finite elements $(Q_i)_{i=1,\dots,M}$ to cover Ω and bilinear basis functions $(\phi_j)_{j=1,\dots N:=n\times m}$ which are centered at the corner points of the finite elements. We denote by $\mathcal{Q} := \operatorname{span}\{Q_i : i = 1, \dots, M\}$. The initial data u^0 is given as discrete values on a rectangular grid of size $n \times m$ and is identified with the function $u^0 = \sum_{i=1}^N u_i^0 \phi_i$.

The minimizer u_{NCBV} (non-convex bounded variation) of the functional $\mathcal{R}(I)$ solves the optimality condition $\partial\mathcal{R}(I)(u_{\mathrm{NCBV}}) = 0$, where $\partial\mathcal{R}(I)$ is the subgradient of $\mathcal{R}(I)$. In the weak form the optimality condition reads as

$$\begin{aligned} \frac{u-u^0}{|\nabla u|}\phi_j + \left(\alpha - \frac{(u-u^0)^2}{2|\nabla u|^2}\right)\frac{\nabla u \nabla \phi_j}{|\nabla u|} &= 0 \quad \text{if } \sqrt{2\alpha}|\nabla u| > |u-u^0|\,, \\ \sqrt{2\alpha}\frac{u-u^0}{|u-u^0|}\phi_j &= 0 \quad \text{if } \sqrt{2\alpha}|\nabla u| \le |u-u^0|\,, \end{aligned} \tag{18}$$

where $j = 1, \dots, N$. The second equation implies that if $\sqrt{2\alpha}|\nabla u| \le |u-u^0|$, then $u(x) = u^0(x)$, from which it follows that $|\nabla u(x)| = 0$. With the abbreviation

$$a(u) = \frac{1}{|\nabla u|} \wedge \frac{\sqrt{2\alpha}}{|u-u^0|}\,, \quad b(u) = \left(\left(\alpha - \frac{|u-u^0|^2}{2|\nabla u|^2}\right) \vee 0\right)\frac{1}{|\nabla u|}$$

equation (18) reads as follows

$$\sum_{i=1}^N \int_\Omega a(u)\phi_i\phi_j\, u_i + b(u)\nabla\phi_i\nabla\phi_j u_i = \int_\Omega a(u)\phi_i\phi_j u_i^0 \quad (j = 1, \dots, N)\,. \tag{19}$$

Let $U := (u_1, \dots, u_N)^T$, $U^0 := (u_1^0, \dots, u_N^0)^T$,

$$M_{ij} := \int_\Omega \phi_i\phi_j \text{ and } L_{ij} := \int_\Omega \nabla\phi_i\nabla\phi_j\,.$$

We approximate $a(u)$ and $b(u)$ by elementwise constant functions $\tilde{a}(U)$ and $\tilde{b}(U)$.

Using this notation and these approximations, (18) reads as

$$\left(\tilde{a}(U_{\mathrm{NCBV}})M + \tilde{b}(U_{\mathrm{NCBV}})L\right) U_{\mathrm{NCBV}} = \tilde{a}(U_{\mathrm{NCBV}})M\,U^0\,. \tag{20}$$

This system is solved applying the fixed point iteration:

$$\tilde{a}(U_{\mathrm{NCBV}}^{(s)})\, M\, U_{\mathrm{NCBV}}^{(s+1)} + \tilde{b}(U_{\mathrm{NCBV}}^{(s)})\, L\, U_{\mathrm{NCBV}}^{(s+1)} = \tilde{a}(U_{\mathrm{NCBV}}^{(s)})\, M\, U_{\mathrm{NCBV}}^{0} \quad (s = 0, 1, 2, \ldots) . \tag{21}$$

The iteration is terminated if a given tolerance tol is reached, i.e., if $|U_{\mathrm{NCBV}}^{(s+1)} - U_{\mathrm{NCBV}}^{(s)}| \leq \mathrm{tol}$ or s exceeds a given limit. In each iteration step, for solving the linear system for $U_{\mathrm{NCBV}}^{(s+1)}$ we use the C(onjugate)G(radient)-method. In order to avoid occurring oscillations, the following modified scheme can be used: For $s = 0, 1, 2, \ldots$, solve

$$a(U_{\mathrm{NCBV}}^{(s)})\, M\, U_{\mathrm{NCBV}}^{(*)} + \tilde{b}(U_{\mathrm{NCBV}}^{(s)})\, L\, U_{\mathrm{NCBV}}^{(*)} = M\, U_{\mathrm{NCBV}}^{(s)} \tag{22}$$

due to the unknown function $U_{\mathrm{NCBV}}^{(*)}$ and using the solution set

$$U_{\mathrm{NCBV}}^{(s+1)} = U_{\mathrm{NCBV}}^{(s)} + \delta^s\, (U_{\mathrm{NCBV}}^{*} - U_{\mathrm{NCBV}}^{(s)}) \quad (s = 0, 1, 2, \ldots) , \tag{23}$$

where $0 < \delta^s \leq 1$ tends to zero for increasing s.

5 Results

In this section we show that the iterated solution of $\mathcal{R}(I)$ gives similar results as solving the MCM equation. We show that $v(T)$, the solution of the MCM equation and $u_{\mathcal{N}}^{\mathcal{N}}$ are almost identical. We recall that $u_k^{\mathcal{N}}$ is the minimizer of the functional $\mathcal{R}(I)$ where u^0 is replaced by $u_{k-1}^{\mathcal{N}}$, $k = 1, \ldots, \mathcal{N}$ and $\alpha = T/\mathcal{N}$.

The MCM equation at time $T = \Delta T \tilde{\mathcal{N}}$ is calculated by solving the system of equations (note that ΔT needs not be identical to α)

$$\tilde{c}(U_{\mathrm{MCM}})(M + \Delta T\, L) U_{\mathrm{MCM}} = \tilde{c}(U_{\mathrm{MCM}}) M U_{\mathrm{MCM}}^{k-1} \quad (k = 1, \ldots, \tilde{\mathcal{N}}) \tag{24}$$

and denoting the solution by U_{MCM}^k. A vector U_{MCM} is associated with the function $u_{\mathrm{MCM}} = \sum_{i=1}^N (U_{\mathrm{MCM}})_i \phi_i$ from which an approximation $\tilde{c}(U_{\mathrm{MCM}})$ for $c(u) = \frac{1}{|\nabla u|}$ is determined that is piecewise constant on the finite elements. I.e., $\tilde{c}(U_{\mathrm{MCM}})|_{Q_{ij}} = c(u_{\mathrm{MCM}})(p_{ij})$, where p_{ij} is the midpoint of cell Q_{ij}. The implemented FE-Method for solving the Mean Curvature Motion essentially follows [28].

For fixed k, we again use a fixed point iteration to solve (24):

$$\tilde{c}(U_{\mathrm{MCM}}^{(s)})(M + \Delta T\, L) U_{\mathrm{MCM}}^{(s+1)} = \tilde{c}(U_{\mathrm{MCM}}^{(s)}) M U_{\mathrm{MCM}}^{k} \quad (s = 0, 1, \ldots) . \tag{25}$$

If $\|U_{\mathrm{MCM}}^{(s+1)} - U_{\mathrm{MCM}}^{(s)}\| < \mathrm{tol}$ the iteration is terminated and $U_{\mathrm{MCM}}^{k+1} := U_{\mathrm{MCM}}^{(s+1)}$.

In the following we present two numerical comparisons of regularization, i.e., minimizing the functional (10), iterative regularization, and solving the MCM equation (4).

5. Radmoser, E., Scherzer, O., Weickert, J.: Scale-space properties of nonstationary iterative regularization methods. Journal of Visual Communication and Image Representation **11** (2000) 96–114
6. Mrázek, P., Weickert, J., Steidl, G.: Correspondences between wavelet shrinkage and nonlinear diffusion. In: [30]. (2003) 101–116
7. Steidl, G., Weickert, J.: Relations between soft wavelet shrinkage and total variation denoising. In: [31]. (2002) 198–205
8. Evans, L.C.: Regularity for fully nonlinear elliptic equations and motion by mean curvature. In: Viscosity solutions and applications (Montecatini Terme, 1995). Volume 1660 of Lecture Notes in Math. Springer, Berlin (1997) 98–133
9. Alvarez, L., Guichard, F., Lions, P.L., Morel, J.M.: Axioms and fundamental equations of image processing. Arch. Ration. Mech. Anal. **123** (1993) 199–257
10. Scherzer, O.: Explicit versus implicit relative error regularization on the space of functions of bounded variation. In: [32]. (2002) 171–198
11. Lenzen, F., Scherzer, O.: Tikhonov type regularization methods: history and recent progress. In: Proceeding Eccomas 2004. (2004)
12. Grasmair, M., Scherzer, O.: Relaxation of non-convex singular functionals. (2004) submitted.
13. Scherzer, O.: A posteriori error estimates for nonlinear ill–posed problems. Nonlinear Analysis **45** (2001) 459–481
14. Osher, S., Rudin, L.I.: Feature-oriented image enhancement using shock filters. SIAM J. Numer. Anal. **27** (1990) 919–940
15. Rudin, L., Osher, S., Fatemi, E.: Nonlinear total variation based noise removal algorithms. Physica D **60** (1992) 259–268
16. Evans, L., Gariepy, R.: Measure Theory and Fine Properties of Functions. CRC–Press, Boca Raton (1992)
17. Ambrosio, L., Fusco, N., Pallara, D.: Functions of bounded variation and free discontinuity problems. The Clarendon Press Oxford University Press, New York (2000)
18. Temam, R.: Problèmes mathématiques en plasticité. Volume 12 of Méthodes Mathématiques de l'Informatique [Mathematical Methods of Information Science]. Gauthier-Villars, Montrouge (1983)
19. Rudin, W.: Real and Complex Analysis. McGraw-Hill, New York (1987) 3rd edition.
20. Dacorogna, B.: Direct Methods in the Calculus of Variations. Springer–Verlag, Berlin (1989)
21. Roberts, A.W., Varberg, D.E.: Convex functions. Academic Press [A subsidiary of Harcourt Brace Jovanovich, Publishers], New York-London (1973) Pure and Applied Mathematics, Vol. 57.
22. Grasmair, M., Scherzer, O.: Relaxation of nonlocal variational functionals. (2004) submitted.
23. Bouchitté, G., Fonseca, I., Mascarenhas, L.: A global method for relaxation. Arch. Rational Mech. Anal. **145** (1998) 51–98
24. Buttazzo, G., Dal Maso, G.: Γ-limits of integral functionals. J. Analyse Math. **37** (1980) 145–185
25. Aviles, P., Giga, Y.: Variational integrals on mappings of bounded variation and their lower semicontinuity. Arch. Ration. Mech. Anal. **115** (1991) 201–255
26. Fonseca, I., Müller, S.: Quasi-convex integrands and lower semicontinuity in L^1. SIAM J. Math. Anal. **23** (1992) 1081–1098
27. Fonseca, I., Müller, S.: Relaxation of quasiconvex functionals in $BV(\Omega, R^p)$ for integrands $f(x, u, \nabla u)$. Arch. Ration. Mech. Anal. **123** (1993) 1–49

28. Deckelnick, K., Dziuk, G.: Convergence of a finite element method for non-parametric mean curvature flow. Numer. Math. **72** (1995) 197–222
29. Nielsen, M., Johansen, P., Olsen, O., Weickert, J., eds.: Scale-Space Theories in Computer Vision. Lecture Notes in Computer Science Vol. 1683, Springer Verlag (1999) Proceedings of the Second International Conference, Scale-Space'99, Corfu, Greece, 1999.
30. Griffin, L., Lillholm, M., eds.: Scale-Space Methods in Computer Vision. Lecture Notes in Computer Science Vol. 2695, Springer Verlag (2003) Proceedings of the 4th International Conference, Scale-Space 2003, Isle of Skye, UK, June 2003.
31. Van Gool, L., ed.: Pattern Recognition. Springer, Berlin (2002) Lecture Notes in Computer Science, Vol. 2449.
32. Nashed, M., Scherzer, O., eds.: Interactions on Inverse Problems and Imaging. Volume 313. AMS (2002) Contemporary Mathematics.

Tree Edit Distances from Singularity Theory

Ole Fogh Olsen

The IT University of Copenhagen,
Rued Langgaardsvej 7, DK-2300 Copenhagen, Denmark
fogh@itu.dk

Abstract. An representation based on the singularity structure of the gradient magnitude over scale is used as the atoms in a space of images. This representation is summarized as a rooted tree. The generic transitions of the functional of the scale space images are analysed and listed for the scale parameter and one free parameter. A distance measure between images is deduced soly from these generic transistions. The singular transitions are translated into the language of the tree transitions such that one generic transition corresponds to one unit edit operation of the tree structure. The distance between two images is the size of the smallest set of edit operations necessary to transform the corresponding tree representations into each other.

1 Introduction

The content of an image manifests itself at multiple, a prior unknown levels of scale or resolution. This has been addressed by the computer vision community in a principled fashion by the so-called scale space theories or multi scale schemes[11, 13]. Scale space theory ensures an image representation invariant to rotation, translation and scaling (or invariant to other groups of transformations[24]) and provides a regularization of the original image to a differentiable output which makes the vast toolbox of differential geometry available. In its simplest form, a scale space image is a continuum of increasingly blurred images also refered to as the Gaussian scale space due to the generating kernel. In this paper we will not consider the vast amount of alternative scale space schemes only the Gaussian scale space.

This machinery has opened for the creation of a range of feature detectors (interest points) defined and detected by (semi-)algebraic expressions of derivatives possibly automatically tuned to the appropriate scale[4, 5, 18]. The framework is mathematically well founded in a principled way allowing for derivation and analysis of properties of the system [25, 3, 22]. Recent research has investigated the geometry of scale space images [2] for instance the trajectories of extrema[6]. The geometry of scale space images relates the details present at low scale (high resolution) to the coarse overall objects on the high scale (low resolution) [19] and offers the opportunity to analyse information over scale also denoted deep structure analysis.

R. Kimmel, N. Sochen, J. Weickert (Eds.): Scale-Space 2005, LNCS 3459, pp. 316–326, 2005.

Information of objects over a range of scales can be represented as graphs or trees in the algorithmic data structure sense. Such a scale space tree of features provides an invariant representation as above but also an representation which is invariant to minor changes in the configurations of the singularities in the scale space image more precisely invariant to a small local diffeomorphism. Hence it represents the topology of the singularity paths over scale. Preliminary investigations [26, 20, 14] show that it is feasible to construct scale space trees based on different kinds of singularities also for 3D data sets.

Distance measures between the scale space trees can be established to assess the distance between the content in two different images. Standard graph or tree matching algorithms do not per default provide good distance measure between scale space trees, simply because an atomic tree transform in the algorithmic sense can correspond to a large series of scale space image transformations and vice versa. Schemes have to be developed for finding the appropriate atomic transformations from the image analysis point of view and find their algorithmic counterpart for implementation purposes.

For this, we propose the generic tree transformations to be deduced from the generic transitions of the singular paths. The necessity for only generic transitions is obvious. All imaginable transitions will give rise to a combinatoric explosion of possible transitions but according to singularity theory only the generic cases will occur in almost all cases (loosely spoken the non-generic cases occur with probability zero) and are very limited in number. The presence of non-generic structures[15, 12] can for instance indicated symmetric structures which are unlikely to occur in all most all images. Genericity is always stated in relation to a base set of functions. In this paper it is the set of solutions to the heat equation.

The idea is inspired by the successful line of work within shape analysis, specifically within medial representations/skeletonisations of shapes [7, 8] and symmetry set representation [7, 8, 16]. In this area, classical singularity theory has been applied and extended to determine the relation between geometric fiducial points on the outline of the shape and the central points in the medial representation. Next step has been to derive the generic transitions for the fiducial points for general warping of the shape and relate these results to the corresponding changes in the medial representation [9]. The counterpart in the scale space tree approach is to establish the generic transitions of a scale space image when the original image is changed. These transitions will be translated into the language of algorithmic tree transitions[21].

We present an extended annotated scale space tree detected from the multi scale structure of the squared gradient magnitude. We will derive the list of possible transitions for the singularities of the gradient squared under the parameters of scale and an extra control parameter. These transitions will be use to deduce corresponding tree transitions which will form the basis for an image matching scheme.

2 A Multiscale Gradient Magnitude Tree

The gradient magnitude from a Gaussian scale space image has previously been suggested as the underlying representation for a semi-automatic segmentation [23, 17, 1]. For this, a tree structure was constructed based on the generic transitions of an image evolving under the heat equation, the complete list of these transitions was derived and analysed in previous work[22]. It was shown that the fold and the cusp catastrophes occur generically in Gaussian scale space for the squared gradient magnitude.

The possible transitions through the sampled scale space is shown schematic in figure 1. Also included in the figure are the dual regions to the minima namely the catchment basins or Voronoi areas. The events (annihilation, merging, creation, splitting) are named after the interaction between the saddle and the minimum (or minima). In the cases of annihilation (b) and merging (c) two minima and a saddle are reduced to one minimum, corresponding to a disappearing border between the two segments. The cases of creation (d) and splitting (e) are the reverse events where an emerging saddle corresponds to the appearing of a border between the segments (dual to the two minima). A line in the figure from a segment to a segment indicates a edge in the corresponding tree. Hence, the parent-child edges in the tree are in all cases indirectly given by the saddle connecting the involved minima.

The squared gradient magnitude will contain several global minima with a value of zero. They coincide and correspond to the singularities of the image itself. The squared gradient magnitude will also have local minima which are just all non-global minima. In terms of image geometry a local minima of the gradient magnitude corresponds to a point where the second order structure of the image has a degeneracy in the direction of the image gradient. In other words in the point in the gradient direction the image looks like $x^3 + x$

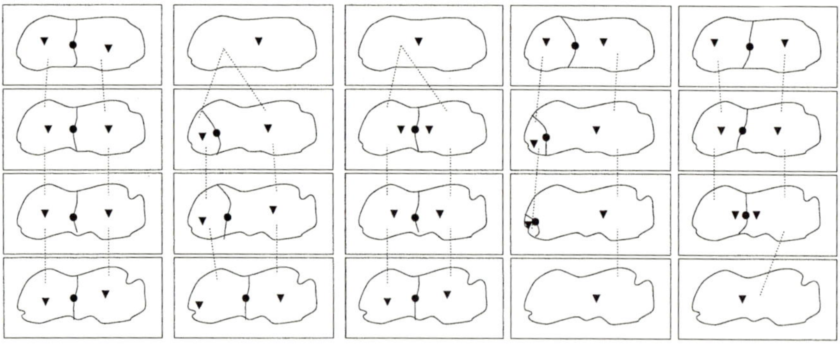

(a) No change (b) Annihilation fold local (c) Merging cusp global-to-local (d) Creation fold local (e) Splitting cusp local-to-global

Fig. 1. Generic events of the gradient magnitude. Scale increases upwards in the figure. Minima and saddles are symbolised with triangles and circles, respectively

The annihilation or creation of one minimum with one saddle will always involve a local minimum. The merging will involve the joining of two global minima and saddle into one local minimum. The splitting event will destroy one local minimum and introduce one saddle and two global minima at a higher scale. These facts originate from the simple fact that global minima for the gradient magnitude squared correspond to singularities in the original image and the cusp event in the gradient magnitude image will correspond to fold event in the original image. Hence global minima in the gradient magnitude will always interact in pairs never alone or uneven numbers.

This extra information will be used to annotated the tree structure with the type of the different minima. Nodes in the tree correspond to the minima of the squared gradient magnitude of the scale space image. Each level of the tree corresponds to a sampled scale in the scale space image. The direct correspondence between levels and scale discretisation is not a necessity for the representation. Nodes with only one child and one parent can just be collapsed with cost zero. This will avoid the undesired growth of the tree in limit of finer and finer scale discretisation. The nodes in the tree are connected according to the derived possible transitions of the minima. The nodes are annotated as local or global minima. This more rich tree structure limits later on the amount of possible matches inbetween trees. That is, for a given part of the first tree less possible matches exist in the second tree. Of course this also results in a more detailed list of possible transitions.

The full tree syntax is presented in figure 2. In figure 1 the difference between transitions involving local and global minima was not illustrated. This is included in figure 2. Circles denoted internal nodes in a tree, small and big circles indicate respective local and global minima. In case (a) there is no change of either a local or global minimum. In case (b) a local minimum is annihilated and its corresponding node will be connected to its neighbour which is either a local or global minimum. In case (c) two global minima are merging into one local minimm. In case (d) a local minimum is created somewhere in the tree.

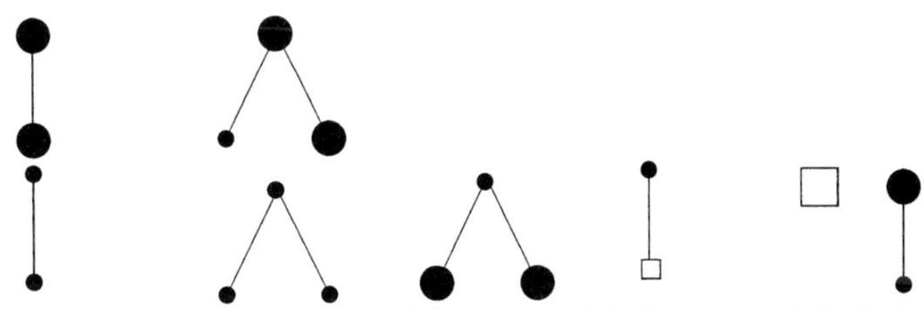

(a) No change (b) Annihilation fold (c) Merging cusp (d) Creation fold (e) Splitting cusp

Fig. 2. The tree syntax corresponding to the generic scale transitions in figure 1. Circles are internal nodes. Squares are leaves. Big symbols correspond to global minima and small symbols correspond to local

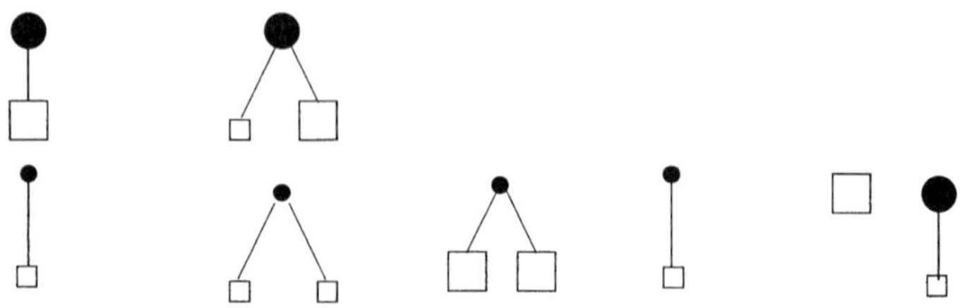

(a) No change (b) Annihilation fold (c) Merging cusp (d) Creation fold (e) Splitting cusp

Fig. 3. The tree syntax as in figure 2 instantiated on the lowest level in the tree

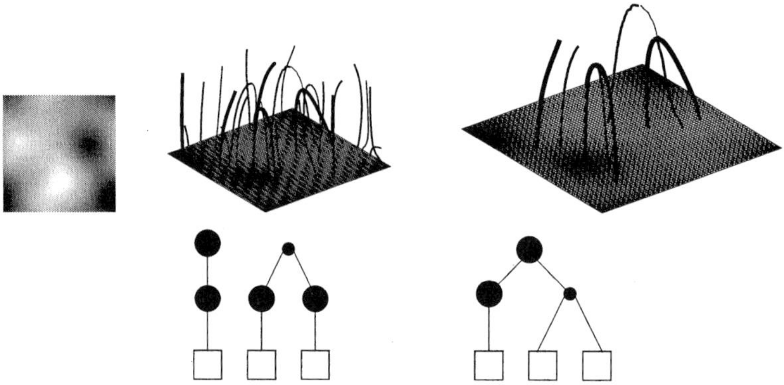

Fig. 4. Top left: An image, top middle: singularity paths for the image (maximum=2 point line, saddle=3 point line,global minimum=fat line, local minimum=smallest line), top right: a subset of the paths. Bottom: the two groups in the subset of singularity paths correspond to two disjoint parts of the tree. Note that only the largest and smallest paths correspond directly to nodes and edges in the tree. Bottom left: the first group is a merging of two global minima into one local minima plus a path with no change. Bottom right: two global minima (leaves) merge into one local minima which annhilates and is connected to the neighbouring global minimum. Please note that the latter global minimum is for simplicity of the figure only depicted in the tree syntax and among all the paths in the middel not in the subset of paths shown to the top right

In case (e) a local minimum splittes into two global minima. Hence the local minimum is linked to a global minimum which will have a global minimum as neighbour. In this paper we only consider the framework of trees not graphs. Therefore in case (e) the local minimum is only connected to one of the global minima. The representation of a splitting as a child node with two parents would ruin the tree structure and introduce the more general graph structure. Such a representation has been discussed as a interesting and relevant alternative in previous publication by the author and others [10, 19].

Each of the cases and the sub cases in figure 2 can occur on the lowest level (the root is in the top) of the tree. In such a case the lowest circles in each case will be transform to leaves and denoted with a square instead. This is illustrated in figure 3

In figure 4 is an example of an image and the corresponding singularity paths. For simplicity a subset of these has been selected and the corresponding tree structure for this subset is depicted. Note how the subtree are constructed by combine the subtrees from figure 2 and for the lowest level in tree the subtrees from figure 2 are used.

3 Tree Transformations

When one image is warped into another this will of course also change the corresponding multi scale trees from one to another. Because the trees are deduced and builded from the catastrophes (an abrut change of structure) in the singu-

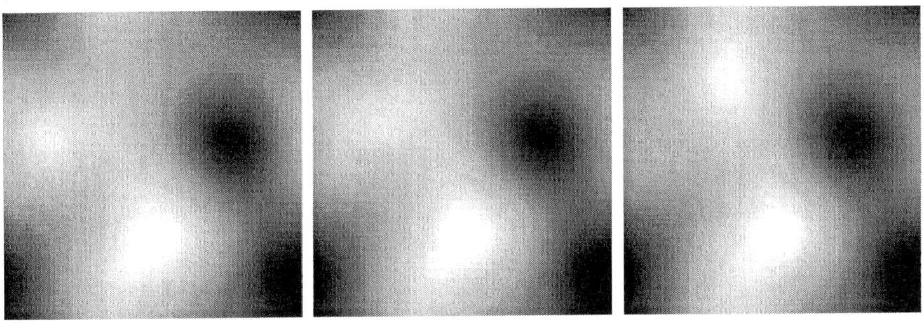

Fig. 5. A constructed image sequence made by taken a random image and adding a one-pixel size peak in three different locations and then blurr the sum. In left frame the peak is close the to left border of the image then moving along a straight line ending close to top part of the image in the last right frame

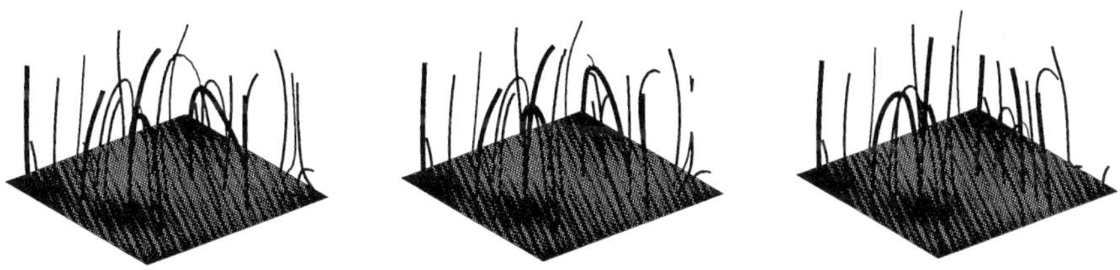

Fig. 6. Frames plus all singular paths. Lines of width 2 denote maxima of gradient magnitude (G). Lines of width 3 denote saddles of G. Thin lines with width 1 are local minima. Fat lines are global minima

Fig. 7. A subset of the singular paths in figure 6. Two series of transitions are depicted. Intermediate size indicates saddles. Thin line corresponds to local minimum. Fat line denotes global minimum

lar paths the transitions of the trees will correspond to the sudden change of the paths known as higher order events in singularity theory or catastrophes. In terms of singularity theory a family of multi scale images is a family of functions controlled by two parameters (scale and another one) from this the possible transitions can be analysed and the corresponding transitions on the tree structure can be deduced.

In the following we provide an example of the connection between transitions of paths and tree transitions before giving the full list of tree transitions. In figure 5 is depicted a constructed image sequence. In figure 6 are the constructed frames plus their corresponding singular paths.

In figure 6 several series of transitions occur. In figure 7 only a subset is presented for clarity. The subset consists of two independent series of transitions; one in the background and one in the foreground.

In the foreground series the connection between three global minima is changed. A "no change" path and a merge between two global minima interact and changes connections such that the "no change" path afterwards merges with the middle global path and the far right path becomes a "no change" paths. Another phrasing would be that the middle global minima swaps its relation to its nearest neighbour; it swaps from being the detail of one structure to being a detail of the neighbouring structure.

The background series involves two changes first the annihilation in the top is resolved from the left frame to the middle frame, secondly the merge disappear from the middle frame to the right frame. In the presented scale range it corresponds to an extended lifetime over scale of the involved structure. The first event the disappearing of the annihilation corresponds to the structure persists further over scale instead of becoming a detail of a larger structure. The second event has a similar interpretation since the global minimum persists over a long scale range extend beyond the visualised levels of scale.

In figure 8 is shown the tree structure corresponding to the singular paths in figure 7. As is illustrated the minimal singular paths corresponds directly to the derived tree structure. Nodes correspond to the sampled scale levels and the edges in the tree are derived from catastrophe points on the singular paths.

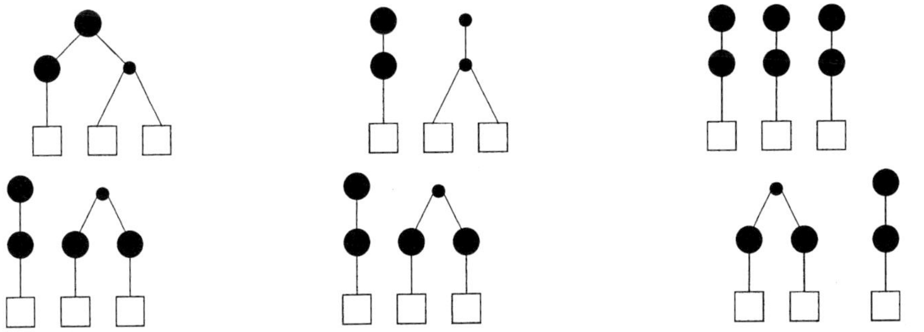

Fig. 8. The tree structures corresponding to the minimum paths in figure 7. The top row corresponds the series of transitions in background of figure 7. The bottom row corresponds to the foreground series of transitions

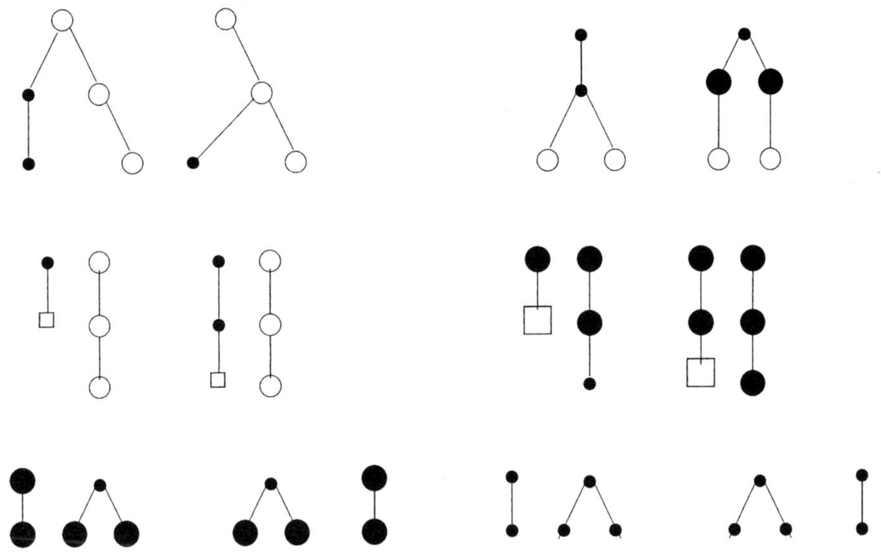

Fig. 9. The four types of tree transitions. The transitions can occur from left to right and from right to left. Open circles indicate either a local or global minimum annotation. These can be expanded according to the tree syntax provided in the previous sections

3.1 Generic Tree Transformations

Only a very limited number of local tree transformations is possible if one only consider transformations correspond to the generic transitions of the singular paths. In figure 9 is listed the generic edit operations on the tree structures.

In the following section we will derive and explain how exactly these transitions are derived from the catastrophes.

4 Co-dimension One Transitions in Scaled Families of the Gradient Magnitude

The co dimension one transitions are introduced by an extra control parameter such that the family of Gaussian scale space images depends on two spatial variables, the scale parameter and the control parameter. One can think of the control parameter as describing a path through the underlying set of images. When traveling through the family of scale space images using this extra degree of freedom then for all most all image instances there will only be the usual catastrophes on the singular paths but for specific fixed values of the control parameter the singular paths will exhibit extraordinary catastrophes (which are non-generic for a single scale space image) but are generic for a family. These are the transitions between two set of generic paths. These higher order catastrophes correspond to the collision of an extra singular path through a "ordinary" catastrophe point. Simple events can also occur namely the resolving of a catastrophe or the shifting of it to a high or low location in scale.

In the following it can be useful to compare the transitions with the tree syntax in figure 2. An annihilation event for the gradient magnitude can be shifted towards higher or lower scale. This will correspond to the tree transition illustrated in figure 9 top row left. A creation event can also be shifted in the scale direction which will result in the tree transition depicted in figure 9 middle row left.

A shift for the merge event will result in tree transition shown in figure 9 top row right. As in the other subfigures the top and bottom nodes stay fixed in the tree; hence the transition is fully depicted in the figures.

The splitting event can make a shift in accordance with the illustration in figure 9 middle row right. Please note the difference to the middle row left where there is no restriction on the neighboring nodes.

The cusp catastrophe with two global minima and a saddle (corresponding to ordinary fold catastrophe for the image involving an extremum and a saddle) can collide with another global minimum path and swap the connectivity between the three involved global minima. This corresponds to the transition in figure 9 bottom left.

The fold catastrophe can collide with another local minimum path (this will in the transition moment correspond to the cusp catastrophe involving local minima). This will also result in the swap of the connectivity between the involved minima. This corresponds to the transition given in figure 9 bottom right.

This concludes the list of possible transitions and their counterpart in terms of tree transitions.

5 Conclusion

An image representation based on multi scale singularity tree has been proposed. The syntax of the resulting multi scale tree has been presented. The possible transitions of the multi scale trees have been listed as the basis of image matching scheme based on edit operation distance of trees. In order to do this, the codimension one transitions of the scale space images have been derived and translated to tree transitions. It remains to apply this matching scheme to ensembles of real world images and evaluate its practical performance. It might also be possible to weight the different edit operations with a cost proportional to their probability of occurence. This is the objective for future work.

References

1. Erik Dam, Peter Johansen, Ole Fogh Olsen, Andreas Thomsen, Tron Darvann, Andy B. Dobrzeniecki, Nuno V. Hermann, Noriyuki Kitai, Sven Kreiborg, Per Larsen, Martin Lillholm, and Mads Nielsen. Interactive multi-scale segmentation in clinical use. In *European Congress of Radiology 2000*, March 2000. Abstract and video accepted for CompuRAD.
2. James Damon. Local morse theory for solutions to the heat equation and gaussian blurring. *Journal of Differential Equations*, 115(2), January 1995.
3. James Damon. Properties of ridges and cores for two-dimensional images. *Journal of Mathematical Imaging and Vision*, 10:163–174, 1999.
4. L. M. J. Florack, B. M. ter Haar Romeny, J. J. Koenderink, and M. A. Viergever. Cartesian differential invariants in scale-space. *JMIV*, 3(4):327–348, November 1993.
5. Luc Florack, Bart ter Haar Romeny, Jan Koenderink, and Max Viergever. General intensity transformations and differential invariants. *Journal of Mathematical Imaging and Vision*, 4(2):171–187, 1994.
6. Luc M J Florack and Arjan Kuijper. The topological structure of scale-space images. *Journal of Mathematical Imaging and Vision*, 12:65–79, 2000.
7. P. J. Giblin and B. B. Kimia. On the local form and transitions of symmetry sets, and medial axes, and shocks in 2D. In *Proc. of ICCV*, pages 385–391, Greece, September 1999. IEEE Computer Society.
8. P. J. Giblin and B. B. Kimia. On the local form of symmetry sets, and medial axes, and shocks in 3D. In *Proc. of CVPR*, pages 566–573. IEEE Computer Society, 2000.
9. P. J. Giblin and B. B. Kimia. Transitions of the 3D medial axis under a one-parameter family of deformations. In *ECCV*, pages 718–724, 2002.
10. Lewis D. Griffin. *Descriptions of Image Structure.* PhD thesis, Uni. of London, 1995.
11. T. Iijima. Basic theory on normalization of a pattern (in case of typical one-dimensional pattern). *Bulletin of Electrotechnical Laboratory*, 26:368–388, 1962. (in Japanese).
12. Peter Johansen, Mads Nielsen, and Ole Fogh Olsen. Singular points in one-dimensional gaussian scale space. Journal of Mathematical Imaging and Vision, 2000.
13. Jan J. Koenderink. The structure of images. *Biological Cybernetics*, 50:363–370, 1984.
14. Arjan Kuijper and L.M.J. Florack. Using catastrophe theory to derive trees from images. *Journal of Mathematical Imaging and Vision*, 2004.
15. Arjan Kuijper and Luc Florack. The relevance of non-generic events in scale space models. *International Journal of Computer Vision*, 2004.

16. Arjan Kuijper, Ole Fogh Olsen, and Peter Giblin. Data structures from shapes using the symmetry set. In *SIAM conference on Imaging Science*, 2004.
17. M. Letteboer, O.F. Olsen, E. Dam, P. Willems, M. Viergever, and W.J. Niessen. Segmentation of tumors in mr brain images using an interactive multi-scale watershed algorithm. *Academic Radiology*, 2004.
18. Tony Lindeberg. *Scale–Space Theory in Computer Vision.* The Kluwer International Series in Engineering and Computer Science. Kluwer Academic Publishers, Boston, USA, 1994.
19. Ole Fogh Olsen. *Generic Image Structure.* PhD thesis, Department of Computer Science, University of Copenhagen, Universitetsparken 1, DK-2200 Copenhagen East, Denmark, March 2000. Technical Report DIKU-2000/04.
20. Ole Fogh Olsen. The scale structure of the gradient magnitude. Technical report, IT University of Copenhagen, Rued langgaards Vej 7, DK-2300 Copenhagen, Denmark, 2003. ITU Technical report : TR-2003-29.
21. Ole Fogh Olsen. Inducing edit distance from generic transitions. In *SIAM conference on imaging science*, 2004.
22. Ole Fogh Olsen and Mads Nielsen. Generic events for the gradient squared with application to multi-scale segmentation. In *Scale-Space Theory in Computer Vision, Proc. 1st International Conference*, volume 1252 of *Lecture Notes in Computer Science*, pages 101–112, Utrecht, The Netherlands, July 1997.
23. Ole Fogh Olsen and Mads Nielsen. Multi-scale gradient magnitude watershed segmentation. In *ICIAP'97 - 9th International Conference on Image Analysis and Processing*, volume 1310 of *Lecture Notes in Computer Science*, pages 6–13, Florence, Italy, September 1997.
24. Peter Olver, Guillermo Sapiro, and Allen Tannenbaum. Differential invariance signatures and flows in computer vision: a symmetry group approach. In Bart ter Haar Romeny, editor, *Geometry-Driven Diffusion in Computer Vision*, chapter 11. Kluwer Academic Publishers, 1994.
25. J.H. Rieger. Generic evolutions of edges on families of diffused greyvalue surfaces. *Journal of Mathematical Imaging and Vision*, 5(3):207–217, 1995.
26. K. Somchaipeng, J. Sporring, S. Kreiborg, and P. Johansen. Software for extracting 3d mssts. Technical report, EU project IST-2001-35443, Deliverable No. 8 http://www.itu.dk/Internet/sw1953.asp.

The Stochastic Structure of Images

Jan-Mark Geusebroek*

ISLA, Informatics Institute, Faculty of Science,
University of Amsterdam, Kruislaan 403,
1098 SJ Amsterdam, The Netherlands
mark@science.uva.nl

Abstract. As resolving power increases, image features evolve in an iterative fashion; large scale features will persist, and finer and finer scale features are resolved at each increase in resolution. As such, the observation process is shown to overwhelm natural image statistics. Observation by a linear imaging process imposes natural image statistics to be of random multiplicative nature, rather than additive. The scaling behavior of the random process is driven by the gradient structure in the scene irradiance. From the general structure of multiplicative processes, image statistics are proven to follow a sequential fragmentation process. From these theoretical results, analytical forms for the distributions of image derivative filter responses and gradient magnitude are derived.

1 Introduction

The observation of scenes is dominated by coincidence. Coincidence stems from the fact that we perceive a two-dimensional projection of the physical world, the projection affected by occlusion, reflection, and clutter. Coincidence due to the viewpoint causing an accidental background. Coincidence due to the lighting conditions and accidental reflection characteristics. If the viewpoint, view, light, and the background are arbitrarily chosen, this does not imply the resulting image has completely arbitrary statistics. Objects having uniform visual characteristics impose structure to the scene. Shadow and shading effects, although accidental, are spatially correlated. Composition of the scene, by human or nature alike, brings structure to the scene. All these effects cause images to be covered by the laws of correlated spatial disorder.

Axiom 1. *The spatial structure of the irradiance from natural scenes is dominated by correlated disorder.*

From our experiments, we have reasons to believe that a large fraction of recorded images is covered by the axiom [1]. Hence, natural scenes are largely non regular. Repetitions of structures are considered a different phenomenon.

* This work is sponsored by the Netherlands Organization for Scientific Research (NWO).

R. Kimmel, N. Sochen, J. Weickert (Eds.): Scale-Space 2005, LNCS 3459, pp. 327–338, 2005.

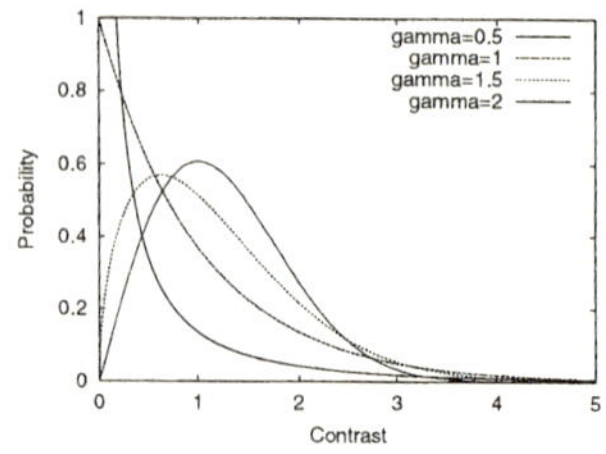

Fig. 2. Illustration of the Weibull distribution $p(r)$ for various values of γ

Solving the sequential fragmentation equation Eq. (14) for a power-law distribution yields

$$p(r) = c_1 \left(\frac{r}{\beta}\right)^{\gamma-1} \int_r^{\infty} p(r')\, dr' \tag{15}$$

which is solved by

$$p(r) = c_1 \left(\frac{r}{\beta}\right)^{\gamma-1} e^{-\frac{1}{\gamma}\left(\frac{r}{\beta}\right)^{\gamma}} \ , \tag{16}$$

and where c_1 normalizes $p(r)$ to yield a Weibull probability density function,

$$c_1 = \frac{1}{\int_0^{\infty} \left(\frac{r}{\beta}\right)^{\gamma-1} e^{-\frac{1}{\gamma}\left(\frac{r}{\beta}\right)^{\gamma}}} = \frac{1}{\beta} \ . \tag{17}$$

The value of the shape parameter γ ranges from 0 to 2 for the distribution to be positive semi-definite [26] (see Fig. 2).

Corollary 3. *The gradient magnitude $|\nabla E|$ in natural images follows a Weibull probability density,*

$$p(r) = \frac{1}{\beta} \left(\frac{r}{\beta}\right)^{\gamma-1} e^{-\frac{1}{\gamma}\left(\frac{r}{\beta}\right)^{\gamma}} \ , \quad 0 < \gamma \leq 2 \ .$$

So far, we considered an isotropic distribution of responses, resulting in the Weibull distribution Corollary 3 of the one-dimensional gradient magnitude. To proceed with the joint density of a two-dimensional response distribution, $p(x, y)$, consider the sum of the orthogonal derivative magnitudes $r = |x| + |y|$ for an isotropic random field. Note that at a later stage results will be generalized to hold for the gradient magnitude $r' = \sqrt{x^2 + y^2}$. Due to isotropy, the gradient magnitude response distribution r will be identical to the one-dimensional case Corollary 3. The marginal statistics for the response of the derivative operator in the x-direction, $p_x(x)$, is obtained by integration over the y-variable,

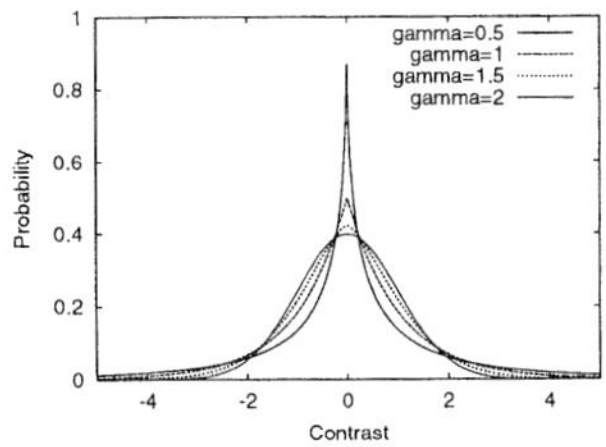

Fig. 3. Illustration of the marginal distribution $p_x(x)$ for various values of γ

$$p_x(x) = \int_{-\infty}^{+\infty} \frac{1}{\beta} \left(\frac{|x| + |y|}{\beta} \right)^{\gamma-1} e^{-\frac{1}{\gamma}\left(\frac{|x|+|y|}{\beta}\right)^{\gamma}} dy \tag{18}$$

$$= -2e^{-\frac{1}{\gamma}\left(\frac{|x|+|y|}{\beta}\right)^{\gamma}} \Big|_0^{\infty} , \tag{19}$$

which, after normalization, leads to an integrated form of the Weibull distribution (see Fig. 3).

Corollary 4. *The response to the directional derivative $\nabla_x E$ in natural images follows an integrated Weibull probability density,*

$$p_x(x) = c_2 e^{-\frac{1}{\gamma}\left|\frac{x}{\beta}\right|^{\gamma}} .$$

The normalization factor is given by

$$c_2 = \frac{\gamma}{2\gamma^{\frac{1}{\gamma}} \beta \Gamma(1/\gamma)} ,$$

where $\Gamma(\alpha)$ represents the complete Gamma function,

$$\Gamma(\alpha) = \int_0^{\infty} t^{\alpha-1} e^{-t} dt .$$

Note that any power transformation of the form $x' = x^{\alpha}$ does not affect the marginal distribution $p_x(x)$, other than a reparameterization $\gamma' = \gamma/\alpha$, $\beta' = \beta^{\alpha}$. Hence, the gradient magnitude $\nabla E^2 = \nabla_x E^2 + \nabla_y E^2$ is of the form $r = |x| + |y|$, which will be Weibull distributed according to Corollary 3. Furthermore, the response to the directional derivative $\nabla_w E$ in any direction w will be distributed according to Corollary 4. Finally, non-linear gamma transforms applied to the intensity at any fixed resolution in the scaling cascade will not affect the results, other than a reparameterization of the Weibull parameters γ and β as discussed above.

5 Conclusions

This paper shows the linear imaging process to impose the stochastic structure of images. The observation by a linear convolution over the incident light diffuses the once emitted random field, the diffusion being driven by the gradient

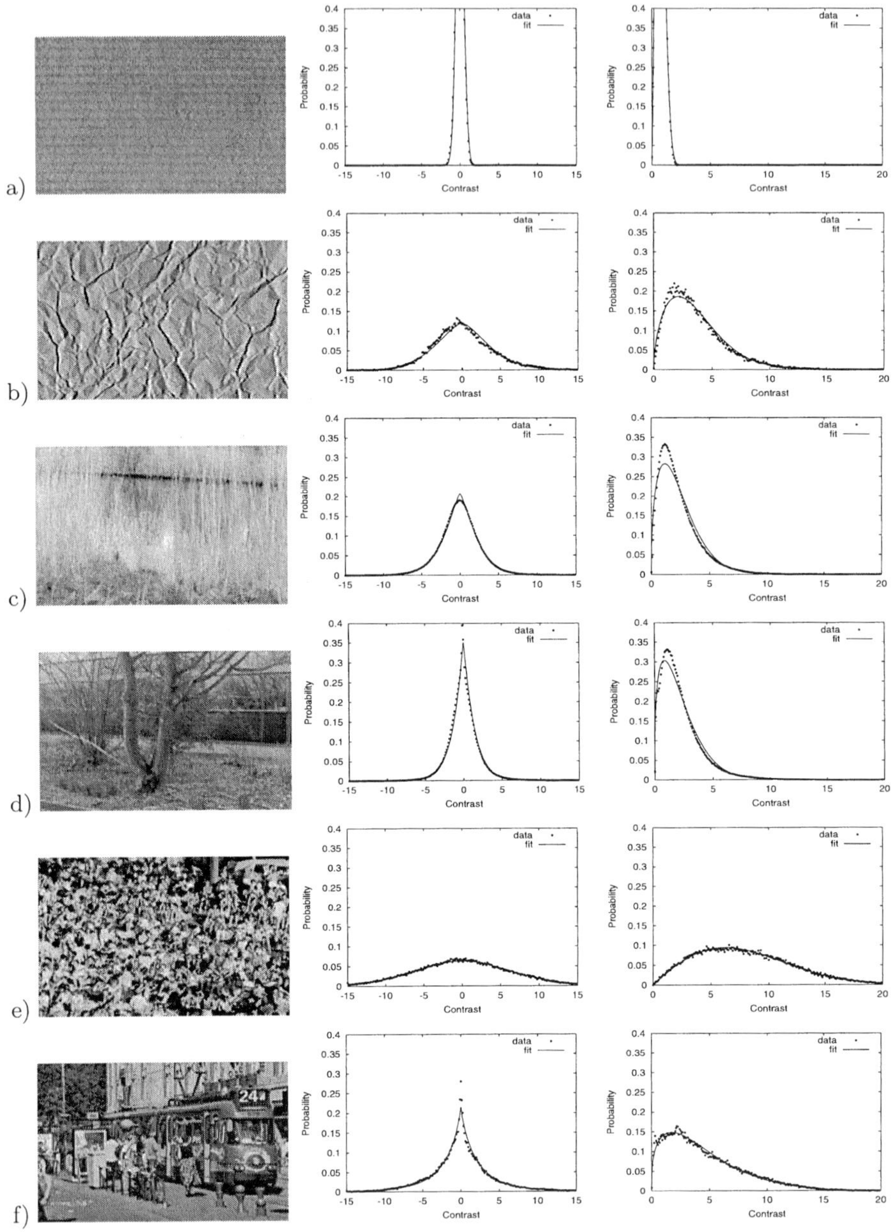

Fig. 4. Examples of the fit of the various probability densities to the statistics of natural images. The middle column shows the distribution of the derivative operator ($\sigma = 3$ pixels) in the x-direction of the original image (left column), the right column shows the gradient magnitude distribution. **a.** corduroy and **b.** crumpled paper from the Curet dataset [27]; **c.** image 271 and **d.** 167 from the van Hateren collection [28]; **e.** image 848012 and **f.** 673021 from the Corel Gallery collection

structure in the projected intensity. The multiplicative process for correlated intensity fluctuations combined with the additive stochastic laws for uncorrelated fluctuations cause image statistics to follow the laws of sequential fragmentation. From these theoretical results, analytical expressions for the statistics of gradient magnitude and derivative filter responses are derived.

A major point is the percentage of images covered by the posed statistical laws. The question essentially addresses the validity of Axiom 1. For our experiments we have reasons to believe that this is a large fraction of recorded images [1]. See Fig. 4 for some examples.

The statistical characteristics of multiplicative processes are dominated by the extreme events that are on the tail of the distribution [22]. For a multiplicative process, the distribution of the product and the behavior of the moments are crucially sensitive to extreme events. When increasing imaging resolution, progressively more extreme events become accessible. This increased access to the tails of the distribution will manifest itself in the sporadic appearance of exceptional realizations that will cause the observed statistics to fluctuate as a function of resolution. Consequently, a multi-scale approach is essential for a general vision system to probe the statistical structure of images.

The stochastic structure of any image is affected by the linear observation kernel. More precisely, the power-law and Weibull distribution of image statistics are imposed by the observation instrument. The occurrence of these general distributions are not a property of the objective world, that is, the structure of natural scenes other than randomness. The structure of the scene underlying the observation only influences the Weibull parameters γ and β. In this contribution, I have derived the exact influence of the linear observation instrument on the observed image statistics.

References

1. Geusebroek, J.M., Smeulders, A.W.M.: Fragmentation in the vision of scenes. In: Proc. 9th Int. Conf. Comput. Vision. Volume 1., IEEE Computer Society (2003) 130–135
2. Field, D.J.: Relations between the statistics of natural images and the response properties of cortical cells. J. Opt. Soc. Am. A **4** (1987) 2370–2393
3. Mallat, S.G.: A theory for multiresolution signal decomposition: The wavelet representation. IEEE Trans. Pattern Anal. Machine Intell. **11** (1989) 674–693
4. Ruderman, D.L., Bialek, W.: Statistics of natural images: Scaling in the woods. Phys. Rev. Let. **73** (1994) 814–817
5. van der Schaaf, A., van Hateren, J.H.: Modelling the power spectra of natural images: Statistics and information. Vision Res. **36** (1996) 2759–2770
6. Simoncelli, E.P.: Modeling the joint statistics of images in the wavelet domain. In: Proc. SPIE. Volume 3813., SPIE (1999) 188–195
7. Sigman, M., Cecchi, G.A., Gilbert, C.D., Magnasco, M.O.: On a common circle: Natural scenes and Gestalt rules. Proc. Natl. Acad. Sci. USA **98** (2001) 1935–1940
8. Schwartz, O., Simoncelli, E.P.: Natural signal statistics and sensory gain control. Nature Neurosci. **4** (2001) 819–825

9. Grenander, U., Srivastava, A.: Probability models for clutter in natural images. IEEE Trans. Pattern Anal. Machine Intell. **23** (2001) 424–429
10. Lee, A.B., Mumford, D., Huang, J.: Occlusion models for natural images: A statistical study of a scale-invariant dead leaves model. Int. J. Comput. Vision **41** (2001) 35–59
11. Srivastava, A., Liu, X., Grenander, U.: Universal analytical forms for modeling image probabilities. IEEE Trans. Pattern Anal. Machine Intell. **24** (2002) 1200–1214
12. Geusebroek, J.M., Smeulders, A.W.M.: A physical explanation for natural image statistics. In Chantler, M., ed.: Proceedings of the 2nd International Workshop on Texture Analysis and Synthesis (Texture 2002), Heriot-Watt University (2002) 47–52
13. Srivastava, A., Lee, A.B., Simoncelli, E.P., Zhu, S.C.: On advances in statistical modeling of natural images. J. Math. Imaging Vision **18** (2003) 17–33
14. Koenderink, J.J.: The structure of images. Biol. Cybern. **50** (1984) 363–370
15. Havlin, S., Selinger, R.B., Schwartz, M., Stanley, H.E., Bunde, A.: Random multiplicative processes and transport in structures with correlated spatial disorder. Phys. Rev. Let. **61** (1988) 1438–1441
16. Castaing, B., Dubrulle, B.: Fully developed turbulence: A unifying point of view. J. Phys. II. (France) **5** (1995) 895–899
17. Arneodo, A., Bacry, E., Manneville, S., Muzy, J.F.: Analysis of random cascades using space-scale correlation functions. Phys. Rev. Let. **80** (1998) 708–711
18. Brown, W.K.: A theory of sequential fragmentation and its astronomical applications. J. Astrophys. Astr. **10** (1989) 89–112
19. Papoulis, A., Pillai, S.U.: Probability, Random Variables, and Stochastic Processes. 4 edn. McGraw-Hill, New-York (2002)
20. Benzi, R., Biferale, L., Crisanti, A., Paladin, G., Vergassola, M., Vulpiani, A.: A random process for the construction of multiaffine fields. Physica D **65** (1993) 352–358
21. Havlin, S., Schwartz, M., Selinger, R.B., Bunde, A., Stanley, H.E.: Universality classes for diffusion in the presence of correlated spatial disorder. Phys. Rev. A **40** (1989) 1717–1719
22. Redner, S.: Random multiplicative processes: An elementary tutorial. Am. J. Phys. **58** (1990) 267–273
23. Levy, M., Solomon, S.: Power laws are logaritmic boltzmann laws. Int. J. Mod. Phys. C **7** (1996) 595–601
24. Sornette, D., Cont, R.: Convergent multiplicative processes repelled from zero: Power-laws and truncated power-laws. J. Phys. I. (France) **7** (1997) 431–444
25. Feller, W.: An Introduction to Probability Theory and its Applications. Volume 2. John Wiley and Sons, New York (1971)
26. Gnedenko, B.V., Kolmogorov, A.N.: Limit Distributions for Sums of Independent Random Variables. Addison-Wesley, Boston (1968)
27. Dana, K.J., van Ginneken, B., Nayar, S.K., Koenderink, J.J.: Reflectance and texture of real world surfaces. ACM Trans Graphics **18** (1999) 1–34
28. van Hateren, J.H., van der Schaaf, A.: Independent component filters of natural images compared with simple cells in primary visual cortex. Proc. R. Soc. Lond. B **265** (1998) 359–366

Skeletons of 3D Shapes*

Jayant Shah

Mathematics Department, Northeastern University, Boston MA
shah@neu.edu

Abstract. A new method for determining skeletons of 3D shapes is described. It is a combination of the approach based on the "grass-fire" technique and Zhu's approach based on first finding portions of the shape where its width is approximately constant. The method specifically does not require presmoothing of the shape and is robust in the presence of noise. In an appendix, a method based on variational calculus is formulated for determining pruned, smoothed shape skeletons by minimizing a functional.

1 Introduction

Although a large number of papers have appeared on ways to determine shape skeletons, until recently [3, 7, 8, 11, 18], these papers have dealt with 2D shapes [4, 6, 10, 12, 15, 16, 17, 19]. Extension of these techniques to 3D shapes is non-trivial since skeletons for 3D shapes consist of intersecting 2-dimensional surfaces [7]. This paper presents a generalization of the method developed in [17] to 3D shapes. The emphasis is on developing a computationally robust method for determining shape skeletons in the presence of noise and the inevitable numerical inaccuracies inherent in computation on a discrete grid.

The usual definition of the skeleton of a 3D shape is that it is the locus of the centers of maximal spheres contained in the shape. If the radii of these spheres are recorded at the corresponding points on the skeleton, the shape can be fully recovered as an envelope of the spheres centered on the skeleton with radii recorded on the skeleton. Construction of the skeleton by drawing maximal spheres at all points within the shape is clearly impractical and determining the tangency between a sphere and a noisy shape boundary is problematic. An alternative is the "grassfire" technique in which the shape is imagined to be filled with dry grass and a fire is started at the shape boundary. The time of arrival of the grassfire front at a point equals the distance ρ of that point from the shape boundary. The shape skeleton is the locus of points where fronts from two or more directions meet. Alternatively, it is the discontinuity locus of the gradient of the distance function ρ. When the colliding fronts come from opposite directions, the point of their collision can be determined fairly accurately. As the angle between the front normals decreases, it becomes numerically more and more difficult to

* This work was supported by NIH Grant I-R01-NS34189-08.

R. Kimmel, N. Sochen, J. Weickert (Eds.): Scale-Space 2005, LNCS 3459, pp. 339–350, 2005.

detect their collision. It is even more difficult to locate the singular points of the skeleton where more than two fronts come together. Another difficulty is that the skeleton preserves the noise present in the shape boundary in the form of numerous extraneous branches which must be pruned. It is also necessary to identify and prune branches corresponding to shape features which are deemed irrelevant for the task at hand.

An alternative to the approach described above is the one proposed by Zhu [19] for 2D shapes. A 2D shape is viewed as a collection of ribbons which are glued together. A ribbon is a part of the shape where the shape width is approximately constant, so that, after some smoothing, the opposite boundaries of the ribbon are nearly parallel. Chords may then be found which are approximately orthogonal to the smoothed boundary. Zhu minimizes a functional designed to determine optimal chords which are as closely normal to the smoothed boundary as possible. The functional also incorporates terms such as a penalty for parts of the shape in which proper chords cannot be found and a penalty for having many short ribbon segments instead of a few long ones. The relative importance of these desirable properties depends on the values of the parameters in the functional. The centers of the chords define the medial axes of the ribbons which are then extended to form junctions in an optimal way. Note that these junctions are not necessarily the points where the maximal discs touch the shape boundary at two or more points. Instead, their location is governed by criteria such as a minimal number of junctions and angles subtended by the branches at the junctions. Parts of the shape which are neither ribbon-like nor associated with junctions are ignored (pruned). It is not clear how to determine a good initial approximation to start the minimization process, how to tune the numerous parameters involved in the functional and how to generalize the method to 3D shapes.

The approach proposed in this paper is a combination of the two approaches described above and consists of three steps. The "gray skeleton" of a shape is defined in §2 by associating each point inside and outside the shape with a numerical value which is a monotonic function of the angle at which the fronts intersect. One half of this angle is what is called the object angle in the literature [3,5]. We define this angle to be zero at points away from the skeleton. The first important step is an accurate calculation of the object angle at every point based on the observation that normals to the fronts may be determined by using fairly large neighborhoods. An alternative method for calculating the object angle based on the divergence theorem is given in [5]. Corresponding to Zhu's chords, we now have a set of boundary points associated with each point P on the gray skeleton where the front normals at P intersect the shape boundary.

Gray skeletons contain numerous points associated with noise and its pruning constitutes the second step which is described in §3. Parts of the shape with slowly varying width are determined by thresholding the gray skeleton. Then, branches of the thresholded skeleton are extended along the gray-skeleton in such a way that they join up without picking up extraneous branches caused by noise. The method is illustrated by means of an example in §4. An alternative approach to pruning the gray skeleton by contracting the shape boundary towards the gray skeleton homotopically is described in [16].

Due to noise, the boundary of the skeleton obtained by the method described above may have a tattered appearance. The skeleton may also have gaps and short isolated branches. Regularization of the skeleton is the third step. In §5, we propose an approach based on minimization of a functional. The 2D version of this functional has been implemented in [17]; however, its 3D version involves regularization of the skeleton boundary which is a space curve and has yet not been implemented.

Zhu's functional includes a term which penalizes wiggles in the skeleton. In the appendix, we present an approach based on variational calculus for smoothing (and pruning) shape skeletons by minimizing a functional which is analogous to functionals used for segmenting gray-scale images.

2 Gray Skeleton

In this section, we define the gray skeleton Γ of a shape from which the topological skeleton K is extracted. A shape is simply an open bounded set D; its boundary will be denoted by ∂D. The topological shape skeleton K is assumed to have the regularity properties usually assumed in practice. For example, its dimension is at most 2. We start with the distance function

$$\rho(P) = \max_{X \in \partial D} dist(X, P) \tag{1}$$

which may be computed quickly, for example, by Sethian's fast marching algorithm [1] which solves the eikonal equation $\|\nabla \rho\| = 1$ by the "upwind" finite difference scheme. We estimate the angle between the normals to the intersecting fronts exploiting the property that the gradient lines of ρ are straight lines except where they cross the shape skeleton [14]. The gradient line of ρ passing through a point P off the shape skeleton connects P to the point Q on the shape boundary nearest to P. The vector $\overrightarrow{QP}$ is the radius of the maximal sphere at P. It is normal to the fronts advancing to P. If P is a point of K where there are exactly two boundary points, Q^+ and Q^-, nearest to P, the maximal sphere centered at P touches the shape boundary only at Q^+ and Q^-. Exactly two fronts intersect at P and the angle between their velocity vectors is equal to the angle between the vectors $\overrightarrow{Q^+P}$ and $\overrightarrow{Q^-P}$ (Fig. 1). Let φ denote one half of the angle between the vectors $\overrightarrow{Q^+P}$ and $\overrightarrow{Q^-P}$; it takes values between 0 and $\frac{\pi}{2}$. If the shape boundary is smooth at Q^+, then the vector $\overrightarrow{Q^+P}$ is orthogonal to the shape boundary and φ is the angle between the chord $\overrightarrow{Q^+Q^-}$ and the tangent plane at Q^+.

It is shown in [7] that the bisector of the angle between $\overrightarrow{Q^+P}$ and $\overrightarrow{Q^-P}$ is tangent to the shape skeleton at P. Since $\|\nabla^{\pm}\rho\| = 1$ where $\nabla^{\pm}\rho$ denotes the gradients in the directions $\overrightarrow{Q^{\pm}P}$, we have

$$\cos\varphi = \frac{d\rho}{ds} \tag{2}$$

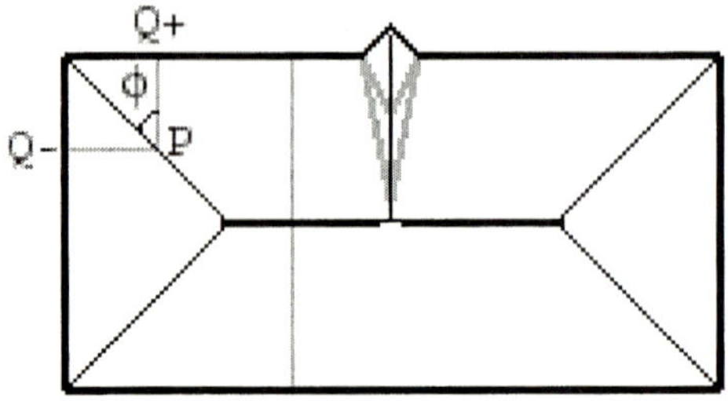

Fig. 1. Geometry of the distance function

where ds denotes the infinitesimal arc-length along the skeleton in the direction of the bisector. It is the projection of the gradients $\nabla^{\pm}\rho$ onto the plane tangent to the skeleton at P. Thus, the larger the value of φ, the slower the rate of change in the width of the shape. We also have

$$\sin\varphi = \frac{1}{2}\left\|\nabla^{+}\rho - \nabla^{-}\rho\right\| \tag{3}$$

which is one half the jump in $\nabla\rho$ across the shape skeleton. The larger the object angle φ, the larger the jump in $\nabla\rho$, and the easier it is to detect the corresponding portion of the shape skeleton.

Let $\varphi = 0$ in the complement of $\partial D \cup K$ where $\nabla\rho$ is continuously differentiable. We define the gray skeleton, Γ, of the shape by letting its value equal to $\sin\varphi$. The gray skeleton is defined everywhere except on the set J of points of K where the maximal sphere touches the shape boundary at more than two points. (J is assumed to have dimension ≤ 1. Our strategy is to determine K/J from the gray skeleton in the complement of $\partial D \cup J$ and then extend it over parts of J which lie in the closure of K/J . This strategy still leaves out special cases such as circular cylinders and balls where the closure of K/J does not contain all of J. Therefore, it would be useful to extend the definition of the gray skeleton to all of K. An elegant method to define such an extension due to Dimitrov, Damon and Siddiqi [5] is based on the notion of average flux. These authors use the divergence theorem and calculate the average flux as a limit of the surface integral of $\nabla\rho$ over spherical neighborhoods as the neighborhood size shrinks to zero. A possible alternative method for calculating the average flux is to integrate the laplacian of ρ in the sense of distributions. Only the singular part of the laplacian (including its poles) contributes to the average flux and since the singular part depends only on the object angles, it may be computed accurately using large neighborhoods. In this paper, the purpose of calculating Γ is to extract K from it and it is sufficient to use a simple approximation of Γ at a points in J. At a point P in J. we pick two points, Q^{+} and Q^{-}, among the set of boundary points nearest to P such that the object angle φ determined by the vectors $\overrightarrow{Q^{+}P}$ and $\overrightarrow{Q^{-}P}$ is the largest possible and set $\Gamma(P) = \sin\varphi$. Note that the set J may be quite large because there may be numerous intersecting branches produced by noise. The suggested method amounts to either ignoring or lumping together contribution of less important branches.

To estimate φ, we need to determine the gradient directions $\overrightarrow{Q^+P}$ and $\overrightarrow{Q^-P}$ which are the directions in which the directional derivative of ρ is maximum and equals 1. We compute the directional derivatives of ρ in all *inward* radial directions at every point P in a given shape and determine the directions in which it is nearly equal to 1. On a discrete grid there are only finitely many radial directions available depending upon the size of the neighborhood N which determines the numerical accuracy of φ.Since the gradient lines are straight lines and $||\nabla\rho|| = 1$, we may use fairly large neighborhoods. The algorithm in detail is as follows:

Step 1: Scan the boundary ∂N of a neighborhood N of P. If R is a point belonging to ∂N, the derivative of ρ in the direction $\overrightarrow{RP}$ is

$$D_R(\rho) = \frac{\rho(P) - \rho(R)}{\left|\overrightarrow{RP}\right|} \tag{4}$$

Step 2: Determine the local maxima of $D_R(\rho)$ along ∂N. Among these maxima, choose the ones which are approximately equal to 1 within the tolerance determined by the size of N.

Step 3: If there is a single gradient direction at a point, we set $\varphi = 0$. If there are two or more maxima chosen in *Step 2*, calculate the object angle between rays corresponding to all possible pairs maxima and choose the largest value.

3 Pruning

A straightforward approach to pruning is by thresholding. A high threshold results in a set of disconnected skeleton branches correspond to the shape parts with slowly varying width. If the threshold is set too low, the skeleton will include branches due to noise. The following algorithm gets around this difficulty by extending the branches of the thresholded skeleton along the gray skeleton into the thicker parts of the shape, possibly forming junctions, without picking up extraneous branches belonging to noise.

Step 1: Choose two thresholds $\overline{\theta}$ and $\underline{\theta}$ for angle φ with $\overline{\theta} > \underline{\theta}$.

Step 2: Threshold the gray skeleton by $\overline{\theta}$. This eliminates irrelevant branches so that what remains corresponds to significant parts of the shape.

Step 3: Extend the branches of the thresholded skeleton in the direction of increasing ρ provided that φ remains greater than $\underline{\theta}$ throughout. The effect is to extend the skeleton branches towards the more blobby or thicker parts of the shape. As the branch is extended, it may encounter junctions with noise (or protrusion) related branches, but *these branches are not followed since they ascend from the junctions towards the shape boundary.* If the value of $\overline{\theta}$ is set too high, some numerical difficulties may be encountered during this step due to inaccuracies in the values of ρ and the discreteness of the grid. If the value of $\overline{\theta}$ is

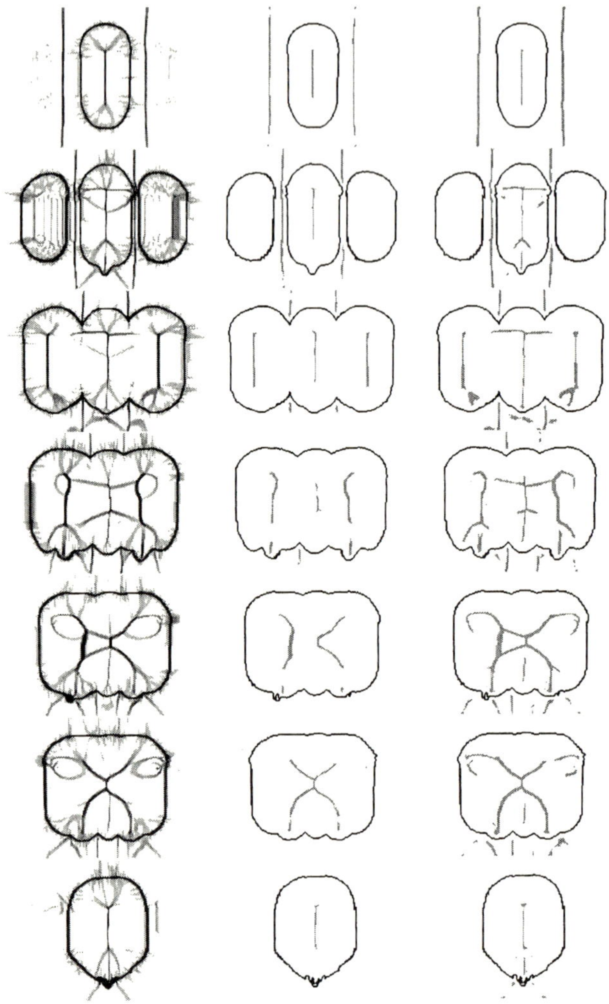

Fig. 4. Gray skeleton and pruned skeletons: Horizontal sections

shape and a single component outside. Notice that as the shape get less cylinder-like (small φ), the gray skeleton gets fuzzier reflecting the difficulty in locating the skeleton accurately.

5 Regularization of the Skeleton Boundary

As mentioned in the Introduction, the boundary of the skeleton obtained by the method described above may have a tattered appearance due to noise in the data. It may also have gaps and and short isolated branches. To address these defects, we define a regularizing functional

$$E(K) = \int_K (c - \alpha) + \beta |\partial K| \tag{5}$$

where K is a subset of the gray skeleton Γ representing the pruned skeleton, $c = \cos\varphi$ is the cost function defined on the gray skeleton, α is a cost threshold, ∂K is the boundary of K and $|\partial K|$ is its length. If $\beta = 0$, E is minimized by setting K equal to the set of points of the gray skeleton G where $c \leq \alpha$. If $\beta > 0$, we may locally shorten ∂K even if it means extending K along G to include voxels where $c > \alpha$. Thus, minimization of E requires the techniques of curve evolution on a surface. Details of this approach are being worked out.

6 Appendix

We present an approach based on variational calculus to determine a pruned, smoothed skeleton directly by minimizing a functional. We exploit a basic identity that the distance function satisfies [9]. Let

$$u = \nabla(\frac{1}{2}\rho^2) \tag{6}$$

where ρ is the distance function. Since $\|\nabla\rho\| = 1$ and $u = \rho\nabla\rho$, $\|u\| = \rho$. It follows that u satisfies the identity

$$u = \nabla(\frac{1}{2}u \cdot u) \tag{7}$$

We have the following converse due to Gomes and Faugeras [9]:

Proposition: Suppose $u : R^n \to R^n$ satisfies Identity (6) almost everywhere and u is continuous at all points of the set $M = u^{-1}(0)$. Then, $u = \nabla(\frac{1}{2}\rho^2)$ where ρ is the distance from M.

Taking into account that the identity fails where u is discontinuous, namely, on the shape skeleton, we define the functional

$$\int_{R-K} \alpha \left\| \frac{1}{2}\nabla(u \cdot u) - u \right\|^2 + |K| \tag{8}$$

subject to the condition that $u = 0$ on the shape boundary. Here R is an open subset of R^n containing the shape, K is the discontinuity locus of u and $|K|$ is its volume (length, area, etc). The problem with this functional is that $\left\|\frac{1}{2}\nabla(u \cdot u) - u\right\|^2 = O(\rho^2)$ and hence, the penalty for violationg Identity (6) depends on its distance from ρ. To remedy this, we consider the normalized functional

$$E(u, K) = \int_{R-K} \alpha \left\| \nabla \|u\| - \frac{u}{\|u\|} \right\|^2 + |K| \tag{9}$$

Since minimization of E balances the cost of the modifying or removing a segment of K against the cost of violating Identity (6), the result is a pruned and smoothed skeleton K. The functional does not attempt to regularize the boundary of K; inclusion of such a term would make the functional considerably more difficult to implement. To apply the method of gradient descent, we use the Ambrosio-Tortorelli appoximation of E:

$$E_\lambda(u,v) = \int_R \left\{ \alpha \left\| \nabla \|u\| - \frac{u}{\|u\|} \right\|^2 (1-v)^2 + \frac{\lambda}{2} \|\nabla v\|^2 + \frac{v^2}{2\lambda} \right\} \tag{10}$$

The corresponding gradient descent equations are:

$$\frac{\partial u}{\partial t} = \beta u + (1-v)\nabla \|u\| \tag{11}$$

$$\frac{\partial v}{\partial t} = \nabla \cdot \nabla v - \frac{v}{\lambda^2} + \frac{2\alpha}{\lambda}(1-v) \left\| \nabla \|u\| - \frac{u}{\|u\|} \right\|^2 \tag{12}$$

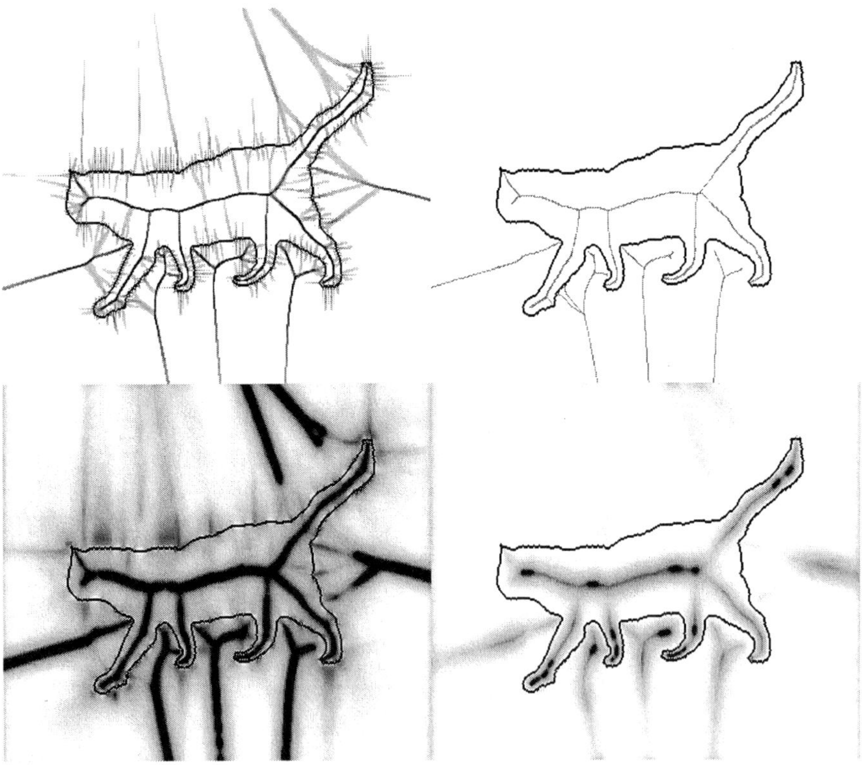

Fig. 5. Top row: Gray skeleton and pruned skeleton Bottom row: Function v with two different values of α

where

$$\beta = (1-v)\left\{\nabla \cdot \nabla \|u\| - \frac{u}{\|u\|}\right\} - 2\left(\nabla \|u\| - \frac{u}{\|u\|}\right) \cdot \nabla v \tag{13}$$

We illustrate this approach for determining shape skeletons by means of an example of a 2D shape shown in Fig. 4. The top row shows the gray skeleton and the skeleton obtained by pruning the gray skeleton. The bottom row depicts the function v obtained using Eqs. 10 and 11 with two different values of α. Just as in the case of segmentation functionals, two objections may be raised against this approach: First, we have much less control over pruning and smoothing of the skeleton than when we use the method of gray skeletons; second, function v is spread out and we need a ridge finding method to locate the skeleton precisely by following the ridges of v. Nonetheless, it is of interest to formulate the problems of determining shape skeletons and of segmenting images within the same variational framework.

References

1. D. Adalsteinsson and J.A. Sethian: "The fast construction of extension velocities in level set methods", J. of Computational Physics, 1999, pp. 2-22.
2. L. Ambrosio and V.M. Tortorelli: "On the Approximation of Functionals depending on Jumps by Quadratic, Elliptic Functionals", Boll. Un. Mat. Ital. (1992).
3. N. Amenta, S. Choi and R. Kolluri: "The Power Crust", Proc. of the sixth ACM Symposium on Solid Modeling, pp.249-260, 2001.
4. S. Bouix and K. Siddiqi: "Divergence-based medial surfaces", Sixth European Conference on Computer Vision, Dublin, Ireland, June, 2000.
5. P. Dimitrov, J.N. Damon and K. Siddiqi: "Flux Invariants for Shape", Conference on Computer Vision and Pattern Recognition (CVPR'03), 2003.
6. P. Giblin and B.B. Kimia: "On local form and transitions of symmetry sets and medial axes and shocks in 2D", International Conference on Computer Vision, 1999, pp.385-391.
7. P. Giblin and B.B. Kimia: "A formal classification of 3D medial axis points and their local geometry', IEEE Trans. Pattern Analysis and Machine Intelligence, 2003.
8. J.D. Furst, S.M. Pizer, D.M. Eberly: "Marching Cores: A Method for Extracting Cores from 3D Medical Images", Proc. of the Workshop on Mathematical Methods in Biomedical Image Analysis (MMBIA), pp. 124-130, June, 1996.
9. J. Gomes and O. Faugeras: "Using the Vector Distance Functions to Evolve Manifolds of Arbitrary Codimension", Third International Conference on Scale Space and Morphology in Computer Vision, 2001.
10. B.B. Kimia, A.R. Tannenbaum, and S.W. Zucker: "Shapes, shocks and deformations, I", International Journal of Computer Vision, 15(3), 1995.
11. F. Leymarie and B.B. Kimia: "Symmetry-based representation of 3D data", International Conference on Image Processing, October, 2001, pp.581-584.
12. T.-L. Liu, D. Geiger and R.V. Kohn: "Representation and self-similarity of shapes", International Conference on Computer Vision, Mumbai, India, 1998.
13. G. Malandain and S. Fernandez-Vidal: "Euclidean Skeletons", Image and Vision Computing, v.16, pp.317-327, 1998.

14. C. Mantegazza and A.C. Mennucci: "Hamilton-Jacobi equations and distance functions on Riemannian manifolds", Appl. Math. Opt. v.47, no.1, 2003, pp.1-25.
15. J. Siddiqi, A. Shokoufandeh, S. Dickinson and S. Zucker: "Shock graphs and shape matching", International Journal of Computer Vision, 35(1), September, 1999, pp. 13-32.
16. K. Siddiqi, S. Bouix, A. Tannenbaum and S. Zucker: "The hamilton-jacobi skeleton", International Conference on Computer Vision, 1999, pp. 828-834.
17. J. Shah: "Gray Skeletons and Segmentation of Shapes", to appear in CVIU.
18. S. Tari and J. Shah: "Local symmetries of shapes in arbitrary dimension", Sixth International Conference on Computer Vision, 1998.
19. S.C. Zhu: "Stochastic jump-diffusion process for computing medial axes in markov random fields", IEEE Trans. Pattern Analysis and Machine Intelligence, 21(11), 1999.

Scale-Space Generation via Uncertainty Principles

Chen Sagiv[1], Nir A. Sochen[1], and Yehoshua Y. Zeevi[2]

[1] Department of Applied Mathematics,
University of Tel Aviv, Ramat-Aviv, Tel-Aviv 69978, Israel
{chensagi, sochen}@post.tau.ac.il
[2] Department of Electrical engineering,
Technion - Israel Institute of Technology,
Technion City, Haifa 32000, Israel
zeevi@ee.technion.ac.il

Abstract. This study is concerned with the uncertainty principles which are related to the Weyl-Heisenberg, the $SIM(2)$ and the Affine groups. A general theorem which associates an uncertainty principle to a pair of self-adjoint operators was previously used in finding the minimizers of the uncertainty principles related to various groups, e.g., the one and two-dimensional Weyl-Heisenberg groups, the one-dimensional Affine group, and the two-dimensional similitude group of $\mathbb{R}^2$, $SIM(2) = \mathbb{R}^2 \times (\mathbb{R}^+ \times SO(2))$. In this study the relationship between the affine group in two dimensions and the $SIM(2)$ group is investigated in terms of the uncertainty minimizers. Moreover, we present scale space properties of a minimizer of the $SIM(2)$ group.

1 Introduction

The 2D Gabor function and Gabor-Morlet wavelets are commonly used in computer vision. Mostly in relation to texture analysis, synthesis and segmentation. The use of these functions is usually motivated by the fact that the Gaussian window minimizes the uncertainty and attains the maximal possible accuracy in both the spatial and frequency domains. In fact, the Gabor transform is a representation of the Weyl-Heisenberg group while the 2D Gabor-Morlet transform is a representation of the 2D affine group or of subgroups thereof. Since both the 2D Gabor-Morlet wavelet transform and the multi-window Gabor transform involve rotation and scaling (and potentially few more transformations) in addition to the usual translation and frequency modulations, it makes sense to look for a window shape that maximizes the accuracy in all attributes. This study explores this question and shows that this aim can only be partially attained.

The Gaussian function appears as a pivot in scale-space theory as well, where its successive applications to images produce coarser resolution images. It is shown, in fact, that the family of Gaussian functions posses semi-group properties with respect to the width of the Gaussian. This raises the question whether

R. Kimmel, N. Sochen, J. Weickert (Eds.): Scale-Space 2005, LNCS 3459, pp. 351–362, 2005.

families of functions that minimize the uncertainty for other groups of transformations posses the same characteristic. It is shown in this study that this is in fact true for the cases that we consider. This is an intriguing fact whose full significance is not yet fully understood.

In this study we consider the results obtained for the similitude group [1, 3] and apply them to the affine group in two dimensions. Moreover, we explore the scale-space nature of the minimizer derived by Ali, Antoine and Gazeau [1] and find that their solution has smoothing and edge detection attributes which can produce scale-space representation of images.

The rest of this paper is organized as follows: First, we review the uncertainty principle theorem for self-adjoint operators and point out related works. We then apply it to the Weyl-Heisenberg group and the affine group in one and two dimensions. We conclude by pointing out the scale-space properties of the minimizers obtained.

2 Background and Related Work

The uncertainty principle is a fundamental concept in quantum mechanics as well as in signal and information theory. In quantum mechanics, the Heisenberg uncertainty principle states that the position and momentum of a particle cannot be simultaneously known. In signal and information theory, Gabor [5] showed that there exists a trade off between time resolution and frequency resolution for one-dimensional signals, and that there is a lower bound on their product. These results were extended to consideration of images [9].

A special attention has been given to the functions which attain the lower bound of the inequality defined by the uncertainty principle. It is used to define the canonical coherent states for quantum systems in physics. In signal processing it was discussed, inter alia, by Gabor. He showed that Gaussian-modulated complex exponentials provide the best trade-off for time resolution and frequency resolution. These are equivalent to a family of canonical coherent states generated by the Weyl-Heisenberg group.

A general theorem which is well known in quantum mechanics and harmonic analysis [4] relates an uncertainty principle to any two self-adjoint operators and provides a mechanism for deriving a minimizing function for the uncertainty equation: two self-adjoint operators, A and B obey the uncertainty relation:

$$\Delta A_f \Delta B_f \geq \frac{1}{2}|\langle [A, B] \rangle| \quad \forall f, \tag{1}$$

where $\Delta A_f, \Delta B_f$ denote the variances of A and B respectively with respect to the signal f. The triangular parnthesis mean an average over the signal i.e. $\langle X \rangle = \int f^* X f$. The mean of the action of an operator P on a function f is denoted as: $\mu_P = \langle P \rangle$ and the commutator $[A, B]$ is given by: $[A, B] := AB - BA$. A function f is said to have minimal uncertainty if the inequality turns into an equality. This happens if and only if there exists a $\lambda \in i\mathbb{R}$ such that

$$(A - \mu_A) f = \lambda (B - \mu_B) f. \tag{2}$$

Thus, any two self-adjoint operators, whose commutator does not vanish, lead to an uncertainty principle. Moreover, the constraint for equality, together with a realization of the operators in differential form, lead to a set of partial differential equations. The solution is the function which minimizes the uncertainty for the relevant operators.

Both windowed Fourier and wavelet transforms are related to group theory, as both can be derived from square integrable group representations [6]. The windowed Fourier transform is related to the Weyl-Heisenberg group, and the wavelet transform is related to the affine group. The general uncertainty theorem [4] stated above provides a tool for obtaining uncertainty principles using the infinitesimal generators of the groups' representations. In the case of the Weyl-Heisenberg group, the canonical functions which minimize the corresponding uncertainty relation are Gaussian functions. The canonical functions which minimize the uncertainty relations for the affine group in one dimension and for the similitude group in two dimensions were the subject of previous studies, among them is the study of Dhalke and Maass [3] and that of Ali, Antoine and Gazeau [1].

Dahlke and Maass [3], as well as Ali, Antoine and Gazeau [1] studied the uncertainty principle for a sub-group of the affine group, the similitude group of $\mathbb{R}^2$, $SIM(2) = \mathbb{R}^2 \times (\mathbb{R}^+ \times SO(2))$, which is related to the wavelet transform. Dahlke and Maass [3] have included commutators with elements of the enveloping algebra, i.e. polynomials in the generators of the algebra, and managed to find the $2D$ isotropic Mexican hat. Ali, Antoine and Gazeau [1] derived a possible minimizer in the frequency domain for some fixed direction. Their solution is a real wavelet which is confined to some convex cone in the positive half plane of the frequency space and is exponentially decreasing inside.

3 The Weyl-Heisenberg Group

The uncertainty principle related to the Weyl-Heisenberg group has a tremendous importance in two main fields; in quantum mechanics, the uncertainty principle prohibits the observer from exactly knowing the location and momentum of a particle. In signal processing, the uncertainty principle provides a limit on the localization of the signal in both time (spatial) and frequency domains.

Let G be the Weyl-Heisenberg group,

$$G := \{(\omega, b, \tau) | b, \omega \in \mathbb{R}, \tau \in \mathbb{C}, |\tau| = 1\} \tag{3}$$

with group law

$$(\omega, b, \tau) \circ (\omega', b', \tau') = (\omega + \omega', b + b', \tau\tau' e^{i\frac{(\omega b' - \omega' b)}{2}}). \tag{4}$$

Let π be a representation of the group's action on $L^2(\mathbb{R})$; then, the coefficients generated by $\langle f, \pi(x)\psi\rangle$ are known as the windowed Fourier transform of the function f, with ψ being the window function. The windowed Fourier transform is defined by:

$$\langle f, \pi(x)\psi \rangle = (G_\psi f)(\omega, b) = \int f(x)\psi(x-b)e^{-i\omega x}dx \tag{5}$$

The Fourier transform is a profound tool in signal processing. The Gaussian window function $\psi(x) = e^{-\frac{x^2}{2}}$ has an important role in the windowed Fourier analysis as it minimizes the Weyl-Heisenberg uncertainty principle. Next, we review the derivation of the uncertainty principles for the Weyl-Heisenberg group in one and two dimensions using the uncertainty principle theorem. The reader may find the classical proofs of the uncertainty principle for the Weyl-Heisenberg group in the work of Gabor [5] for one-dimensional signals and in the work of Daugman [2] for two-dimensional signals.

3.1 The One Dimensional Case

The unitary irreducible representation of the Weyl-Heisenberg group in $L^2(R)$ can be defined by: $[U(\omega, b)f](x) := e^{i\omega x}f(x-b)$. The following infinitesimal generators of the group are then given by:

$$(T_\omega f)(x) := i\frac{\partial}{\partial\omega}[U(\omega,b)f](x)|_{\omega=0,b=0} = -xf \tag{6}$$

$$(T_b f)(x) := i\frac{\partial}{\partial b}[U(\omega,b)f](x)|_{\omega=0,b=0} = -i\frac{d}{dx}f \tag{7}$$

The one-dimensional uncertainty principle for the Weyl-Heisenberg group can be derived using the general uncertainty principle.

Corollary: [4] Let $A = T_\omega$ and $B = T_b$ be the infinitesimal operators of the Weyl-Heisenberg group: $A = -x$, $B = -i\frac{\partial}{\partial x}$. If $f \in L^2(R)$ and $a = \mu_A, b = \mu_B \in R$ we have: $\|(A-a)f\|_2\|(B-b)f\|_2 \geq \frac{1}{4}\|f\|_2$, with equality being obtained iff

$$f(x) = ce^{2\pi ibx}e^{-\pi r(x-a)^2} \tag{8}$$

for some $c \in \mathbb{C}$, $r \in \mathbb{R}_+$.

3.2 The Two-Dimensional Case

The unitary irreducible representation of the Weyl-Heisenberg group in $L^2(R^2)$ in two dimensions is given by: $[\widetilde{U}(\omega_1,\omega_2,b1,b2,\tau)f](x,y) = \tau e^{i(\omega_1 x+i\omega_2 y)}f(\overrightarrow{u}-\overrightarrow{b})$, where $\overrightarrow{u} = (x,y)$, $\overrightarrow{b} = (b_1, b2)$. The following infinitesimal generators of the group can be defined as:

$$(T_{\overrightarrow{\omega}} f)(\overrightarrow{u}) := i\frac{\partial}{\partial\overrightarrow{\omega}}[Uf](\overrightarrow{u})|_{\overrightarrow{\omega}=0,\overrightarrow{b}=0} = -\overrightarrow{u}f \tag{9}$$

$$(T_{\overrightarrow{b}} f)(\overrightarrow{u}) := i\frac{\partial}{\partial\overrightarrow{b}}[Uf](\overrightarrow{u})|_{\overrightarrow{\omega}=0,\overrightarrow{b}=0} = -i\nabla f \tag{10}$$

The only non-vanishing commutators of these four operators are:

$$[T_{w_k}, T_{b_k}] = -i \quad , \quad k = 1, 2 \; . \tag{11}$$

Thus, an uncertainty principle can be obtained for translations in the space and frequency domains. This can be solved for each dimension separately. It is interesting to note that using the Weyl-Heisenberg group, there is no coupling between the x and y components. Thus attaining a certain accuracy in the x component does not affect the degree of accuracy of the y component. If we derive the minimization equation, we simply get the same result for the one-dimensional analysis for both x and y. The separability of the Weyl-Heisenberg group results in separable gaussian functions as the minimizers of the combined uncertainty. This is, in fact, an inherent property of the Gaussian functions.

4 The Affine Group

Let G be the affine group, and let U be its canonical left action on $L^2(R)$; the coefficients generated by $\langle f, U(x)\psi \rangle$ are known, in the one-dimensional case, as the wavelet transform of a function f, with ψ as a mother wavelet, or template. The wavelet transform is defined by:

$$(W_\psi f)(a,b) = \int_R f(x)|a|^{-\frac{1}{2}}\overline{\psi(\frac{x-b}{a})}dx \tag{12}$$

4.1 The One-Dimensional Case

Let A be the affine group,

$$A := \{(a,b)|(a,b) \in R^2, a \neq 0\} \tag{13}$$

with group law

$$(a,b) \circ (a',b') = (aa', ab' + b). \tag{14}$$

A unitary group representation obtained by the action of $U(A)$ on $f(x)$ is given by:

$$[U(a,b)f](x) = |a|^{-\frac{1}{2}} f\left(\frac{x-b}{a}\right) \tag{15}$$

In preparation for our extension to two dimensions and other groups, we quote the main results presented in the work of Dahlke and Maass [3] for the one dimensional affine group. First, the self-adjoint infinitesimal operators are calculated by computing the derivatives of the representation at the identity element:

$$\begin{aligned} T_a &= -i(\frac{1}{2} - x\frac{\partial}{\partial x}) \\ T_b &= -i\frac{\partial}{\partial x}. \end{aligned} \tag{16}$$

Using these operators, the affine uncertainty principle is given, and the following differential equation can be solved to obtain the uncertainty minimizer:

$$(T_a - \mu_a)f = \lambda(T_b - \mu_b)f, \tag{17}$$

parameter t_1+t_2. This is all very well known of course. The interesting question is whether the minimizer of the uncertainty relations of other groups depends on parameters such that it forms a semi-group with respect to convolution. We consider here the minimizers of the uncertainties related to the $SIM(2)$ group and to the affine group.

The solution offered by Dahlke and Maass is scale-space by nature. The minimizer that they found is the Mexican hat function: $\psi(x,y) = \beta(1-\beta r^2)e^{-\beta r^2}$, where $r := \sqrt{x^2+y^2}$. Its Fourier transform is $\pi^2 k^2 e^{-\frac{\pi^2 k^2}{\beta}}$. Clearly, if we define $\beta = 1/t$ then the semi-group property is trivially satisfied, with t as the semi-group parameter. Note that this is a scale-space of *edge detector* and not of the image as usual. It is in fact an element of the jet-space of the traditional Gaussian scale-space.

The rest of this section is devoted to exploring the scale-space nature of the minimizer given by Ali, Antoine and Gazeau for the uncertainty related to the $SIM(2)$ group [1]. Their solution is given in the frequency space (k_x, k_y). It is a

Fig. 1. A one-dimensional rectangular pulse function

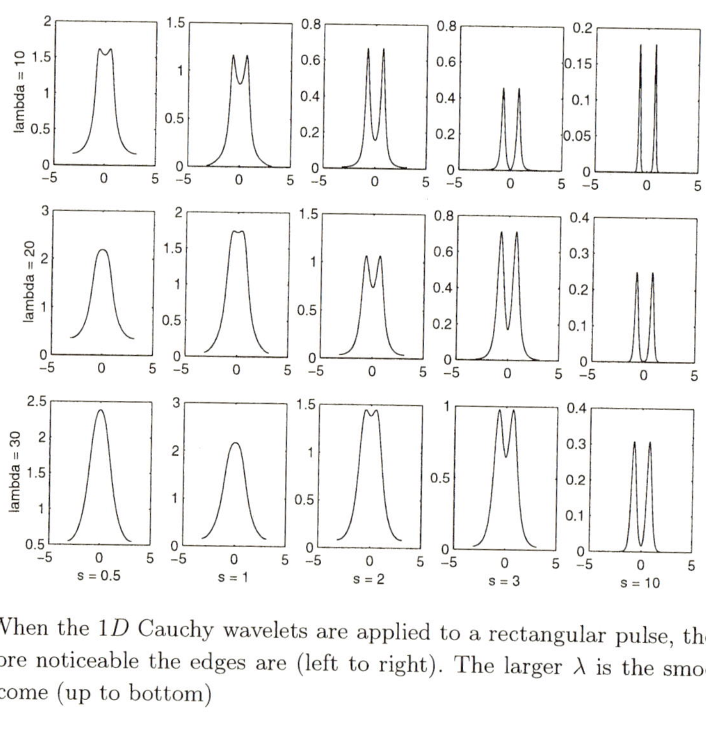

Fig. 2. When the $1D$ Cauchy wavelets are applied to a rectangular pulse, the larger s is the more noticeable the edges are (left to right). The larger λ is the smoother the edges become (up to bottom)

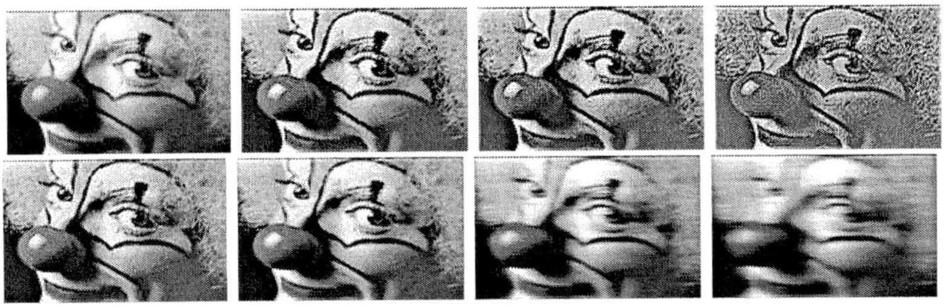

Fig. 3. 1st row: For a constant value of $\lambda = 0.00001$, increasing the value of s, $0.01, 0.2, 0.5, 1$ (left to right), results in edge enhancement. 2nd row: For a constant value of $s = 0.2$, increasing the value of λ is increased: $0.001, 0.01, 0.05, 0.1$ (left to right) results in a effect of motion-blurring in the x-direction

function which vanishes outside some convex cone in the half-plane $k_x > 0$ and is exponentially decreasing inside:

$$\hat{\psi}_{s\lambda}(\boldsymbol{k}) = c|\boldsymbol{k}|^s e^{-\lambda k_x}, \tag{30}$$

where $s > 0$ and $\lambda > 0$. It is quite obvious, from the mere definition of the function, that successive applications of the filters with two values of either s or λ correspond to a single application of an effective parameter: $\hat{\psi}_{s_1\lambda_1}\hat{\psi}_{s_2\lambda_2} = \hat{\psi}_{(s_1+s_2)(\lambda_1+\lambda_2)}$. Moreover, this function has the following properties: The portion $|\boldsymbol{k}|^s = (k_x^2 + k_y^2)^{\frac{s}{2}}$ in frequency space is the transformation (up to a sign) of the Laplacian operator in the spatial space :$\Delta^{\frac{s}{2}}$, and thus can be considered as an edge enhancement operator. The portion $e^{-\lambda k_x}$ can be considered as a directional smoothing operator.

We look first at the one-dimensional equivalent of the solution of Ali, Antoine and Gazeau [1], which is known as the Cauchy wavelets [7,8]: $\hat{\psi}(\xi) = c\xi^s e^{-\lambda\xi}$ for $\xi \geq 0$ and $\hat{\psi}(\xi) = 0$ for $\xi < 0$, and $s > 0$. Their application to a rectangular pulse function (Fig. 1) provides the following results: as s increases, the edges become more evident, thus the edge is enhanced, while as λ increases, the signal becomes smoother (Fig. 2).

We next apply the two-dimensional minimizer filter to a test image of a clown, symmetrizing the filters as follows: $\hat{\psi}(\hat{k}) = c|\boldsymbol{k}|^s e^{-\lambda|k_x|}$. When the value of λ is kept constant, increasing s results in a progressive edge enhancement (Fig. 3 1st row). When the value of s is kept constant and the value of λ is increased, there is a motion blurring effect in the x-direction (Fig. 3 2nd row).

6 Discussion and Conclusions

In this work we study the possibility of designing a window shape that is optimal with respect to all the possible parameters of the two-dimensional affine transform. The study is based on minimizing the uncertainty relations that are inherent in the non-commutative affine symmetry. We generalized ideas and techniques that were used by Dahlke and Maass [3] and Ali, Antoine and Gazeau [1] for lower dimensional groups.

Our study shows that there is no function that minimizes the uncertainty with respect to all parameters of the affine transformations. We were able to show, though, the existence of an L^2 window that minimizes a subset of the commutation relations.

Moreover, the scale-space properties of the minimizer offered by Ali, Antoine and Gazeau, are considered. We find that the two-parameter minimizer family is a semi-group with respect to each parameter and that modifying the function's parameters results in either edge enhancement or motion-like blurring.

Our preliminary results point to the need to further explore the scale-space attributes of uncertainty minimizers.

References

1. S.T. Ali, J.P. Antoine and J.P. Gazeau, "Coherent States, Wavelets and Their Generalizations", Springer-Verlag, 2000.
2. J.G. Daugman, "Uncertainty relation for resolution in space, spatial frequency, and orientation optimized by two-dimensinal visual cortical filters", *J. Opt. Soc. Amer.* 2(7), 1985, 1160-1169.
3. S. Dahlke and P. Maass, "The Affine Uncertainty Principle in One and Two Dimensions", *Comput. Math. Appl.*, 30(3-6), 1995, 293-305.
4. G. Folland, "Harmonic Analysis in Phase Space", *Princeton University Press*, Princeton, NJ, USA, 1989.
5. D. Gabor, "Theory of communication",*J. IEEE* , 93, 1946, 429-459.
6. A. Grossmann, J. Morlet and T. Paul, "Transorms associated to square integrable group representations, II. Examples",*Ann. Inst. H. Poincaré*, 45, 1986, 293-309.
7. J.R. Klauder, "Path Integrals for Affine Variables", in: Functional Integration, Theory and Applications, 101-119, J.P. Antoine and E. Tirapegui (Eds.), Plenum Press, New York and London, 1980.
8. Th. Paul and K. Seip, "Wavelets in Quantum Mechanics", in: Wavelets and their Applications, 303-322, M.B. Ruskai, G. Beylkin, R. Coifman, I. Daubechies, S. Mallat, Y. Meyer and L. Raphael (Eds.), Jones and Bartlett, Boston, 1992.
9. M. Porat and Y.Y. Zeevi, "The generalized Gabor scheme of image representation in biological and machine vision", *IEEE Transactions on PAMI*, 10(4), 1988, 452-468.

Scale Invariant Texture Analysis Using Multi-scale Local Autocorrelation Features

Yousun Kang, Ken'ichi Morooka, and Hiroshi Nagahashi

Imaging Science and Engineering Laboratory,
Tokyo Institute of Technology, Yokohama, 226-8503 Japan
{yskang, morooka, longb}@isl.titech.ac.jp

Abstract. We have developed a new framework for scale invariant texture analysis using multi-scale local autocorrelation features. The multi-scale features are made of concatenated feature vectors of different scales, which are calculated from higher-order local autocorrelation functions. To classify different types of textures among the given test images, a linear discriminant classifier (LDA) is employed in the multi-scale feature space. The scale rate of test patterns in their reduced subspace can also be estimated by principal component analysis (PCA). This subspace represents the scale variation of each scale step by principal components of a training texture image. Experimental results show that the proposed method is effective in not only scale invariant texture classification including estimation of scale rate, but also scale invariant segmentation of 2D image for scene analysis.

1 Introduction

Texture analysis plays an important role in the interpretation and understanding of real-world images so that it is a useful research area in computer vision and pattern recognition. Texture analysis has been applied to many practicable vision systems such as industrial inspection, remote sensing, biomedical imaging, ground classification, and segmentation of satellite or aerial imagery. Recent applications show the potential of natural-scene analysis by utilizing texture segmentation and texture labeling [1], [2].

One of the major problems in texture analysis is that the textures in the real world are often not uniform because of changes in orientation, scale or other visual appearance. Especially, several researches on scale and rotation invariant properties of texture images have been reported in recent years [3], [4]. Our research is focused on the scale invariant property of a texture image in a statistical feature space. The statistical features extracted from co-occurrence or autocorrelation functions represent a spatial distribution of gray values. Texture patterns can be represented by a large number of feature vectors in high-dimensional feature space. Classifiers are also designed to reduce the computational complexity of an enormous amount of feature vectors. We should therefore make an effort

R. Kimmel, N. Sochen, J. Weickert (Eds.): Scale-Space 2005, LNCS 3459, pp. 363–373, 2005.

to develop robust invariant features that can be extracted and classified with a low-computational complexity in statistical approaches [5].

In this paper, we propose a theoretically and computationally simple framework in which it is possible to discriminate a large range of scaled texture images efficiently in spite of gray-scale variations. The proposed multi-scale features are utilized in scale invariant texture classification and scale invariant segmentation for scene analysis including estimation of scale rate. To classify the different texture types, K-class *Fisher criterion* [6] for a multi-class classification problem is employed in their discriminant space. We separate feature vectors of texture patterns of the same type from a linear discriminant space. The separated discriminant space consists of feature vectors of different scale rate although they are extracted from texture patterns of the same type. According to the scale variation, thus, texture patterns can also be classified by using principal components analysis.

Scale variation of texture patterns is related to texture gradient, which is an important cue for depth perception comparable to binocular disparity [7]. Texture gradient can be represented by gradual scale variation of texture patterns in a 2D scene. The proposed method is applied to scale invariant texture segmentation of a 2D scene. In addition, we can evaluate relative distance of a 2D scene image by estimating the scale rate of a training texture pattern. The experimental results show the effectiveness of the proposed method for scale invariant texture classification and segmentation including the estimation of scale rate. In the extensive application of our proposed method, we will make it possible to perceive a surface orientation and a relative distance in a natural 2D scene.

2 Multi-scale Local Autocorrelation Features

2.1 Multi-scale Features

We assume that texture is a kind of repetitive pattern over a certain ranges of scale. An image texture can appear in different ways according to the scale of observation so that the scale concept and the notion of multi-scale representation are of crucial importance in texture analysis [8]. A number of approaches to multi-scale representations, which are more or less related to scale space theory such as pyramids, wavelets and multigrid methods have been developed [9]. As for our method, the pyramid representation of gray-level is used to build the scale space, which combines a sub-sampling operation with different levels of spatial resolution.

Figure 1 illustrates a method of multi-scale feature extraction from a training texture image. Local windows of $(n \times n)$ size are sampled from all over the edge of texture pattern in randomly. According to the scale step i, we extract m number of sample windows, thus, we can get $(m \times i)$ feature vectors from a training texture image with d-dimensionality. The rate at which the scale step increases i is kept constant to make a linear scale space. We concatenate feature vectors $(\boldsymbol{S}_{i-1}, \boldsymbol{S}_i, \boldsymbol{S}_{i+1})$ of different scale to a single vector $\boldsymbol{Ms}_i$ of $(d \times 3)$ dimension. A multi-scale feature consists of three feature vectors that give three consecutive scales.

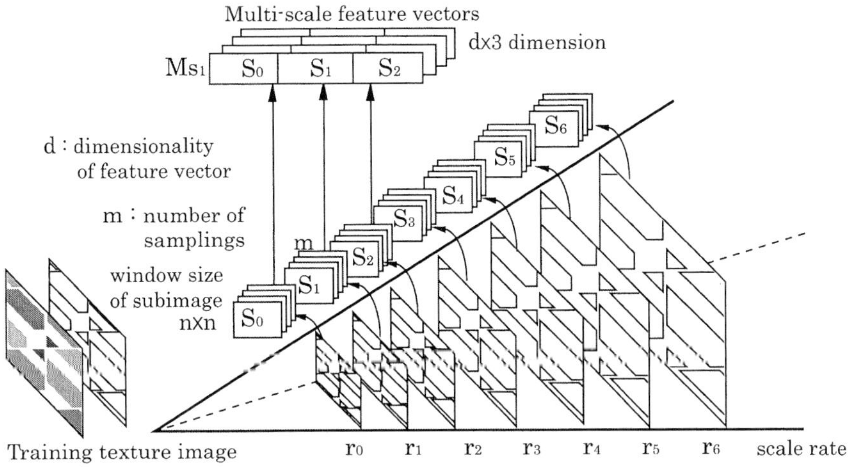

Fig. 1. Extraction of multi-scale feature vectors from scale space

2.2 Higher Order Local Autocorrelation

Various methods for extracting scale invariant features have been applied in texture analysis and image recognition (e.g., a Gabor filter and Wavelets and so on). The autocorrelation functions possess uniqueness property for even orders, and they have advantages of being shift-invariant and computational low cost. The autocorrelation function has been used in a wide range of applications: face detection [10], pattern recognition [11], and scale invariant image recognition [12].

A local autocorrelation function can be used to assess the amount of regularity as well as the fineness/coarseness of a texture image. An important property of many textures is the repetitive nature of the placement of texture elements in the image. Due to advantages of low cost and repetitive nature of texture image, a higher order local autocorrelation function is employed in feature extraction module.

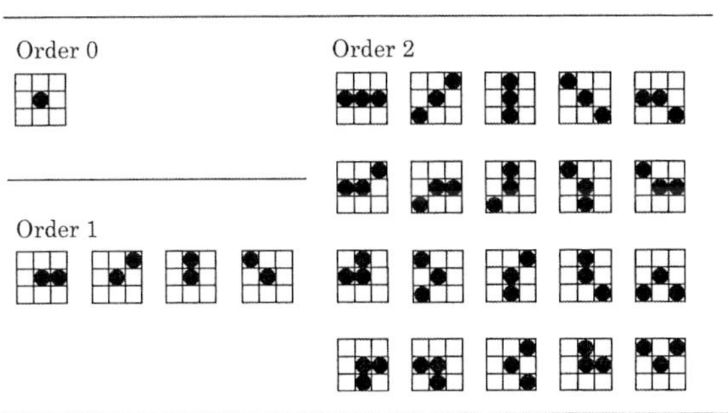

Fig. 2. Local mask patterns for higher-order local autocorrelation features

Higher order autocorrelation functions are defined by

$$r_x^n(a_1, a_2, \cdots, a_n) = \int_D f(x)f(x + a_1) \cdots f(x + a_n)dx, \tag{1}$$

where n denotes the order of the autocorrelation function, x is the image coordinate vector, and a_i are the displacement vectors. A function $f(x)$ stands for the image intensity on the retinal plane D [13]. Considering computational cost, we limit the order n to 2. Using the mask patterns shown in Fig. 2, the feature extraction module computes 25 local autocorrelation coefficients from a texture image. For each mask pattern, a product is calculated by multiplying pixels in the masks.

3 Discriminant Space for Classification

To get more effective feature set, we perform an orthogonal transformation on all training feature vectors by using the $Karhunen - Loève$ (KL) transformation [14]. The orthogonal transformation produces a new sequence of uncorrelated texture images on the higher auto-correlation feature space so that feature vectors have a set of the most independent output components. A feature extraction module usually includes a process which determines an appropriate subspace of dimensionality d' in the original feature space of dimensionality $d(d' \leq d)$. The $Karhunen - Loève$ method also chooses a dimensionality reducing linear projection, but we do not reduce the dimensionality of feature space to preserve the original features.

To classify each texture image represented by its feature vector, we use the *Fisher criterion* for K-class classification associated with *linear discriminant analysis* (LDA). The task of our classification is to assign test texture patterns that have the same type of texture but different scales to one of the classes with the same type of texture patterns. This is called scale-invariant texture classification. The Fisher method is the simplest and most popular approach in linear discriminant analysis (LDA). The method is further generalized by Rao [15] into the multiple class problem. The Fisher method projects high-dimensional data onto a subspace to maximize the separation of inter-classes. For a K-class problem with $K > 2$, a transformation matrix from d-dimensional feature space to m-dimensional space ($m \leq d$) is determined such that the *Fisher criterion* of total scatter versus average within-class scatter is maximized [16]. The within-class scatter matrix, $\boldsymbol{S_W}$, and the between-class scatter matrix, $\boldsymbol{S_B}$, are written as follows :

$$\boldsymbol{S_W} = \sum_{i=1}^{K} \Big(P(\omega_i) \frac{1}{n_i} \sum_{\boldsymbol{x} \in \omega_i} (\boldsymbol{x} - \boldsymbol{m}_i)(\boldsymbol{x} - \boldsymbol{m}_i)^T \Big), \tag{2}$$

$$\boldsymbol{S_B} = \sum_{i=1}^{K} P(\omega_i)(\boldsymbol{m}_i - \boldsymbol{m})(\boldsymbol{m}_i - \boldsymbol{m})^T, \tag{3}$$

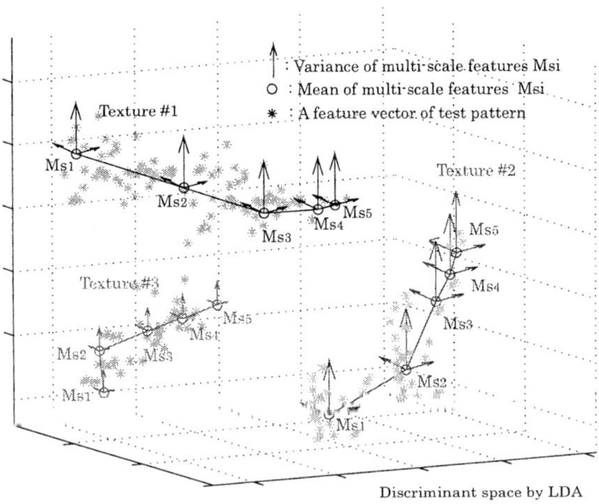

Fig. 3. Feature vectors of test texture patterns and prototypes represented by means of multi-scale features in the linear discriminant space

where $\boldsymbol{x}$ is a feature vector, $\boldsymbol{m}_i$ is the mean vector of class ω_i, $\boldsymbol{m}$ is the overall mean of all classes, K is the number of classes, n_i is the number of patterns in each class ω_i, and $P(\omega_i)$ is its prior probability. We assume that n_i reflects the prior probability, where n is the number of all patterns, so that a prior probability can be represented as $P(\omega_i) = n_i/n$. We can, thus, obtain the transformation matrix $\boldsymbol{A}$, which maximizes J_F, so-called *Fisher criterion*:

$$J_F(\boldsymbol{A}){=}tr(\frac{\boldsymbol{A}^T\boldsymbol{S_B}\boldsymbol{A}}{\boldsymbol{A}^T\boldsymbol{S_W}\boldsymbol{A}}). \tag{4}$$

Solving this optimization problem results in finding the eigenvectors of $\boldsymbol{S_W^{-1}}\boldsymbol{S_B}$ corresponding to the d largest eigenvalues. As long as $d \geq (K-1)$, no information is lost when the classes are normally distributed. We can project the d-dimensional features into $(K-1)$-dimensional space or more reduced dimensional space related to the cumulative proportion of eigenvalues. The objective of dimensionality reduction below $(K-1)$ is to find a subspace in which a projection of the class means preserves distance such that the class separability is maintained as good as possible [17].

In the linear discriminant space built by the K-class Fisher criterion, re-projected feature vectors of training data constitute spatial distributions in accordance with scale variation. Figure 3 shows means and variances of the distributions of feature vectors in limited 3-dimensional discriminant space for visualization. We use a LDA classifier based on the minimum Euclidean distance to classify test texture patterns. Note that LDA using a few prototypes per class is the simplest and the most practical classifier.

We make prototypes by using the means of multi-scale features extracted from the same type of training texture patterns at each scale step. A class consists of

i prototypes of multi-scale features $\boldsymbol{Ms}_i$ and i represents the scale step of multi-scale features. We then compared these i prototypes with test texture patterns.

4 Subspaces for Estimation

For the estimation of scale rate of texture patterns, we make separate subspaces by using Principal Component Analysis. The subspaces are composed of the same type of training texture patterns, but with different scale, and can represent scale variation of a training texture type. The PCA is a standard technique for extracting the structure from a high-dimensional data set. It reduces the redundancy contained within the feature vectors by creating a new series of feature vectors in which the axes of the new coordinate system point in the direction of decreasing variance. Suppose that a feature vector and its dimensionality d can be written as $\{\boldsymbol{x} = (x_1, x_2, \ldots x_d)^T \in R^d\}$. We extracted feature vectors from higher local autocorrelation function with 25-dimensionality as described in section 2.2. The number of feature vectors from a training texture image is $(m \times i)$, where m is the number of sampling windows and i is a number of scale steps of feature vectors, which are represented by $\{\boldsymbol{x}^{(k)}\}(i = 1, \ldots m \times i)$.

To begin the PCA transformation of feature vectors within a class, the covariance matrix $\boldsymbol{C_x}$ of the all data set of feature vectors can be defined as

$$\boldsymbol{C_x} = \frac{1}{m \times i - 1} \sum_{k=1}^{m \times i} \{(\boldsymbol{x}^{(k)} - \overline{\boldsymbol{x}})(\boldsymbol{x}^{(k)} - \overline{\boldsymbol{x}})^T\} \tag{5}$$

where $\overline{\boldsymbol{x}}$ is the mean of $\boldsymbol{x}^{(k)}$. Using the covariance matrix, we obtain the eigenvalues λ are obtained from $|\boldsymbol{C_x} - \lambda I| = 0$, where I is an identity matrix. The eigenvectors $\boldsymbol{e}$ define the axes of the components and are obtained from

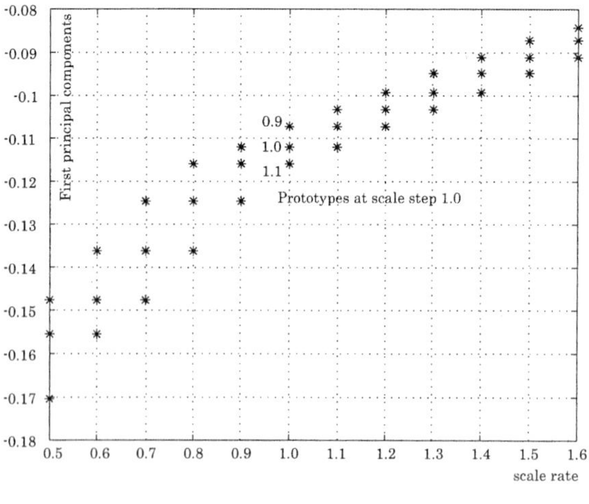

Fig. 4. An example of prototypes from a training texture image in their subspace

$(\boldsymbol{C_x} - \lambda I)\boldsymbol{e} = 0$. The eigenvector of the largest eigenvalue is computed for the new first component of reduced subspace. New first components have the largest percentage of the total variance in original feature space, so the first components reflect a big change of feature vectors according to the scale variation.

We compute means of first principal components at each scale step, and then concatenate mean vectors of a different scale to a single prototype like multi-scale features. Therefore, a prototype consists of three means of first principal components at each scale step including a pre-scale step and a post-scale step. Figure 4 shows an example of prototypes from a training texture image in their subspace. To estimate the scale rate of classified test texture patterns, the prototypes are compared with test texture patterns based on euclidean distance.

5 Experimental Result

The effectiveness of the proposed method for scale invariant texture analysis has been well tested. We performed two major experiments: scale invariant texture classification including the estimation of scale rate, and scale invariant texture segmentation of an artificial 2D image for scene analysis.

5.1 Scale Invariant Texture Classification

We used sixteen natural texture images from the Brodatz album [18] (as shown in Fig. 5) for the experiments on scale invariant texture classification. All textures are originally gray-scale images with 256 levels, and the size of a texture image is 640 × 640 pixels. To get the training texture patterns, we randomly extracted 500

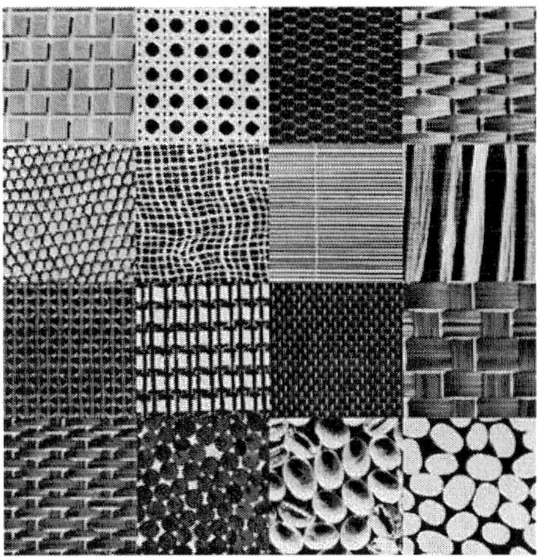

Fig. 5. The sixteen classes of textures from the Brodatz album. **Row 1**:D1, D101, D34, D56. **Row 2**:D22, D103, D49, D51. **Row 3**: D52, D20, D6, D64. **Row 4**: D65, D66, D74, D75

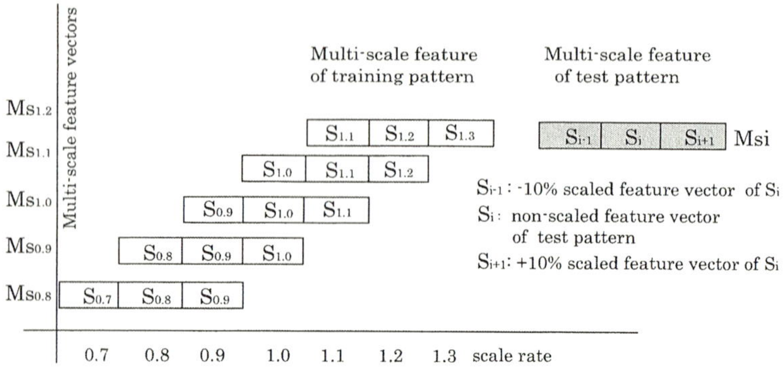

Fig. 6. Multi-scale features of training and test patterns

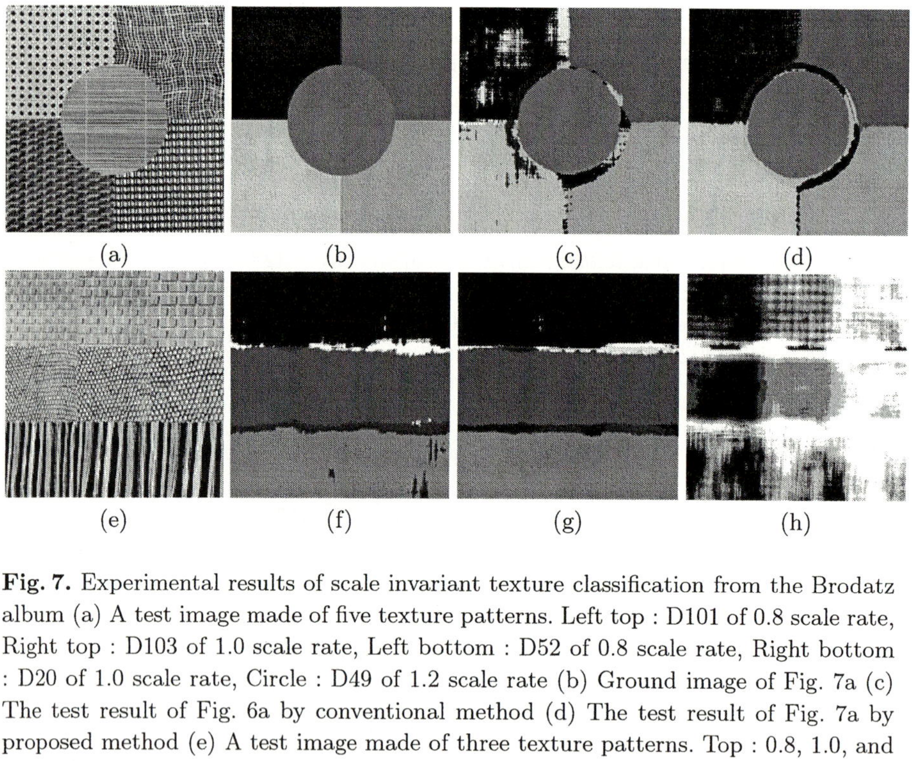

Fig. 7. Experimental results of scale invariant texture classification from the Brodatz album (a) A test image made of five texture patterns. Left top : D101 of 0.8 scale rate, Right top : D103 of 1.0 scale rate, Left bottom : D52 of 0.8 scale rate, Right bottom : D20 of 1.0 scale rate, Circle : D49 of 1.2 scale rate (b) Ground image of Fig. 7a (c) The test result of Fig. 6a by conventional method (d) The test result of Fig. 7a by proposed method (e) A test image made of three texture patterns. Top : 0.8, 1.0, and 1.2 scaled image of D1, Middle : 0.8, 1.0, and 1.2 scaled image of D22, Bottom : 0.8, 1.0, and 1.2 scaled image of D52 (f) Test result of Fig. 7e by conventional method (g) The test result of Fig. 7e by proposed method (h) The test result of scale estimation of Fig. 7e by proposed method

sub-samples of 90 × 90 pixels from each texture image, with seven different scale steps (0.7, 0.8, 0.9, 1.0, 1.1, 1.2, and 1.3). We then made the multi-scale features

Table 1. The error rate of test texture images from Brodatz album

	Conventional method	Proposed method
Five texture patterns (Fig. 7a)	9.70%	7.35%
Nine texture patterns (Fig. 7e)	7.96%	6.97%

with five scale steps (0.8, 0.9, 1.0, 1.1, and 1.2) represented by 75 dimensionality feature vectors from 56,000 sample patterns (500 × 7 × 16). When we make a multi-scale feature of a test texture pattern, that pattern is scaled into 90% and 110% patterns. Three feature vectors of the two scaled and original patterns are then concatenated into one vector as shown in Fig. 6.

The size of the test images in Fig. 7 are 900 × 900 and the sources are shown in Figs. 7(a) and (e). To make the multi-scale features for test texture patterns, we sequentially extract a sub-region of 90 × 90 pixels with three different scale steps from a test image. Only 25-dimensional feature vectors extracted from higher-order local autocorrelation functions are used in a conventional method as a feature extraction module.

A conventional method includes a classifier, which regards a class as all training texture patterns of all scale steps. This conventional method can not estimate the scale rate of texture patterns. We obtained the scale invariant texture classification results without post-processing as shown in Figs. 7(d) and (g). For comparison, the results obtained from conventional method also presented in Figs. 7(c) and (f). The error rates of proposed method were lower than that of conventional method (as shown in Table 1). In addition, Fig. 7(h) shows the result of scale estimation of classified texture images in their subspace.

(a) (b) (c) (d)

Fig. 8. Experimental results of scale invariant texture segmentation (a) a road test image (b) the result of estimation of relative distance using scale invariant texture segmentation in Fig. 8a (c) a avenue test image (d) the result of estimation of relative distance using scale invariant texture segmentation in Fig. 8c

5.2 Scale Invariant Texture Segmentation

Tsutsui [19] reported on the neural correlates for perception of 3D surface orientation from texture gradient. The scale invariant property of a texture image is also related to texture gradient for perception of 3D surface orientation. Texture gradient can be represented by gradual scale variation of a texture pattern in the 2D scene image. The proposed method is applied to scale invariant segmentation of 2D image. Several experiments were performed via two kinds of artificial 2D images as shown in Fig. 8. The road image in Fig. 8(a) has only two classes of block and ground texture patterns with 12 scale steps (50% - 170%). Figure 8(b) shows the segmentation result including estimation of relative distance using a block-training texture image.

The avenue image in Fig. 8(c) has four classes which consist of leave, trunk, grass, and ground with five scale steps of multi-scale features (70%-120%). The test texture patterns are extracted at a 10% differential scale rate to make multi-scale features. The result of scale invariant segmentation including estimation of relative distance is shown in Fig. 8(d). In our future research, we will apply multi-scale feature classification to perceive 3D surface orientation of natural scene analysis.

6 Conclusion

We have developed a new framework for scale invariant texture analysis using multi-scale local autocorrelation features. The proposed multi-scale features are extracted from higher order local autocorrelation functions and feature vectors of different scale are concatenated to a single feature vector. To classify different texture types from the given test image, a linear discriminant classifier (LDA) was employed. The scale rate of test texture patterns in their reduced subspace can also be estimated by using principal component analysis (PCA). Experimental results show that the proposed method is effective in not only scale invariant texture classification including estimation of scale rate, but also scale invariant segmentation of 2D image for scene analysis. We plan to apply this method to the perception of 3D surface orientation and relative distance estimation for robot eye.

Acknowledgment

The first author has been supported by the Postdoctoral Fellowships for Foreign Researchers of the Japan Society for the Promotion of Science (JSPS).

References

1. M. Varma and A. Zisserman, Estimating Illumination Direction from Textured Images, Proc. Int. Conf. Computer Vision and Pattern Recognition (CVPR'03), vol.1, pp.179-186, 2004.

2. M. Turtinen and M. Pietikainen, Visual Training and Classification of Outdoor Textured Scene Images, The 3rd international workshop on texture analysis and synthesis, pp.101-106, 2003.
3. C. Pun and M. Lee, Log-polar wavelet energy signatures for rotation and scale invariant texture classification, IEEE Trans. PAMI (Pattern Analysis and Machine Intelligence), vol.25, no.5, pp.590-602, 2003.
4. T. Ojala, M. Pietikainen and T. Maenpaa, Multiresolution gray-scale and rotation invariant texture classification with local binary patterns, IEEE Trans. PAMI, vol.24, no.7, pp.971-987, 2002.
5. A.K. Jain, P. Duin and J. Mao, Statistical Pattern Recognition: A Review, IEEE Trans. PAMI, vol.22, no.1, pp.4-37, 2000.
6. R.A. Fisher, The Statistical Utilization of Multiple Measurements, Annals of Eugenics, vol.8, pp.376-386, 1938.
7. J. J. Gibson, Perception of the Visual World, HoughtonMifflin, Boston, 1950.
8. T. Lindeberg, Scale-space theory: A basic tool for analysing structures at different scales, J. of Applied Statistics, vol.21(2), pp.224-270, 1994.
9. T. Lindeberg, Scale-space for discrete signals, IEEE Trans. PAMI, vol.12, no.3, pp.234-254, 1990
10. F. Goudail, E. Lange, T. Iwamoto, K. Kyuma and N. Otsu, Face Recognition System Using Local Autocorrelations and Multiscale Integration, IEEE Trans. PAMI, vol.18, no.10, pp. 1024-1028, 1996
11. V. Popovici and J. Thiran, Higher Order Autocorrelations for Pattern Classification, International Conference on Image Processing (ICIP), pp.724-727, 2001.
12. M. Kreutz, B. Volpel, and H. Janssen, Scale-invariant image recognition based on higher-order autocorrelation features, Pattern Recognition, vol.29, no.1, pp.19-26, 1996.
13. J. McLaughlin and J. Raviv, Nth-Order Autocorrelations in Pattern Recognition, Information and Control, vol.12, pp.121-142, 1968.
14. J. Fraleigh and R. Beauregard, Linear Algebra, Addison-Wesley, 1995.
15. C.R. Rao, The Utilization of Multiple Measurements in Problems of Biological Classification, J. Royal Statistical Soc. ser. B, vol.10, pp.159-203, 1948.
16. K. Fukunaga, Introduction to Statistical Pattern Recognition, Academic Press, 1990.
17. M. Loog, R. Duin and R. Haeb-Umbach, Multiclass Linear Dimension Reduction by Weighted Pairwise Fisher Criteria, IEEE Trans. PAMI, vol.23, no.7, pp.762-766, 2001.
18. P. Brodatz, Textures:A Photographic Album for Artist and Designers, Dover, 1966.
19. K. Tsutsui, H. Sakato, T. Naganuma and M. Taira, Neural Correlates for Perception of 3D surface Orientation from Texture Gradient, SCIENCE, vol.298, pp.409-412, 2002.

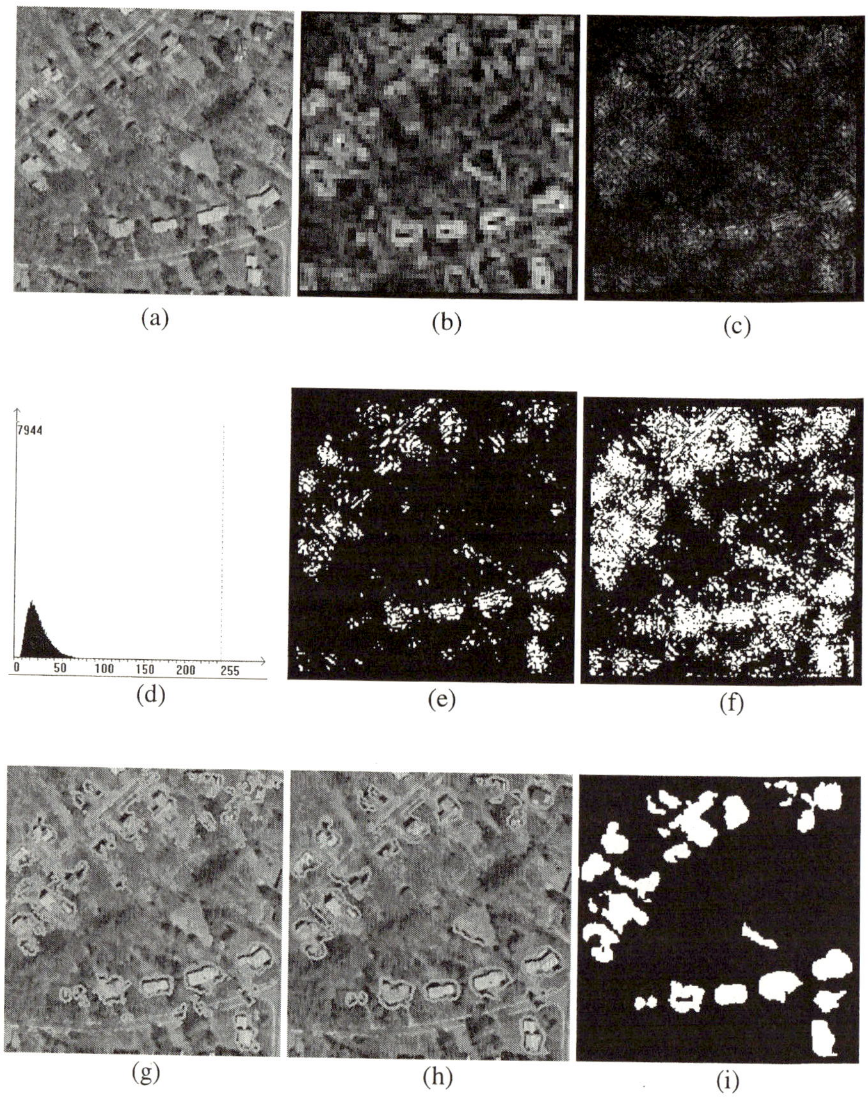

Fig. 1. Comparison of segmentation results. (a) Original 8-bit gray aerial image. (b) Texture edge image. (c) Fractal error image. (d) Histogram of the fractal error image. (e) Threshold=59. White: man-made. (f) OSTU threshold image. White: man-made. (g) Segmentation results only depend on the fractal error feature. (h) Our method results. (i)White: man-made areas

We tested the man-made feature segmentation algorithm using many aerial images. Fig.1 (c) is the fractal error image (normalized from 0 to 255). The lighter areas in the fractal error image correspond to pixels with higher fractal errors; the darker areas correspond to lower fractal errors. Thus, lighter areas in the resultant image (e),(f) tend to indicate man-made objects. In Fig.1 (e) we set the threshold 59 in order to get the minimum error compare with the man-made areas by hand-labeled segmentation.

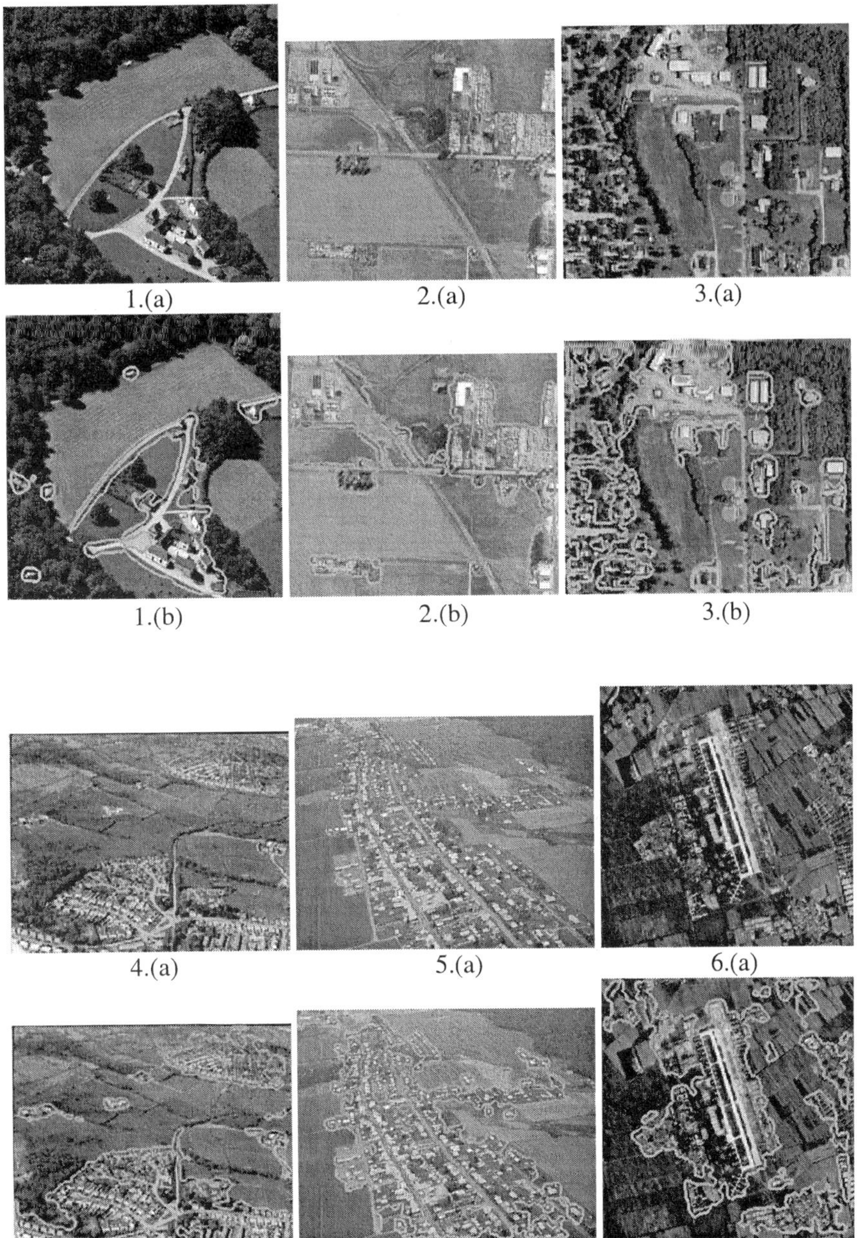

1.(a) 2.(a) 3.(a)

1.(b) 2.(b) 3.(b)

4.(a) 5.(a) 6.(a)

4.(b) 5.(b) 6.(b)

Fig. 2. Segmentation results

From Fig. 1 (e) and (f) we can see that it's very difficult to set a proper threshold. The segmentation error varies greatly according to the threshold. The border of the man-made area zigzagged through the image. Using the fractal error constraint to evolving

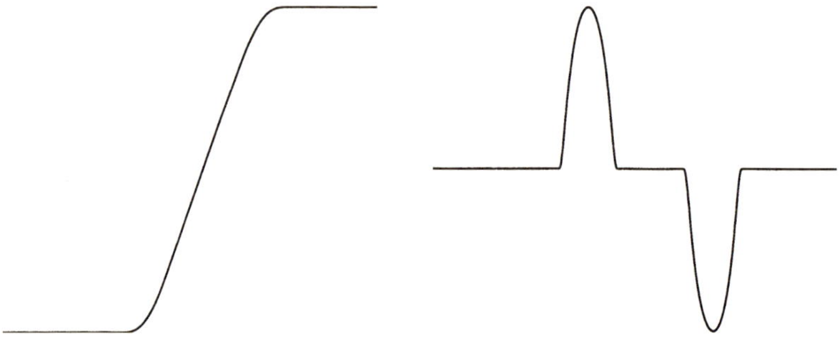

Fig. 4. Edge profile (left) with second derivative (right)

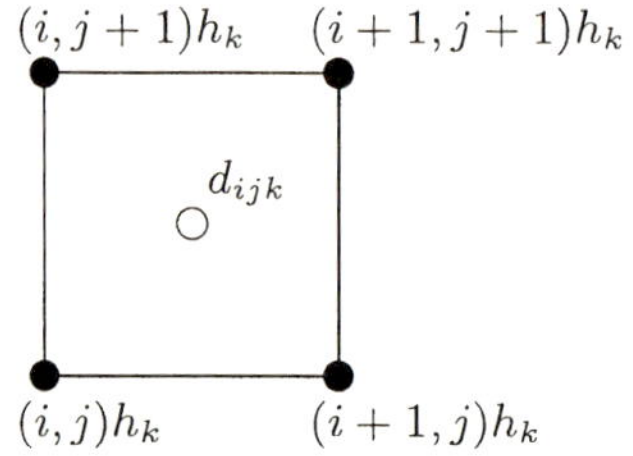

Fig. 5. Cell $C_{ijk} \subset \Omega_k$ with vertices

Generalization. We now have to generalize to two space dimensions. We let L_n be the operator stemming from a discretization by the bilinear finite element method of the Poisson operator $-\Delta$. It can be represented by the 3×3 stencil (or mask)

$$L_n \quad \sim \quad \begin{bmatrix} -1 & -1 & -1 \\ -1 & +8 & -1 \\ -1 & -1 & -1 \end{bmatrix}. \tag{19}$$

Both the original operator $-\Delta$ and its above approximation are invariant to rotation. If P_k, $k = 1, \ldots, n$ are prolongations by means of bilinear interpolation then at the coarser grids all L_k produced by (9) turn out to be represented by the stencil (19) as well (but associated with subsequently coarser grids $S(\Omega_k)$, $0 \leq k < n$), see [24].

Fusion and Finite Elements. Considering the definition (12) of E_k we have to ensure that at each level k the f_k resulting from the recombination (17) remains in the range of L_k or else $E_k f_k$ cannot be applied. We achieve this by composing f_k in a finite element manner. The horizontal diffusion operator and vertical diffusion operator are treated separately. Only the contribution of the horizontal operator is described, the contribution of the vertical operator is the analogue. Consider the cell $C_{ijk} \subset \Omega_k$ defined by four indices as indicated in Figure 5.

This cell yields contributions to the stencils of L_k at the four corners, e.g. at gridpoints (i, j) and $(i+1, j)$ it contributes the respective stencils

$$d_{ijk}\frac{1}{6}\begin{bmatrix} 0 & 1 & -1 \\ 0 & 2 & -2 \\ 0 & 0 & 0 \end{bmatrix} \quad \text{and} \quad d_{ijk}\frac{1}{6}\begin{bmatrix} -1 & 1 & 0 \\ -2 & 2 & 0 \\ 0 & 0 & 0 \end{bmatrix}$$

where $d_{ijk} \in \mathbb{R}$ is the diffusion coefficient located at the center of the cell (for now $d_{ijk} = 1$). Such stencils, together with their horizontally mirrored counterparts, add up to stencil (19). When the above stencils are applied on an image grid-function we observe that the contributions at the pixels (i, j) and $(i+1, j)$ have the same amplitude but opposite sign, hence their sum vanishes.

When fusing a set of m images, for each image grid-function we compute per cell C_{ijk} the contribution, then choose the one from the set of m that is largest

Fig. 6. Top: out-of-focus input images with focus on the right-hand side (left), and with focus on the left-hand side (right). Bottom: fusion of out-of-focus images (left), detail (right)

in absolute value and add this value to the value at pixel (i, j) and the same value but with opposite sign to the value at pixel $(i + 1, j)$. After scanning all cells, the resulting recombined f_k has the desired property.

4.3 Example Fusion Problem

We apply the fusion algorithm of Section 4.2 to two out-of-focus input images, see the top row of Figure 6, the result is to be seen at the bottom row. The quality matches the one obtained by use of the Laplacian pyramid [5] as multiresolution scheme (result not shown).

5 Concluding Remarks

A new multiresolution scheme has been proposed, based on an image transform by a discretized elliptic partial differential operator and use of a multigrid operator, leading to a pyramidal representation. It is shown how this scheme can be applied for image fusion. A single experiment has been added to demonstrate its usefulness. More experiments and an comparison with established methods are in preparation.

The Poisson case as described is just a special case. The framework of the multigrid image transform and multigrid fusion algorithm remains valid if we use the Laplace operator with varying diffusion coefficients instead. An application thereof can be envisaged if we involve segmentation. This is a topic for future research.

References

1. Acton, S.T.: Multigrid Anisotropic Diffusion, IEEE Transactions on Image Processing, Vol. 7, No. 3 (1998)
2. Aubert, G., Kornprobst, P.: Mathematical Problems in Image Processing, Partial Differential Equations and the Calculus of Variations. Applied Mathematical Sciences 147. Springer Verlag, New York (2002)
3. Botta, E.F.F., Dekker, K., Notay, Y., van der Ploeg, A., Vuik, C., Wubs, F.W., de Zeeuw, P.M.: How fast the Laplace equation was solved in 1995. J. Applied Numerical Mathematics 24 (1997) 439–455
4. Brandt, A.: Multi-level adaptive techniques (MLAT) for partial differential equations: ideas and software. In: Rice, J. (ed.): Mathematical Software. Academic Press, New York (1977) 277–318
5. Burt, P.J., Adelson, E.H.: The Laplacian pyramid as a compact image code. IEEE Transactions on Communications, **31** 4 (1983) 532–540
6. Burt, P.J., Kolczynski, R.J.: Enhanced Image Capture through Fusion. Proceedings Fourth International Conference on Computer Vision, Berlin, IEEE (1993)
7. Canny, J.: A Computational Approach to Edge Detection. IEEE Transactions on Pattern Analysis and Machine Intelligence Vol. 8 No. 6 (1986) 679–698
8. Hackbusch, W.: Multi-Grid Methods and Applications. Springer, Berlin (1985)

9. Kimmel, R., Yavneh, I.: An algebraic multigrid approach for image analysis, SIAM J. Sci. Comput. Vol. 24, No. 4 (2003) 1218–1231
10. Li, H., Manjunath, B.S., Mitra, S.K.: Multisensor image fusion using the wavelet transform. Graphical Models and Image Processing **57** 3 (1995) 235–245
11. Mallat, S.: A theory for multiresolution signal decomposition: the wavelet representation. IEEE Pattern Analysis and Machine Intelligence Vol. 11, No. 7 (1989) 674–693
12. Osher, S., Sethian, J.A.: Fronts propagating with curvature dependent speed: algorithms based on the Hamilton-Jacobi formulation. J. of Computational Physics **79** (1988) 12–49
13. Piella, G.: A general framework for multiresolution image fusion: from pixels to regions. Information Fusion **9** (2003) 259–280
14. Piella, G.: Adaptive Wavelets and their Applications to Image Fusion and Compression. Ph.D. Thesis, CWI & University of Amsterdam (2003)
15. Sethian, J.A.: Level Set Methods: Evolving interfaces in geometry, fluid mechanics, computer vision, and materials science. Cambridge University Press (1996)
16. Shapira, Y.: Matrix-Based Multigrid: Theory and Applications. Kluwer Academic Publishers, Boston (2003)
17. Simoncelli, E.P., Freeman, W.T.: The steerable pyramid: a flexible architecture for multi-scale derivative computation. Proceedings of the IEEE International Conference on Image Processing, IEEE Signal Processing Society (1995) 444—447
18. Sweldens, W.: The lifting scheme: A construction of second generation wavelets. SIAM J. Math. Anal. **29** 2 (1997) 511–546
19. Terzopoulos, D.: Image Analysis using Multigrid Relaxation Methods, IEEE Transactions on Pattern Analysis and Machine Intelligence Vol. 8, No. 2 (1986) 129–139
20. Toet, A.: Hierarchical image fusion. Machine Vision and Applications **3** 1 (1990) 1–11
21. Trottenberg, U., Oosterlee, C.W., Schüller, A.: Multigrid. Academic Press, London (2001)
22. Weickert, J.: Anisotropic Diffusion in Image Processing. Teubner-Verlag, Stuttgart (1998)
23. Wesseling, P.: An Introduction to Multigrid Methods. John Wiley & Sons Ltd., Chichester (1991)
24. de Zeeuw, P.M.: Matrix-dependent prolongations and restrictions in a blackbox multigrid solver. J. Comput. Appl. Math. **33** (1990) 1–27

A Total Variation Motion Adaptive Deinterlacing Scheme

Sune Keller, François Lauze, and Mads Nielsen

The IT University of Copenhagen,
Rued Langgaardsvej 7, 2300 Kbh. S, Denmark
{sunebio, francois, malte}@itu.dk
http://www.itu.dk/image/index.html

Abstract. We propose a new way of deinterlacing using a total variation scheme. Starting by the Bayesian inference formulation of total variation we do MAP by rewriting the problem into PDEs that can be solved by simple numerical schemes. Normally deinterlacing schemes are developed ad hoc with online hardware implementation directly at eye, sometimes with some frequency analysis as only theoretical base. Our belief is that mathematically well based image models are needed to do optimal deinterlacing and by our work presented here, we hope to prove it. Comparing the output of our scheme with those of ten known deinterlacing schemes shows very promising results.

1 Introduction

Interlaced scan has been use since the birth of television in the 1930's and is the scanning format used in the television standards PAL and NTSC. Interlacing is separates a full frame image into two parts called fields, one containing all horizontally odd numbered line and the other containing all the even lines. When recorded in interlaced scan the fields are separated in time and two neighboring fields cannot be merged to one full frame without problems.

Interlacing saves bandwidth and lowers the cost of cameras and CRTs as it is possible to combine a high rate of fields per second (to avoid large area flicker in the image) with a relatively high vertical resolution. This looked fine to the human visual system (HVS) in the early days of television but as screen size grew and television sets produced brighter images, interlacing artefacts started to show.

Interlacing artefacts have many names and are often mixed up when described, as they can be described both from a frequency analysis point of view and by their visual appearance. They are by visual appearance

- Line crawl due to vertical motion in the image and the time difference between the two fields composing one frame.
- Serration of edges due to horizontal motion in the image and the time difference between the two fields composing a frame. It happens to edges at all orientations except those close to and at horizontal orientation.

R. Kimmel, N. Sochen, J. Weickert (Eds.): Scale-Space 2005, LNCS 3459, pp. 408–418, 2005.

- Interline flicker due to fine stationary details appearing only in either odd or even fields of the image as they are to small (that is of to high a vertical frequency) to be sampled in both even and odd fields.

A further discussion of frequency analysis and aliasing in interlaced image sequences can be seen in [2] and [11].

One way of reducing the effect of these artifacts in terms of visibility to the human eye is to interpolate new fields and raise the field rate as done in 100 Hz TV sets [7] and [2]. Another way is to convert the interlaced sequence to a progressive sequence by interpolation of image information in the missing lines of the fields to make a full frame of each field. This conversion is called deinterlacing. Progressive scan is used in all PC monitors, in projectors, and in flat panel displays (LCDs and Plasmas). So as many new displays for television are progressive and as PC and television are merged (set top boxes for digital television, DVDs, television tuners for PCs, and video editing on PCs) there is obviously a big need for deinterlacing. Deinterlacing is difficult, as turning e.g. 50 fields per second into 50 frames per second requires a doubling of the amount of image data without introducing new artefacts to annoy the human visual perception. We propose a new scheme for deinterlacing developed from techniques used in image and image sequence inpainting, and we have implemented ten known and widely used deinterlacing schemes to compare it with. Our scheme uses Total Variation (TV) based in a Bayesian framework and do MAP by minimizing an energy functional. This is accomplished by deriving and solving solving corresponding Partial Differential Equations (PDEs) obtained through the calculus of variations. This is in contrast to many known deinterlacers that have been developed ad hoc (and in a heuristic way) with online hardware implementation directly in eye. Therefore they are often simplified to keep hardware costs down. We start with a theoretically well-based offline design that by further development could end up as online hardware. Section 2 will describe the other deinterlacing schemes implemented, section 3 will describe our proposed scheme, section 4 shows the results and in section 5 we draw our conclusions.

2 Standard Deinterlacing

To measure the performance of our deinterlacing scheme, we have implemented ten other schemes known from literature and/or available software and hardware ([2]).

Line Doubling (LDB) is very simple. Every interpolated horizontal line is a repetition of the previous existing line ([15] and [17]). Line Averaging (LAV) is a vertical average of the above and below pixels, since they are both known ([2], [15] and [17]). Field Insertion (FI), a.k.a. merging or weaving, fills in the blanks with neighbouring lines in time and is essentially a temporal version of LDB. The result is very similar to the image seen on an interlaced display ([2] and [17]). Field averaging (FAV) is a temporal version of LAV ([17]), while Vertical Temporal interpolation (VT) is a simple 50/50 combination of LAV and FAV ([17]). Many more advanced but not significantly better VT filters have been suggested, e.g. by BBC Research ([16]). All schemes mentioned so far are fixed, linear filters, whereas the next five are nonlinear and adapt to certain conditions in their local neighborhood and chose one of several possible interpolations depending on the local image content to yield better results.

Median filtering (Med) is a real classic in image processing and is used for deinterlacing in many variations ([2], [3], [8], [14], [13] and [15]). We have chosen a 3-tap vertical temporal version from [2] although we use the forward temporal neighbor instead of the backwards. Motion adaptive deinterlacing (MA) can be done in a countless number of ways and we have chosen a version suggested in [13] and [14]. It does simple Motion detection and takes advantage of the qualities of simpler schemes under different conditions: FAV in presence of no motion, Median filtering when motion is slow and LAV when fast motion is detected. Thresholds classify the motion. Weighted Vertical Temporal deinterlacing (wVT) is a simpler way of doing motion adaptation than the previous mentioned scheme, MA, and gives, instead of a hard switching between schemes, a smooth weighted transition between temporal and vertical interpolation. The scheme is described in detail in [9]. Edge Adaptive deinterlacing (EA) has been suggested in several forms, e.g. in [6], [9] and [15]. We have chosen a scheme that based on Summed Absolute Differences (SAD) selects a direction of interpolation as described in [15], although we have modified it to detect the best of five directions, 0°, $\pm 26^o$ and $\pm 45^o$ from vertical. Successive Approximation (SA) is the second level of approximation in [9] although the its edge adaptive scheme working on the first deinterlaced approximation has been swapped with the EA scheme that works directly on the interpolated original and thereby taking the successiveness out of the scheme but in the same instance also removing the possibility of error propagation.

Med is a simple adaptive scheme, EA adopts to the orientation of edges while MA, wVT and SA are Motion adaptive.

3 Total Variation Deinterlacing

In this section we introduce a novel deinterlacing scheme based on Total Variation minimization. We first proceed in a Bayesian fashion and deduce a variational formulation through MAP estimation in continuous settings following [10]. We then compute the associated Euler-Lagrange equations and their associated gradient descent formulations. The discretization of the latter will provide our numerical schemes. We first introduce the notations used in the sequel. Ω will denote the spatio-temporal domain of the progressive sequence, $F \subset \Omega$ the domain of the known fields, u_0 will denote the interlaced sequence, and by abuse of notations, it will also denote the known data on F.

3.1 Bayesian Framework

Let u denote a progressive sequence and u_0 the known sequence of interlaced fields. According to Bayes' Theorem

$$p(u|u_0) \propto p(u_0|u)p(u) \ . \tag{1}$$

The term on the left hand side is the a posteriori to be maximized (MAP) and the first term on the right hand the side is a model term and the second is a prior on image sequences. For the model term we choose a simple Dirac distribution $p(u_0|u) = \delta((u - u_0)|_F)$ because we wish to keep the existing pixels unchanged.

We have investigated two distributions for the prior term p(u). First, by viewing the image sequence u as a 3D volume, we set

$$p(u) \propto e^{-\lambda \sum_x |\nabla_3 u(x)|} \tag{2}$$

with x running over all the pixels in the sequence, $\nabla_3 u$ a discrete spatio-temporal gradient and λ a positive constant.

Nevertheless, it is somewhat unnatural to treat an image sequence as a 3D volume. We introduce therefore a simple model that separates spatial and temporal dimensions and we assume independence of the spatial and temporal distributions. Our image prior thus becomes

$$p(u) = p(u_s, u_t) = p(u_s)p(u_t) \tag{3}$$

where $p(u_s)$ refers to the spatial distribution of images and $p(u_t)$ to the temporal correlation between frames. For the spatial prior we use

$$p(u_s) \propto e^{-\lambda \sum_x |\nabla u(x)|} \tag{4}$$

with x again running over all the pixels in the sequence, ∇u a discrete *spatial* gradient and λ a positive constant. This has proven a robust model, well studied in the computer vision community; see for instance [1], [4] or [12]. The temporal prior

$$p(u_t) \propto e^{-\mu \sum_x |\partial_t u(x)|} \tag{5}$$

where $\partial_t u$ denotes the time-derivative of u and introduces the motion adaptive aspect of our algorithm, μ being a positive constant.

3.2 Variational Formulation - Euler-Lagrange Equations

Following [10] in order to compute the Maximum A Posteriori (MAP) solution, u, for our problems, we take the $-\log$ of each term to reformulate it as a minimization problem. Instead of using the $|\cdot|$ function which is non differentiable at the origin, we replace it by the approximation $\psi(s^2) = \sqrt{s^2 + \varepsilon^2}$, with $\varepsilon = 0.1$ or 0.01 in our experiments. From (2), with this modification we obtain u as the solution of

$$\operatorname*{Arg\,min}_u \int_\Omega \psi(|\nabla_3 u|)dx, \quad u = u_0|_F \ . \tag{6}$$

From standard calculus of variations and the fact that $\psi'(s)/s = 1/\psi(s)$, its Euler-Lagrange equation is

$$-\mathrm{div}\left(\frac{\nabla_3 u}{\psi(|\nabla_3 u|)}\right) = 0, \quad u = u_0|_F \tag{7}$$

Table 1. MSE from deinterlacing the four sequences `Person`, `C&T`, `BS` and `BSNM`. 3D and 2D+1D are the two versions of our scheme with the number of iterations given after the name. Clearly our schemes give the best results of all on the natural image sequences `Person` and `C&T`

Scheme		`Person`	`C&T`	`BS`	`BSNM`
LDB		17.90	79.72	1623.6	755.8
LAV		5.53	26.31	924.6	678.4
FI		22.26	472.25	3935.1	0
FAV		9.03	284.22	2514.4	0
VT		5.62	94.34	1146.8	169.6
Med		9.27	65.72	1154.1	363.4
MA		5.53	28.18	*840.6*	363.4
wVT		5.07	67.97	1056.0	0
EA		8.77	32.87	957.1	210.1
SA		5.36	48.79	862.4	82.0
3D TV	2	5.02	27.00	1066.2	666.8
3D TV	50	*4.85*	56.29	1078.3	461.8
2D+1D TV	2	4.97	*26.06*	919.2	666.2
2D+1D TV	20	*4.86*	*26.23*	890.9	567.9
2D+1D TV	200	5.11	32.89	*804.4*	242.5

We also give a subjective evaluation as the final judge of the result is the human visual system. A discussion of how to determine the quality of deinterlacing is given in [2]. Table 1 gives the objective results.

On `C&T` and `BS` it is seen from the MSE's that in presence of large motion, our scheme offers only little improvement, and only for the motion adaptive 2D+1D version, where the spatial and temporal gradients are separated. The 3D version suffers from having a spatio-temporal gradient. Over time (in terms of number of iterations) the 2D+1D improves a lot on the `BS` but not on `C&T`, which contains the larger motions of the two. Although the 3D does not perform to good overall, it actually improves the per-frame MSE in 23 of the 98 frames in `C&T`. In presence of none or only small motion, our scheme wins as it can be seen from the MSE's on the `BSNM` and `Person`. After only 2 iterations a 10 % improvement is seen on the MSE of `Person` and it increases with the number of iterations. Taking SA as initial guess instead of LAV on `BSNM` gave a 9 % improvement in MSE after 20 iterations of 2D+1D.

The results for 2D+1D after 200 iterations show that convergence in MSE stops for the two natural sequences. This is due to the smoothing of the TV prior and noise in the original sequence. Further studies showed that the lowest values in MSE was reached after 40-60 iterations.

Subjectively our scheme produces the best visible results on all four sequences but `BSNM`. `BSNM` is fully stationary, so the temporal schemes give 100 % perfect results on it. On `BS` and `C&T` the improvement is moderate, but on `Person` the results from our scheme are clearly the best. After two iterations we already

Fig. 2. From top to bottom: deinterlacing with MA, wVT and 2D+1D TV. Only zoom-ins of the full frame is shown here

se a good subjective result for the 2D+1D scheme, but after 20 iterations (and 50 for the 3D) the results are really good, even making us doubt which is the deinterlaced when comparing to the progressive original.

Figure 2 top, middle and bottom also illustrate the potential of our method. The sequence used for the illustration shows a pan of Christianborg Castle in Copenhagen and as it only exists as interlaced, no MSE's can be calculated. The top picture shows the result of the wVT scheme while the middle one shows the result of the MA scheme. Serious artifacts are visible for both schemes, serration for wVT and erroneous detection and interpolation for MA. The bottom picture shows the result of our 2D+1D scheme after 20 iterations, and clearly asserts the quality improvement obtained with our scheme.

LAV in itself is, given its simplicity, a remarkably well performing deinterlacer as the results in table 1 indicates, but as figure 3 shows, the 2D+1D scheme is able to improve the quality of deinterlacing significantly after 20 iterations. Note in the ornaments how the details have been sharpened and the jagged edges have been removed. The sequence used, `Church`, is stationary, shot with the camera on a camera mounting but rather noisy do to low lighting.

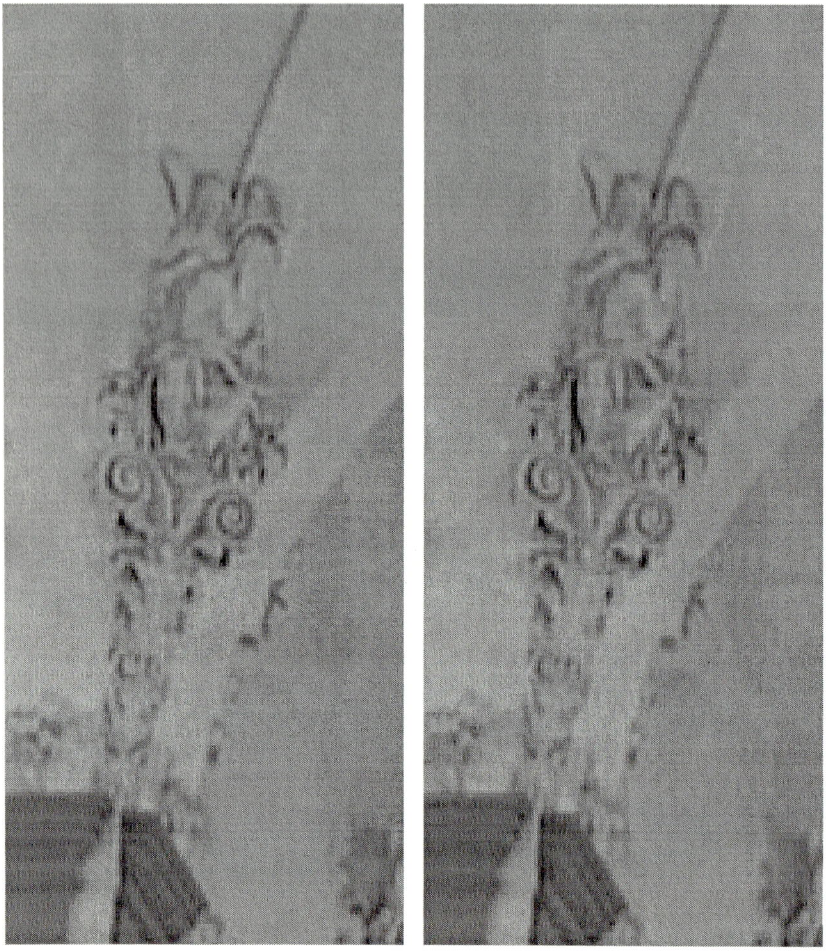

Fig. 3. Deinterlacing of the sequence `Church`. Left: 2D+1D TV after 20 iterations. Right: LAV used as initialization for 2D+1D TV. Clearly 2D+1D TV improves on the LAV initialization. Only zoom-ins of the full frame is shown here. Notice the in particular the arm of the upper angel and the helixes (spirals) in the middle

Further investigations on the 2D+1D schemes has also shown that the number of iterations to obtain a certain quality of the result can be reduced by a factor of three to six by increasing the time step in the gradient descent without the loss of stability. The number of operations and complexity per iteration of the 2D+1D schemes are the same as for the most complex of the ten known deinterlacing schemes, SA. This together with the increase in time step and good results after a few iterations gives rise to our believes that an online hardware implementation of 2D+1D TV MA Deinterlacing is possible.

5 Conclusion

We have shown that a technique so far used for inpainting can be redeveloped to do deinterlacing and further on our Total Variation Motion Adaptive Deinterlacing outperforms ten known fixed or adaptive deinterlacers. Our deinterlacer is still in its youth and its potential not yet fully explored. The quality of results and the computational complexity both indicate that hardware implementation can reach high quality results in realtime.

TV deinterlacing is a novel approach and introduces a whole new theoretical framework for video processing and the results advocate the further exploration of the ideas presented here.

References

1. Aubert, G. Kornprobst, P.: Mathematical Problems in Image Processing: Partial Differential Equations and the Calculus of Variations. Applied Mathematical Sciences, Vol. 147. Springer-Verlag, Berlin Heidelberg New York (2002)
2. Bellers, E. B., De Haan, G.: Deinterlacing: A Key Technology for Scan Rate Conversion, Elsevier, Amsterdam (2000)
3. Capodiferro, L.: Interlaced to Progressive Conversion by Median Filtering. In: Chiarglione, L. (ed.): Signal Processing of HDTV, II, Elsevier Science Publishers, Amsterdam (1990)
4. Chan, T. F., Shen, J.: Mathematical Models for Local Nontexture Inpaintings. SIAM Journal on Applied Mathematics, vol. 62, no. 3 (2002) 1019-1043
5. Chanas, L.: Mthodes variationelles pour la restauration de squences d'images fortement dgrades. Application aux images blouies par laser. Ph.D. Thesis, University of Cergy-Pontoise (2001)
6. Doyle, T., Looymans, M.: Progressive Scan Conversion Using Edge Information. In: Chiarglione, L. (ed.): Signal Processing of HDTV, II, Elsevier Science Publishers, Amsterdam (1990)
7. Dreier, H.-J.: Line Flicker Reduction by Adaptive Signal Processing. In: Chiarglione, L. (ed.): Signal Processing of HDTV, II, Elsevier Science Publishers, Amsterdam (1990)
8. Haavisto, P., Juhola, J., Neuvo, Y.: Scan Rate Up-Conversion Using Adaptive Weighted Median Filtering. In: Chiarglione, L. (ed.): Signal Processing of HDTV, II, Elsevier Science Publishers, Amsterdam (1990)
9. Kovacevic, J., Safranek, R. J., Yeh, E. M.: Deinterlacing by Successive Approximation. IEEE Trans. Image Proc., Vol. 6, no. 2 (1997)
10. Mumford, D.: Bayesian Rationale For The Variational Formulation. In: ter Haar Romeny, B. M. (ed.): Geometry-Driven Diffusion In Computer Vision, Kluwer Academic Publishers, Dordrecht (1994) 135–146
11. Pigeon, S., Guillotel, P.: Advantages and Drawbacks of Interlaced and Progressive Scanning Formats. CEC RACE/HAMLET Deliverable no R2110/WP2/DS/R/004/b1 (1995)
12. Rudin, L., Osher, S., Fatemi, E.: Non Linear Total Variation Based Noise Removal Algorithms. Physica D. 60. (1992) 259-286
13. Skarabot, A. Ramponi, G., Buriola, L.: FPGA architecture for a videowall image processor. Proc. SPIE Intern. Symp. Electronic Imaging 2001 (2001)

14. Skarabot, A., Ramponi, G., Toffoli, D.: Image sequence processing for videowall visualization. Proc. IST/SPIE 12th Annual Intern. Symp. Electronic Imaging 2000 (2000)
15. Tekalp, A. M.: Digital Video Processing. Prentice Hall, Upper Saddle River, NJ, (1995)
16. Thomas, G. A.: A Comparison of Motion-Compensated Interlace-to-Progressive Conversion Methods. Signal Process.: Image Commun., vol. 12, no. 3 (1998)
17. Wang, Y. Ostermann, J., Zhang, Y.Q.: Video Processing and Communications. Prentice Hall, Upper Saddle River, NJ (2002)

A Geometric Formulation of Gradient Descent for Variational Problems with Moving Surfaces

Jan Erik Solem and Niels Chr. Overgaard

Applied Mathematics Group, School of Technology and Society,
Malmö University, Sweden
{jes, nco}@ts.mah.se

Abstract. Developments within the computer vision community have led to the formulation of many interesting problems in a variational setting. This paper introduces the manifold of admissible surfaces and a scalar product on its tangent spaces. This makes it possible to properly define gradients and gradient descent procedures for variational problems involving m-surfaces. These concepts lead to a geometric understanding of current state of the art evolution methods and steepest descent evolution equations. By geometric reasoning, common procedures within the variational level set framework are explained and justified. Concrete computations for a general class of functionals are presented and applied to common variational problems for curves and surfaces.

Keywords: variational methods, level set methods, differential geometry, functionals, surface and curve evolution, geometric flows, gradient.

1 Introduction

Formulating problems in a variational setting has become a common procedure for solving problems in computer vision and image analysis. This paper will focus on variational formulations involving surfaces of codimension one, e.g. curves in images, surfaces in space and their generalization to higher dimensions. Some useful examples are the alignment of curves to image edges [1], segmentation in medical images [2], surface fitting to 3D data [3], variational stereo reconstruction [4], the popular geodesic active contours [5] which find shortest curves using derived metrics and minimal surface approaches [6, 7]. In all these examples a curve- or surface motion is derived from a functional and using this motion the curve or surface is evolved until some interesting local minimum is found.

It is the purpose of this paper to analyze this process, simplify it, and give a proper definition of the gradient descent procedure. This will explain and justify some commonly used procedures for these types of variational problems.

Let us first recall how gradient descent works in the finite dimensional setting. To find the minimum of a differentiable function $f : \mathbf{R}^m \to \mathbf{R}$, pick a point $\mathrm{x}_0 \in \mathbf{R}^m$ and solve the initial value problem

$$\dot{\mathrm{x}}(t) = -\nabla f(\mathrm{x}(t)), \qquad \mathrm{x}(0) = \mathrm{x}_0 \ , \tag{1}$$

R. Kimmel, N. Sochen, J. Weickert (Eds.): Scale-Space 2005, LNCS 3459, pp. 419–430, 2005.

We require the level set function to satisfy the condition $\nabla\phi(\mathbf{x}) \neq 0$ for all $\mathbf{x} \in \Gamma$. This is always possible because Γ is a regular surface. The sets $\{\mathbf{x} : \phi(\mathbf{x}) < 0\}$ and $\{\mathbf{x} : \phi(\mathbf{x}) > 0\}$ are called the *inside* and the *outside* of Γ, respectively. Using this convention, the outward unit normal $\mathbf{n}$ and the mean curvature κ of Γ is given by, cf. [9],

$$\mathbf{n} = \frac{\nabla\phi}{|\nabla\phi|} \quad \text{and} \quad \kappa = \nabla \cdot \frac{\nabla\phi}{|\nabla\phi|} \ . \tag{5}$$

The latter requires $k \geq 2$ thus we will assume $k = 2$ in the rest of this paper.

Using the representation above we define a *regular surface evolution* or *regular motion* by adding a time dependence to the level set function, $\phi : \mathbf{R}^{m+1} \times I \to \mathbf{R}$ for some open (time-) interval $I \subset \mathbf{R}$, where ϕ is C^2 with respect to the space variables and C^1 with respect to t. Moreover we require that[2] $\nabla\phi \neq 0$ on $\Gamma(t) = \{\mathbf{x} : \phi(\mathbf{x}, t) = 0\}$. The function $\phi(\mathbf{x}, t)$ then represents the regular surface evolution $t \mapsto \Gamma(t)$.

Let $t \mapsto \Gamma(t)$ be such a regular surface evolution. Suppose that a point, or particle, moves along with the evolving surface. If the motion of the particle is given by the curve $\boldsymbol{\alpha} : I \to \mathbf{R}^{m+1}$ then the identity $\phi(\boldsymbol{\alpha}(t), t) \equiv 0$ holds, and differentiation with respect to time yields

$$\frac{\partial\phi}{\partial t} + \dot{\boldsymbol{\alpha}} \cdot \nabla\phi = 0 \quad \text{or} \quad \frac{\partial\phi}{\partial t} + v_n|\nabla\phi| = 0 \ , \tag{6}$$

where $\dot{} = d/dt$ and $v_n = \dot{\boldsymbol{\alpha}} \cdot \mathbf{n}$ is the normal component of the particle velocity. This is the well-known *level set equation.*

3 Deriving the Equations of Motion: Common Approach

As a motivation for the work in Sections 4, 5 and 6, we give a brief review of the standard procedure for deriving the evolution equation for Γ in the variational level set formulation. It is our goal to explain this procedure.

A common approach is to use a parametrization of the m-surface to derive the normal velocity, cf. e.g. [1, 2, 4]. The drawbacks of this approach are that unless $m = 1$, i.e. planar curves, the computations are tedious and also they do not easily generalize to arbitrary dimensions. An elegant way of deriving motion equations was proposed in [7] using the method of moving frames. Another alternative, which will be reviewed here, is to derive the equation of motion directly on the level set function ϕ as in e.g. [10, 11].

To illustrate the procedure we will use a simple example. If Γ is represented by a level set function ϕ, the area functional $E(\Gamma) = \int_\Gamma 1\, d\sigma$ can be written[3]

$$E(\phi) = \int_{\mathbf{R}^{m+1}} |\nabla\phi|\delta(\phi)\, d\mathbf{x} = \int_{\mathbf{R}^{m+1}} F(\phi, \nabla\phi)\, d\mathbf{x} \ ,$$

[2] $\nabla\phi$ is the gradient with respect to the spatial directions, i.e. the $\mathbf{x}$-variable.

[3] It is often convenient to write $E(\phi)$ instead of $E(\Gamma)$, as we shall see later.

where $|\nabla\phi|\delta(\phi)\,d\mathbf{x}$ is the surface measure, see e.g. Hörmander [12–Thm. 6.1.5]. Standard formulas from calculus of variations give the following necessary condition for extremals [13],

$$\frac{\partial F}{\partial \phi} - \sum_{i=1}^{m+1} \frac{\partial}{\partial x_i}\left(\frac{\partial F}{\partial \phi_{x_i}}\right) = 0 \ ,$$

which gives the simple Euler-Lagrange equation $\kappa\delta(\phi) = 0$ for the problem. The standard procedure is to evolve an initial surface using either of the PDEs

$$\frac{\partial \phi}{\partial t} = \pm\kappa\delta(\phi) \ , \tag{7}$$

until a steady state is reached, which is then an extremal of E. First one checks whether the plus or the minus sign in (7) gives a descent direction, since this information is not contained in the Euler-Lagrange equation (here plus is the correct sign). Then (7) is made into an equation of the form (6), thus extending the motion to all level sets of ϕ, by replacing $\delta(\phi)$ with $|\nabla\phi|$, which yields

$$\frac{\partial \phi}{\partial t} = \kappa|\nabla\phi| \ .$$

It has been shown [10] that this procedure of replacing $\delta(\phi)$ by $|\nabla\phi|$ indeed gives a descent PDE and we will present a geometric explanation of this in Section 6.

4 Gradient Interpretation

In this section a manifold M of m-surfaces $\Gamma \subset \mathbf{R}^{m+1}$ is defined, and the tangent space of the manifold is determined at any $\Gamma \in M$. Then a scalar product on the tangent spaces is introduced, allowing us to give a gradient interpretation of the Gâteaux derivative (or differential) dE of a given functional $E : M \to \mathbf{R}$.

Let Γ_0 be a closed, connected, m-dimensional surface in $\mathbf{R}^{m+1}$. The manifold M is defined as the set of surfaces Γ' which can be obtained from Γ_0 by some regular surface evolution $t \mapsto \Gamma(t)$. This means that there should exist a level set function $\phi : \mathbf{R}^{m+1} \times I \to \mathbf{R}$, such that if we set $\Gamma(t) = \{\mathbf{x} : \phi(\mathbf{x}, t)\}$ then $\Gamma(0) = \Gamma_0$ and $\Gamma(t) = \Gamma'$, for some $t > 0$.

If we agree to write $\Gamma' \sim \Gamma$ if Γ' can be obtained from Γ by a regular surface evolution, then clearly $\Gamma \sim \Gamma'$ (symmetry) because we can always reverse the time in the evolution. Also $\Gamma \sim \Gamma$ (reflexivity). Finally it can be shown that if $\Gamma' \sim \Gamma$ and $\Gamma'' \sim \Gamma'$, then $\Gamma'' \sim \Gamma$ (transitivity). Thus $\sim$ is an equivalence relation on surfaces in $\mathbf{R}^{m+1}$ and $M = \{\Gamma' : \Gamma' \sim \Gamma_0\}$ is an equivalence class of $\sim$. In particular, any $\Gamma \in M$ can be used in the definition of M.

Our next aim is to determine the tangent space of M at the "point" $\Gamma_0 \in M$. Before we can do that, we need to clarify what we mean by a *tangent vector* of M at Γ_0. For this purpose, take a point $\mathbf{x}_0 \in \Gamma_0$, and let $\phi = \phi(\mathbf{x}, t)$ be a level set function which defines a regular surface evolution $\Gamma(t) = \{\mathbf{x} : \phi(\mathbf{x}, t) = 0\}$ for $-\epsilon < t < \epsilon$ with $\Gamma(0) = \Gamma_0$.

5 Gradient Descent for Functionals Defined on Surfaces

In this section the gradient is computed, and the corresponding gradient descent motion constructed, for a general class of functionals defined on surfaces. The functionals under consideration may depend on surface points and normals, and the class is general enough to cover most applications in computer vision. Concrete examples are found in Section 7.

The result in Theorem 1 below is not new in itself, it was derived recently by Goldlücke and Magnor [7] using Cartan's aesthetically pleasing method of moving frames. Our reason for rederiving this result is to illustrate the theory in Section 4, and to show that the computations can also be carried out in the level set framework, by relying on distribution theory.

Let Γ be an m-surface in $\mathbf{R}^{m+1}$ and

$$E(\Gamma) = \int_\Gamma g(\mathbf{x}, \mathbf{n})\, d\sigma \ , \tag{12}$$

be an energy functional with $g : \mathbf{R}^{m+1} \times S^m \to \mathbf{R}$. Representing Γ as the zero set of $\phi(\mathbf{x})$ gives,

$$E(\phi) = \int g\left(\mathbf{x}, \frac{\nabla\phi}{|\nabla\phi|}\right) |\nabla\phi|\delta(\phi)\, d\mathbf{x} \ .$$

Consider the variation $\phi^s = \phi + s\psi$ of ϕ. This gives the normal velocity $v = -\psi/|\nabla\phi|$ at $s = 0$. The following results are needed

$$\frac{d}{ds}(|\nabla\phi^s|)\Big|_{s=0} = \frac{\nabla\phi}{|\nabla\phi|} \cdot \nabla\psi \ , \tag{13}$$

$$\frac{d}{ds}\left(\frac{\nabla\phi^s}{|\nabla\phi^s|}\right)\Big|_{s=0} = \frac{\nabla\psi}{|\nabla\phi|} - \left(\frac{\nabla\phi}{|\nabla\phi|} \cdot \frac{\nabla\psi}{|\nabla\phi|}\right) \frac{\nabla\phi}{|\nabla\phi|} \ , \tag{14}$$

and

$$\frac{d}{ds}(\delta(\phi^s))\Big|_{s=0} = \delta'(\phi)\psi \ . \tag{15}$$

Taking the Gâteaux derivative with $v = -\psi/|\nabla\phi|$ gives

$$dE(\Gamma)v = \frac{d}{ds}E(\phi + s\psi)\Big|_{s=0} = \int \frac{d}{ds}\left[g\left(\mathbf{x}, \frac{\nabla\phi^s}{|\nabla\phi^s|}\right)|\nabla\phi^s|\delta(\phi^s)\right] d\mathbf{x}\Big|_{s=0} \ .$$

Let us use the notation $g_\mathbf{n} = \nabla_{S^m} g$, for the gradient on the unit sphere S^m, as defined in (3). Since $g : \mathbf{R}^{m+1} \times S^m \to \mathbf{R}$ we have $g_\mathbf{n} \in T_\mathbf{n} S^m$, hence $g_\mathbf{n} \cdot \mathbf{n} = 0$. This means that the derivative $\frac{d}{ds} g\left(\mathbf{x}, \frac{\nabla\phi^s}{|\nabla\phi^s|}\right)$ is simply the tangential component

$$\frac{d}{ds} g\left(\mathbf{x}, \frac{\nabla\phi^s}{|\nabla\phi^s|}\right) = g_\mathbf{n} \cdot \left[\frac{\nabla\psi}{|\nabla\phi|} - \left(\frac{\nabla\phi}{|\nabla\phi|} \cdot \frac{\nabla\psi}{|\nabla\phi|}\right)\frac{\nabla\phi}{|\nabla\phi|}\right] = g_\mathbf{n} \cdot \frac{\nabla\psi}{|\nabla\phi|} \ ,$$

where we have used (14). The Gâteaux derivative is then simply

$$dE(\Gamma)v = \int g_\mathbf{n} \cdot \nabla\psi\, \delta(\phi)\, d\mathbf{x} + \int g \left[\frac{\nabla\phi}{|\nabla\phi|} \cdot \nabla\psi\right] \delta(\phi)\, d\mathbf{x} + \int g\, |\nabla\phi|\delta'(\phi)\psi\, d\mathbf{x} \ .$$

Integration by parts on $\nabla\psi$ gives

$$
\begin{aligned}
dE(\Gamma)v &= \int (-\psi)\nabla\cdot[g_{\mathbf{n}}\delta(\phi)]\,d\mathbf{x} + \int(-\psi)\nabla\cdot\left[g\frac{\nabla\phi}{|\nabla\phi|}\delta(\phi)\right]d\mathbf{x} \\
&\quad - \int(-\psi)g\,|\nabla\phi|\delta'(\phi)\,d\mathbf{x} \\
&= \int(-\psi)\Big\{(\nabla\cdot g_{\mathbf{n}})\,\delta(\phi) + g_{\mathbf{n}}\cdot\nabla\phi\,\delta'(\phi) + \nabla\cdot\left[g\,\frac{\nabla\phi}{|\nabla\phi|}\right]\delta(\phi) \\
&\qquad + g\,\frac{\nabla\phi}{|\nabla\phi|}\cdot\nabla\phi\,\delta'(\phi) - g\,|\nabla\phi|\delta'(\phi)\Big\}\,d\mathbf{x}\ .
\end{aligned}
$$

Since $g_{\mathbf{n}}\cdot\mathbf{n}=0$ implies $g_{\mathbf{n}}\cdot\nabla\phi=0$ and the two last terms cancel we get

$$
dE(\Gamma)v = \int\Big(\frac{-\psi}{|\nabla\phi|}\Big)\nabla\cdot\left[g_{\mathbf{n}} + g\,\frac{\nabla\phi}{|\nabla\phi|}\right]|\nabla\phi|\delta(\phi)\,d\mathbf{x} = \langle v, \nabla\cdot[g_{\mathbf{n}} + g\,\mathbf{n}]\rangle_\Gamma,
$$

The results of the derivations above are summarized in the following theorem.

Theorem 1. *Let $g = g(\mathbf{x},\mathbf{n})$ be continuously differentiable with respect to $\mathbf{x}$ and twice continuously differentiable with respect to $\mathbf{n}$. Then the functional E defined by (12) has the differential*

$$
dE(\Gamma)v = \langle\,\nabla\cdot[g_{\mathbf{n}} + g\,\mathbf{n}]\,, v\,\rangle_\Gamma\ ,
$$

for $v\in T_\Gamma M$, in particular $\nabla_M E = \nabla\cdot[g_{\mathbf{n}}+g\,\mathbf{n}]$.

Corollary 1. *The gradient descent evolution for the minimization of (12), when formulated in terms of a level set function ϕ, is*

$$
\frac{\partial\phi}{\partial t} = (\nabla\cdot[g_{\mathbf{n}}+g\,\mathbf{n}])|\nabla\phi|\ .
$$

Proof. The normal velocity of the evolution $\Gamma(t)$ given by $\phi(\mathbf{x},t)$ is $\dot\Gamma(t) = -(\partial\phi/\partial t)/|\nabla\phi|$, and $-\nabla_M E = -\nabla\cdot[g_{\mathbf{n}}+g\,\mathbf{n}]$. In view of (11) in Definition 2 the desired result follows.

This result is in agreement with [7] and [4]. Let us use the notation $\nabla E = \nabla_M E$ from now on.

6 Explaining the "Common Approach"

As pointed out in Section 3, a common approach, which is found in nearly all cases where the the Euler-Lagrange equation contains the factor $\delta(\phi)$, is to replace this factor by $|\nabla\phi|$ with the motivation that it does not affect the steady state solution, but only the speed of descent, cf. [10]. This section offers a geometric explanation and justification for this procedure. In order not to obscure the argument with technicalities, the explanation is carried out for the surface area functional $E(\Gamma) = \int_\Gamma 1\,d\sigma$, already considered in Section 3.

Using the formulas (13) and (15) from the previous section, the Gâteaux derivative of E is found to be

$$dE(\Gamma)v = \frac{d}{ds}E(\phi + s\psi)\Big|_{s=0} = \int \left[\frac{\nabla\phi}{|\nabla\phi|}\cdot\nabla\psi\right]\delta(\phi)\,d\mathbf{x} + \int |\nabla\phi|\delta'(\phi)\psi\,d\mathbf{x}$$

$$= \int(-\psi)\left(\nabla\cdot\frac{\nabla\phi}{|\nabla\phi|}\right)\delta(\phi)\,d\mathbf{x} = \int(-\psi)\kappa\;\delta(\phi)\,d\mathbf{x}\ , \tag{16}$$

where $v = -\psi/|\nabla\phi|$ is the normal velocity of the surface evolution $\phi^s(\mathbf{x}) = \phi(\mathbf{x}) + s\psi(\mathbf{x})$. If the integrand on the right hand side of (16) is multiplied by $|\nabla\phi|/|\nabla\phi|$, then the numerator of this quote can be absorbed in the surface measure $d\sigma = |\nabla\phi|\delta(\phi)\,d\mathbf{x}$, and the the denominator together with $-\psi$ forms the normal velocity v. Thus the differential becomes

$$dE(\Gamma)v = \int(\frac{-\psi}{|\nabla\phi|})\kappa\,d\sigma = \int_\Gamma v\,\kappa\,d\sigma = \langle v,\kappa\rangle_\Gamma\ . \tag{17}$$

So the gradient is $\nabla E = \kappa$. The gradient descent is obtained by setting the normal velocity of the moving surface $\phi(\mathbf{x},t) = 0$ equal to minus the gradient, that is $-(\partial\phi/\partial t)/|\nabla\phi| = -\nabla E = -\kappa$ or $\partial\phi/\partial t = \kappa|\nabla\phi|$, which agrees with the result in Section 3. This amounts to the same as replacing $\delta(\phi)$ with $|\nabla\phi|$ in the Euler-Lagrange equation and using this as the right hand side of the evolution equation, but here there is a geometric meaning behind the reasoning.

This extension of the motion by replacing $\delta(\phi)$ with $|\nabla\phi|$ is sometimes motivated by the fact that it makes the equation of motion morphological, i.e. it does not depend on the slope of ϕ, cf. [10]. This means that for a monotone and increasing function $f : \mathbf{R} \to \mathbf{R}$ with $f(0) = 0$ the equation of motion (6) has the same solution if ϕ is replaced by the new level set function $f(\phi(\mathbf{x}))$. This is no surprise since it follows directly from the analysis of the normal velocity in Section 4, in particular Lemma 1.

7 Examples

This section shows how the framework introduced in the previous sections can be applied to concrete examples commonly used in computer vision and image analysis applications.

Flux. The functional representing the outward flux of a vector field $\mathbf{v} : \mathbf{R}^{m+1} \to \mathbf{R}^{m+1}$ through the surface Γ, as in [2] and [1],

$$E(\Gamma) = \int_\Gamma \mathbf{v}(\mathbf{x})\cdot\mathbf{n}\,d\sigma\ ,$$

where $g(\mathbf{x},\mathbf{n}) = \mathbf{v}(\mathbf{x})\cdot\mathbf{n}$, has gradient

$$\nabla E = \nabla\cdot[g_{\mathbf{n}} + g\,\mathbf{n}] = \nabla\cdot[\mathbf{v} - (\mathbf{v}\cdot\mathbf{n})\,\mathbf{n} + (\mathbf{v}\cdot\mathbf{n})\,\mathbf{n}] = \nabla\cdot\mathbf{v}\ ,$$

and gradient evolution equation equal to $\partial\phi/\partial t = (\nabla\cdot\mathbf{v})|\nabla\phi|$.

Geodesic Active Contours and Surface Potential. Let $I : \mathbf{R}^2 \to \mathbf{R}$ denote an image. To find a smooth curve Γ that fits the edges in the image the following functional has been proposed

$$E(\Gamma) = \int_\Gamma f(|\nabla I|)\, d\sigma \ , \tag{18}$$

for some function $f : \mathbf{R}_+ \to \mathbf{R}_+$. This elegant approach is called the geodesic active contours [5] since extremal curves are a geodesics on a Riemannian manifold whose metric is determined by f.

Here $g(\mathbf{x}, \mathbf{n}) = f(|\nabla I(\mathbf{x})|)$ does not depend on $\mathbf{n}$ so the gradient is $\nabla E = \nabla \cdot (f(|\nabla I|)\mathbf{n})$, which gives the gradient descent motion,

$$\frac{\partial \phi}{\partial t} = \Big[\nabla \cdot (f(|\nabla I|)\mathbf{n})\Big] |\nabla \phi| = \Big[f(|\nabla I|)\, \kappa + \nabla f(|\nabla I|) \cdot \mathbf{n}\Big] |\nabla \phi| \ .$$

This approach extends naturally to higher dimensional images, cf. [6]. For instance, the gradient of surface potential functional $E(\Gamma) = \int_\Gamma d(\mathbf{x})\, d\sigma$, where e.g. $d(\mathbf{x}) : \mathbf{R}^{m+1} \to \mathbf{R}$ is the distance potential in [3], is $\nabla E = \nabla d \cdot \mathbf{n} + d\kappa$.

Average Potential. Consider the case of the surface potential above. Sometimes it is desirable to minimize (maximize) the *average* potential value over Γ thereby removing the area dependence. This can be formulated as

$$E(\Gamma) = \frac{\int_\Gamma d(\mathbf{x})\, d\sigma}{\int_\Gamma d\sigma} = \frac{F(\Gamma)}{G(\Gamma)} \ ,$$

as has been done in e.g. [14] and [1]. It was shown above that $\nabla F = \nabla d \cdot \mathbf{n} + d\kappa$ and $\nabla G = \kappa$, so by the quotient rule the gradient is simply

$$\nabla E = \frac{(\nabla d \cdot \mathbf{n} + d\kappa) G(\Gamma) - F(\Gamma)\kappa}{G(\Gamma)^2} = \frac{\nabla d \cdot \mathbf{n} + \kappa(d - E(\Gamma))}{G(\Gamma)} \ ,$$

Notice that the numerator can be used as a descent direction because $G(\Gamma) > 0$.

Volume Potential. The analysis above has been focused on functionals involving surface integrals. A final example shows that it is possible to use the same framework for volume integrals. Let Ω denote the interior of Γ, $d(\mathbf{x}) : \mathbf{R}^{m+1} \to \mathbf{R}$ be a potential function and define the energy as the volume integral

$$E(\Gamma) = \int_\Omega d(\mathbf{x})\, d\mathbf{x} = \int d(\mathbf{x})(1 - H(\phi))\, d\mathbf{x} \ ,$$

where $H(\cdot)$ is the Heaviside function. The Gâteaux derivative of E is

$$\begin{aligned} dE(\Gamma)v &= \frac{d}{ds} E(\phi + s\psi)\Big|_{s=0} = \int \frac{d}{ds}(1 - H(\phi + s\psi))\Big|_{s=0} d(\mathbf{x})\, d\mathbf{x} \\ &= \int (-\psi)\delta(\phi) d(\mathbf{x})\, d\mathbf{x} = \int (\frac{-\psi}{|\nabla \phi|}) d(\mathbf{x}) |\nabla \phi| \delta(\phi)\, d\mathbf{x} \\ &= \int_\Gamma v\, d(\mathbf{x})\, d\sigma = \langle v, d(\mathbf{x}) \rangle_\Gamma \ , \end{aligned} \tag{19}$$

which shows that the gradient is $\nabla E = d(\mathbf{x})$ and the gradient descent motion $\partial \phi / \partial t = d(\mathbf{x}) |\nabla \phi|$. This is a classical result, cf. e.g. [15].

8 Conclusions

In this paper we introduced a scalar product on the manifold of admissible surfaces which made it possible to define the gradient of m-surface functionals in arbitrary dimensions and proper gradient descent procedures. Using distribution theory and a level set representation we derived gradient descent motion equations for a general class of functionals. This class covers most variational surface formulations in computer vision and image analysis. The derivations were performed using geometric quantities such as normal velocity and gradient and are therefore valid for any choice of surface representation. Finally, we used this framework to explain common procedures in the variational level set domain, and apply the framework to some useful examples.

References

1. Kimmel, R., Bruckstein, A.: Regularized Laplacian zero crossings as optimal edge integrators. Int. Journal of Computer Vision **53** (2003) 225–243
2. Vasilevskiy, A., Siddiqi, K.: Flux maximizing geometric flows. IEEE Trans. on PAMI **24** (2002) 1565–1578
3. Zhao, H., Osher, S., Merriman, B., Kang, M.: Implicit and non-parametric shape reconstruction from unorganized points using a variational level set method. In: Computer Vision and Image Understanding. (2000) 295–319
4. Faugeras, O., Keriven, R.: Variational principles, surface evolution, PDEs, level set methods, and the stereo problem. IEEE Transactions on Image Processing **7** (1998) 336–344
5. Caselles, V., Kimmel, R., Sapiro, G.: Geodesic active contours. Int. Journal of Computer Vision (1997)
6. Caselles, V., Kimmel, R., Sapiro, G., Sbert, C.: Minimal surfaces based object segmentation. IEEE Transactions on Pattern Analysis and Machine Intelligence **19** (1997) 394–398
7. Goldlücke, B., Magnor, M.: Weighted minimal hypersurfaces and their applications in computer vision. In: European Conference on Computer Vision. Volume 2., Prague, Czech Republic (2004) 366–378
8. do Carmo, M.: Differential Geometry of Curves and Surfaces. Prentice-Hall (1976)
9. Thorpe, J.A.: Elementary Topics in Differential Geometry. Springer-Verlag (1985)
10. Zhao, H., Chan, T., Merriman, B., Osher, S.: A variational level set approach to multiphase motion. J. Computational Physics 127 (1996) 179–195
11. Bertalmio, M., Cheng, L.T., Osher, S., Sapiro, G.: Variational problems and partial differential equations on implicit surfaces. J. Comput. Phys. **174** (2001) 759–780
12. Hörmander, L.: The Analysis of Linear Partial Differential Operators I. Springer, Berlin (1990)
13. Gelfand, I., Fomin, S.: Calculus of Variations. Dover Publications (2000)
14. Fua, P., Leclerc, Y.: Model driven edge detection. MVA **3** (1990) 45–56
15. Paragios, N., Deriche, R.: Geodesic active regions: A new paradigm to deal with frame partition problems in computer vision. International Journal of Visual Communication and Image Representation (2000)

On Image Reconstruction from Multiscale Top Points

Frans Kanters[1], Martin Lillholm[2], Remco Duits[1], Bart Janssen[1], Bram Platel[1], Luc Florack[1], and Bart ter Haar Romeny[1]

[1] Eindhoven University of Technology,
Den Dolech 2, Postbus 513, 5600 MB Eindhoven, The Netherlands
{F.M.W.Kanters, R.Duits, B.J.Janssen, B.Platel, L.M.J.Florack, B.M.terHaarRomeny}@tue.nl

[2] IT University of Copenhagen, Rued Langgaards Vej 7,
2300 Copenhagen S, Denmark
grumse@itu.dk

Abstract. Image reconstruction from a fiducial collection of scale space interest points and attributes (e.g. in terms of image derivatives) can be used to make the amount of information contained in them explicit. Previous work by various authors includes both linear and non-linear image reconstruction schemes. In this paper, the authors present new results on image reconstruction using a top point representation of an image. A hierarchical ordering of top points based on a stability measure is presented, comparable to feature strength presented in various other works. By taking this into account our results show improved reconstructions from top points compared to previous work. The proposed top point representation is compared with previously proposed representations based on alternative feature sets, such as blobs using two reconstruction schemes (one linear, one non-linear). The stability of the reconstruction from the proposed top point representation under noise is also considered.

1 Introduction

Nielsen, Lillholm and Griffin presented a linear minimal variance reconstruction scheme to reconstruct an image given a set of scale space interest points and the local N-jet in those points [7,5]. Based on that reconstruction algorithm, Kanters et al. [3] presented a closed form solution for the Gram matrix of that linear framework. The prior used in these linear frameworks is however not sufficient to create visually attractive reconstructions if not enough constraints are used. To overcome this problem Nielsen and Lillholm proposed a prior based on natural image statistics, the Brownian reconstruction [7,5]. Recently a generalization of the linear reconstruction framework is proposed by Janssen et al. which — in a different way — also tries to overcome this problem while maintaining linearity [2].

These algorithms have been used with various *feature points* (collections of scale space interest points as input for the reconstruction algorithm), but only very little results are presented using so-called top points. In this paper, we present new results in reconstructions from top points using the Brownian reconstruction algorithm as well as the reconstruction scheme proposed by Janssen. Reconstructions from top points are

R. Kimmel, N. Sochen, J. Weickert (Eds.): Scale-Space 2005, LNCS 3459, pp. 431–442, 2005.

compared with results obtained from previously proposed image representations based on alternative feature sets such as blobs [6] in section 4. Furthermore, a hierarchical ordering of top points is introduced in section 3, based on a stability measure. Lillholm and Nielsen showed that for blobs it makes sense to use an ordering by *feature strength* for reconstruction, to select the most salient feature points [5]. In section 4, we present results comparing our ordering of top points with the ordering of blobs by feature strength. Finally the stability of the reconstruction from top points is considered in section 4.

2 Reconstruction Schemes

To reconstruct an image from a set of points and measurements in these points, two different reconstruction schemes are used. These schemes are projections on the intersection of a set of measurement filters. Both methods try to minimize an "energy" function while maintaining the same measurements in the points of the reconstructed image (constraints). The first method is proposed by Lillholm and Nielsen [5] and minimizes a global property while the second method proposed by Janssen et. al. [2] uses a more general inner product to calculate the measurements and minimizes the norm of that inner product space.

For the first method a measurement c_i in a certain point i is defined as an $\mathbb{L}_2$ inner product between a given filter ϕ_i and the original image I:

$$c_i = \langle \phi_i | I \rangle_{\mathbb{L}_2} = \int_{\Omega} \phi_i(x) I(x) dx \tag{1}$$

The following functional should be minimized to get the reconstruction $\hat{I}$:

$$S[\hat{I}] = \Psi + \sum_i \lambda_i \langle I - \hat{I} | \phi_i \rangle_{\mathbb{L}_2} \tag{2}$$

where Ψ is some prior which should be minimized and the other terms are the constraints. For the remainder of this paper we use a prior known as the Tikhonov regularizer [9] which results in:

$$S[\hat{I}] = \int_{\Omega} |\nabla \hat{I}|^2 dx + \sum_i \lambda_i \langle I - \hat{I} | \phi_i \rangle_{\mathbb{L}_2} \tag{3}$$

This prior is based on a Brownian motion image model and therefore the reconstruction is referred to as Brownian reconstruction. The solution $\hat{I}$ of this non linear minimization problem is found using an iterative algorithm, cf. Lillholm and Nielsen [5].

For the second method the definition of the inner product is generalized. For a positive, symmetric operator A we can define the inner product:

$$\langle f|g \rangle_A = \langle f|g \rangle_{\mathbb{L}_2} + \langle Af|Ag \rangle_{\mathbb{L}_2} = \langle f|(I + A^{\dagger}A)g \rangle_{\mathbb{L}_2} \tag{4}$$

Given a filter ψ_i and minimizing the norm of the A-inner product space, the functional to minimize becomes:

$$S[\hat{I}] = \frac{1}{2} \langle \hat{I} | \hat{I} \rangle_A + \sum_i \lambda_i \langle I - \hat{I} | \psi_i \rangle_A \tag{5}$$

Note that ψ_i is chosen in such a way that: $c_i = \langle\phi_i|I\rangle_{\mathbb{L}_2} = \langle\psi_i|I\rangle_A$. The solution $\hat{I}$ of this *linear* minimization problem boils down to an orthogonal projection on the intersection of the measurement filters, cf. Janssen et al. [2]. For the remainder of this paper, the definition of A proposed by Janssen et al. is used:

$$A = \gamma\sqrt{-\Delta} \tag{6}$$

Which results in minimizing:

$$S[\hat{I}] = \frac{1}{2}\int_\Omega \hat{I}^2 dx + \gamma^2 \int_\Omega |\nabla \hat{I}|^2 dx + \sum_i \lambda_i \langle I - \hat{I}|\phi_i\rangle_{\mathbb{L}_2} \tag{7}$$

In which again the Tikhonov regularizer can be identified, among with another term in the prior. Note that there is a free parameter γ which has an optimum value dependent on the image, number of points and reconstruction order [2]. Since the reconstruction scheme is based on a Sobolev type inner product it is referred to as a STIP (Sobolev Type Inner Product) reconstruction for the remainder of this paper.

Both reconstruction methods try to "complete" the reconstructed image in areas where no constraints are present by some regularization. However, the first method minimizes a global constraint and is implemented in an iterative way (but with a nice statistical background for the prior) while the second method uses an alternative inner product which results in a linear system.

3 Feature Points and Their Feature Strength

This section describes the various feature points used for image reconstruction with the previously described algorithms. Note that in principle every point in scale space can be used but it is shown by various authors that some feature points are more suitable than others [5, 3].

Blobs. The first feature points used in this paper are blobs as proposed by Lindeberg [6]. Blobs have been successfully used as feature points in the work by Nielsen and Lillholm [7]. They are defined by:

$$\max_{x,\sigma}(\sigma^2 \Delta I) \tag{8}$$

The feature strength of blobs is defined as: $|\Delta I|$.

Top Points. Top points have previously been used as feature points in work by various authors [7, 5, 3]. A top point is a critical point at which the determinant of the Hessian is zero:

$$\begin{cases} \nabla I = 0 \\ \det(\mathbf{H}) = 0 \end{cases} \tag{9}$$

Defining the feature strength of a top point is not as straight forward as defining it for blobs. We assume that feature strength can be linked to the stability of a top point.

Instabilities of critical points arise in areas in the image that consist of almost constant intensity (genericity implies that flat plateaux do not occur in the image). On the other hand, critical points are more stable in an area with structure. The amount of structure contained in a *spatial* area around a critical point can be quantified by the *total (quadratic) variation* (TV) norm over that area [1]. By using a spatial Taylor series around a considered critical point the TV-norm simplifies to Eqn. (10) which is referred to as the *differential TV-norm* [8].

$$\text{diff tv} = \sigma^4 Tr(\mathbf{H}^2) \tag{10}$$

in which $\mathbf{H}$ denotes the Hessian matrix and σ denotes the scale at which the stability of the critical point, or top point in this context, is calculated.

Besides the top points of the original image, top points of the gradient magnitude of the image, the Laplacian of the image or higher order differential invariants of the image also yield sparse sets of points in scale space. These can also be used as feature point sets for our reconstruction algorithms.

Scale Space Saddle Points. Scale space saddle points are critical points in scale space, for which not only the spatial derivatives, but also the scale derivative is zero:

$$\begin{cases} \nabla I = 0 \\ \Delta I = 0 \end{cases} \tag{11}$$

Scale space saddle points have been used for (pre)segmentation by Kuiper and Florack [4]. Since they are critical points just like the top points, the same measure of stability, Eqn. (10), can be used.

Figure 1 shows all the different points for an image of Lena's eye projected on the original image. In each feature point a number of measurements can be used to constrain the reconstruction. In this paper, we use measurements in the form of an inner product of the image with a derivative of a Gaussian filter, resulting in a derivative of the image at a certain spatial position and scale. Since the image can be locally well described by a Taylor expansion, a natural choice of measurements is the *local N-jet* (Complete set of derivatives from 0-th order up to N-th order) in each feature point.

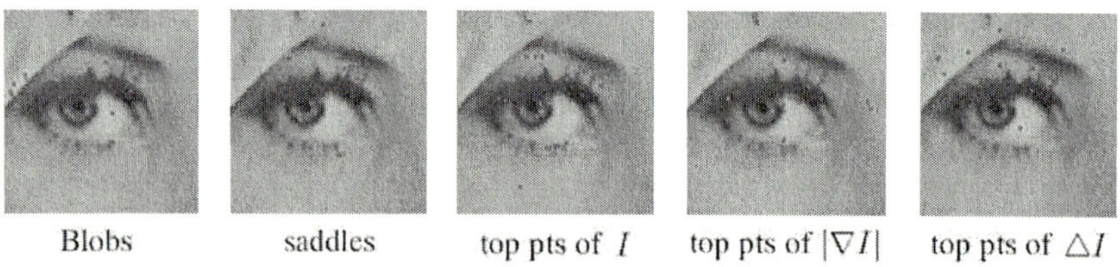

Fig. 1. Examples of the different points used for reconstruction, projected on the lena64 image. From left to right: Blobs, scale space saddles, top points of image I, top points of $|\nabla I|$ and top points of $\triangle I$. For all feature points, the 19 strongest (relative to their respective feature strength) points are shown

4 Experimental Results

Experiments are done only using two images. The first one is a sub image of the famous Lena image and has dimensions of 64x64 pixels. The second image is a magnetic resonance image of the human brain with dimensions of 128x128 pixels. This image has a large homogeneous black background and therefore is not considered a *generic* image. It is, however, a real life image and for future medical applications it is important to investigate reconstructions of this image. Figure 2 shows the two images. In all cases up to fourth order derivatives (4-jet) were used as measurements in each point unless stated differently. Note that for real quantitative conclusions the experiments should be run on a larger number of different images.

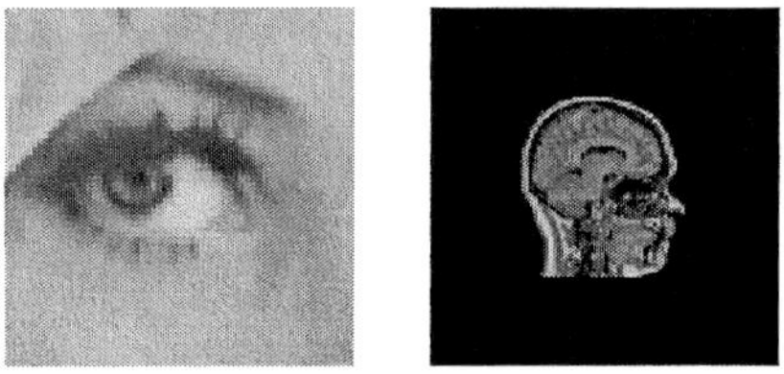

Fig. 2. Our test images lena64 and MR128

4.1 Ordering of the Feature Points

The first experiment is to show the influence of the ordering of the feature points on the reconstruction quality. To measure the reconstruction quality the *Root Mean Square Error* (RMSE) is used on the normalized images:

$$\text{RMSE} = \frac{1}{\sqrt{MN}}\sqrt{\sum (I - \hat{I})^2} \tag{12}$$

with I the original image, $\hat{I}$ the reconstructed image and M and N the dimensions of the images. Reconstructions were done from top points, ordered by their differential TV norm and from blobs, ordered by their feature strength using both reconstruction algorithms. Figure 3 shows the resulting RMSE for the lena image. Figure 4 shows some examples of the reconstructions for the lena64 image. For the STIP reconstruction $\gamma = 32$ is used.

In general one can observe that the brownian reconstructions outperform the STIP reconstructions both in terms of RMSE as well as visually. Comparing reconstructions from a low number of feature points a human observer would prefer reconstructions from top points, since more visually attractive information is present. However, a large clear geometric feature is missing (The ridge to the left of the eye), which is present in the blob reconstruction. The RMSE for the Brownian reconstruction shows almost no difference between the top points and the blobs for a low number of points since apparently the same amount of information is missing. This shows that in case of different feature points the RMSE does not always reflect the visual quality of the reconstruction. For

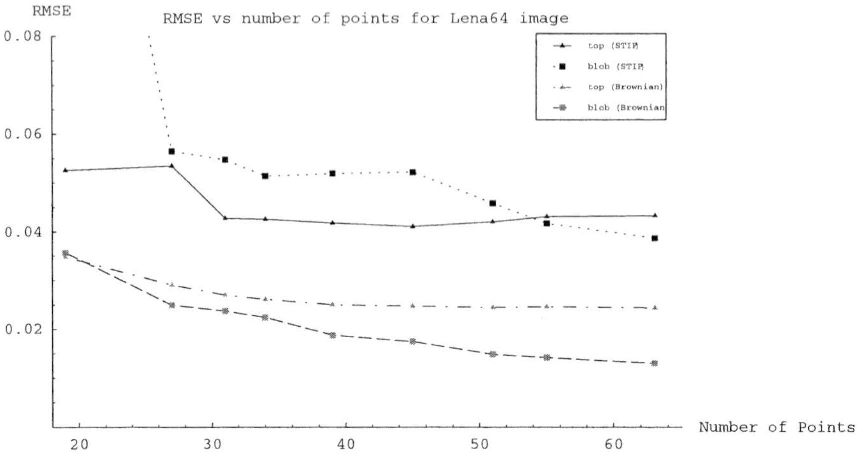

Fig. 3. RMSE of reconstructed Lena64 image with respect to number of reconstruction points

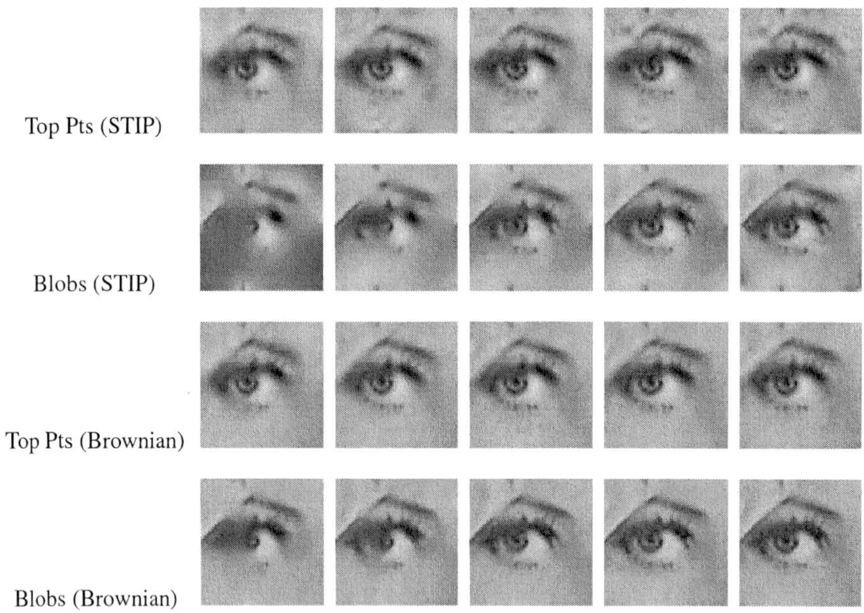

Fig. 4. Examples of reconstructed Lena64 image with respect to number of reconstruction points. The top row is reconstructed from top points using STIP reconstruction, The second row is reconstructed from blob points using STIP reconstruction, the third and fourth row are reconstructions using Brownian reconstruction from the same number of resp. top points and blobs. The columns from left to right show reconstructions from resp. 19,31,39,51 and 63 feature points

a higher number of feature points the top points are outperformed by the blobs both in RMSE as visually. From figure 4 one can conclude that the ordering of the top points is performing well. The reconstruction from the 20 most stable top points is visually very close to the reconstruction from all 63 top points, which is not the case for the blobs.

The experiment is repeated for the MR128 image. Figure 5 shows the resulting RMSE for the MR128 image and figure 6 shows some examples of the reconstructions for the MR128 image. Again for a low number of feature points the top points seem to perform better than blobs in terms of visual quality, while for a high number of feature points it is the other way around. Both reconstruction methods have a slightly higher RMSE due to intensity "leakage" into the black background which can also be seen in figure 6. As can be expected this problem is less present in the STIP reconstruction, since it is suppressed by the first term of Eqn. (7). Also from figure 6 one can conclude that the ordering of top points is performing well since most of the visually interesting image information is present in the reconstruction from the 55 most stable top points.

4.2 Robustness of the Reconstruction from Top Points

An experiment is done to investigate the robustness of the reconstruction from top points under noise. This experiment is to simulate detection errors by adding random noise to the position of the top points after detection but before calculating the measurements. Figure 7 shows the RMSE for noise on the top point positions up to 2 pixels for reconstructions from resp. 20, 30 and 40 top points (ordered by their feature strength). All reconstructions were made using the STIP algorithm with $\gamma = 32$ using up to second order derivatives. Figure 8 shows some examples of the experiments done with 30 points.

Figure 7 shows that for a small perturbation on the top point positions, the RMSE does not change much, which is also reflected in the visual quality of the images in figure 8. For perturbations up to 2 pixels the RMSE stays below 0.09 and the visual quality of the reconstructions is still very reasonable. From figure 7 one can also conclude that for small perturbations reconstructions from 20 points have a slightly higher RMSE than reconstructions from 30 or 40 points, as expected. However, the reconstructions from 20 points have an almost constant RMSE up to perturbations of 1 pixel, while reconstructions from 30 or 40 points only have a constant RMSE for perturbations up to approximately 0.6 pixels. This shows that the proposed ordering is strongly related to the stability of the points. Note that in some cases the random noise on the position can result in near dependant equations, which can result in some artefacts as is shown in figure 8. Also note that there seems to be a local maximum in the RMSE for perturbations of approximately 1-1.25 pixels. This is related to the spatial distribution of the top points in the image and the previously mentioned problem of dependant equations.

4.3 Influence of Points Used for the Reconstruction

This experiment is to compare reconstructions from different feature points. Table 1 shows the results. The first two columns show reconstructions using the Brownian reconstruction, the last two columns show reconstructions using the STIP reconstruction. For the blobs, the points with the highest feature strength were used. For the top points

problem there, is the seamless integration of the partial embeddings into a single continuous mapping. Hence, we are motivated to extend the current scheme to better handle complex meshes. It can also be used as a component of a partitioning based scheme.

For textures such as the soccer ball shown in Fig. 2, we denote σ, the intrinsic scale of the texture image. σ is the largest distance that should be preserved by the embedding, while larger distances can be distorted. For example, for the soccer ball, σ is the width of the pentagon.

This approach is related to local regression [21], where functions are estimated locally by polynomials and the local scale is given by the scale parameter σ. Following the local regression formulations in [21] the weights were computed using the Nadaraya-Watson kernel

$$K_\sigma(x) = D\left(\frac{|x|}{\sigma}\right)$$

with

$$D(t) = \begin{cases} \frac{3}{4}\left(1 - t^2\right) & \text{if } |t| \leq 1 \\ 0 & \text{otherwise.} \end{cases}$$

Equation (4.2) is reduced to

$$\varepsilon = \sum_{i,j} w_{ij}\varepsilon_{ij}^2 = \sum_{|D_G(i,j)|<\sigma} K_\sigma\left(D_G(i,j)\right)\varepsilon_{ij}^2. \tag{4.6}$$

Note that by using the weighted MDS (WMDS), the scheme preserves only the distances $\left|D_G\left(\theta^i, \theta^j\right)\right| < \sigma$. Thus, there is no need to store all of the relative distances and compared to the unweighted MDS and [14] the matrix of distances becomes sparse. This makes the weighted MDS more suitable for embedding large meshes.

Intrinsic Scale Estimation. For a given spherical texture map the intrinsic scale can be estimated using spherical harmonics. This is a set of basis function defined on the surface of the sphere analogous to Fourier analysis. Spherical harmonics have been used in graphics to efficiently represent the bidirectional reflection distribution function

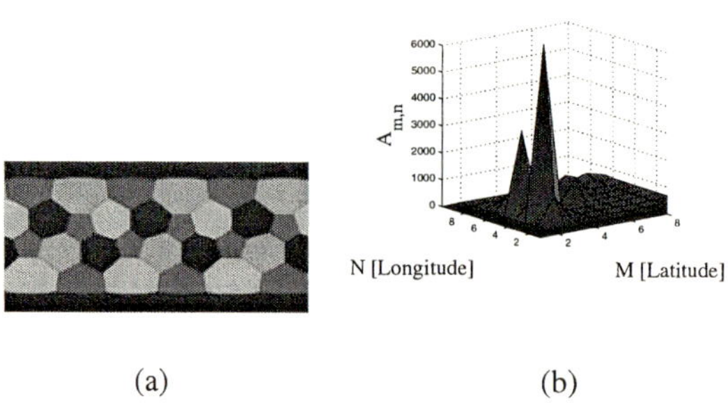

Fig. 3. Computing the intrinsic scale of a texture using spherical harmonics. The spherical texture map used in Fig. 2 is given in (a) in spherical coordinates. The spherical harmonics coefficients are shown in (b). The maxima is detected in $(m = 2, n = 4)$, corresponding to 2 cycles in the latitude (vertical axis of (a)) and 4 cycles in the longtitude (horizontal axis of (a))

(BRDF) of different materials. Thus by computing the spherical harmonics coefficients, one can estimate the dominant spherical frequencies corresponding to the intrinsic scale σ. Such an example is given in Fig. 3 where the spherical harmonics coefficients of Fig. 3a are given in Fig. 3b.

We note that the Weighted MDS can also be applied to planar least squares MDS, where the intrinsic scale σ can be estimated by the Fourier transform of a given texture map and detecting the dominant frequencies.

5 Experimental Results

The proposed technique was tested using several meshes. We start by comparing the proposed scheme to the MDS based planar embedding given in [14]. The 'face' mesh (6000 vertices) has a half-sphere-like topology and the 'David' mesh (10000 vertices) is used to compare the embeddings of a full-sphere-like object. We show the applicability of the scheme to medical visualization by annotating and texture mapping the 'cortex' mesh (6000 vertices) and a segment of it. Finally, we apply the WMDS to texture mapping the 'Stanford bunny' (6000 vertices) and show the improvement compared to the regular MDS.

Given θ^i (the spherical embedding of a vertex V_i), the color associated to V_i was determined by sampling the spherical texture image I at θ^i. Subpixel values of θ^i were handled by bilinear interpolation of I. The visualizations of the texture mapped meshes were produced by the VTK mesh viewer [27]. The multi-resolution MDS scheme was used in all of the simulations, with three resolution scales. At each scale the computation continued until $\Delta\varepsilon$, the reduction of the embedding error, became less than 10^{-7}.

Figure 4 and Table 1 show the visualizations artifacts and the embedding error of the 'face' mesh, respectively. This is a half-sphere-like object with low curvature, except for the nose area which is a high curvature structure. The embedding was computed using the regular (unweighted) MDS. Figures 4a and 4b, show similar visual quality, except for the nose area, where the planar MDS shows a larger distortion. Thus, both embedding have a similar average error.

(a) Spherical MDS (b) Cartesian MDS

Fig. 4. The Spherical and Planar MDS applied to the 'face' mesh (6000 vertices). Note the embedding error of the cartesian MDS around the region of the nose

Table 1. Embedding error comparison between cartesian and spherical embeddings

Name	Vertices#	Embedding type	Embedding error
Face	3000	spherical	$3.1 \cdot 10^{-5}$
Face	3000	cartesian	$3.4 \cdot 10^{-5}$
Face	6000	spherical	$1.15 \cdot 10^{-5}$
Face	6000	cartesian	$1.25 \cdot 10^{-5}$

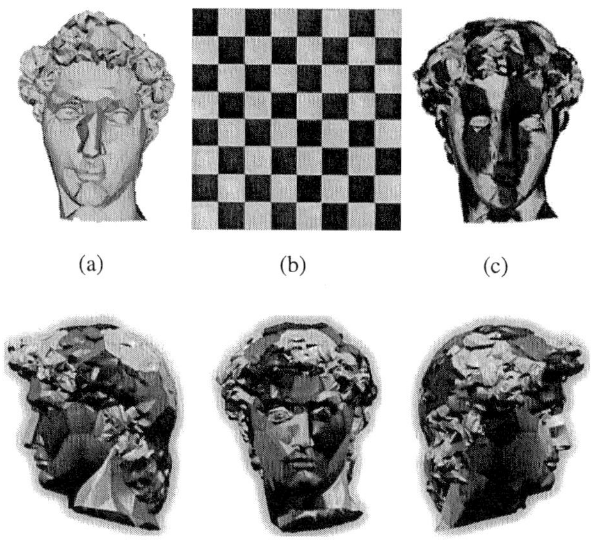

(a) (b) (c)

(d) (e) (f)

Fig. 5. Texture mapping of a sphere-like mesh. (a) The head of the David mesh (1000 vertices). (b) The planar chess texture used for the planar texture mapping in (c). (c) The results of a LSMDS based planar texture mapping. (d), (e) and (f) Texture mapping results using the spherical MDS

A significant improvement is evident in the embedding of the full-sphere-like 'David' mesh given in Fig. 5. The planar embedding results in significant embedding error, while the spherical MDS gives reasonable results.

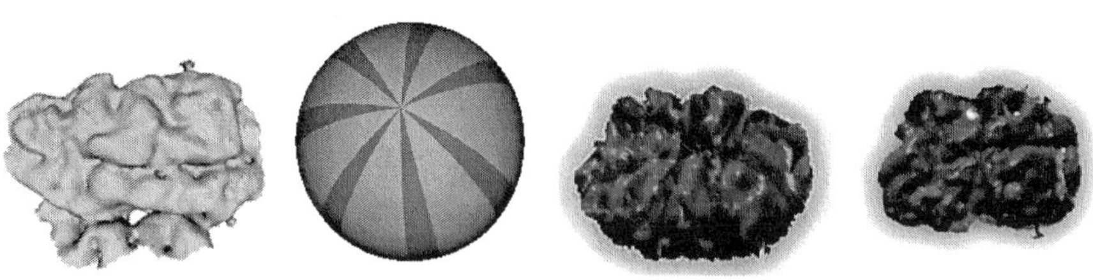

(a) (b) (c) (d)

Fig. 6. Annotation of parts of the Cortex

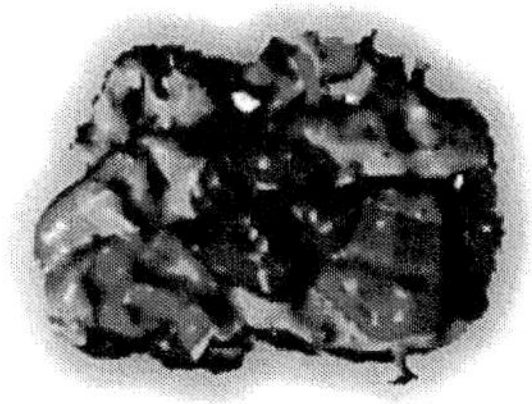

Fig. 7. Coloring part of the cortex. Note the accuracy with which the pentagon patterns are preserved

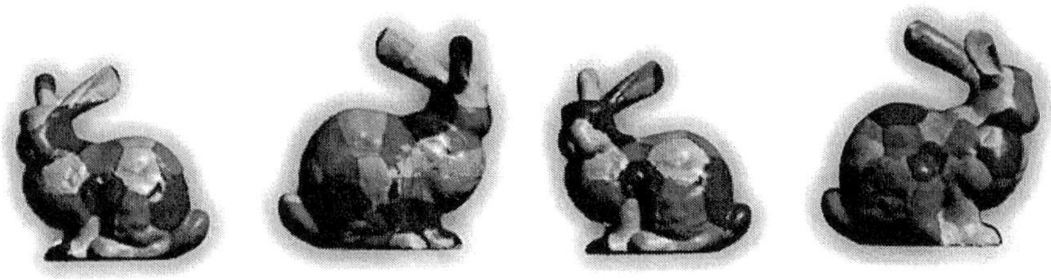

(a) (b) (c) (d)

Fig. 8. Local spherical MDS texture mapping results. (a) and (b) are the results of the regular spherical MDS while (c) and (d) are both sides of the Stanford bunny textured mapped using the local spherical MDS. Note the improved mapping in (b)

Figures 6 and 7 apply the spherical MDS to the unfolding of the brain cortex for medical visualization. We were able to annotate the cortex in Fig. 6 and color it in Fig. 7. Note the lack of geometrical distortions in these figures.

Finally, we compare the results of the spherical and weighted spherical MDS using the 'Stanford bunny'. This object is far from a sphere and has points of negative and positive mean curvature. Texture mapping it using the spherical MDS, results in significant embedding errors, as depicted in Figs. 8a and 8b. There is no embedding that minimizes the difference in both the local and global distances. The results of the WMDS applied with $\sigma = 0.1$ are demonstrated in Figs. 8c and 8d, that depict the improved texture mapping. Fig. 8e depicts the texture mapping of the earth's texture.

These results asses the validity of the mapping (bijectivity). For patches with high isoperimetric distortion (sphere-like shapes), classical geodesic-MDS often gives wrong results. This is not evident in our results. Note that the measure we optimize is a global

Table 2. Timing results for the single scale and multiscale MDS. The multi-scale scheme significantly reduces the computation time

Mesh size [vertices#]	1000	2000	3000	6000	10000
Single scale	4.4s	11.4s	32.7s	166.8s	420.0s
Multiscale	2.0s	5.2s	25.2s	67.1s	217.5s

one. The triangulation is a particular numerical representation of the geometry. While mapping a sphere-like surface onto a plane one could experience face flips. However, choosing the sphere as a target simplified map, such effects are reduced drastically. Using least squares scaling one could restrict the triangles to preserve their orientation on the sphere. However, we did not find such a restriction a necessity in our applications.

The timing results for the single and multiscale MDS are given table 2. The algorithms were implemented using non-optimized C++ and the multi-scale MDS implementation uses the same code as the single scale scheme for the computations within each resolution scale. The 'Stanford bunny' (6000 vertices) was used for the timing measurements on a 2.8GHz PC. Evidently, the multi-scale scheme improved the algorithm's timing by 50-100 percent.

6 Summary

An efficient and accurate method for embedding surfaces onto a sphere was presented. The method is based on the fast marching on triangulated domains algorithm followed by multi-dimensional scaling, and was shown to provide improved visual results compared to planar flattening. Furthermore, we presented a weighted MDS formulation which allows us to better handle objects with non-sphere-like geometry. Finally, we derived a fast multi-scale optimization scheme for the numerical solution of the problem. In future work we will study the use of the proposed scheme in a partitioning based embedding scheme. The main challenge is to device a partitioning scheme that will allow a continuous mapping between the embedded partitions.

References

1. Arad, M.: Isometric texture mapping for free-form surfaces. Computer Graphics Forum **16** (1997) 247–256 ISSN 1067-7055.
2. Azariadis, P.N., Aspragathos, N.A.: On using planar developments to perform texture mapping on arbitrarily curved surfaces. Computers and Graphics **24** (2000) 539–554
3. Neyret, F., Cani, M.P.: Pattern-based texturing revisited. SIGGRAPH (1999) 235–242.
4. Praun, E., Finkelstein, A., Hoppe, H.: Lapped textures. In: Proceedings of the 27th annual conference on Computer graphics and interactive techniques, ACM Press/Addison-Wesley Publishing Co. (2000) 465–470
5. Gu, X., Wang, Y., Chan, T.F., Thompson, P.M., Yau, S.T.: Genus zero surface conformal mapping and its application to brain surface mapping. IEEE Transaction on Medical Imaging **23** (2004)
6. Haker, S., Angenent, S., Tannenbaum, A., Kikinis, R., Sapiro, G., Halle, M.: Conformal surface parameterization for texture mapping. IEEE Trans. on Visualization and Computer Graphics **6** (2000) 181–189
7. Sheffer, A., de Sturler, E.: Parameterization of faceted surfaces for meshing using angle based flattening. Engineering with Computers **17** (2001) 326–337
8. Eck, M., DeRose, T., Duchamp, T., Hoppe, H., Lounsbery, M., Stuetzle, W.: Multiresolution analysis of arbitrary meshes. Computer Graphics **29** (1995) 173–182
9. Maillot, J., Yahia, H., Verroust, A.: Interactive texture mapping. In: Proceedings of the 20th annual conference on Computer graphics and interactive techniques, ACM Press (1993) 27–34

10. Sheffer, A.: Spanning tree seams for reducing parameterization distortion of triangulated surfaces. In: Proceedings of the Shape Modeling International 2002 (SMI'02), IEEE Computer Society (2002) 61
11. Wolfson, E., Schwartz, E.L.: Computing minimal distances on polyhedral surfaces. 11 (1989) 1001–1005
12. Schwartz, E.L., Shaw, A., Wolfson, E.: A numerical solution to the generalized mapmaker's problem: Flattening nonconvex polyhedral surfaces. 11 (1989) 1005–1008
13. Grossman, R., Kiryati, N., Kimmel, R.: Computational surface flattening: A voxel-based approach. 24 (2002) 433–441
14. Zigelman, G., Kimmel, R., Kiryati, N.: Texture mapping using surface flattening via MDS. IEEE Trans. on Visualization and Computer Graphics **8** (2002) 198–207
15. Wandell, B.A., Chial, S., Backus, B.: Visualization and measurements of the cortical surface. Journal of Cognitive Neuroscience (2000)
16. Cox, M., Cox, T.: Multidimensional Scaling. Chapman and Hall (1994)
17. Borg, I., Groenen, P.: Modern Multidimensional Scaling - Theory and Applications. Springer (1997)
18. Kruskal, J.B., Wish, M.: Multidimensional Scaling. Sage (1978)
19. Kimmel, R., Sethian, J.: Computing geodesic paths on manifolds. Proc. of National Academy of Science **95** (1998) 8431–8435
20. Elad, A., Kimmel, R.: Spherical flattening of the cortex surface. In Malladi, R., ed.: Geometric Methods in Biomedical Image Processing, Springer (2002) 77–90
21. Hastie, T., Ribshirani, R., Friedman, J.H.: The elements of statistical learning: data mining, inference and prediction. Springer (2002)
22. Tenenbaum, J.B., de Silva, V., Langford, J.C.: A global geometric framework for nonlinear dimensionality reduction. Science **290** (2000) 2319–2323
23. Sethian, J.: A review of the theory, algorithms, and applications of level set method for propagating surfaces. Acta Numerica, Cambridge University Press (1996)
24. Gill, P.: Practical Optimization. Academic Press (1982)
25. Mann, S., Picard, R.: Virtual bellows: constructing high quality stills from video, Austin, TX, IEEE International Conference Image Processing (1994) 363–367
26. Melax, S.: A simple, fast and effective polygon reduction algorithm. Game Developer Journal (1998)
27. Schroeder, W., Martin, K., Lorensen, B.: The Visualization Toolkit: An Object-Oriented Approach to 3D Graphics. Prentice Hall (1997)

On Similarity-Invariant Fairness Measures

Alexander Brook[1], Alfred M. Bruckstein[2], and Ron Kimmel[2]

[1] Dept. of Mathematics, Technion, Israel
abrook@technion.ac.il
[2] Dept. of Computer Science, Technion, Israel

Abstract. After introducing the basic principles behind the similarity-invariant smoothness measures for curves and surfaces, with references to the relevant literature, we discuss the ramifications of scale-invariance in various problems of image processing and analysis, and point out some unanswered questions.

1 Introduction

In this paper we consider variational measures of fairness which are invariant under similarity transformations and reparametrizations, that is integrals of the form

$$J(\gamma) = \int_\gamma F(p, C, C'_p, C''_{pp}, \dots)\, dp$$

(and analogous integrals for surfaces) invariant under

- change of parametrization $p = p(\bar{p})$, where p is a smooth monotonically increasing function,
- translation $C = \bar{C} + v$, v being a constant vector,
- rotation $C = U\bar{C}$, for any orthonormal matrix U,
- scaling $C = \alpha\bar{C}$, for any nonzero (usually positive) scalar α.

The combination of the first three properties is usually called Euclidean invariance. Most image processing theories and algorithms are Euclidean invariant, so we will focus on scale invariance.

In this paper we discuss the different (but often confused) meanings of invariance and review the previous work on fairness measures for curves and surfaces, trying to produce an exhaustive list of functionals that are such measures. We then consider the need for (and the possibility of) scale invariance in image processing and image analysis (in particular edge detection and integration).

2 Invariance of a Measure Versus Invariance of a Solution

An important distinction, which is often overlooked in works dealing with invariance of variational solutions, is the difference between the invariance of the

R. Kimmel, N. Sochen, J. Weickert (Eds.): Scale-Space 2005, LNCS 3459, pp. 456–467, 2005.

solution to a minimization problem and the invariance of the functional being minimized.

There are two possible uses for a fairness measure. One is the general question "Is this a nice curve?" or "Which of these two curves looks better?" (see [1–p. 30] for a discussion of definitions of fairness). Another is the specific question "What is the best curve satisfying the given conditions?" Most frequently, this is used in contour completion. The basic setting is of two short segments in the plane that are to be connected in the smoothest way possible. Such segments are usually treated as 1-jets, that is, the only information extracted from them is the position and the tangent.

The invariance of the functional is important if we use the functional as a universal measure of smoothness; we wish to give each curve a number that will be a measure of its smoothness, and this number should be scale-invariant.

On the other hand, if we are only interested in solving a specific problem of finding the most fair curve that connects two given segments, for example, we are only dealing with the minimizers of the functional. It is possible that the minimizers will be scale-invariant while the functional is not.

This phenomenon is remarked upon in [1–p. 72]:

> Minimal Variation Curves are invariant under rigid body transformation and uniform scaling. The *value* of the Minimal Variation Curve functional, however, changes with a change of scale.

Let us consider a very simple example. Say, we want to connect two points by the shortest possible curve. The corresponding functional is just the length of the curve, $\int_\gamma ds$, and the solution is, of course, the straight segment connecting the two points. Both the functional and the solution are invariant under rigid motions.

However, the solution is also scale invariant. Indeed, if we dilate the plane, the straight segment will remain the shortest possible line even after dilation. And yet the functional that defines this curve is not scale invariant, of course!

How does this happen? If we dilate the space by λ, the length functional is multiplied by λ:

$$\int_\gamma ds \quad \mapsto \quad \lambda \int_\gamma ds,$$

and so is the Euler-Lagrange equation:

$$C'' = 0 \quad \mapsto \quad \lambda C'' = 0,$$

but these two equations are equivalent. Thus, a functional which is not invariant under certain group (dilations) possesses an Euler-Lagrange equation which is invariant under this group.

This non-equivalence is mentioned in [2–p. 255] and [3–p. 236]. In both cases, after the theorem that says "every symmetry of the functional is a symmetry of the Euler-Lagrange equation", it is mentioned that the converse is not true and that "the most common examples" are those of scaling transformations. Also, the above theorem (if the functional is invariant then the Euler-Lagrange equation is invariant) is not a trivial one, due to the fact that generally, not all of the Euler-Lagrange equation's solutions are minimizers of the functional.

We do not know whether these are the only examples, or whether it is possible to give a complete classification of such counterexamples. We hope to answer these questions elsewhere.

3 Smoothness Measures for Curves

3.1 Basic Invariants

Classical Similarity Invariants. First of all, if we drop the requirement of invariance to scale, we are left with parametrization and Euclidean invariance. It is well known and easily established (see [4–p. 81] and [5–8.1.1]) that Euclidean and reparametrization invariant functionals must have the form

$$\int_\gamma F(\kappa, \kappa_s, \kappa_{ss}, \dots) ds,$$

where s is the arclength parameter, and κ is the curvature. (For a curve in space we will have to add torsion to the list, but space curves are rarely, if ever, used in image processing.)

From these functionals we want to select those that are invariant to scaling. If we replace C with αC, $\kappa^{(i)}$ becomes $\kappa^{(i)}/\alpha^{i+1}$. Substituting these to F, we see that we are looking for functions F that are homogeneous (of order -1, so that $F\,ds$ is invariant) with respect to α (the same idea is used in [6]). Let us see what this leaves us. One possibility is just to write down the list of these functions:

$$\kappa, \quad \frac{\kappa_s}{\kappa} = -\frac{d}{ds}(\ln|\kappa|), \quad \frac{\kappa_s^2}{\kappa^3}, \quad \frac{\kappa_{ss}}{\kappa^2}, \quad \frac{\kappa_{ss}}{\kappa_s} = -\frac{d}{ds}(\ln|\kappa_s|) \tag{1}$$

and so on.

We can go on with this list, by considering higher order derivatives and higher powers, none of which is a good idea. High order derivatives are numerically challenging; PDEs of order higher than four are nearly impossible to implement. High powers are unreasonable in a smoothness measure because of their high sensitivity to noise (also known as sensitivity to outliers or non-robustness).

We can bring some order to the list (1) with the help of invariants theory. The invariant arclength and curvature for the similarity group are given in [3] as

$$\frac{u_{xx}\,dx}{1+u_x^2} \quad \text{and} \quad \frac{(1+u_x^2)u_{xxx} - 3u_x u_{xx}^2}{u_{xx}^2},$$

which upon inspection turn out to be $d\theta = \kappa\,ds$ (notice that the turn angle θ itself is not even Euclidean invariant!) and $\mu = \kappa_s/\kappa^2$. Appropriately, both quantities are homogeneous of degree 0 in the sense described above. Thus, the invariant measures are $\int \Phi(\mu, \frac{d\mu}{d\theta}, \dots) d\theta$, which generates the same list as above; in particular,

$$d\mu = \left(\frac{\kappa_{ss}}{\kappa^2} - \frac{\kappa_s^2}{\kappa^3}\right) d\theta.$$

Winding Number and Total Absolute Curvature. The integral $\frac{1}{2\pi}\int_\gamma \kappa\, ds$ is called the winding number, for it equals $\frac{1}{2\pi}\int_\gamma d\theta$, which is the number of the turns the (closed) curve makes. The integral $\int \kappa\, ds$ on an open piece of a curve is the turn angle. This in particular means that κ is a null Lagrangian, that is, an attempt to write down the Euler-Lagrange equation for the winding number will result in $0 = 0$. In any case, the winding number is obviously unsuitable as a smoothness measure.

We can consider another similarity invariant, the total absolute curvature $\int_\gamma |\kappa|\, ds$. The excess of this functional above 2π shows how much the curve wiggles before closing upon itself ([7] calls it "angular total variation"). This quantity was a subject of some research; see the review [8]. The relevant result is that $\int_\gamma |\kappa|\, ds \geqslant 2\pi$ for any closed curve, and the equality holds if and only if γ is a convex curve. Thus, in principle, it is a fairness measure, and the corresponding flow should bring any closed curve to a convex one, which is a desirable behavior for a short time. However, $|\kappa|$ is also a null Lagrangian, except at the inflection points.

The article [9] gives an algorithm for minimizing $\int_\gamma (a + b|\kappa|)\, ds$ by a direct construction of the minimum, which is a polygonal path. This approach will not work for $a = 0$.

Other Invariants. First of all, there are two functionals that are integrals of full differentials: the similarity arclength $\int d\theta$ (the turn angle) and $\int \mu\, d\theta = \int \frac{\kappa_s}{\kappa} ds = \int d\ln|\kappa|$. Since they result in 0 for any closed curve, they are unsuitable for our needs.

What is left is $\int \mu^2\, d\theta = \int \frac{\kappa_s^2}{\kappa^3} ds$. It is unclear whether it has any meaning in familiar terms.

3.2 The Idea of Weiss and Others

So far, the situation is far from satisfying. What we want is something that is invariant, numerically feasible and has a clear meaning. Being unable to achieve this goal within the framework of standard variational problems, researchers turned to other possibilities, namely, to non-local functionals. The following idea was first published in [10], then in [11] and [12].

If we have a functional J that is Euclidean invariant, parametrization invariant, and homogeneous with respect to scaling (α from Sect. 3.1), then $(\int_\gamma ds)^p J$ is similarity invariant for some value of p. The most popular example is the scale-invariant elastica (called Minimal Energy Curve in [12]) $\int_\gamma ds \int_\gamma \kappa^2\, ds$.

3.3 Scale-Invariant Minimal Energy Curve (Elastica)

The functional $\int_\gamma \kappa^2\, ds$ gives the bending energy of an elastic rod. The usual boundary conditions are derivatives at the endpoints, which is the common setting of the contour interpolation problem. Sometimes the constant length constraint $\int_\gamma ds = L$ is added.

The paper [11] introduces length-constrained elastica and scale-invariant elastica, and provides the Euler-Lagrange equations in terms of the turn angle. Another measure suggested there is the weighted functional $L \int_0^L W(s)\kappa^2(s)\,ds$; the article does not mention the now obvious possibility to select an image dependent weight W and use this for segmentation.

The article [12] introduces all four functionals (Minimal Energy Curve and Minimal Variation Curve and their scale-invariant counterparts) in a quest for a fair and stable curve interpolation. The main rationale for the scale-invariant measures is that they are more stable. The minimization is by quintic Hermite splines (Hermite splines readily accommodate the constraints provided by the boundary segments). A different discretization approach to these functionals is offered in [13].

In [14] an area constraint is added to the usual length constraint on elastica:

$$\alpha \int \kappa^2\,dl + \mu\left(\int dl - L\right) + \sigma\left(\iint_{\text{interior}} dx - A\right)$$

(notice that the constraints are not squared). The authors are mainly interested in the dependence of the minimizers' behavior on the Lagrange multipliers μ and σ. Their basic approach is through the phase plane of the Euler-Lagrange equations for κ, that is the plane κ, κ'; the basic ideas were present in [15], but here they are taken much further.

Non-closed elastica with fixed endpoints or fixed endpoints and end directions is the subject of [16]. Existence is shown and explicit formulas are given.

There is a standing conjecture of De Giorgi (mentioned in [17]) that $\int \kappa^2\,dl$ can be approximated in the sense of Γ-convergence by

$$\frac{1}{\varepsilon}\int_\Omega \left(2\varepsilon\Delta z - \frac{W'(z)}{4\varepsilon}\right)^2 dx\,dy,$$

where $W(z) = (1 - z^2)^2$, and the curve is the zero level set of $z(x, y)$. Some results on this can be found in [18].

3.4 Scale-Invariant Minimal Variation Curve

The functional $\left(\int_\gamma ds\right)^3 \int_\gamma \kappa_s^2\,ds$ is proposed in [12] as a scale-invariant version of the minimal variation curve $\int_\gamma \kappa_s^2\,ds$ and later studied numerically in [13]. The generated curves are usually more smooth than elasticae, of course, and the scale-invariant version is more stable then the usual Minimal Variation Curve [12].

A recent article [19] deals with the problem of curve completion. The authors propose a scale-invariant completion model which they think is scale-invariant Minimal Variation Curve; unfortunately, they are wrong. Their solution, namely Euler spiral, satisfies $\kappa_{ss} = 0$; the authors erroneously claim that this is the Euler-Lagrange equation of the Minimal Variation Curve functional $\int_\gamma \kappa_s^2\,ds$. The correct equation $\kappa^{(4)} + \kappa^2\ddot{\kappa} - \frac{1}{2}\kappa\dot{\kappa}^2 = 0$ is given in [20], and is also contained in [21], derived by the methods from [15]. Note that since the equation $\kappa_{ss} = 0$

is homogeneous it is indeed, obviously, scale-invariant. This is similar to the situation we have described in Sect. 2.

In [22] the energy $\int(\alpha\kappa_s^2+\beta)\,ds$ arises as the mode of a certain type of random walk, that attempts to model curve completion.

3.5 Other Possibilities

One possibility to go in a different (though not necessarily more promising) direction is to use the area inside the curve instead of the length, for example $[\text{area}(\gamma)]^{1/2}\int_\gamma \kappa^2\,ds$. The most obvious invariant using the area is the isoperimetric ratio $\frac{\text{length}^2}{\text{area}}$. Of course, it is only suited for closed curves [23].

Also, if we want a solution to a specific problem, we might as well consider flows (PDEs) that do not come from an Euler-Lagrange equation of a functional. For some examples see [19, 24, 25].

4 Smoothness Measures for Surfaces

4.1 Integrals of Curvatures

Just as with curves, we are led to the study of integrals of powers of curvatures, namely of the Gaussian curvature K and the mean curvature H.

K is a null Lagrangian; the Gauss-Bonnet theorem states that on a closed surface $\int_S K = 2\pi\chi(S)$, where χ the Euler characteristic of the surface.

The integrals $\int_S K^2$ and $\int_S(K^2+\lambda)$ are treated in [25]. The integral of the mean curvature vector is zero on a closed surface: $\int_S H\hat{n} = \mathbf{0}$ [26]. The integral of H itself is widely studied in integral geometry (e.g. [27]) and gives, in a sense, the mean width of the body enclosed by a surface. It is unlikely that it will make a good smoothness measure.

The most widely studied is the Willmore functional $\int_S H^2$, which is similarity invariant. The book [28] gives some basic results on this functional, including the Euler-Lagrange equation, and provides an introduction to the literature. In particular, there are works on numerical minimization of the Willmore functional.

4.2 Other Invariant Measures

Besides Willmore functional $\int_S H^2$ there are other similarity invariant smoothness measures for surfaces.

The paper [12] proposes the following "minimal variation surface" scale-invariant functional:

$$\int dS \int \left[\left(\frac{\partial\kappa_1}{\partial\hat{e}_1}\right)^2 + \left(\frac{\partial\kappa_2}{\partial\hat{e}_2}\right)^2\right] dS.$$

Here $\frac{\partial\kappa_1}{\partial\hat{e}_1}$ is a directional derivative of a principal curvature in the corresponding principal direction (see [1] for details).

The isoperimetric ratio can be generalized to closed surfaces as $\frac{\text{area}^3}{\text{volume}^2}$ [23].

5 Smoothness Measures for Images and Scale-Invariant Image Processing

5.1 Scale-Invariant Smoothness Measures for Images

If we accept the "morphological" point of view, and regard an image as a collection of isophotes, we can generate a smoothness measure on images from any smoothness measure on curves, by integrating over the intensity range: $\int_{\mathbb{R}} f(I^{-1}(u))\,du$, where $I : \Omega \to \mathbb{R}$ is the image. Using the coarea formula we obtain for $f(\gamma) = \int_\gamma ds$ and for $f(\gamma) = \int_\gamma |\kappa|^p\,ds$

$$\int_{\mathbb{R}} \int_{I^{-1}(u)} du = \int_\Omega |\nabla u|\,dx\,dy,$$

$$\int_{\mathbb{R}} \int_{I^{-1}(u)} |\kappa|^p\,du = \int_\Omega |\kappa|^p |\nabla u|\,dx\,dy = \int_\Omega \left|\nabla \cdot \frac{\nabla u}{|\nabla u|}\right|^p |\nabla u|\,dx\,dy.$$

These formulas appear in [9] and [29] as a justification of a functional

$$\int_\Omega (a + b|\kappa|^p)|\nabla u|\,dx\,dy,$$

used to solve the problem of disocclusion. To obtain a scale-invariant functional, we have to multiply the functionals instead of adding them, arriving at

$$\int_\Omega |\nabla u|\,dx\,dy \cdot \int_\Omega \kappa^2 |\nabla u|\,dx\,dy \qquad (2)$$

and

$$\left(\int_\Omega |\nabla u|\,dx\,dy\right)^3 \cdot \int_\Omega \kappa_s^2 |\nabla u|\,dx\,dy$$

(note that $\kappa_s = \nabla\left(\nabla \cdot \frac{\nabla u}{|\nabla u|}\right) \cdot \frac{(u_y, -u_x)}{|\nabla u|}$).

5.2 Do we Want Image Processing to be Scale Invariant?

It seems difficult to really justify the need for scale invariance in image processing. Our main objection to the idea from [11] that

> a small circle should be considered as smooth as a large one

is that in an image, the ultimately small one-pixel circle is most probably due to noise, and most image restoration techniques rely on this distinction in scale between noise and signal. Thus, scale-invariant image processing will be very noise-sensitive, if not unstable.

We did not check this objection numerically. It might ultimately turn out that the regularizing effect of a discretization will overcome these problems somehow.

Also, it is possible to achieve partial scale invariance by building an image processing algorithm on a pyramid, that is, on a range of scales; in this case, we can decide explicitly on the minimal scale at which we "stop the invariance". But complete scale invariance means inability to detect noise.

The only kind of image processing that probably needs scale invariance is inpainting (disocclusion) [30, 31]. As noted (in a somewhat different context) in [31]:

> In application, an average image often contains objects of a large dynamic range of scales. Hence in most inpainting problems, it is commonly found that "slim" objects are broken by the inpainting domains even though the domains themselves are small (to human observers). A good inpainting scheme should encourage the connection of those broken slim objects.

Still, so far there are no inpainting algorithms that are completely invariant.

5.3 What is Similarity-Invariant Image Processing?

What condition should an image processing algorithm satisfy to be called "scale-invariant"? There are at least two answers to this question, both unsatisfactory in our opinion.

The first possibility is to demand that the processing results on a scaled version of an image will be the same as the results on the image itself, up to scaling. This is the usual commutativity demand for invariance. The only problem with this criterion is that it is too easily satisfied—we just scale all images to a fixed size before processing. Usually this is done even without actually scaling the image, which will introduce errors, but by selecting the image size as the unit of measurement [32, 33].

Another variant on basically the same principle is to ask that an object will be processed the same, regardless of its size in a given picture. This, again, is customarily done by selecting some characteristic dimension of the object as the unit length for the algorithm [34, 35].

On the other hand, we can ask for an algorithm that will act invariantly on each object in the image, regardless of the scale, this without trying to actually measure the object as part of the algorithm. This is what we see in most Euclidean invariant algorithms, so it seems to be a reasonable requirement. For example, if two identical chairs are seen in an image from different distances, we would like to have in the processed image two corresponding parts, differing only by scale, and that without any attempt on the part of the algorithm to identify the objects and their dimensions.

If we want to have a variational model to have this property, the functional must be local. In particular, we can not allow products of non-trivial integrals, like in (2). We are not aware of any image processing or image analysis algorithm, which is scale-invariant in this sense.

2. Olver, P.J.: Applications of Lie groups to differential equations. Second edn. Volume 107 of Graduate Texts in Mathematics. Springer-Verlag, New York (1993)
3. Olver, P.J.: Equivalence, invariants, and symmetry. Cambridge University Press (1995)
4. Giaquinta, M., Hildebrandt, S.: Calculus of variations. I. Springer-Verlag (1996)
5. Giaquinta, M., Hildebrandt, S.: Calculus of variations. II. Springer-Verlag (1996)
6. Bruckstein, A.M., Katzir, N., Lindenbaum, M., Porat, M.: Similarity-invariant signatures for partially occluded planar shapes. Int. J. Comp. Vision **7** (1992) 271–285
7. Ambrosio, L., Masnou, S.: A direct variational approach to a problem arising in image reconstruction. Interfaces Free Bound. **5** (2003) 63–81
8. Willmore, T.J.: Tight immersions and total absolute curvature. Bull. London Math. Soc. **3** (1971) 129–151
9. Masnou, S.: Disocclusion: a variational approach using level lines. IEEE Trans. Image Processing **11** (2002) 68–76
10. Weiss, I.: 3D shape representation by contours. Comp. Vis. Graphics Image Proc. **41** (1988) 80–100
11. Bruckstein, A.M., Netravali, A.N.: On minimal energy trajectories. Comp. Vision Graphics Image Proc. **49** (1990) 283–296
12. Moreton, H.P., Séquin, C.H.: Scale-invariant minimum-cost curves: fair and robust design implements. Computer Graphics Forum **12** (1993) C–473–C–484 Eurographics '93 (R. J. Hubbold and R. Juan, eds).
13. Bruckstein, A.M., Holt, R.J., Netravali, A.N.: Discrete elastica. Applicable Analysis **78** (2001) 453–485
14. Arreaga, G., Capovilla, R., Chryssomalakos, C., Guven, J.: Area-constrained planar elastica. Phys. Rev. E **65** (2002)
15. Griffiths, P.A.: Exterior differential systems and the calculus of variations. Volume 25 of Progress in Mathematics. Birkhäuser Boston (1983)
16. Linnér, A.: Existence of free nonclosed Euler-Bernoulli elastica. Nonlinear Anal. **21** (1993) 575–593
17. Esedoglu, S., Shen, J.: Digital inpainting based on the Mumford-Shah-Euler image model. European J. Appl. Math. **13** (2002) 353–370
18. Loreti, P., March, R.: Propagation of fronts in a nonlinear fourth order equation. European J. Appl. Math. **11** (2000) 203–213
19. Kimia, B.B., Frankel, I., Popescu, A.M.: Euler spiral for shape completion. Int. J. Comp. Vision **54** (2003) 159–182
20. Anderson, I.M.: The variational bicomplex. Academic Press (to appear)
21. Cheung, W.S.: C^∞-invariants on loop spaces. Proc. Amer. Math. Soc. **100** (1987) 322–328
22. August, J., Zucker, S.W.: Sketches with curvature: the curve indicator random field and Markov processes. IEEE Trans. Pattern Analysis and Machine Intelligence **25** (2003) 387–400
23. Delingette, H.: On smoothness measures of active contours and surfaces. In: IEEE Workshop on Variational and Level Set Methods (VLSM'01). (2001) 43–50
24. Chopp, D.L., Sethian, J.A.: Motion by intrinsic Laplacian of curvature. Interfaces Free Bound. **1** (1999) 107–123
25. Polden, A.: Curves and surfaces of least total curvature and fourth-order flows. Dissertation, Universität Tübingen (1996)
26. Chen, B.Y.: Mean curvature vector of a submanifold. In: Differential geometry (Proc. Sympos. Pure Math., Vol. XXVII, Stanford Univ., Stanford, Calif., 1973), Part 1. Amer. Math. Soc. (1975) 119–123

27. Santaló, L.A.: Integral geometry and geometric probability. Volume 1 of Encyclopedia of Mathematics and its Applications. Addison-Wesley Publishing Co., Reading, Mass.-London-Amsterdam (1976)
28. Willmore, T.J.: Riemannian geometry. Clarendon Press, Oxford University Press (1993)
29. Chan, T.F., Kang, S.H., Shen, J.: Euler's elastica and curvature-based inpainting. SIAM J. Appl. Math. **63** (2002) 564–592
30. Bertalmio, M., Sapiro, G., Caselles, V., Ballester, C.: Image inpainting. In: Proceedings Of The 27th Conf. on Computer graphics (SIGGRAPH). (2000) 417 – 424
31. Chan, T.F., Shen, J.: Non-texture inpainting by curvature-driven diffusions (2001)
32. Li, S.Z.: Similarity invariants for 3D space curve matching. In: Proceedings of the First Asian Conf. on Computer Vision. (1993) 454–457
33. Werman, M., Weinshall, D.: Similarity and affine invariant distance between point sets. IEEE Trans. Pattern Analysis and Machine Intelligence **17** (1995) 810–814
34. Lisani, J.L., Moisan, L., Monasse, P., Morel, J.M.: Affine invariant mathematical morphology applied to a generic shape recognition algorithm. In: Proceedings of the Int. Symposium on Mathematical Morphology. (2000)
35. Doermann, D., Rivlin, E., Weiss, I.: Logo recognition. Technical Report CS-TR-3145, University of Maryland, College Park (1993)
36. Alter, T.D., Basri, R.: Extracting salient curves from images: An analysis of the saliency network. Int. J. Computer Vision **27** (1998) 51–69
37. Sha'ashua, A., Ullman, S.: Structural saliency: the detection of globally salient structures using a locally connected network. In: Proceedings of the Second Int. Conf. on Computer Vision. (1988) 321–327
38. Tang, C.K., Lee, M.S., Medioni, G.: Tensor voting. In Boyer, K.L., Sarkar, S., eds.: Perceptual Organization for Artificial Vision Systems. Kluwer (2000)
39. Sharon, E., Brandt, A., Basri, R.: Completion energies and scale. IEEE Trans. Pattern Anal. Mach. Intelligence **22** (2000) 1117–1131
40. Kimmel, R., Bruckstein, A.M.: Regularized Laplacian zero crossings as optimal edge integrators. Int. J. Comp. Vision **53** (2003) 225–243
41. Ma, T., Tagare, H.D.: Consistency and stability of active contours with euclidean and non-euclidean arc lengths. IEEE Trans. Image Processing **8** (1999) 1549–1559

On α Kernels, Lévy Processes, and Natural Image Statistics

Kim Steenstrup Pedersen[1], Remco Duits[2], and Mads Nielsen[1]

[1] Image Analysis Group, Department of Innovation,
IT University of Copenhagen, Denmark
kimstp@itu.dk, malte@itu.dk

[2] Department of Biomedical Engineering,
Eindhoven University of Technology, The Netherlands
R.Duits@tue.nl

Abstract. The probability distribution on the set of naturally occurring images is sparse with most of the probability mass on a small subset of all possible images, hence not all images are equally likely to be seen in nature. This can indirectly be observed by studying the marginal statistics of filter responses on natural images. Intensity differences, or equivalently responses of linear filters, of natural images have a spiky distribution with heavy tails, which puts a large proportion of the probability mass on small intensity differences, but at the same time giving a reasonable probability on large differences. This is due to the fact that images consist mostly of smooth regions separated by discontinuous boundaries. We propose to model natural images as stochastic Lévy processes with α kernel distributed intensity differences. We will argue that the scale invariant α kernels of the recently proposed α scale space theory provides a promising model of the intensity difference distribution (or in general linear filter responses) in conjunction with the Lévy process model of natural images.

1 Introduction

By a natural image we mean any image that we may encounter on our walk through life. This include both scenes of man-made objects as well as nature scenes. In Fig. 1, we give two examples of natural images.

One of the most striking statistical properties of natural images is the occurrence of scale invariant (or self similar) statistics [1, 2, 3]. Besides this, results show that natural images are non-Gaussian and that the probability distribution on the space of images is sparse (see e.g. [4]). Statistical models of images taking into account the statistical properties of natural images have proven to be useful in various application areas within image analysis and computer vision, e.g. [5, 6].

Mumford and Gidas [7] suggest that infinitely divisible distributions are the natural tool for stochastic models of images. They suggest the random wavelet

R. Kimmel, N. Sochen, J. Weickert (Eds.): Scale-Space 2005, LNCS 3459, pp. 468–479, 2005.

Fig. 1. Two examples of natural images

expansion, an image formation model under which an image is a linear combination of randomly positioned and scaled random wavelets or object profiles. The model consists of a Gaussian part and a part governed by a compound Poisson process (see e.g. [8] for a def.), hence the model form a stochastic Lévy process (see e.g. [8]). In this paper, we propose a simplified version of this model in which the Lévy measure of the compound Poisson process is the Lévy measure of the α kernel distribution and we ignore the Gaussian part of the model. We suggest that images can be modelled as sample functions from the Lévy process with α kernel distribution on $\mathbb{R}$. The α kernel distribution is interesting, for among other reasons, because it is an α-stable infinitely divisible distribution. Hence, the Lévy process $\{X_t : t \geq 0\}$ is self-similar with $\{X_{at} : t \geq 0\} \stackrel{d}{=} \{a^{1/(2\alpha)} X_t : t \geq 0\}$, where $\stackrel{d}{=}$ denotes equality in law (identically distributed). α kernels are also the only non-trivial symmetric and α-stable distributions on $\mathbb{R}$ with $0 < \alpha \leq 1$.

The α kernel distribution of the image Lévy process is connected to the marginal statistics of intensity differences or in general linear filter responses.

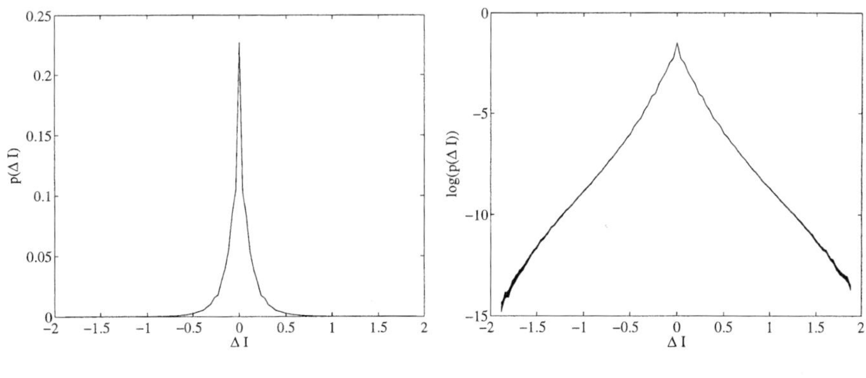

Fig. 2. Histogram of intensity differences $P(\Delta I)$ and log-histogram of intensity differences $\log(P(\Delta I))$. Each curve represent one of the 50 bootstrap histograms of $\Delta_x I$ and $\Delta_y I$ combined (see Sec. 4 for definition of ΔI, $\Delta_x I$, and $\Delta_y I$)

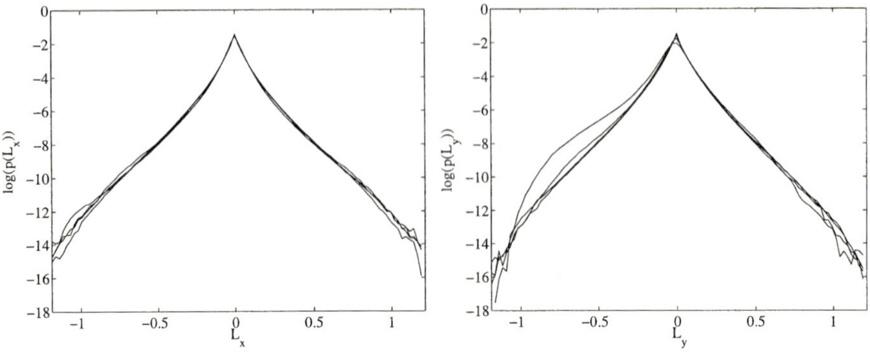

Fig. 3. Log-histograms $\log(P(\cdot))$ of scale space derivatives (L_x, L_y) measured at four different scales, $s^{1/2} \in [2, 32]$

The marginal statistics of natural images tells us something about the structure of the probability distribution on the set of naturally occurring images. For instance the distribution of intensity differences $\Delta_x I(i,j) = I(i,j+1) - I(i,j)$, an approximation to derivatives of the image function, is a spiky distribution with heavy tails (see Fig. 2 and Fig. 3). This first of all shows us the non-Gaussian nature of natural images. Furthermore, the spike at $\Delta_x I = 0$ tells us that smooth regions occur often in images, but at the same time the heavy tails tell us that abrupt changes occur frequently. One interpretation is that the smooth regions have boundaries at which abrupt changes occur.

The marginal statistics of natural images in the form of the distribution of intensity differences, and in general linear filter responses, have been studied and parametric models of the distributions have been proposed. These models include the generalised Laplace distribution proposed by Huang and Mumford [9] and the Bessel K forms proposed by Grenander and Srivastava [10, 11]. Contrary to the generalised Laplace distribution the Bessel K forms [10] are based on an image formation model. This model is called the transported generator model, under which an image is formed by a linear combination of randomly transformed 2D object profiles. This model leads to the Bessel K forms as the distribution of the marginal filter statistics [11].

We propose to use the α kernel distributions as a parametric model of the distribution of linear filter responses on natural images. The underlying Lévy process model accounts for this marginal statistics.

α kernels were first introduced by Pauwels et al. [12] as a family of scale invariant scale space kernels which include the Gaussian kernel as a member. Recently, this family of kernels have been studied by Duits et al. [13, 14] who introduced α scale space theory.

It is interesting to note the connection between Lévy processes and α scale space theory, since the α scale space evolution equation is the pseudo differential equation corresponding to an α-stable Lévy process. This is seen from the the generator $-(-\Delta)^\alpha$ of the α scale spaces. In particular, for $\alpha = 1$ the diffusion equation is the Kolmogorov equation of the well-known Wiener process or Brownian motion. The case $\alpha = 1/2$, leads to the Poisson scale space [15, 14], which is a harmonic extension to the upper plane. This directly follows from the factorisation $u_{ss} + \Delta u = (\partial_s - \sqrt{-\Delta})(\partial_s + \sqrt{-\Delta})u = 0$.

Finally, we would like to mention the scale invariant dead leaves model of natural images [16] which has some resemblance to the random wavelet expansion proposed by Mumford and Gidas [7]. In the dead leaves model images are build by throwing (substituting) random objects on to the image plane. The model can generate images which has similar marginal statistics as natural images. One of the differences is that in the dead-leaves model it is not assumed that images are formed by an additive process, as is the case for the random wavelet expansion, but by substitution. The drawback of the dead leaves model is that it is difficult to operationalise.

The structure of the paper is as follows: In Sec. 2, we introduce α kernels as probability distributions. We proceed in Sec. 3 by introducing Lévy processes and proposing that Lévy processes with α kernel distribution can be used as a statistical model of natural images. In Sec. 4, we argue by showing empirical results that α kernels can be used as parametric models of the statistics of linear filter responses. We end in Sec. 5 with concluding remarks.

2 α Kernels as Probability Distributions

Pauwels et al. [12] prove that from the four axioms, scale invariance, the recursivity principle (also known as the semi-group property), and rotation and translation invariance, one can deduce a family of scale space kernels — α kernels — parameterised by the α parameter and a scale parameter. This family of kernels form the basis of the recently developed α scale space theory by Duits et al. [13, 14].

The recursivity principle (or semi-group property) of a kernel function $K(x, s)$ can be stated as

$$\forall s, t \geq 0 : (K(\cdot, s) * K(\cdot, t))(x) = K(x, s + t) \tag{1}$$

where $*$ denotes convolution.

A kernel function $K(x, s)$ is scale invariant, if there exist a parent kernel ϕ such that, for any s, $K(x, s)$ can be expressed in terms of a rescaling of ϕ

$$K(x, s) = \frac{1}{\psi(s)} \phi\left(\frac{x}{\psi(s)}\right) , \tag{2}$$

where $\psi : \mathbb{R}_+ \to \mathbb{R}_+$ is a rescaling function.

Besides these two conditions Pauwels et al. [12] suggest that a scale space kernel should fulfil the following conditions: rotation and translation invariance, mass preservation, i.e. $\int_{\mathbb{R}} K(x, s)\, dx = 1$, $K(x, s)$ should be even, i.e. $K(x, s) = K(-x, s)$, and $K(x, s)$ should be integrable and continuous in both x and s. Kernels fulfilling these conditions we will call the α kernels.

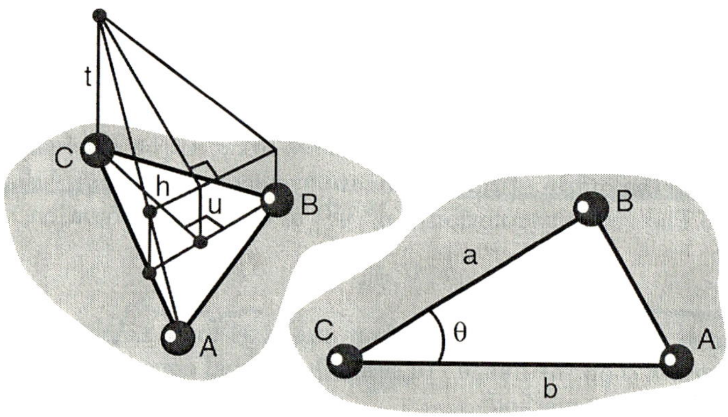

Fig. 3. The numerical stencil used to update $\phi(C)$ according to $\phi(A)$ and $\phi(B)$

- If $u < t$ and $a\cos\theta < \frac{b(t-u)}{t} < \frac{a}{\cos\theta}$, then $\phi(C) = \min\{\phi(C), t + \phi(A)\}$. Else, $\phi(C) = \min\{\phi(C), b + \phi(A), a + \phi(B)\}$.

The numerical scheme described in the previous paragraph enables the update of a pixel according to two of its neighbors. In order to use this scheme to generate the entire distance map the following algorithm [12] is used.

Initialization:

1. The pixel at the origin (for which the kernel is constructed) is defined as *Accepted* and given an initial value of zero.
2. All the other pixels are defined as *Far* and given the value infinity.

Iterations:

1. *Far* 'neighbors' of *Accepted* pixels are defined as *Close.*
2. The values of the *Close* pixels are updated according to the numerical scheme.
3. The *Close* pixel with the minimal value becomes an *Accepted* pixel.
4. If there remain any *Far* pixels, return to step 1.

We use the term 'neighbors' to describe pixels that belong to the same numerical stencil. These pixels are not necessarily neighboring pixels in the image. We find these 'neighbors' during the pre-processing stage described previously.

The complexity of the algorithm is upper bounded by $O(n \log n)$, where n is the number of pixels in the image. The $\log n$ results from using a min-heap data structure for sorting the *Close* pixels [12]. Since There is no need to use all the pixels in the image in order to update one pixel (the value of the kernel for most of these pixels is negligible), we can bound in advance the neighborhood in which the eikonal equation is solved. Thus, we decrease substantially the size of the heap used for the fast marching and enhance its efficiency.

sion equation is the Kolmogorov equation of the well-known Wiener process or Brownian motion. The case $\alpha = 1/2$, leads to the Poisson scale space [15, 14], which is a harmonic extension to the upper plane. This directly follows from the factorisation $u_{ss} + \Delta u = (\partial_s - \sqrt{-\Delta})(\partial_s + \sqrt{-\Delta})u = 0$.

Finally, we would like to mention the scale invariant dead leaves model of natural images [16] which has some resemblance to the random wavelet expansion proposed by Mumford and Gidas [7]. In the dead leaves model images are build by throwing (substituting) random objects on to the image plane. The model can generate images which has similar marginal statistics as natural images. One of the differences is that in the dead-leaves model it is not assumed that images are formed by an additive process, as is the case for the random wavelet expansion, but by substitution. The drawback of the dead leaves model is that it is difficult to operationalise.

The structure of the paper is as follows: In Sec. 2, we introduce α kernels as probability distributions. We proceed in Sec. 3 by introducing Lévy processes and proposing that Lévy processes with α kernel distribution can be used as a statistical model of natural images. In Sec. 4, we argue by showing empirical results that α kernels can be used as parametric models of the statistics of linear filter responses. We end in Sec. 5 with concluding remarks.

2 α Kernels as Probability Distributions

Pauwels et al. [12] prove that from the four axioms, scale invariance, the recursivity principle (also known as the semi-group property), and rotation and translation invariance, one can deduce a family of scale space kernels — α kernels — parameterised by the α parameter and a scale parameter. This family of kernels form the basis of the recently developed α scale space theory by Duits et al. [13, 14].

The recursivity principle (or semi-group property) of a kernel function $K(x, s)$ can be stated as

$$\forall s, t \geq 0 : (K(\cdot, s) * K(\cdot, t))(x) = K(x, s + t) \tag{1}$$

where $*$ denotes convolution.

A kernel function $K(x, s)$ is scale invariant, if there exist a parent kernel ϕ such that, for any s, $K(x, s)$ can be expressed in terms of a rescaling of ϕ

$$K(x, s) = \frac{1}{\psi(s)} \phi\left(\frac{x}{\psi(s)}\right) , \tag{2}$$

where $\psi : \mathbb{R}_+ \to \mathbb{R}_+$ is a rescaling function.

Besides these two conditions Pauwels et al. [12] suggest that a scale space kernel should fulfil the following conditions: rotation and translation invariance, mass preservation, i.e. $\int_{\mathbb{R}} K(x, s)\, dx = 1$, $K(x, s)$ should be even, i.e. $K(x, s) = K(-x, s)$, and $K(x, s)$ should be integrable and continuous in both x and s. Kernels fulfilling these conditions we will call the α kernels.

Pauwels et al. [12] prove that, given the above stated conditions, the Fourier transform of the α kernels $K^{(\alpha)}(x,s)$ have the form

$$\widehat{K^{(\alpha)}}(\omega,s) = e^{-s|\omega|^{2\alpha}} \ , \ \omega \in \mathbb{R} \ , \ 0 < \alpha \leq 1 \ , \ s > 0 \ . \tag{3}$$

Unfortunately, an analytical expression for $K^{(\alpha)}(x,s)$ only exists for $\alpha = 1/2$ and $\alpha = 1$. Note that the case $\alpha = 1/2$ correspond to the Poisson scale space filter [17] and $\alpha = 1$ correspond to the Gaussian filter of linear scale space [18].

The α kernels, as defined in (3), are everywhere positive for $0 < \alpha \leq 1$ and normalised such that $\int_{\mathbb{R}} K^{(\alpha)}(x,s)\,dx = 1$ due to mass preservation. The α kernels thereby fulfils the properties of a probability density function. From now on we write $P^{(\alpha)}(x,s) = K^{(\alpha)}(x,s)$ for the parametric α kernel probability density on $\mathbb{R}$ in order to make clear that we use the α kernels as probability distributions and not as filters. The Fourier transform of the kernel $K^{(\alpha)}(x,s)$ given in (3) is the characteristic function $\widehat{P^{(\alpha)}}(\omega,s)$ of the distribution $P^{(\alpha)}(x,s)$.

It is interesting to note that for $\alpha = 1/2$, (3) is the characteristic function of the Cauchy distribution. For $\alpha = 1$, (3) corresponds to the characteristic function of the Gaussian distribution.

As probability density functions, the α kernels have various interesting properties. The most interesting is that α kernels are so-called infinitely divisible distributions. The following definition is taken from Sato [8] (page 31, Def. 7.1).

Definition 1. *A probability measure μ on $\mathbb{R}^d$ is infinitely divisible if, for any positive integer n, there is a probability measure μ_n on $\mathbb{R}^d$ such that $\mu = \mu_n^{n*}$. Where μ^{n*} denotes n-fold convolution, $\mu^{n*} = \underbrace{\mu * \cdots * \mu}_{n}$.*

The α kernels $P^{(\alpha)}(x,s)$ are infinitely divisible distributions, because we can easily find a distribution, namely the α kernels themselves $P^{(\alpha)}(x,s/n)$, that by n-fold convolution gives the α kernels $P^{(\alpha)}(x,s) = (P^{(\alpha)}(x,s/n))^{n*}$. This follows from

$$\widehat{P^{(\alpha)}}(x,s) = e^{-s|\omega|^{2\alpha}} = \underbrace{\widehat{P^{(\alpha)}}(x,s/n) \cdot \ \cdots \ \cdot \widehat{P^{(\alpha)}}(x,s/n)}_{n} = e^{-|\omega|^{2\alpha}\sum_{i=1}^{n} s/n} \ . \tag{4}$$

Other examples of infinitely divisible distributions are the Gaussian, Cauchy, and Dirac δ distributions on $\mathbb{R}^d$. The Poisson distribution on $\mathbb{R}$ is another example (see e.g. [8] for more examples). We find it interesting to note that the Laplace distribution on $\mathbb{R}^d$, $p(x) = 1/Ze^{-|x/s|}$, is infinitely divisible. To our knowledge there is no proof or disproof of the infinitely divisibility of the generalised Laplace distribution $p(x) = 1/Ze^{-|x/s|^{\beta}}$, except for $\beta = 1$ and $\beta = 2$.

To make explicit the connection between infinitely divisibility and the sum of independent random variables we state the following definition:

Definition 2. *A random variable X is said to be infinite divisible if for all $n \in \mathbb{N}$ there exists a random variable X_n such that $X \stackrel{d}{=} \sum_{i=1}^{n} X_{n,i}$, where $\{X_{n,i}\}_{i=1,\ldots n}$ are independent and distributed as X_n.*

Notice that according to this definition the distribution function must satisfy $F(x) = F_n^{n*}(x)$, where F_n is the probability distribution function of X_n (and F of X).

To see that this definition coincides with the definition of infinitely divisible probability measures, definition 1, we need to apply the well known theorem, which states that if two random variables X and Y are independent and distributed by resp. f and g then $X+Y$ is distributed ($\sim$) by $f*g$ (see e.g. [19]), i.e.

$$\{X+Y\} \sim f*g \ . \tag{5}$$

From this it follows that definition 2 coincides with definition 1.

Infinitely divisibility is tightly related to the class of stochastic processes known as Lévy processes and self similarity, a generalisation of scale invariance, of stochastic processes. We will discuss this further in Sec. 3.

We would also like to mention that the α kernels do not have finite variance for $0 < \alpha < 1$. If the α kernels had finite variance, then the infinitely divisibility of the α kernels and the central limit theorem implies that the α kernels must be Gaussian. Hence, variance is not a good measure of scale.

3 α Kernels, Lévy Processes, and Image Models

We will now argue for the use of stochastic Lévy processes with α kernel distribution as a model of natural images. A stochastic process is usually defined as a family of random variables $\{X_t : t \geq 0\}$, where X_t is a random variable on $\mathbb{R}^d$. Here we are interested in the case $d = 1$. The definition of Lévy processes is as follows (Sato [8], page 3, Def. 1.6):

Definition 3. *A stochastic process $\{X_t : t \geq 0\}$ on $\mathbb{R}^d$ is a Lévy process (in law) if the following conditions are satisfied:*

1. *For any choice of $n \geq 1$ and $0 \leq t_0 < t_1 < \cdots < t_n$, the random variables $X_{t_0}, X_{t_1} - X_{t_0}, X_{t_2} - X_{t_1}, \ldots, X_{t_n} - X_{t_{n-1}}$ are independent.*
2. *$X_0 = 0$ almost surely.*
3. *The distribution of $X_{s+t} - X_s$ does not depend on s.*
4. *It is stochastically continuous, i.e.*

$$\lim_{s \to t} P[\|X_s - X_t\| > \epsilon] = 0 \ , \ \forall \epsilon > 0 \ .$$

The following theorem states the connection between Lévy processes and infinitely divisible distributions.

Theorem 1. *1. If $\{X_t : t \geq 0\}$ is a Lévy process in law on $\mathbb{R}^d$, then, for any $t \geq 0$, the distribution P_{X_t} of X_t is an infinitely divisible distribution and, letting $P_{X_1} = \mu$, we have $P_{X_t} = \mu^t$.*

2. Conversely, if μ is an infinitely divisible distribution on $\mathbb{R}^d$, then there is a Lévy process in law $\{X_t : t \geq 0\}$ such that $P_{X_1} = \mu$.

3. *If* $\{X_t\}$ *and* $\{X'_t\}$ *are Lévy processes in law on* $\mathbb{R}^d$ *such that* $P_{X_1} = P_{X'_1}$*, then* $\{X_t\}$ *and* $\{X'_t\}$ *are identical in law.*

(The proof can be found in Sato [8], page 35, Theorem 7.10).

We want to use stochastic processes as a model of images, hence we need to think of the parameter $t \in \mathbb{R}$ as the parameterisation of an oriented[1] straight line in the domain of the image, i.e. the plane $\mathbb{R}^2$. Then X_t is the intensity of the image as we move along the line. The above stated definition and theorem then applies to any non-overlapping oriented straight lines in the plane. Hence, all non-overlapping oriented straight lines in a Lévy process image are 1 dimensional Lévy processes X_t.

Brownian motion is an example of a Lévy process, which has a Gaussian distribution. Brownian images have been investigated as stochastic models of images, e.g. [7, 20].

In the next section, we model the distribution of increments $X_{t_k} - X_{t_{k-1}}$ by the α kernels. We suggest the Lévy process with α kernel distributed increments $X_{t_k} - X_{t_{k-1}}$ as a simple non-Gaussian stochastic model of natural images. But what is the distribution of X_t? The answer is that it is also α kernel distributed. To see this, first realise given n steps on an oriented line through the stochastic process we have

$$X_t = (X_{t_1} - X_{t_0}) + \cdots + (X_{t_n} - X_{t_{n-1}}) \ . \tag{6}$$

From (5) it follows that a sum X_t of n α kernel distributed $P^{(\alpha)}$ independent random variables $(X_{t_k} - X_{t_{k-1}})$ has a distribution determined by the n-fold convolution of $P^{(\alpha)}$, i.e. $P(X_t, s') = (P^{(\alpha)}(X_{t_k} - X_{t_{k-1}}, s))^{n*}$, which as we proofed in (4) is also an α kernel distribution $P(X_t, s') = P^{(\alpha)}(X_t, ns)$.

One of the important properties of the Lévy process with α kernel distribution is that it is self similar, a generalisation of scale invariance.

Self similarity is related to the concept of stability of the distribution of the Lévy process [8]. An infinitely divisible probability measure μ is called stable if, for any $a > 0$, there are $b > 0$ and $c \in \mathbb{R}^d$ such that

$$\hat{\mu}(\omega)^a = \hat{\mu}(b\omega)e^{i\langle c,\omega\rangle} \ . \tag{7}$$

If $c = 0$, then the measure μ is called strictly stable. A stochastic process $\{X_t : t \geq 0\}$ is called broad sense self similar if, for any $a > 0$, there are $b > 0$ and a function $c(t) : [0, \infty) \to \mathbb{R}^d$ such that

$$\{X_{at} : t \geq 0\} \stackrel{d}{=} \{bX_t + c(t) : t \geq 0\} \ . \tag{8}$$

If $c(t) = 0$, the process $\{X_t : t \geq 0\}$ is called self similar. A Lévy process is self similar or broad sense self similar if and only if, its distribution is respectively strictly stable or stable (a proof can be found in Sato [8], page 71, Prop. 13.5).

[1] Assuming an orientation means that we consider the increments to be independent. Otherwise, we would be able to go back following our own footsteps and the increments will no longer be independent.

If $b = a^{1/(2\alpha)}$, the Lévy process $\{X_t : t \geq 0\}$ is called α-stable. α-stable processes has various important properties, see Sato [8] and the theorem below.

α kernel distributions are strictly stable and α-stable. Hence, the Lévy process with α kernel distribution is self similar with $b = a^{1/(2\alpha)}$ and $c(t) = 0$.

It is interesting to note that in the literature on Lévy processes and infinitely divisible distributions, e.g. [8], the α kernel distribution is also pointed out as being special, as the following theorem reveals.

Theorem 2. *A non-trivial probability measure μ on $\mathbb{R}^d$ is rotation invariant (symmetric for $d = 1$) and α-stable with $0 < \alpha \leq 1$ if and only if*

$$\hat{\mu}(\omega) = e^{-s|\omega|^{2\alpha}}$$

with $s > 0$.

(For a proof see Sato [8], page 86, first part of Theorem 14.14)

α stability is connected with the generalisation of the well-known central limit theorem. Recall that the central limit theorem states that given n independently identically distributed random variables X_i with finite mean and variance, then the distribution of the sum $\sum_{i=1}^{n} X_i$ converges to the Gaussian distribution in the limit $n \to \infty$ (see e.g. for a proof [21]).

But the sum of independently identically Cauchy distributed random variables (with infinite variance) will converge to a Cauchy distribution. There also exist examples of sums of independent random variables identically distributed with infinite variance that will converge to a Gaussian distribution (for example $f(x) = 2|x|^{-3} \log|x|$ for $|x| \geq 1$ and $f(x) = 0$ for $|x| \leq 1$).

The generalisation of the central limit theorem states that if a sum of independently identically distributed random variables converges, it converges to a stable distribution. In particularly, a convergent sum of identically (by say f) distributed symmetric random variables will converge to a stable symmetrically distributed (i.e. distributed by α-kernel) random variable. By considering asymptotics on $\int_{-x}^{x} y^2 f(y)\text{d}y$ it is possible to characterise the unique limiting distribution (see Feller [19], chapter IX.8).

The image formation model we propose here is essentially a simplification of the random wavelet expansion model proposed by Mumford and Gidas [7]. To see this we first have to introduce the Lévy-Khintchine representation, which states that the characteristic function of any infinitely divisible distribution μ must have a form dependent on a constant part, a Gaussian part and a measure ν known as the Lévy measure. For μ on $\mathbb{R}$ this boils down to

$$\hat{\mu}(\omega) = e^{i\gamma\omega - \sigma^2\omega/2 + \hat{\nu}(\omega)} \tag{9}$$

where ν must satisfy $\nu(\{0\}) = 0$ and have finite variance on the closed unit ball $\int_{\mathbb{R}} min\{|x|^2, 1)\}\nu(dx) < \infty$. Here $\hat{\nu}$ is essentially the Fourier transform of ν (see [8], page 37, Theorem 8.1, for further details and proof of this theorem).

Mumford and Gidas [7] propose that images I are formed by adding independent random images I_i, $I = I_0 + I_1 + \sum_{i \geq 2} I_i$, where I_0 is constant and I_1 is a Gaussian random image. The random images I_i for $i \geq 2$ are samples from a compound Poisson process governed by some distribution ν. This model implies a Lévy process with a distribution having the characteristic function given in (9) and ν as Lévy measure.

What we suggest in this paper is to choose the Lévy measure ν to be the Lévy measure of α kernels, that is $\hat{\nu}(\omega) = -s|\omega|^{2\alpha}$. Here we also ignore the constant and Gaussian part, hence our model essentially boils down to a compound Poisson process with $\hat{\nu}(\omega) = -s|\omega|^{2\alpha}$. But the missing constant and Gaussian parts could be added to the α kernel based model with the drawback that we will lose the symmetry and α stability property of the distribution. We believe that the Lévy process with α kernel distribution is an interesting model because of the α-stability of the distribution which implies the special form of self-similarity, $\{X_{at} : t \geq 0\} \stackrel{d}{=} \{a^{1/\alpha} X_t : t \geq 0\}$. Self-similarity or scale invariance is apparently an important property of natural images [1, 2, 3, 4].

4 Statistics of Linear Filter Responses

We will now argue that α kernels $P^{(\alpha)}(x, s)$ can be used as a parametric model of the distribution of linear filter responses on natural images. We do not attempt to argue that α kernels fit the marginal statistics of natural images better than for instance the generalised Laplace distribution [9] or the Bessel K forms [10, 11]. Our goal is simply to point out the existence of the α kernels as parametric models of this statistics. As we argued in the previous sections, the α kernels have a lot to offer as basis for stochastic models of natural images, because of the connection with Lévy processes and infinitely divisibility.

In the following experiments, we do all processing on the log intensities $\log(I(i, j) + 1)$, where a discrete image is considered to be a matrix of real numbers $I : \mathbb{Z}^2 \to \mathbb{R}$. The choice of log intensities was pointed out by Koenderink [22] as the natural choice for image processing. By intensity difference we mean $\Delta_x I(i, j) = \log(I(i, j+1)+1) - \log(I(i, j)+1)$, $\Delta_y I(i, j) = \log(I(i+1, j)+1) - \log(I(i, j) + 1)$, and write $\Delta I = \{\Delta_x I, \Delta_y I\}$ for the combined data set.

As examples of linear filters we use scale normalised scale space derivatives, see e.g. Florack [23],

$$L_{x^n y^m}(x, y, s) = s^{\frac{n+m}{2}} \frac{\partial^{n+m}}{\partial x^n \partial y^m} G(x, y, s) * f(x, y) , \tag{10}$$

where $G(x, y, s) = \frac{1}{4\pi s} e^{-(x^2+y^2)/(4s)}$ is the Gaussian kernel and s is the scale parameter.

The intensity differences histograms can be seen in Fig. 2 and examples of histograms for scale space derivatives can be found in Fig. 3.

We picked approximately 1600 images from the van Hateren natural image database [24] and we bootstrapped 50 histograms by taking 50.000 samples at random from each image followed by 50 times sampling with replacement of 50.000 points from this per image set. The bootstrap histograms gives a measure

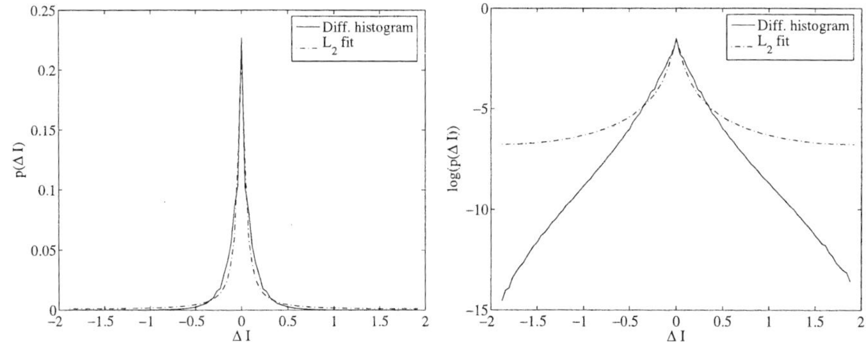

Fig. 4. Graphs of $p(\Delta I)$ and $\log p(\Delta I)$. The dash dotted graph represents the fitted α kernel distribution with the bootstrap mean α and s, $\alpha = 0.376 \pm 8 \cdot 10^{-5}$, $s = 0.187 \pm 7 \cdot 10^{-5}$

of the robustness of the estimated parameter values when we fit α kernels to the histograms, i.e. we can calculate the parameter bootstrap mean and standard deviation error.

In order to fit the α kernels to the histograms we use the L_2 norm as a measure of difference between two distributions. We do the minimisation in the Fourier domain,

$$L_2(\widehat{H}, \widehat{P^{(\alpha)}}) = \frac{1}{N} \sum_{i=1}^{N} |\widehat{H}(\omega_i) - \widehat{P^{(\alpha)}}(\omega_i, s)|^2 , \tag{11}$$

where $\widehat{H}(\omega_i)$ denotes the discrete Fourier transform of the histogram. The minimisation is implemented by using the Nelder-Mead simplex direct search method as implemented in Matlab. The results of fitting the α kernels to intensity difference histograms can be seen in Fig. 4. The fitted α kernels match the central

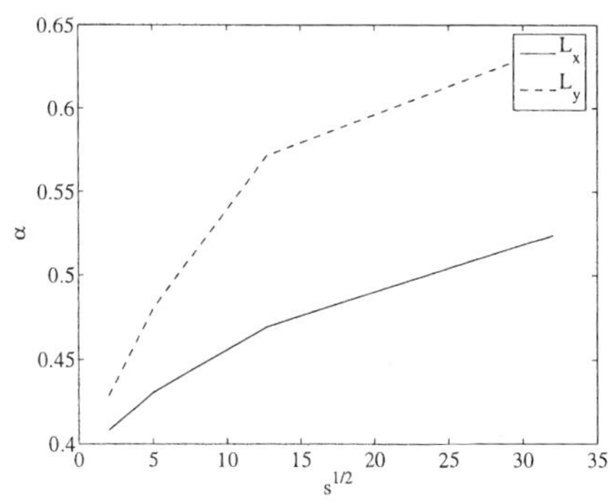

Fig. 5. Estimated bootstrap mean α values of L_x and L_y across scale s

part of the distribution fairly well, but there are problems with the tails because of lack of sufficient data.

The α kernels can also be fitted to histograms of linear filter responses, and Fig. 5 shows estimated α values as a function of scale for L_x and L_y scale space derivatives. What we observe is that when the scale goes up the estimated α values increase as well. This means that convolving with a Gaussian kernel alters the statistics (i.e. the shape of the distribution) of the image Lévy process. One may argue that we could choose another scale space filter that would preserve the statistics of the Lévy process and this choice would be the α kernel corresponding to the α kernel distribution of the image process.

5 Conclusion

We propose to use the Lévy processes with α kernel distribution as a non-Gaussian stochastic model of natural images. This model explains the marginal statistics of natural images. The model is essentially a simplification of the random wavelet expansion model proposed by Mumford and Gidas [7]. Furthermore, we have argued that α kernels can be used as parametric models of the marginal linear filter statistics of natural images. Hence, the increments of the natural image stochastic process are α kernel distributed.

The image formation model we propose is simpler, but however crude, compared to the model suggested by Mumford and Gidas [7], but nevertheless we believe that this model can be useful, especially because of its simplicity and that we are capable of analysing its properties. Future applications will tell whether this is correct or not.

Lévy processes have independent increments, but one can easily argue that this is not true for images, because objects or homogeneous regions introduce dependencies between pairs of pixels inside of the same object. Furthermore, using a Lévy process implies that images are formed by an additive process, which might not be a correct image formation model (substitution seems more appropriate). These are limitations of the model.

For future work, we would like to study the probability distribution of the model in the scale space jet. Here it is interesting to notice that the statistical properties of wavelet decompositions of α-stable processes have been studied by among others Pesquet-Popescu [25]. Pesquet-Popescu proves that the wavelet coefficients of an α-stable process also form an α-stable process (Prop. 4 in [25]). We find it interesting to extend the results reported in [25] to linear scale space on images, i.e. extend from compact support test functions to the infinite support Gaussian test functions. Furthermore, it would be helpful if we could draw samples from our stochastic image model. This might require the use of Markov Chain Monte Carlo methods.

References

1. Field, D.J.: Relations between the statistics of natural images and the response properties of cortical cells. J. Optic. Soc. of Am. **4** (1987) 2379–2394

2. Ruderman, D.L., Bialek, W.: Statistics of natural images: Scaling in the woods. Physical Review Letters **73** (1994) 814–817
3. Pedersen, K.S., Lee, A.B.: Toward a full probability model of edges in natural images. In: Proceedings of 7th ECCV. LNCS 2350 (2002) 328–342
4. Srivastava, A., Lee, A.B., Simoncelli, E.P., Zhu, S.: On advances in statistical modeling of natural images. Journal of Mathematical Imaging and Vision **18** (2003) 17–33
5. Lillholm, M., Nielsen, M., Griffin, L.D.: Feature-based image analysis. International Journal of Computer Vision **52** (2003) 73–95
6. Pedersen, K.S., Lillholm, M.: Brownian images: A generic background model. In: Proceedings of Workshop on Statistical Learning in Computer Vision. (2004)
7. Mumford, D., Gidas, B.: Stochastic models for generic images. Quarterly of Applied Mathematics **59** (2001) 85–111
8. Sato, K.: Lévy Processes and Infinitely Divisible Distributions. Cambridge University Press (1999)
9. Huang, J., Mumford, D.: Statistics of natural images and models. In: Proc. of IEEE Conf. on Computer Vision and Pattern Recognition. (1999)
10. Grenander, U., Srivastava, A.: Probability models for clutter in natural images. IEEE Transaction on Pattern Analysis and Machine Intelligence **23** (2001) 424–429
11. Srivastava, A., Liu, X., Grenander, U.: Universal analytical forms for modeling image probabilities. IEEE Trans. on PAMI **24** (2002) 1200–1214
12. Pauwels, E.J., van Gool, L.J., Fiddelaers, P., Moons, T.: An extended class of scale-invariant and recursive scale space filters. IEEE T-PAMI **17** (1995) 691–701
13. Duits, R., Felsberg, M., Florack, L.: α scale spaces on a bounded domain. In Griffin, L., ed.: Proceedings of Scale-Space'03. (2003)
14. Duits, R., Florack, L., de Graaf, J., ter Haar Romeny, B.: On the axioms of scale space theory. Journal of Mathematical Imaging and Vision **20** (2004) 267–298
15. Felsberg, M., Sommer, G.: Scale adaptive filtering derived from the Laplace equation. In Radig, B., Florczyk, S., eds.: 23. DAGM Symposium Mustererkennung, München. LNCS vol. 2191, Springer (2001) 124–131
16. Lee, A.B., Mumford, D., Huang, J.: Occlusion models for natural images: A statistical study of a scale-invariant dead leaves model. International Journal of Computer Vision **41** (2001) 35–59
17. Felsberg, M., Sommer, G.: The monogenic scale-space: A unified approach to phase-based image processing in scale-space. JMIV **21** (2004) 5–26
18. Koenderink, J.J.: The structure of images. Biological Cybernetics **50** (1984) 363–370
19. Feller, W.: An Introduction to Probability Theory and its Applications, volume 2. John Wiley and sons, Inc., New York, London, Sydney (1966)
20. Pedersen, K.S.: Properties of brownian image models in scale-space. In: Proceedings of the 4th Scale-Space conference. LNCS 2695 (2003) 281–296
21. Tjur, T.: Probability Based on Radon Measures. John-Wiley (1980)
22. Koenderink, J.J.: Image processing done right. In: Proceedings of the ECCV. LNCS 2350 (2002) 158–172
23. Florack, L.: Image Structure. Computational Imaging and Vision. Kluwer Academic Publishers, Dordrecht (1997)
24. van Hateren, J.H., van der Schaaf, A.: Independent component filters of natural images compared with simple cells in primary visual cortex. Proc. R. Soc. Lond. Series B **265** (1998) 359 – 366
25. Pesquet-Popescu, B.: Statistical properties of the wavelet decomposition of certain non-gaussian self-similar processes. Signal Processing **75** (1999) 303–322

An Analysis of Variational Alignment of Curves in Images

Niels Chr. Overgaard and Jan Erik Solem

Applied Mathematics Group,
School of Technology and Society,
Malmö University, Sweden
{nco, jes}@ts.mah.se

Abstract. In this paper a common variational formulation for alignment of curves to vector fields is analyzed. This variational approach is often used to solve the problem of aligning curves to edges in images by choosing the vector field to be the image gradient. The main contribution of this paper is an analysis of the Gâteaux derivative and the descent motion of the corresponding alignment functional, improving on earlier research in this area. Several intermediate results are proved and finally a theorem concerning necessary conditions for extremals of the alignment functional is derived. The analysis of the evolution is performed using a level set formulation and results from distribution theory.

Keywords: variational methods, level set methods, edge integration, curve evolution, alignment.

1 Introduction

In this paper we study the properties of an alignment functional, proposed by Kimmel and Bruckstein, which is frequently used for variational alignment of curves and surfaces in image analysis and computer vision. The focus will be on the two-dimensional case. A common problem is to align curves to the edges in an image where edges are found at locations with high image gradient. This is a fundamental problem within many applications such as e.g. image segmentation. This paper will deal with some unresolved issues of this particular alignment functional. These issues will be described below, and the analysis will result in theorems about ascent directions and a criterium, which is stronger than the results found in the literature, for a local maximum of this particular and common case.

The paper is neither concerned with edge detection in the low level image analysis sense, nor with edge detection using *area based* segmentation techniques [1, 2], recently analyzed in [3]. Instead, the paper is in the spirit of the *active contours* or *snakes* [4] and the *geodesic active contours* [5, 6] in that curves are aligned to edges using a curve integral measure. In the latter case the resulting curve will be a geodesic in a Riemannian space with a specified metric. The

R. Kimmel, N. Sochen, J. Weickert (Eds.): Scale-Space 2005, LNCS 3459, pp. 480–491, 2005.

in-depth analysis of such functionals in [7] leads to the functional which is the topic of this paper. The work presented here builds directly on the treatment of this special alignment functional in [8] which is further developed in [9].

1.1 Variational Alignment

Let $\mathbf{v} : \mathbf{R}^2 \to \mathbf{R}^2$ be a continuously differentiable vector field, and Γ be a regular curve in the plane. We shall study the alignment functional given by the following curve integral,

$$E_A(\Gamma) = \int_\Gamma |\mathbf{v} \cdot \mathbf{n}| \, d\sigma \ , \tag{1}$$

where $d\sigma$ is the Euclidian curve length element on Γ, and $\mathbf{n}$ is a unit normal vector field on Γ. This functional was used in [8, 7] and analyzed in [9]. The functional E_A is defined entirely in terms of geometric quantities, and its value is therefore independent of the way we choose to represent Γ. In this paper we use the level set method as a theoretical tool, and the calculations for E_A are all carried out in this framework, using results from distribution theory.

One aim with this paper is to discuss the use of the PDE

$$\frac{\partial \phi}{\partial t} = \operatorname{sign}_\varepsilon(\mathbf{v} \cdot \mathbf{n})(\nabla \cdot \mathbf{v})|\nabla \phi| \ , \tag{2}$$

as an ascent motion[1] for the problem of finding a curve Γ^* such that

$$E_A(\Gamma^*) = \max_\Gamma E_A(\Gamma) \ . \tag{3}$$

Here $\nabla \cdot \mathbf{v} = \partial v_1/\partial x_1 + \partial v_2/\partial x_2$ is the divergence of $\mathbf{v}$ and $\operatorname{sign}_\varepsilon : \mathbf{R} \to [-1, 1]$ is any continuous, nondecreasing approximation of the sign-function satisfying

$$\operatorname{sign}_\varepsilon(0) = 0 \quad \text{and} \quad \operatorname{sign}_\varepsilon(x) = \operatorname{sign}(x) \quad \text{for } |x| > \varepsilon \ . \tag{4}$$

We shall also touch upon the minimization problem for (1) in Section 4, since it has some relevance even for the problem in (3).

In [7] the following result, which we state as a theorem, regarding extremals of (1) was proved.

Theorem 1 (Kimmel-Bruckstein). *Let Γ be a local maximum of (1). If $\mathbf{v} \cdot \mathbf{n}$ does not change sign along Γ, then $(\nabla \cdot \mathbf{v})(\mathbf{x}) = 0$ for all $\mathbf{x} \in \Gamma$.*

If $\mathbf{v} = \nabla u$, where $u : \mathbf{R}^2 \to \mathbf{R}$ is an image, and the curve Γ is a local maximum for E_A which satisfies the hypothesis of the theorem, then $\nabla \cdot \mathbf{v} = \nabla \cdot (\nabla u) = \Delta u = 0$ on Γ. This means that the extremal curves are Marr-Hildreth edges, cf. [7]. In Section 8 we prove a result in which the hypothesis about $\mathbf{v} \cdot \mathbf{n}$ is not needed, hence it strengthens the above theorem.

---

[1] In our context it is more natural to consider the max-problem for E_A instead of the min-problem for $-E_A$, thus we speak of "ascent" instead of "descent".

1.2 Contribution of the Paper

The contributions of this paper are described in Sections 6 to 8. In addition to the written analysis, the contributions are stated and proved in Lemma 1 and Theorem 2, 3, 4 and 5. In this paper we compute the correct Gâteaux derivative of (1) and show that there are cases where the functional is not differentiable. We also show that the evolution equation derived in [7], i.e. the PDE (2), is an ascent direction under the weak assumption that the sign-function satisfies (4). This justifies the derivations of Kimmel and Bruckstein [7] even without the assumption of constant sign on Γ. Finally we derive a necessary condition for local maxima of (1) which is stronger than Theorem 1.

2 Level Set Representation

A time dependent curve $\Gamma(t)$ is implicitly represented as the zero level set of a function $\phi(\mathbf{x},t) : \mathbf{R}^2 \times I \to \mathbf{R}$ as $\Gamma(t) = \{\mathbf{x}\ ;\ \phi(\mathbf{x},t) = 0\}$, where $I \subset \mathbf{R}$ is some open (time-) interval. The sets $\{\mathbf{x} : \phi(\mathbf{x},t) < 0\}$ and $\{\mathbf{x} : \phi(\mathbf{x},t) > 0\}$ are called the *inside* and *outside* of Γ, respectively. Using this convention the outward unit normal $\mathbf{n}$ and the (mean) curvature κ are given in terms of the level set functions by the formulas, cf. [10]:

$$\mathbf{n} = \frac{\nabla\phi}{|\nabla\phi|} \quad \text{and} \quad \kappa = \nabla \cdot \frac{\nabla\phi}{|\nabla\phi|} \ . \tag{5}$$

Suppose a particle, whose motion is described by the parametrized curve $\mathbf{x} = \mathbf{x}(t)$, follows the moving curve $\Gamma(t)$ at all times $t \in I$, then $\phi(\mathbf{x}(t),t) \equiv 0$, and differentiation of this identity with respect to t gives

$$\frac{\partial\phi}{\partial t} + \mathbf{u} \cdot \nabla\phi = 0 \ \Leftrightarrow \ \frac{\partial\phi}{\partial t} + u_n|\nabla\phi| = 0 \ , \tag{6}$$

where $\mathbf{u} = d\mathbf{x}(t)/dt$ and $u_n = \mathbf{u} \cdot \mathbf{n}$ is the normal component of the particle velocity. This is the so-called *level set equation.* By solving (6) from some initial value, Γ will evolve according to the velocity $\mathbf{u}$. More details can be found in e.g. [11].

3 A Simple Example: Edges in Images

In this section we will consider the example of $\mathbf{v} = \nabla u$ for aligning curves to edges in images. We will refer to this concrete example to illustrate some of the ideas presented in the following sections such as the singular curve in Section 7. In addition to the alignment term in the functional, in practice a regularizing term is needed to avoid aligning the curve to image noise, cf. [8]. The following functional is minimized

$$E(\Gamma) = \alpha \int_\Gamma 1 \, d\sigma - E_A(\Gamma) = \int_\Gamma (\alpha - |\nabla u \cdot \mathbf{n}|) \, d\sigma \ , \tag{7}$$

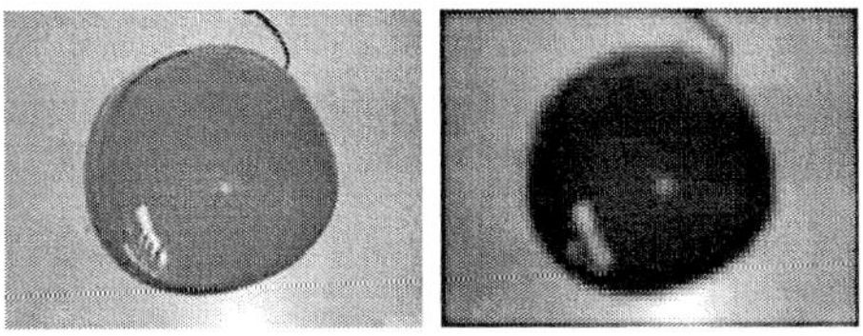

(a) (b)

Fig. 1. Image used for illustrating the curve motion (8). (a) original image. (b) image smoothed with a Gaussian filter and down-sampled

where the first term is the curve length weighted with a constant $\alpha > 0$ and the second term maximizes the alignment of the curve normal to the image gradient. This means that the curve is attracted to regions with large gradient magnitude $|\nabla u|$ and oriented so that $\mathbf{n}$ or $-\mathbf{n}$ is in the gradient direction. An initial curve is evolved using the following PDE

$$\frac{\partial \phi}{\partial t} = (\alpha \kappa - \text{sign}_\varepsilon(\nabla u \cdot \mathbf{n}) \Delta u)|\nabla \phi| \ . \tag{8}$$

To illustrate the curve motion we used an image of a lamp seen from above, see Figure 1. The image was smoothed using a Gaussian filter and then down-sampled. The smoothening is applied in order to eliminate alignment to noise and also to determines the scale of the edges. The function ϕ is at all times kept to be a signed distance function. The curve evolution is shown in the top row of Figure 4 overlaid on the images.

4 Incidence Sets

A curve Γ is clearly a (global) minimum for the alignment functional E_A if the vector field $\mathbf{v}$ is everywhere tangent to Γ. If this is the case we say that Γ is *incident with* $\mathbf{v}$. If $\mathbf{v}(\mathbf{x})$ is tangent to Γ for all points $\mathbf{x}$ belonging to some segment I of Γ we say that Γ is *locally incident with* $\mathbf{v}$. The following analysis shows that it is important to know whether Γ has this property or not.

Let $\mathbf{v}$ be a continuous vector field, and assume that Γ is a curve which is locally incident with $\mathbf{v}$ on a segment $I \subset \Gamma$, and assume further that $\mathbf{v} \neq 0$ in I. Consider a deformation Γ' of Γ such that $\Gamma' = \Gamma$ outside I, see Figure 2, and let I' denote the part of the deformed curve which corresponds to the segment I. The change in the value of the functional is

$$E_A(\Gamma') - E_A(\Gamma) = \int_{I'} |\mathbf{v} \cdot \mathbf{n}|\, d\sigma - \underbrace{\int_I |\mathbf{v} \cdot \mathbf{n}|\, d\sigma}_{=0} > 0 \quad \text{(strict inequality)}, \tag{9}$$

since the continuity implies that $\mathbf{v}$ has almost the same magnitude and direction in a neighbourhood of the deformation, and because of the $|\cdot|$-sign in the definition of E_A. It follows that any perturbation of Γ in I will increase the value of E_A, which proves

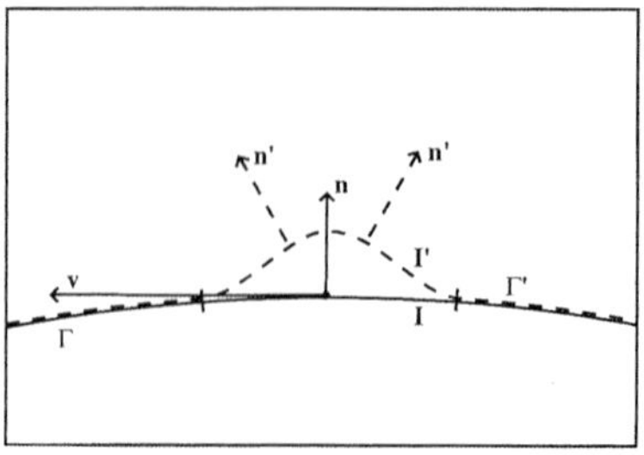

Fig. 2. The case where $\mathbf{v} \cdot \mathbf{n} = 0$ on a segment I of the curve Γ

Lemma 1. *If a curve Γ is locally incident with* $\mathbf{v}$ *on some segment I, and* $\mathbf{v}$ *does not vanish on some neighbourhood of I, then Γ cannot be a local maximum of (1), in particular Γ is not a solution to (3).*

Let Γ be represented by the level set function ϕ, and pick a smooth function ψ such that Γ' is represented by $\phi(\mathbf{x})+t\psi(\mathbf{x})$ when $t = 1$. A more careful analysis of the difference in (9), which lack of space prevents us from giving here, shows that unless $\mathbf{v}$ satisfies certain strict conditions near I, then $E_A(\phi+t\psi) - E_A(\phi) = c|t|$ for some constant $c > 0$. This implies that if Γ is locally incident with $\mathbf{v}$, then the functional need not be (Gâteaux-) differentiable at Γ. It is therefore important to avoid curves which are locally incident to $\mathbf{v}$, if we are to apply differential calculus to solve the problem (3).

5 Mathematical Results for Curve Measures

In this section we briefly recall some results from distribution theory concerning the composition of a distribution with a smooth function. We also establish a lemma showing that a certain product of two distributions makes sense.

First a bit of useful terminology from differential calculus: Let $U \subset \mathbf{R}^n$ be an open set and $f : U \to \mathbf{R}^m$ a (sufficiently) smooth function. We use the symbol $f'(\mathbf{x})$, $\mathbf{x} \in U$, to denote the *Jacobian matrix* $(\partial f_i(\mathbf{x})/\partial x_j)$. A point $\mathbf{y} \in \mathbf{R}^m$ is called a *regular value of* f if, for all $\mathbf{x}$ such that $f(\mathbf{x}) = \mathbf{y}$, the matrix $f'(\mathbf{x})$ has full rank. (If $m = 1$ this means that $f'(\mathbf{x}) = \nabla f(\mathbf{x})$ does not vanish on the set where $f(\mathbf{x}) = \mathbf{y}$, and if $m = n$ then $\det f'(\mathbf{x}) \neq 0$ on the solution set.)

Let $f : U \to \mathbf{R}^m$ be as above and $u \in \mathcal{D}'(\mathbf{R}^m)$ a given distribution. It is a standard result in distribution theory, that if all $\mathbf{y} \in \text{supp}(u)$ (=*the support of* u) are regular values of f, then the composition $u \circ f = u(f)$, of u by f, is a well-defined distribution in $\mathcal{D}'(U)$. In this paper we need two particular instances of this result involving the Dirac distribution $\delta \in \mathcal{D}'(\mathbf{R}^m)$.

(a) If $f : U \to \mathbf{R}$ $(m = 1)$ is C^1 and zero is a regular value for f, then $\delta(f) \in \mathcal{D}'(U)$ is a distribution with support on the set $\Sigma = \{\mathbf{x} : f(\mathbf{x}) = 0\}$ and

$$\delta(f) = \frac{d\sigma}{|\nabla f|}, \tag{10}$$

where $d\sigma$ is the Euclidean surface measure on Σ, here considered as a distribution in U with $\mathrm{supp}(d\sigma) \subset \Sigma$.

(b) If $f : U \to \mathbf{R}^n$ $(n = m)$ is C^1 and $\mathbf{0} = (0, \ldots, 0)$ is a regular value for f, then the composition of $\delta \in \mathcal{D}'(\mathbf{R}^n)$ by f is

$$\delta(f) = \sum_{\substack{\mathbf{p} \in U: \\ f(\mathbf{p}) = 0}} \frac{\delta_{\mathbf{p}}}{|\det f'(\mathbf{p})|}, \tag{11}$$

where $\delta_{\mathbf{p}} \in \mathcal{D}'(U)$ is the Dirac distribution situated at $\mathbf{p} \in U$.

From now on we focus on the two-dimensional case. If $\mathbf{u} = (u_1, u_2) \in \mathbf{R}^2$ then the vector obtained by a counter-clockwise rotation of $\pi/2$ radians is

$$\hat{\mathbf{u}} = (-u_2, u_1). \tag{12}$$

Also, if $\mathbf{u}, \mathbf{v} \in \mathbf{R}^2$, we define the anti-symmetric scalar product

$$\mathbf{u} \wedge \mathbf{v} = \hat{\mathbf{u}} \cdot \mathbf{v}.$$

Suppose $U \subset \mathbf{R}^2$ is an open connected set (for instance the image domain) and $g, h : U \to \mathbf{R}$ is a pair of continuously differentiable functions. Set $A = \{\mathbf{x} \in U : g(\mathbf{x}) = 0\}$ and $B = \{\mathbf{x} \in U : h(\mathbf{x}) = 0\}$. If we assume zero to be a regular value of both g and h, then, by the implicit function theorem, A and B are well-defined differentiable curves in U, see Figure 3(a).

Consider also the mapping $G : U \to \mathbf{R}^2$ defined by $G(\mathbf{x}) = (g(\mathbf{x}), h(\mathbf{x}))$. If $(0,0)$ is a regular value for G, that is, if $\det G'(\mathbf{p}) = \nabla g(\mathbf{p}) \wedge \nabla h(\mathbf{p}) \neq 0$ for all $\mathbf{p} \in A \cap B$, then the curves A and B will intersect non-tangentially or *transversally* at $\mathbf{p} \in A \cap B$, as indicated in Figure 3(a). In particular, $A \cap B$ consists of isolated points.

An orientation of A is defined the following way. If we go along A, in the direction of the orientation, then $g(\mathbf{x}) < 0$ for points $\mathbf{x}$ on the left-hand side of A. B is given an orientation in a similar manner. The orientations are indicated in Figure 3(a) by arrows on the curves.

Using these orientations we can speak of an "index" of an intersection $\mathbf{p} \in A \cap B$. The index is the function $\mathrm{ind}(g, h, \cdot) : A \cap B \to \{-1, +1\}$ given by: $\mathrm{ind}(g, h, \mathbf{p}) = +1$ if h changes sign from negative to positive when we go along A in the direction of the orientation, and $\mathrm{ind}(g, h, \mathbf{p}) = -1$ if h passes form positive values to negative ones. (Our hypothesis of transversal intersections guarantees that these are the only two possible cases, hence the index is well defined.) There is a neat analytical expression for the index. Let $\mathbf{n}_g = \nabla g / |\nabla g|$ and $\mathbf{n}_h = \nabla h / |\nabla h|$ denote the unit normal fields on A and B, respectively. If $\mathbf{p} \in A \cap B$ then the vector $\hat{\mathbf{n}}_g(\mathbf{p})$ is tangent to A and points along the orientation, so it is easy check that

$$\mathrm{ind}(g, h, \mathbf{p}) = \frac{\mathbf{n}_g(\mathbf{p}) \wedge \mathbf{n}_h(\mathbf{p})}{|\mathbf{n}_g(\mathbf{p}) \wedge \mathbf{n}_h(\mathbf{p})|}. \tag{13}$$

The index depends on the order of g, h, in fact $\mathrm{ind}(h, g, \mathbf{p}) = -\mathrm{ind}(g, h, \mathbf{p})$. When the order is understood we simply write $\mathrm{ind}(\mathbf{p})$.

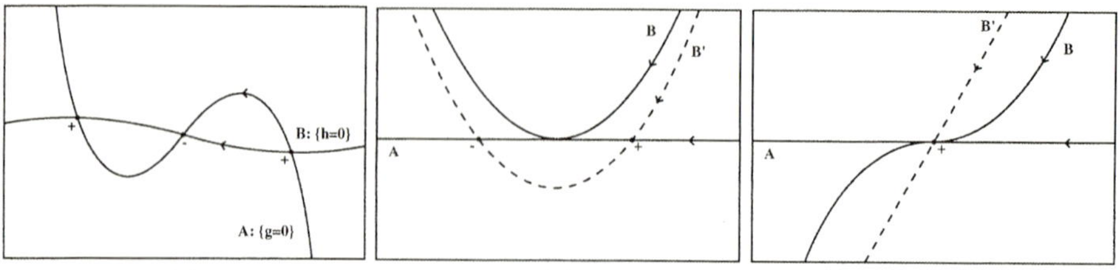

(a) (b) (c)

Fig. 3. (a) Transversal intersections of A and B. (b) The continuity argument for non-transversal intersection without crossing and (c) with crossing

Theorem 2. *Assume that* $g, h : U \to \mathbf{R}$ *are continuously differentiable functions with zero as a regular value, and such that the associated curves* $A = \{\mathbf{x} : g(\mathbf{x}) = 0\}$ *and* $B = \{\mathbf{x} : h(\mathbf{x}) = 0\}$ *intersect transversally. Then the product of the distributions* $\delta(g), \delta(h) \in \mathcal{D}'(U)$ *is well-defined, and*

$$(\nabla g \wedge \nabla h)\delta(g)\delta(h) = \sum_{\mathbf{p} \in A \cap B} \operatorname{ind}(\mathbf{p})\, \delta_{\mathbf{p}}, \tag{14}$$

where $\operatorname{ind}(\mathbf{p}) = \operatorname{ind}(g, h, \mathbf{p})$.

Proof. The product $\delta(g)\delta(h)$ is to be interpreted as a tensor product of Dirac distributions in the (g, h)-plane, hence we have $\delta(g)\delta(h) = \delta(g, h) \in \mathcal{D}'(\mathbf{R}^2)$. In order to express this product in the (x, y)-plane we need to compute the composition by the mapping $G(x, y) = (g(x, y), h(x, y))$ from U to $\mathbf{R}^2$. Thus it follows from (11) that

$$\delta(g, h) = \delta(G) = \sum_{\mathbf{p} \in G^{-1}(\mathbf{0})} \frac{\delta_{\mathbf{p}}}{|\det G'(\mathbf{p})|} = \sum_{\mathbf{p} \in A \cap B} \frac{\delta_{\mathbf{p}}}{|\nabla g(\mathbf{p}) \wedge \nabla h(\mathbf{p})|}.$$

If we multiply this identity by the continuous function $(\nabla g \wedge \nabla h)$, and apply the analytic expression (13) for the index, then the assertion of the theorem follows.

The result of Theorem 2 can be extended to non-transversal (but discrete) intersections of A and B by a continuity argument, which we sketch briefly.

Consider the situation in Figure 3(b), where A and B meet in a point $\mathbf{p}$ without crossing (h does not change sign along A). Let B' be a small perturbation of B. Then there are two possibilities. Either B' does not intersect any more, in which case there is no contribution to (14), or B' intersects A in two nearby points with opposite indices, as in Figure 3(b). The two contributions to (14) are equal in size but have opposite signs, so they cancel (in the sense of distributions) as B' approaches B. We conclude that the formula (14) holds if we agree to set $\operatorname{ind}(\mathbf{p}) = 0$ if A and B meet without crossing.

Suppose instead that A and B meet non-transversally, but they cross, as in Figure 3(c). Approximating B by a sequence of curves, like B', which intersect A

transversally, we see that all of these curves give exactly the same contribution to (14), so passing to the limit, it follows that (14) must hold for B as well.

We shall therefore feel free to use the result of Theorem 2 in all cases where A and B do not coincide along a common segment.

6 The Gâteaux Derivative of E_A: Formal Calculations

In this section formal calculations show that the differential (or Gâteaux derivative) of the alignment functional (1) contains a term with the second order (distributional) derivative of the $|\cdot|$-function. This term is analyzed in Section 8.

First we rewrite the (1) as

$$E_A(\Gamma) = \int_\Gamma |\mathbf{v}\cdot\mathbf{n}|\,d\sigma = \int_{\mathbf{R}^2} |\mathbf{v}\cdot\frac{\nabla\phi}{|\nabla\phi|}||\nabla\phi|\delta(\phi)\,d\mathbf{x} = \int_{\mathbf{R}^2} |\mathbf{v}\cdot\nabla\phi|\delta(\phi)\,d\mathbf{x}\ .$$

Let Γ is represented by the level set function ϕ, and consider a variation of Γ given by $\phi^s = \phi + s\psi$. Then the normal velocity of the moving curves (at $s = 0$) is $v = -\psi/|\nabla\phi|$. The differential is given by

$$\begin{aligned} dE(\Gamma)v &= \frac{d}{ds}E(\phi+s\psi)\Big|_{s=0} = \int_{\mathbf{R}^2}\frac{d}{ds}\left[|\mathbf{v}\cdot\nabla\phi^s|\delta(\phi^s)\right]\,d\mathbf{x}\Big|_{s=0} \\ &= \int_{\mathbf{R}^2} \operatorname{sign}(\mathbf{v}\cdot\nabla\phi)(\mathbf{v}\cdot\nabla\psi)\delta(\phi) + |\mathbf{v}\cdot\nabla\phi|\delta'(\phi)\psi\,d\mathbf{x}\ . \end{aligned} \tag{15}$$

Integrating by parts on $\nabla\psi$ yields

$$\begin{aligned} dE(\Gamma)v &= \int_{\mathbf{R}^2}(-\psi)\nabla\cdot[\mathbf{v}\operatorname{sign}(\mathbf{v}\cdot\nabla\phi)]\delta(\phi) + (-\psi)\operatorname{sign}(\mathbf{v}\cdot\nabla\phi)(\mathbf{v}\cdot\nabla\phi)\delta'(\phi)\,d\mathbf{x} \\ &\quad - \int_{\mathbf{R}^2}(-\psi)|\mathbf{v}\cdot\nabla\phi|\delta'(\phi)\,d\mathbf{x} \\ &= \int_{\mathbf{R}^2}\frac{-\psi}{|\nabla\phi|}\nabla\cdot[\mathbf{v}\operatorname{sign}(\mathbf{v}\cdot\nabla\phi)]|\nabla\phi|\delta(\phi)\,d\mathbf{x}. \end{aligned} \tag{16}$$

If Γ is a local maximum for E_A then $dE_A(\Gamma)v = 0$ for all normal velocities $v = -\psi/|\nabla\phi|$, hence the Euler-Lagrange equation for the problem (3) becomes

$$\operatorname{sign}(\mathbf{v}\cdot\mathbf{n})(\nabla\cdot\mathbf{v})|\nabla\phi|\delta(\phi) + \underbrace{2\delta(\mathbf{v}\cdot\mathbf{n})(\nabla(\mathbf{v}\cdot\mathbf{n})\cdot\mathbf{v})|\nabla\phi|\delta(\phi)}_{(*)} = 0\ . \tag{17}$$

From the result above we have shown that there is an additional term in the Euler-Lagrange equation, containing the second order (distributional) derivative of the $|\cdot|$-function, which is not present in (2). In many applications the second order term $(*)$ is neglected. Can this be justified and motivated mathematically? Or in other words, is (2) still a descent direction for (1) if this term is neglected. To answer this we need to consider the properties of $(*)$.

It should be noted that a part of this problem was analyzed in [7] and [9] under the assumption that $\mathbf{v}\cdot\mathbf{n}$ does not change sign along the curve. For the case of aligning curves to edges in images this is a reasonable assumption at the edges, e.g. a dark object on a bright background will have $\mathbf{v}$ pointing outward along the edge if the vector field is the image gradient $\mathbf{v} = \nabla u$. However, away from the edges this assumption is perhaps too strong.

One has to be cautious when analyzing the influence of this second order term in the level set framework. The reason for this is that both $\mathbf{v}$ and $\mathbf{n}$ are defined in $\mathbf{R}^2$ (or some subset of $\mathbf{R}^2$) but geometrically, the values of $\mathbf{n}$ are only meaningful at Γ. This means that the influence of $(*)$ outside Γ can not be considered a property of the alignment functional (1). Instead this is a property of the representation chosen. This is not a problem as long as one keeps this in mind throughout the analysis.

7 The Singular Curve

In this section we will show that the zero set of $\mathbf{v} \cdot \mathbf{n}$ can be interpreted as an image curve. If $\mathbf{v} \cdot \mathbf{n}$ has constant sign, then the alignment functional is

$$E_A(\Gamma) = \int_\Gamma |\mathbf{v} \cdot \mathbf{n}|\, d\sigma = \text{sign}(\mathbf{v} \cdot \mathbf{n}) \int_\Gamma \mathbf{v} \cdot \mathbf{n}\, d\sigma = \text{sign}(\mathbf{v} \cdot \mathbf{n}) \int_\Omega \nabla \cdot \mathbf{v}\, d\mathbf{x} \ ,$$

where Ω is the interior of Γ, so the corresponding motion PDE is exactly (2). Together with the analysis above this indicates that the difficulties arise where the zero set of $\mathbf{v} \cdot \mathbf{n}$ intersects Γ. This incidence set will play an important role in the analysis of $(*)$.

Let $f : \mathbf{R}^2 \to \mathbf{R}$ be defined by $f(\mathbf{x}) = \mathbf{v}(\mathbf{x}) \cdot \mathbf{n}(\mathbf{x})$ and set $\Sigma = \{\mathbf{x}\ ;\ f(\mathbf{x}) = 0\}$. Here the extension of the curve normal $\mathbf{n}$ is determined by keeping ϕ as a signed distance function, i.e. $|\nabla \phi| = 1$. The curve Σ is called the *singular curve* and may consist of several closed components. Figure 4 shows the singular curve as Γ evolves for the example in Section 3. Although the properties of Σ away from

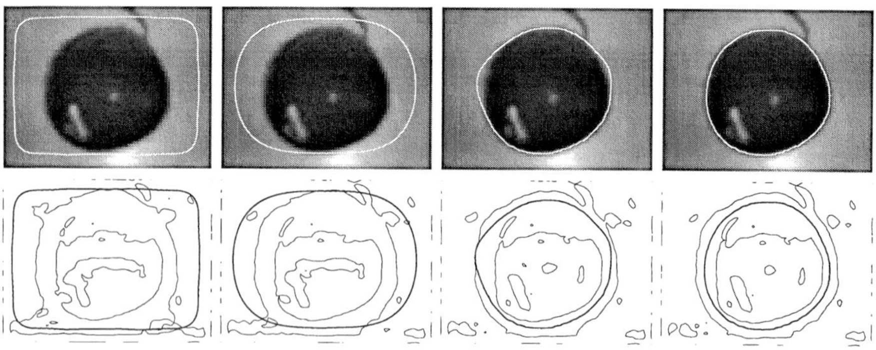

(a) (b) (c) (d)

Fig. 4. The curve evolution in the image using (8). (top) curve evolution overlaid on the image. (bottom) the curve Γ (thick) together with the singular curve Σ

Γ are in fact arbitrary (another extension of $\mathbf{n}$ could have been chosen), Figure 4 shows that there are several places where Σ and Γ coincide and intersect.

8 The Gâteaux Derivative of E_A: Final Analysis

In this section we show how to compute the differential of E_A, that is, we show how to analyze the term $(*)$ in Section 6. In this analysis the singular curve Σ, which was introduced in the previous section, will play a crucial rôle. In order to state the following result, let $\langle u, \varphi \rangle_\Gamma$ denote the value of the distribution $u \in \mathcal{D}'(\Gamma)$ on Γ applied on the test function $\varphi \in C^\infty(\Gamma)$. Also, let $\mathbf{t} = \hat{\mathbf{n}}$ denote the unit tangent field on Γ. We now come to the main result of this paper:

Theorem 3. *If Γ is a curve which is not incident with $\mathbf{v}$ (cf. Section 4) then the functional E_A defined by (1) is differentiable at Γ with the differential*

$$dE_A(\Gamma)v = \int_\Gamma \operatorname{sign}(\mathbf{v} \cdot \mathbf{n})(\nabla \cdot \mathbf{v})\, v\, d\sigma + \left\langle 2 \sum_{\mathbf{p} \in \Gamma \cap \Sigma} (\mathbf{v} \cdot \mathbf{t}) \operatorname{ind}(\mathbf{p}) \delta_\mathbf{p}, v \right\rangle_\Gamma \tag{18}$$

for all normal velocities v at Γ.

Proof. The assertion of the theorem follows from the result (16) of our formal calculations in Section 6, if we can show that

$$2\delta(\mathbf{v} \cdot \mathbf{n})(\nabla(\mathbf{v} \cdot \mathbf{n}) \cdot \mathbf{v})|\nabla\phi|\delta(\phi) \ , \tag{19}$$

i.e. the term $(*)$ in (17), is well-defined and equals the bracket in equation (18).

First, observe that if $\mathbf{p} \in \Gamma \cap \Sigma$, then $\mathbf{v}(\mathbf{p})$ is tangent to Γ, so $\mathbf{v} = (\mathbf{t} \cdot \mathbf{v})\mathbf{t}$ at $\mathbf{p}$. Therefore the scalar product $(\nabla(\mathbf{v} \cdot \mathbf{n}) \cdot \mathbf{v})|\nabla\phi|$ in (19) can be rewritten as

$$(\nabla f \cdot \mathbf{v})|\nabla\phi| = (\mathbf{t} \cdot \mathbf{v})(\nabla f \cdot \mathbf{t})|\nabla\phi| = (\mathbf{t} \cdot \mathbf{v})(\nabla f \cdot \hat{\mathbf{n}})|\nabla\phi| = (\mathbf{t} \cdot \mathbf{v})(\nabla f \wedge \nabla\phi)$$

Since Γ is not incident with $\mathbf{v}$ it follows from Lemma 2 (with $g = f$ and $h = \phi$) that (19) equals

$$\left\langle 2 \sum_{\mathbf{p} \in \Gamma \cap \Sigma} (\mathbf{v} \cdot \mathbf{t}) \operatorname{ind}(\mathbf{p}) \delta_\mathbf{p}, v \right\rangle_\Gamma \ ,$$

as claimed, and the proof is complete.

The following result shows that the hypothesis of Theorem 1 will be automatically satisfied in most cases.

Theorem 4. *If Γ is a local maximum of (1), in particular if Γ solves (3), then the vector field $\mathbf{v}$ must satisfy*

$$(\nabla \cdot \mathbf{v})(\mathbf{x}) = 0 \qquad \textit{for all } \mathbf{x} \in \Gamma, \tag{20}$$

and at least one of the following two alternatives hold: Either
(i) $\mathbf{v} \cdot \mathbf{n}$ *does not change sign on Γ, or*
(ii) $\Gamma \cap \Sigma$ *is non-empty and if $\mathbf{p} \in \Gamma \cap \Sigma$, with* $\operatorname{ind}(\mathbf{p}) \neq 0$, *then* $\mathbf{v}(\mathbf{p}) = \mathbf{0}$.

Proof. It follows from Lemma 1 that Γ is not locally incident with $\mathbf{v}$, so E_A is differentiable at Γ, by Theorem 3, and since Γ is a local maximum, $dE_A(\Gamma)v = 0$ for all admissible normal velocities v at Γ.

We claim that each term in (18) must be zero independently. If not, then $\mathbf{v}(\mathbf{p}) \cdot \mathbf{n}(\mathbf{p}) \operatorname{ind}(\mathbf{p})\delta_{\mathbf{p}} \neq 0$ for at least one $\mathbf{p} \in \Gamma \cap \Sigma$. Choose a non-negative normal velocity $v \in C^\infty(\Gamma)$, with $v(\mathbf{p}) = 1$, which is zero except in the vicinity of $\mathbf{p}$. If the support of v is sufficiently concentrated around $\mathbf{p}$, then the contribution from the first term of (18) can be made arbitrarily small, whereas that from the second term is constant and equal to $\mathbf{v}(\mathbf{p}) \cdot \mathbf{n}(\mathbf{p}) \operatorname{ind}(\mathbf{p})$. This contradicts the necessary condition $dE_A(\Gamma)v = 0$, hence our claim is true, in particular (20) holds.

Now, if (i) is not true, then $\Gamma \cap \Sigma$ must necessarily be non-empty. Since each term of (18) is zero, we see that $\mathbf{v}(\mathbf{p}) \cdot \mathbf{n}(\mathbf{p}) \operatorname{ind}(\mathbf{p})\delta_{\mathbf{p}} = 0$ for each $\mathbf{p} \in \Gamma \cap \Sigma$, but this can be the case only if either $\operatorname{ind}(\mathbf{p}) = 0$ or $\mathbf{v}(\mathbf{p}) \cdot \mathbf{t}(\mathbf{p}) = 0$. This proves that (ii) holds because we know from earlier that $\mathbf{v}$ is tangent to Γ at $\mathbf{p} \in \Gamma \cap \Sigma$.

Theorem 3 can also be used to show the following result regarding ascent directions for (3).

Theorem 5. *The curve evolution with the normal velocity*

$$v(\Gamma) = \operatorname{sign}_\varepsilon(\mathbf{v} \cdot \mathbf{n})(\nabla \cdot \mathbf{v}) \ ,$$

where $\operatorname{sign}_\varepsilon$ *is defined as in (4), is an ascent direction for the problem (3).*

Proof. We only need to show that the differential of E_A applied to $v(\Gamma)$ is non-negative. Using Theorem 3 and the fact that $\operatorname{sign}_\varepsilon(0) = 0$, we find that

$$\begin{aligned}
dE(\Gamma)v(\Gamma) &= \int_\Gamma \operatorname{sign}(\mathbf{v} \cdot \mathbf{n})(\nabla \cdot \mathbf{v})\, v(\Gamma)\, d\sigma + \Big\langle 2 \sum_{\mathbf{p} \in \Gamma \cap \Sigma} (\mathbf{v} \cdot \mathbf{t}) \operatorname{ind}(\mathbf{p})\delta_{\mathbf{p}}, v(\Gamma) \Big\rangle_\Gamma \\
&= \int_\Gamma \operatorname{sign}(\mathbf{v} \cdot \mathbf{n})(\nabla \cdot \mathbf{v}) \operatorname{sign}_\varepsilon(\mathbf{v} \cdot \mathbf{n})(\nabla \cdot \mathbf{v})\, d\sigma \\
&\quad \underbrace{+ \Big\langle 2 \sum_{\mathbf{p} \in \Gamma \cap \Sigma} (\mathbf{v} \cdot \mathbf{t}) \operatorname{ind}(\mathbf{p})\delta_{\mathbf{p}}\, , - \operatorname{sign}_\varepsilon(\mathbf{v} \cdot \mathbf{n})(\nabla \cdot \mathbf{v}) \Big\rangle_\Gamma}_{=0} \\
&= \int_\Gamma |\nabla \cdot \mathbf{v}|^2 \, |\operatorname{sign}_\varepsilon(\mathbf{v} \cdot \mathbf{n})|\, d\sigma \ \geq 0 \ ,
\end{aligned}$$

which proves the claim.

This shows that the equation of motion (2) used in [8, 7], is indeed an ascent direction even without assuming constant sign of $\mathbf{v} \cdot \mathbf{n}$ along Γ.

9 Conclusions

In this paper we have extended earlier derivations [9, 8, 7] of the gradient motion for aligning curves to vector fields using a particular energy functional. We derived the Gâteaux derivative of this functional and showed that there are cases where it is not differentiable. We also showed that the evolution equation (2) is an ascent direction under some weak assumptions, justifying and extending the applications of earlier work, e.g. [7]. Finally we derived necessary conditions for local maxima of the alignment functional, improving on earlier results.

References

1. Mumford, D., Shah, J.: Optimal approximation by piecewise smooth functions and associated variational problems. Comm. Pure Appl. Math. **42** (1989) 577–685
2. Chan, T., Vese, L.: Active contours without edges. IEEE Trans. Image Processing **10** (2001) 266–277
3. Brook, A., Kimmel, R., Sochen, N.: Variational restoration and edge detection for color images. J. Math. Imaging Vision **18** (2003) 247–268
4. Kass, M., Witkin, A., Terzopoulos, D.: Snakes: Active contour models. Int. J. Computer Vision **1** (1987) 321–331
5. Kichenassamy, S., Kumar, A., Olver, P., Tannenbaum, A., Yezzi, A.: Gradient flows and geometric active contour models. In: ICCV. (1995) 810–815
6. Caselles, V., Kimmel, R., Sapiro, G.: Geodesic active contours. Int. Journal of Computer Vision (1997)
7. Kimmel, R., Bruckstein, A.: Regularized Laplacian zero crossings as optimal edge integrators. Int. Journal of Computer Vision **53** (2003) 225–243
8. Kimmel, R.: Fast edge integration. In Paragios, N., Osher, S., eds.: Geometric Level Set Methods in Imaging, Vision and Graphics, Springer-Verlag (2003)
9. Desolneux, A., Moisan, L., Morel, J.M.: Variational snake theory. In Paragios, N., Osher, S., eds.: Geometric Level Set Methods in Imaging, Vision and Graphics, Springer-Verlag (2003)
10. Thorpe, J.A.: Elementary Topics in Differential Geometry. Springer-Verlag (1985)
11. Osher, S.J., Fedkiw, R.P.: Level Set Methods and Dynamic Implicit Surfaces. Springer Verlag (2002)

Enhancing Images Painted on Manifolds

Alon Spira and Ron Kimmel

Department of Computer Science, Technion, Israel
{salon, ron}@cs.technion.ac.il

Abstract. The fields of image processing, computer vision and computer graphics have concentrated traditionally on regular 2D images. Recently, images painted on 2D manifolds are becoming more popular and are used in face recognition, volumetric medical image processing, 3D computer graphics, and many other applications. The need has risen to regularize this type of images.

Various manifold representations are the input for these applications. Among the main representations are triangulated manifolds and parametric manifolds. We extend the short time image enhancing Beltrami kernel from 2D images to these manifold representations. This approach suits also other manifold representations that can be easily converted to triangulated manifolds, such as implicit manifolds and point clouds.

The arbitrary time step enabled by the use of the kernel filtering approach offers a tradeoff between the accuracy of the flow and its execution time. The numerical scheme used to construct the kernel makes the method applicable to all types of manifolds, including open manifolds and self intersecting manifolds. The calculations are done on the 2D manifold itself and are not affected by the complexity of the manifold or the dimension of the space in which it is embedded. The method is demonstrated on images painted on synthetic manifolds and is used to selectively smooth face images. Incorporating the geometrical information of the face manifolds in the regularization process yields improved results.

1 Introduction

The Beltrami framework [5, 17] enables state of the art image regularization. It produces a spectrum of image enhancing algorithms ranging from the L_2 linear diffusion to the L_1 non-linear flows. Apart from regular 2D images, the framework was used for textures, video, and volumetric data [6], non-Euclidean color spaces [18], and orientation diffusion [8]. A detailed review can be found in [23].

The recent increase in applications using images painted on 2D manifolds, requires the development of computational tools for regularizing such images. An approach based on harmonic maps was developed to enhance images painted on implicit manifolds [2, 3, 9]. In this approach the manifold is the zero set of a level set function [10] defined in the space in which the manifold is embedded. As noted before [19], this approach has three main drawbacks: the need to extend the manifold to the embedding space, performing the calculations there

R. Kimmel, N. Sochen, J. Weickert (Eds.): Scale-Space 2005, LNCS 3459, pp. 492–502, 2005.

(which might be computationally prohibitive for spaces with more than three dimensions) and the method's applicability only to manifolds represented by a level set and thus excluding more general manifolds, such as open manifolds and self intersecting ones.

Sochen et. al. [15, 14] extended the Beltrami flow for images painted on explicit and implicit manifolds. They have also shown the Beltrami flow to be a generalization of the approach discussed in the previous paragraph. Still, the explicit numerical schemes used to implement the Beltrami flow require an upper bound on the time step used and might result in many iterations. Furthermore, the method was not extended to triangulated manifolds, which are common in many applications.

Recently, Bajaj et. al. [1] and Clarenz et. al. [4] presented combined regularizations of triangulated manifolds and the images painted on them. Both the manifolds and the images undergo anisotropic diffusions. The numerical scheme in [1] consists of Loop's subdivision while [4] uses a finite element discretization in space. Both use semi-implicit finite difference discretizations in time.

A short time kernel for the Beltrami flow on regular 2D images was presented in [22]. It followed the introduction of a short time kernel for 1D non-linear diffusion [16] and an approximation for the 2D Beltrami operator [13]. These kernels implement the flows by 'convolving' the signals with the kernels, similar to the implementation of the heat equation by a convolution with a gaussian kernel. The numerical implementation of the kernels enables an arbitrary time step that gives a tradeoff between the accuracy of the flow and its execution time.

We present here an extension of the short time Beltrami kernel to images painted on manifolds. This kernel enjoys several important advantages,

- Efficiency achieved by performing the calculations on the 2D manifold itself.
- Flexibility through the tradeoff enabled by the selection of an arbitrary time step.
- Robustness by the applicability of the kernel to all possible 2D manifolds, including open manifolds and self intersecting ones.
- Simplicity due to the applicability of the method to all popular manifold representations including triangulated manifolds, parametric manifolds, implicit manifolds and point clouds. The difference in the implementation of the method to the various manifold representations lies only in the preprocessing stage.

In order to compute the short time kernel we need to calculate geodesic distances between pixels in the image. For images painted on parametric manifolds we use fast marching [11, 12, 24] on parametric manifolds [20, 21]. For images painted on triangulated manifolds we use fast marching on triangulated manifolds [7].

This paper is organized as follows. The first section describes the Beltrami flow for images painted on manifolds. In Section 2 it is implemented by the short time kernel. Section 3 describes the calculation of geodesic distance maps on images. In Section 4 the kernel is demonstrated on images painted on synthetic manifolds and is used to regularize face images. The conclusions are in Section 5.

2 The Beltrami Flow

According to the Beltrami framework [5, 17] the image is represented by $\{X^1, X^2, \ldots, X^M, I^1, I^2, \ldots, I^N\}$, with X^i the spatial coordinates and I^j the intensity components. The following derivation will assume color images painted on parametric manifolds embedded in $\mathbb{R}^3$, where we have $M = 3, N = 3$ and the image is $\{x(u^1, u^2), y(u^1, u^2), z(u^1, u^2), I^1(u^1, u^2), I^2(u^1, u^2), I^3(u^1, u^2)\}$. For other values of M and N the derivation is virtually identical.

If we choose the embedding space to be Euclidean, its metric h_{ij} is represented by the diagonal matrix H, with ones in the first M rows and β^2 in the next N. β is the relative scale between the spatial coordinates and the intensity components. The metric elements g_{ij} of the image are derived from the metric elements h_{ij} and the embedding by the pullback procedure

$$G = (g_{ij}) = \begin{pmatrix} 1 + \beta^2 \sum_a (I_1^a)^2 & \beta^2 \sum_a I_1^a I_2^a \\ \beta^2 \sum_a I_1^a I_2^a & 1 + \beta^2 \sum_a (I_2^a)^2 \end{pmatrix}, \tag{1}$$

with I_j^i the derivative of I^i with respect to u^j.

The Beltrami flow is obtained by minimizing the area of the image

$$S = \iint \sqrt{g} du_1 du_2, \tag{2}$$

with respect to the embedding, where $g = \det(G) = g_{11}g_{22} - g_{12}^2$. The corresponding Euler-Lagrange equations as a gradient descent process are

$$I_t^i = \Delta_g I^i, \tag{3}$$

where Δ_g is the Laplace-Beltrami operator which is the extension of the Laplacian to manifolds.

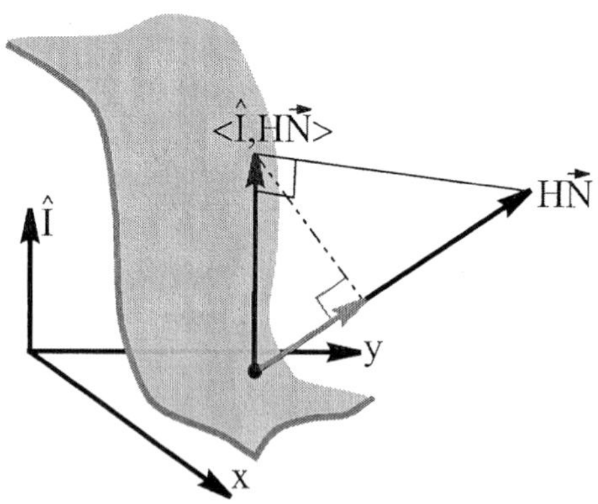

Fig. 1. In the Beltrami flow for 2D regular gray level images the image surface moves according to the intensity component of the mean curvature flow. Geometrically, only the projection of this movement on the normal to the surface matters

For gray level images we have

$$I_t = \Delta_g I = H\langle \hat{I}, \boldsymbol{N} \rangle,$$

i.e., the image surface moves according to the intensity component of the mean curvature flow, see Fig. 1.

3 A Short Time Kernel for the Beltrami Flow

As shown in [22], a kernel exists for the Beltrami flow on 2D regular images. This is similar to the simpler case of linear diffusion, where applying the PDE

$$I_t = \Delta I \tag{4}$$

to the 2D regular image $I(u^1, u^2, t_0)$ for the duration t is equivalent to convolving the image with a Gaussian kernel

$$\begin{aligned} I(u^1, u^2, t_0 + t) &= \iint I(\tilde{u}^1, \tilde{u}^2, t_0) K(|u^1 - \tilde{u}^1|, |u^2 - \tilde{u}^2|; t) d\tilde{u}^1 d\tilde{u}^2 = \\ &= I(u^1, u^2, t_0) * K(u^1, u^2; t) \ , \end{aligned} \tag{5}$$

where the kernel is given by

$$K(u^1, u^2; t) = \frac{1}{4\pi t} \exp\left(-\frac{(u^1)^2 + (u^2)^2}{4t} \right) \ . \tag{6}$$

Because of the non-linearity of the Beltrami flow (the Beltrami operator depends on the data I), the Beltrami kernel is a short time kernel, that if used iteratively, has an equivalent effect to that of the Beltrami flow.

The main idea behind the kernel is presented in Fig 2. For the Gaussian kernel the amplitude of the filtered image at a specific pixel is the sum of the neighboring pixels' amplitudes weighted according to their distance along the coordinate axis. For the nonlinear Beltrami kernel the weighting is according to the geodesic distance on the image itself. The Beltrami kernel 'resides' on the image while for the linear kernel the Gaussian 'resides' on the coordinate axis. This is the reason why linear diffusion blurs the image while the Beltrami flow removes the noise but keeps the edges intact.

In each iteration of the Beltrami kernel we use

$$I^i(u^1, u^2, t_0 + t) = \iint I^i(\tilde{u}^1, \tilde{u}^2, t_0) K(u^1, u^2, \tilde{u}^1, \tilde{u}^2; t) d\tilde{u}^1 d\tilde{u}^2 \ , \tag{7}$$

with the kernel

$$K(u^1, u^2, \tilde{u}^1, \tilde{u}^2; t) = \frac{H_0}{t} \exp\left(-\frac{\left(\int_{(u^1,u^2)}^{(\tilde{u}^1,\tilde{u}^2)} ds \right)^2}{4t} \right)$$

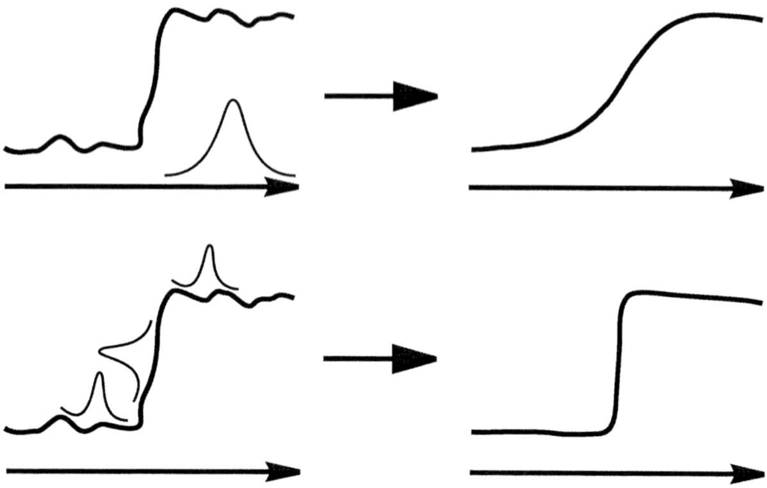

Fig. 2. Filtering an image with a linear Gaussian kernel (top) and a nonlinear Beltrami kernel (bottom)

$$= \frac{H_0}{t} \exp\left(-\frac{d_g^2\left((u^1,u^2),(\tilde{u}^1,\tilde{u}^2)\right)}{4t}\right), \tag{8}$$

where ds is an arc-length element on the image, and $d_g(p_1, p_2)$ is the geodesic distance between two pixels p_1 and p_2. For the full derivation of the kernel see [22]. The derivation of the short time kernel for the Beltrami flow on images painted on manifolds is the same. The difference lies in the calculation of geodesic distances on this kind of images, which will be detailed in the next section.

The resulting update step for the Beltrami kernel is

$$I^i(u^1,u^2,t_0{+}t) = \frac{H_0}{t} \iint_{(\tilde{u}^1,\tilde{u}^2)\in N(u^1,u^2)} I^i(\tilde{u}^1,\tilde{u}^2,t_0) \exp\left(-\frac{\left(\int_{(u^1,u^2)}^{(\tilde{u}^1,\tilde{u}^2)} ds\right)^2}{4t}\right) d\tilde{u}^1 d\tilde{u}^2, \tag{9}$$

with $N(u^1, u^2)$ the neighborhood of the pixel (u^1, u^2), where the value of the kernel is above a certain threshold. Due to the monotone nature of the fast marching algorithm used in the next section for the solution of the eikonal equation, once a pixel is reached, where the value of the kernel is smaller than the threshold, the algorithm can stop and thereby naturally bound the numerical support of the kernel. The value of the kernel for the remaining pixels is negligible. Therefore, the eikonal equation is solved only in a small neighborhood of each pixel. H_0 is taken such that integration over the kernel in the neighborhood $N(u^1, u^2)$ of the pixel equals one.

4 Solving the Eikonal Equation on Images Painted on Manifolds

As shown in the previous section, the construction of the kernel for a pixel requires the calculation of the geodesic distances between the pixel and its neighbors. We place the origin of the coordinate system of the image ($u^1 = u^2 = 0$) at the pixel. The viscosity solution $\phi(u^1, u^2)$ of the eikonal equation

$$\|\nabla_g \phi\| = 1, \tag{10}$$

is the required geodesic distance map from the pixel to its neighbors, where $\nabla_g \phi$ is the gradient of ϕ on the image. To solve the eikonal equation on the image we use the fast marching method.

Regular 2D images are parametric manifolds, where the metric g_{ij} is given for every point. Therefore, calculating the geodesic distances needed for implementing the kernel to these images [22] was done by an extension of the fast marching method [11, 12, 24] to parametric manifolds [20, 21]. The same method is used here for images painted on parametric manifolds. For images painted on triangulated manifolds we use fast marching on triangulated manifolds [7]. Since the embedding space in our case has at least four dimensions (gray scale images painted on manifolds), calculating the distances explicitly on the 2D image is advantageous.

The original fast marching method solves the eikonal equation in an orthogonal coordinate system. In this case, the numerical support for the update of a grid point consists of one or two points out of its four neighbors. For images, where $g_{12} \neq 0$, we get a non-orthogonal coordinate system on the image. The numerical support should include non-neighboring grid points (pixels). For parametric manifolds the method uses the metric of the image at each pixel in order to find the pixels used for the numerical scheme. In the case of triangulated manifolds, the triangulation is given in advance and it determines the numerical support for each pixel.

The updated pixel together with the two other pixels in its numerical support constitute the vertices of a triangle. This triangle is the numerical stencil for updating the pixel. If the triangle is obtuse, it should be split and replaced by two acute triangles. For parametric manifolds the splitting is done according to the metric at the updated pixel, see [20, 21]. For triangulated manifolds an "unfolding" scheme is used, see [7].

After this pre-processing stage, all the triangles in the numerical grid are acute, as in Fig. 3. The figure shows the method by which the vertex (pixel) C is updated according to the vertices A and B. The objective is to find t such that $\frac{t-u}{h} = 1$ and use it to calculate $\phi(C)$ based on $\phi(A)$ and $\phi(B)$.

The numerical scheme according to [7] is

- $u = \phi(B) - \phi(A)$.
- Solve the quadratic equation

$$(a^2 + b^2 - 2ab\cos\theta)t^2 + 2bu(a\cos\theta - b)t + b^2(u^2 - a^2\sin^2\theta) = 0.$$

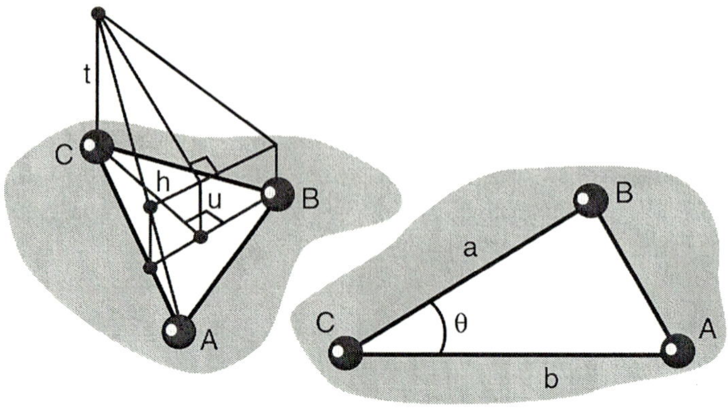

Fig. 3. The numerical stencil used to update $\phi(C)$ according to $\phi(A)$ and $\phi(B)$

– If $u < t$ and $a \cos\theta < \frac{b(t-u)}{t} < \frac{a}{\cos\theta}$, then $\phi(C) = \min\{\phi(C), t + \phi(A)\}$. Else, $\phi(C) = \min\{\phi(C), b + \phi(A), a + \phi(B)\}$.

The numerical scheme described in the previous paragraph enables the update of a pixel according to two of its neighbors. In order to use this scheme to generate the entire distance map the following algorithm [12] is used.

Initialization:

1. The pixel at the origin (for which the kernel is constructed) is defined as *Accepted* and given an initial value of zero.
2. All the other pixels are defined as *Far* and given the value infinity.

Iterations:

1. *Far* 'neighbors' of *Accepted* pixels are defined as *Close.*
2. The values of the *Close* pixels are updated according to the numerical scheme.
3. The *Close* pixel with the minimal value becomes an *Accepted* pixel.
4. If there remain any *Far* pixels, return to step 1.

We use the term 'neighbors' to describe pixels that belong to the same numerical stencil. These pixels are not necessarily neighboring pixels in the image. We find these 'neighbors' during the pre-processing stage described previously.

The complexity of the algorithm is upper bounded by $O(n \log n)$, where n is the number of pixels in the image. The $\log n$ results from using a min-heap data structure for sorting the *Close* pixels [12]. Since There is no need to use all the pixels in the image in order to update one pixel (the value of the kernel for most of these pixels is negligible), we can bound in advance the neighborhood in which the eikonal equation is solved. Thus, we decrease substantially the size of the heap used for the fast marching and enhance its efficiency.

5 Simulations and Results

We first demonstrate the effect of the manifold on the resulting enhanced image. In Fig. 4 a texture image is painted on a flat plane (a regular 2D image) as well as on the manifold $\{x, y, \sin(2\pi x)\sin(2\pi y)\}$. Both images are enhanced using the short time Beltrami kernel. While the texture in the regular 2D image is smoothed evenly, the degree of smoothing in the image painted on the manifold differs according to the geometry of the manifold. In planar areas of the manifold (its peaks and troughs) the smoothing is the same as in the regular image, but on the slopes the spatial extent of the kernel is smaller and there is less smoothing, as can be expected from Fig. 2.

Figure 5 shows the difference between enhancing a color face image as a regular 2D image and enhancing it as an image painted on the face manifold. In both cases one iteration of the kernel with a time step of $t = 0.5$ was applied, only grid points with a kernel value above 0.01 were used for the filtering and the fast

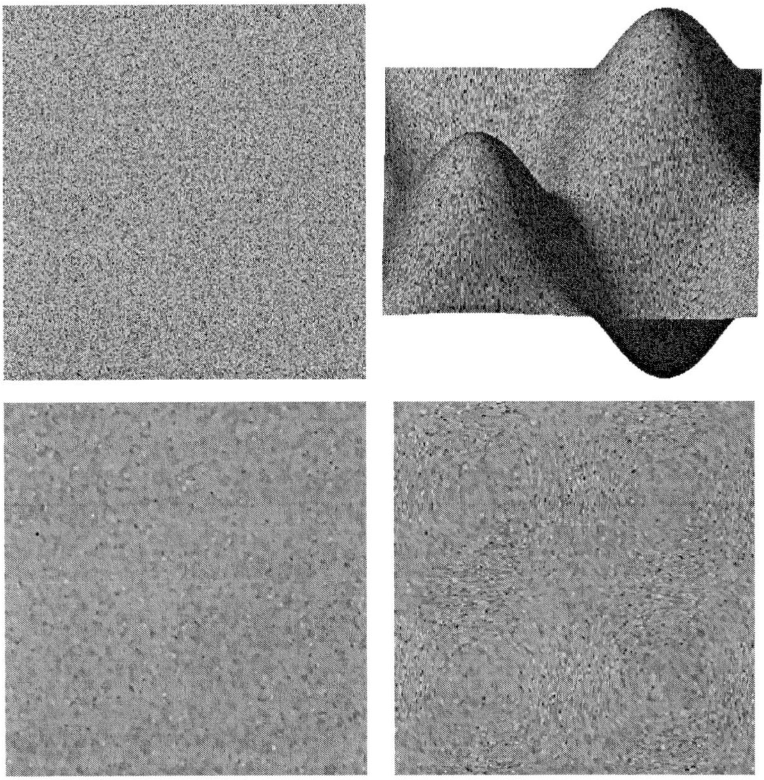

Fig. 4. The effect of the manifold on the enhanced images. The regular 2D image is on the left and the texture painted on the manifold is on the right. The original textures are on the top and the enhanced textures on the bottom

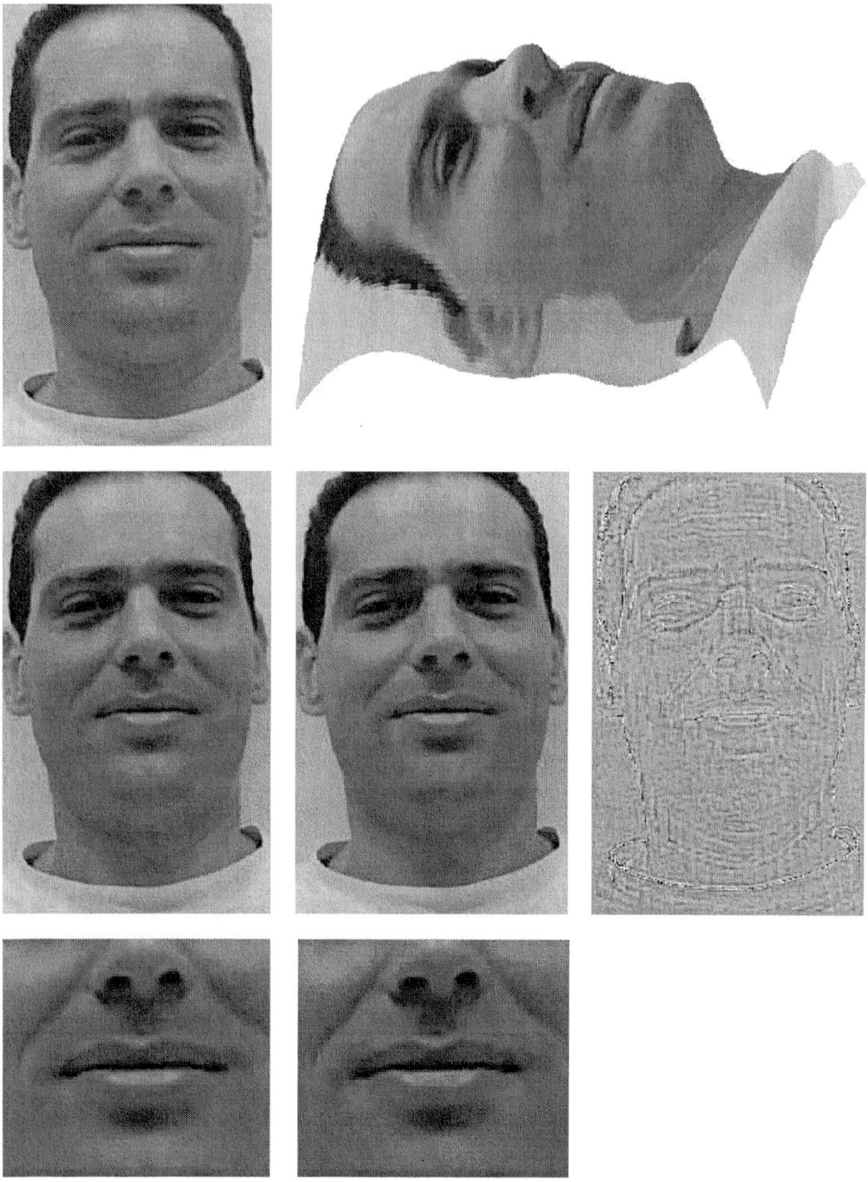

Fig. 5. Enhancing a color face image with the Beltrami kernel. On the left the image is treated as a regular 2D image and on the right as an image painted on the face manifold. On the top are the original images and in the middle and the bottom are the enhanced images. The image in the middle right is the difference between the images on its left

marching was restricted to a neighborhood of 7×7 around each updated pixel. An average of 20 pixels were used in the kernel as a result of these parameters. A comparison between the output images shows that a kernel that takes into account the geometry of the manifold smoothes more in flat regions such as the forehead and smoothes less in edges of the manifold such as the lips. The overall effect is a more selective smoothing and a better looking output image.

6 Conclusions

We have presented a short time kernel for the Beltrami flow for images painted on manifolds. Incorporating the metric of the manifold in the flow produces better results in applications such as face image regularization. The numerical implementation of the kernel handles every possible manifold represented by virtually every common manifold representation. It enjoys low computational complexity further enhanced by an arbitrary time step that enables trading accuracy for a shorter execution time. All these attributes make the Beltrami kernel a highly practical tool for real life graphics and 3D image processing applications.

References

1. C. Bajaj and G. Xu. Anisotropic diffusion of surfaces and functions on surfaces. *ACM Transactions on Graphics*, 22:4–32, January 2003.
2. M. Bertalmio, L. Cheng, S. Osher, and G. Sapiro. Variational problems and partial differential equations on implicit surfaces. *Journal of Computational Physics*, 174:759–780, October 2001.
3. M. Bertalmio, F. Memoli, L. Cheng, G. Sapiro, and S. Osher. Variational problems and partial differential equations on implicit surfaces: Bye bye triangulated surfaces? *Geometric Level Set Methods in Imaging, Vision, and Graphics, S. Osher, N. Paragios, eds., Springer, New York*, pages 381–398, 2003.
4. U. Clarenz, U. Diewald, and M. Rumpf. Processing textured surfaces via anisotropic geometric diffusion. *IEEE Transactions on Image Processing*, 13(2):248–261, 2004.
5. R. Kimmel, R. Malladi, and N. Sochen. Image processing via the beltrami operator. In *Proc. of 3-rd Asian Conf. on Computer Vision*, Hong Kong, January 1998.
6. R. Kimmel, R. Malladi, and N. Sochen. Images as embedding maps and minimal surfaces: Movies, color, texture, and volumetric medical images. *International Journal of Computer Vision*, 39(2):111–129, 2000.
7. R. Kimmel and J. Sethian. Computing geodesic paths on manifolds. *Proceedings of National Academy of Sciences*, 95(15):8431–8435, July 1998.
8. R. Kimmel and N. Sochen. Orientation diffusion or how to comb a porcupine? *special issue on PDEs in Image Processing, Computer Vision, and Computer Graphics, Journal of Visual Communication and Image Representation*, 13:238–248, 2002.
9. F. Memoli, G. Sapiro, and S. Osher. Solving variational problems and partial differential equations mapping into general target manifolds. *Journal of Computational Physics*, 195(1):263–292, 2004.

10. S. Osher and J. Sethian. Fronts propagation with curvature dependent speed: Algorithms based on hamilton-jacobi formulations. *J. Comput. Phys.*, 79:12–49, 1988.
11. J. Sethian. A fast marching level set method for monotonically advancing fronts. *Proceedings of National Academy of Sciences*, 93(4):1591–1595, 1996.
12. J. Sethian. *Level Set Methods and Fast Marching Methods.* Cambridge university press, 1996.
13. N. Sochen. Stochastic processes in vision: From langevin to beltrami. In *Proc. of International Conference on Computer Vision*, Vancouver, Canada, July 2001.
14. N. Sochen, R. Deriche, and L. Lopez-Perez. The beltrami flow over implicit manifolds. In *Proc. of 9th International Conference on Computer Vision*, Nice, October 2003.
15. N. Sochen, R. Deriche, and L. Lopez-Perez. The beltrami flow over manifolds. Technical Report TR-4897, INRIA Sophia-Antipolis, Sophia Antipolis, France, 2003.
16. N. Sochen, R. Kimmel, and A. Bruckstein. Diffusions and confusions in signal and image processing. *Journal of Mathematical Imaging and Vision*, 14(3):195–209, 2001.
17. N. Sochen, R. Kimmel, and R. Malladi. A general framework for low level vision. *IEEE Trans. on Image Processing*, 7(3):310–318, 1998.
18. N. Sochen and Y. Y. Zeevi. Representation of colored images by manifolds embedded in higher dimensional non-euclidean space. In *Proc. of ICIP98*, pages 166–170, Chicago, IL, January 1998.
19. A. Spira and R. Kimmel. Geodesic curvature flow on parametric surfaces. In *Curve and Surface Design: Saint-Malo 2002*, pages 365–373, Saint-Malo, France, June 2002.
20. A. Spira and R. Kimmel. An efficient solution to the eikonal equation on parametric manifolds. In *INTERPHASE 2003 meeting, Isaac Newton Institute for Mathematical Sciences, 2003 Preprints, Preprint no. NI03045-CPD*, UK, June 2003.
21. A. Spira and R. Kimmel. An efficient solution to the eikonal equation on parametric manifolds. *Interfaces and Free Boundaries*, 6(3):315–327, September 2004.
22. A. Spira, R. Kimmel, and N. Sochen. Efficient beltrami flow using a short time kernel. In *Proc. of Scale Spcace 2003, Lecture Notes in Computer Science (vol. 2695)*, pages 511–522, Isle of Skye, Scotland, UK, June 2003.
23. A. Spira, N. Sochen, and R. Kimmel. Geometric filters, diffusion flows, and kernels in image processing. *Handbook of Geometric Computing- Applications in Pattern Recogntion, Computer Vision, Neuralcomputing, and Robotics, Bayro Corrochano Eduardo (Ed.), Springer Verlag, Heidelberg*, pages 203–230, February 2005.
24. J. Tsitsiklis. Efficient algorithms for globally optimal trajectories. *IEEE Trans. on Automatic Control*, 40(9):1528–1538, 1995.

A Two-Step Area Based Method for Automatic Tight Segmentation of Zona Pellucida in HMC Images of Human Embryos*

Adam Karlsson, Niels Chr. Overgaard, and Anders Heyden

Applied Mathematics Group,
School of Technology and Society, Malmö University, Sweden
{Adam.Karlsson, nco, ah}@ts.mah.se

Abstract. An important prognostic parameter for assessing the success of an in vitro fertilization treatment is the variation in thickness of the zona pellucida. Zona pellucida, the envelope of the human embryo, is usually visualized using Hoffman modulation contrast microscopy (HMC). This paper considers automatic segmentation of zona pellucida in HMC images of human embryos. There are two subproblems: (a) the embryo should be separated from the background and (b) the zona should be separated from the rest of the embryo. (a) is solved using a robust formulation of a classical area based method and (b) is solved using a probabilistic method. Both solutions are set in a variational framework using a novel image model for the zona. This variational framework is adapted to handle images in which large artefacts are covered with masks. Since the zona has a simple topology we focus on parametric models and a representation by trigonometric sums is considered.

1 Introduction

In the Nordic countries about 2% of all childbirths are the results of assisted reproductive technology. This has resulted in a significant increase of multiple pregnancies. Multiple pregnancies are associated with an increased risk of complications, resulting in unnecessary suffering for the offspring and increased health care costs. To reduce the number of multiple pregnancies a single embryo transfer would be necessary in many of the in vitro fertilization (IVF) treatments. This raises the need for more accurate and reliable criteria for assessing the potential of an embryo to give rise to a normal pregnancy resulting in a healthy child. It has recently been shown that the zona pellucida thickness variation (ZPTV) is a very good predictor for the development potential. ZPTV might reflect other qualities of the embryo than the older morphological criteria [1, 2]. The hypothesis is that ZPTV could represent the ability of hatching.

* This work has been supported by the Swedish Research Council,project 621-2001-358 and by IVF-Kliniken i Öresund. The image material has been collected by personnel at IVF-Kliniken i Öresund, to whom we express our gratitude.

R. Kimmel, N. Sochen, J. Weickert (Eds.): Scale-Space 2005, LNCS 3459, pp. 503–514, 2005.

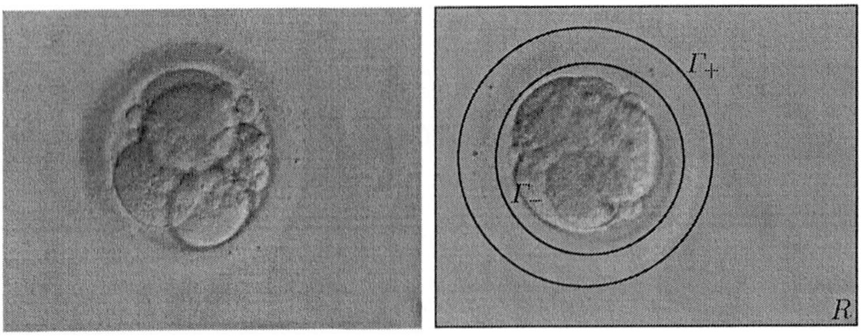

Fig. 1. Form left to right: a) An HMC image of a human embryo. The zona pellucida is the non cellular envelope. b) The curves Γ_+ and Γ_- defining the regions R^+ and R^-

Fig. 1a shows an image of a human embryo. The zona pellucida is the non cellular layer enveloping the embryo. The image is taken using a video camera mounted on a Hoffman Modulation Contrast (HMC) microscope. The HMC technique is a light microscopy contrast technique which generates contrast from phase gradients, originating from differences in the refractive index of the specimen [3]. Since this technique requires no staining it is well suited for studies of living specimen. The HMC microscope uses a modulator with a given orientation. This results in phase differences being most clearly visible in one direction and invisible in the orthogonal direction.

This paper considers the problem of automatically segmenting the zona in HMC images, which is the crucial step in an automated procedure for ZPTV measurements. An automated system would be time-saving and the measurements obtained would be independent from the system operator. Today it is hard to achieve reproducible measurements among different technicians.

An appealing method for segmentation problems is to use some sort of active contour model [4]. However, most active contour models rely on the use of edge maps and in our case, since the zona is diffuse, it is difficult to develop good edge maps. Instead, we consider area based segmentation methods.

Relation to Previous Work. The variational framework used has similarities to the *minimal partitioning problem* of Mumford and Shah [5], studied in [6, 7] by Chan and Vese, and to its probabilistic counterpart developed by Paragios, Rousson and Deriche [8, 9]. In [10] the characteristics of HMC images are studied, in [11] modelling of human embryos is considered and in [12] an approach to blastomere detection is presented. None of these papers address automatic segmentation. There are commercially available tools for assisting laboratory technicians in measuring embryo parameters [13]. To our knowledge these systems use little automation and require much user interaction. The work most closely related to the present paper is the authors' [14], in which the first steps were taken towards an automated procedure for segmentation of the zona.

Contribution of the Paper. The contributions of this paper are found in Sections 3 to 8. First of all, the application *automatic segmentation of HMC images* *of human embryos*, must be considered novel, see above. A new image model for the zona, which improves the one in [14], is developed, cf. Section 3, and used again, as an expectation prior, in the probabilistic area based method in Section 4. A two-step iterative strategy for optimizing the segmentation functionals is presented in Section 7. Some of the images contain large artefacts, which ruins the segmentation results. To solve this problem, a modification of the functionals is introduced in Section 5. This modification allows the application of the method on images where artefacts are covered by masks. Smaller artefacts are handled automatically using a robust formulation of the classical area-based method, see Section 3. Curves are represented by trigonometric polynomials. In Section 6 the Gâteaux derivatives of the functionals are computed in terms of the (Fourier-) coefficients of these trigonometric sums. The formulas, which we believe are new, can be used in more general active contour problems.

Problem Formulation. The segmentation of the zona can be divided into two subproblems: (a) The segmentation of the embryo from the background, i.e. finding the *outer circumference* of the zona (cf. Section 3) and (b) the segmentation of the zona from the interior of the embryo, i.e. finding the *inner circumference* of the zona (cf. Section 4). Since the segmentations are to be used for measurements (of ZPTV), and not for tasks like e.g. tracking of objects, it is critical that the localization of the outer and inner circumferences are precise. That is, the segmentation must be *tight*.

When describing the segmentation procedures for (a) and (b) we assume that we have already found an initialization consisting of two simple closed curves Γ_+ and Γ_-, cf. Fig. 1(b), such that Γ_+ lies in the background and surrounds Γ_- and the embryo, and Γ_- lies in the zona and surrounds the interior of the embryo. How such an initialization is obtained is described in Section 7.

2 Area Based Segmentation

We are going to use two different variants of area based variational segmentation, both of which are covered by the general description given in this section.

Functionals for Area Based Segmentation. Let $I : R \to [-1,1]$ be an image. A pair of subsets (D_0, D_1) of R is called a *decomposition of R* (into two parts) if D_0 and D_1 have no interior points in common and $R = D_0 \cup D_1$. The common boundary of D_0 and D_1 is denoted Γ. It is sometimes easier to refer to a decomposition (D_0, D_1) by referring to Γ.

A *two-phase segmentation* of I is a decomposition (D_0^*, D_1^*) of the image domain R which is optimal with respect to some criterium. One of the subdomains, D_1^* say, is often referred to as the *object*. D_0^* is then called the *background*.

An area-based two-phase segmentation method consists of two ingredients:
(a) A pair of *image models* $I_0, I_1 : R \to [-1,1]$. Each image model depends on points (pixels) $\mathbf{x}$ in the image domain, and a number of parameters $\boldsymbol{\mu}_i = (\mu_i^1, \ldots, \mu_i^{p_i}) \in \mathbf{R}^{p_i}$, that is $I_i = I_i(\mathbf{x}, \boldsymbol{\mu}_i)$, $i = 0, 1$.
(b) A pair of *penalty functions* $V_0, V_1 : \mathbf{R} \to \mathbf{R}_+$.

The *segmentation functional* is the total penalty over the entire image:

$$J[(D_0, D_1), (\boldsymbol{\mu}_0, \boldsymbol{\mu}_1)] = \int_{D_0} V_0(I(\mathbf{x}) - I_0(\mathbf{x}, \boldsymbol{\mu}_0))\, d\mathbf{x} + \int_{D_1} V_1(I(\mathbf{x}) - I_1(\mathbf{x}, \boldsymbol{\mu}_1))\, d\mathbf{x}. \tag{1}$$

A segmentation in the area based framework is a decomposition $R = D_0^* \cup D_1^*$ and a choice of parameters $\boldsymbol{\mu}_0^*, \boldsymbol{\mu}_1^*$ which are optimal in the sense that

$$J[(D_0^*, D_1^*), (\boldsymbol{\mu}_0^*, \boldsymbol{\mu}_1^*)] = \min_{(D_0, D_1), (\boldsymbol{\mu}_0, \boldsymbol{\mu}_1)} J[(D_0, D_1), (\boldsymbol{\mu}_0, \boldsymbol{\mu}_1)]. \tag{2}$$

If the penalty functions depend on parameters, $V_i = V_i(\cdot, \boldsymbol{\sigma}_i)$, where $\boldsymbol{\sigma}_i \in \mathbf{R}^{q_i}$, $i = 0, 1$, then $\boldsymbol{\sigma}_0$ and $\boldsymbol{\sigma}_1$ must be included in the variational problem (2).

Minimizing the Segmentation Functional. In order to solve the minimization problem (2) we use the identity

$$\min_{(D_0, D_1), (\boldsymbol{\mu}_0, \boldsymbol{\mu}_1)} J[(D_0, D_1), (\boldsymbol{\mu}_0, \boldsymbol{\mu}_1)] = \min_{(D_0, D_1)} \left\{ \min_{(\boldsymbol{\mu}_0, \boldsymbol{\mu}_1)} J[(D_0, D_1), (\boldsymbol{\mu}_0, \boldsymbol{\mu}_1)] \right\}.$$

The inner optimization problem can be solve for $R = D_0 \cup D_1$ fixed. Clearly

$$\min_{(\boldsymbol{\mu}_0, \boldsymbol{\mu}_1)} J = \min_{\boldsymbol{\mu}_0} \int_{D_0} V_0(I - I_0(\cdot, \boldsymbol{\mu}_0))\, d\mathbf{x} + \min_{\boldsymbol{\mu}_1} \int_{D_1} V_1(I - I_1(\cdot, \boldsymbol{\mu}_1))\, d\mathbf{x}. \tag{3}$$

If the parameter dependence $\boldsymbol{\mu}_i \mapsto I_i(\cdot, \boldsymbol{\mu}_i)$ is linear, which it is in our case, then reasonable assumptions on the penalty functions V_0, V_1 ensure that there exist *unique* parameters $\boldsymbol{\mu}_0^* = \boldsymbol{\mu}_0^*(D_0)$ and $\boldsymbol{\mu}_1^* = \boldsymbol{\mu}_1^*(D_1)$ which solve (3). For quadratic penalty functions the optimal parameters are given by explicit formulas, as we shall see in the following sections.

Next, introduce the *reduced functional* depending only on the decomposition:

$$\hat{J}[(D_0, D_1)] = J[(D_0, D_1), (\boldsymbol{\mu}_0^*(D_0), \boldsymbol{\mu}_1^*(D_1))].$$

Then (2) becomes the problem of finding a decomposition D_0^*, D_1^* such that

$$\hat{J}[(D_0^*, D_1^*)] = \min_{(D_0, D_1)} \hat{J}[(D_0, D_1)]. \tag{4}$$

To compute the Gâteaux derivative (or differential) of $\hat{J}$ at Γ, take a differentiable function $\nu : \Gamma \to \mathbf{R}$ defined on the curve, and define a curve evolution by $\Gamma(t) = \mathbf{x} + t\nu(\mathbf{x})\mathbf{n}(\mathbf{x})$, for $\mathbf{x} \in \Gamma$. Here $\mathbf{n}$ denotes the outward unit normal field on Γ. Clearly $\Gamma(0) = \Gamma$ and the *normal velocity* of the curve evolution at $t = 0$ is $\dot{\Gamma}(0) = \nu$. The differential of $\hat{J}$ is given by

$$d\hat{J}(\Gamma; \nu) = \frac{d}{dt} \hat{J}(\Gamma(t))\Big|_{t=0} = \int_{\Gamma} \left\{ V_1(I - I_1(\cdot, \boldsymbol{\mu}_1^*)) - V_0(I - I_0(\cdot, \boldsymbol{\mu}_0^*)) \right\} \nu\, ds,$$

where ds is the curve element on Γ. The *gradient of* $\hat{J}$ is defined as the *real valued* function on Γ given by

$$\nabla \hat{J}(\Gamma; \mathbf{x}) = V_1(I(\mathbf{x}) - I_1(\mathbf{x}, \boldsymbol{\mu}_1^*)) - V_0(I(\mathbf{x}) - I_0(\mathbf{x}, \boldsymbol{\mu}_0^*)), \qquad (\mathbf{x} \in \Gamma), \tag{5}$$

where $\boldsymbol{\mu}_i^* = \boldsymbol{\mu}_i^*(D_i)$ are the optimal parameters.

Given an initial decomposition Γ_0, a (local) minimizer of $\hat{J}$ (and a solution of (4)) can be found by solving the *gradient descent motion problem*,

$$\dot{\Gamma}(t) = -\nabla \hat{J}(\Gamma(t)), \qquad \Gamma(0) = \Gamma_0, \tag{6}$$

where $\dot{\Gamma}(t)$ denotes the normal velocity of the curve evolution.

3 Finding the Outer Circumference

This section describes the segmentation functional for the outer circumference of the zona. This functional is inspired by the one in [6], but differs from it by the introduction of a new, more elaborate, image model and the use of truncated penalty functions, which leads to a more robust penalization procedure.

The segmentation functional is defined on the restriction of the original image to the set $R^+ = \{\mathbf{x} \in R : \mathbf{x} \text{ is outside } \Gamma_-\}$, see Fig. 1b. We seek a decomposition of R^+ into D_1, the zona, and D_0, the background.

The intensity of the background is essentially constant, so we set

$$I_0^+(\mathbf{x}, \mu_0) = \mu_0. \tag{7}$$

The model, I_1^+, for the zona is much more interesting. The angular dependence of the intensity, which is so conspicuous in HMC images, see Fig. 1a, is built into the model by the introduction of the distance function $d_+(\mathbf{x}) = \text{dist}(\mathbf{x}, \Gamma_-)$:

$$I_1^+(\mathbf{x}, \mu_1, \mathbf{w}) = \mu_1 + \mathbf{w}^T \nabla d_+(\mathbf{x}), \tag{8}$$

where $\mu_1 \in \mathbf{R}$ and $\mathbf{w} \in \mathbf{R}^2$ are parameters. The outward unit vector field ∇d_+ describes the angular dependence in the images, and $\mathbf{w}$ corresponds to the direction and amplitude of this angular dependence.

Many of the images we study contain small artefacts, in the form of spermatozoa and cell fragments. This makes the commonly used penalty function $\frac{1}{2}|\cdot|^2$ unsuitable, because it is sensitive to outliers. Instead, we consider a *robust formulation* in which the penalty functions are truncated:

$$V_i(t) = \frac{1}{2}\min(t^2, \tau_i^2) \qquad (i = 0, 1), \tag{9}$$

where $\tau_0, \tau_1 > 0$ are fixed parameters.

Plugging (7), (8) and (9) into equation (1) in Section 2 we get the segmentation functional for the outer circumference:

$$\begin{aligned} J_+(\Gamma, \mu_0, \mu_1, \mathbf{w}) = &\frac{1}{2}\int_{D_0} \min\left[(I(\mathbf{x}) - \mu_0)^2, \tau_0^2\right] d\mathbf{x} \\ &+ \frac{1}{2}\int_{D_1} \min\left[(I(\mathbf{x}) - \mu_1 - \mathbf{w}^T \nabla d_+(\mathbf{x}))^2, \tau_1^2\right] d\mathbf{x}. \end{aligned} \tag{10}$$

To find the optimal parameters μ_0^*, μ_1^* and $\mathbf{w}^*$ for a fixed decomposition $R^+ = D_0 \cup D_1$, we differentiate under the integral signs in (10), and get the identities

$$\mu_0^* = |D_0'(\mu_0^*)|^{-1} \int_{D_0'(\mu_0^*)} I(\mathbf{x})\, d\mathbf{x} \tag{11}$$

and

$$\mu_1^* = |D_1'(\mu_1^*, \mathbf{w}^*)|^{-1} \int_{D_1'(\mu_1^*,\mathbf{w}^*)} I(\mathbf{x})\, d\mathbf{x},$$

$$\mathbf{w}^* = \left[\int_{D_1'(\mu_1^*,\mathbf{w}^*)} \nabla d_+(\mathbf{x})(\nabla d_+(\mathbf{x}))^T\, d\mathbf{x}\right]^{-1} \int_{D_1'(\mu_1^*,\mathbf{w}_1^*)} I(\mathbf{x})\nabla d_+(\mathbf{x})\, d\mathbf{x}, \tag{12}$$

where $D_0'(\mu_0^*) = \{\mathbf{x} \in D_0 : |I(\mathbf{x}) - I_0^+(\mathbf{x}, \mu_0^*)| < \tau_0\}$ and $D_1'(\mu_0^*, \mathbf{w}^*) = \{\mathbf{x} \in D_1 : |I(\mathbf{x}) - I_1^+(\mathbf{x}, \mu_0^*, \mathbf{w}^*)| < \tau_1\}$. Equations (11) and (12) are fixed point problems for the optimal parameters μ_0^*, μ_1^* and $\mathbf{w}^*$ and can in principle be solved using fixed point iterations. If $\tau_0 = \tau_1 = \infty$ then $V_i(\cdot) = \frac{1}{2}|\cdot|^2$ and the solution is

$$\mu_0^* = |D_0|^{-1} \int_{D_0} I(\mathbf{x})\, d\mathbf{x}, \qquad \mu_1^* = |D_1|^{-1} \int_{D_1} I(\mathbf{x})\, d\mathbf{x}, \tag{13}$$

$$\mathbf{w}^* = \left[\int_{D_1} \nabla d_+(\mathbf{x})(\nabla d_+(\mathbf{x}))^T\, d\mathbf{x}\right]^{-1} \int_{D_1} I(\mathbf{x})\nabla d_+(\mathbf{x})\, d\mathbf{x}. \tag{14}$$

In our implementation (11) and (12) are not solved exactly. Instead (13) and (14) are used as predictors of the optimal parameter values, after which one iteration using (11) and (12) is performed.

From (5) it follows that the gradient of $\hat{J}_+$ is

$$\nabla \hat{J}_+(\Gamma) = \tfrac{1}{2}\min\left[(I - \mu_1^* - \nabla d_+^T \mathbf{w}^*)^2, \tau_1^2\right] - \tfrac{1}{2}\min\left[(I - \mu_0^*)^2, \tau_0^2\right], \tag{15}$$

where μ_0^*, μ_1^* and $\mathbf{w}^*$ solve (11) and (12).

4 Finding the Inner Circumference

For finding the inner circumference a segmentation functional is again defined on a restriction of the image, the set $R^- = \{\mathbf{x} \in R : \mathbf{x} \text{ is inside } \Gamma_+\}$, see Fig. 1b. For an optimal placement of Γ_+, R^- would correspond to the embryo. We seek a decomposition of R^- into D_0, the interior of the embryo, and D_1, the zona.

We use a probabilistic method similar to that of Rousson and Deriche in [9]. This is a Bayesian approach in which the likelihood of a decomposition (D_0, D_1) is maximized given an observation of the image I. The method assumes that the gray levels of the pixels are independent random variables which are normally distributed. More precisely, if a pixel $\mathbf{x} \in R$ is assigned to D_i, it is assumed that $I(\mathbf{x}) \sim \mathcal{N}(I_i^-(\mathbf{x}), \sigma_i)$ for $i = 0, 1$. The means I_i^- are the models of our framework (cf. Section 2) used as expectation priors.

We have yet to define a model which describes the structure in the interior of the embryo. Instead, we use a constant image model

$$I_0(\mathbf{x}, \mu_0) = \mu_0. \tag{16}$$

The thesis being, that our zona model and the variance differences are sufficient to model the difference between the interior and the zona. For the zona we use

$$I_1^-(\mathbf{x}, \mu_1, \mathbf{w}) = \mu_1 + \mathbf{w}^T \nabla d_-(\mathbf{x}) \tag{17}$$

where $d_-(\mathbf{x}) = -\operatorname{dist}(\mathbf{x}, \Gamma_+)$. The minus sign has been introduced for convenience, since then $\nabla d_+(\mathbf{x}) \approx \nabla d_-(\mathbf{x})$ in the zona, which gives the same interpretation of $\mathbf{w}$ in both zona models (8) and (17).

The segmentation functional, which is essentially minus the logarithm of the likelihood functions, is

$$\begin{aligned} J_-(\Gamma, \boldsymbol{\mu}, \mathbf{w}, \boldsymbol{\sigma}) = & \frac{1}{2} \int_{D_0} \log \sigma_0^2 + \frac{|I(\mathbf{x}) - \mu_0|^2}{\sigma_0^2} \, d\mathbf{x} \\ & + \frac{1}{2} \int_{D_1} \log \sigma_1^2 + \frac{|I(\mathbf{x}) - \mu_1 - \mathbf{w}^T \nabla d_-(\mathbf{x})|^2}{\sigma_1^2} \, d\mathbf{x}. \end{aligned} \tag{18}$$

This is the same as (1) with the image models (16) and (17), and the parameter dependent penalty functions

$$V_i(t, \sigma_i) = \frac{1}{2} \left[\log \sigma_i^2 + \frac{|t|^2}{\sigma_i^2} \right] \qquad (i = 0, 1).$$

The optimal values for the parameters μ_0, μ_1 and $\mathbf{w}$ are found using the formulas (13) and (14) of Section 3 with $d_-(\mathbf{x})$ replacing $d_+(\mathbf{x})$ in (14) and the optimum for the standard deviations are clearly given by

$$\sigma_i^* = \left[\frac{1}{|D_i|} \int_{D_i} |I - I_i|^2 \, d\mathbf{x} \right]^{1/2} \qquad (i = 0, 1).$$

The gradient of $\hat{J}_-$ is

$$\nabla \hat{J}_-(\Gamma) = \log \frac{\sigma_1^*}{\sigma_0^*} + \frac{1}{2} \frac{|I - \mu_1^* - \nabla d_-^T \mathbf{w}^*|^2}{{\sigma_1^*}^2} - \frac{1}{2} \frac{|I - \mu_0^*|^2}{{\sigma_0^*}^2}. \tag{19}$$

5 Supervised Segmentation Using Masks

In this section we generalize the formulation given in Section 2 to include *handling of occlusion.* If it can be a priori established that parts of an image scene belongs to neither the background D_0 nor the object D_1, it would be preferable to exclude these parts from the calculations. In our application such regions consist of large artefacts occluding parts of the image scene.

For this purpose we introduce a new fixed set M, called a *mask*, and decompose the image domain R into four regions $D_0 \setminus M$, $D_0 \cap M$, $D_1 \setminus M$ and $D_1 \cap M$. For the regions where image information is retrievable, $D_0 \setminus M$ and $D_1 \setminus M$, we use the penalty functions V_0 and V_1, as before. For the regions were no information is given, each pixel is assigned a penalty which is equal to the average penalty over the corresponding region with known information, i.e. for $\mathbf{x} \in D_i \cap M$ the penalty is $\fint_{D_i \setminus M} V_i(\mathbf{x}, \boldsymbol{\mu}_i)\, d\mathbf{x}$. The total penalty over the image domain becomes

$$J[(D_0, D_1), (\boldsymbol{\mu}_0, \boldsymbol{\mu}_1)] = |D_0| \fint_{D_0 \setminus M} V_0(\mathbf{x}, \boldsymbol{\mu}_0)\, d\mathbf{x} + |D_1| \fint_{D_1 \setminus M} V_1(\mathbf{x}, \boldsymbol{\mu}_1)\, d\mathbf{x}.$$

Under the same assumptions as in Section 2 the minimum over the parameters $(\boldsymbol{\mu}_0, \boldsymbol{\mu}_1)$ can be solved for explicitly. For our segmentation problem we can use the formulas (11), (12), (13) and (14) if D_0 is replaced by $D_0 \setminus M$ and D_1 by $D_1 \setminus M$, to find the optimal values for the parameters.

Set $\hat{J}_i(D_i) = |D_i| \fint_{D_i \setminus M} V_i(\mathbf{x}, \boldsymbol{\mu}_i^*)\, d\mathbf{x}$ and define $\hat{J}(D_0, D_1) = \hat{J}_0(D_0) + \hat{J}_1(D_1)$. The differential of $\hat{J}$ is computed using the product and quotient rule:

$$d\hat{J}(\Gamma; \nu) = \int_{\Gamma \cap M^c} \left\{ V_1(\cdot, \boldsymbol{\mu}_1^*) - V_0(\cdot, \boldsymbol{\mu}_0^*) + \frac{\hat{J}_0(D_0)}{|D_0 \setminus M|} - \frac{\hat{J}_1(D_1)}{|D_1 \setminus M|} \right\} \nu\, ds + \\ + \int_{\Gamma} \left\{ \frac{\hat{J}_1(D_1)}{|D_1|} - \frac{\hat{J}_0(D_0)}{|D_0|} \right\} \nu\, ds \qquad (20)$$

where M^c is the the complement of M with respect to R.

6 Representation by Trigonometric Sums

Here we introduce a trigonometric representation of the curve Γ:

$$\Gamma(\theta) = \Gamma(\theta; \mathbf{a}, \mathbf{A}_1, \ldots, \mathbf{A}_N) = \mathbf{a} + \sum_{k=1}^{N} \mathbf{A}_k \begin{bmatrix} \cos(k\theta) \\ \sin(k\theta) \end{bmatrix}, \qquad \theta \in [0, 2\pi], \qquad (21)$$

where $N \geq 1$ is fixed, $\mathbf{a} \in \mathbf{R}^2$ and $\mathbf{A}_1, \ldots, \mathbf{A}_N$ are 2-by-2 matrices. From the differential (2) of the reduced functional $\hat{J}$, we calculate the gradient with respect to the parameters of this representation. An outward orthogonal field $\hat{\Gamma}$ of Γ is constructed by making a 90 degree clockwise rotation of the tangent vector $\Gamma'(\theta)$, that is:

$$\hat{\Gamma}(\theta) = \sum_{k=1}^{N} k \mathbf{P} \mathbf{A}_k \mathbf{P}^T \begin{bmatrix} \cos(k\theta) \\ \sin(k\theta) \end{bmatrix}, \qquad (22)$$

were $\mathbf{P} = \left[\begin{smallmatrix} 0 & 1 \\ -1 & 0 \end{smallmatrix}\right]$. Recall formula (2) for the differential of $\hat{J}$ and consider an arbitrary small displacement $\delta\Gamma$ of Γ. This displacement gives a normal velocity $\nu = \mathbf{n}^T \delta\Gamma$ and since $\mathbf{n}\, ds = \hat{\Gamma}\, d\theta$ we, with $G(\theta) = (\nabla \hat{J})(\Gamma(\theta))$, get:

$$d\hat{J}(\Gamma;\delta\Gamma) = \int_0^{2\pi} G(\theta)\delta\Gamma^T \hat{\Gamma}(\theta)\,d\theta. \tag{23}$$

We now seek an expression for the gradient $\nabla\hat{J} = (\nabla_{\mathbf{a}}\hat{J}, \nabla_{\mathbf{A}_1}\hat{J}, \ldots, \nabla_{\mathbf{A}_N}\hat{J})$ of $\hat{J}$ with respect to the parameters $\mathbf{a}$ and $\mathbf{A}_j$. Suppose the $\mathbf{A}_j$ are kept fixed and consider the variation $\mathbf{a} \to \mathbf{a} + \delta\mathbf{a}$, where $\delta\mathbf{a} \in \mathbf{R}^2$. Then it is easy to see that $\delta\Gamma = \delta\mathbf{a}$, so $d\hat{J}(\Gamma;\delta\Gamma) = \delta\mathbf{a}^T \int_0^{2\pi} G(\theta)\hat{\Gamma}(\theta)\,d\theta$. Thus

$$\nabla_{\mathbf{a}}\hat{J}(\Gamma) = \int_0^{2\pi} G(\theta)\hat{\Gamma}(\theta)\,d\theta. \tag{24}$$

Now, if we make a small variation of $\mathbf{A}_j$, $\mathbf{A}_j \to \mathbf{A}_j + \delta\mathbf{A}$ then $\delta\Gamma = \delta\mathbf{A}\begin{bmatrix}\cos(j\theta)\\ \sin(j\theta)\end{bmatrix}$ so by equation (23) $d\hat{J}(\Gamma;\delta\Gamma) = \int_0^{2\pi} G(\theta)\begin{bmatrix}\cos(j\theta) & \sin(j\theta)\end{bmatrix}\delta\mathbf{A}^T\hat{\Gamma}\,d\theta$.

The expression $\mathrm{tr}(\mathbf{B}^T\mathbf{A})$, where $\mathbf{A}, \mathbf{B} \in \mathbf{R}^{2\times 2}$, defines a scalar product on the vector space of 2-by-2-matrices. Hence, the linear functional $\Lambda : \mathbf{A} \mapsto \mathbf{u}^T\mathbf{A}\mathbf{v}$, where $\mathbf{u}, \mathbf{v} \in \mathbf{R}^2$ are fixed vectors, can be written in the form $\Lambda(\mathbf{A}) = \mathrm{tr}(\mathbf{B}^T\mathbf{A})$ with $\mathbf{B} = \mathbf{u}\mathbf{v}^T$. It follows that $d\hat{J}(\Gamma;\delta\Gamma) = \mathrm{tr}(\mathbf{B}^T\delta\mathbf{A})$, where $\mathbf{B} = \nabla_{\mathbf{A}_j}\hat{J}$ and

$$\nabla_{\mathbf{A}_j}(\Gamma)\hat{J} = \int_0^{2\pi} G(\theta)\hat{\Gamma}\begin{bmatrix}\cos(j\theta) & \sin(j\theta)\end{bmatrix}\,d\theta. \tag{25}$$

Introducing the Fourier coefficients of $G(\theta)$, $\begin{bmatrix}\alpha_k\\ \beta_k\end{bmatrix} = \frac{1}{\pi}\int_0^{2\pi} G(\theta)\begin{bmatrix}\cos(k\theta)\\ \sin(k\theta)\end{bmatrix}\,d\theta$, $k = 0, \ldots, 2N$, we may rewrite the formulas (24) and (25) as:

$$\nabla_{\mathbf{a}}\hat{J}(\Gamma) = \pi\sum_{k=1}^{N} k\mathbf{P}\mathbf{A}_k\mathbf{P}^T\begin{bmatrix}\alpha_k\\ \beta_k\end{bmatrix}$$

$$\nabla_{\mathbf{A}_j}\hat{J}(\Gamma) = \frac{\pi}{2}\sum_{k=1}^{N} k\mathbf{P}\mathbf{A}_k\mathbf{P}^T\begin{bmatrix}\alpha_{k+j}+\alpha_{k-j} & \beta_{k+j}-\beta_{k-j}\\ \beta_{k+j}+\beta_{k-j} & \alpha_{k-j}-\alpha_{k+j}\end{bmatrix}.$$

Given an initial curve $\Gamma = \Gamma(\cdot;\mathbf{a}^0, \mathbf{A}_1^0, \ldots, \mathbf{A}_N^0)$, a gradient descent curve evolution, cf. (6), is obtained by solving the initial value problem:

$$\begin{cases}\dot{\mathbf{a}} = -\nabla_{\mathbf{a}}\hat{J}(\Gamma), & \mathbf{a}(0) = \mathbf{a}^0,\\ \dot{\mathbf{A}}_j = -\nabla_{\mathbf{A}_j}\hat{J}(\Gamma), & \mathbf{A}_j(0) = \mathbf{A}_j^0,\ j = 1, \ldots, N,\end{cases} \tag{26}$$

where $\Gamma = \Gamma(\cdot;\mathbf{a}, \mathbf{A}_1, \ldots, \mathbf{A}_N)$.

7 Implementation

The images are normalized by subtraction of a linear intensity model adapted to the image in a very rough segmentation of the background. This background estimate is found using a fixed threshold and mathematical morphology. For practical purposes we therefore use $\mu_0 = 0$ in (7). This procedure also gives an initial ellipse E surrounding the embryo.

To find the segmentation of the zona we seek the solutions of (10) and (18). Using trigonometric representations the solutions are found by solving (26). This is done repeatedly as described last in this section under *The Two-Step Method.* Before this process can start we need the initial curves Γ_+ and Γ_-.

Initialization. The curve Γ_+ is found from the ellipse E by using the curve evolution $t \mapsto \Gamma(t)$, with $\Gamma(0) = E$ and

$$\dot{\Gamma}(t) = -\nabla \hat{J}_+(\Gamma(t)),$$

where $\dot{\Gamma}(t)$ is the normal velocity of the moving curve. $\nabla \hat{J}_+(\Gamma)$ is the gradient of $\hat{J}_+$ given in equation (15) using $\tau_0 = \tau_1 = \infty$, except for the fact that $\mu_1^*, \mathbf{w}^*$ are computed by (13) and (14) using an annular region A instead of D_1. The annular region $A = A(t)$, which has a fixed width and $\Gamma(t)$ as its outer boundary, actually moves along with the evolving curve.

Now, Γ_+ is defined as $\Gamma(+\infty)$, i.e. the curve that the evolution converges to. Notice that the procedure does not (necessarily) come from a minimization problem (because of A). $\Gamma(t)$ is, nevertheless, a well-defined curve evolution which is evolved using (26). The curve Γ_- is found by scaling Γ_+. We use $\Gamma_- = \alpha\Gamma_+$, where $\alpha = 0.8$.

The Two-Step Method. Given the initializations Γ_+, Γ_- the inner segmentation problem is solved (cf. Section 4) using (26) with with $N = 1$ and Γ_- as initial curve. This yields a segmentation Γ_-^* for the inner circumference. A new pair of initializers Γ_+, Γ_-^* is then formed and the outer segmentation problem (cf. Section 3) is solved using $N = 1$ and with Γ_+ as initial curve. This gives a segmentation Γ_+^* for the outer circumference. The pair Γ_+^*, Γ_-^* is a segmentation of the zona with $N = 1$.

If a segmentation of the zona with $N = 2$ is wanted, the above procedure is repeated with the initializers Γ_+^*, Γ_-^* and $N = 2$. This can be done any number of times giving segmentations for $N > 2$.

8 Experiments

The proposed method has been on 60 images. The images have been manually collected and were not primarily intended for automatic measurements. Therefore, the image quality, with respect to e.g. illumination and focusing, is far from optimal. Furthermore, some of the images contain small or large artefacts.

The evaluation was performed as outlined in Section 7. To represent the outer circumference two matrices were used ($N = 2$) and the inner circumference was described using three matrices ($N = 3$). The truncation parameters in the robust penalty function (9) were set to $\tau_1 = \tau_2 = 10$. Out of the 60 images 12 were considered to have large artefacts. For these, manually specified masks and the modified method described in Section 5 were used. Fig. 2 shows some of the results obtained using the proposed method on embryos with small or no artefacts. Notice that the small artefacts in the bottom two images do not disturb the method. In Fig. 3a result obtained using a mask is shown together

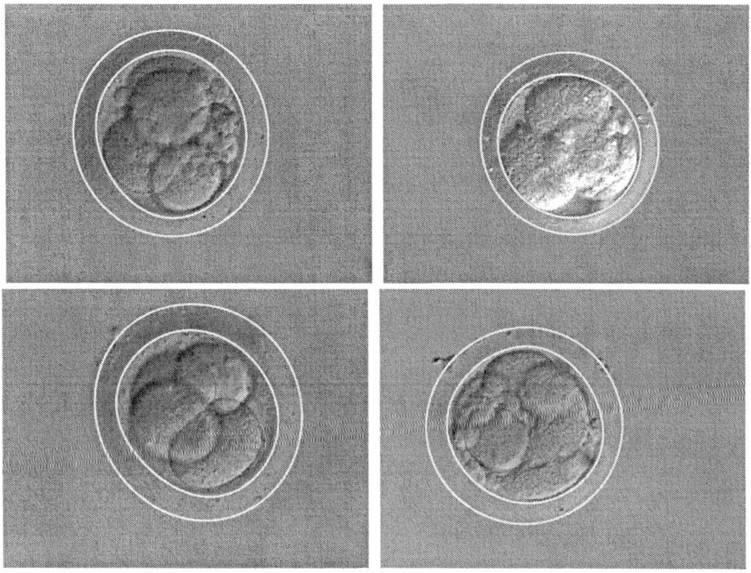

Fig. 2. Segmentation results using the proposed method on embryos with small or no artefacts. The top left image is the same as shown in Fig. 1a

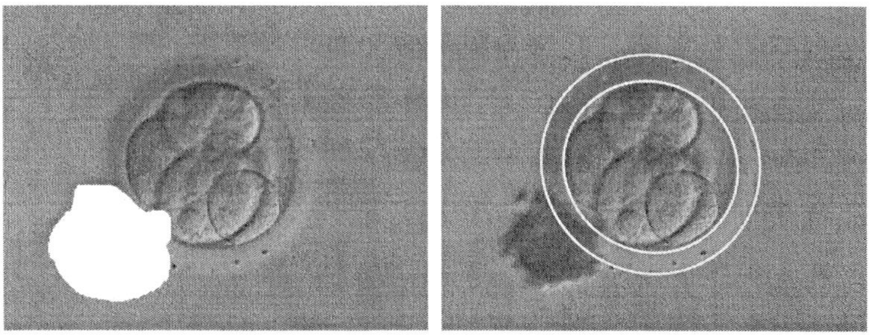

Fig. 3. Segmentation result on an image with a large artefact using a manually specified mask and the modified method of Section 5

with the mask. Out of the 48 images with small or no artefacts, 44 were correctly segmented and 8 of the 12 images with large artefacts were correctly segmented. This corresponds to success rates of 92% and 75%, respectively, and an overall success rate of 88%.

9 Conclusion

We have presented an area based method for segmentation of zona pellucida in HMC images of human embryos. The method is robust and does not depend on edges in the images. Instead an image model is used, which takes advantage of the characteristic appearance of HMC images. Considering that the quality of the images used for evaluation are far from optimal, the results of the proposed method are very good. The few unsuccessful results can be traced to poor image quality. The results indicate that if an automated image acquiring system is used, assuring more stable image quality, we have a segmentation method which is tight and stable enough for automatic measurements of the zona pellucida.

References

1. Palmstierna, M., Murkes, D., Csemiczky, G., Andersson, O., Wramsby, H.: Zona pellucida thickness variation and occurrence of visible mononucleated blastomeres in preembryos are associated with a high pregnancy rate in IVF treatment. Journal of Assisted Reproduction and Genetics **15** (1998) 70–75
2. Gabrielsen, A., Bhatnager, P.R., Petersen, K., Lindenberg, S.: Influence of zona pellucida thickness of human embryos on clinical pregnancy outcome following in vitro fertilization treatment. Journal of Assisted Reproduction and Genetics **17** (2000) 323–328
3. Hoffman, R., Gross, L.: The modulation contrast microscope. Nature **254** (1975) 586–588
4. Kass, M., Witkin, A., Terzopoulos, D.: Snakes: Active contour models. Int. J. Computer Vision **1** (1987) 321–331
5. Mumford, D., Shah, J.: Optimal approximation by piecewise smooth functions and associated variational problems. Comm. Pure Appl. Math. **42** (1989) 577–685
6. Chan, T., Vese, L.: Active contours without edges. IEEE Transactions on Image Processing **10** (2001) 266–277
7. Chan, T., Vese, L.: Image segmentation using level sets and the piecewise constant Mumford-Shah model. CAM Report 00-14, UCLA (2000)
8. Paragios, N., Deriche, R.: Geodesic active regions: A new framework to deal with frame partition problems in computer vision. Journal of Visual Communication and Image Representation (2002) 249–268
9. Rousson, M., Deriche, R.: A variational framework for active and adaptive segmentation of vector valued images. Technical Report 4512, INRIA (2002)
10. Holm Olsen, N., Sporring, J., Nielsen, M., Hnida, C., Ziebe, S.: Reconstructing the optical thickness from Hoffman modulation contrast images. In: 13th Scandinavian Conference on Image Analysis. (2003) 526–533
11. Damgaard Pedersen, U., Fogh Olsen, O., Holm Olsen, N.: A multiphase variational level set approach for modelling human embryos. In: IEEE Proc. Workshop on Variational, Geometric and Level Set Methods in Computer Vision. (2003) 25–32
12. Sporring, J., Holm Olsen, N., Nielsen, M.: Detection and localization of random signals. In: Proc. Scale Space Methods in Computer Vision. (2003) 785–797
13. Image House A/S Medical: (http://www.ihmedical.dk/)
14. Karlsson, A., Overgaard, N.C., Heyden, A.: Automatic segmentation of zona pellucida in HMC images of human embryos. In: Proc. International Conference on Pattern Recognition. Volume 3. (2004) 518–521

Relations Between Higher Order TV Regularization and Support Vector Regression*

G. Steidl[1], S. Didas[2], and J. Neumann[1]

[1] Faculty of Mathematics and Computer Science, D7, 27,
University of Mannheim, 68131 Mannheim, Germany
{steidl, jneumann}@uni-mannheim.de
http://www.kiwi.math.uni-mannheim.de

[2] Mathematical Image Analysis Group,
Faculty of Mathematics and Computer Science, Building 27,
Saarland University, 66123 Saarbrücken, Germany
didas@mia.uni-saarland.de
http://www.mia.uni-saarland.de

Abstract. We study the connection between higher order total variation (TV) regularization and support vector regression (SVR) with spline kernels in a one-dimensional discrete setting. We prove that the contact problem arising in the tube formulation of the TV minimization problem is equivalent to the SVR problem. Since the SVR problem can be solved by standard quadratic programming methods this provides us with an algorithm for the solution of the contact problem even for higher order derivatives. Our numerical experiments illustrate the approach for various orders of derivatives and show its close relation to corresponding nonlinear diffusion and diffusion–reaction equations.

1 Introduction

In this paper, we are interested in constructing a function u that minimizes the functional

$$\int_0^1 (u(x) - f(x))^2 + 2\lambda |u^{(m)}(x)| \, \mathrm{d}x. \tag{1}$$

More precisely, we are concerned with a discrete version of (1), where the functions are only considered at equispaced points.

For $m = 1$ and arbitrary space dimensions, we are in the classical Rudin–Osher–Fatemi setting [16] applied in image denoising and segmentation. Several numerical solution algorithms were proposed, see, e.g., [24] and references therein. A quite interesting method uses the tube formulation of (1). In one space dimension, the tube approach is known as a non–parametric regression model in statistics [10]. A generalization to the two-dimensional setting was proposed in

* This joint research was supported by the Deutsche Forschungsgemeinschaft within the projects WE 2602/2-2 and SCHN 457/5-2. This is gratefully acknowledged.

R. Kimmel, N. Sochen, J. Weickert (Eds.): Scale-Space 2005, LNCS 3459, pp. 515–527, 2005.

[7]. The heart of the tube method consists in the solution of a contact problem within a tube of width depending on the regularization parameter $\lambda > 0$. For $m = 1$, this contact problem can be solved efficiently by the so–called 'taut string algorithm' [4] in one dimension, but becomes harder in higher dimensions [7].

In recent years, there has been a growing interest in higher order variational methods [13, 17, 3, 26, 9, 8]. In particular, a tube approach for $m \geq 2$ was addressed for one dimension in [10] and for higher dimensions based on Meyer's G–norm [12] in [14].

In this paper, we will see that the contact problem can be tackled by solving a simple quadratic optimization problem, namely a so–called *support vector regression* (SVR) problem. SVR methods became very popular in machine learning during the last years, see [23]. The SVR approach also approximates a given function within a tube, but by minimizing a different cost functional. The SVR solution is always contained in a previously determined reproducing kernel Hilbert space. We will prove that in our discrete setting the solution of the contact problem corresponding to (1) coincides with the SVR solution in an appropriately chosen reproducing kernel Hilbert space. This space is a discrete variant of the Sobolev space W_0^m which has a reproducing kernel determined by splines of degree $2m - 1$. We remark that similar results can be obtained by applying the dual approach of Chambolle [2] to our setting. This is discussed in [20]. In this paper, we want to emphasize the spline point of view.

Our paper is organized as follows. We start by developing the tube formulation and SVR with spline kernels in a discrete setting in Sections 2 and 3, respectively. In Section 4, we prove the equivalence of the SVR problem and the key part of the tube algorithm, the contact problem. To prepare a numerical comparison, a discretization of corresponding partial differential equations (PDEs) is provided in Section 5. Our denoising experiments in Section 6 demonstrate properties of our method for various orders of derivatives and show the relation of the variational approach to the numerical solution of corresponding diffusion and diffusion–reaction equations. The paper is concluded with Section 7.

2 Tube Characterization of TV Regularization Functionals with Higher Order Derivatives

In the following, we are concerned with discrete functions defined on some subsets of the integers. As a discrete version of the m–th derivative we choose the m–th finite difference

$$D^m u(j) := \sum_{k=0}^{m} (-1)^{k+m} \binom{m}{k} u\left(j - \left\lfloor \frac{m}{2} \right\rfloor + k\right), \tag{2}$$

where $\lfloor x \rfloor$ denotes the largest integer $\leq x$. Given the values $f(j)$, $j = 1, \ldots, n$, we are interested in finding a discrete function u that minimizes the functional

$$J(u) := \sum_{j=1}^{n} (u(j) - f(j))^2 + 2\lambda \sum_{j=1}^{n} |D^m u(j)|, \tag{3}$$

where we suppose the boundary conditions

$$D^m u(j) := 0, \qquad j = -\left\lfloor \frac{m-1}{2} \right\rfloor, \ldots, \left\lfloor \frac{m}{2} \right\rfloor ; n - \left\lfloor \frac{m-1}{2} \right\rfloor, \ldots, n + \left\lfloor \frac{m}{2} \right\rfloor . \quad (4)$$

The boundary conditions for $j = 1, \ldots, \lfloor \frac{m}{2} \rfloor ; n - \lfloor \frac{m-1}{2} \rfloor, \ldots, n$ imply that the second sum in (3) runs indeed only from $\lfloor \frac{m}{2} \rfloor + 1$ to $n - \lfloor \frac{m+1}{2} \rfloor$. The other boundary conditions are imposed to keep the summation index in the following derivation simple. We remark that these boundary conditions are equivalent to $D^k u(0) = D^k u(n) = 0$, $k = m, \ldots, 2m - 1$.

Since the functional J is strictly convex, our problem has a unique solution. A necessary and sufficient condition for u to be the minimizer of (3) is that the zero vector is an element of the subgradient $\partial J(u)$, i.e., for $j = 1, \ldots, n$, the following inclusions must be fulfilled:

$$0 \in u(j) - f(j) + \lambda \sum_{k=0}^{m} (-1)^k \binom{m}{k} \frac{D^m u(j - \lfloor \frac{m+1}{2} \rfloor + k)}{|D^m u(j - \lfloor \frac{m+1}{2} \rfloor + k)|}, \quad (5)$$

where $y/|y| := [-1, 1]$ if $y = 0$, and where the same quotient $D^m u(\cdot)/|D^m u(\cdot)|$ in different inclusions denotes the same numbers in $[-1, 1]$. Moreover, the summands corresponding to our boundary conditions (4) are zero.

We want to find linear combinations of the right–hand sides of (5) such that most of the terms in the sum vanish. For this, we introduce a discrete equivalent to the m–th power function by $k^{(m)} := 1$ for $m = 0$ and $k^{(m)} := k(k+1)\ldots(k+m-1)$ for $m \geq 1$ and a discrete version of the m–th antiderivative of a function f by

$$F_f(j) := \sum_{k=1}^{j} \frac{(j+1-k)^{(m-1)}}{(m-1)!} f(k), \qquad j = 1, \ldots, n. \quad (6)$$

Then we obtain the following proposition which can be considered as a discrete counterpart of a result in [10].

Proposition 1 (Tube Characterization of TV Minimization). *The function u is a solution of (3) if and only if F_u fulfills the conditions*

$$F_u(j) \in F_f(j) - (-1)^m \lambda \frac{D^m u(j + \lfloor \frac{m}{2} \rfloor)}{|D^m u(j + \lfloor \frac{m}{2} \rfloor)|}, \qquad j = 1, \ldots, n - m \quad (7)$$

and $F_u(n-j) = F_f(n-j)$, $j = 0, \ldots, m-1$.

The basic idea of the proof is the following: For $j \in \{1, \ldots, n\}$ and $k = 1, \ldots, j$, we multiply the k–th inclusion in (5) by $(j+1-k)^{(m-1)}/(m-1)!$, add the corresponding j expressions and transfer $F_u(j)$ to the opposite side. By (6) and setting $F_u(j) := 0$, $j = -(m-1), \ldots, 0$, we obtain u from given F_u by

$$u(j) = \sum_{k=0}^{m} (-1)^k \binom{m}{k} F_u(j-k), \qquad j = 1, \ldots, n. \quad (8)$$

so that by (11), the support vectors j can be also characterized by $D^{2m}U(j) \neq 0$.

We summarize the properties of the SVR solution:

(S1) $U(j) = 0$ for $j = -(m-1), \ldots, 0$,
(S2) $U(n-j) = F(n-j)$ for $j = 0, \ldots, m-1$,
(S3) U lies in a tube around F of width 2λ, i.e.,
$|F(j) - U(j)| \le \lambda$ for $j = 1, \ldots, n-m$.
(S4) Let $\tilde{\Lambda} := \{j \in \{1, \ldots, n\} : D^{2m}U(j) \neq 0\}$.
If $j \in \tilde{\Lambda}$, then $U(j)$ contacts the boundary of the tube,
where j are the support vectors obtained by solving (14).

Comparing these properties with (T1) – (T4), we see that for $F = F_f$ only the fourth condition differs.

Finally, we remark that the SVR solution U can be considered as sparse approximation of F. In particular, by [6], U (without the last m equality constraints) is also the solution of the unconstrained minimization problem

$$\|F - U\|^2_{\mathcal{W}_0^m} + 2\lambda \|\boldsymbol{c}\|_{\ell_1} \to \min .$$

4 Equivalence of Tube and SVR Solution

In the following, we set $F := F_f$ in (14) and show that the solution U of (14) coincides with the solution of the contact problem (T1) – (T4). Since by the reproducing kernel property $\boldsymbol{c}'\boldsymbol{K}\boldsymbol{c} = \|U\|^2_{\mathcal{W}_0^m}$, we can use (10) to rewrite (14) as

$$\mathcal{E}(U) := \sum_{j=-\lfloor \frac{m-1}{2} \rfloor}^{n-\lfloor \frac{m+1}{2} \rfloor} (D^m U(j))^2 \to \min, \tag{16}$$

subject to
$$|U(j) - F_f(j)| \le \lambda, \quad j = 1, \ldots, n-m,$$
$$U(n-j) = F_f(n-j), \quad j = 0, \ldots, m-1.$$

In particular, for $m = 1$, we minimize just the sum of the squared lengths

$$\sum_{j=0}^{n-1} (U(j+1) - U(j))^2 \to \min,$$

instead of the lengths in (9). However, by the following proposition both problems are equivalent.

Proposition 2 (Equivalence of Contact and SVR Problem). *The solution of the contact problem* (T1) – (T4) *coincides with the solution of the SVR problem* (16).

Proof. Let U be the solution of (16). Assume that $j \in \tilde{\Lambda}$ is an upper contact point that does not fulfill (T4), i.e., $(-1)^m D^{2m} U(j) > 0$. (Similar arguments can be used for lower contact points.) By (2), this means that

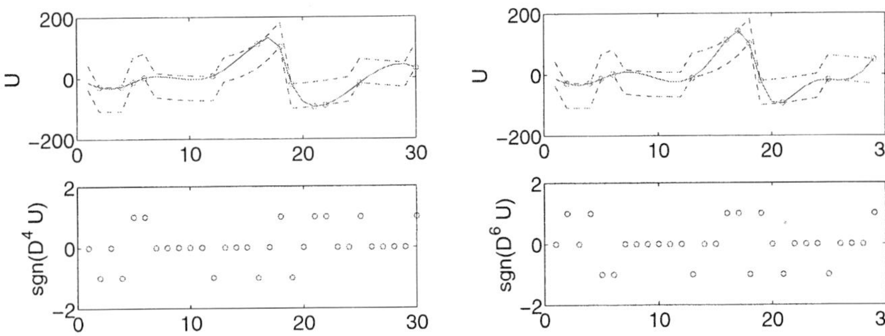

Fig. 1. Property (T4) of U computed by (16) for $m = 2$ (left) and $m = 3$ (right). Top: U (solid line) with tube (dashed line) and contact points. Bottom: Sign of $D^{2m}U$

$$U(j) > W(j) := -(-1)^m \sum_{\substack{k=0 \\ k \neq m}}^{2m} \binom{2m}{k} (-1)^k U(j-m+k) \Big/ \binom{2m}{m}.$$

By definition $W(j)$ lies on the discrete spline of degree $2m-1$ through $U(j \pm k)$, $k = 1, \ldots, m$. Now we define a function V which is equal to U except at j, where

$$V(j) := \begin{cases} W(j), & \text{if } W(j) > U(j) - 2\lambda, \\ U(j) - 2\lambda & \text{otherwise.} \end{cases}$$

Obviously, V fulfills the constraints of (16) and

$$U(j) > V(j) \geq W(j). \tag{17}$$

We show that $\mathcal{E}(V) < \mathcal{E}(U)$. This contradicts the choice of U as minimizer of (16) and we are done. Replacing D^m in $\mathcal{E}$ by (2) and regarding that $V(i) = U(i)$ for $i \neq j$, we obtain after some technical computations that

$$\mathcal{E}(U) - \mathcal{E}(V) = \binom{2m}{m} \big(U(j) - V(j)\big) \big(U(j) + V(j) - 2W(j)\big).$$

Now we have by (17) that $\mathcal{E}(U) - \mathcal{E}(V) > 0$. This completes the proof. □

Fig. 1 demonstrates property (T4) for the solution U of (16).

5 Parabolic PDEs with Higher Order Derivatives

Regularization methods are closely related to parabolic PDEs by the Euler–Lagrange equation, see, e.g.,[19]. To allow for a comparison of our tube–SVR method with PDE approaches we shortly describe their relations. For this, we consider the slightly modified version of (1) suggested in [1, 3]

$$\int_0^1 (u(x) - f(x))^2 + 2\lambda\,\varphi\left((u^{(m)})^2\right)\,\mathrm{d}x$$

with $\varphi(s^2) := \left(\varepsilon^2 + s^2\right)^{\frac{1}{2}}$. A minimizer u of this functional necessarily satisfies the Euler–Lagrange equation

$$\frac{u-f}{\lambda} = (-1)^{m+1}\frac{\partial^m}{\partial x^m}\left(2\,\varphi'\left((u^{(m)})^2\right)u^{(m)}\right) \tag{18}$$

with natural boundary conditions $u^{(k)}(0) = u^{(k)}(1) = 0$, $k = m,\dots,2m-1$, see [5]. These boundary conditions are in agreement with our boundary conditions (4). Introducing an additional time variable t, the left–hand side of equation (18) can be understood as fully implicit time discretization of the *diffusion equation*

$$\frac{\partial u}{\partial t} = (-1)^{m+1}\frac{\partial^m}{\partial x^m}\left(\frac{u^{(m)}}{\sqrt{(u^{(m)})^2+\varepsilon^2}}\right) \tag{19}$$

with natural boundary conditions, initial value f and stopping time λ. To solve (19) we use finite differences for the derivatives in space and an explicit Euler scheme in time. This leads to the following iterative scheme:

$$\begin{aligned}
u^0(j) &:= f(j),\\
v^k(j) &:= \frac{D^m u^k(j)}{\sqrt{\left(D^m u^k(j)\right)^2+\varepsilon^2}},\\
u^{k+1}(j) &:= u^k(j) - \tau\sum_{l=0}^{m}(-1)^l\binom{m}{l}\,v^k\left(j-\left\lfloor\frac{m+1}{2}\right\rfloor+l\right),
\end{aligned}$$

where we set $D^m u^k(j) := 0$ for $j = 0,\dots,\lfloor\frac{m}{2}\rfloor; n-\lfloor\frac{m-1}{2}\rfloor,\dots,n$. This scheme satisfies stability in the Euclidean norm if the time step size τ is chosen sufficiently small, namely $\tau \le \frac{\varepsilon}{2^{2m-1}}$. In comparisons with regularization methods we use the regularization parameter λ as stopping time, i.e., we iterate until $k = \frac{\lambda}{\tau}$.

Alternatively, we can also approximate a solution of (18) by solving the *diffusion–reaction equation*

$$\frac{\partial u}{\partial t} = \frac{u-f}{\lambda} + (-1)^m\frac{\partial^m}{\partial x^m}\left(2\,\varphi'\left((u^{(m)})^2\right)u^{(m)}\right). \tag{20}$$

A discretization of this equation can be obtained in a similar way to the one for the diffusion equation. It should be noted that the steady state of (20) for $t\to\infty$ yields a solution of (18) while the diffusion approach (19) leads in general only to an approximation of the solution. Only for the classical setting $m = 1$ without additional ε–regularization, it was shown in [21, 15] that the analytical solution of the space discrete diffusion equation (19) is equivalent to the solution of the optimization problem (1). For a space continuous version we refer to [22]. Even for first order derivatives this is a very special property of the TV regularization function φ.

6 Denoising Experiments

In this section, we show by denoising experiments that our tube–SVR approach works well even in comparison with corresponding PDE methods and demonstrate the influence of higher order derivatives.

As examples we have used the signals shown in Fig. 2. The first signal is piecewise polynomial, and Gaussian noise with standard deviation 10 was added. The other one consists of piecewise sine signals and the noise standard deviation is 1. First, we have determined the optimal parameters λ for our tube–SVR denoising method with respect to the maximal signal–to–noise–ratio (SNR) defined by $\mathrm{SNR}(g,u) := 10\log_{10}\left(\frac{\|g\|_2^2}{\|g-u\|_2^2}\right)$ with original signal g. We have applied the tube–SVR method described at the end of Section 2, where the contact problem was solved by applying the Matlab quadratic programming routine to (16). This routine is based on an active set method. The results are contained in Tab. 1.

Then we compared the quality of the results obtained by our tube–SVR approach and by the PDE methods for various orders of derivatives m. In our PDE experiments we have used a regularization parameter $\varepsilon = 10^{-4}$ and for each order the maximal time step size. One should be aware of the influence of the parameter ε for both PDE methods and the number of iterations for the diffusion–reaction method. For smaller values of ε one could even obtain better results at the cost of a higher number of iterations. Figs. 3 and 4 show the denoising results. Since one can visually not distinguish between the tube–SVR and the diffusion–reaction results we have only plotted

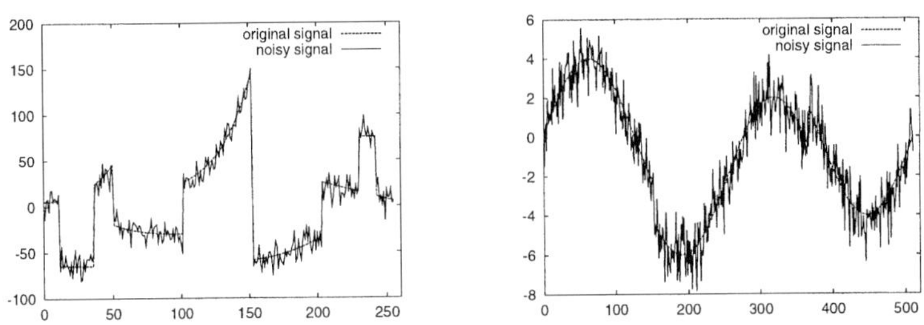

Fig. 2. Test signals. Left: Piecewise polynomial signal, 256 pixels, SNR 14.74dB. Right: Piecewise sine signal, 512 pixels, SNR 10.25dB

Table 1. Optimal parameters λ and SNR values for tube–SVR denoising

	Polynomial signal		Sine signal	
Order m	λ	SNR (dB)	λ	SNR (dB)
1	15	21.34	3	20.03
2	5	18.45	16	21.96
3	2	17.59	174	21.91

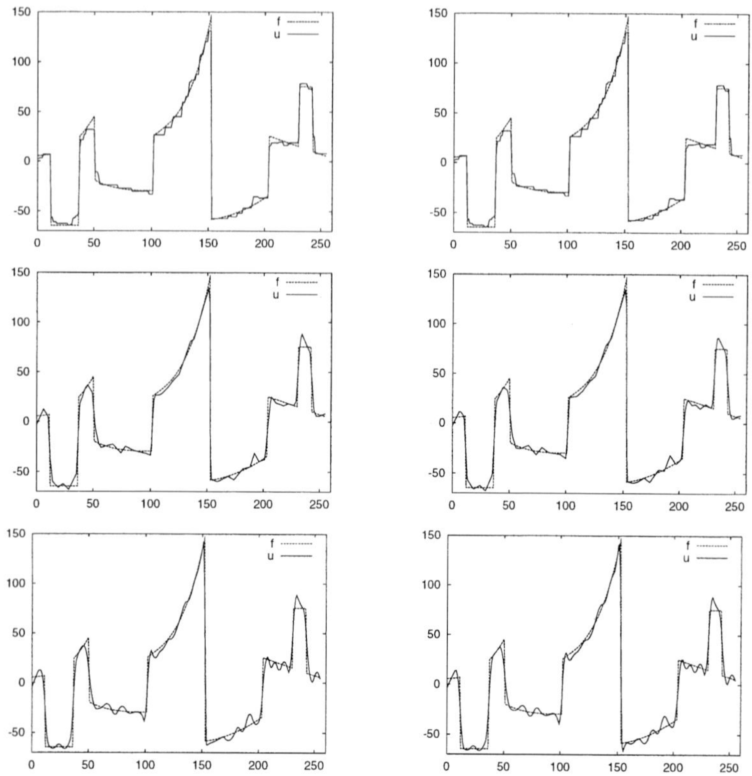

Fig. 3. Denoising results for the piecewise polynomial signal with $\lambda = 15$. Left: Tube–SVR method. Right: Diffusion method. Top: First order. Middle: Second order. Bottom: Third order

Table 2. Difference between tube–SVR method and diffusion/diffusion–reaction approach

Order m	Polynomial signal			Sine signal		
	Diffusion–reaction		Diffusion	Diffusion–reaction		Diffusion
	Iterations	l_∞–norm	l_∞–norm	Iterations	l_∞–norm	l_∞–norm
1	10^7	$6.2 \cdot 10^{-4}$	$1.2 \cdot 10^{-2}$	10^7	$9.0 \cdot 10^{-4}$	$8.5 \cdot 10^{-3}$
2	10^8	$8.2 \cdot 10^{-4}$	7.2	10^8	$1.3 \cdot 10^{-2}$	$1.9 \cdot 10^{-1}$
3	10^8	$6.0 \cdot 10^{-4}$	5.1	$5 \cdot 10^8$	$1.1 \cdot 10^{-1}$	$1.0 \cdot 10^{-1}$

the diffusion results in the PDE part. However, the diffusion results look also very similar except for slight smoothing effects for $m = 2$. To affirm this impression numerically, Tab. 2 shows the maximal absolute differences between the results of our tube–SVR method and the diffusion/diffusion–reaction methods.

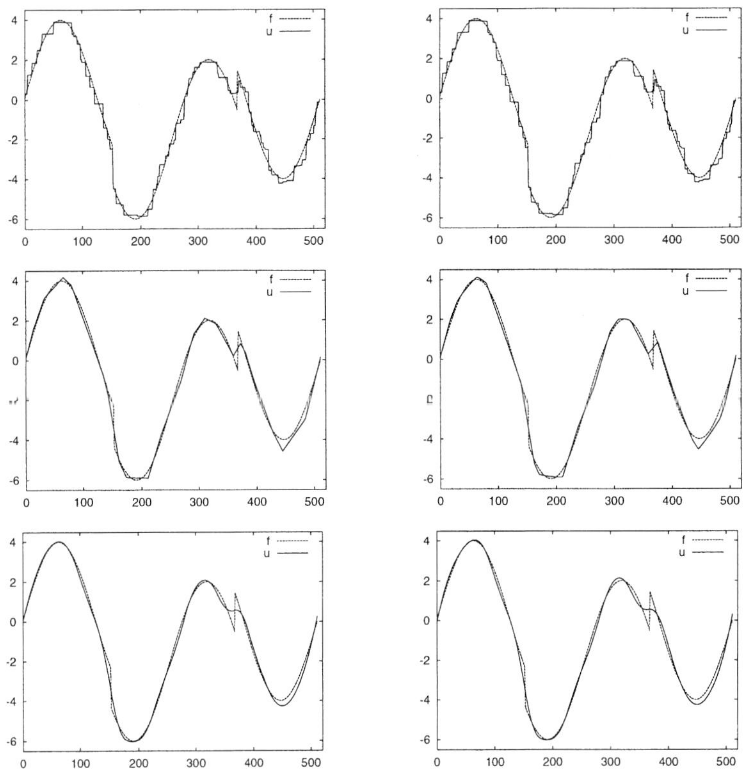

Fig. 4. Denoising results for the piecewise sine signal. Left: Tube–SVR method. Right: Diffusion method. Top: First order, $\lambda = 3$. Middle: Second order, $\lambda = 16$. Bottom: Third order, $\lambda = 174$

7 Conclusions

We have proved that the contact problem arising in the tube formulation of the minimization problem with ℓ_2 data term and TV regularization term with higher order derivatives can be formulated as SVR problem with discrete spline kernels. Therefore the problem is closely related to spline interpolation with variable knots. The results can also be considered from a different point of view, namely by applying Chambolle's dual approach to our setting, see [20]. This will be the basis for handling higher space dimensions. In our denoising experiments we have also incorporated corresponding nonlinear diffusion and diffusion–reaction equations with higher order derivatives which lead to similar results.

References

1. R. Acar and C. R. Vogel. Analysis of bounded variation penalty methods for ill–posed problems. *Inverse Problems*, 10:1217–1229, 1994.

2. A. Chambolle. An algorithm for total variation minimization and applications. *Journal of Mathematical Imaging and Vision*, 20:89–97, 2004.
3. T. F. Chan, A. Marquina, and P. Mulet. High-order total variation-based image restoration. *SIAM Journal on Scientific Computing*, 22(2):503–516, 2000.
4. P. L. Davies and A. Kovac. Local extremes, runs, strings and multiresolution. *Annals of Statistics*, 29:1–65, 2001.
5. S. Didas. Higher order variational methods for noise removal in signals and images. Diploma Thesis, Dept. of Math., Saarland University, Saarbrücken, Germany, 2004.
6. F. Girosi. An equivalence between sparse approximation and support vector machines. *Neural Computation*, 10(6):1455–1480, 1998.
7. W. Hinterberger, M. Hintermüller, K. Kunisch, M. von Oehsen, and O. Scherzer. Tube methods for BV regularization. *Journal of Mathematical Imaging and Vision*, 19:223–238, 2003.
8. M. Loog. Support blob machines: The sparsification of linear scale-space. In T. Pajdla and J. Matas, eds., *Computer Vision – ECCV 2004, Part IV*, volume 3024 of *Lecture Notes in Computer Science*, 14–24, Springer, Berlin, 2004.
9. M. Lysaker, A. Lundervold, and X.-C. Tai. Noise removal using fourth-order partial differential equations with applications to medical magnetic resonance images in space and time. Technical Report CAM-02-44, Department of Mathematics, University of California, 2002.
10. E. Mammen and S. van de Geer. Locally adaptive regression splines. *Annals of Statistics*, 25(1):387–413, 1997.
11. O. L. Mangasarian and L. L. Schumaker. Best summation formulae and discrete splines via mathematical programming. *SIAM Journal on Numerical Analysis*, 10(3):448–459, 1973.
12. Y. Meyer. *Oscillating Patterns in Image Processing and Nonlinear Evolution Equations*, volume 22 of *University Lecture Series*. AMS, Providence, 2001.
13. M. Nielsen, L. Florack, and R. Deriche. Regularization, scale-space and edge detection filters. *Journal of Mathematical Imaging and Vision*, 7:291–307, 1997.
14. A. Obereder, S. Osher, and O. Scherzer. On the use of dual norms in bounded variation type regularization. Technical report, Department of Computer Science, University of Innsbruck, 2004.
15. I. Pollak, A. S. Willsky and Y. Huang. Nonlinear evolution equations as fast and exact solvers of estimation problems. *IEEE Trans. on Signal Processing*, in print.
16. L. I. Rudin, S. Osher, and E. Fatemi. Nonlinear total variation based noise removal algorithms. *Physica D*, 60:259–268, 1992.
17. O. Scherzer. Denoising with higher order derivatives of bounded variation and an application to parameter estimation. *Computing*, 60:1–27, 1998.
18. O. Scherzer. Taut string algorithm and regularization programs with G-norm. *Journal of Mathematical Imaging and Vision*, in print.
19. O. Scherzer and J. Weickert. Relations between regularization and diffusion filtering. *Journal of Mathematical Imaging and Vision*, 12:43–63, 2000.
20. G. Steidl. A note on the dual treatment of higher order regularization functionals. Technical Report 278, Institute of Mathematics, University of Mannheim, 2004.
21. G. Steidl, J. Weickert, T. Brox, P. Mrázek, and M. Welk. On the equivalence of soft wavelet shrinkage, total variation diffusion, total variation regularization, and SIDEs. *SIAM Journal on Numerical Analysis*, 42(2):686–713, 2004.

22. D. M. Strong. *Adaptive Total Variation Minimizing Image Restoration.* PhD thesis, Department of Mathematics, University of California, 1997.
23. V. N. Vapnik. *Statistical Learning Theory.* John Wiley and Sons, Inc., 1998.
24. C. R. Vogel. *Computational Methods for Inverse Problems.* SIAM, Philadelphia, 2002.
25. G. Wahba. *Spline Models for Observational Data.* SIAM, Philadelphia, 1990.
26. Y.-L. You and M. Kaveh. Fourth-order partial differential equations for noise removal. *IEEE Transactions on Image Processing*, 9(10):1723–1730, 2000.

Perfusion Analysis of Nonlinear Liver Ultrasound Images Based on Nonlinear Matrix Diffusion

A. Kissi[1], S. Cormier[2], L. Pourcelot[1], A. Bleuzen[1], and F. Tranquart[1]

[1] INSERM unité 619, CHRU Bretonneau,
2 bis boulevard Tonnellé 37044 Tours, France
kissi_a@med.univ-reims.fr, tranquart@med.univ-tours.fr
[2] CRESTIC-L.E.R.I., Université de Reims Champagne-Ardenne,
rue des Crayères B.P. 1035, 51687 Reims Cedex 2 Reims, France
stephane.cormier@univ-reims.fr

Abstract. Doppler has been used for many years for cardiovascular exploration in order to visualize vessel walls as well as anatomical or functional diseases. The use of ultrasound contrast agents makes it possible to improve ultrasonic information.

Recently, nonlinear imaging has emerged as a powerful tool for characterizing pathologies by studying their perfusion. In this paper, we present a new method for estimating the perfusion parameter over a sliding window in order to accurately characterize liver lesions from two-dimensional nonlinear ultrasound images. This method is inspired by the Lucas and Kanade Algorithm coupled with coherence enhancing diffusion in order to suppress the speckle and transparent motions due to the presence of contrast agents.

1 Introduction

Doppler echography is an imaging modality that enables to visualize blood flow and perfusion. When it is coupled with ultrasound contrast agents, it allows to ensure a high quality of perfusion study which is of great interest for effective diagnosis. Several methods have been proposed to quantify perfusion especially in contrast echocardiography. They are based on optical flow methods with promising results.

However, images are buried in speckle noise, depending on contrast agents density and their nonlinear behavior. Furthermore, there are transparent motions due to the presence of contrast agents. In this study, we present a new method to characterize liver lesions by perfusion quantification. It is based on Lucas and Kanade algorithm for optical flow which estimates perfusion parameters over a sliding window.

This paper is organized as follow: Section 1 describes ultrasound contrast agents and nonlinear imaging. Sections 2 and 3 presents our method which is divided into two steps, first of all a coherence enhancing diffusion and then the

R. Kimmel, N. Sochen, J. Weickert (Eds.): Scale-Space 2005, LNCS 3459, pp. 528–535, 2005.

Lucas Kanade algorithm. And finally, section 4 shows the results of our method in liver lesions. To the best of our knowledge, there are no published works about perfusion analysis with optical flow in nonlinear liver ultrasound imaging.

2 Nonlinear Ultrasound Imaging

Conventional Doppler imaging is able to image vascular flow. It is thus best seen as a tool for imaging the macrocirculation, rather than the microcirculation (small or deep vessels with minimal flow). Despite this, it is considerably useful in tumor imaging, as many benign lesions are hypovascular compared with adjacent tissues, while most malignant lesions show hypervascularity. It is the simplest approach for a qualitative visual assessment of Doppler blood flow.

To image microcirculation as macrocirculation, a common approach is to use an injection of intravenous ultrasound contrast agents for boosting detected signal from small blood vessels. These agents are encapsulated microbubbles which diameter (3μm) is much below than the red blood cells one. The particular behavior of the bubbles makes that, in presence of ultrasound wave, they can be expanded more easily than they can be compressed: this phenomenon is described as "non-linearity". They vibrate particularly strongly at moderate acoustic intensity, used for diagnostic ultrasound imaging. This makes them several thousand times more reflective than blood cells and thus enhances reflected ultrasound signals. The mean frequency of reception is twice the mean frequency of emission, and thus the corresponding non-linear signal from bubbles is only detected, since solid tissues weakly resonate [3].

At high acoustic intensity, the bubbles vibrate and are destroyed emitting transient harmonics.

Nonlinear imaging uses the nonlinear response from bubbles for image formation. Signal intensity depends on the density of contrast agent. As bubbles are strictly intravascular, it enables to image exclusively the blood supply for an organ.

The perfusion of an organ provides information on the relative vascularization of a suspected lesion, so it can be characterized. The contrast enhancement can be quantified to give an index of the tissue perfusion. The liver is filled according to the following equation:

$$SI(t) = SI_{max}(1 - \exp(-\beta t)) \tag{1}$$

where SI_{max} represents the maximum signal reached within the dedicated region of interest, β the perfusion frequency. The lesion is characterized by its perfusion index R such as $R = 1/\beta$. This index R representats the rapidity of the enhancement in a given area.

Injection of microbubbles contributes to a reduction of artifacts acoustically generated in the image (side lobe effects, multiple scattering, on-axis reverberations). But in the same time, as conventional echographic images, harmonic images are buried in noise, called speckle depending on contrast agents density and their nonlinear response [12], [11]. However, because bubbles response depends nonlinearly on incident amplitude, the speckle formation becomes nonlinear too, inducing marked limitation in quantifying contrast enhancement.

3 Optical Flow

The optical flow is used to track the pathology during a perfusion sequence. According to the theory, let $I_A(x)$ and $I_B(x)$ be intensity values. When there is motion between these images, we have $I_A(x) = I_B(x+v)$, where v is the motion vector. The injection of contrast agents disturbs motion estimates by the creation of additive transparent signals [7]. We obtain: $I_A(x) = I_B(x+v) + \epsilon(x+v)$.

To obtain a good optical flow estimation, ϵ has to converge towards 0. In angiography, Ferrant et al. in [2] has proposed to match away from contrast agent region to avoid transparent motions. In our study, contrast agents spread all over the region of interest. Thus, this previous method cannot be applied so we choose to develop a preprocessing smoothing step to eliminate the speckle and the transparent motions.

3.1 Image Filtering

As proved in [1] and [8], the coherence enhancing diffusion enables to eliminate the speckle. The basic idea of this process is to smooth along the orientation of image structures. In order to represent the orientation of structures, a tensor is used. The structure tensor, introduced in the 80's, provides a consistent representation of both local orientation and type of structure. It can be represented by a matrix function, depending on the image gradient, according to the following equation:

$$J_\rho(\nabla I_\sigma) = L_\sigma * (\nabla I_\sigma \nabla {I_\sigma}^t) \tag{2}$$

where L_σ denotes a gaussian with standard deviation σ (window size over which the orientation information is averaged), and $I_\sigma = L_\sigma * I$. The structure tensor can be decomposed by its eigenvalues μ_i and eigenvectors ω_i:

$$J_\rho(\nabla I_\sigma) = \begin{pmatrix} \omega_1 & \omega_2 \end{pmatrix} \begin{pmatrix} \mu_1 & 0 \\ 0 & \mu_2 \end{pmatrix} \begin{pmatrix} \omega_1^t \\ \omega_2^t \end{pmatrix} \tag{3}$$

The eigensystem of the structure tensor carries the orientation representation. The eigenvalues λ_i reflects the type of structure, depending on the gray level variations, and the eigenvectors ω_i, the orientation of this structure.

The structure tensor allows us to separate the image into constant areas, corners and straight edges according to the number of non-zero eigenvalues. The parameter used to measure the spread of the eigenvalues is the coherence, noted φ. It is measured by:

$$\varphi = (\mu_1 - \mu_2)^2 \tag{4}$$

The diffusion tensor D uses the same eigenvectors as the structure tensor in order to apply smoothing in directions given by the structure image [14]. It adapts its eigenvalues to enhance the coherence. Thus, the eigenvalues are related to the anisotropy of the image $(\mu_1 - \mu_2)$ through the Tuckey's biweight robust estimator. It preserves sharp boundaries and improves automatic stopping of the diffusion in the gradient direction. Hence, the eigenvalues are chosen as:

$$\lambda_1 = \begin{cases} \alpha * (1 - \frac{\varphi}{s^2}) & \text{if} (\mu_1 - \mu_2)^2 < s^2 \\ 0 & \text{else} \end{cases} \qquad (5)$$
$$\lambda_2 = \alpha$$

For the discretization, an explicit model introduced in [15] is used. It uses rotational optimal filters.

3.2 Optical Flow Estimation

Optical flow estimation is of major interest in computer vision. It has been applied in biomedical imaging as in the assessment of myocardial deformation [9]. In ultrasound contrast imaging, optical flow has been used in echocardiography to guide snakes for studying the right ventricule hemodynamics and to analyse the intramyocardial perfusion with promising results [5].

The Lucas and Kanade algorithm for optical flow estimates the local motion by assuming that it is constant all over the window. The following equation has to be resolved:

$$f_x u + f_y v + f_z = 0 \qquad (6)$$

where f_x, f_y, f_z are partial derivates and (u, v) the parameters of optical flow. The equation of structure tensor in (2) can also be used to calculate motion estimates as proved by Weickert and Brox in [13]. Furthermore, the coupling between the nonlinear anisotropic diffusion (such as edge or coherence enhancing diffusion) and optical flow can improve the optical flow estimation in noisy sequences [10]. Thus, we use the structure tensor in the denoising step for estimating optical. The parameter ρ gives the window size over which the optical flow is averaged. Then, it allows to eliminate transparent motions. We obtain the following equation:

$$\begin{pmatrix} L_\rho * I_x^2 & L_\rho * I_x I_y \\ L_\rho * I_x I_y & L_\rho * I_y^2 \end{pmatrix} \begin{pmatrix} u \\ v \end{pmatrix} = \begin{pmatrix} -L_\rho * I_x I_z \\ -L_\rho * I_y I_z \end{pmatrix} \qquad (7)$$

In order to have (u, v), we use a broad Gaussian filter to avoid the aperture problem and to obtain dense optical flow. In our study, we use $\rho = 5$.

4 Results

To illustrate the performance of the process described, two applications on focal liver lesions are proposed below. Fifteen sequences of 160 non compressed images have been obtained, using a Siemens Acuson Sequoia 512 with a 4C1 probe, in CPS mode, after Sonovue injection (Bracco SpA imaging, Italy) with an MI = 0.21.

In our experiments, the frame rate is high enough to obtain a gradual perfusion and constant brightness variations. Haussecker and Fleet have proposed in [6] to use a physical model of the time varying brightness in image sequence in order to estimate the optical flow. In contrast liver harmonic imaging, the perfusion may differ between patients as well as image zones.

The first shortcoming of ultrasound images is its noise. For conventionnal images, noise is mainly defined by the speckle [4]. For the contrast harmonic imaging, the definition is more complex since it behavior changes as function of their size, transmition frequency and strongly on the concentration. In our case, at the early stage of perfusion, the concentration of bubbles is still too low and we might approximate the image speckle noise by conventionnal tissue speckle. However, at late stages of perfusion, the scattering of bubbles becomes dominant and nonlinear. Hence, the conventionnal speckle definition is not applicable. To the best of our knowledge, there is not published paper about bubble noise characterization and this issue is out of the scope of this manuscript.

In our study, the diffusion parameters have been set on the maximum of perfusion and used during the sequence. As the spatial local anisotropy or coherence differed in healthy versus pathologic tissues, a nonlinear anisotropic diffusion was applied for smoothing and emphasizing structural features with $\sigma = 5$.

Motion ambiguities may arise when there is an insufficient representation of spatial information. This holds in regions with a specular reflection or in homogeneous regions of weak scattering. The local coherence, measured by $(\mu_1 - \mu_2)^2$, help us to adapt locally the smoothing according to the type of tissue and thus overcome this shortcoming. When the tissue is "homogeneous", μ_1 and μ_2 are close and the diffusion becomes isotropic. On the other hand, when the tissue is fully structured, the diffusion becomes anisotropic. These regions correspond to $(\mu_1 - \mu_2)^2 > s^2$ [1]. In our study, the stopping criterion s has been set $\sqrt{2}$.

We used a sliding window over wich we calculate the perfusion parameters. It followed the liver movements which are associated to the breath respiration. They are a combination of three translations (lateral, anterio-posterior, cranio-

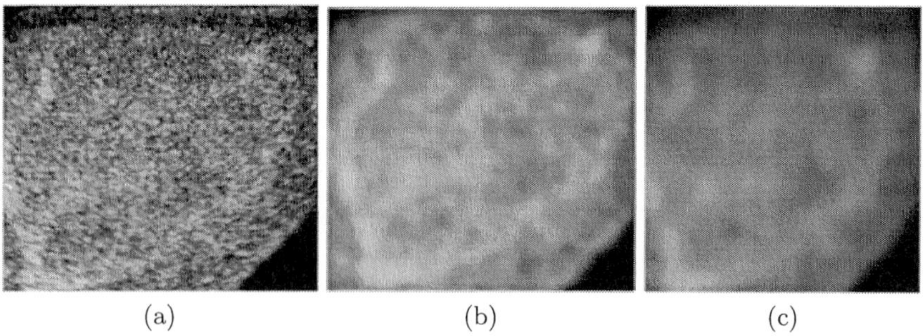

(a) (b) (c)

Fig. 1. Exemple of a hypervascular metastasis taken at the maximum of perfusion : (a) the original image, (b) and (c) are the smoothed image with respectivelly $s = 0.5$ and $s = \sqrt{2}$

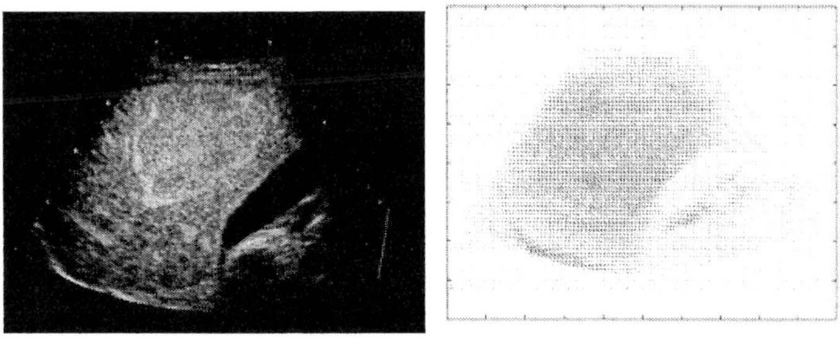

(a) (b)

Fig. 2. Exemple of the flow for a hypervascular metastasis : (a) the original image, (b) the flow obtained

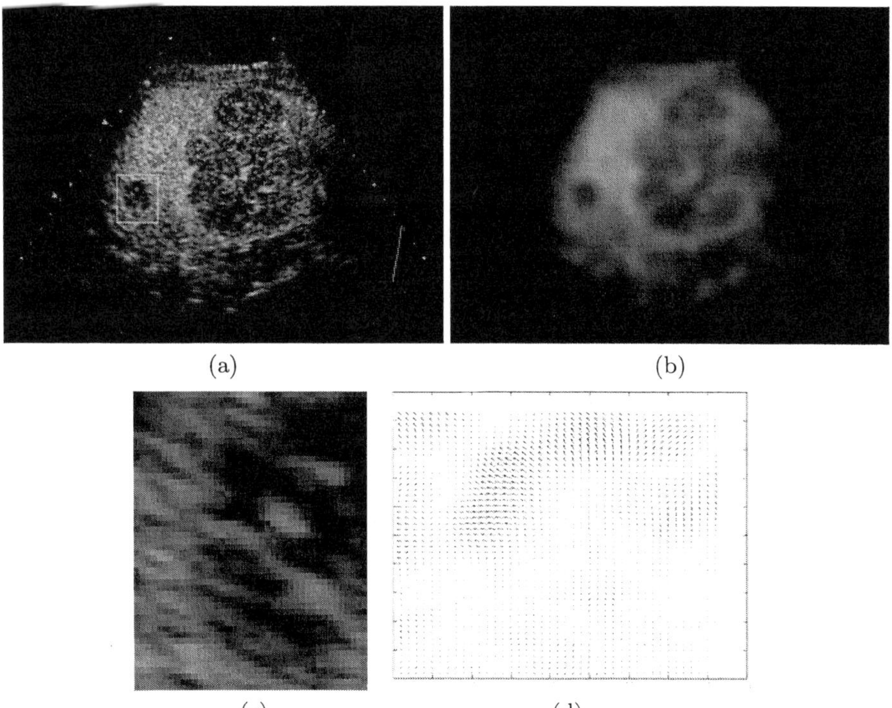

(a) (b) (c) (d)

Fig. 3. Exemple of two hypovascular metastasis :(a) is the original image, (b) its smoothed version and (d) the optical flow over a window around the little metastasis in figure (a)

caudal). Over the window, we calculate the optical flow and the mean grey level during the sequence. An example of two successive images and their optical flow is given in figures 2. We obtained the curve given by the figure 3.

Stabilised Nonlinear Inverse Diffusion for Approximating Hyperbolic PDEs

Michael Breuß[1], Thomas Brox[2], Thomas Sonar[1], and Joachim Weickert[2]

[1] Technical University Brunswick, Computational Mathematics,
Pockelsstraße 14, 38106 Brunswick, Germany
{m.breuss, t.sonar}@tu-bs.de
www.icm.tu-bs.de

[2] Mathematical Image Analysis Group, Faculty of Math. and Computer Science,
Saarland University, Building 27, 66041 Saarbrücken, Germany
{brox, weickert}@mia.uni-saarland.de
www.mia.uni-saarland.de

Abstract. Stabilised backward diffusion processes have shown their use for a number of image enhancement tasks. The goal of this paper is to show that they are also highly useful for designing shock capturing numerical schemes for hyperbolic conservation laws. We propose and investigate a novel flux corrected transport (FCT) type algorithm. It is composed of an advection step capturing the flow dynamics, and a stabilised nonlinear backward diffusion step in order to improve the resolution properties of the scheme. In contrast to classical FCT procedures, we base our method on an analysis of the discrete viscosity form. This analysis shows that nonlinear backward diffusion is necessary. We employ a slope limiting type approach where the antidiffusive flux determined by the viscosity form is controlled by a limiter that prohibits oscillations. Numerical experiments confirm the high accuracy and shock capturing properties of the resulting scheme. This shows the fruitful interaction of PDE-based image processing ideas and numerical analysis.

1 Introduction

Starting with Rudin's Ph.D. thesis in 1987 [18], many ideas from computational fluid dynamics and the numerics of hyperbolic conservation laws have entered the field of image processing. Because problems of fluid dynamics and hyperbolic conservation laws involve the formation of shocks, sophisticated numerical methods such as total variation diminishing (TVD) and essentially non-oscillatory (ENO) schemes had to be devised to give a sharp resolution at shocks and to avoid visible numerical oscillations [8, 9, 13]. On the image processing side, image edges carry important information and may be regarded as shocks as well. Often edges are blurred, so it is natural to apply shock-enhancing concepts from computational fluid dynamics. This has led to PDE formulations of shock filters [14] and to stabilised linear backward diffusion [15]; see Fig. 1 for an example. In case that noise is present as well, one aims at preserving or enhancing edges, while simultaneously smoothing at more homogeneous regions. Combinations of shock filtering with mean curvature motion [1] and in particular nonlinear diffusion

R. Kimmel, N. Sochen, J. Weickert (Eds.): Scale-Space 2005, LNCS 3459, pp. 536–547, 2005.

Fig. 1. Left: Original image with blurred edges. **Right:** Image sharpened with stabilized linear backward diffusion [15]

filtering [16] are suitable concepts to achieve this goal. Interesting variants of nonlinear diffusion include stabilised inverse diffusion equations (SIDEs) [17] and so-called forward-and-backward (FAB) diffusion [7] that explicitly uses negative diffusivities in a certain gradient range. Last but not least, total variation minimisation [19] has been proposed as a variational framework for discontinuity-preserving denoising.

While numerical ideas for hyperbolic conservation laws had undoubtedly a strong impact on modern image analysis, fertilisation in the inverse direction – where image processing methods are applied to improve the numerics of hyperbolic conservation laws – have started only recently: in [10, 11] different variants of numerical schemes are proposed that combine the second-order Lax-Wendroff scheme with anisotropic diffusion filtering with a diffusion tensor [23]; see also [22] for related ideas. All these strategies start with a hyperbolic scheme that gives sharp shock resolution, but suffers from oscillations in the shock areas. Anisotropic diffusion regards such oscillations as noise at edges that can be removed by smoothing along the edge.

On the other hand, there are also monotone first order schemes for hyperbolic conservation laws that do *not* produce oscillations at shocks. Unfortunately, they suffer from strong blurring effects (dissipation) since they involve a significant amount of numerical diffusion (viscosity) to achieve their favourable stability properties. However, if one takes such a scheme as starting point, an interesting question would be if there are useful ideas inspired from edge-enhancing PDE-based image processing that allow to sharpen these shocks. Ideally they should also turn the first order basis method having low accuracy into a higher accurate second-order method without introducing oscillations. For simplicity we focus on the one-dimensional scalar case. We will see that the resulting methods can be regarded to belong to the class of flux-corrected transport (FCT) schemes [2], but in contrast to classical FCT schemes they offer the advantage that they are also applicable to the important class of *nonlinear* conservation laws. It turns out that the appropriate sharpening process from image processing must be a stabilised nonlinear inverse diffusion step. It resembles the stabilised linear inverse diffusion filter that has been proposed by Osher and Rudin [15] for deblurring images.

Our paper is organised as follows: in Section 2 we describe a classical Upwind scheme and analyse its intrinsic numerical diffusion, while Section 3 gives an introduction to FCT schemes. This analysis forms the basis for our novel image-processing inspired

FCT scheme that we present in Section 4. Its stability properties are analysed in Section 5. Section 6 presents experiments with linear and nonlinear test scenarios where we compare our method with a state-of-the-art shock capturing scheme: a TVD method with van Leer flux limiter. We conclude our paper with a summary in Section 7.

2 The Classical Upwind Scheme and Its Numerical Diffusion

In this paper, we deal with the numerical approximation of hyperbolic conservation laws of type

$$u_t + (f(u))_x = 0, \tag{1}$$

where $u := u(x,t)$ is a scalar-valued function of a one-dimensional space variable x and time t, subscripts denote partial derivatives, and the flux function f is supposed to satisfy $f'(.) \geq 0$.

The underlying method for our novel FCT technique in the next section is the classical *Upwind scheme*

$$U_j^{k+1} = U_j^k - \lambda \left(f_j^k - f_{j-1}^k \right). \tag{2}$$

Thereby, we use as in the following the notation U for discrete data in contrast to the sought solution u, and we denote the ratio of mesh parameters as $\lambda = \Delta t / \Delta x$. The upper index k in U_j^k denotes as usual the temporal level $k\Delta t$ while analogously the lower index j denotes the spatial mesh point $j\Delta x$. For shortness of notation, we abbreviate

$$f_j^k := f\left(U_j^k\right).$$

Unless stated otherwise, we consider all occurring methods to be stable under the usual CFL condition, see e.g. [13] for details concerning this notion.

One desirable property of the Upwind scheme consists of the fact that the scheme does not produce numerical oscillations:

Proposition 1 (Extrema Diminishing Properties of the Upwind Scheme). *The Upwind scheme (2) is a generalised monotone scheme in the sense of LeFloch and Liu [12], i.e., it is a local extremum diminishing (LED) scheme, while it also does not introduce new extrema during a computation, i.e., it diminishes also the number of extrema (NED property).*

Proof. The validity of the assertion follows since under a CFL condition

$$U_j^{k+1} \in \operatorname{conv}\left(U_{j-1}^k, U_j^k\right)$$

always follows, where conv denotes the convex hull. (Compare the *data compatibility* notion due to Roe [20].) ♣

Unfortunately, the Upwind scheme also has a severe disadvantage: it suffers from undesirable blurring effects (dissipation). To quantify these viscous artifacts we write the scheme (2) in its *viscous form*, i.e.,

$$\underbrace{U_j^{k+1} = U_j^k - \frac{\lambda}{2}\left(f_{j+1}^k - f_{j-1}^k\right)}_{\text{(A)}} + \underbrace{\frac{Q_j^{+,k}}{2}\left(U_{j+1}^k - U_j^k\right) - \frac{Q_j^{-,k}}{2}\left(U_j^k - U_{j-1}^k\right)}_{\text{(B)}}. \tag{3}$$

The underlying idea behind this decomposition is to consider part (A) as a second order approximation of (1) in space (and first order in time), while part (B) is (in leading order) the discrete counterpart of the numerical diffusion incorporated in the method (2).

One easily verifies that (2) and (3) can be made identical by choosing viscosity coefficients Q_j^+ and Q_j^- that satisfy

$$Q_j^{+,k} = \lambda \frac{f_{j+1}^k - f_j^k}{U_{j+1}^k - U_j^k} \quad \text{and} \quad Q_j^{-,k} = \lambda \frac{f_j^k - f_{j-1}^k}{U_j^k - U_{j-1}^k}. \tag{4}$$

for $U_{l+1}^k \neq U_l^k$, $l \in \{j, j-1\}$. Note that our assumption $f'(.) \geq 0$ ensures that the viscosities $Q_j^{\pm}$ are nonnegative. Since the viscosities are proportional to the diffusion coefficients it follows that forward diffusion takes place. This numerical diffusion is responsible for the undesirable blurring effects that are observed with this first order method. We observe that, in spite of the simplicity of the Upwind scheme, an inherent diffusion process with nonlinear (!) viscosities $Q_j^{\pm}$ is involved. These nonlinear viscosities are inversely proportional to the derivative of u. In this respect they closely resemble the diffusivities in TV denoising of images [19].

3 FCT Schemes

A common method to compensate for the before mentioned blurring artifacts is the *flux corrected transport* (FCT) algorithm of Boris and Book [2]: a numerical scheme with much numerical diffusion used as a predictor for the evolution is corrected by an antidiffusive step. This principle is used as a basis of many successful FCT type algorithms, see especially [24] and the references therein.

The classical FCT approach as described in [2] is motivated by the method of the *modified equation*: the numerical diffusion incorporated in the predictor step is computed by means of the differential advection-diffusion-equation that the viscous predictor scheme actually approximates with second order accuracy. The resulting diffusion coefficient is annihilated by the antidiffusive step while a limiting procedure ensures that no oscillations develop. For more information on the modified equation, see e.g. the books [8, 13]. The described strategy can be refined by considering an analysis of wave coefficients in the linear case, ensuring especially for linear advection problems a high approximation quality; see [3, 4].

In order to describe the classical FCT method based on the Upwind scheme we use the following data notions:

- $U_j^{k+1/2}$ for the data obtained with the Upwind scheme starting from U_j^k
- U_j^{k+1} for the data obtained after the antidiffusive step.

Let us define the abbreviate notion

$$\Delta U_{j+1/2}^k := U_{j+1}^k - U_j^k. \tag{5}$$

Then the traditional FCT approach amounts to an *antidiffusion step* realised via

$$U_j^{k+1} = U_j^{k+1/2} - g_{j+1/2} + g_{j-1/2} \tag{6}$$

where the fluxes g are chosen in a fashion such that the following directive holds:

Construction Principle 1 (Boris and Book [2]).
"No antidiffusive flux transfer of mass can push the density value at any grid point beyond the density value at neighboring points."

The traditional FCT scheme realises this principle by setting

$$g_{j+1/2} := \mathrm{minmod}\left(\Delta U_{j-1/2}^{k+1/2},\, \eta_{j+1/2}\Delta U_{j+1/2}^{k+1/2},\, \Delta U_{j+3/2}^{k+1/2}\right) \tag{7}$$

where

$$\mathrm{minmod(a, b, c)} := \mathrm{sgn(b)} \max\left(0, \min(\mathrm{sgn(b)a}, |\mathrm{b}|, \mathrm{sgn(b)c})\right) \tag{8}$$

and

$$\eta_{j+1/2} := \frac{\lambda}{2}\bar{a}\,(1 - \lambda\bar{a}), \tag{9}$$

with $\bar{a}$ determined by

$$\bar{a} := \max_{U \in \mathrm{conv}\left(U_j^{k+1/2},\, U_{j+1}^{k+1/2}\right)} |f'(U)|.$$

Note that (9) is equivalent to Δt times the discrete version of the viscous term of the modified equation obtained by using a local linearisation of (1).

4 A New FCT Scheme with Nonlinear Inverse Diffusion

Let us now introduce a novel variant of FCT schemes that incorporates image processing ideas on stabilised inverse diffusion. In contrast to the previous section, our considerations are based solely on the viscosity form (3). This is a new feature of possible FCT algorithms. Our method of derivation can be advantageous concerning a rigorous analysis of the combined method, especially with respect to the nonlinear case.

A naive step to achieve inverse diffusion would consist of applying a direct antidiffusion process to the predicted data $U^{k+1/2}$ from the Upwind scheme by setting

$$\tilde{g}_{j+1/2} := \frac{1}{2}Q_j^{+,k+1/2}\Delta U_{j+1/2}^{k+1/2}. \tag{10}$$

It is immediately clear that such an antidiffusive step without a direct minmod-type stabilisation as used in (7) may introduce many oscillations. Thus, we limit the antidiffusive flux $\tilde{g}$ from (10) by

$$g_{j+1/2} := \mathrm{minmod}\left(\tilde{g}_{j-1/2}, \tilde{g}_{j+1/2}, \tilde{g}_{j+3/2}\right) \tag{11}$$

using the minmod function (8).

The discrete inverse diffusion employed here is similar to an image enhancement algorithm due to Osher and Rudin [15]. However, while the filter of Osher and Rudin is the stabilised inverse filter to *linear* diffusion, we extend this algorithm to be the stabilised inverse filter to *nonlinear* diffusion.

5 Stability Analysis

We are now ready to prove the following stability assertion.

Lemma 1 (Local Extremum Principle).
Let

$$\text{sign}\left(\Delta U_{j+1/2}^{k+1/2}\right) = \text{sign}\left(\Delta U_{j-1/2}^{k+1/2}\right) \neq 0 \tag{12}$$

hold. Then the FCT scheme defined by

$$U_j^{k+1} = U_j^{k+1/2} - g_{j+1/2} + g_{j-1/2} \tag{13}$$

using g from (11) satisfies locally a discrete minimum–maximum principle.

Proof. The aim is to show that

$$U_j^{k+1} \in \text{conv}\left(U_{j-1}^{k+1/2}, U_j^{k+1/2}, U_{j+1}^{k+1/2}\right)$$

holds. We only consider the situation defined by

$$\left|U_j^{k+1/2} - U_{j-1}^{k+1/2}\right| \leq \left|U_{j+1}^{k+1/2} - U_j^{k+1/2}\right|, \tag{14}$$

the other case can be treated analogously.

For simplicity, we omit the superscript $k + 1/2$ in the following computations. The idea is, starting from (13), to derive the estimate

$$\left|-g_{j+1/2} + g_{j-1/2}\right| \leq |U_j - U_{j-1}|$$

since then the sought convex hull condition is satisfied. Thus, we compute using the Lipschitz continuity of f and a corresponding Lipschitz constant L

$$\begin{aligned}
&\left|-g_{j+1/2} + g_{j-1/2}\right| \\
&\leq \left|g_{j+1/2}\right| + \left|g_{j-1/2}\right| \\
&\overset{(11)}{\leq} \left|\frac{1}{2}Q_j^+ \Delta U_{j-1/2}\right| + \left|\frac{1}{2}Q_j^+ \Delta U_{j-1/2}\right| \\
&\overset{(4)}{=} \lambda \left|\frac{f_j - f_{j-1}}{U_j - U_{j-1}}\right| \left|U_j^k - U_{j-1}^k\right| \\
&= \lambda \left|f_j - f_{j-1}\right| \\
&\leq \lambda L \left|U_j - U_{j-1}\right| \\
&\leq \left|U_j - U_{j-1}\right|
\end{aligned}$$

by the CFL condition. ♣

Because of the properties of the minmod function, the core of the proof also works without the assumption (12). Thus we can give directly

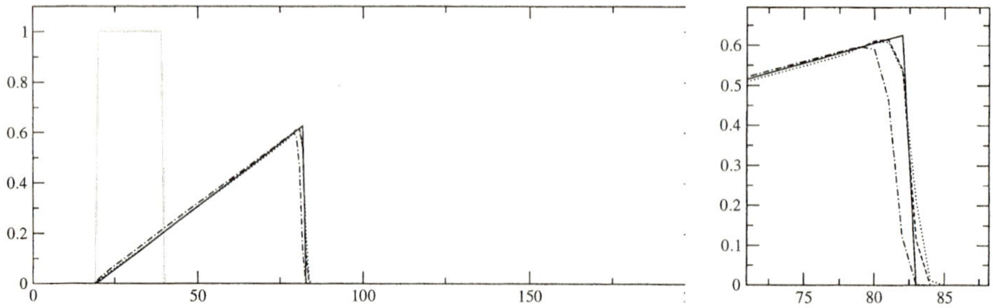

Fig. 3. Burgers equation. **Left:** Grey line: initial condition. Solid line: correct solution after $t = 100$. Dotted line: Upwind with antidiffusion. Dashed line: TVD method. Dash-dotted line: Upwind without antidiffusion. **Right:** Zoom

It can be observed that the proposed FCT-like scheme yields approximately the same accuracy as the TVD method, which supports the order test. When compared to the Upwind scheme without antidiffusion, the antidiffusion step clearly reveals its impact.

Burgers' Equation

A nonlinear test problem is concerned with Burgers' equation

$$u_t + \left(\frac{1}{2}u^2\right)_x = 0$$

supplemented by the initial condition

$$u_0(x) = \begin{cases} 1: & 20 \leq x < 40 \\ 0: & \text{else} \end{cases} .$$

This square wave decays to an N-wave like every solution of Burgers' equation. Thus the example has a profound meaning, see e.g. [6, 13] for discussions.

The computational parameters are the same as in the linear example before, and we compute the solution at $t = 100$. We compare again the numerical solutions obtained by the Upwind scheme with and without antidiffusion, as well as the TVD method. The corresponding numerical solutions together with the exact solution are displayed in Fig. 3. For better comparison, a detailed cutout of the region around the shock is depicted beside. Again it can be seen that the proposed scheme yields results very close to those of the TVD method. Note that in this nonlinear case, the classical FCT scheme is not applicable anymore without additional considerations.

Buckley-Leverett Equation

A second nonlinear and even non-convex test problem is based on the Buckley-Leverett equation

$$u_t + \left(\frac{u^2}{u^2 + \frac{1}{2}(1-u)^2}\right)_x = 0$$

supplemented by a Riemann problem as initial condition:

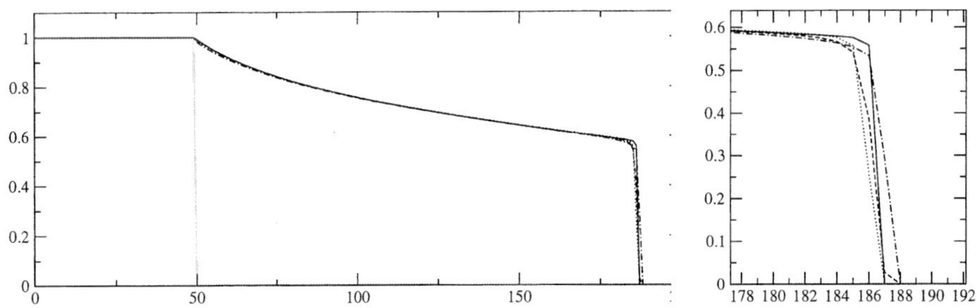

Fig. 4. Buckley-Leverett equation. **Left:** Grey line: initial condition. Solid line: correct solution after $t = 100$. Dotted line: Upwind with antidiffusion. Dashed line: TVD method. Dash-dotted line: Upwind without antidiffusion. **Right:** Zoom

$$u_0(x) = \begin{cases} 1 : & 0 \leq x < 50 \\ 0 : & \text{else} \end{cases} .$$

With the same settings as in the tests before, we obtain the numerical solutions depicted in Fig. 4. The outcome is similar to the experiments before: while the scheme with nonlinear antidiffusion is very close to the TVD scheme, there is quite some difference to the Upwind scheme without antidiffusion, although this difference is smaller than in the other experiments.

7 Conclusions

We have presented a novel FCT-type algorithm for hyperbolic conservation laws. It incorporates stabilised nonlinear inverse diffusion in order to improve the shock resolution of a first-order Upwind scheme. The nonlinear inverse diffusion step is inspired from a stabilised linear inverse diffision filter that has been proposed by Osher and Rudin for deblurring images. In contrast to classical FCT methods, our scheme arises naturally from the viscosity form of the basic scheme. As a consequence it also applies to the important class of nonlinear problems, even if the flux function is nonconvex. A theoretical analysis has shown that the novel scheme satisfies a global extremum principle and other desirable stability properties, while experiments with linear and nonlinear test scenarios indicate that it has approximately second order accuracy properties. It gives far better results than its underlying Upwind scheme and – in spite of its simplicity – it is even competitive to modern TVD methods for shock capturing approximations of hyperbolic PDEs. Its simplicity accounts for expectations that the method may be better accessible to theoretical analysis than TVD methods. In our ongoing work we further intend to analyse generalisations to the higher dimensional case as well as to systems of conservation laws.

Our work has shown that the connection between numerical schemes for hyperbolic conservation laws and image enhancement methods is not a one-way road: in the meantime, many PDE-based image enhancement techniques have reached a degree of maturity such that they may be highly useful for a number of problems outside the field of image analysis. It is our goal to investigate more of these fascinating interdisciplinary connections in the future.

Acknowledgements

The authors gratefully acknowledge the financial support of their work by the *Deutsche Forschungsgemeinschaft (DFG)* under the grants No. SO 363/9-1 and WE 2602/1-2.

References

1. Alvarez L. and Mazorra L. (1994): Signal and image restoration using shock filters and anisotropic diffusion. SIAM J. Num. Math. **31**, pp. 590–605
2. Boris, J.P. and Book, D.L. (1973): Flux corrected transport. I. SHASTA, a fluid transport algorithm that works. J. Comp. Phys. **11**, No. 1, pp. 38-69
3. Boris, J.P., Book, D.L. and Hain, K. (1975): Flux corrected transport II: Generalizations of the method. J. Comp. Phys. **18**, pp. 248-283
4. Boris, J.P. and Book, D.L. (1976): Flux corrected transport. III. Minimal error FCT algorithms. J. Comp. Phys. **20**, pp. 397-431
5. Breuß, M. (2004): The correct use of the Lax-Friedrichs method. RAIRO Math. Models Num. Anal. **38**, No. 3, pp. 519-540
6. Evans, L. (1998): Partial Differential Equations. American Mathematical Society
7. Gilboa G., Sochen, N.A., and Zeevi Y.Y. (2002): Forward-and-backward diffusion processes for adaptive image enhancement and denoising. IEEE Trans. Image Proc. **11**, No. 7, pp. 689-703
8. Godlewski, E. and Raviart, P.-A. (1991): Hyperbolic Systems of Conservation Laws. Ellipses, Edition Marketing
9. Godlewski, E. and Raviart, P.-A. (1996): Numerical Approximation of Hyperbolic Systems of Conservation Laws. Springer, New York
10. Grahs, T., Meister, A., and Sonar T. (2002): Image processing for numerical approximations of conservation laws: nonlinear anisotropic artificial dissipation. SIAM J. Sci. Comp. **23**, No. 5, pp. 1439-1455
11. Grahs, T. and Sonar T. (2002): Entropy-controlled artificial anisotropic diffusion for the numerical solution of conservation laws based on algorithms from image processing. J. Visual Commun. Image Repr. **13**, No. 1/2, pp. 176-194
12. LeFloch, P.G. and Liu, J.-G. (1999): Generalized monotone schemes, discrete paths of extrema, and discrete entropy conditions. Math. Comp., **68**, No. 227, pp. 1025-1055
13. LeVeque, R.J. (1992): Numerical Methods for Conservation Laws. Birkhäuser, 2nd Edition
14. Osher, S. and Rudin, L. (1990): Feature-oriented image enhancement using shock filters. SIAM J. Num. Anal. **27**, pp. 919-940
15. Osher, S. and Rudin, L. (1991): Shocks and other nonlinear filtering applied to image processing. SPIE Vol. 1567: Applications of Digital Image Processing XIV, pp. 414-425
16. Perona, P. and Malik, J. (1990): Scale space and edge detection using anisotropic diffusion. IEEE Trans. Pattern Anal. Mach. Intell., **12**, pp. 629–639
17. Pollak, I., Willsky, A.S., and Krim, H. (2000): Image segmentation and edge enhancement with stabilized inverse diffusion equations. IEEE Trans. Image Proc. **9**, No. 2, pp. 256-266
18. Rudin, L.I. (1987): Images, Numerical Analysis of Singularities and Shock Filters. Ph.D. thesis, California Institute of Technology, Pasadena, CA

19. Rudin, L.I., Osher, S., and Fatemi, E. (1992): Nonlinear total variation based noise removal algorithms. Physica D **60**, pp. 259–268
20. Roe, P.L. (1981): Numerical algorithms for the linear wave equation. Technical Report 81047, Royal Aircraft Establishment, Bedford, UK
21. Sweby, P.K. (1984): High resolution schemes using flux limiters for hyperbolic conservation laws. SIAM J. Num. Anal. **21**, pp. 995-1011
22. Wei, G.W. (2002): Shock capturing by anisotropic diffusion oscillation reduction. Comp. Phys. Commun. **144**, pp. 317-342
23. Weickert, J. (1998): Anisotropic Diffusion in Image Processing. Teubner, Stuttgart.
24. Zalesak, S.T. (1997): Introduction to "Flux corrected transport. I. SHASTA, a fluid transport algorithm that works". J. Comp. Phys. **135**, pp. 170-171

Sparse Finite Element Level-Sets for Anisotropic Boundary Detection in 3D Images*

Martin Weber[1], Andrew Blake[2], and Roberto Cipolla[1]

[1] Department of Engineering, University of Cambridge, UK
{mw232, cipolla}@eng.cam.ac.uk
http://mi.eng.cam.ac.uk/research/vision/
[2] Microsoft Research, Cambridge, UK
ablake@microsoft.com

Abstract. Level-Set methods have been successfully applied to 2D and 3D boundary detection problems. The geodesic active contour model has been particularly successful. Several algorithms for the discretisation have been proposed and the banded approach has been used to improve efficiency, which is crucial in 3D boundary detection. In this paper we propose a new scheme to numerically represent and evolve surfaces in 3D. With the new scheme, efficiency and accuracy are further improved. For the representation, space is partitioned into tetrahedra and finite elements are used to define the level-set function. Extreme sparsity is obtained by maintaining data only for tetrahedra that contain the zero level-set. We formulate the evolution PDE in weak form and incorporate a normalisation term. We obtain a stable scheme with consistent sub-grid accuracy without having to rely on any re-initialisation procedure. Boundary detection is performed using an anisotropic extension of the isotropic geodesic model. With the sparse representation, the anisotropic model is computationally feasible. We present experimental results on volumetric data sets including images with a significant amount of noise.

1 Introduction

Boundary surface detection in noisy 3D images is a vital ingredient for the analysis of medical scans, such as 3D ultrasound. In principle, one can attempt to form a surface by combining the contours of slices through the volumetric image. However, noisy images require a 3D method to facilitate the exploitation of the full spatial context of the problem. Geodesic [1, 2, 3, 4] surface detection has been proposed and Boykov and Kolmogorov [5] have recently presented an efficient discrete algorithm using *graph cuts*, including the anisotropic case. However, greater accuracy can be achieved in the continuous setting when a consistent sub-grid definition is being employed. *Deformable models* [6] have been used to estimate curve and surface models from image data. *Level-set methods* [7, 8, 9, 10, 1, 2] represent the surface implicitly, are topologically flexible and overcome

* This work was supported by the EPSRC, the Cambridge European Trust and DAAD.

R. Kimmel, N. Sochen, J. Weickert (Eds.): Scale-Space 2005, LNCS 3459, pp. 548–560, 2005.

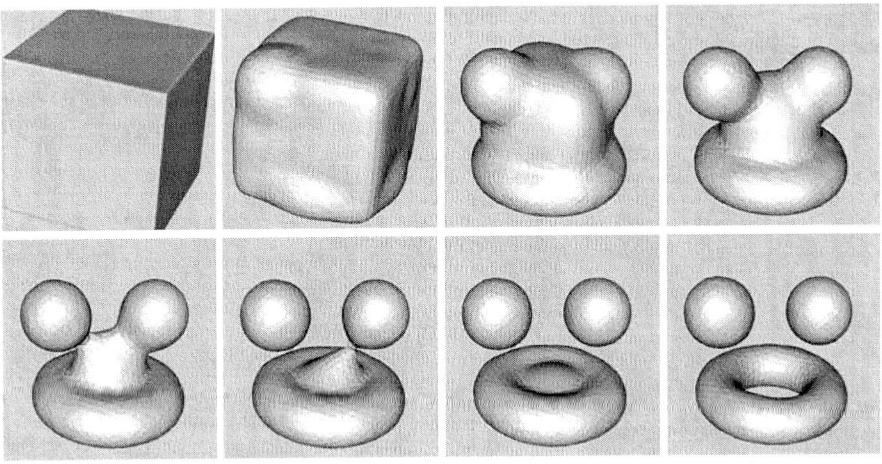

Fig. 1. Topological changes: stages of a geodesic evolution in the novel finite element scheme (starting from the top left). Three synthetic objects are to be detected. The evolution is initialised as a cube and splits automatically into the components. The generation of the hole inside the torus (bottom right) is an example of a topological change where the number of components remains unchanged (resolution $40 \times 40 \times 40$)

other problems of parametric models, such as self-intersection. However, level-set implementations in 3D are computationally involved. Preusser and Rumpf [11] have proposed a level-set framework with cubical finite elements. However, their numerical representation is not sparse (all levels are evolved) which requires a computational power exceeding that of current PCs.

Level set-methods [7, 8, 9, 10] introduce a level-set function ϕ, to represent the interface Γ implicitly as the zero level-set: $\Gamma := \phi^{-1}(0)$ where $\phi : \mathbb{R}^3 \rightarrow \mathbb{R}$ is a *Lipschitz-continuous*, real valued function. The implicit representation links ϕ (as the introduced *analytic entity*) with the *geometric entity* Γ: $\phi \mapsto \Gamma(\phi)$ and allows for changes in the topology during the evolution (Figure 1). Furthermore, it was pointed out [12] that this relationship can be made one-to-one by imposing the signed distance constraint. That is, the constraint fixes the gauge freedom[1]. The conceptual advantage is then that ϕ is (up to a sign) uniquely determined by Γ and that one can also write $\Gamma \mapsto \phi(\Gamma)$. In this way ϕ gets the intrinsic geometric meaning as the distance function for Γ.

1.1 Differential Minimisation and Level-Set Evolution

For the evolution, one introduces an evolution parameter $t \in R$ and ϕ becomes *time*-dependent[2]. One starts with an initial function $\phi(0, .)$ and prescribes an

[1] the freedom of using any multiple $\psi \phi$ in the place of ϕ with *any* positive valued function $\psi > 0$.

[2] One refers to the parameter t as *time* although it is not related to physical time.

evolution $\phi(t,.)$. In Section 2, we introduce a cost-functional C and define the evolution via gradient descent. The resulting evolution equation is a PDE:

$$\frac{d\phi}{dt} = \beta \tag{1}$$

where at the interface Γ, β is the differential of the cost: $\beta|_\Gamma := -\frac{\delta C}{\delta \phi}$ and is defined elsewhere in such a way as to maintain the signed distance constraint [12]. The signed distance constraint is well known for its desirable conceptual and numerical properties. Where ϕ is differentiable, we have $|\nabla\phi(x)| = 1$ and, for $x \in \Gamma$ the particularly simple expressions for the *normal* $N(x) = \nabla\phi(x) \in S^2$ and *mean curvature* $\kappa(x) = \Delta\phi(x) \in \mathbb{R}$ along Γ.

1.2 Previous Level-Set Methods and Numerical Problems

In the following sections, u denotes the numerical representation of the level-set function ϕ. The major issues for the numerical implementation of (1) concern efficiency, stability and accuracy. Improved efficiency and stability was obtained using sparse finite element level-sets, as we reported in [13]. Here we extend the 2D scheme to 3D and include anisotropic cost-functionals. One advantage of level-set methods over discrete methods is the potential for sub-grid accuracy.

Consistent Sub-Grid Accuracy: sub-grid accuracy requires the definition of the level-set function inside each cell of the grid. Figure 2 details the standard way in which this can be achieved [12].

Unfortunately, this sub-grid definition has several problems:

- **mixed polynomial degree:** the standard representation (see Figure 2) singles out the directions of the coordinate axes (along which u is piece-wise affine). In the interior of the cell it is generally a cubic polynomial.
- **tiling ambiguities:** the graphical output of the implicit interface is not straightforward. To obtain a facet approximation, one employs so called *iso-contour algorithms* ([10] p.425) such as the *marching cube algorithm*, in which case ambiguities in the *tiling* have to be resolved.
- **sub-grid definition cannot be used for the evolution:** the definition of *curvature motion* [8] inside a cell is problematic. To see this, let us assume that we have an initial signed distance function ϕ. Then, curvature motion is simply given by the Laplacian: $\beta = \Delta\phi$. However, the Laplacian of the numerical representation u vanishes identically: $\Delta u(x) = 0$ for any x inside the cell, independent of any node-values on the grid. This is related to the fact that u is affine along coordinate directions (alternatively, one can verify in Figure 2 that $\Delta e_0 = 0$, like any other nodal basis function). Therefore, the sub-grid definition cannot be used when the interface is evolved. Instead, one [8] computes differential operators (like the Laplacian) with finite difference operators on *stencils* that contain several cells. However, it is not entirely clear how the evolution in terms of finite difference operators is related to the sub-grid localisation of interfaces as defined in Figure 2.

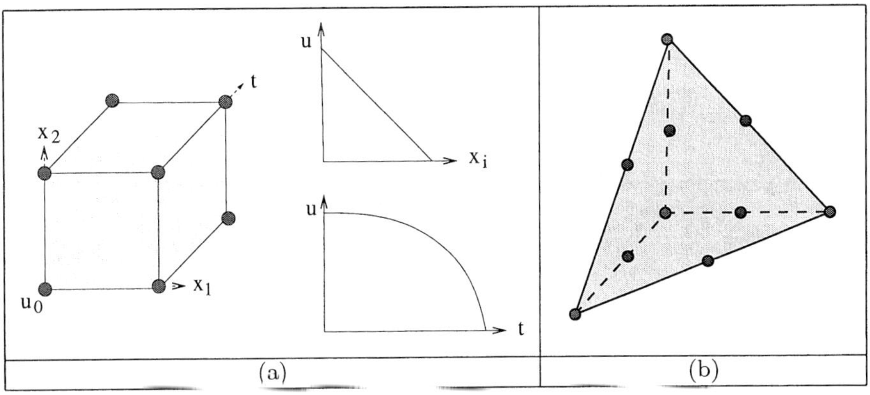

Fig. 2. Sub-grid definition: The figure illustrates the sub-grid definition of a level-set function u in terms of (a) cubical cells and (b) tetrahedral elements. (a) in the case of cubical cells, a product of functions that are affine in the direction of the coordinate-axes is used. One can express u as linear combination $u = \sum_{i=0}^{7} u_i e_i$ of nodal basis functions e_i. For instance, the basis function for node 0 is: $e_0(x) = (1-x_1)(1-x_2)(1-x_3)$ which evaluates to 1 at node 0 and vanishes on all other nodes. Unfortunately, the e_i (and hence u) are of mixed polynomial degree: along all coordinate axes they are affine but on the diagonal they are cubic polynomials. (b) tetrahedral elements are defined by the 4 nodes located at the vertices (1st degree case) and, 6 additional nodes at the edge-midpoints in the 2nd degree case. A general 1st or 2nd degree polynomial is prescribed by the values u_i at the nodes. Coordinate axes are not treated as special directions here

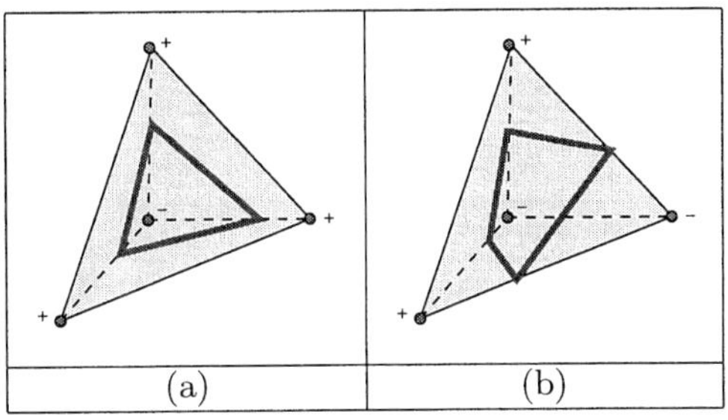

Fig. 3. Implicit facet in 3D: depending on the signs of the node-values, the implicit facet is either (a) a triangle (if one sign differs from the other three) or (b) a planar quadrangle (if there are two pairs of different signs)

These problems are resolved when simplices are used instead of cubical cells:

- **fixed degree:** inside each element, u is a polynomial of fixed degree.
- **simple graphical output:** with first degree simplex-elements, the interface inside each element is always a planar facet (Figure 3). Hence, no tiling ambiguity occurs and the output of the interface is straightforward.

– **consistent use of the sub-grid definition in the evolution:** the evolution equations are treated in weak formulation, which allows us to define the evolution equation consistently with the defined sub-grid accuracy.

2 Anisotropic Interface Optimisation

By adopting an interface optimisation formulation, we incorporate the prior knowledge/demand that the interface is a continuous boundary Γ. As is well known, this implies robustness to noise. The task is to find local minima of the cost

$$C = \int_{\Gamma} \gamma \tag{2}$$

which is a two-dimensional *Lebesgue integral* of a density $\gamma > 0$ (specified below) over the interface Γ. In the anisotropic case, γ depends not just on location but also orientation. Global minima of (2) are trivial (sets of measure zero) unless additional constraints are imposed [5]. We follow here the strategy of interface evolution [7, 8] by deforming an initial interface to obtain local minima.

2.1 Edge Indication Function

We denote the real-valued volumetric image by $f : \mathbb{R}^3 \to \mathbb{R}$. Edges are related to the gradient of f. In order to *quantify* the strength of edges, a function g is introduced. In the literature [14], g is referred to as *edge indication function.* The particular functional form of g is not crucial. It is however essential that $g \geq 0$ and usually g is monotonous with respect to feature gradient magnitude. We also adopt the convention to normalise g to have values in the unit interval $[0; 1]$. We let $g(x)$ depend on the gradient magnitude $|\nabla f(x)|$ and choose [15]:

$$g := 1 - \exp\left(-\tfrac{a}{|\nabla f_\sigma|^q}\right) \tag{3}$$

where f_σ denotes the smoothed version of the image f obtained by convolution of f with a Gaussian of width σ. Furthermore, a and q are real, positive constants that determine the response of the edge/feature detection.

In order to complete the definition of the edge detector function g, one can adjust the parameters a and q in (3) automatically to the image-data. For instance we determine a, q in such a way that the average gradient magnitude $\langle|\nabla f_\sigma|\rangle$ over the image results in $g = \frac{1}{2}$ and that the slope of g with respect to the gradient magnitude equals $-1/\langle|\nabla f_\sigma|\rangle$ at this point. These are sensible choices which are invariant to the affine value-transformations of the image [16].

2.2 An Anisotropic Density

In this section we define a cost $C(\Gamma)$ for the boundary detection by specifying the density γ in (2). One viable density γ is obtained by simply choosing the edge indication function g itself: $\gamma(x, N) = g(x)$. In this case, one refers to the density as *(isotropic) Riemannian metric* [17,9] since the cost $C(\Gamma) = \int_\Gamma g$ is the *Riemannian area.* The attraction to the edges/features in the image is geometrically obtained by driving the evolution towards g-minimal surfaces.

This is the geodesic model, which is capable of detecting boundaries even in the presence of noise. However, the isotropic model is known to be sub-optimal. While the edge-strength is taken into account, the orientation information provided by ∇f_σ is not utilised. We can further improve the performance of the geodesic boundary detection by including an orientation-dependent term. Formally, this means that we consider *anisotropic* densities γ which depend non-trivially on the orientation of the interface N.

Orientation-dependent terms have been used previously: Kimmel and Bruckstein [18] detect boundary-curves in 2D images using *Laplacian zero crossings.* However, their formulation requires additional regularising functionals and parameters which are needed to determine their relative strength. Instead, the formulation presented here is based on a single cost-functional that is to be minimised. Boykov and Kolmogorov [5] consider an anisotropic extension of the geodesic model in their discrete setting. In our continuous formulation, the anisotropic extension is conceptually straightforward with consistent sub-grid accuracy and computationally feasible due to the extreme sparsity of the representation.

We consider the general quadratic expression

$$\gamma = \tfrac{1}{2}\langle N, N\rangle_G \tag{4}$$

where G is a matrix-valued function and $\langle N, N\rangle_G := N^\top G N$. We use the edge detector g to define G:

$$G := \mathbf{1} + \frac{g-1}{|\nabla f_\sigma|^2}\,\nabla f_\sigma \otimes \nabla f_\sigma \tag{5}$$

where $v \otimes v := v\,v^\top$. This choice is motivated by the following properties:

- **alignment case** ($\nabla f_\sigma \propto N$)**:** we obtain $G\,N = g\,N$, which is equivalent to geodesic motion.
- **non-alignment case** ($df_\sigma\,N = 0$)**:** we obtain $G\,N = N$, equivalent to *curvature motion.*
- **weak gradient case** ($\nabla f_\sigma \approx 0$)**:** since $g \to 1$, we obtain $G \to \mathbf{1}$ which results in *curvature motion.*

When compared to the isotropic geodesic motion, there is an additional aligning force. Figure 4 demonstrates the orienting force for the special case where $g = 0$. In this case (5) becomes $G = \mathbf{1} - \frac{1}{|\nabla f|^2}\nabla f \otimes \nabla f$. The following proposition (proved in [16]) quantifies the aligning property :

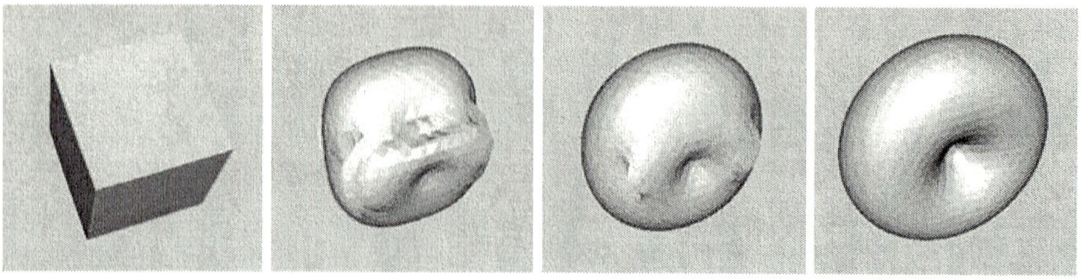

Fig. 4. Orienting motion: the figure illustrates the orienting property of quadratic motion $\gamma = \frac{1}{2}\langle N, N\rangle_G$. Orientation is prescribed here by selecting a distance map f and setting $G := \mathbf{1} - \nabla f \otimes \nabla f$. In this example, f is the distance from a circle. The interface is oriented by the evolution without being contracted or attracted to any features such as edges ($|\nabla f| \equiv 1$). Rather, any initial interface is progressively aligned to the orientation prescribed by f. A temporary local 'folding' of the interface can be observed. This is related to the fact that the cost functional does not discriminate between N and $-N$; however, the cost of transitions between the two possible alignments means that eventually there is one consistent alignment

---

Proposition 1. *Difference between the anisotropic and geodesic cost-functionals:*

The difference between the cost functionals is given by $\Delta C := 2\int_\Gamma \gamma - \int_\Gamma g = \int_\Gamma \Delta\gamma$ (here Δ denotes differences, not the Lapacian) with the difference-density

$$\begin{aligned}\Delta\gamma &= \langle N, N\rangle_G - g \\ &= (1-g)\,(1-a)\end{aligned} \tag{6}$$

with the aligning term $a := \left\langle N, \frac{1}{|\nabla f_\sigma|}\nabla f_\sigma\right\rangle^2 \in [0;1]$. Consequently,

- $\Delta\gamma \geq 0$: hence an additional cost is present in the anisotropic density.
- $\Delta\gamma \propto (1-g)$: hence the orienting power is most pronounced at edges and vanishes in feature-free regions.
- $\Delta\gamma \propto (1-a)$: if the interface is aligned with the feature ($a \approx 1$) there is no additional cost; when the alignment is worst (i.e. orthogonal, $a \approx 0$) the orienting force is maximal.

2.3 Gradient Descent for the Anisotropic Scheme

Surface evolution is defined as the gradient descent of the cost (2) with density (4). Using *variational calculus*, one can derive [16] the following result:

Proposition 2. *Gradient descent of quadratic densities*

The normal motion $\beta|_\Gamma = -\frac{\delta C}{\delta\phi}$ for the gradient descent is given by

$$\beta|_\Gamma = \operatorname{div} V \tag{7}$$

$$V = (G - \gamma\,\mathbf{1})\,N \tag{8}$$

where N is the surface normal. We call V the *descent-generating vector field.* The geodesic model is a special case in which $G = 2\,g\mathbf{1}$ and hence $V = g\,N$. Equation (7) has also been used by Vasilevskiy and Siddiqi in their flux maximising finite difference scheme [19]. Here, the vector filed V derives from (8) and we exploit equation (7) in our weak scheme. The fact that such a vector field exists is of prime importance to our scheme since it enables a straightforward application of Gauss's integral theorem for the weak formulation of Section 4.

3 Sparse Finite Element Representation

The new numerical representation u of the level-set function employs a sparse finite element complex [13]. For its definition in 3D, we have to partition space into tetrahedra.

3.1 Partitioning of Three-Space into Tetrahedra

Unfortunately, unlike in the 2D case, it is not possible to partition $\mathbb{R}^3$ into standard 3-simplices. However, the methodology of [13] is not confined to standard simplices. Here, we obtain a convenient partitioning $\mathcal{M}$ of space as *Delaunay tetrahedrisation* [20] of the vertex-set $(2\,\mathbb{Z}^3) \cup ((1,1,1)^\top + (2\,\mathbb{Z}^3))$. A particularly convenient feature of this mesh is that all tetrahedra of the mesh have the same shape. Each element has two even $(2\mathbb{Z}^3)$ and two odd $(2\mathbb{Z}^3 + (1,1,1)^\top)$ vertices. For instance, the four vertices $\{(0\,0\,0)^\top, (2\,0\,0)^\top, (1\,{-1}\,1)^\top, (1\,1\,1)^\top\}$ span a tetrahedron of the mesh $\mathcal{M}$.

As in the 2D case [13], we restrict the actual numerical representation to the sparse simplicial sub-complex which consists only of those elements which contain the zero level-set. We call this minimal set the *active complex* $\mathcal{A} \subset \mathcal{M}$.

3.2 Finite Elements for Functional Representations

The evolution equations of Section 4 involve the continuous functionals u and G_{ij} which have to be represented numerically. This is achieved with consistent sub-grid definition by using standard finite element methods [21]. We briefly describe the technique for u (every component of G is represented similarly).

u is a linear combination of nodal-basis functions: $u(x) = \sum_j u_j\, e_j(x)$. The nodes are located at the vertices that are contained in the active complex $\mathcal{A}$. In the second degree case ($p = 2$), additional nodes are inserted as in Figure 2b. Each node defines a nodal basis function e_j. Inside a tetrahedral element $T \in \mathcal{A}$, e_j is defined as the unique [21] polynomial of degree p which evaluates to 1 at the node and vanishes on all other nodes: $e_j(x_i) = \delta_{ij}$. Note that e_j is a globally defined continuous function over the area Ω covered by the active complex $\mathcal{A}$.

The fact that the integral over u is a linear map allows us to integrate efficiently by using linear combinations of pre-computed integrals over (products of) basis functions. In fact, this type of integration can be performed analytically [16] and we only need to store one real value u_j at each node. For first degree elements, each active element $T \in \mathcal{A}$ contains one planar facet (Figure 3).

4 Surface Evolution with Sparse Finite Elements

We now show how a stable evolution can be defined to realise the gradient descent expression (7) in terms of the sparse representation. The sparse representation is updated in two stages:

- first, a differential update corresponding to a time-step Δt is performed. This alters the values of nodes u_j.
- secondly, the active complex $\mathcal{A}$ is updated to restore the minimal containing property. Elements that no longer contain the zero level-set are deleted while neighbouring elements get activated if the zero-level set has moved into their domain. As in the 2D case, the criteria that control the activation and removal of elements are simply obtained[3] and functional extrapolation is used to initialise newly activated nodes.

For the algorithmic details of this process, we refer to [13, 16]. The differential update with *time-step* Δt: $u(t+\Delta t,.) = u(t,.) + v(t,.), v = \Delta t\, \beta$ combines two equations, which we formulate in *weak form (Petrov Galerkin)*:

- the normalisation of the level-set function is maintained by demanding $\langle \nabla u, \nabla u \rangle - 1 = 0$. This is formulated in the weak sense by

$$z_1^i = \int_\Omega e_i \left(\langle \nabla(u+v), \nabla(u+v) \rangle - 1 \right) \tag{9}$$

- the interface motion $v - \Delta t\, \beta|_\Gamma = 0$ is formulated as

$$z_2^i = \int_\Omega e_i \left(v - \Delta t\ \beta \right) \tag{10}$$

Note that $\langle \nabla u, \nabla u \rangle - 1 = 0$ (and likewise (9)) has a *flat direction* since it is invariant to any level-shift $u \to u + c,\ c \in \mathbb{R}$. While this would complicate the use of the equation in isolation[4], it is not a problem when the equation is combined with the interface motion equation (10).

If there are n active nodes, we have $2\,n$ equations and determine the update v as the least-square solution to $|z_1|^2 + |z_2|^2$. This is a sparse, banded problem which we solve [13] by the *conjugate gradient method* [22].

The interface speed $\beta_\Gamma = \mathrm{div} V$ depends on second order derivatives, since V depends on $N = \nabla u$. In *strong form*, this causes the same problems as the finite-difference approach (Section 1.2) since the desired second order derivatives cannot be obtained from the sub-grid definition inside the elements and are ill defined at element-transitions. This problem is resolved in the *weak* formulation. Instead of having second order derivatives which are ill-defined at element transitions, we encounter boundary integrals (which we compute efficiently [16]).

---

3 This reduces to the task of determining if a boundary-face of $\mathcal{A}$ (a triangle) contains any part of the zero level-set. For $p = 1$ this simply depends on whether all signs of u_j at the nodes of the triangle agree.

4 This equation is used in some re-initialisation schemes [8].

By Proposition 2, the gradient descent has a descent-generating vector field $\beta = \operatorname{div} V$, hence we can apply Gauss's theorem and perform partial integration. This allows us to eliminate second order derivative operators in (10):

$$\int_{\Omega} e_i \beta = \int_{\partial\Omega} e_i \langle V, \mathcal{V} \rangle - \int_{\Omega} de_i \, V \tag{11}$$

where $\mathcal{V}$ denotes the normal along the boundary $\partial\Omega$. The right hand side does not contain any second order derivatives of the level-set function.

5 Experimental Results

In this section, we apply the new method to real medical image data. Figures 5 and 6 illustrate results of the boundary detection (white-grey matter boundary)

(a) (b) (c)

Fig. 5. Cortical brain structure: (a) displays a slice through the volumetric MRI data. (b) and (c) show views of the VRML model created by the described method

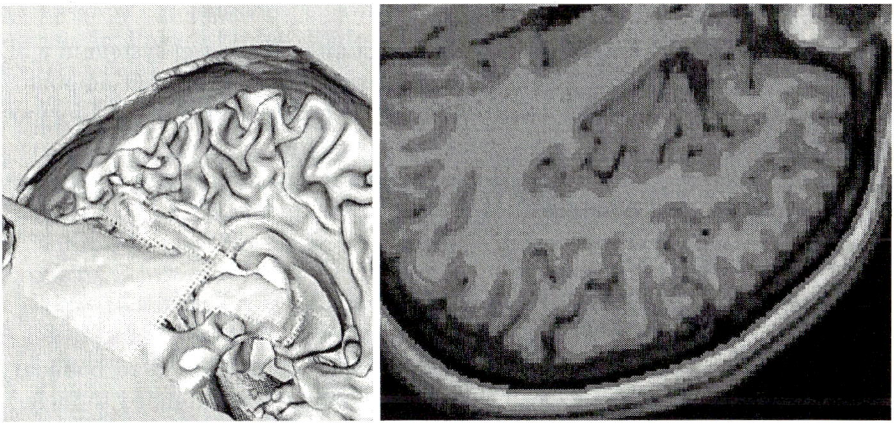

(a) (b)

Fig. 6. MRI-brain scan: the cortical brain structure is detected from an MRI-image, using the quadratic cost. The evolution was initialised by a sphere located in the area of white matter and inflated with the aid of a balloon force ($c = -0.1$). (a) shows a projective view of the obtained surface and in (b) a section of the 3D interface is superimposed on the corresponding slice of the input-image

15. Weickert, J., ter Haar Romeny, B., Viergever, M.: Efficient and reliable schemes for nonlinear diffusion filtering. IEEE Trans. Image Processing **7** (1998) 398–410
16. Weber, M.: Curve and Surface Reconstruction from Images and Sparse Finite Element Level-Sets. PhD thesis, Department of Engineering, University of Cambridge (2004)
17. Do Carmo, M.: Differential Geometry of Curves and Surfaces. Prentice-Hall (1976)
18. Kimmel, R., Bruckstein, A.: Regularized Laplacian zero crossings as optimal edge integrators. International Journal of Computer Vision **53** (2003) 225–243
19. Vasilevskiy, A., Siddiqi, K.: Flux maximizing geometric flows. In: Proc. IEEE Int. Conf. on Computer Vision, Vancover, Canada (2001) 149–154
20. Edelsbrunner, H.: Geometry and Topology for Mesh Generation. Cambridge University Press, Cambridge (2001)
21. Zienkiewicz, O., Morgan, K.: Finite Elements & Approximation. John Wiley & Sons, New York (1983)
22. Schwarz, H.: Numerische Mathematik. Teubner, Stuttgart (1993)

A Scale Space Method for Volume Preserving Image Registration

Eldad Haber[1] and Jan Modersitzki[2,*]

[1] Dept. of Mathematics and Computer Science,
Emory University, Atlanta GA 30322
haber@mathcs.emory.edu

[2] Inst. of Mathematics, University of Lübeck, Germany
modersitzki@math.uni-luebeck.de

Abstract. Image registration is an inherently ill-posed problem. Therefore one typically aims to provide as much information about the underlying application as possible. In particular for tumor monitoring, volume preservation of the wanted deformation is a central point. Based on [9], we propose a new scale space approach to volume preserving image registration. The main advantage of the new approach is that the constraints appear linearly and therefore the system matrices resembles Stokes matrices, which appears in computational fluid dynamics.

We present the scale space framework, a composition based numerical approach and its implementation. Finally, we demonstrate the outstanding features of this idea by a real life example.

1 Introduction

Image registration is one of the fundamental tasks in today's image processing and in particular in medical imaging; see, e.g., [11, 4, 15] and references therein. The objective of image registration is to make images which are taken at different times, from different perspectives, and/or from different devices to be more alike. Loosely, the goal of image registration is to find a "*reasonable*" deformation such that the "*distance*" between a reference image R and a deformed version of a template image T becomes small.

Image registration is an ill-posed problem (cf., e.g., [12]) and therefore need to be regularized. Different types of regularizers can be used to specify the meaning of reasonable. However, for particular applications, one may want to provide additional information. Typical examples include the knowledge of the location of anatomical landmarks or markers in the images and/or additional physical properties of the deformation field.

A situation of particular clinical interest is the analysis of pairs of images acquired before and after contrast administration; see, e.g., [13] and references

* Jan Modersitzki was supported by the US National Institute of Health under Grant NIHR01 HL 068904.

R. Kimmel, N. Sochen, J. Weickert (Eds.): Scale-Space 2005, LNCS 3459, pp. 561–572, 2005.

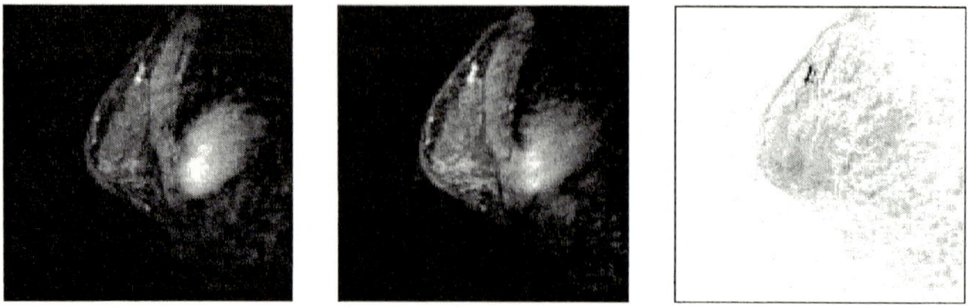

Fig. 1. MRI's of a female breast, LEFT: during the wash-in phase, MIDDLE: during the wash-out phase, and RIGHT: difference image

therein. As a typical example, Figure 1 shows two different magnetic resonance images (MRIs) of a female breast as they are used routinely for tumor monitoring (images from Bruce Daniel, Lucas Center for Magnetic Resonance Spectroscopy and Imaging, Stanford University). The first image shows an MRI section taken during the so-called wash-in phase of a radiopaque marker and the second image shows the analogous section during the so-called wash-out phase. A comparison of these two images indicates a suspicious region in the upper part of the images. This region can be detected easily if the images have been registered: tissue located at a certain position in the wash-in image is related to the tissue at the same position in the wash-out phase. Generally, however, a quantitative analysis is a delicate matter since observable differences are not only related to contrast uptake but also due to motion of the patient, like, for example, breathing or heart beat. Therefore, image registration becomes an inevitable task. However, as pointed out by Rohlfing et al. [13], there is a substantial difficulty with the registration of pre- and post-contrast images. Bright regions seem to enlarge during the so-called wash-in phase. This enhancement is due to contrast uptake but not to movement of the patient. Therefore, the geometry of these regions should not be changed by the registration procedure. Most importantly, for this particular application, a distance minimizing registration can produce unrealistic results. A critical feature of the wanted deformation is that it has to preserve tissue volume, particularly the tumor volume. From a clinical point of view, a volume change is unacceptable even if it results in a much smaller image distance.

A general approach to constrained image registration has been proposed in [9]. Focussing on volume preserving registration, a variational approach with a Tikhonov-type regularization has been presented. The approach leads to a nonlinear constrained optimization problem which is solved by a Sequential Quadratic Programming (SQP). At each iteration a large system of linear equations has to be solved and the iteration matrix changes as the iteration proceeds. As it is typical for constrained optimization, the system is a so-called KKT system, where for this application the off-diagonal blocks are nonlinear. In this paper, we explore a scale space type regularization applied to the constrained problem. We replace the Tikhonov regularization by a dynamical process. The regularization of the problem is obtained by performing a finite number of steps of the discretized process. We demonstrate that this approach has unique properties when applied to the problem of volume preserving image registration. A major advantage of this new scale space approach is that the linear systems to be solved at each iteration simplify. Particularly, the off-diagonals become linear. Therefore, one can use a wealth of efficient algorithms developed for computational fluid dynamics.

This paper is organized as follows. In Section 2 we discuss the scale space regularization and formulate the system of equations it leads to. In Section 3 we discuss discretization issues of the problem. In Section 4 we present a numerical scheme for the solution of the discrete scale space problem. In Section 5 we apply the new approach and demonstrate the effectiveness of our algorithm on a realistic example.

2 Scale Space Regularization for Image Registration

In the first part we explain our notation and a straightforward Tikhonov regularization. We introduce the building blocks of our implementation. Here, for ease of presentation, we focus on the sum of squared differences (or L_2 Norm) as a distance measure (or misfit) and the elastic potential as a regularizer (or displacement semi-norm). However, other distance measures or regularizer can be used as well; see [8, 12].

In the second part we discuss our scale space Tikhonov regularization. The main observation is that a composition type approach for the transformation does allow for locally linear constraints. Therefore, a numerical treatment of the scale space approach has severe advantages.

2.1 General Notations and Tikhonov Regularization

With $d \in \mathbb{N}$ we denote the spatial dimension of the given images $R, T : \mathbb{R}^d \to \mathbb{R}$ which are assumed to be sufficiently smooth. Thus, $T(\boldsymbol{x})$ gives a gray value at a spatial position $\boldsymbol{x}$. Without loss of generality, we assume that the supports of the images are contained in a bounded domain $\Omega := (0,1)^d$ and in particular $R(\boldsymbol{x}) = T(\boldsymbol{x}) = 0$ for $\boldsymbol{x} \notin \Omega$.

Our goal is to find a "*reasonable*" transformation φ such that the "*distance*" between the reference image and the deformed template image is small. As usual in image registration, we set $\varphi(\boldsymbol{x}) = \boldsymbol{x} + \boldsymbol{u}(\boldsymbol{x})$ and work the displacement $\boldsymbol{u} = (u^1, \ldots, u^d)^T : \mathbb{R}^d \to \mathbb{R}^d$ rather than the transformation φ. However, we will also use φ whenever it makes notation shorter.

It is well-known that the registration problem is ill-posed and therefore needs to be regularized [12]. Tikhonov-type regularization is commonly used to balance between image distance and regularity of $\boldsymbol{u}$. A mathematical formulation of the regularized and constrained problem thus reads:

$$\text{minimize} \quad \mathcal{D}(T, R; \boldsymbol{u}) + \tfrac{\alpha}{2}\|\mathcal{B}\boldsymbol{u}\|^2 \tag{1a}$$

$$\text{subject to} \quad \mathcal{C}(\boldsymbol{u}) = 0, \tag{1b}$$

where $\mathcal{D}$ measures image distance, $\|\mathcal{B}\boldsymbol{u}\|^2$ is a quadratic regularizer, $\mathcal{B}$ a partial differential operator, $\alpha > 0$ is a regularization parameter that compromises between similarity and regularity, and $\mathcal{C}$ are the volume preserving constraints. Problem (1) is usually solved for a sequence of decreasing α's. For each α of this sequence, we obtain a smaller misfit and a generally larger displacement semi-norm.

For ease of presentation, we focus on the Sum of Square Differences (SSD) as a distance measure,

$$\mathcal{D}(\boldsymbol{u}) := \mathcal{D}(T, R; \boldsymbol{u}) := \tfrac{1}{2}\|T(\boldsymbol{x}+\boldsymbol{u}) - R\|^2, \tag{2}$$

and the elastic potential with Lamé constants λ and μ as regularizer. Hence,

$$\begin{aligned} \mathcal{S}[\boldsymbol{u}] &= \int_\Omega \langle \mathcal{B}\boldsymbol{u}, \mathcal{B}\boldsymbol{u}\rangle \; d\boldsymbol{x} && (3)\\ &:= \int_\Omega \frac{\lambda+\mu}{2}\|\nabla\cdot\boldsymbol{u}\|^2 + \frac{\mu}{2}\sum_{i=1}^{d}\|\nabla\boldsymbol{u}_i\|^2 \; d\boldsymbol{x}. \end{aligned}$$

For the purpose of this paper, a transformation is volume preserving if

$$\det(\nabla\varphi(\boldsymbol{x})) = 1 \quad \text{for all} \quad \boldsymbol{x} \in \Omega;$$

see [9] for an extended discussion. Our definition implies that volume preserving maps also preserve orientation, which is an additional desirable feature in medical registration. With $I_d \in \mathbb{R}^{d,d}$ denoting the d-by-d identity matrix, our pointwise constraint thus becomes

$$\mathcal{C}(\boldsymbol{u}) := \det(I_d + \nabla\boldsymbol{u}) - 1 = 0. \tag{4}$$

Introducing a Lagrange multiplier p, the Lagrangian of (1) is

$$\mathcal{L}(\boldsymbol{u}, p) = \tfrac{1}{2}\|T(\boldsymbol{u}) - R\|^2 + \frac{\alpha}{2}\|\mathcal{B}\boldsymbol{u}\|^2 + \int_\Omega \mathcal{C}(\boldsymbol{u})\cdot p \; d\boldsymbol{x}$$

and the continuous Euler-Lagrange equations for (1) are

$$\begin{aligned} 0 &= (\nabla T(\boldsymbol{u}))^\top (T(\boldsymbol{u}) - R) + \alpha\mathcal{B}^*\mathcal{B}\boldsymbol{u} \\ &\quad - \nabla\cdot\left[\,\det(I_d+\nabla\boldsymbol{u})(I_d+\nabla\boldsymbol{u})^{-\top}\cdot p\,\right], && (5a)\\ 0 &= \det(I_d + \nabla\boldsymbol{u}) - 1, && (5b)\\ 0 &= \boldsymbol{n}\cdot\nabla\boldsymbol{u}_i; \qquad i = 1..d && (5c) \end{aligned}$$

see [9]. Here, ∇T is the gradient of T and $\mathcal{B}^*$ the adjoint of $\mathcal{B}$. The system (5) is a highly coupled system of nonlinear partial differential equations (PDE). The differential operator $\mathcal{B}^*\mathcal{B}$ in (5a) is a linear, elliptic operator. The last term in (5a) is related to the derivative of the constraints which also show up in (5b). It is not easy to show either existence or uniqueness of a solution of the PDE (5).

For the purpose of this paper, we therefore assume existence of a solution and remark that proving its existence is a subject of further research.

After freezing the coefficients of the linearized system (5), a Local Fourier Analysis can be used to show ellipticity of the system for small displacements $\boldsymbol{u}$; see, e.g. [8]. However, for large displacements ellipticity is crucial and numerical difficulties may arise; see [15]. This motivates us to use a modified Iterative Tikhonov Regularization approach where only small displacements are considered and a sequence of elliptic problems has to be solved.

2.2 A Modified Iterated Tikhonov regularization

We now explore a modified iterated Tikhonov regularization approach and show that it has some favorable numerical properties.

We assume that at some stage k we have an approximate solution $\varphi_k(\boldsymbol{x}) = \boldsymbol{x} + \boldsymbol{u}_k(\boldsymbol{x})$ which is volume preserving and therefore obeys Eq. (4). We can also associate to this solution an image distance $\mathcal{D}(\boldsymbol{u}_k)$ and a displacement norm $\|\mathcal{B}\boldsymbol{u}_k\|^2$. We seek a method to update φ_k such that the new transformation φ_{k+1} gives a smaller image distance and an equal or larger displacement norm. Using the common iterated Tikhonov regularization techniques, one would compute a linear perturbation $\boldsymbol{v}$ and set

$$\varphi_{k+1} = \varphi_k + \boldsymbol{v}_k \quad \text{or} \quad \boldsymbol{u}_{k+1} = \boldsymbol{u}_k + \boldsymbol{v}_k. \tag{6}$$

However, this straightforward approach does not lead to a simplification of the equations or to an improvement of the numerics. The main problem is that the volume preserving constraint is nonlinear and therefore

$$\mathcal{C}(\boldsymbol{u}_k + \boldsymbol{v}_k) \neq \mathcal{C}(\boldsymbol{u}_k) + \mathcal{C}(\boldsymbol{v}_k).$$

A better approach is to update the whole φ_k in terms of a composition. This can be done by computing a new function $\psi_k(\boldsymbol{x}) = \boldsymbol{x} + \boldsymbol{v}_k(\boldsymbol{x})$ and setting

$$\varphi_{k+1}(\boldsymbol{x}) = \varphi_k(\psi_k(\boldsymbol{x})) \quad \text{or} \quad \boldsymbol{u}_{k+1}(\boldsymbol{x}) = \boldsymbol{u}_k(\boldsymbol{x} + \boldsymbol{v}_k(\boldsymbol{x})) + \boldsymbol{v}_k(\boldsymbol{x}).$$

The following Lemma is a key observation for the development of an efficient algorithm.

Lemma 1. *The set of volume preserving mappings with the composition is a non-commutative group.*

Proof. Since the set of continuous functions forms a group with respect to composition, the statement follows from

$$\det(\nabla[\varphi(\psi(\boldsymbol{x}))]) = \det(\nabla\varphi(\psi(\boldsymbol{x})) \cdot \nabla\psi(\boldsymbol{x})) = \det(\nabla\varphi) \cdot \det(\nabla\psi) = 1,$$

where φ and ψ are assumed to be volume preserving. ∎

The above Lemma allows us to derive efficient algorithms. At stage k, we define $T_k(\boldsymbol{x}) := T(\varphi_k(\boldsymbol{x}))$. We are than seeking for a (small) update $\boldsymbol{v}$ as a solution of

$$\text{minimize } \tfrac{1}{2}\|T_k(\boldsymbol{x}+\boldsymbol{v}) - R\|^2 + \tfrac{\alpha}{2}\|\mathcal{B}\boldsymbol{v}\|^2 \tag{7a}$$

$$\text{subject to } \mathcal{C}(\boldsymbol{v}) = 0. \tag{7b}$$

Note that for a minimizer $\boldsymbol{v}_k$ of problem (7) we have

$$T_k(\boldsymbol{x}+\boldsymbol{v}_k) = T(\varphi_k(\boldsymbol{x}+\boldsymbol{v}_k)) = T(\varphi_k(\psi_k(\boldsymbol{x}))).$$

In contrast to the straightforward linear approach (6), the new approach is based on composition. This approach is related to the Euler coordinates framework as used in fluid registration [1], see also [12–§10.4.2]. In other words, we change from material to spatial coordinates but drop the material derivative within the regularizer. Thus, the new scale space approach give different solutions than the Iterative Tikhonov regularization. In the next section we demonstrate that the solution of the sequence of problems (7) can be computed very effectively using a Newton-Multigrid method.

3 Discretization

There are two main approaches for the discretization of the registration problems (1) and (7)). In the first so-called *optimize-discretize* approach one forms the objective function, then differentiates to obtain the continuous Euler-Lagrange equations, which are finally discretized and solved numerically; see, e.g., [10, 2, 12]. In the second so-called *discretize-optimize* approach one directly discretizes the problem and then solves a finite but typically high-dimensional optimization problem; see, e.g., [7]. The advantage of the latter approach is that standard optimization methods can be used. We therefore prefer the discretize-optimize approach. However, in order to take advantage of efficient optimization techniques, all parts of the discrete problem need to be continuously differentiable.

Choosing a stable discretization method for a system of partial differential equations (PDE's) with mixed derivatives is a non-trivial matter. As proposed in [9], we use staggered grids (cf. Figure 2) which are very common for stable discretizations of the related problem of incompressible fluid flow (see, e.g., [5]) and electromagnetics (see, e.g., [17, 7]) where operators such as the gradient, curl, and divergence are discretized.

In this section we briefly summarize the discretization we use. Further discussion and details are given in [9].

3.1 Discretizing $\boldsymbol{u}$ and $\mathcal{S}$

We assume that our discrete images have $m_1 \times \ldots \times m_d$ pixels, where $d = 2, 3$ is the image dimensionality. For ease of presentation, we also assume that each pixel

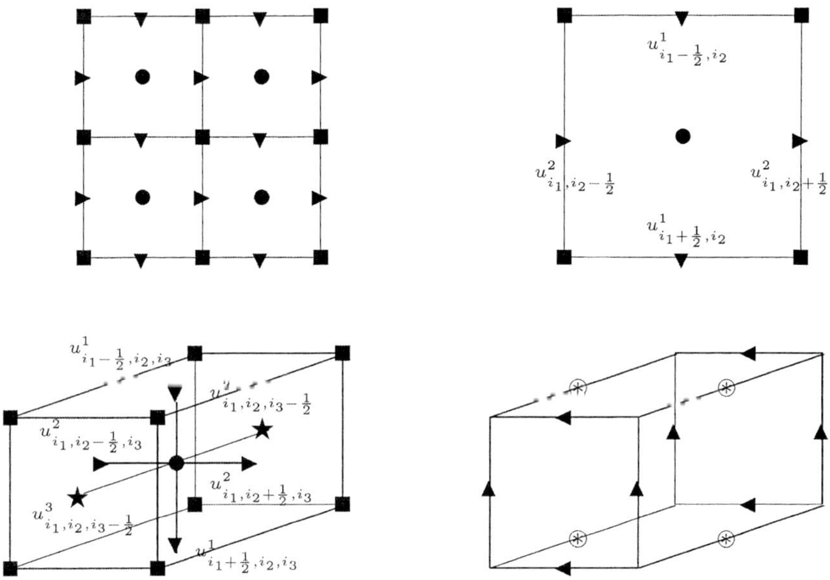

Fig. 2. Staggered grids for dimensions $d = 2, 3$, nodal ■, cell centered •, face staggered grids (▼ in x_1-, ► in x_2-, and ★ in x_3-direction), and edge staggered grids (▲ in x_1-, ◄ in x_2-, and ⊛ in x_3-direction). TOP LEFT: $d = 2$, four pixels, TOP RIGHT: $d = 2$ pixel (i_1, i_2) with grids, BOTTOM LEFT: $d = 3$, voxel (i_1, i_2, i_3) with face staggered grids and positions of u^1, u^2, u^3, BOTTOM RIGHT: $d = 3$, edge staggered grids

is square or cubic where each side has length h. In our description we allow for half step indices. As usual in image processing, we identify pixels/voxels with cell centered grid points $x_{j,k,\ell}$, which are therefore labeled with full integers indices. Given a pixel/voxel $x_{j,k,\ell}$, their faces are numbered with a half index, $x_{j\pm\frac{1}{2},k,\ell}$, $x_{j,k\pm\frac{1}{2},\ell}$, and $x_{j,k,\ell\pm\frac{1}{2}}$, and we discretize the ith component u^i of $\boldsymbol{u}$ on the ith face for every pixel/voxel. With some abuse of notation, we denote the discrete analog of the continuous vector field by $\boldsymbol{u} = (u^1, \ldots, u^d)^\top$, where u^i denotes the grid function which is approximated on the face-staggered grid.

If needed, the derivatives $\partial_j u^k$ are approximated by the short (central) differences,

$$\partial_j u^k \approx \partial_j^h u^k := \tfrac{1}{h}(u^k_{\ldots,i_j+\frac{1}{2},\ldots} - u^k_{\ldots,i_j-\frac{1}{2},\ldots}). \tag{8}$$

Note that no boundary conditions are needed to approximate derivatives in the normal directions ($\partial_j u^j$). For the tangential directions ($\partial_j u^k$, $j \neq k$) we imposed Neumann boundary conditions.

Since many regularizers are phrased in terms of the more complex differential operators ∇ and divergence $\nabla\cdot$, we introduce the notation ∇^h and $\nabla^h\cdot$ for the discrete analogs. Using these discrete analogs the elastic potential $\mathcal{S}$ is discretized as

$$\|\boldsymbol{B}\boldsymbol{u}\|^2 := \frac{\lambda+\mu}{2}\| \nabla^h \cdot \boldsymbol{u}\|^2 + \frac{\mu}{2}\sum_{i=1}^{d} \| \nabla^h \boldsymbol{u}_i\|^2$$

In the course of the registration process we require derivatives. Upon differentiation of the regularizer we obtain the Navier-Lamé operator

$$\frac{1}{2}\frac{\partial\|\boldsymbol{B}\boldsymbol{u}\|^2}{\partial \boldsymbol{u}} = (\lambda+\mu)(\nabla^h\cdot)^\top \nabla^h \cdot \boldsymbol{u} + \mu\Delta^h \boldsymbol{u} =: \boldsymbol{A}\boldsymbol{u}.$$

where Δ^h is the vector Laplacian on a staggered grid

3.2 Discretizing T and $\mathcal{D}$

We are heading for fast and efficient optimization scheme and therefore differentiability does play a key role. Thus, although computationally superior, d-linear image approximations can not be used, since they are not continuously differentiable.

If we require a continuously differentiable objective function we require to have a continuous image model. Since the images are typically noisy but derivatives are needed we use a smoothing B-spline to approximate the image where the smoothing parameter is chosen using the Generalized Cross Validation method [6]. For data interpolation using B-splines see [16]. Since the grid is regular, we can quickly evaluate the spline coefficients using a cosine transform. The continuous smooth approximation is denoted by T^{spline}; see [9] for details.

Given the staggered grid representation of $\boldsymbol{u}$ we use averaging operators P_j for the transfer to the cell centered positions, we set

$$T(\boldsymbol{u}) := T^{\text{spline}}(x^1 + P_1 u^1, \ldots, x^d + P_d u^d),$$

see [9] for details. We denote the Jacobian of T by

$$T_{\boldsymbol{u}} := \frac{\partial T}{\partial \boldsymbol{u}}(\boldsymbol{u}) = \left(\text{diag}(P_1^\top \partial_1 T), \ldots \text{diag}(P_d^\top \partial_d T)\right),$$

where the partial derivatives $\partial_j T$ are evaluated at the spatial positions $(x^1 + P_1 u^1, \ldots, x^d + P_d u^d)$. Using a spline approximation for T, $T_{\boldsymbol{u}}$ becomes a sparse matrix with only four non-zero diagonals.

Our discretization of the SSD (2) is straightforward,

$$\boldsymbol{D}(\boldsymbol{u}) := \tfrac{1}{2}\|T(\boldsymbol{u}) - R\|_2^2 \quad \text{and thus} \quad \boldsymbol{D}_{\boldsymbol{u}}(\boldsymbol{u}) = T_{\boldsymbol{u}}(\boldsymbol{u})^\top (T(\boldsymbol{u}) - R).$$

3.3 Discretizing $\mathcal{C}$

In our discretization of the volume preserving constraints we note that derivatives of every field to every direction are needed and that they need to be centered at the same location. We shortly review the work in [9]. To simplify the discussion we discuss the 2D case only, the extension to 3D is lengthy but straightforward.

In 2D the constraint reads

$$\mathcal{C}(\boldsymbol{u}) = \nabla \cdot \boldsymbol{u} + u_x^1 u_y^2 - u_y^1 u_x^2 = u_x^1 + u_y^2 + u_x^1 u_y^2 - u_y^1 u_x^2.$$

Using the short difference (8) the derivatives u_x^1 and u_y^2 naturally centered at the cell center. To discretize derivatives in the tangential direction we simply use long differences.

4 The Discrete Scale Space Process

After discretization we obtain a discrete scale space process. At each iteration we require to approximately solve the discrete system

$$\text{minimize} \quad \tfrac{1}{2}\|T_k(\boldsymbol{v}) - R\|^2 + \tfrac{\alpha}{2}\|\boldsymbol{B}\boldsymbol{v}\|^2 \tag{9a}$$

$$\text{subject to} \quad \boldsymbol{C}(\boldsymbol{v}) = 0. \tag{9b}$$

Common to other iterative regularization methods (cf., e.g., [3]), we use a single Gauss-Newton step to approximate the solution of the discrete system (9) and immediately update T_k. After linearization we obtained the following quadratic constrained optimization problem

$$\begin{aligned} \text{minimize} \quad & \tfrac{1}{2}\|T_k + (\nabla^h T_k)\boldsymbol{v} - R\|^2 + \tfrac{\alpha}{2}\|\boldsymbol{B}\boldsymbol{v}\|^2 \\ \text{subject to} \quad & \nabla^h \cdot \boldsymbol{v} = 0. \end{aligned}$$

The main advantage of the scale space approach is that the constraints in (9) are linearized with respect to $\boldsymbol{v}$. Therefore, the nonlinear parts vanish and we end up with linear constraints. The Gauss-Newton step is given by the solution of

$$\begin{pmatrix} \boldsymbol{M} + \alpha \boldsymbol{A} & (\nabla^h \cdot)^\top \\ \nabla^h \cdot & \boldsymbol{0} \end{pmatrix} \begin{pmatrix} \boldsymbol{v} \\ \boldsymbol{p} \end{pmatrix} = \begin{pmatrix} (\nabla^h T_k)^\top (T_k - R) \\ \boldsymbol{0} \end{pmatrix}, \tag{10}$$

where $\boldsymbol{p}$ is a discrete Lagrange multiplier and $\boldsymbol{M}$ is a diagonal positive semi-definite matrix which approximates $(\nabla^h T_k)^\top \nabla^h T_k$. The system (10) is similar to the Stokes system in Fluid Dynamics. It is well known that staggered discretization is crucial in order to obtain a stable system; see, e.g., [5]. For Fluid Dynamic problems, effective numerical methods where developed; see, e.g., [14, 15]). In particular, multigrid solvers are among the most efficient schemes to solve problem (10).

The system (10) is solved numerically and $\boldsymbol{v}$ is updated. However, since $\boldsymbol{v}$ is a solution for the linearized problem, this update may not be volume preserving. In order to guarantee the volume preservation of our numerical solution, we explicitly project this solution onto the constraint. As explained in [9], the projection is computed by solving

$$\boldsymbol{C}(\boldsymbol{v} + \boldsymbol{w}) = 0$$

for the correction $\boldsymbol{w}$. To be precise, we compute a least squares solution of the linearized problem

$$\boldsymbol{C}(\boldsymbol{v}+\boldsymbol{w}) \approx \boldsymbol{C}(\boldsymbol{v}) + \boldsymbol{C}_v \boldsymbol{w} = 0, \tag{11}$$

where $\boldsymbol{C}_v$ is the derivative of $\boldsymbol{C}$. This system is underdetermined because we have less equations than unknowns. The above process can be thought of as an orthogonal projection to the volume preserving constraint.

The above algorithm is summarized in Algorithm 1.

Algorithm 1
A scale space approach to volume preserving image registration.
$\boldsymbol{u} \leftarrow \mathrm{SPIR}(\alpha, \boldsymbol{u})$;

set $k=0$, $T_k(\boldsymbol{x}) = T(\boldsymbol{x})$, $\boldsymbol{\varphi}_k(\boldsymbol{x}) = \boldsymbol{x}$;
while not stop **do**
  compute image distance, displacement semi-norm, and $\boldsymbol{M}$;
  set up system (10) and solve for $\boldsymbol{v}$;
  project to the constraint by solving (11) for $\boldsymbol{w}$;
  set $\boldsymbol{\psi}_k(\boldsymbol{x}) = \boldsymbol{x} + \boldsymbol{v} + \boldsymbol{w}$ and $\boldsymbol{\varphi}_{k+1}(\boldsymbol{x}) = \boldsymbol{\varphi}_k(\boldsymbol{\psi}_k(\boldsymbol{x}))$;
  compute $T_{k+1}(\boldsymbol{x}) = T(\boldsymbol{\varphi}_{k+1}(\boldsymbol{x}))$;
  update $k \leftarrow k+1$;
end while

5 A Numerical Example and Discussion

To demonstrate the effectiveness of our new scale space approach we apply the algorithm to the breast images in Figure 1. The images where analyzed in [9] using the straighforward Tikhonov regularization discussed in Section 2.1 and we use them for comparison here. The magnetic resonance images are noisy and only slightly shifted and therefore it is expected that a small number of scale space iterations is needed to achieve a sufficient level of image similarity. Indeed, setting the regularization parameter of each iteration to $\alpha = 10^{-4}$ it takes nine iterations to obtain a relative image distance of 0.87. The history of the iteration is presented in Table 1.

The Tikhonov regularization approach presented in [9] takes ten iterations with a fixed regularization parameter $\alpha = 10^{-5}$ to achieve about the same image distance and about the same displacement semi-norm. For this particular example, the images and the displacements obtained using the new scale space approach and those obtained from the Tikhonov regularization method are visually identical; see Figure 3.

There are two main advantages to the scale approach over the Traditional Tikhonov regularization. First, every iteration of the scale space method is significatively simpler compared with the Tikhonov regularization approach. The main advantage is that the iteration matrix is fixed and therefore efficient methods can be designed for the solution of the linear system. In particular, the

Table 1. The relative image distance $\boldsymbol{D}(\boldsymbol{u}_k)/\boldsymbol{D}(\mathbf{0})$ and the displacement seminorm versus iteration k for the scale space iteration

iteration	1	2	3	4	5	6	7	8	9
distance	0.98	0.96	0.91	0.90	0.90	0.89	0.88	0.88	0.87
$10^3 \cdot \lVert \boldsymbol{Bu} \rVert^2$	1.2	1.4	1.8	1.7	1.8	1.8	1.9	2.0	2.0

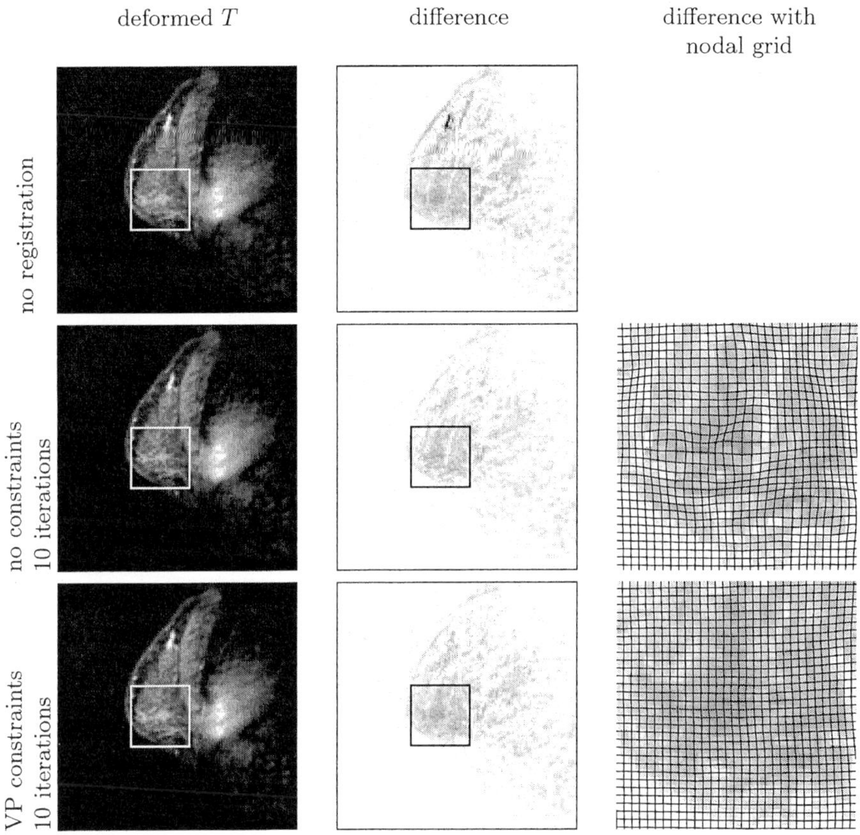

Fig. 3. Registration results for the images of Fig. 1. LEFT COLUMN deformed template images T_k, MIDDLE COLUMN difference image $|R - T_k|$ with a region of interest (ROI), RIGHT COLUMN ROI with nodal grid, vertices connected by straight lines ; ROW 1: no registration, ROW 2: no constraints ten iterations, and ROW 3: volume preserving constraints ten iterations

system is similar to the Stokes system and we intend to explore the use of multigrid methods for its solution in a consecutive paper. The second advantage is that we need not search for the regularization parameter. A rough choice is sufficient and the SSD is reduced by the scale space iteration.

References

1. Gary Edward Christensen, *Deformable shape models for anatomy*, Ph.D. thesis, Sever Institute of Technology, Washington University, 1994.
2. U. Clarenz, M. Droske, and M. Rumpf, *Towards fast non–rigid registration*, Inverse Problems, Image Analysis and Medical Imaging, AMS Special Session Interaction of Inverse Problems and Image Analysis, vol. 313, AMS, 2002, pp. 67–84.
3. H.W. Engl, M. Hanke, and A. Neubauer, *Regularization of inverse problems*, Kluwer, 1996.
4. J. M. Fitzpatrick, D. L. G. Hill, and C. R. Maurer Jr., *Image registration, handbook of medical imaging*, Volume 2: Medical Image Processing and Analysis, SPIE (2000), 447–513.
5. C.A.J. Fletcher, *Computational techniques for fluid dynamics*, vol. II, Springer-Verlag, 1988.
6. G. Golub, M. Heath, and G. Wahba, *Generalized cross-validation as a method for choosing a good ridge parameter*, Technometrics **21** (1979), 215–223.
7. E. Haber and U. Ascher, *Fast finite volume simulation of 3D electromagnetic problems with highly discontinuous coefficients*, SIAM J. Scient. Comput. **22** (2001), 1943–1961.
8. E. Haber and J. Modersitzki, *Multilevel methods for image registration*, SIAM J. on Scientific Computing **To Appear** (2004), xxx.
9. ______, *Numerical solutions of volume preserving image registration*, Inverse Problems **20** (2004), 1621–1638.
10. Stefan Henn and Kristian Witsch, *Iterative multigrid regularization techniques for image matching*, SIAM J. on Scientific Comp. (2001), 1077–1093.
11. C.R. Maurer and J.M. Fitzpatrick, *A review of medical image registration*, In: Interactive Image-Guided Neurosurgery, American Association of Neurological Surgeons (Park Ridge, IL), Aug 1993, pp. 17–44.
12. J. Modersitzki, *Numerical methods for image registration*, Oxford, 2004.
13. Torsten Rohlfing, Calvin R. Maurer, Jr. David A. Bluemke, and Michael A. Jacobs, *Volume-preserving nonrigid registration of MR breast images using free-form deformation with an incompressibility constraint*, IEEE Transactions on Medical Imaging **22** (2003), 730–741.
14. D. Silvester, H. Elman, D. Kay, and A. Wathen, *Efficient preconditioning of the linearized Navier-Stokes equations*, J. Comp. Appl. Math. **128** (2001), 261–279.
15. U. Trottenberg, C. Oosterlee, and A. Schuller, *Multigrid*, Academic Press, 2001.
16. G. Wahba, *Spline models for observational data*, SIAM, Philadelphia, 1990.
17. K.S. Yee, *Numerical solution of initial boundary value problems involving Maxwell's equations in isotropic media*, IEEE Trans. on antennas and propagation **14** (1966), 302–307.

Piecewise Constant Level Set Methods and Image Segmentation

Johan Lie[1], Marius Lysaker[2], and Xue-Cheng Tai[1]

[1] Department of Mathematics, University of Bergen,
Joh. Brunsgate 12, N-5008 Bergen, Norway
{johanl, tai}@mi.uib.no
http://www.mi.uib.no/BBG
http://www.mi.uib.no/~tai

[2] Simula Research Laboratory, M. Linges v 17,
Fornebu P.O.Box 134, N-1325 Lysaker, Norway
mariul@simula.no

Abstract. In this work we discuss variants of a PDE based level set method. Traditionally interfaces are represented by the zero level set of continuous level set functions. We instead use piecewise constant level set functions, and let interfaces be represented by discontinuities. Some of the properties of the standard level set function are preserved in the proposed method. Using the methods for interface problems, we minimize a smooth locally convex functional under a constraint. We show numerical results using the methods for image segmentation.

1 Introduction

The level set method was proposed by Osher and Sethian in [19] as a versatile tool for tracing interfaces separating a domain Ω into subdomains. Interfaces are treated as the zero level set of higher dimensional functions. Moving the interfaces can implicitly be done by evolving level set functions instead of explicitly moving the interfaces. Applications of the level set method include image analysis, reservoir simulation, inverse problems, computer vision and optimal shape design [5, 4, 10, 26, 18, 22, 24]. In this work, we discuss some variants of the level set method. The primary concern for our approach is to remove the connection between the level set functions and the signed distance function and thus remove some of the computational difficulties associated with the calculation of the Eikonal equation. Another motivation is to avoid numerical problems associated with the Heaviside and Delta functions used in some level set formulations [5, 25]. The third concern of this approach is to develop fast algorithms for level set methods. Due to the fact that the functional and the constraints for this approach are rather smooth, it is possible to apply Newton types of iterations to construct fast algorithms for the proposed model. For experimental purposes, we have used gradient type of methods here, and we restrict ourselves to image segmentation.

R. Kimmel, N. Sochen, J. Weickert (Eds.): Scale-Space 2005, LNCS 3459, pp. 573–584, 2005.

For a given digital image $u_0 : \Omega \to \mathbb{R}$, the aim is to separate Ω into a set of subdomains Ω_i such that $\Omega = \cup_{i=1}^n \Omega_i$ and u_0 is nearly a constant in each Ω_i. Having determined the partition of Ω into a set of subdomains Ω_i, one can do further modelling on each domain independently and automatically. One general image segmentation model was proposed by Mumford and Shah in [16]. For numerical approximations, see [2]. Using this model, the image u_0 is decomposed into $\Omega = \cup_i \Omega_i \cup \Gamma$, where Γ is a curve separating the different domains. Inside each Ω_i, u_0 is approximated by a smooth function. The optimal partition of Ω is found by minimizing the Mumford-Shah functional. Chan and Vese [5, 25] minimized the piecewise constant Mumford-Shah functional using level set functions. Motivated by the Chan-Vese approach, we will in this article solve the segmentation problem in a different way, i.e. by introducing a piecewise constant level set function ϕ. Instead of using the zero level of a function to represent the interface between subdomains, the interfaces are implicitly represented by the discontinuities of a set of characteristic functions ψ_i. Note that both the Chan-Vese model and our model can be extended to shape recognition using the framework of [7, 6].

The rest of this article is structured as follows. Our framework and the minimization functional used for image segmentation is formulated in §2. The segmentation problem is formulated as a minimization problem with a smooth cost functional under a constraint. We are minimizing the piecewise constant Mumford-Shah functional associated with special level set models. In §3 and §4 we explain our two variants of the level set method for image segmentation in more detail. Both sections include algorithms and numerical results. We conclude with a brief discussion. For a more detailed treatment of the methods, including more numerical results we refer the reader to [14, 15]. We also refer the reader to [23, 21, 9, 11, 7] for some related methods.

2 A Framework for Subdomain Representation

In this section a framework for representing subdomains of Ω is developed. Each subdomain Ω_i is associated with a basis function ψ_i, such that $\psi_i = 1$ in Ω_i and zero elsewhere. The basis functions are constructed using one or several level set functions $\{\phi_j\}_{j=1}^l$. Two different realizations of the basis functions ψ_i are shown in §3 and §4. Each ψ_i is compactly supported in Ω_i. Thus we can construct a piecewise constant function u by a weighted sum of the characteristic functions. If we let $\mathbf{c} = \{c_i\}_{i=1}^n$ be a set of real scalars, we can represent a piecewise constant function u taking these n distinct constant values by

$$u = \sum_{i=1}^{n} c_i \psi_i. \tag{1}$$

Let η be Gaussian noise, and $u_0 = u + \eta$. If u_0 is almost equal to a constant in n subdomains it can be approximated by (1) provided we optimally choose $\mathbf{c}$ and $\{\psi_i\}_{i=1}^n$. This is done by solving a minimization problem subject to a constraint corresponding to the choice of basis functions. The constraint controls the structure of possible solutions. We return to the specific constraints in §3 and §4.

The simple structure of the characteristic functions gives us the opportunity to measure the lengths of curves surrounding Ω_i and the area of each region Ω_i by

$$|\partial\Omega_i| = \int_\Omega |\nabla\psi_i| dx, \qquad \text{and} \qquad |\Omega_i| = \int_\Omega \psi_i dx. \tag{2}$$

Here we note that $|\partial\Omega_i|$ is the Total Variation (TV)-norm of ψ_i [20].

The above framework can be used as a tool for image segmentation. Let u_0 be an image to be segmented. We want to construct a piecewise constant function u which approximates u_0 in a proper sense. The segmentation can be formulated as a minimization of the following functional

$$F(\phi, \mathbf{c}) = \frac{1}{2}\int_\Omega |u - u_0|^2 dx + \beta \sum_{i=1}^{n} \int_\Omega |\nabla\psi_i| dx, \tag{3}$$

where u is on the form (1). The first term of F is a least squares fidelity term, measuring the closeness of u to u_0. The second term is a regularizer term measuring the length of all the curves separating the subdomains. We here note the similarity between functional (3) and the functional used by Chan and Vese in [25], where u is represented using Heaviside functions. For uniquely classifying each point in the image, we need to introduce a constraint $K(\phi) = 0$ (c.f. §3 §4) and solve the constrained optimization problem

$$\min_{\mathbf{c},\phi} \; F(\mathbf{c}, \phi) \quad \text{subject to} \quad K(\phi) = 0. \tag{4}$$

This problem is solved using the augmented Lagrangian method [1, 17]. A minimizer of F corresponds to a saddle-point of the augmented Lagrangian functional

$$L(\mathbf{c}, \phi, \lambda) = F(\mathbf{c}, \phi) + \int_\Omega \lambda K(\phi)\, dx + \frac{r}{2}\int_\Omega |K(\phi)|^2 dx, \tag{5}$$

where λ is a function defined on the same domain as ϕ called the Lagrangian multiplier, $r \in \mathbf{R}^+$ is a penalty parameter being characteristic for the augmented Lagrangian method. r can either be chosen to be a fixed small number during the iterative process or it can be increased during iterations to accellerate convergence [17]. We have used both the augmented Lagrangian method and the standard Lagrangian method. Numerical experiments indicate that the augmented Lagrangian method is best suited for this minimization problem. At a saddle-point of (5) we must have

$$\frac{\partial L}{\partial \phi} = 0, \quad \frac{\partial L}{\partial c_i} = 0 \text{ and } \frac{\partial L}{\partial \lambda} = 0. \tag{6}$$

Essentially we minimize L w.r.t $\mathbf{c}$ and ϕ, and maximize L w.r.t λ. In §3 and §4 we introduce iterative algorithms to find the saddle-points in (6) coming from two different level set formulations. The rest of the current section is devoted to calculations being common for both approaches.

Since u is linear with respect to $\mathbf{c}$, we see that L is quadratic with respect to $\mathbf{c}$. Thus the minimization problem w.r.t $\mathbf{c}$ can be solved exactly. Note that

$$\frac{\partial L}{\partial c_i} = \int_\Omega \frac{\partial L}{\partial u}\frac{\partial u}{\partial c_i} = \int_\Omega (u - u_0)\psi_i \, dx, \qquad \text{for } i = 1, 2, \ldots n. \tag{7}$$

Therefore, the minimizer satisfies a linear system of equations $A\mathbf{c}^k = b$ in the following form

$$\sum_{j=1}^{n} \int_\Omega (\psi_j \psi_i) c_i^k \, dx = \int_\Omega u_0 \psi_i \, dx, \qquad \text{for } i = 1, 2, \ldots n. \tag{8}$$

In the above $\psi_j = \psi_j(\phi^k)$, $\psi_i = \psi_i(\phi^k)$ and thus, $\mathbf{c}^k = \{c_i^k\}_{i=1}^n$ depends on ϕ^k. We form the $(n \times n)$ matrix A and vector b and solve the equation $Ac^{\ k} = b$ using an exact solver. The minimization with respect to ϕ will be solved by the following gradient method

$$\phi^{new} = \phi^{old} - \Delta t \frac{\partial L}{\partial \phi}(\mathbf{c}, \phi^{old}, \lambda), \tag{9}$$

where Δt is a small positive number determined by trial and error. For a given $\mathbf{c}$ and λ, we need to iterate many times in order to find the minimizer with respect to ϕ. This can be interpreted as a scalespace method with t as the scale.

3 Piecewise Constant Level Set Method Using a Polynomial Approach

We first present the polynomial piecewise constant level set method (PCLSM). Assume that we need to find n regions $\{\Omega_i\}_{i=1}^n$ which form a portion of Ω. In order to identify the regions, we want to find a piecewise constant function taking the values

$$\phi = i \text{ in } \Omega_i, \quad i = 1, 2, \ldots, n. \tag{10}$$

With this approach we just need one function to identify all the phases in Ω. The basis functions ψ_i associated with ϕ are defined in the following form

$$\psi_i = \frac{1}{\alpha_i} \prod_{\substack{j=1 \\ j \neq i}}^{n} (\phi - j) \text{ and } \alpha_i = \prod_{\substack{k=1 \\ k \neq i}}^{n} (i - k). \tag{11}$$

It is clear that the function u given by (1) is a piecewise constant function and $u = c_i$ in Ω_i if ϕ is as given in (10). The function u is a polynomial of order $n{-}1$ in ϕ. Each ψ_i is expressed as a product of linear factors of the form $(\phi - j)$, with the ith factor omitted. Thereupon $\psi_i(\mathbf{x}){=}1$ for $\mathbf{x} \in \Omega_i$, and $\psi_i(\mathbf{x} = 0$

elsewhere as long as (10) holds. To ensure that equation (1) gives us a unique representation of u, i.e. at convergence different values of ϕ should correspond to different function values $u(\phi)$ in (1), we introduce

$$K(\phi) = (\phi - 1)(\phi - 2) \cdots (\phi - n) = \prod_{i=1}^{n} (\phi - i). \tag{12}$$

If a given function $\phi : \Omega \mapsto R$ satisfies

$$K(\phi) = 0, \tag{13}$$

there exists a unique $i \in \{1, 2, \ldots, n\}$ for every $x \in \Omega$ such that $\phi(x) = i$. Thus, each point $x \in \Omega$ can belong to one and only one phase if $K(\phi) = 0$. The constraint (13) is used to guarantee that there is no vacuum and overlap between the different phases. In [27] some other constraints for the classical level set methods were used to avoid vacuum and overlap.

Following the framework in §2, we will use the basis functions (11), the constraint (12) and the representation (1) of u. To find a minimizer of (5), we need to find the saddle point where $\frac{\partial L}{\partial \phi} = 0$ and $\frac{\partial L}{\partial \lambda} = 0$. Remember that $\frac{\partial L}{\partial c_i}$ is zero if $\{c_i\}_{i=1}^n$ are computed from (8). We use the Uzawa-type Algorithm 1 to find a saddle point of $L(\mathbf{c}, \phi, \lambda)$. The algorithm has a linear convergence rate and its convergence has been analyzed in [13] and used in [4, 3].

Algorithm 1. Choose initial values for ϕ^0 and λ^0. For $k = 1, 2, \ldots$, do:

– Find $\mathbf{c}^k$ from

$$L(\mathbf{c}^k, \phi^{k-1}, \lambda^{k-1}) = \min_{\mathbf{c}} L(\mathbf{c}, \phi^{k-1}, \lambda^{k-1}). \tag{14}$$

– Use (1) to update $u = \sum_{i=1}^n c_i^k \psi_i(\phi^{k-1})$.
– Find ϕ^k from

$$L(\mathbf{c}^k, \phi^k, \lambda^{k-1}) = \min_{\phi} L(\mathbf{c}^k, \phi, \lambda^{k-1}). \tag{15}$$

– Use (1) to update $u = \sum_{i=1}^n c_i^k \psi_i(\phi^k)$.
– Update the Lagrange-multiplier by

$$\lambda^k = \lambda^{k-1} + rK(\phi^k). \tag{16}$$

– If not converged: Set k=k+1 and go to step 1.

To compute $\frac{dL}{d\phi}$ we utilize the chain rule to get

$$\frac{\partial L}{\partial \phi} = (u - u_0)\frac{\partial u}{\partial \phi} - \beta \sum_{i=1}^{n} \nabla \cdot \left(\frac{\nabla \psi_i}{|\nabla \psi_i|} \right) \frac{\partial \psi_i}{\partial \phi} + \lambda \frac{\partial K}{\partial \phi} + rK \frac{\partial K}{\partial \phi}. \tag{17}$$

It is easy to get $\partial u/\partial \phi$, $\partial \psi_i/\partial \phi$ and $\partial K/\partial \phi$ from (1), (11) and (12). We use the gradient method (9) to solve (15). We do a fixed number of iterations, for example 400 iterations or stop the iteration after the L_2 norm of gradient has been reduced by 10%.

Remark 1. The updating for the constant values in (14) is ill-posed. A small perturbation of the ϕ function produces a large perturbation for the c_i values. Due to this reason, we have tried out a variant of Algorithm 1. In each iteration we alternate between (15) and (16), while (14) is only carried out if $\|K(\phi^{new})\|_{L^2} < \frac{1}{10}\|K(\phi^{old})\|_{L^2}$. Here, ϕ^{old} denotes the value of ϕ when (14) was carried out the last time and ϕ^{new} denotes the current value of ϕ. If we use such a strategy, we can do just one or a few iterations for the gradient scheme (9) and Algorithm 1 is still convergent. This strategy is particular efficient when the amount of noise is high.

Remark 2. In Algorithm 1, we give initial values for ϕ and λ. We first minimize with the constant values, and then minimize with the level set function. The multiplier is updated in the end of each iteration. In situations where good initial values for $\mathbf{c}$ are available, an alternative variant of Algorithm 1 may be used, i.e. we first minimize with the level set function followed by a minimization for the constant values and then update the multiplier.

Remark 3. There is no particular reason why we chose to use integers as roots of the polynomial. Perhaps a better choice would be to use a Chebyshev polynomial with its roots as buildingblocks for the characteristic functions ψ_i. This could for example be accomplished by interchanging the integer values in (10) by the roots of a Chebyshev polynomial of degree n defined on the interval $[-1, 1]$ [12].

$$z_i = \cos\left(\frac{\pi(i-\frac{1}{2})}{n}\right) \text{ in } \Omega_i, \quad i = 1, 2, \ldots, n. \tag{18}$$

The exact same framework could then be used to construct a set of characteristic functions $\{\psi_i\}_{i=1}^n$ by

$$\psi_i = \frac{1}{\alpha_i}\prod_{\substack{j=1\\ j\neq i}}^{n}(\phi - z_j) \text{ and } \alpha_i = \prod_{\substack{k=1\\ k\neq i}}^{n}(i - z_k). \tag{19}$$

The corresponding constraint would then be

$$K(\phi) = (\phi - z_1)(\phi - z_2)\cdots(\phi - z_n) = \prod_{i=1}^{n}(\phi - z_i). \tag{20}$$

In this work we have not implemented this approach.

3.1 Numerical Experiments Using the Polynomial Approach

In this section we validate the piecewise constant level set method with numerical examples. We only consider 2-D images and restrict ourself to gray-scale images, although the model permits any dimension and can be extended to vector-valued images as well. Our results will be compared with the related works [5, 25]. To complicate the segmentation process we typically expose the original image with

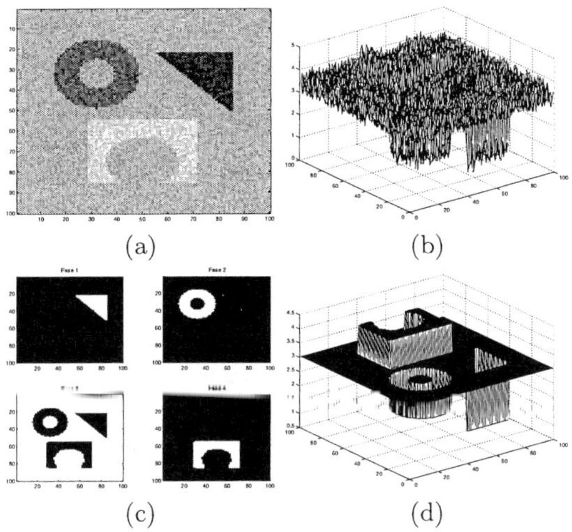

Fig. 1. (a) Observed image u_0 (SNR$\approx$ 5.2). (b) Initial level set function ϕ. (c) Each separate phase $\phi = 1 \vee 2 \vee 3 \vee 4$ are depicted as a bright region. (d) At convergence ϕ approaches 4 constant values

Gaussian distributed noise and use the polluted image as the observation data u_0. To demonstrate a 4-phase segmentation we begin with a noisy synthetic image containing 3 objects (and the background) as shown in Figure 1(a). This is the same image as Chan and Vese used to examine their multiphase algorithm [5, 25]. The observation data u_0 is given in Figure 1(a) and the only assumption we make is that a 4-phase model should be utilized to find the segmentation. In Figure 1(d) the ϕ function is depicted at convergence. The function ϕ approaches the predetermined constants $\phi = 1 \vee 2 \vee 3 \vee 4$. Each of these constants represents one unique phase as seen in Figure 1(c). Our result is in accordance with what Chan and Vese reported in [5, 25].

In many applications the number of objects to detect are not known a priori. A robust and reliable algorithm should find the correct segmentation even when the exact number of phases is not known. By introducing a model with

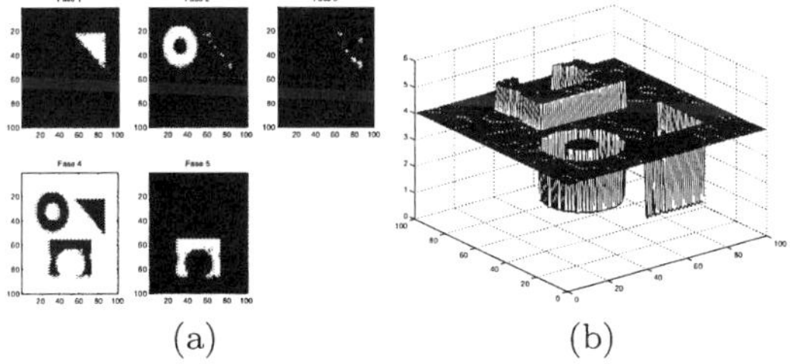

Fig. 2. (a) Each phase $\phi = 1 \vee 2 \vee 3 \vee 4 \vee 5$ is depicted as a bright region. (b) At convergence ϕ approaches 4 constant values

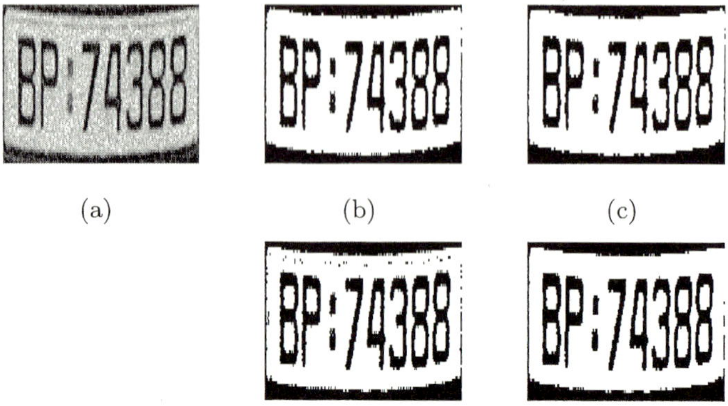

(a) (b) (c) (d) (e)

Fig. 3. Character and number segmentation from a car plate using our method (b),(c) and CV (d),(e)

more phases than one actually needs, we can find the correct segmentation if all superfluous phases are empty when the algorithm has converged. To see if our algorithm can handle such a case we again use Figure 1(a) as the observation image and utilize a 5-phase model. Our results are reported in Figure 2. One of the 5 phases must be empty if a 5-phase model is used to find a 4-phase segmentation. Due to the high noise level some pixels can easily be misclassified and contribute to the phase that should be empty. (The regularization parameter β is not tuned in order to get rid of this misclassification.) The level set function shown in Figure 2(b) approaches the constants $\phi = 1 \vee 2 \vee 4 \vee 5$, except from the few misclassified pixels where $\phi = 3$ as seen in Figure 2(a). By comparing Figure 1(c) (where a 4-phase model is used) and Figure 2(a) (where a 5-phase model is used), we observe only small changes in the segmented phases, except from the extra nonempty phase $\phi = 3$ in Figure 2(a).

Below we proceed with one example using a real image. We want to demonstrate that PPCLSM (polynomial PCLSM) can be uses to extract characters or numbers from images. We use an image of a license plate. To evaluate the segmentation process, the Chan/Vese method [5, 25], for short (CVM), is examined using the same input image. Both algorithms are processed using two different regularization parameters. With the amount of noise in Fig. 3 (a), both PPCLSM and CVM miss some details along the edges of the characters and numbers. To compare the result of (CV) and (PPCLSM), both algorithms were terminated after 500 iterations. The regularization parameters used in the experiment are (b) $\beta = 0.1$, (c) $\beta = 3$, (d) $\beta = 1$ and (e) $\beta = 10^4$. Thus we observe that both our method and CVM are quite robust to noise, and the choice of regularization parameters. The processing time is almost equal for both methods in this example. We have done no efforts for code optimization. The complexity of the implementation is also almost equal for both methods.

4 The Binary Approach for PCLSM

We will now introduce an alternative realization of the characteristic functions in (1). Using the following approach, we can represent a maximum of 2^N subdomains, using N level set functions $\{\phi_i\}_{i=1}^N$. To simplify notation, we form the vector $\boldsymbol{\phi} = \{\phi_1, \phi_2, \dots, \phi_N\}$. Let us first assume that the interface Γ is enclosing $\Omega_1 \subset \Omega$. By standard level set methods the interior of Ω_1 is represented by points $\boldsymbol{x} : \phi(\boldsymbol{x}) > 0$, and the exterior of Ω_1 is represented by points $\boldsymbol{x} : \phi(\boldsymbol{x}) < 0$. We instead let $\phi(\boldsymbol{x}) = 1$ if $\boldsymbol{x}$ is an interior point of Ω_1 and $\phi(\boldsymbol{x}) = -1$ if $\boldsymbol{x}$ is an exterior point of Ω_1. As proposed, Γ is implicitly defined as the discontinuity of ϕ. Representing four subdomains is done analogously as in [25, 7] by

$$u(\boldsymbol{x}) = \begin{cases} c_1, \text{ if } \phi_1(\boldsymbol{x}) = 1, \phi_2(\boldsymbol{x}) = 1, \\ c_2, \text{ if } \phi_1(\boldsymbol{x}) = 1, \phi_2(\boldsymbol{x}) = -1, \\ c_3, \text{ if } \phi_1(\boldsymbol{x}) = -1, \phi_2(\boldsymbol{x}) = 1, \\ c_4, \text{ if } \phi_1(\boldsymbol{x}) = -1, \phi_2(\boldsymbol{x}) = -1. \end{cases}$$

Thus, a piecewise constant function taking four different constant values can be written

$$\begin{aligned} u = &\frac{c_1}{4}(\phi_1+1)(\phi_2+1) - \frac{c_2}{4}(\phi_1+1)(\phi_2-1) \\ &- \frac{c_3}{4}(\phi_1-1)(\phi_2+1) + \frac{c_4}{4}(\phi_1-1)(\phi_2-1). \end{aligned} \tag{21}$$

Using (21), we can form the set of basis functions ψ_i as in the following

$$u = c_1 \underbrace{\frac{1}{4}(\phi_1+1)(\phi_2+1)}_{\psi_1} + c_2 \underbrace{(-1)\frac{1}{4}(\phi_1+1)(\phi_2-1)}_{\psi_2} + \dots, \tag{22}$$

and we can write: $u = \sum_{i=1}^4 c_i \psi_i$. The set of ψ_i for the multiple subdomain case is constructed by generalization of this. For $i = 1, 2, \dots, 2^N$, let $(b_1^{i-1}, b_2^{i-1}, \dots, b_N^{i-1})$ be the binary representation of $i-1$, where $b_j^{i-1} = 0 \vee 1$. Let $s(i) = \sum_{j=1}^N b_j^{i-1}$, and write ψ_i and u as

$$\psi_i = \tfrac{(-1)^{s(i)}}{2^N} \prod_{j=1}^N (\phi_j + 1 - 2b_j^{i-1}) \quad \text{and} \quad u = \sum_{i=1}^{2^N} c_i \psi_i. \tag{23}$$

It is now easy to see that these basis functions have the properties needed for the framework in §2. Using this representation for the basis functions, we need N constraints, one constraint K_i to each of the level set functions ϕ_i. We use the constraints: $K_i(\phi_i) = \phi_i^2 - 1 \; \forall i$. Setting $K_i(\phi_i) = 0$ implies ϕ_i can only take the values ± 1 at convergence.

Having determined the choice of basis functions $\{\psi_i\}_{i=1}^n$ and the representation of u by (23), we find the saddle point of L by the augmented Lagrangian method. This means that we must minimize L w.r.t $\boldsymbol{\phi}$ and $\boldsymbol{c}$, and maximize L w.r.t $\boldsymbol{\lambda}$, which has the same dimension as $\boldsymbol{\phi}$.

We minimize L w.r.t $\boldsymbol{\phi}$ by using the gradient method (9) for all the N level set functions. The gradients for the level set functions are given as

$$\frac{\partial L}{\partial \phi_i} = (u - u_0)\sum_{j=1}^{2^N} c_j \frac{\partial \psi_j}{\partial \phi_i} - \beta \sum_{j=1}^{2^N} \nabla \cdot \left(\frac{\nabla \psi_j}{|\nabla \psi_j|}\right)\frac{\partial \psi_j}{\partial \phi_i} + 2\lambda_i \phi_i + 2r(\phi_i^2 - 1)\phi_i. \tag{24}$$

The constraints K_i are independent of the constant values c_i and thus the same formula (8) can be used to update the c_i values.

Similar to the algorithm used for the polynomial approach for the PCLSM, we use the following algorithm to find a saddle point for the binary approach for the PCLSM.

Algorithm 2. Choose initial values for $\boldsymbol{\phi}^0$ and $\boldsymbol{\lambda}^0$. For $k = 1, 2, \ldots$, do:

– Update $\boldsymbol{\phi}^k$ by (9), to approximately solve

$$L(\mathbf{c}^{k-1}, \boldsymbol{\phi}^k, \boldsymbol{\lambda}^{k-1}) = \min_{\boldsymbol{\phi}} L(\mathbf{c}^{k-1}, \boldsymbol{\phi}, \boldsymbol{\lambda}^{k-1}). \tag{25}$$

– Construct $u(\mathbf{c}^{k-1}, \boldsymbol{\phi}^k)$ by $u = \sum_{i=1}^{2^N} c_i^{k-1} \psi_i^k$.
– Update $\mathbf{c}^k$ by (8), to solve

$$L(\mathbf{c}^k, \boldsymbol{\phi}^k, \boldsymbol{\lambda}^{k-1}) = \min_{\mathbf{c}} L(\mathbf{c}, \boldsymbol{\phi}^k, \boldsymbol{\lambda}^{k-1}). \tag{26}$$

– Update the multiplier by

$$\boldsymbol{\lambda}^k = \boldsymbol{\lambda}^{k-1} + r\boldsymbol{K}(\boldsymbol{\phi}^k). \tag{27}$$

– If not converged: Set k=k+1 and go to step 1.

Remark 4. Remarks 1 and 2 after Algorithm 1 also apply to algorithm 2.

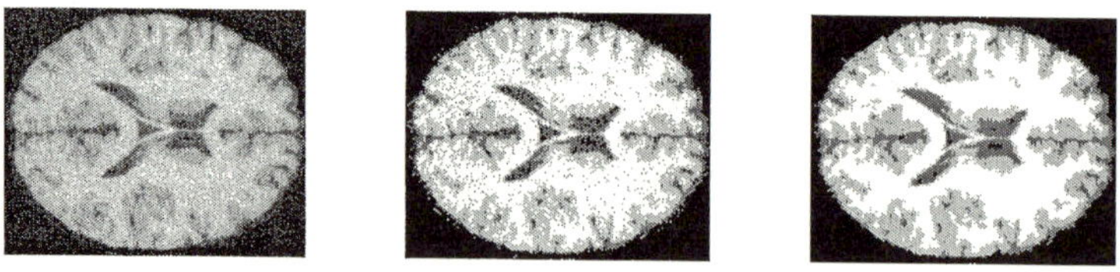

(a) (b) (c)

Fig. 4. In this example, the image (a) is first segmented using the isodata method (b). Then the result is further processed using our method, and the final result is shown in (c)

4.1 Numerical Experiments Using the Binary Approach

Now we will present one of the numerical results achieved using the binary piecewise constant level set formulation (BPCLSM). For more numerical results, see [15]. We will here show a segmentation of a Magnetic Resonance image of a brain using two level set functions. The goal is to partition the image into three different tissue classes in addition to the background. To accelerate the convergence of our method, we first preprocess the image using a simple tresholding, the isodata method [8]. In (b), we show the result of the isodata segmentation of u_0 [8]. We here observe that main structures are preserved, but also highly oscillating patterns occur. We use the results from the isodata segmentation to construct initial values for ϕ_1 and ϕ_2, run our algorithm with these initial values, and end up with the image depicted in Fig. 4 (c). Observe that the main structures are still very well preserved, but most of the (unwanted) highly oscillating patterns are removed. By initializing the algorithm in this way, we both accelerate the convergence of our algorithm, and in addition more or less avoid the problem of local minimizers. In the example, we used the parameters $\beta = 5 \cdot 10^{-3}$, $\mu = 5$, and 1000 iterations.

5 Conclusion

In this article we have discussed a framework for subdomain identification. We have also pointed out two methods for image segmentation using this framework. Recently, there has been done work to extend the method to 3-D data sets. Work is also done to incorporate a Newton-type of iteration for improving the convergence properties of the method. The PPCLSM is favourable in terms of computational complexity and memory requirements, and in terms of handling cases where a priori information of the number of subdomains is lacking. Molding BPCLSM into existing software for level set methods is possibly easier than PPCLSM because of similarities in the machinery of standard level set methods.

References

1. D. P. Bertsekas, *Constrained optimization and Lagrange multiplier methods*, Academic Press Inc., 1982.
2. A. Chambolle, *Image segmentation by variational methods: Mumford and Shah functional and the discrete approximations*, SIAM J. Appl. Math., 55 (1995), pp. 827–863.
3. T. F. Chan and X.-C. Tai, *Identification of discontinuous coefficients in elliptic problems using total variation regularization*, SIAM J. Sci. Comput., 25 (2003), pp. 881–904.
4. T. F. Chan and X.-C. Tai, *Level set and total variation regularization for elliptic inverse problems with discontinuous coefficients*, J. Comput. Phys., 193 (2003), pp. 40–66.
5. T. F. Chan and L. A. Vese, *Active contours without edges*, IEEE Trans. Im. Proc., 10 (2001), pp. 266–277.

6. D. Cremers, T. Kohlberger, and C. Schnörr, *Shape statistics in kernel space for variational image segmentation*, Patt. Recogn., 36 (2003), pp. 1929–1943.
7. D. Cremers, N. Sochen, , and C. Schnörr, *Multiphase dynamic labeling for variational recognition-driven image segmentation*, in ECCV 2004, LNCS 3024, 2004, pp. 74–86.
8. F. R. Dias Velasco, *Thresholding using the ISODATA clustering algorithm*, IEEE Trans. Systems Man Cybernet., 10 (1980), pp. 771–774.
9. S. Esedoglu and Y.-H. R. Tsai, *Threshold dynamics for the piecewise constant mumford-shah functional*, UCLA-CAM Report 04-63, (2004).
10. R. P. Fedkiw, G. Sapiro, and C.-W. Shu, *Shock capturing, level sets, and PDE based methods in computer vision and image processing: a review of Osher's contributions*, J. Comput. Phys., 185 (2003), pp. 309–341.
11. F. Gibou and R. Fedkiw, *A fast hybrid k-means level set algorithm for segmentation*, Stanford Tech. Rep., 2002, (in review)., (2002).
12. M. T. Heath, *Scientific computing: an introductory survey.*, 2001.
13. K. Kunisch and X.-C. Tai, *Sequential and parallel splitting methods for bilinear control problems in Hilbert spaces*, SIAM J. Numer. Anal., 34 (1997), pp. 91–118.
14. J. Lie, M. Lysaker, and X.-C. Tai, *A variant of the level set method and applications to image segmentation*, UCLA, CAM-report, 03-50, (2003).
15. J. Lie, M. Lysaker, and X.-C. Tai, *A binary level set model and some applications for mumford-shah image segmentation*, UCLA, CAM-report, 04-31, (2004).
16. D. Mumford and J. Shah, *Optimal approximation by piecewise smooth functions and associated variational problems*, Comm. Pure Appl. Math, 42 (1989), p. 577685.
17. J. Nocedal and S. J. Wright, *Numerical optimization*, Springer Series in Operations Research, 1999.
18. S. Osher and R. Fedkiw, *Level Set Methods and Dynamic Implicit Surfaces*, vol. 153 of Appl. Math. Sci., Springer, 2003.
19. S. Osher and J. A. Sethian, *Fronts propagating with curvature-dependent speed: algorithms based on Hamilton-Jacobi formulations*, J. Comput. Phys., 79 (1988), pp. 12–49.
20. L. Rudin, S. Osher, and E. Fatemi, *Nonlinear total variation based noise removal algorithm*, Physica D., 60 (1992), pp. 259–268.
21. C. Samson, L. Blanc-Feraud, G. Aubert, and J. Zerubia, *A level set model for image classification*, IJCV, 40 (2000), pp. 187–198.
22. J. A. Sethian, *Level set methods and fast marching methods*,Cambridge University Press, second ed., 1999.
23. B. Song and T. F. Chan, *A fast algorithm for level set based optimization*, Tech. Rep. CAM 02-68, UCLA, 2002.
24. X.-C. Tai and T. F. Chan, *A survey on multiple set methods with applications for identifying piecewise constant functions*, Int. J. Num. Anal. and Mod., 1 (2004), pp. 25–48.
25. L. A. Vese and T. F. Chan, *A multiphase level set framework for image segmentation using the mumford and shah model*, Int. J. of Comp. Vis., 50 (2002), pp. 271–293.
26. J. Weickert and G. Kühne, *Fast methods for implicit active contour models*, in Geometric level set methods in imaging, vision, and graphics, Springer, 2003, pp. 43–57.
27. H.-K. Zhao, T. Chan, B. Merriman, and S. Osher, *A variational level set approach to multiphase motion*, J. Comput. Phys., 127 (1996), pp. 179–195.

PDE-Based Deconvolution with Forward-Backward Diffusivities and Diffusion Tensors

Martin Welk, David Theis, Thomas Brox, and Joachim Weickert

Mathematical Image Analysis Group,
Faculty of Mathematics and Computer Science, Bldg. 27,
Saarland University, 66041 Saarbrücken, Germany
{welk, theis, brox, weickert}@mia.uni-saarland.de
http://www.mia.uni-saarland.de

Abstract. Deblurring with a spatially invariant kernel of arbitrary shape is a frequent problem in image processing. We address this task by studying nonconvex variational functionals that lead to diffusion-reaction equations of Perona–Malik type. Further we consider novel deblurring PDEs with anisotropic diffusion tensors. In order to improve deblurring quality we propose a continuation strategy in which the diffusion weight is reduced during the process. To evaluate our methods, we compare them to two established techniques: Wiener filtering which is regarded as the best linear filter, and a total variation based deconvolution which is the most widespread deblurring PDE. The experiments confirm the favourable performance of our methods, both visually and in terms of signal-to-noise ratio.

1 Introduction

In many application contexts, image acquisition leads to blurred images. Blurring is caused e.g. by motion of objects and/or camera during the recording, from defocussing or from specific errors in the optics of the camera. It is therefore desirable to devise methods how to sharpen – to deblur – images. A variety of different approaches has been proposed in the literature which differ greatly in the assumptions made about the image and the blurring process.

An important case of image blurring is convolution with a fixed kernel. This type of space-invariant blurring is especially found when defocussing and optical errors or translatory motion of the camera have caused the blurring.

Under ideal conditions, convolution could be reverted by convolving with an inverse kernel that could, e.g., be computed via the Fourier domain. However, taking the reciprocal of the Fourier transform of some kernel always leads to an unbounded function that needs to be cut at some frequency; moreover, for other than very simple kernels (like Gaussians), zeroes occur which introduce poles into the inverse. For these reasons, a pseudo-inverse is used. By a slight extension of this idea, one obtains Wiener filtering which relies essentially on a regularisation

R. Kimmel, N. Sochen, J. Weickert (Eds.): Scale-Space 2005, LNCS 3459, pp. 585–597, 2005.

of the inversion operation [16, 7]. These linear approaches give reasonable results but the cut-off of high frequencies introduces characteristic oscillatory artifacts (Gibbs-like phenomena) that cannot be avoided by any linear approach [2].

In the case of blurring by Gaussian convolution the special relation between Gaussian convolution and linear diffusion also makes methods based on stabilised inverse diffusion a possible choice [9].

Another class of methods aims at so-called blind deconvolution in which the kernel is not known a priori but is reconstructed simultaneously with the sharpened image. Obviously, assumptions on the image are inevitable here, otherwise the problem is principally underdetermined. A favourable way to encode a-priori assumptions on the image is the use of a variational framework [17]. Variational blind deconvolution has been investigated in [1]. This approach combines deconvolution with segmentation. The convolution kernels are restricted to Gaussians. In the second part of [1], the combination of deconvolution with a known (not necessarily Gaussian) kernel with segmentation is discussed.

The ill-posedness of all deblurring problems makes it reasonable to involve any available a-priori knowledge in the reconstruction process. Methods for deblurring with known kernels are therefore not made superfluous by blind deconvolution techniques; their better understanding can even support the development of blind deconvolution methods.

In this article, we describe PDE-based approaches for deblurring in case of convolution with a space-invariant kernel. We do not make specific assumptions on symmetry of the kernel, instead, our method is designed to work even for fairly irregularly shaped kernels. Our approach is motivated first by a variational model for deconvolution. The involved variational problem is solved via a diffusion–reaction equation where the diffusivity is linked to the regulariser. While total variation (TV) regularisation is a common choice in the literature, we investigate the nonconvex regularisation which leads to forward-backward diffusivity. By generalising to diffusion-reaction PDEs which are no longer associated with variational formulations, we can also include anisotropic diffusion tensors in our study. Experiments show that improvements over established deblurring techniques can be achieved by these methods. Since the non-uniqueness of steady states plays an important role, the treatment of the diffusion weight is a central issue. We present a strategy to avoid unwanted solutions.

We also discuss problems occurring at image boundaries which are caused by the admission of kernels without particular symmetries. Note that many deconvolution methods discussed in the literature restrict the shape of the kernel, e.g. to Gaussians with principal axes parallel to the image boundaries, or motion blurs in directions parallel to the image boundaries. The specific symmetries of such kernels make the treatment of boundaries considerably easier.

We proceed as follows. In Section 2, we discuss linear deconvolution methods. Section 3 then describes a variational deblurring model and derives our basic deblurring PDE. Boundary conditions and the choice of the regularisation parameter are discussed separately. Section 4 is dedicated to extensions of the basic model. These include the continuation strategy and the introduction of anisotropic diffusivities. Experiments are presented in Section 5.

Related Work. Deblurring problems have attracted the attention of computer vision researchers for a long time, and numerous publications exist on this and related topics. Variational approaches to deconvolution with total variation (TV) regularisation have been investigated e.g. by Marquina and Osher [8] for non-blind, and Chan and Wong [6] for blind deconvolution, or recently Bar, Sochen, and Kiryati [1] which addresses both classes of problems. TV regularisation in image restoration has earlier been studied by Rudin, Osher, and Fatemi [12], see also [10]. A variational approach to blind deconvolution with more general regularisation has been presented by You and Kaveh [17].

Research on existence, uniqueness, and stability of solutions for these and related problems can be found in the work by Bertero, Poggio, and Torre [3]. Continuation strategies have been considered for non-convex variational problems in visual reconstruction by Blake and Zisserman [4] and more specifically in the context of total-variation based denoising by Chan, Chan and Zhou [5].

2 Linear Models

We assume that we have an image f which is the result of convolving the original (sharp) image u with some kernel h and superposing some additive noise n,

$$f(x,y) = (u*h)(x,y) + n(x,y) \ .$$

Assume first that the noise n can be neglected. By Fourier transform, the equation then goes into

$$\hat{f} = \hat{u} \cdot \hat{h} \ .$$

If h is known, one could in principle divide $\hat{f}$ by $\hat{h}$ to restore $\hat{u}$ and thereby u. However, this *inverse filtering* procedure faces the problem that in general $\hat{h}$ will possess zeroes. These represent frequencies which are deleted by blurring with h and must therefore not be present in a noise-free blurred image. But still $\hat{h}$ is close to zero in the vicinity of its zeroes, and, even if it has no zeroes, for high frequencies. But in frequency ranges where $|\hat{h}|$ is small, even minimal amounts of noise are tremendously amplified, rendering the procedure extremely unstable.

The simplest approach to handle this difficulty is the *pseudo-inverse filtering* which eliminates frequencies for which $\hat{h}$ is smaller than some threshold H. A more advanced regularisation of inverse filtering is *Wiener filtering* [16] which replaces $\hat{h}^{-1}$ by $\hat{h}^{-1}\,|\hat{h}|^2/(|\hat{h}|^2 + H^2)$ such that we obtain

$$\hat{u} = \frac{1}{\hat{h}} \cdot \frac{|\hat{h}|^2}{|\hat{h}|^2 + H^2} \cdot \hat{f} \ .$$

This filter displays better stability than pseudoinverse filtering. It has properties of a band-pass and is therefore even well-suited to deal with moderate noise.

All deconvolution methods described up to here are linear methods which allow for efficient implementations via Fourier transforms. However, all of them display characteristic, shadow-like artifacts particularly near edges and also near the image boundaries depending on the boundary conditions used. As proven in [2–p. 119pp.], linear methods cannot avoid these artifacts. Further improvements can therefore be achieved only with nonlinear methods.

3 The Basic Deblurring PDE

3.1 Variational Motivation

Deconvolution of an image can be achieved by minimising the energy functional

$$E(u) = \int\limits_{\Omega} (h * u - f)^2 + \alpha \Psi(|\nabla u|^2)\,\mathrm{d}x\,. \tag{1}$$

The first summand in the integral – the data term – is the square error of the reconstruction of the blurred image from the deblurred image candidate. This data term arises naturally in the deconvolution context and is also used in the variational blind deconvolution models in [17, 1]. The second summand – the smoothness term or regulariser – uses a monotonically increasing function $\Psi : \mathbb{R}_0^+ \to \mathbb{R}$ to enforce the smoothness of the deconvolved image.

Note that an unregularised energy consisting only of the data term already has the original image as a global minimiser. Unfortunately, this minimum is by far not unique since the data term is in general not strictly convex. If the Fourier transform of h has zeroes, then contributions of the corresponding frequencies may be added to u without changing the data term. Even if $\hat{h}$ has no zeroes, it is very small for e.g. high frequencies. Contributions from such frequencies hardly influence the data term. Hence, the data term cannot effectively suppress artifacts like those encountered for linear methods. The smoothness term is needed to reduce these ambiguities. In the case of a strictly convex regulariser Ψ, the energy E as a whole might even be convex and the minimum therefore unique.

Solutions of our variational problem satisfy the Euler-Lagrange equation

$$0 = \tilde{h} * (h * u - f) - \alpha \,\mathrm{div}\left(\Psi'(|\nabla u|^2)\nabla u\right)\,.$$

Here $\tilde{h}$ denotes the mirror-kernel $\tilde{h}(x, y) := h(-x, -y)$. A gradient descent leading for $t \to \infty$ to a minimiser of E is given by

$$\partial_t u = -\tilde{h} * (h * u - f) + \alpha \,\mathrm{div}\left(g(|\nabla u|^2)\nabla u\right)\,, \tag{2}$$

a diffusion–reaction equation where the diffusion term with diffusivity $g(s^2) = \Psi'(s^2)$ is related to the regulariser in the energy functional. This PDE can be solved numerically, in the simplest case by an explicit discretisation.

3.2 Choice of the Diffusivity

An important point in determining the properties of the deconvolution process is the choice of the diffusivity g. The simplest case, the constant diffusivity

$g(s^2) = 1$ which corresponds to Tikhonov regularisation $\Psi(s^2) = s^2$, tends towards an over-smoothed deblurring result because high gradients at edges of the reconstructed image are penalised over-proportionally. Moreover, in this case the whole deconvolution method is again linear and suffers from the artifacts described in the previous section.

Total variation (TV) diffusivity $g(s^2) = 1/|s|$, mostly in its regularised form $g(s^2) = 1/\sqrt{s^2+\varepsilon^2}$, is a popular choice (see [6, 8, 1]), particularly since it enforces piecewise constant results and therefore encourages sharp edges in the image. We therefore include TV diffusivity in our experiments.

Another interesting choice in isotropic nonlinear diffusion models is the Perona–Malik diffusivity $g(s^2) = (1 + s^2/\lambda^2)^{-1}$ that is related to the nonconvex regulariser $\Psi(s^2) = \lambda^2 \ln(1 + s^2/\lambda^2)$, see [11, 14]. Note that the smoothness energy $\Psi(|\nabla u|^2)$ is no longer convex in this case. It is therefore expected that depending on the initial conditions different solutions are obtained.

To reduce the noise sensitivity of isotropic Perona–Malik diffusion (see [14, 13]) it can be stabilised by using a Gaussian-smoothed gradient ∇u_σ in the diffusivity argument, turning the diffusion expression into $\operatorname{div}(g(|\nabla u_\sigma|^2)\,\nabla u)$. Though this stabilised Perona–Malik diffusivity can easily be used in our diffusion–reaction equations (which in this case cease to be gradient descents for energy functionals), experiments indicate that it bears no clear advantages in this case.

3.3 Boundary Conditions

For solving the diffusion–reaction equations of type (2), suitable boundary conditions must be specified. In many diffusion-based image processing methods, reflecting Neumann boundary conditions work well because they guarantee conservation properties as well as a continuous extension of the image at its boundary. Periodic boundary conditions for a rectangular domain lead instead to a wrap-around of image information between opposite boundaries; moreover, they introduce discontinuities which often entail artifacts in the processed image.

Unfortunately, the usage of reflecting boundary conditions for deconvolution with space-invariant kernels is bound to fail if the kernel is not symmetric w.r.t. the image boundary directions because reflected parts of the image would be blurred with a reflected kernel, violating the model assumptions. Since periodic boundary conditions are compatible with any shift-invariant blur, without imposing symmetry constraints on the kernel, we use periodic boundary conditions or modifications of them.

A chief disadvantage of periodic boundary conditions are the discontinuities introduced at the image boundaries. These lead to strong artifacts near the image boundaries. To mitigate these artifacts as well as the undesired wrap-around of image information in the deblurring with periodic boundary conditions, the image can be extended continuously to a larger image with equal grey-values at opposing boundaries. Periodic boundary conditions then no longer introduce discontinuities, and the wrap-around influences mostly the amended parts of the image. Since the assumptions of our deblurring model are still violated near the image boundaries, boundary artifacts are reduced but not perfectly eliminated.

For quality measurements in our experiments, we therefore arrange a special setting. We start by extending the sharp test image via horizontal and vertical reflection to quadruple size. Periodic extension of this larger image is equivalent to reflecting extension of the original image. Now the large image is blurred in a periodic setting (i.e. with the left boundary wrapping over into the right one etc.) with the irregular kernel. The resulting image has lost the symmetry of the original larger image. In deblurring this image, we use again periodic boundary conditions. While this approach cannot be used in real deblurring applications where the blurring process is not subject to our control, its advantage is that it admits a measurement of the deblurring quality, e.g. in terms of signal-to-noise ratio, without including discontinuity and boundary artifacts which would dominate the total result otherwise. By doing so, we ensure that the model assumptions are met everywhere, at the cost of making this boundary treatment unsuitable for naturally blurred images.

3.4 Choice of the Diffusion Weight

The extreme ill-posedness of the problem makes the choice of the diffusion weight a difficult problem. We discuss this for the deblurring processes which minimise a variational functional. In absence of noise, the non-regularised energy consisting only of the data term is minimised by the correct solution. However, the data term is insensitive to certain perturbations (those being annihilated by convolution with h), preventing in general the solution to be unique. The diffusion–reaction equation (2) in this case turns into a fixed-point equation without diffusion part.

Assume now the energy is made convex by a suitable regulariser such that a unique solution exists. Even if the weight of the regularisation (and thus of the diffusion) is very small, it is practically only the regularisation term which chooses the solution among those which cannot be discriminated by the smoothness term. As a consequence, even a small diffusion weight can drive the solution far away from the true unblurred image, leading to a deblurring result which heavily depends on the type of regulariser (hence, diffusivity) used.

Particularly with nonconvex regularisers and the corresponding forward–backward diffusivities, the existence of multiple steady states of the deblurring process constitutes another issue. The solution which is really obtained depends heavily on the initial conditions. When using the blurred image as initialisation with small diffusion weight, similar artifacts as for linear deblurring methods evolve. Large diffusion weights, on the other hand, induce an over-smoothing and loss of small-scale details in the image.

3.5 Numerical Implementation

In order to solve equation (2) numerically, finite difference discretisations are used for the diffusion term as well as for the left-hand side. The simplest way to do so is to use a forward difference for $\partial_t u$ and central differences from the old time step for the diffusion term. Denoting by τ the time step size and by Δx, Δy the spatial step sizes in x, y direction, we are led to the explicit scheme

$$\begin{aligned}\frac{u_{ij}^{k+1} - u_{ij}^k}{\tau} &= -R_{ij}^k + \tfrac{\alpha}{2\Delta x}\big((g_{i+1,j}^k + g_{ij}^k)(u_{i+1,j}^k - u_{ij}^k) - (g_{ij}^k + g_{i-1,j}^k)(u_{ij}^k - u_{i-1,j}^k)\big) \\ &\quad + \tfrac{\alpha}{2\Delta y}\big((g_{i,j+1}^k + g_{ij}^k)(u_{i,j+1}^k - u_{ij}^k) - (g_{ij}^k + g_{i,j-1}^k)(u_{ij}^k - u_{i,j-1}^k)\big)\end{aligned}$$

where the diffusivity $g(|\nabla u|^2)$ (or stabilised $g(|\nabla u_\sigma|^2)$) is discretised by

$$g_{ij}^k = \Psi'\left(\left(\frac{u_{i+1,j}^k - u_{i-1,j}^k}{\Delta x}\right)^2 + \left(\frac{u_{i,j+1}^k - u_{i,j-1}^k}{\Delta y}\right)^2\right)$$

and R_{ij}^k discretises the reaction term $\tilde{h} * (h * u - f)$ at pixel (i, j) and time $k\tau$.

The discretisation of the reaction term poses a difficulty. Since it contains convolutions, its direct computation in each time step would be extremely costsome, even taking into account that $\tilde{h} * h$ can be precomputed once for all time steps. Note that R_{ij}^k is computed via the Fourier domain. Though this still requires one DFT (or FFT for suitable image size) and one inverse transform per time step, computing time is considerably reduced for kernels with large support.

4 Extensions

4.1 Continuation Strategy for Optimisation

It has been explicated that the deblurring model is ill-posed, i.e., it reveals not only several local optima, but may even have multiple global optimum solutions that do not depend continuously on the initial data. Consequently, the gradient descent often does not yield the original image as solution, but some other steady state which can be significantly different. In most cases it contains a rather huge amount of oscillatory structures not present in the original data.

A remedy for this ill-posedness has been the supplement of a regulariser to the energy functional. This regulariser introduces the a-priori knowledge that smooth solutions should be preferred. However, despite the usefulness of non-quadratic regularisers which allow for discontinuities in the solution, the negative consequence of the regularisation is a result that is smoother than the original image, since not the complete amount of blurring is reversed by the process.

Actually, there is no regularisation necessary for the model to yield the original image as an optimum solution of the energy. Even the opposite is true: just in the case *without* regularisation the model has the original image as one of the optimum solutions. The regularisation must only be introduced in order to guide the gradient descent towards one out of several optima that shows the least oscillatory behaviour.

Since the regularisation primarily serves as guidance for the optimisation process, the proposed alternative approach continuously reduces the amount of regularisation during optimization. Instead of considering only one energy functional with a fixed amount of regularisation, a cascade of functionals is taken into account. Starting with a rather large amount of regularisation, this amount is reduced from one member of the cascade to the next, and finally leads to the energy functional without any regularisation. The first members of the cascade prefer smooth solutions and therefore may yield good initialisations close to the smoothest optimum of the next version with less regularisation. On the other hand, the later members of the cascade tend more and more to solve the original deblurring problem without regularisation and therefore yield sharper results. This way, one finally runs into an optimum of the functional without regularisation, yet choosing a specific optimum that is smooth. In most cases this optimum is not exactly the original image (it is often still too smooth), yet it is supposed to be closer to this image than the solutions one obtains without this continuation strategy, i.e. either with a fixed amount of smoothness or without any regularisation.

4.2 Anisotropic Model

An improved reconstruction of edges can further be achieved by substituting the isotropic diffusivity g with an anisotropic diffusion tensor $D(\nabla u_\sigma)$. In our model, we use $D(\nabla u_\sigma) = g(\nabla u_\sigma \nabla u_\sigma^{\mathrm{T}})$ where the Perona–Malik diffusivity g is applied to the symmetric matrix $\nabla u_\sigma \nabla u_\sigma^{\mathrm{T}}$, as usual, by letting g act on the eigenvalues and leaving the eigenvectors unchanged [14]. The resulting equation

$$0 = \tilde{h} * (h * u - f) - \alpha \operatorname{div}\left(D(\nabla u_\sigma)\nabla u\right)$$

is not the gradient descent for an energy because of the smoothed gradient. However, this smoothing is inevitable in order to have true anisotropy.

5 Experiments

To illustrate and validate the methods described in the preceding sections, we show experimental results obtained with two test images and two different convolution kernels, Fig. 1. One is a banana-shaped blob with irregularly distributed

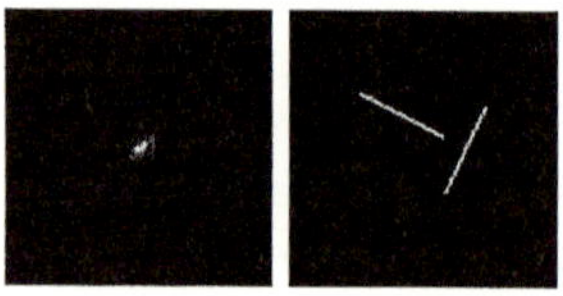

Fig. 1. Convolution kernels. **Left:** Banana-shaped kernel. **Right:** Discontinuous kernel consisting of two line-shaped components

Fig. 2. Left to right: Photograph, 480 × 640 pixels. – Blurred with banana-shaped kernel. – Deblurred by diffusion–reaction method with Perona–Malik diffusivity, $\lambda = 5$, $\alpha = 0.001$, 1000 iterations. – Photograph blurred with discontinuous kernel. – Deblurred by diffusion–reaction method with Perona–Malik diffusivity, $\lambda = 1$, the continuation strategy was used with two steps for the diffusion weight: 2400 iterations with $\alpha = 0.01$ followed by 2400 iterations with $\alpha = 0$

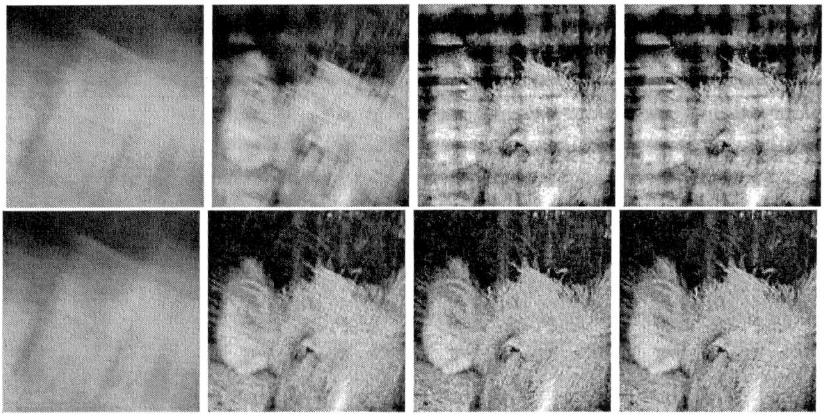

Fig. 3. Deblurring of a detail of the photograph from Fig. 2 with different boundary treatment. **Left to right:** Photograph detail blurred with discontinuous kernel. – Linear deblurring by Wiener filter. – Diffusion–reaction deblurring with TV diffusivity. – Diffusion–reaction deblurring with Perona–Malik diffusivity. **Top row:** Continuous periodic extrapolation of the blurred image (realistic method). While details are well reconstructed, shadow-like boundary artifacts affect the overall quality. **Bottom row:** Same with special setting to suppress boundary artifacts. The image was extended by reflection to four times its original size *before* blurring. This quadruple-size blurred image was then blurred and deblurred with periodic continuation

intensity. This comes close to the blurring of photographs taken with bad illumination and moving camera and objects. The second convolution kernel is discontinuous; it is assembled from two line-shaped parts which are similar to motion blurs. It has been selected as an example of a very challenging kernel.

The first test image used in Figs. 2 and 3 is a photograph with many small-scale details. In fact, this is a colour image to which our diffusion–reaction equa-

Fig. 4. Top left: Grey-value test image. **Top middle:** Blurred with discontinuous kernel. **Top right:** Linear deblurring by Wiener filter, boundary treatment by continuous extrapolation. **Bottom left:** Diffusion–reaction deblurring with TV diffusivity and continuation strategy (2 levels). **Bottom middle:** Same with Perona–Malik diffusivity. **Bottom right:** Same with anisotropic diffusion tensor

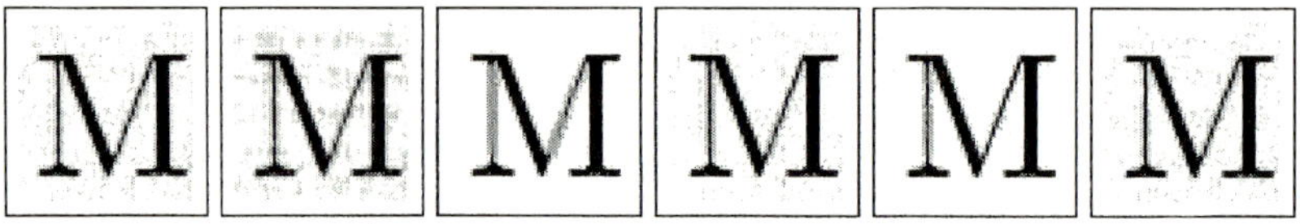

Fig. 5. Detail from deblurred grey-value images, with boundary artifacts suppressed by special test setting. **Left to right:** Linear filtering. – Unregularised variational model. – Perona–Malik, constant diffusion weight. – Perona–Malik followed by nonregularised iteration (two-step continuation strategy). – Perona–Malik, continuation strategy with 10 steps. – Anisotropic diffusion–reaction, continuation strategy with 10 steps

tions were adapted in the straightforward way with channel coupling. This procedure is well-established in nonlinear diffusion literature [14]. The second test image used in Figs. 4 and 5 is a grey-value image of three print letters. It differs from the first image by its composition of fairly homogeneous regions.

In Fig. 2 we blur the first test image with both kernels and restore it by diffusion–reaction deblurring with Perona–Malik diffusivity. For the discontinuous kernel, we also use the continuation strategy in a simple form with one positive α followed by a fixed-point iteration with $\alpha = 0$. Excellent deblurring quality is achieved for the banana kernel (despite its irregularity) while for the discontinuous kernel some shadow-like boundary artifacts are observed.

In Fig. 3, a more detailed comparison of deblurring algorithms is presented for a detail from the photograph blurred with the discontinuous kernel, including Wiener filter as an example of linear deblurring, diffusion–reaction filtering with TV, and Perona–Malik diffusivity. Here we also demonstrate the use of our special test setting to avoid boundary artifacts in quality measurements.

Results for the grey-value test image are shown in Fig. 4. Here, we concentrate on the discontinuous kernel. Besides the methods mentioned above we show also diffusion–reaction deblurring with anisotropic diffusion tensor which performs particularly well for this type of strongly segmented images.

Table 1. SNR (dB) for deblurring with the discontinuous kernel. First values: specific test setting for boundary conditions, second values: with continuous extrapolation

	Wiener filtering	Diffusion–reaction, Perona–Malik	Diffusion–reaction, TV diffusivity
Letters	15.6 / 7.1	18.4 / 7.3	17.1 / 7.2
– with contin. strategy		19.3 / 7.4	19.2 / 7.1
Photograph detail	15.9 / 9.4	14.9 / 6.0	14.3 / 5.9

Fig. 5 shows a detail of our grey-value test image to demonstrate the improvements made by anisotropic diffusion tensors and continuation strategy. Pure Perona–Malik deblurring reduces oscillatory artifacts quite well but smears thin lines while the fixed-point iteration with $\alpha = 0$ restores many details but generates artifacts similar to those of linear deconvolution. The continuation strategy combines a better restoration of details with a reasonable suppression of artifacts. Even in its simplest form with two steps it bears a clear improvement; more steps lead to further enhancement. The sharpness of edges is further improved by using the anisotropic diffusion tensor.

In Table 1 we compile measurements of the signal-to-noise ratio (SNR)

$$\mathrm{SNR}(v,u) = 10\,\log_{10}\frac{\mathrm{var}(u)}{\mathrm{var}(u-v)}\ \mathrm{dB}$$

where u is the original image and v the deblurring result. Throughout the measurements Perona–Malik deblurring tends to slightly better SNR than deblurring with TV diffusivity. However, not always do SNR measurements reflect sufficiently well visual judgement. For the photograph, e.g., the Wiener filter performs better than diffusion–reaction deblurring in terms of SNR. On the other hand, Fig. 3 clearly reveals the superiority of diffusion–reaction deblurring.

6 Conclusions and Ongoing Work

In this paper we have developed diffusion–reaction based deconvolution methods with forward–backward diffusivities motivated from non-convex regularisation. We have established a continuation strategy for the control of the diffusivity weight that allows to combine the suppression of artifacts provided by large diffusion weights with the good reconstruction of details that is typically achieved with small diffusion weights. We have further extended our algorithm by introducing an anisotropic diffusion tensor which allows for a further enhancement of edges in the deblurring process. The favourable performance of the algorithms even for severely blurred images and irregularly shaped kernels has been demonstrated visually and by signal-to-noise ratio measurements.

Ongoing work is dedicated to improvements in the numerical efficiency of our deblurring algorithms, e.g. by utilising additive operator splitting schemes (see [15]). We also work on improving the treatment of boundaries to reduce artifacts even in the restoration of severely blurred images. Because of the observed discrepancies between SNR measurements and visual quality judgement, the development of more adequate quality criteria is a further issue.

References

1. L. Bar, N. Sochen, N. Kiryati. Variational pairing of image segmentation and blind restoration. In T. Pajdla, J. Matas, editors, *Computer Vision – ECCV 2004,* vol. 3022 of *Lecture Notes in Computer Science,* pages 166–177, Berlin, 2004, Springer
2. M. Bertero and P. Boccacci. *Introduction to Inverse Problems in Imaging.* IoP Publishing, Bristol, 1998.
3. M. Bertero, T. A. Poggio, and V. Torre. Ill-posed problems in early vision. *Proceedings of the IEEE*, 76(8):869–889, Aug. 1988.
4. A. Blake and A. Zisserman. *Visual Reconstruction.* MIT Press, Cambridge, MA, 1987.
5. T. Chan, R. Chan, and H. Zhou. A continuation method for total variation denoising problems. In F. Luk, editor, *Proceedings of the SPIE Conference on Advanced Signal Processing Algorithms,* vol. 2563 of *Algorithms, Architectures, and Implementations,* pages 314–325, San Diego, 1995
6. T. F. Chan and C. K. Wong. Total variation blind deconvolution. *IEEE Transactions on Image Processing*, 7:370–375, 1998.
7. R. C. Gonzalez and R. E. Woods. *Digital Image Processing.* Addison–Wesley, Reading, second edition, 2002.
8. A. Marquina, S. Osher. A new time dependent model based on level set motion for nonlinear deblurring and noise removal. In M. Nielsen, P. Johansen, O. F. Olsen, and J. Weickert, editors, *Scale-Space Theories in Computer Vision*, volume 1682 of *Lecture Notes in Computer Science*, pages 429–434. Springer, Berlin, 1999.
9. S. Osher and L. Rudin. Shocks and other nonlinear filtering applied to image processing. In A. G. Tescher, editor, *Applications of Digital Image Processing XIV*, volume 1567 of *Proceedings of SPIE*, pages 414–431. SPIE Press, Bellingham, 1991.
10. S. Osher, L. Rudin. Total variation based image restoration with free local constraints. *Proceedings of the IEEE ICIP*, 31–35, Austin, 1994
11. P. Perona and J. Malik. Scale space and edge detection using anisotropic diffusion. *IEEE Transactions on Pattern Analysis and Machine Intelligence*, 12:629–639, 1990.
12. L. I. Rudin, S. Osher, and E. Fatemi. Nonlinear total variation based noise removal algorithms. *Physica D*, 60:259–268, 1992.
13. J. Weickert. A review of nonlinear diffusion filtering. In B. ter Haar Romeny, L. Florack, J. Koenderink, and M. Viergever, editors, *Scale-Space Theory in Computer Vision*, volume 1252 of *Lecture Notes in Computer Science*, pages 3–28. Springer, Berlin, 1997.
14. J. Weickert. *Anisotropic Diffusion in Image Processing.* Teubner, Stuttgart, 1998.

15. J. Weickert, B. M. ter Haar Romeny, and M. A. Viergever. Efficient and reliable schemes for nonlinear diffusion filtering. *IEEE Transactions on Image Processing*, 7(3):398–410, Mar. 1998.
16. N. Wiener. *Extrapolation, Interpolation and Smoothing of Stationary Time Series with Engineering Applications.* Cambridge (Mass.), 1949, The MIT Press
17. Y.-L. You and M. Kaveh. Anisotropic blind image restoration. In *Proc. 1996 IEEE International Conference on Image Processing*, volume 2, pages 461–464, Lausanne, Switzerland, Sept. 1996.

Denoising of Audio Data by Nonlinear Diffusion

Martin Welk, Achim Bergmeister, and Joachim Weickert

Mathematical Image Analysis Group,
Faculty of Mathematics and Computer Science, Bldg. 27,
Saarland University, 66041 Saarbrücken, Germany
{welk, bergmeister, weickert}@mia.uni-saarland.de
http://www.mia.uni-saarland.de

Abstract. Nonlinear diffusion has long proven its capability for discontinuity-preserving removal of noise in image processing. We investigate the possibility to employ diffusion ideas for the denoising of audio signals. An important difference between image and audio signals is which parts of the signal are considered as useful information and noise. While small-scale oscillations in visual images are noise, they encode essential information in audio data. To adapt diffusion to this setting, we apply it to the coefficients of a wavelet decomposition instead of the audio samples themselves. Experiments demonstrate that the denoising results are surprisingly good in view of the simplicity of our approach. Nonlinear diffusion promises therefore to become a powerful tool also in audio signal processing.

1 Introduction

Degradation of signals by noise is a ubiquitous phenomenon. In practically any field of signal processing the removal of noise therefore is a key problem. In the field of image processing, diffusion processes are among the most effective and theoretically best understood denoising techniques [18]. While linear diffusion is highly effective in removing noise, it blurs indiscriminately all image information and therefore removes, or at least severely degrades, important image features such as edges along with the noise. Nonlinear diffusion processes – isotropic as well as anisotropic – have therefore gained increasing interest in the last 15 years [14, 2, 18, 17]. They allow to treat details of different size and contrast differently. Thus they enable the design of image filters which remove noise effectively while at the same time edges are preserved and in some cases even enhanced.

The question therefore arises naturally whether diffusion filters can also be used to denoise other classes of signals. Audio signals are one class of signals which is of similar importance as images. We want therefore to investigate the possibility of denoising digital audio signals by diffusion processes.

First of all, audio signals are one-dimensional, so the range of applicable techniques is constrained to linear and isotropic nonlinear diffusion. Another observation is that music samples, like images, contain well-localised features that should be preserved in the denoising process. It can therefore be expected that a good filtering process should again be inhomogeneous and therefore nonlinear.

R. Kimmel, N. Sochen, J. Weickert (Eds.): Scale-Space 2005, LNCS 3459, pp. 598–609, 2005.

Direct application of diffusion to sampled audio signals faces a serious problem. Typically, small oscillatory details are the first structures that a diffusion filter removes from a signal. This is well appropriate for image processing; however, in audio signals the most important features consist of oscillations. It needs therefore a re-consideration which features in an audio signal are typically noise and which are useful information.

Targeting at the denoising of sampled music or speech, we see that useful information basically should consist of periodic oscillations with only a moderate number of different frequencies occurring at the same time while noise is supposed to be made up of irregular oscillations which are not concentrated at single frequencies. It seems therefore reasonable to separate useful signal components from noise by a suitable frequency analysis method. The necessity to keep signal components well localised in time motivates us to prefer wavelet decomposition [5, 13] over Fourier analysis.

An established technique for denoising of data in wavelet representations is wavelet shrinkage [6, 7]. Though it is by far not an optimal denoising technique for audio data, it serves for us as a reference because of its simplicity. We will compare the denoising results of our diffusion-based methods with those of wavelet shrinkage working on the same wavelet representations.

A signal restoration approach that is related to ours since it manipulates wavelet coefficients using variational ideas is described in [4]. Further approaches which combine variational and wavelet ideas in a different manner to denoise signals and images can be found in [1, 8, 12].

Section 2 gives an outline of the wavelet methods that will be needed in this paper. In Section 3 we introduce diffusion processes for wavelet coefficients. Application of these filters to digital audio data is illustrated by experiments in Section 4 which are discussed quantitatively and qualitatively. A summary and outlook in Section 5 conclude the paper.

2 Signal Processing with Wavelets

Wavelet methods in signal processing rely on the representation of a signal f with respect to a basis consisting of scaling functions φ_i^j and wavelet functions ψ_i^j. All scaling and wavelet functions are shifted and dilated versions of one scaling function φ with low-pass characteristics and one wavelet function ψ with band-pass properties, i.e.

$$\varphi_i^j(z) = 2^{-j/2}\varphi(2^{-j}z - i), \qquad \psi_i^j(z) = 2^{-j/2}\psi(2^{-j}z - i)\ .$$

If the wavelet functions ψ_i^j and scaling functions $\varphi_i^{j_0}$ form an orthonormal basis, we have

$$f(z) = \sum_{i\in\mathbb{Z}} c_i^{j_0}\varphi_i^{j_0}(z) + \sum_{j=-\infty}^{j_0}\sum_{i\in\mathbb{Z}} d_i^j\psi_i^j(z)$$

where $c_i^j := \langle f, \varphi_i^j\rangle$, $d_i^j := \langle f, \psi_i^j\rangle$ with $\langle\cdot,\cdot\rangle$ being the $L_2(\mathbb{R})$ inner product. Note that this representation uses scaling functions only on the coarsest level j_0 while wavelet functions of level j_0 and all finer levels are used. For a discrete signal, represented by a function f which is constant on $[k, k+1)$ for every integer k, only wavelet levels $j \geq 1$ occur.

2.1 Haar Wavelet Representations

Let us consider the simplest wavelet, the *Haar wavelet* [10] $\psi(z) = \chi_{[0,\frac{1}{2})}(z) - \chi_{[\frac{1}{2},1)}(z)$ with corresponding scaling function $\varphi(z) = \chi_{[0,1)}$, where χ_I is the characteristic function of the interval I. Then, wavelet analysis and synthesis can be efficiently carried out via the *two-scale relations*

$$c_i^j = \frac{c_{2i}^{j-1} + c_{2i+1}^{j-1}}{\sqrt{2}}, \qquad d_i^j = \frac{c_{2i}^{j-1} - c_{2i+1}^{j-1}}{\sqrt{2}}.$$

For a discrete signal as described above, the c_i^0 equal the signal samples.

A wavelet representation of this kind, called *decimated* wavelet decomposition, constitutes a hierarchical subdivision of the domain of definition of the signal, thereby bearing the clear disadvantage of lacking translation invariance. A simple but effective way to overcome this problem in case of a discrete signal of finite length is the *cycle-spinning* procedure [3]. Cycle-spinning essentially means that the decimated wavelet analysis and synthesis of a signal of length N is carried out N times: for the original signal and all $N-1$ possible cyclically shifted copies of it. The analysis step therefore yields N redundant wavelet representations encoding the same signal. In the synthesis step, N concurrent signals are generated. However, if the reconstruction is done with processed wavelet coefficients, these N signals will in general no longer coincide. The final reconstruction result is therefore obtained by averaging the concurrent reconstructions. Filters designed with these transforms are shift-invariant by construction.

2.2 Soft Wavelet Shrinkage

The processing of a signal is typically performed in three steps. First, the *analysis* step in which the given signal is transformed into wavelet representation; second, some operation on the wavelet coefficients; third, the *synthesis* in which the modified coefficients are used to reconstruct the processed version of the signal.

One class of denoising methods widely studied in literature are *wavelet shrinkage* procedures. *Soft wavelet shrinkage* applies the shrinkage function

$$S_\theta(y) = \begin{cases} y - \theta \operatorname{sgn}(y), & |y| > \theta, \\ 0, & |y| \leq \theta \end{cases}$$

to the wavelet coefficients after the analysis step. The modified coefficients $\tilde{d}_i^j := S_\theta(d_i^j)$ are used with the unchanged scaling coefficients c_i^j in the synthesis step.

In [16] it was shown that the denoising quality of shift-invariant soft Haar wavelet shrinkage is improved by using the level-dependent shrinkage parameter $\theta_j = 2^{-j/2}\theta_0$ on level j instead of one uniform parameter θ.

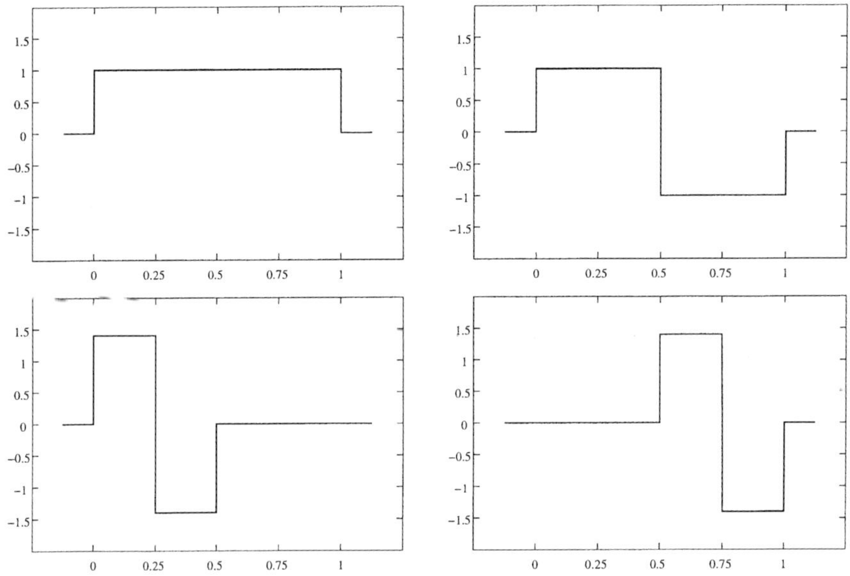

Fig. 1. Basis for a two-level decimated Haar wavelet decomposition. **Top left:** Scaling function φ_0^1. **Top right:** Wavelet function ψ_0^1. **Bottom left:** Wavelet function ψ_0^2. **Bottom right:** Wavelet function ψ_1^2

2.3 Daubechies Wavelets

In image processing, excellent denoising results can be achieved using Haar wavelets, particularly its shift-invariant modification. However, for audio signal processing Haar wavelets are often considered insufficient because they have only one vanishing moment. Daubechies wavelets [5] are often suggested as a better choice in this context. They can be constructed with arbitrarily many vanishing moments. In every case, their wavelet and scaling functions can be found via a recursive procedure. For the simplest Daubechies wavelet (Daubechies-4), one has

$$\varphi(1) = \frac{1+\sqrt{3}}{2}, \qquad \varphi(2) = \frac{1-\sqrt{3}}{2}, \qquad \varphi(k) = 0, \ k \in \mathbb{Z} \setminus \{1,2\},$$

$$\varphi(z) = \frac{1}{4}\Big((1+\sqrt{3})\varphi(2z) + (3+\sqrt{3})\varphi(2z-1) + (3-\sqrt{3})\varphi(2z-2) + (1-\sqrt{3})\varphi(2z-3)\Big)$$

$$\psi(z) = \frac{1}{4}\Big(-(1+\sqrt{3})\varphi(2z) + (3+\sqrt{3})\varphi(2z-1) - (3-\sqrt{3})\varphi(2z-2) + (1-\sqrt{3})\varphi(2z-3)\Big)$$

for all $z = 2^{-j}k$, with integers j, k, and by continuity on the whole real line (see [5, 11]).

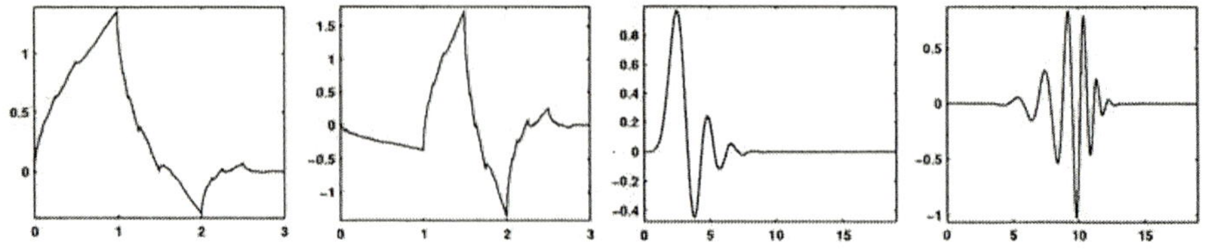

Fig. 2. Left to right: Scaling function for Daubechies-4 wavelet. – Corresponding wavelet function. – Scaling function for Daubechies-20 wavelet. – Corresponding wavelet function. Generated using MATLAB with Wavelet toolbox

Scaling and wavelet functions of the Daubechies-4 and Daubechies-20 wavelets used in our experiments are shown in Fig. 2.

3 Diffusion of Wavelet Coefficients

Let a one-dimensional signal $s(z)$ be given. The partial differential equation

$$v_t = \partial_z(g(v_z^2)\, v_z), \qquad (z,t) \in \mathbb{R} \times (0,+\infty) \tag{1}$$

with initial condition $v(z,0) = s(z)$ describes a one-dimensional diffusion process which embeds the signal $s(z)$ into a family $v(z,t)$ of signals, parametrised with $t \in [0,+\infty)$, which constitute smoothed versions of s. The parameter t can be considered as a diffusion time which, however, must be well distinguished from the signal time z. Here, the *diffusivity* $g(y^2)$ should be a bounded, non-increasing, positive function of its nonnegative argument. For $t \to \infty$, the signal will tend to a constant function. To obtain a smoothed signal, it is therefore necessary to choose a stopping time T which determines the degree of smoothing.

Besides the case $g(y^2) = 1$ of linear diffusion we consider the Perona–Malik diffusivity $g(y^2) = \frac{1}{1+y^2/\lambda^2}$ where λ is a threshold parameter [15]. The purpose of non-linear diffusion with such a decreasing diffusivity is to suppress smoothing at locations with large gradients which are supposed to be edge-like discontinuities holding important information.

A simple explicit discretisation of (1) is given by

$$v_i^{k+1} = v_i^k + \frac{\tau}{h}\left(g((\dot{v}_i^k)^2)\dot{v}_i^k - g((\dot{v}_{i-1}^k)^2)\dot{v}_{i-1}^k\right), \qquad \dot{v}_i^k = \frac{1}{h}(v_{i+1}^k - v_i^k)$$

with step sizes τ for diffusion time h for the signal parameter.

In applying the diffusion equation to the coefficients of a decimated wavelet representation, the coefficients of each wavelet level are considered as one channel, such that diffusion does not transfer amplitudes between different frequency bands. It requires attention that the step size h doubles from each level to the next coarser one. Having this in mind, linear diffusion can be implemented straightforward by

$$[d_i^j]^{k+1} = [d_i^j]^k + \frac{\tau}{2^{2j}}([d_{i+1}^j]^k - 2[d_i^j]^k + [d_{i-1}^j]^k)$$

where $[d_i^j]^k$ denotes the wavelet coefficient d_i^j in the decomposition of the signal in the k-th diffusion-time step, and we have assumed that the temporal resolution in signal time is 1 for the samples, i.e. wavelet level 0.

In nonlinear diffusion of multi-channel signals (like colour images) it is essential that the discontinuities where smoothing is suppressed are localised at equal positions in all channels [9]. To achieve this, one uses a common diffusivity which incorporates gradient information from all channels and steers uniformly the diffusion in all of them. The same argument applies also in our model. Here it is assumed that large differences between neighbouring wavelet coefficients signify boundaries of acoustic events extended in time which should not be blurred.

Consequently, we want to use a common diffusivity also in the nonlinear diffusion of our wavelet coefficients. In computing this common diffusivity in a decimated wavelet representation, it needs to be clarified which neighbour differences should contribute to which diffusivities. Since diffusivities are to steer diffusion *between* neighbouring wavelet coefficients of one level, the proper location where to estimate the diffusivity is the central point between the coordinates of the wavelet coefficients themselves. To determine which neighbour differences should enter a particular diffusivity, we look at the influence zones of the wavelet coefficients, i.e. for each coefficient the group of subsequent samples that it depends on. Then each diffusivity is influenced exactly by the neighbour differences of those wavelet coefficients with influence zones starting or ending at the position of this diffusivity, see Fig. 3. Vice versa, the diffusion between two adjacent wavelet coefficients is regulated only by the diffusivity at the single inter-sample location where the two influence zones meet.

This procedure can directly be motivated as a simple subsampling of the coarser wavelet levels. By writing down each wavelet coefficient $[d_i^j]^k$ of the j-th level 2^{j-1} times, a vector-valued signal with the signal-time resolution of the finest wavelet level 1 arises. The i-th vector in this signal reads

$$([d_i^1]^k, [d^2_{\lfloor i/2 \rfloor}]^k, \ldots, [d^{j_0}_{\lfloor i/2^{j_0-1} \rfloor}]^k)^{\mathrm{T}}$$

The diffusivity between the i-th and $(i+1)$-th vector in this signal is

$$g_{i+1} := g\left(\sum_{j=1}^{j_0}\left([d^j_{\lfloor (i+1)/2^j \rfloor}]^k - [d^j_{\lfloor i/2^j \rfloor}]^k\right)^2\right)$$

where $\lfloor z \rfloor$ denotes the largest integer less or equal z. Here we have weighted all wavelet levels equally; one could also use different weights for the wavelet levels.

After computing the new diffusion time-step, the 2^{j-1} copies of the coefficient $[d_i^j]^k$ are no longer identical; the new value $[d_i^j]^{k+1}$ is then obtained by averaging the concurrent values which amounts exactly to

$$[d_i^j]^{k+1} = [d_i^j]^k + \frac{\tau}{2^{2j}}\left(g_{2^{j-1}(i+1)} \cdot \left([d_{i+1}^j]^k - [d_i^j]^k\right) - g_{2^{j-1}i} \cdot \left([d_i^j]^k - [d_{i-1}^j]^k\right)\right) .$$

In the context of decimated wavelet shrinkage, it can be criticised that diffusivities at different locations have different numbers of influencing coefficient

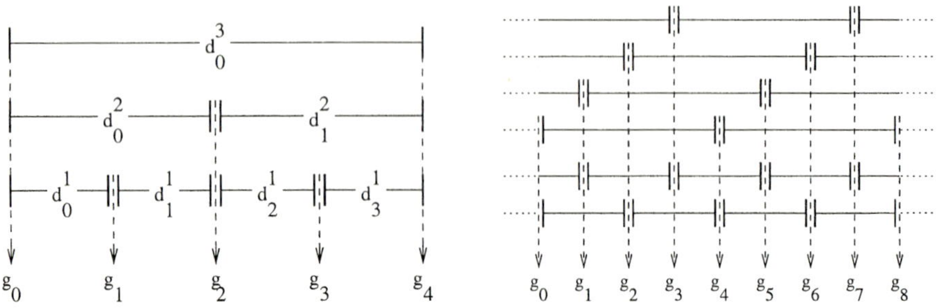

Fig. 3. Influence of differences of neighbouring wavelet coefficients on diffusivities for the coupled nonlinear diffusion. **Left:** Decimated Haar wavelet representation with 3 levels. Wavelet coefficients d_j^k with their influence zones are shown. By $g_0, \ldots, g_4$ subsequent diffusivities are denoted. **Right:** Shift-invariant Haar wavelet representation with 2 levels. Again, $g_0, \ldots, g_8$ are subsequent diffusivities. Variables for wavelet coefficients are omitted

pairs. One could consider compensation factors to remedy this. On the other hand, as soon as we switch to shift-invariant Haar wavelets, the problem disappears. In this situation, if we count identical wavelet coefficients only once, we have for each inter-sample location exactly one pair of coefficients in each wavelet level whose influence zones meet there; compare Fig. 3.

We emphasise that in the shift-invariant setting only coefficients of one wavelet level which are part of the same decimated wavelet decomposition communicate directly in the diffusion process. These are coefficients which have not only the same frequency but also the same phase. Coefficients of the same level and different phase, as well as those of different levels, belong to different channels which are linked only by the channel coupling.

4 Experimental Results

For a first impression of the properties of different wavelet bases in audio processing, we degrade a synthetic 200 Hz sine wave[1] by adding Gaussian noise[2] of 10 % the signal variance, and apply to it soft wavelet shrinkage with non-shift-invariant and shift-invariant Haar wavelets as well as with two Daubechies bases, see Fig. 4. It is evident that the signal shape of the shrinked signal is strongly influenced by the shape of the used wavelets. Audio perception is very sensitive to such details in the wave shape such that the signals denoised with decimated Haar, and Daubechies-4 wavelets do not sound too well. Surprisingly, the quality of shift-invariant Haar wavelet shrinkage is subjectively perceived

[1] The sampling frequency of all our audio data is 44,100 Hz.

[2] The choice of Gaussian noise in the experiments presented here is nothing special. Denoising quality was quite similar when, e.g., white noise or recorded noise from technical sources was added.

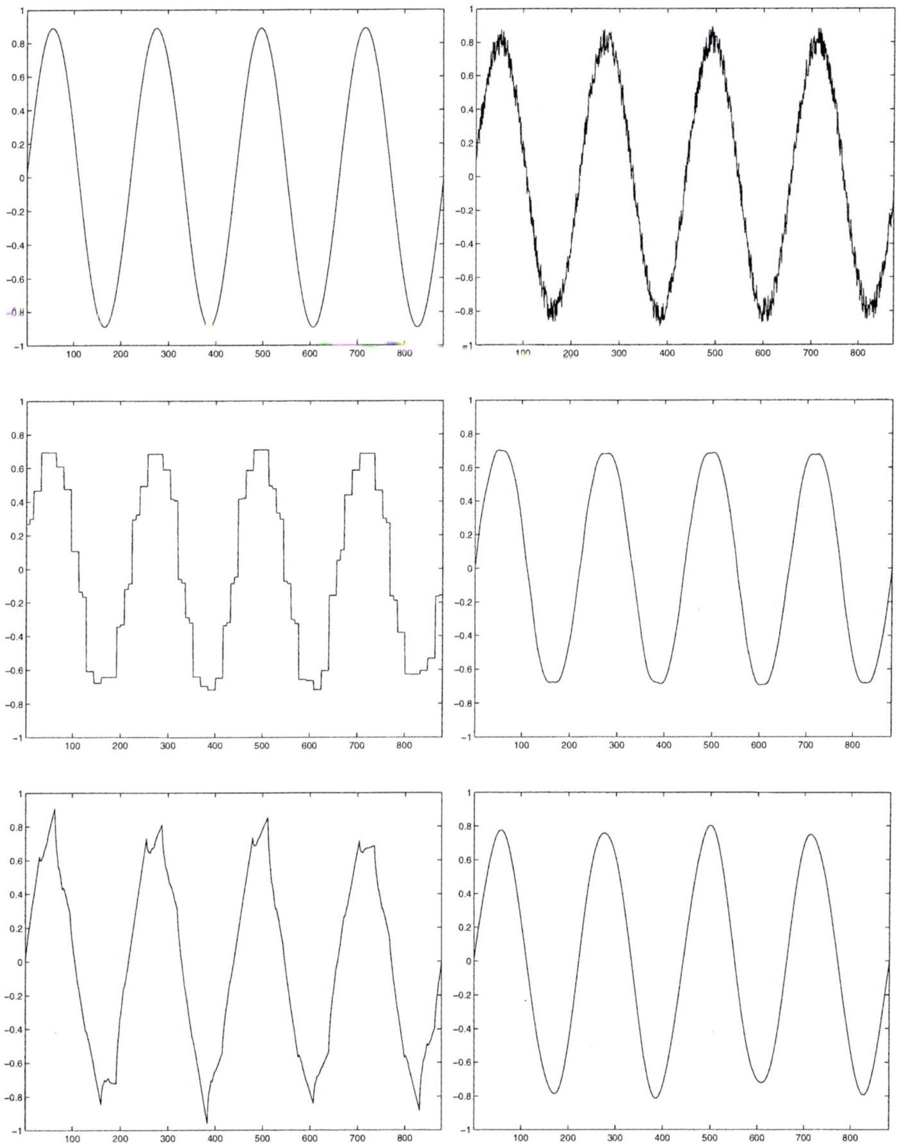

Fig. 4. Soft wavelet shrinkage with different wavelet decompositions, shrinkage parameter always $\theta = 10000$. **Top left:** Original signal (sine wave of 200 Hz). **Top right:** Same with Gaussian noise, noise variance ca. 10 % of signal variance. **Middle left:** Denoised using decimated Haar wavelet shrinkage. **Middle right:** Shift-invariant Haar wavelet shrinkage. **Bottom left:** With Daubechies-4 wavelets. **Bottom right:** Daubechies-20 wavelets

Table 1. SNR measured for different denoising methods

Denoising method	SNR (dB), drum	SNR (dB) instr.
Wavelet shrinkage, shift-invariant Haar	12.50	13.54
Nonlin. diffusion, decimated Haar	12.57	14.13
Nonlin. diffusion, Daubechies-4	12.81	14.49
Nonlin. diffusion, Daubechies-20	13.01	15.05
Nonlin. diffusion, shift-invariant Haar	12.92	13.04
no denoising (noisy signal)	11.16	13.63

superior to that of Daubechies-20 shrinkage although the shift-invariant Haar wavelet shrinkage result shows a slight deformation around the peaks. However, the Daubechies-20 denoised sample displays a jitter of amplitude which is indeed an audible perturbation.

In the further experimental validation of our diffusion denoising model, we use two musical signals: First, a short drum loop; second, a clipping of instrumental music (brass accompanied by strings)[3]. We add Gaussian noise to each of the signals. The noise variance is about 10 % of the signal variance for the drum loop and about 5 % for the instrumental piece. Table 1 compiles results of signal-to-noise ratio (SNR) measurements for selected denoising methods. The SNR is computed as

$$\mathrm{SNR}(u,f) = 10\,\log_{10}\frac{\mathrm{var}(f)}{\mathrm{var}(f-u)}\,\mathrm{dB}$$

where f is the original and u the noisy signal.

According to subjective perception, our nonlinear diffusion method in general leads to better denoising results than soft wavelet shrinkage. The SNR, however, favours in some cases wavelet shrinkage (or even the noisy signal!) which indicates that it might not be an adequate criterion for denoising quality. A characteristic difference between shrinkage and diffusion denoising is shown in Fig. 5. Looking at the right part of the original signal clipping displayed, one notices small high-frequent oscillations which are visible particularly near the extrema of the low-frequent base oscillation. These components are indeed essential for the characteristic timbre of the drum. By removing these signal components along with the noise, wavelet shrinkage compromises the timbre much more than nonlinear diffusion which keeps at least part of these components.

In the remainder of this section, we discuss qualitatively a few more observations made during our experiments. As to the choice of the wavelet basis, auditory impression as well as, in part, the SNR measurements suggest that introducing shift-invariance into Haar wavelet shrinkage raises the quality to a level comparable that of (decimated) Daubechies wavelets. This is observed both for shrinkage and diffusion algorithms.

[3] Audio samples are available via the first author's web page, `http://www.mia.uni-saarland.de/welk`

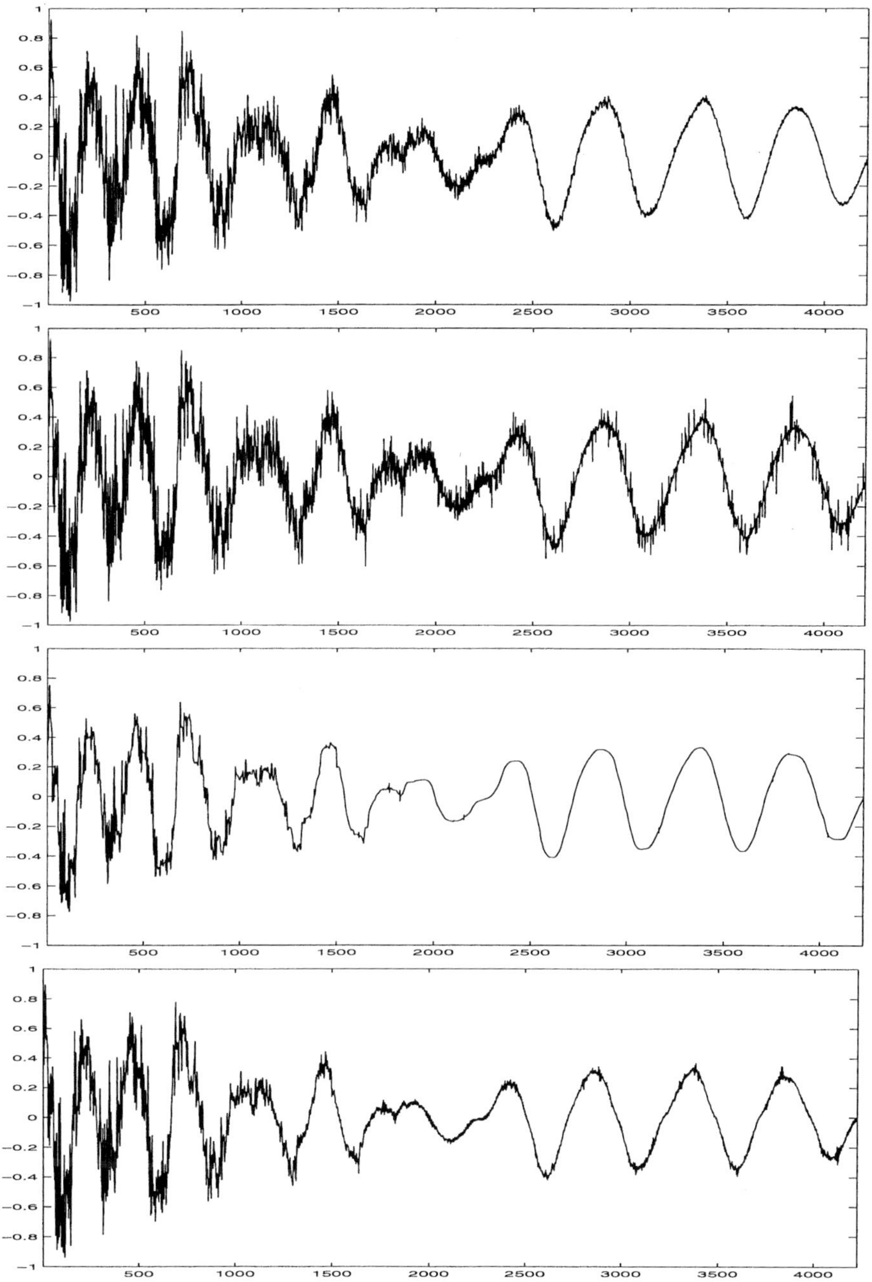

Fig. 5. Denoising of a drum loop signal. Parameters are chosen such that similar degrees of noise reduction are perceived. **Top to bottom:** A drum loop. – Same with Gaussian noise. – Denoised by shift-invariant soft Haar-wavelet shrinkage ($\theta = 5000$). – Denoised by channel-coupled nonlinear diffusion on shift-invariant Haar wavelet decomposition ($t = 100$, $\lambda = 500$, $\sigma = 1.0$)

The number of wavelet levels which are used in our wavelet diffusion process is less important than it appears. Most denoising is achieved just in the finest five wavelet levels; with ten levels, no significant enhancement is encountered. The reason is that the diffusion process is not very effective on coarse levels because of their low resolution and the very small updates which are therefore made per diffusion-time step. Low-frequent (rumbling) noise should therefore be handled by other measures.

Denoising with Haar wavelet methods often creates a characteristic audible artifact. It consists in a slightly rough-ringing noise of specific pitch which is composed of frequencies standing in octave relations to the sampling frequency. This noise gains intensity the longer diffusion acts or the more coefficients are shrunk by wavelet shrinkage. It is also observed that many signals tend to be flattened even by our nonlinear diffusion process for large t. We assume that these two phenomena are two sides of the same medal: By admitting only transfer between wavelet coefficients of equal frequency and phase, our diffusion process tries to keep signal amplitude in the separated frequency and phase components and to avoid extinction. However, this works perfectly only for frequencies which are subdivisions of the sampling frequency by powers of two (and which therefore in some sense are "in resonance" with the wavelet decomposition). Other frequencies are still weakened during the process, inducing the tendency to flatten signals. On the other hand, even some noise is kept in the resonance frequencies and can be perceived with its pitch as soon as the other frequencies are gone.

In agreement with this reasoning, improvements in the algorithm which lead to a better preservation of signal amplitude reduce the sound artifact at the same time. Starting from a simple linear diffusion process on decimated Haar wavelet coefficients, each of the following steps observably reduces both the diminishing of signal amplitude and the appearance of the artifact tone: first, switching to shift-invariant Haar wavelets; second, making diffusion in each channel nonlinear; and third, establishing of the channel coupling.

5 Summary and Outlook

In this paper, we have introduced a method for the denoising of audio signals by nonlinear diffusion. Because of the specifics of audio data compared to images, the diffusion process is not formulated for the digital audio samples but for the wavelet coefficients of a suitable wavelet representation.

Comparisons with wavelet shrinkage techniques reveal a fairly good performance of our method despite its simplicity. Due to the high sensitivity of human auditory perception for even tiny perturbations, the denoising achieved is not satisfactory enough for immediate application Nevertheless, our results clearly indicate that nonlinear diffusion can be successfully adapted to the denoising of audio data.

Ongoing work concentrates on the reduction of artifacts and improvement of the homogeneity of denoising over the frequency range. We are also interested in establishing better quantitative measures for denoising quality that reflect the perceived quality as accurate as possible.

References

1. E. J. Candés and F. Guo. New multiscale transforms, minimum total variation synthesis: Applications to edge-preserving image reconstruction. *Signal Processing*, 82(11):1519–1543, 2002.
2. F. Catté, P.-L. Lions, J.-M. Morel, and T. Coll. Image selective smoothing and edge detection by nonlinear diffusion. *SIAM Journal on Numerical Analysis*, 32:1895–1909, 1992.
3. R. R. Coifman and D. Donoho. Translation invariant denoising. In A. Antoine and G. Oppenheim, editors, *Wavelets in Statistics*, pages 125–150. Springer, New York, 1995.
4. R. R. Coifman and A. Sowa. Combining the calculus of variations and wavelets for image enhancement. *Applied and Computational Harmonic Analysis*, 9(1):1–18, July 2000.
5. I. Daubechies. *Ten Lectures on Wavelets.* SIAM, Philadelphia, 1992.
6. D. L. Donoho. De-noising by soft thresholding. *IEEE Transactions on Information Theory*, 41:613–627, 1995.
7. D. L. Donoho and I. M. Johnstone. Minimax estimation via wavelet shrinkage. *Annals of Statistics*, 26(3):879–921, 1998.
8. S. Durand and J. Froment. Reconstruction of wavelet coefficients using total-variation minimization. *SIAM Journal on Scientific Computing*, 24(5):1754–1767, 2003.
9. G. Gerig, O. Kübler, R. Kikinis, and F. A. Jolesz. Nonlinear anisotropic filtering of MRI data. *IEEE Transactions on Medical Imaging*, 11:221–232, 1992.
10. A. Haar. Zur Theorie der orthogonalen Funktionensysteme. *Mathematische Annalen*, 69:331–371, 1910.
11. A. K. Louis, P. Maass, A. Rieder. *Wavelets: Theory and Applications.* Wiley, New York, 1997.
12. F. Malgouyres. Minimizing the total variation under a general convex constraint for image restoration. *IEEE Transactions on Image Processing*, 11(12):1450–1456, 2002.
13. S. Mallat. *A Wavelet Tour of Signal Processing.* Academic Press, San Diego, second edition, 1999.
14. P. Perona and J. Malik. Scale space and edge detection using anisotropic diffusion. In *Proc. IEEE Computer Society Workshop on Computer Vision*, pages 16–22, Miami Beach, FL, Nov. 1987. IEEE Computer Society Press.
15. P. Perona and J. Malik. Scale space and edge detection using anisotropic diffusion. *IEEE Transactions on Pattern Analysis and Machine Intelligence*, 12:629–639, 1990.
16. G. Steidl, J. Weickert, T. Brox, P. Mrázek, and M. Welk. On the equivalence of soft wavelet shrinkage, total variation diffusion, total variation regularization, and SIDEs. *SIAM Journal on Numerical Analysis*, 42(2):686–713, 2004.
17. J. Weickert. A review of nonlinear diffusion filtering. In B. ter Haar Romeny, L. Florack, J. Koenderink, and M. Viergever, editors, *Scale-Space Theory in Computer Vision*, volume 1252 of *Lecture Notes in Computer Science*, pages 3–28. Springer, Berlin, 1997.
18. J. Weickert. *Anisotropic Diffusion in Image Processing.* Teubner, Stuttgart, 1998.

A Four-Pixel Scheme for Singular Differential Equations

Martin Welk[1], Joachim Weickert[1], and Gabriele Steidl[2]

[1] Mathematical Image Analysis Group,
Faculty of Mathematics and Computer Science, Bldg. 27,
Saarland University, 66041 Saarbrücken, Germany
{welk, weickert}@mia.uni-saarland.de
http://www.mia.uni-saarland.de

[2] Faculty of Mathematics and Computer Science, D7, 27,
University of Mannheim, 68131 Mannheim, Germany
steidl@math.uni-mannheim.de
http://kiwi.math.uni-mannheim.de

Abstract. Singular diffusion equations such as total variation (TV) and balanced forward–backward (BFB) diffusion are appealing: They have a finite extinction time, and experiments show that piecewise constant structures evolve. Unfortunately, their implementation is awkward. The goal of this paper is to introduce a novel class of numerical methods for these equations in the 2D case. They are simple to implement, absolutely stable and do not require any regularisation in order to make the diffusivity bounded. Our schemes are based on analytical solutions for 2×2-pixel images which are combined by means of an additive operator splitting (AOS). We show that they may also be regarded as iterated 2D Haar wavelet shrinkage. Experiments demonstrate the favourable performance of our numerical algorithm.

1 Introduction

Nonlinear diffusion filters [15, 21] constitute an important class of image enhancement methods. Let $\Omega \subset \mathbb{R}^2$ denote our two-dimensional image domain and $f : \Omega \to \mathbb{R}$ an initial greyscale image. Then the idea behind nonlinear diffusion filtering is to consider $f(x)$ as initial condition

$$u(x, 0) = f(x) \qquad \text{on } \Omega \tag{1}$$

of a nonlinear diffusion process

$$\partial_t u = \text{div } (g(|\nabla u|)\, \nabla u) \qquad \text{on } \Omega \times (0, \infty) \tag{2}$$

with suitable boundary conditions, e.g. the reflecting (homogeneous Neumann) boundary conditions

$$\partial_n u = 0 \qquad \text{on } \partial\Omega \times (0, \infty). \tag{3}$$

R. Kimmel, N. Sochen, J. Weickert (Eds.): Scale-Space 2005, LNCS 3459, pp. 610–621, 2005.

Here $\nabla = (\partial_{x_1}, \partial_{x_2})^\top$ denotes the spatial nabla operator and n is the outer normal vector on the image boundary $\partial\Omega$. The resulting solution $u(x,t)$ creates a scale-space family $\{u(x,t) \mid t \geq 0\}$ of processed images, where the diffusion time t serves as scale parameter: Larger values of t give more simplified images $u(x,t)$. In order to preserve (or even enhance) edges and to simultaneously smooth within more homogeneous regions, the diffusivity function $g(|\nabla u|)$ is chosen as a decreasing nonnegative function.

While early proposals for nonlinear diffusion filters use bounded diffusivities [15, 6], more recently there has been a growing interest in unbounded diffusivities that become singular in zero [2, 9, 10, 11, 16]. Experimentally one observes that singular diffusion filters lead to piecewise constant images. This is also in accordance with theoretical results by Nikolova [14] who showed that related discrete variational approaches allow piecewise constant solutions if and only if the regulariser is nondifferentiable in zero.

As a prototype for a class of singular diffusivities we consider the family

$$g(|\nabla u|) = \frac{1}{|\nabla u|^p} \qquad (p \geq 0). \tag{4}$$

These diffusivities offer the advantage that they do not require to tune any image specific contrast parameters. Moreover, they lead to scale invariant filters [1], for which even some analytical results have been established [20].

For $p = 1$ one obtains the *total variation (TV)* diffusion [2, 9], the diffusion filter that corresponds to TV minimisation [18] with a penaliser $\Psi(|\nabla u|^2) = 2\,|\nabla u|$. TV diffusion offers a number of interesting properties such as finite extinction time [3], shape-preserving qualities [4], and equivalence to TV regularisation in 1-D [5, 17]. For $p > 1$ the diffusion not only preserves edges but even enhances them. A diffusivity with $p = 2$ has been considered in [11] for the so-called *balanced forward–backward (BFB)* diffusion filtering.[1]

Although singular diffusion equations have very attractive properties, their numerical implementation is difficult. Explicit finite difference schemes are only stable for time step sizes that are inversely proportional to an upper bound for the diffusivity, while absolutely stable implicit or semi-implicit schemes lead to linear systems of equations with condition numbers that are increasing functions of this bound. As a result, iterative numerical schemes may reveal slow convergence, and in general numerical errors can be amplified. In order to limit all these problems, it is common to regularise the diffusivity function by replacing it by the bounded diffusivity

$$g(|\nabla u|) = \frac{1}{\left(|\nabla u|^2 + \varepsilon^2\right)^{p/2}}\,. \tag{5}$$

In this case, however, one observes that blurring artifacts are introduced and some of the nice theoretical properties of singular nonlinear diffusion filters do no longer hold.

[1] While a complete well-posedness theory exists for $p \leq 1$, some theoretical questions are a topic of ongoing research for the edge-enhancing case $p > 1$.

The goal of the present paper is to address these problems by introducing a novel class of numerical schemes for singular diffusion equations. They are based on an analysis of the dynamical system that results from a space discretisation of singular diffusion filters for images with 2×2 pixels. For this scenario we are able to derive an analytical solution. It serves as a building block in a numerical scheme for general 2-D images, since we can assemble these local analytical solutions by means of an additive operator splitting (AOS) [12, 22] to a global numerical approximation. Our scheme is very simple, it is absolutely stable and reveals good rotation invariance. It does not require regularised diffusivities of type (5). Interestingly, it can also be related to a recently introduced family of shift invariant wavelet shrinkage methods with coupled shrinkage functions [13].

Our paper is organised as follows. In Section 2 we analyse space-discrete singular diffusion filters for 2×2 images, derive their analytical solutions, and relate them to Haar wavelet shrinkage of 2×2 images. These analytical solutions are used in Section 3 for constructing numerical schemes for 2-D images of arbitrary size. We analyse their stability and consistency properties, and show their equivalence to suitable shift invariant wavelet shrinkages. Numerical experiments are presented in Section 4, and the paper is concluded with a summary in Section 5.

Related Work. Relations between one-dimensional discrete TV diffusion, TV regularisation and Haar wavelet shrinkage were investigated in [19]. A main instrument in studying one-dimensional total variation methods were considerations of two-pixel signals. Based on the two-pixel dynamics, a novel scheme for N-pixel TV diffusion could be established. We may regard our present work as a two-dimensional extension. The two-dimensional situation, however, turns out to be significantly more complicated than the one-dimensional scenario. With respect to TV-diminishing flows along the directions of Haar wavelets, our work can also be related to a paper by Coifman and Sowa [8]. A regularisation-free approach to TV regularisation has been proposed by Chambolle [7]. It should be noted that in our paper we consider the parabolic diffusion case instead of the elliptic regularisation setting. Moreover, we do not restrict ourselves to the TV case: Our results hold for any arbitrary singular diffusivity of type (4).

2 Analytical Results for 2×2-Pixel Images

We start by examining the simplest nontrivial 2-D images, namely those having only 2×2 pixels. This will provide the basis of our new numerical scheme for solving singular diffusion equations on $N \times M$-pixel images.

2.1 Nonlinear Diffusion

We consider the diffusion equation (2) for 2×2-pixel images $u = (u_{i,j})_{i,j=1}^{2}$ with periodic boundary conditions and initial image $f = (f_{i,j})_{i,j=1}^{2}$. Due to the periodic boundary conditions every 2×2 cell in the extended image

$$
\begin{array}{c|cc|c}
u_{2,2} & u_{2,1} & u_{2,2} & u_{2,1} \\
\hline
u_{1,2} & u_{1,1} & u_{1,2} & u_{1,1} \\
u_{2,2} & u_{2,1} & u_{2,2} & u_{2,1} \\
\hline
u_{1,2} & u_{1,1} & u_{1,2} & u_{1,1}
\end{array}
$$

contains exactly the same pixels. This is not true for other boundary conditions, e.g., reflecting boundary conditions.

We want to find an appropriate space discretisation of (2) that results in an ordinary system of four differential equations which can be solved analytically. First we notice that in our 2×2 cell there is one distinguished location where the diffusivity can be optimally approximated, namely the midpoint of the cell. Therefore we use only the midpoint diffusivity $g := g(D(u))$ within the whole image, where

$$
\begin{aligned}
D(u) := \frac{1}{2}\Big(&(u_{1,1} - u_{1,2})^2 + (u_{2,1} - u_{2,2})^2 + (u_{1,1} - u_{2,1})^2 + (u_{1,2} - u_{2,2})^2 \\
&+ (u_{1,1} - u_{2,2})^2 + (u_{1,2} - u_{2,1})^2 \Big)^{1/2}
\end{aligned}
\tag{6}
$$

denotes the discretisation of $|\nabla u|$ in the midpoint of the cell, if the grid size $h := 1$ is chosen in both directions. This disretisation is just the average of the two finite difference discretisations of $|\nabla u|$ with respect to the usual directions $x = (1,0)^{\mathrm{T}}$, $y = (0,1)^{\mathrm{T}}$ and with respect to the 45° diagonal directions $\xi = \frac{1}{\sqrt{2}}(1,1)^{\mathrm{T}}$, $\eta = \frac{1}{\sqrt{2}}(1,-1)^{\mathrm{T}}$, respectively. Next, we discretise the remaining gradient and divergence of the right-hand side of $\partial_t u = \operatorname{div}(g \nabla u)$ with respect to the same directions.

With the uniform midpoint diffusivity $g := g(D(u))$, the discretisation related to x and y leads for $i, j = 1, 2$ to

$$
\dot{u}_{i,j} = g \cdot (u_{i+1,j} + u_{i-1,j} + u_{i,j+1} + u_{i,j-1} - 4u_{i,j}),
$$

where the dot denotes time differentiation, and by our boundary conditions to

$$
\begin{aligned}
\dot{u}_{1,1} &= 2g \cdot (u_{1,2} + u_{2,1} - 2u_{1,1}), & \dot{u}_{1,2} &= 2g \cdot (u_{1,1} + u_{2,2} - 2u_{1,2}), \\
\dot{u}_{2,1} &= 2g \cdot (u_{1,1} + u_{2,2} - 2u_{2,1}), & \dot{u}_{2,2} &= 2g \cdot (u_{1,2} + u_{2,1} - 2u_{2,2}).
\end{aligned}
$$

The discretisation with respect to ξ and η results for $i, j = 1, 2$ in

$$
\dot{u}_{i,j} = g \cdot \tfrac{1}{2} (u_{i+1,j+1} + u_{i-1,j-1} + u_{i+1,j-1} + u_{i-1,j+1} - 4u_{i,j})
$$

and by applying the boundary conditions in

$$
\begin{aligned}
\dot{u}_{1,1} &= 2g \cdot (u_{2,2} - u_{1,1}), & \dot{u}_{1,2} &= 2g \cdot (u_{2,1} - u_{1,2}), \\
\dot{u}_{2,1} &= 2g \cdot (u_{1,2} - u_{2,1}), & \dot{u}_{2,2} &= 2g \cdot (u_{1,1} - u_{2,2}).
\end{aligned}
$$

Weighted averaging of both discretisations with weights $\alpha \in [0,1]$ and $1-\alpha$ gives the ordinary system of differential equations

$$\begin{aligned}
\dot{u}_{1,1} &= 2g \cdot (-(1+\alpha)u_{1,1} + \alpha u_{1,2} + \alpha u_{2,1} + (1-\alpha)u_{2,2}) \,, \\
\dot{u}_{1,2} &= 2g \cdot (\alpha u_{1,1} - (1+\alpha)u_{1,2} + (1-\alpha)u_{2,1} + \alpha u_{2,2}) \,, \\
\dot{u}_{2,1} &= 2g \cdot (\alpha u_{1,1} + (1-\alpha)u_{1,2} - (1+\alpha)u_{2,1} + \alpha u_{2,2}) \,, \\
\dot{u}_{2,2} &= 2g \cdot ((1-\alpha)u_{1,1} + \alpha u_{1,2} + \alpha u_{2,1} - (1+\alpha)u_{2,2})
\end{aligned} \tag{7}$$

with initial conditions $u_{i,j}(0) = f_{i,j}$, $i,j = 1,2$. From $\dot{u}_{1,1}+\dot{u}_{1,2}+\dot{u}_{2,1}+\dot{u}_{2,2} = 0$ we see that the average grey value $\mu := \frac{1}{4}(f_{1,1} + f_{1,2} + f_{2,1} + f_{2,2})$ is preserved during the diffusion process.

In this paper, we are mainly interested in the case $\alpha = 1/2$, where system (7) further simplifies to

$$\dot{u}_{i,j} = 4g \cdot (\mu - u_{i,j}), \quad i,j = 1,2, \tag{8}$$

which is a dynamical system with discontinuous right hand side. It is not difficult to verify that this system possesses the unique analytical solution

$$u_{i,j}(t) = \begin{cases} \mu + \left(1 - 4p\,(D(f))^{-p}\,t\right)^{1/p}(f_{i,j} - \mu), & 0 \le t < (D(f))^p/(4p), \\ \mu & t \ge (D(f))^p/(4p). \end{cases} \tag{9}$$

For $p = 1$, particularly, (2) is the TV diffusion equation $\partial_t u = \operatorname{div}(\nabla u/|\nabla u|)$. The analytical solution of our 2×2-pixel version

$$u_{i,j}(t) = \begin{cases} \mu + (1 - 4t/D(f))\,(f_{i,j} - \mu), & 0 \le t < D(f)/4, \\ \mu, & t \ge D(f)/4 \end{cases} \tag{10}$$

shows a linear evolution which can be written in a slightly different form as

$$u_{i,j}(t) = f_{i,j} + (4t/D(f)) \cdot (\mu - f_{i,j}) \min\{1, D(f)/(4t)\}, \qquad i,j = 1,2. \tag{11}$$

For $p = 2$, we obtain the BFB diffusion $\partial_t u = \operatorname{div}\left(\nabla u/|\nabla u|^2\right)$. The analytical solution in the 2×2 setting reads

$$u_{i,j}(t) = \begin{cases} \mu + \sqrt{1 - 8t/(D(f))^2}\,(f_{i,j} - \mu), & 0 \le t < (D(f))^2/8, \\ \mu, & t \ge (D(f))^2/8. \end{cases} \tag{12}$$

2.2 Haar Wavelet Shrinkage

In [19], it was shown that one-dimensional nonlinear diffusion on two-pixel signals coincides with Haar wavelet shrinkage if the shrinkage function is chosen in accordance with the diffusivity and the threshold parameter is equal to the diffusion time. The two-dimensional Haar wavelet transform acts naturally on subsequent 2×2-pixel tiles of an image. Let us choose one such tile, say $f := (f_{i,j})_{i,j=1}^{2}$, and explain how it changes under two-dimensional Haar wavelet shrinkage. One cycle of Haar wavelet shrinkage consists of three steps. In the first step, the *analysis step*, the low and high pass Haar filters are applied to the rows and columns of f. More precisely, f is multiplied from the left and the right by the orthogonal matrix $W := \frac{1}{\sqrt{2}}\begin{pmatrix} 1 & 1 \\ 1 & -1 \end{pmatrix}$. This results in an image $c := (c_{i,j})_{i,j=1}^{2}$ with

$$c_{1,1} = \tfrac{1}{2}(f_{1,1} + f_{1,2} + f_{2,1} + f_{2,2}), \qquad c_{1,2} = \tfrac{1}{2}(f_{1,1} - f_{1,2} + f_{2,1} - f_{2,2}),$$
$$c_{2,1} = \tfrac{1}{2}(f_{1,1} + f_{1,2} - f_{2,1} - f_{2,2}), \qquad c_{2,2} = \tfrac{1}{2}(f_{1,1} - f_{1,2} - f_{2,1} + f_{2,2}).$$

In the second step, the *shrinkage step*, we modify the high-pass coefficients by reducing the absolute values of some or all of them. To this end, we apply a shrinkage function S_θ depending on a threshold parameter θ to the high-pass filtered coefficients, i.e. we compute $S_\theta(c_{1,2})$, $S_\theta(c_{2,1})$, $S_\theta(c_{2,2})$ and leave the low-pass coefficient $c_{1,1}$ as it is. In the third step, the *synthesis step*, we perform just the inverse transform of step 1 on the shrunken image, i.e., since $W^{-1} = W$, we multiply again from the left and the right by W and obtain

$$\begin{aligned}
v_{1,1} &= \mu + \tfrac{1}{2}\left(S_\theta(c_{1,2}) + S_\theta(c_{2,1}) + S_\theta(c_{2,2})\right),\\
v_{1,2} &= \mu + \tfrac{1}{2}\left(-S_\theta(c_{1,2}) + S_\theta(c_{2,1}) - S_\theta(c_{2,2})\right),\\
v_{2,1} &= \mu + \tfrac{1}{2}\left(S_\theta(c_{1,2}) - S_\theta(c_{2,1}) - S_\theta(c_{2,2})\right),\\
v_{2,2} &= \mu + \tfrac{1}{2}\left(-S_\theta(c_{1,2}) - S_\theta(c_{2,1}) + S_\theta(c_{2,2})\right).
\end{aligned}$$

In [13], the authors proposed to choose a diffusion inspired shrinkage function that simultaneously depends on $c_{1,2}$, $c_{2,1}$ and $c_{2,2}$. In contrast to the classical wavelet shrinkage, this results in an improved rotation invariance of the resulting image. We use this knowledge and define our shrinkage function in dependence on

$$D(f) = \left(c_{1,2}^2 + c_{2,1}^2 + c_{2,2}^2\right)^{1/2}.$$

It is straightforward to check that the value $D(f)$ indeed coincides with the one defined in (6). Applying the shrinkage function

$$S_\theta(s; D(f)) := \begin{cases} \left(1 - 4p\,(D(f))^{-p}\,\theta\right)^{1/p} s, & D(f) \geq (4p\,\theta)^{1/p}, \\ 0 & D(f) < (4p\,\theta)^{1/p}. \end{cases} \tag{13}$$

our Haar wavelet shrinkage produces for $i, j = 1, 2$ the values

$$v_{i,j} = \begin{cases} \mu + \left(1 - 4p\,(D(f))^{-p}\,\theta\right)^{1/p}(f_{i,j} - \mu), & D(f) \geq (4p\,\theta)^{1/p}, \\ \mu, & D(f) < (4p\,\theta)^{1/p}. \end{cases} \tag{14}$$

Comparing this equation with (9) we observe that on 2×2 pixels our Haar wavelet shrinkage with shrinkage function (13) coincides with the solution of the nonlinear diffusion equation with diffusivity (4), where the shrinkage parameter θ plays the same role as the diffusion time t.

3 A Numerical Scheme for Images of Arbitrary Size

Now we consider arbitrary $N \times M$-pixel images $u := (u_{i,j})_{i,j=1}^{N,M}$ which are extended to the full planar grid by, e.g., reflecting boundary conditions.

3.1 Numerical Scheme for Nonlinear Diffusion

We discretise the diffusion equation (2) in space again with respect to the x–y and ξ–η directions. Approximating gradient and divergence by finite differences in x and y directions leads to

$$\begin{aligned}\operatorname{div}(g(|\nabla u|)\nabla u)_{i,j} \approx{}& g_{i+\frac{1}{2},j}\cdot(u_{i+1,j}-u_{i,j}) - g_{i-\frac{1}{2},j}\cdot(u_{i,j}-u_{i-1,j})\\ &+ g_{i,j+\frac{1}{2}}\cdot(u_{i,j+1}-u_{i,j}) - g_{i,j-\frac{1}{2}}\cdot(u_{i,j}-u_{i,j-1})\,.\end{aligned}$$

Again we only want to work with the diffusivities at the midpoints $(i+\frac{1}{2},j+\frac{1}{2})$ of the grid cells. Therefore we approximate $g_{i\pm\frac{1}{2},j}$ and $g_{i,j\pm\frac{1}{2}}$ by averaging the values of the neighbouring cell centers, e.g., $g_{i+\frac{1}{2},j} \approx \frac{1}{2}(g_{i+\frac{1}{2},j+\frac{1}{2}} + g_{i+\frac{1}{2},j-\frac{1}{2}})$. We arrive at

$$\begin{aligned}\operatorname{div}(g(|\nabla u|)\nabla u)_{i,j} \approx \tfrac{1}{2}\Big(& g_{i+\frac{1}{2},j+\frac{1}{2}}\cdot(u_{i+1,j}+u_{i,j+1}-2u_{i,j})\\ &+ g_{i+\frac{1}{2},j-\frac{1}{2}}\cdot(u_{i+1,j}+u_{i,j-1}-2u_{i,j})\\ &+ g_{i-\frac{1}{2},j+\frac{1}{2}}\cdot(u_{i-1,j}+u_{i,j+1}-2u_{i,j})\\ &+ g_{i-\frac{1}{2},j-\frac{1}{2}}\cdot(u_{i-1,j}+u_{i,j-1}-2u_{i,j})\Big)\,.\end{aligned}\tag{15}$$

On the other hand, approximation of both gradient and divergence with respect to diagonal directions ξ, η leads to

$$\begin{aligned}&\operatorname{div}(g(|\nabla u|)\nabla u)_{i,j}\\ &\approx \tfrac{1}{2}\Big(g_{i+\frac{1}{2},j+\frac{1}{2}}\cdot(u_{i+1,j+1}-u_{i,j}) + g_{i+\frac{1}{2},j-\frac{1}{2}}\cdot(u_{i+1,j-1}-u_{i,j})\\ &\qquad + g_{i-\frac{1}{2},j+\frac{1}{2}}\cdot(u_{i-1,j+1}-u_{i,j}) + g_{i-\frac{1}{2},j-\frac{1}{2}}\cdot(u_{i-1,j-1}-u_{i,j})\Big)\,.\end{aligned}\tag{16}$$

Weighted averaging of both approximations results in

$$\begin{aligned}&\operatorname{div}(g(|\nabla u|)\nabla u)_{i,j}\\ &\approx \tfrac{1}{2}\Big(g_{i+\frac{1}{2},j+\frac{1}{2}}\cdot(\alpha u_{i+1,j}+\alpha u_{i,j+1}+(1-\alpha)u_{i+1,j+1}-(1+\alpha)u_{i,j})\\ &\quad + g_{i+\frac{1}{2},j-\frac{1}{2}}\cdot(\alpha u_{i+1,j}+\alpha u_{i,j-1}+(1-\alpha)u_{i+1,j-1}-(1+\alpha)u_{i,j})\\ &\quad + g_{i-\frac{1}{2},j+\frac{1}{2}}\cdot(\alpha u_{i-1,j}+\alpha u_{i,j+1}+(1-\alpha)u_{i-1,j+1}-(1+\alpha)u_{i,j})\\ &\quad + g_{i-\frac{1}{2},j-\frac{1}{2}}\cdot(\alpha u_{i-1,j}+\alpha u_{i,j-1}+(1-\alpha)u_{i-1,j-1}-(1+\alpha)u_{i,j})\Big)\,.\end{aligned}\tag{17}$$

In the following, we consider diffusivities g defined by (4) and $\alpha = 1/2$. We set

$$\begin{aligned}\mu^k_{i,j,++} &:= \tfrac{1}{4}(u^k_{i+1,j}+u^k_{i,j+1}+u^k_{i+1,j+1}+u^k_{i,j})\,,\\ \mu^k_{i,j,+-} &:= \tfrac{1}{4}(u^k_{i+1,j}+u^k_{i,j-1}+u^k_{i+1,j-1}+u^k_{i,j})\,,\\ \mu^k_{i,j,-+} &:= \tfrac{1}{4}(u^k_{i-1,j}+u^k_{i,j+1}+u^k_{i-1,j+1}+u^k_{i,j})\,,\\ \mu^k_{i,j,--} &:= \tfrac{1}{4}(u^k_{i-1,j}+u^k_{i,j-1}+u^k_{i-1,j-1}+u^k_{i,j})\,.\end{aligned}$$

Time discretisation via an explicit Euler scheme would yield as fully discretisation of (2) the naive scheme

$$\begin{aligned} u_{i,j}^{k+1} = u_{i,j}^{k} &+ \tau g_{i+\frac{1}{2},j+\frac{1}{2}} \cdot (\mu_{i,j,++}^{k} - u_{i,j}^{k}) + \tau g_{i+\frac{1}{2},j-\frac{1}{2}} \cdot (\mu_{i,j,+-}^{k} - u_{i,j}^{k}) \\ &+ \tau g_{i-\frac{1}{2},j+\frac{1}{2}} \cdot (\mu_{i,j,-+}^{k} - u_{i,j}^{k}) + \tau g_{i-\frac{1}{2},j-\frac{1}{2}} \cdot (\mu_{i,j,--}^{k} - u_{i,j}^{k}). \end{aligned} \tag{18}$$

Here τ denotes the time step size and $u^k = (u_{i,j}^k)_{i,j}$ the approximate solution at pixel (i,j) and time $k\tau$. Unfortunately, due to the singularity of g at zero, this scheme becomes instable with respect to the maximum-minimum principle for arbitrary small time steps if neighbouring pixel values become arbitrary close. We use therefore a different approximation.

The right-hand side of (17) is exactly the average of the four approximations of $\operatorname{div}(g\nabla u)$ in the 2×2-pixel cells that pixel (i,j) belongs to. This inspires the following simple algorithm to compute one time step of a numerical scheme:

For each pixel $(*)$ with coordinates (i,j):

- **Consider the four cells**

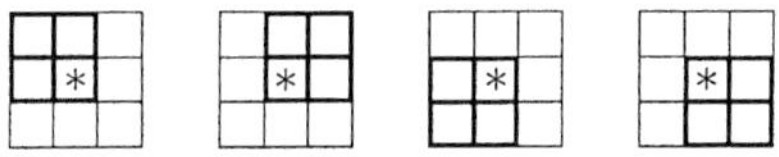

(−−) (+−) (−+) (++)

- **Compute the analytical solutions according to (9). This gives four approximations** (19)

$$u_{i,j,--}^{k+1}, \quad u_{i,j,+-}^{k+1}, \quad u_{i,j,-+}^{k+1}, \quad u_{i,j,++}^{k+1} .$$

- **Average:**

$$u_{i,j}^{k+1} = \tfrac{1}{4}\left(u_{i,j,--}^{k+1} + u_{i,j,+-}^{k+1} + u_{i,j,-+}^{k+1} + u_{i,j,++}^{k+1}\right) .$$

It is worth noting that this averaging scheme is similar to an additive operator splitting (AOS) scheme [12, 22]. One way to look at a usual AOS scheme is that it splits e.g. a two-dimensional dynamical system into two one-dimensional systems, modeling interactions in x and y directions, for which efficient numerical schemes exist. These numerical schemes are then averaged to approximate the 2D solution. Here, we split the dynamical system with right-hand side (17) into four dynamical systems belonging to four-pixel cells each of which can even be solved exactly. Again, an approximation for the solution of the full system is obtained by averaging.

Stability Analysis. The values of the analytical solution (9) at arbitrary times $t \geq 0$ are convex combinations of its initial values. By its construction from the analytical solution (9) the novel scheme (19) therefore satisfies the maximum–minimum principle. Consequently, it is absolutely stable for each τ.

Consistency Analysis. To analyse consistency, let us for simplicity focus on the TV flow, i.e., $p = 1$. Then, by (11) our final scheme reads

$$\begin{aligned} u_{i,j}^{k+1} = u_{i,j}^{k} &+ \tau g_{i+\frac{1}{2},j+\frac{1}{2}} \cdot (\mu_{i,j,++}^{k} - u_{i,j}^{k}) \min\{1, 1/(4\tau g_{i+\frac{1}{2},j+\frac{1}{2}})\} \\ &+ \tau g_{i+\frac{1}{2},j-\frac{1}{2}} \cdot (\mu_{i,j,+-}^{k} - u_{i,j}^{k}) \min\{1, 1/(4\tau g_{i+\frac{1}{2},j-\frac{1}{2}})\} \\ &+ \tau g_{i-\frac{1}{2},j+\frac{1}{2}} \cdot (\mu_{i,j,-+}^{k} - u_{i,j}^{k}) \min\{1, 1/(4\tau g_{i-\frac{1}{2},j+\frac{1}{2}})\} \\ &+ \tau g_{i-\frac{1}{2},j-\frac{1}{2}} \cdot (\mu_{i,j,--}^{k} - u_{i,j}^{k}) \min\{1, 1/(4\tau g_{i-\frac{1}{2},j-\frac{1}{2}})\} . \end{aligned} \tag{20}$$

This scheme can be considered as a stabilisation of the explicit scheme (18). It coincides with (18), and is therefore a consistent approximation for TV diffusion, if each of the four minimum operations on its right-hand side selects the value 1. This consistency condition is fulfilled for

$$0 \le \tau \le \min\{1/(4g_{i+\frac{1}{2},j+\frac{1}{2}}), 1/(4g_{i+\frac{1}{2},j-\frac{1}{2}}), 1/(4g_{i-\frac{1}{2},j+\frac{1}{2}}), 1/(4g_{i-\frac{1}{2},j-\frac{1}{2}})\} .$$

For larger τ it is easy to see that linear diffusion $\partial_t u = \Delta u$ is approximated. This happens in regions where the gradient is already close to zero. In this case, however, the visual differences between linear diffusion and TV diffusion are small.

3.2 Equivalence to Shift and Rotation Invariant Wavelet Shrinkage

Ordinary single scale Haar wavelet shrinkage divides the image into disjoint 2×2-pixel cells and performs Haar wavelet shrinkage on each of these cells as prescribed in Subsection 2.2. Unfortunately, this process is neither shift invariant nor rotation invariant. However, both properties can be achieved with a little more effort by the following procedure:

1. Shift the original image $f_{++} := (f_{i,j})$ one pixel to the right to obtain $f_{-+} := (f_{i-1,j})$, one pixel down to get $f_{+-} := (f_{i,j-1})$ and one pixel to the right and down resulting in $f_{--} := (f_{i-1,j-1})$,
2. Perform wavelet shrinkage (14) on the 2×2 cells of the four images $f_{++}, f_{-+}, f_{+-}, f_{--}$, i.e., four times ordinary Haar wavelet shrinkage.
3. Shift the resulting images back and compute the average.

Obviously, this procedure describes exactly one time step of size $\tau = \theta$ of our novel diffusion scheme (20).

4 Experiments

In Figure 1, we contrast the regularisation-free scheme (20) based on the analytical 2×2-pixel solution for TV diffusion with a standard explicit discretisation. In this scheme, TV diffusivity is approximated by the regularised TV diffusivity $1/\sqrt{|\nabla u|^2 + \varepsilon^2}$. Since the stability condition for explicit schemes imposes to the time step size a bound which is inversely proportional to the upper bound of the

Fig. 1. **Left:** Original image, 93×93 pixels. **Middle:** TV diffusion with standard explicit scheme, where TV diffusivity is regularised with $\varepsilon = 0.01$, $\tau = 0.0025$, 10000 iterations. **Right:** TV diffusion with 2×2-pixel scheme (20) without regularisation of diffusivity, $\tau = 0.1$, 250 iterations

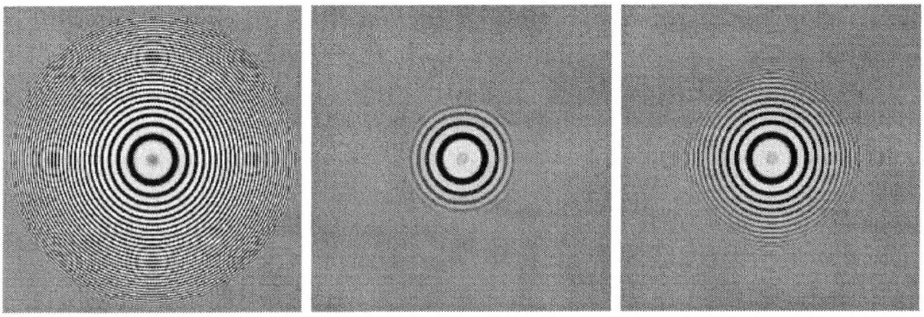

Fig. 2. **Left:** Rotationally symmetric original image, 256×256 pixels. **Middle:** TV diffusion with standard explicit scheme, $\varepsilon = 0.01$, $\tau = 0.0025$, 12000 iterations. **Right:** TV diffusion with 2×2-pixel scheme (20), $\tau = 0.1$, 300 iterations

Fig. 3. **Left:** Original image, 93×93 pixels. **Middle:** Balanced forward–backward diffusion with standard explicit scheme, $\varepsilon = 0.1$, $\tau = 0.0025$, 160000 iterations. **Right:** 2×2-pixel scheme (20), $\tau = 0.1$, 4000 iterations

diffusivity, a high number of iterations is needed for reasonable ε. It can be seen that the 2×2-pixel scheme and the unregularised TV diffusivity which cannot be used in the explicit scheme considerably reduce blurring effects caused by the discretisation.

Figure 2 demonstrates that although the analytically solvable case of the 2×2-pixel cell is not the one with optimal rotational invariance, the rotational invariance is reasonable anyway.

Figure 3 demonstrates balanced forward–backward diffusion. With equal parameters, it can be seen again that the 2×2-pixel scheme looks sharper by preserving finer details. Moreover, it is worth mentioning that we took a time step size that exceeded the largest admissible step size of the explicit scheme by a factor of 40.

5 Conclusion

We have introduced novel numerical schemes for a favourable class of singular nonlinear diffusion equations that includes TV and BFB diffusion. These schemes can be distinguished from other schemes by the fact that they do not require to regularise the diffusivities. They are based on analytical solutions for 4-pixel images. Combining these solutions in an AOS-like manner creates extremely simple algorithms that are absolutely stable in the maximum norm, conditionally consistent and reveal good rotation invariance. Our experiments have shown that they gives sharper results at edges than traditional schemes with regularised diffusivities, even for significantly larger time steps. This more pronounced tendency to create piecewise constant images is particularly suited for singular nonlinear PDEs.

It is our hope that this work will inspire more research on numerical schemes for PDE-based image analysis, in which analytical and numerical concepts are merged.

References

1. L. Alvarez, F. Guichard, P.-L. Lions, and J.-M. Morel. Axioms and fundamental equations in image processing. *Archive for Rational Mechanics and Analysis*, 123:199–257, 1993.
2. F. Andreu, C. Ballester, V. Caselles, and J. M. Mazón. Minimizing total variation flow. *Differential and Integral Equations*, 14(3):321–360, Mar. 2001.
3. F. Andreu, V. Caselles, J. I. Diaz, and J. M. Mazón. Qualitative properties of the total variation flow. *Journal of Functional Analysis*, 188(2):516–547, Feb. 2002.
4. G. Bellettini, V. Caselles, and M. Novaga. The total variation flow in R^N. *Journal of Differential Equations*, 184(2):475–525, 2002.
5. T. Brox, M. Welk, G. Steidl, and J. Weickert. Equivalence results for TV diffusion and TV regularisation. In L. D. Griffin and M. Lillholm, editors, *Scale-Space Methods in Computer Vision*, volume 2695 of *Lecture Notes in Computer Science*, pages 86–100, Berlin, 2003. Springer.
6. F. Catté, P.-L. Lions, J.-M. Morel, and T. Coll. Image selective smoothing and edge detection by nonlinear diffusion. *SIAM Journal on Numerical Analysis*, 32:1895–1909, 1992.
7. A. Chambolle. An algorithm for total variation minimization and applications. *Journal of Mathematical Imaging and Vision*, 20:89–97, 2004.

8. R. R. Coifman and A. Sowa. New methods of controlled total variation reduction for digital functions. *SIAM Journal on Numerical Analysis*, 39(2):480–498, 2001.
9. F. Dibos and G. Koepfler. Global total variation minimization. *SIAM Journal on Numerical Analysis*, 37(2):646–664, 2000.
10. X. Feng and A. Prohl. Analysis of total variation flow and its finite element approximations. Technical Report 1864, Institute of Mathematics and its Applications, University of Minnesota, Minneapolis, MN, July 2002. Submitted to Communications on Pure and Applied Mathematics.
11. S. L. Keeling and R. Stollberger. Nonlinear anisotropic diffusion filters for wide range edge sharpening. *Inverse Problems*, 18:175–190, Jan. 2002.
12. T. Lu, P. Neittaanmäki, and X.-C. Tai. A parallel splitting up method and its application to Navier–Stokes equations. *Applied Mathematics Letters*, 4(2):25–29, 1991.
13. P. Mrázek and J. Weickert. Rotationally invariant wavelet shrinkage. In B. Michaelis and G. Krell, editors, *Pattern Recognition*, volume 2781 of *Lecture Notes in Computer Science*, pages 156–163, Berlin, 2003. Springer.
14. M. Nikolova. Local strong homogeneity of a regularized estimator. *SIAM Journal on Applied Mathematics*, 61(2):633–658, 2000.
15. P. Perona and J. Malik. Scale space and edge detection using anisotropic diffusion. *IEEE Transactions on Pattern Analysis and Machine Intelligence*, 12:629–639, 1990.
16. A. Petrovic, O. Divorra Escoda, and P. Vandergheynst. Multiresolution segmentation of natural images: From linear to non-linear scale-space repesentations. *IEEE Transactions on Image Processing*, 13(8):1104–1114, 2004.
17. I. Pollak, A. S. Willsky, and Y. Huang. Nonlinear evolution equations as fast and exact solvers of estimation problems. *IEEE Transactions on Signal Processing*, 2004. To appear.
18. L. I. Rudin, S. Osher, and E. Fatemi. Nonlinear total variation based noise removal algorithms. *Physica D*, 60:259–268, 1992.
19. G. Steidl, J. Weickert, T. Brox, P. Mrázek, and M. Welk. On the equivalence of soft wavelet shrinkage, total variation diffusion, total variation regularization, and SIDEs. *SIAM Journal on Numerical Analysis*, 42(2):686–713, 2004.
20. V. I. Tsurkov. An analytical model of edge protection under noise suppression by anisotropic diffusion. *Journal of Computer and Systems Sciences International*, 39(3):437–440, 2000.
21. J. Weickert. *Anisotropic Diffusion in Image Processing*. Teubner, Stuttgart, 1998.
22. J. Weickert, B. M. ter Haar Romeny, and M. A. Viergever. Efficient and reliable schemes for nonlinear diffusion filtering. *IEEE Transactions on Image Processing*, 7(3):398–410, Mar. 1998.

Isometric Embedding of Facial Surfaces into $\mathbb{S}^3$

Alexander M. Bronstein[1], Michael M. Bronstein[2], and Ron Kimmel[2]

[1] Department of Electrical Engineering,
alexbron@ieee.org
[2] Department of Computer Science,
bronstein@ieee.org, ron@cs.technion.ac.il
Technion - Israel Institute of Technology,
32000 Haifa, Israel

Abstract. The problem of isometry-invariant representation and comparison of surfaces is of cardinal importance in pattern recognition applications dealing with deformable objects. Particularly, in three-dimensional face recognition treating facial expressions as isometries of the facial surface allows to perform robust recognition insensitive to expressions.

Isometry-invariant representation of surfaces can be constructed by isometrically embedding them into some convenient space, and carrying out the comparison in that space. Presented here is a discussion on isometric embedding into $\mathbb{S}^3$, which appears to be superior over the previously used Euclidean space in sense of the representation accuracy.

1 Introduction

The problem of isometry-invariant representation of surfaces arises in numerous pattern recognition applications dealing with deformable objects. Particularly, in three-dimensional face recognition, it was shown that facial expressions can be modelled as isometric transformations of the facial surface [1, 2]. Under this assumption, the problem of expression invariant face recognition is reduced to finding similarity between isometric surfaces.

Figure 1 illustrates the problem of isometric surface matching. The first row (a)-(c) shows three isometric transformations of the same hand (assume that the fingers do not touch each other, such that the topology is preserved), which, with a bit of imagination, look like a grenade, a dog and a cobra (d)-(f). In other words, from the point of view of their *extrinsic* geometry[1], isometric surfaces can look completely different, while being just instances of the same surface.

---

[1] Formally, *intrinsic* geometry refers to all the properties of the manifold expressed in terms of the metric (first fundamental form), and *extrinsic* geometry refers to properties expressed in terms of the second fundamental form. We use these terms in a broader sense, by which extrinsic geometry defines the properties that describe the way the manifold is immersed in the ambient space.

R. Kimmel, N. Sochen, J. Weickert (Eds.): Scale-Space 2005, LNCS 3459, pp. 622–631, 2005.

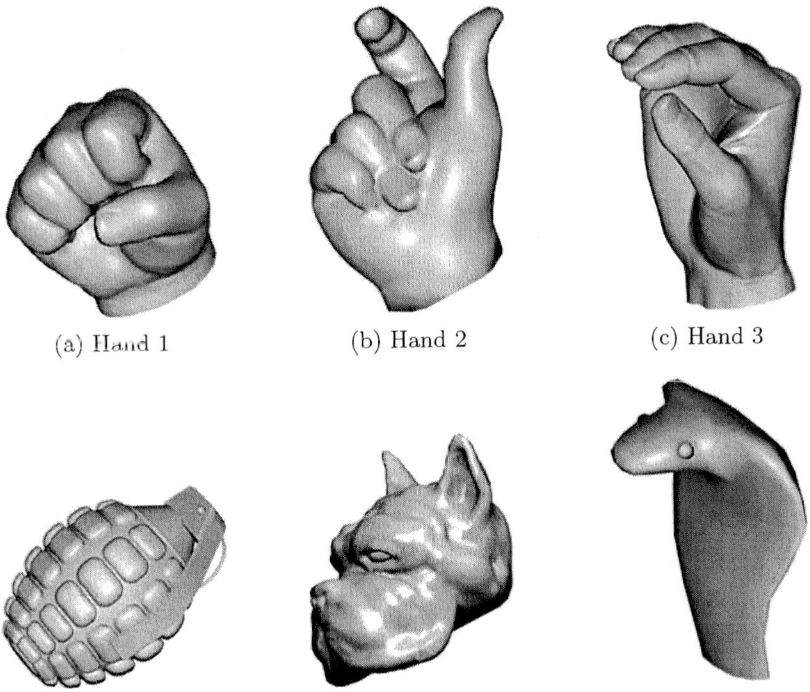

(a) Hand 1 (b) Hand 2 (c) Hand 3

(d) Grenade (e) Dog (f) Cobra

Fig. 1. Illustration of the deformable surface matching problem. (a) - (c): isometries of a hand. (d) - (f): different objects that resemble the hands from extrinsic geometry point of view

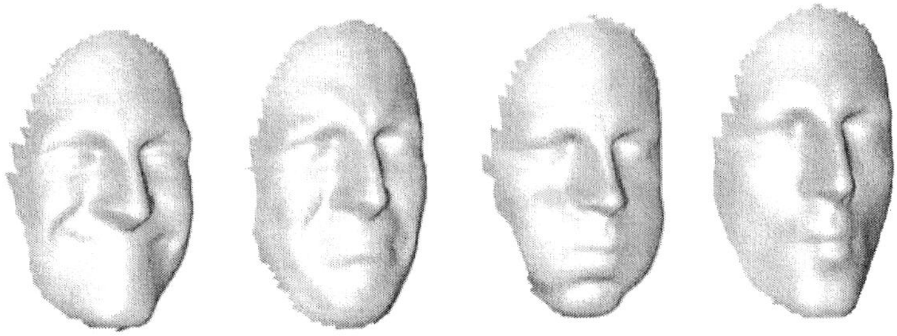

Fig. 2. Several facial expressions of the same person that can be modelled as isometries. Face data shown by courtesy of Eyal Gordon

Formally, given two complete compact smooth Riemannian manifolds $(\mathcal{S}, g)$ and $(\mathcal{Q}, h)$, the diffeomorphism $f : (\mathcal{S}, g) \rightarrow (\mathcal{Q}, h)$ is called an *isometry* if $f{*}h = g$, where $f{*}h$ denotes pullback of the metric. As the result, all the *intrinsic* geometric properties of the surface are preserved. Equivalently, an isometry can be defined as a diffeomorphism preserving the geodesic distances, that is

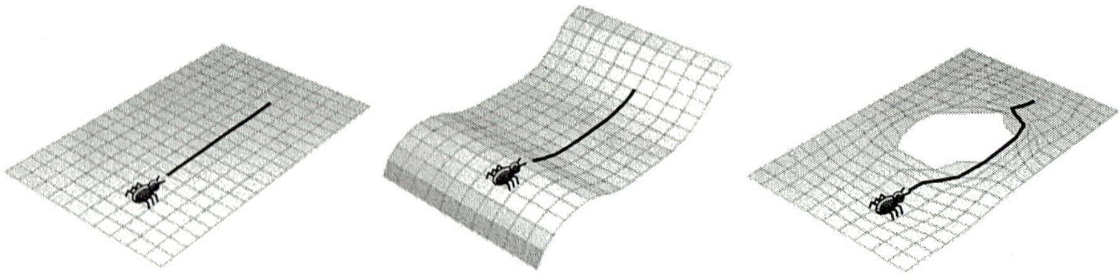

(a) Original surface

(b) Isometric transformation

(c) Non-isometric transformation

Fig. 3. Illustration of isometric (b) and non-isometric (c) transformations of a surface (a). Isometries do not change the intrinsic geometry of the surface, such that an imaginable creature living on the surface does not feel the transformation

$$d_{\mathcal{S}}(\xi_1, \xi_2) = d_{\mathcal{Q}}(\eta_1, \eta_2) \quad \forall \xi_1, \xi_2 \in \mathcal{S},\ \eta_1, \eta_2 \in \mathcal{Q}. \tag{1}$$

where $d_{\mathcal{S}}$ and $d_{\mathcal{Q}}$ denote the geodesic distances induced by g and h, respectively.

2 Bending-Invariant Canonical Forms

Let $(\mathcal{S}, h)$ and $(\mathcal{Q}, h)$ be two-dimensional Riemannian manifolds (surfaces) related by an isometry $f(\mathcal{S}) = \mathcal{Q}$. In the context of face recognition $\mathcal{S}$ and $\mathcal{Q}$ are different expressions of the same face. Since geodesic distances are preserved under an isometry, they are suitable candidates for an isometry-invariant representation of the surface.

However, we should remember that the surfaces $\mathcal{S}$ and $\mathcal{Q}$ are sampled and therefore in practice we have two *finite metric spaces* $(\{\xi_1, ..., \xi_{N_S}\}, \mathbf{D}_{\mathcal{S}})$ and $(\{\eta_1, ..., \eta_{N_Q}\}, \mathbf{D}_{\mathcal{Q}})$, respectively. The matrices $\mathbf{D}_{\mathcal{S}} = (d_{\mathcal{S}}(\xi_i, \xi_j))$ and $\mathbf{D}_{\mathcal{Q}} = (d_{\mathcal{Q}}(\eta_i, \eta_j))$ denote the mutual geodesic distances between the points in $\mathcal{S}$ and $\mathcal{Q}$. There is neither guarantee that $\mathcal{S}$ and $\mathcal{Q}$ are sampled at the same points, nor that the number of samples of the two surfaces is the same ($N_{\mathcal{S}} \neq N_{\mathcal{Q}}$). Moreover, even if the samples are the same, they can be ordered arbitrarily. This ambiguity makes impractical the use of $\mathbf{D}$ itself as an invariant representation.

2.1 Isometric Embedding

An alternative proposed in [3] is to avoid dealing explicitly with the matrix of geodesic distances and represent the Riemannian surface as a subset of some convenient m-dimensional space $\mathcal{S}'^m$, such that the original intrinsic geometry is preserved. We call such a procedure *isometric embedding*. This embedding allows to get rid of the extrinsic geometry, which no more exists in the new space. As a consequence, the resulting representation is identical for all isometric transformations of the surface. Another advantage is related to the fact that a general Riemannian metric is usually inconvenient to work with. The embedding space, on the other hand, can be chosen completely to our discretion. That is, the embedding replaces a complicate geometric structure by a convenient one.

Isometric embedding is a mapping between two finite metric spaces

$$\varphi : (\{\xi_1, ..., \xi_N\} \subset \mathcal{S}, \mathbf{D}) \rightarrow (\{\xi'_1, ..., \xi'_N\} \subset \mathcal{S}'^m, \mathbf{D}') \quad , \tag{2}$$

such that

$$d'_{ij} = d_{ij} \;\; \forall i, j = 1, ..., N. \tag{3}$$

The matrices $\mathbf{D} = (d_{ij}) = (d(\xi_i, \xi_j))$ and $\mathbf{D}' = (d'_{ij}) = (d'(\xi'_i, \xi'_j))$ denote the mutual geodesic distances between the points in the original and the embedding space, respectively. Following Elad and Kimmel, the image of $\{\xi_1, ..., \xi_N\}$ under φ is called the *canonical form* of $(\mathcal{S}, g)$ [3].

In general, such isometric embedding does not exist, and therefore one has to bear in mind that the canonical form is an *approximate* representation of the discrete surface. It is possible to find optimal canonical forms in sense of some metric distortion criterion. Also, the canonical form is not defined uniquely, but up to any transformation in the embedding space that does not alter the distances (e.g. in an Euclidean space, such transformations are translations, rotations and reflections). Yet this ambiguity is much easier to cope with compared to the vast degrees of freedom in the matrix $\mathbf{D}$.

2.2 The Choice of the Embedding Space

An important question is how to choose the embedding space. First, the geometry of the embedding space is important. Popular choices include spaces with flat [4, 3, 5, 6], spherical [7] or hyperbolic [8] geometry. This choice should be dictated by the convenience of a specific space and the resulting embedding error, which, in turn, depends on the embedding error definition.

Secondly, the dimension m of the embedding space must be chosen in such a way that the codimension of $\varphi(\mathcal{S})$ in $\mathcal{S}'^m$ is at least 1. The reason is made clear if we limit our manifolds to be graphs of functions (in our case - functions of two variables). Sampling of a graph $z(x, y)$ produces a set of points, which when embedded into $\mathbb{R}^2$ (or some other two-dimensional manifold), reflect the sampling pattern while the intrinsic geometry is captured mainly by the deformation of the boundaries of that function. Increasing the sampling rate will not enhance the intrinsic geometry captured by the embedded space. It would mainly indicate the x, y parametric shadow, or numerical support in $\mathbb{R}^2$ of the sampled function. On the other hand, when embedded into $\mathbb{R}^3$ (or other higher-dimensional manifolds), the sample points will lie along some two-dimensional submanifold of $\mathbb{R}^3$, and increasing the sampling rate would better capture the geometry of this submanifold.

Embedding with codimension zero (e.g. embedding of a surface in a plane) is useful when the manifold is endowed with some additional property, for example, texture. Such embedding can be thought of as an intrinsic parameterization of the manifold and has been explored in the context of medical visualization [4], texture mapping [6] and registration of facial images [9].

The main focus of present paper is embedding into a three-dimensional sphere $\mathbb{S}^3$. This space appears to be more suitable for embedding of facial surfaces than the Euclidean space used beforehand in [3, 1, 2]. Embedding into a two-dimensional sphere $\mathbb{S}^2$ was employed by Elad and Kimmel for visualization of convoluted brain cortical surface [7]. Here, we present embedding into $\mathbb{S}^3$ as a more accurate representation of facial surfaces.

3 Embedding into $\mathbb{S}^3$

A unit[2] m-dimensional sphere can be represented as the geometric locations of all unit vectors in $\mathbb{R}^{m+1}$

$$\mathbb{S}^m = \left\{ \mathbf{x} \in \mathbb{R}^{m+1} : \|\mathbf{x}\|_2 = 1 \right\}. \tag{4}$$

For every point on $\mathbb{S}^m$, there exists a correspondence between the parameterization coordinates $\xi^1, ..., \xi^m$ and the unit vector in $\mathbb{R}^{m+1}$. Given two points $\boldsymbol{\xi}_i, \boldsymbol{\xi}_j$ on the sphere (corresponding to unit vectors $\mathbf{x}_i, \mathbf{x}_j \in \mathbb{R}^{m+1}$), the geodesic distance between them is the great circle arc length, given by

$$d_{\mathbb{S}^m}(\boldsymbol{\xi}_i, \boldsymbol{\xi}_j) = \cos^{-1}(\langle \mathbf{x}_i, \mathbf{x}_j \rangle). \tag{5}$$

Specifically, $\mathbb{S}^3$ can be parameterized as

$$\begin{aligned} x^1(\boldsymbol{\xi}) &= \cos\xi^1 \cos\xi^2 \cos\xi^3, \\ x^2(\boldsymbol{\xi}) &= \cos\xi^1 \sin\xi^2 \cos\xi^3, \\ x^3(\boldsymbol{\xi}) &= \sin\xi^1 \cos\xi^3, \\ x^4(\boldsymbol{\xi}) &= \sin\xi^3. \end{aligned} \tag{6}$$

where $\boldsymbol{\xi} \in [0, \pi] \times [0, 2\pi] \times [0, \pi]$. The geodesic distance is explicitly expressed as

$$\begin{aligned} d_{\mathbb{S}^3}(\boldsymbol{\xi}_i, \boldsymbol{\xi}_j) = \cos^{-1}(& \cos\xi_i^1 \cos\xi_i^3 \cos\xi_j^1 \cos\xi_j^3 \cos(\xi_i^2 - \xi_j^2) + \\ & \cos\xi_i^3 \cos\xi_j^3 \sin\xi_i^1 \sin\xi_j^1 + \sin\xi_i^1 \sin\xi_j^3). \end{aligned} \tag{7}$$

We use the normalized *weighted stress* [10] as the embedding error criterion

$$\epsilon(\Xi'; \mathbf{D}, \mathbf{W}) = \frac{\sum_{i<j} w_{ij} (d'_{ij}(\Xi') - d_{ij})^2}{\sum_{i<j} d'^2_{ij}(\Xi')} = \frac{A}{B}, \tag{8}$$

where $\Xi' = (\xi'^i_j)$ is a $3 \times N$ matrix representing the parametric coordinates of the canonical form points, and $\mathbf{D} = (d_{ij})$ and $\mathbf{W} = (w_{ij})$ are $N \times N$ matrices of geodesic distances and weights, respectively. The weight are chosen as $w_{ij} = 1$; choosing $w_{ij} \propto d_{ij}^{-2}$ gives the *relative stress* [10].

[2] Without loss of generality, we discuss a unit sphere. The sphere radius is taken into account by scaling the distances.

The stress $\epsilon(\Xi';\mathbf{D},\mathbf{W})$ is minimized w.r.t. Ξ using the BFGS quasi-Newton algorithm (medium-scale optimization implemented in MATLAB function `fminunc`). The gradient of $\epsilon(\Xi';\mathbf{D},\mathbf{W})$ w.r.t. Ξ' is given by

$$\frac{\partial}{\partial \xi_k'^l}\epsilon(\Xi';\mathbf{D},\mathbf{W}) = B^{-2}\left(B\frac{\partial}{\partial \xi_k'^l}A - A\frac{\partial}{\partial \xi_k'^l}B\right), \tag{9}$$

where

$$\frac{\partial}{\partial \xi_k'^l}A = 2\sum_i w_{ik}(d'_{ij} - d_{ij})\frac{\partial}{\partial \xi_k'^l}d'_{ik}, \tag{10}$$
$$\frac{\partial}{\partial \xi_k'^l}B = 2\sum_i d'_{ik}\frac{\partial}{\partial \xi_k'^l}d'_{ik},$$

and

$$\frac{\partial}{\partial \xi_k'^l}d'_{ik} = (1 - C_{ik}^2)^{-1/2}\frac{\partial}{\partial \xi_k'^l}C_{ik},$$
$$C_{ik} = \cos\xi_i'^1\cos\xi_i'^3\cos\xi_k'^1\cos\xi_k'^3\cos(\xi_i'^2 - \xi_k'^2) + \cos\xi_i'^3\cos\xi_k'^3\sin\xi_i'^1\sin\xi_k'^1 + \sin\xi_i'^1\sin\xi_k'^1 \tag{11}$$

4 Numerical Results

The spherical embedding approach was tested on a set of human faces. Facial surfaces were acquired using a structured light 3D scanner [11] and underwent preprocessing by cropping, smoothing and subsampling (see details in [2]). The final surfaces contained about 750 points. An efficient modification of the parametric Fast Marching algorithm [12, 13, 14] was used to measure geodesic distances on the discrete surfaces.

Comparison of canonical form was performed in $\mathbb{R}^4$ using the moments signatures. A $pqrs$ moment of the canonical form is defined as

$$\mu_{pqrs}^{\mathbf{X}'} = \sum_{i=1}^{N}\left(x_i'^1\right)^p\left(x_i'^2\right)^q\left(x_i'^3\right)^r\left(x_i'^4\right)^s, \tag{12}$$

where $\mathbf{X}' = x_j'^i$ denotes the $4\times N$ matrix of $\mathbb{R}^4$ coordinates of the canonical form points, corresponding to the parametric coordinates ξ_j^i.

Canonical forms were first aligned using an Euclidean transformation, by eliminating the first-order moments $m_{1000}, m_{0100}, m_{0010}, m_{0001}$ and the mixed second-order moments $m_{1100}, m_{1010}, m_{1001}, m_{0110}, m_{0101}, m_{0011}$. The axes were reordered according to the second order moments, making the projection onto the first axis x^1 have the largest variance, and onto the fourth axis x^4 the smallest variance. This operation resolves the translation and rotation ambiguities for non-trivial objects. Next, reflections were applied to each axis x^k in order to set

$$\sum_{i=1}^{N}\text{sign}\left(x_i'^k\right) \geq 0. \tag{13}$$

This operation resolves the reflection ambiguity.

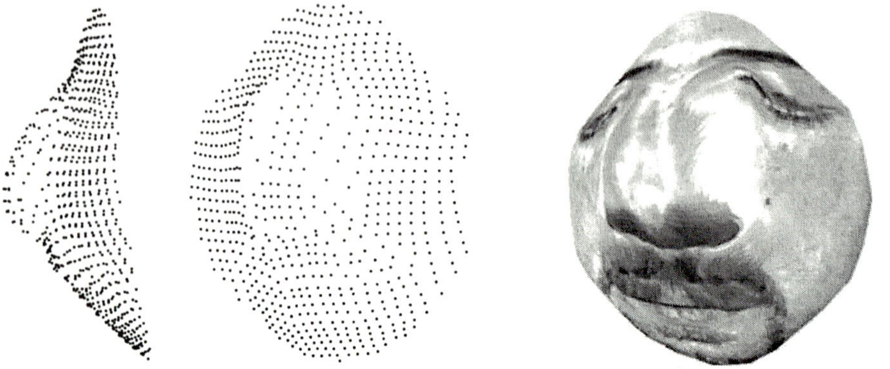

Fig. 4. Maximum-variance projection onto $\mathbb{R}^3$ of a facial surface embedded into $\mathbb{S}^3$ with radius $R = 10cm$

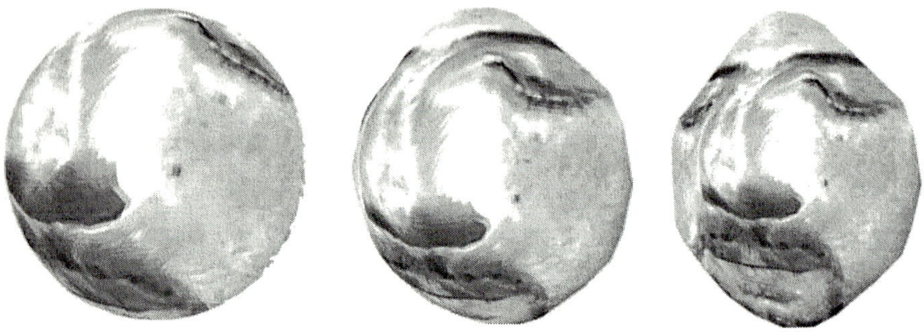

(a) $R = 5cm$ (b) $R = 7.5cm$ (c) $R = 15cm$

Fig. 5. Maximum-variance projections onto $\mathbb{R}^3$ of a facial surface embedded into $\mathbb{S}^3$ with different radii

Next, a signature of moments up to the fifth order was computed. The distance between two canonical forms $\mathbf{X}'$ and $\mathbf{Y}'$ was computed according to the standard Euclidean distance between their moments signatures:

$$d^2_{mom}(\mathbf{X}', \mathbf{Y}') = \sum_{p+q+r+s=2,..,5} \left(\mu^{\mathbf{X}'}_{pqrs} - \mu^{\mathbf{Y}'}_{pqrs} \right)^2 . \tag{14}$$

Figure 4 depicts the maximum-variance projection onto $\mathbb{R}^3$ of a canonical form obtained by embedding into a sphere of radius $R = 10cm$. Figure 5 depicts the embedding of the same face into spheres of different radii.

Figure 6 shows the embedding error ϵ as a function of the sphere radius. The minimum error is obtained around $R = 7.5cm$ and then increases asymptotically, as R grows to infinity. The asymptote corresponds to embedding into $\mathbb{R}^3$.

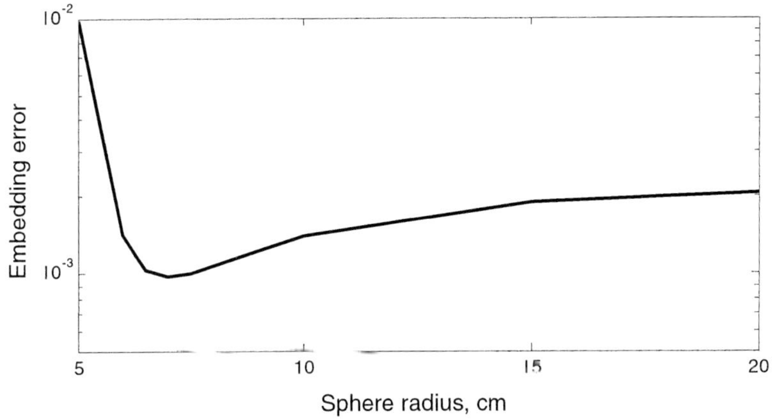

Fig. 6. Embedding error as the function of the sphere radius in cm. The asymptote $R \to \infty$ corresponds to embedding into $\mathbb{R}^3$

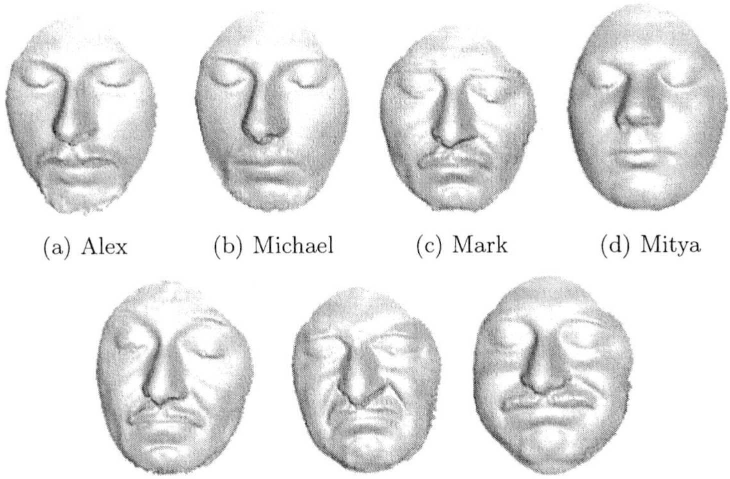

(a) Alex (b) Michael (c) Mark (d) Mitya

(e) Surprise (f) Disgust (g) Inflate

Fig. 7. (a) - (d) Faces used in the experiment. Alex and Michael are identical twins. (e) - (g) Representative facial expressions of subject Mark

Therefore, spherical embedding allows to obtain more than twice lower embedding error.

Finally, Figures 2–8 show a toy "face recognition" experiment that was performed on a set of 33 faces of four subjects in the presence of extreme facial expressions (Figure 2). Figure 8 depicts a two-dimensional visualization of the similarities (in sense of d_{mom}) between the faces.

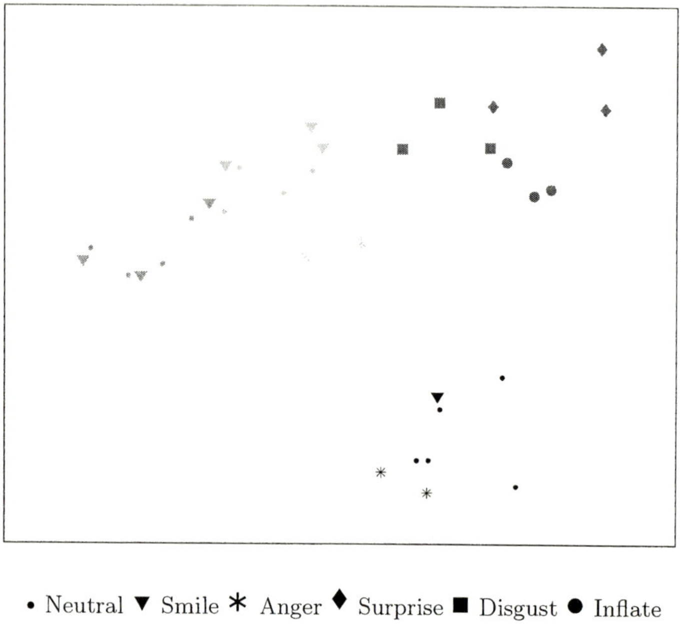

Fig. 8. Two-dimensional visualization of dissimilarities between faces. Colors represent different subjects. Symbols represent different facial expressions

5 Conclusions

We presented the three-dimensional sphere $\mathbb{S}^3$ as an alternative to the Euclidean space used beforehand in [3, 1, 2] to construct expression-invariant representation of human faces. Using $\mathbb{S}^3$ results in smaller embedding error, and leads to a more accurate representation.

References

1. A. Bronstein, M. Bronstein, and R. Kimmel. Expression-invariant 3D face recognition. In Proc. Audio and Video-based Biometric Person Authentication, pages 62–69, 2003.
2. A. Bronstein, M. Bronstein, and R. Kimmel. Three-dimensional face recognition. Technical Report CIS-2004-04, Dept. of Computer Science, Technion, Israel, 2004.
3. A. Elad and R. Kimmel. Bending invariant representations for surfaces. In Proc. CVPR, pages 168–174, 2001.
4. E. L. Schwartz, A. Shaw, and E. Wolfson. A numerical solution to the generalized mapmaker's problem: attening nonconvex polyhedral surfaces. IEEE Trans. PAMI, 11:1005–1008, 1989.
5. R. Grossman, N. Kiryati, and R. Kimmel. Computational surface attening: a voxel-based approach. IEEE Trans. PAMI, 24(4):433–441, 2002.

6. G. Zigelman, R. Kimmel, and N. Kiryati. Texture mapping using surface attening via multi-dimensional scaling. IEEE Trans. Visualization and computer graphics, 9(2):198–207, 2002.
7. A. Elad and R. Kimmel. Spherical attening of the cortex surface. In R. Malladi, editor, Geometric methods in bio-medical image processing, volume 2191, pages 77–89. Springer-Verlag, Berlin Heidelberg New York, 2002.
8. J.Walter and H. Ritter. On interactive visualization of high-dimensional data using the hyperbolic plane. In Proc. ACM SIGKDD Int. Conf. Knowledge Discovery and Data Mining, 2002.
9. A. Bronstein, M. Bronstein, E. Gordon, and R. Kimmel. Fusion of 3D and 2D information in face recognition. In Proc. ICIP, 2004. to appear.
10. I. Borg and P. Groenen. Modern multidimensional scaling - theory and applications. Springer-Verlag, Berlin Heidelberg New York, 1997.
11. A. Bronstein, M. Bronstein, E. Gordon, and R. Kimmel. High-resolution structured light range scanner with automatic calibration. Technical Report CIS-2003-06, Dept. of Computer Science, Technion, Israel, 2003.
12. J. A. Sethian. A review of the theory, algorithms, and applications of level set method for propagating surfaces. Acta numerica, pages 309–395, 1996.
13. R. Kimmel and J. A. Sethian. Computing geodesic on manifolds. In Proc. US National Academy of Science, volume 95, pages 8431–8435, 1998.
14. A. Spira and R. Kimmel. An efficient solution to the eikonal equation on parametric manifolds. Interfaces and Free Boundaries, 2004. To appear.

Author Index

Lecture Notes in Computer Science

For information about Vols. 1–3353

please contact your bookseller or Springer

Vol. 3459: R. Kimmel, N. Sochen, J. Weickert (Eds.), Scale Space and PDE Methods in Computer Vision. XI, 634 pages. 2005.

Vol. 3456: H. Rust, Operational Semantics for Timed Systems. XII, 223 pages. 2005.

Vol. 3455: H. Treharne, S. King, M. Henson, S. Schneider (Eds.), ZB 2005: Formal Specification and Development in Z and B. XV, 493 pages. 2005.

Vol. 3453: L. Zhou, B.C. Ooi, X. Meng (Eds.), Database Systems for Advanced Applications. XXVII, 929 pages. 2005.

Vol. 3452: F. Baader, A. Voronkov (Eds.), Logic for Programming, Artificial Intelligence, and Reasoning. XI, 562 pages. 2005. (Subseries LNAI).

Vol. 3450: D. Hutter, M. Ullmann (Eds.), Security in Pervasive Computing. XI, 239 pages. 2005.

Vol. 3449: F. Rothlauf, J. Branke, S. Cagnoni, D.W. Corne, R. Drechsler, Y. Jin, P. Machado, E. Marchiori, J. Romero, G.D. Smith, G. Squillero (Eds.), Applications on Evolutionary Computing. XX, 631 pages. 2005.

Vol. 3448: G.R. Raidl, J. Gottlieb (Eds.), Evolutionary Computation in Combinatorial Optimization. XI, 271 pages. 2005.

Vol. 3447: M. Keijzer, A. Tettamanzi, P. Collet, J.v. Hemert, M. Tomassini (Eds.), Genetic Programming. XIII, 382 pages. 2005.

Vol. 3444: M. Sagiv (Ed.), Programming Languages and Systems. XIII, 439 pages. 2005.

Vol. 3443: R. Bodik (Ed.), Compiler Construction. XI, 305 pages. 2005.

Vol. 3442: M. Cerioli (Ed.), Fundamental Approaches to Software Engineering. XIII, 373 pages. 2005.

Vol. 3441: V. Sassone (Ed.), Foundations of Software Science and Computational Structures. XVIII, 521 pages. 2005.

Vol. 3440: N. Halbwachs, L.D. Zuck (Eds.), Tools and Algorithms for the Construction and Analysis of Systems. XVII, 588 pages. 2005.

Vol. 3439: R.H. Deng, F. Bao, H. Pang, J. Zhou (Eds.), Information Security Practice and Experience. XII, 424 pages. 2005.

Vol. 3436: B. Bouyssounouse, J. Sifakis (Eds.), Embedded Systems Design. XV, 492 pages. 2005.

Vol. 3434: L. Brun, M. Vento (Eds.), Graph-Based Representations in Pattern Recognition. XII, 384 pages. 2005.

Vol. 3433: S. Bhalla (Ed.), Databases in Networked Information Systems. VII, 319 pages. 2005.

Vol. 3432: M. Beigl, P. Lukowicz (Eds.), Systems Aspects in Organic and Pervasive Computing - ARCS 2005. X, 265 pages. 2005.

Vol. 3431: C. Dovrolis (Ed.), Passive and Active Network Measurement. XII, 374 pages. 2005.

Vol. 3429: E. Andres, G. Damiand, P. Lienhardt (Eds.), Discrete Geometry for Computer Imagery. X, 428 pages. 2005.

Vol. 3427: G. Kotsis, O. Spaniol, Wireless Systems and Mobility in Next Generation Internet. VIII, 249 pages. 2005.

Vol. 3423: J.L. Fiadeiro, P.D. Mosses, F. Orejas (Eds.), Recent Trends in Algebraic Development Techniques. VIII, 271 pages. 2005.

Vol. 3422: R.T. Mittermeir (Ed.), From Computer Literacy to Informatics Fundamentals. X, 203 pages. 2005.

Vol. 3421: P. Lorenz, P. Dini (Eds.), Networking - ICN 2005, Part II. XXXV, 1153 pages. 2005.

Vol. 3420: P. Lorenz, P. Dini (Eds.), Networking - ICN 2005, Part I. XXXV, 933 pages. 2005.

Vol. 3419: B. Faltings, A. Petcu, F. Fages, F. Rossi (Eds.), Constraint Satisfaction and Constraint Logic Programming. X, 217 pages. 2005. (Subseries LNAI).

Vol. 3418: U. Brandes, T. Erlebach (Eds.), Network Analysis. XII, 471 pages. 2005.

Vol. 3416: M. Böhlen, J. Gamper, W. Polasek, M.A. Wimmer (Eds.), E-Government: Towards Electronic Democracy. XIII, 311 pages. 2005. (Subseries LNAI).

Vol. 3415: P. Davidsson, B. Logan, K. Takadama (Eds.), Multi-Agent and Multi-Agent-Based Simulation. X, 265 pages. 2005. (Subseries LNAI).

Vol. 3414: M. Morari, L. Thiele (Eds.), Hybrid Systems: Computation and Control. XII, 684 pages. 2005.

Vol. 3412: X. Franch, D. Port (Eds.), COTS-Based Software Systems. XVI, 312 pages. 2005.

Vol. 3411: S.H. Myaeng, M. Zhou, K.-F. Wong, H.-J. Zhang (Eds.), Information Retrieval Technology. XIII, 337 pages. 2005.

Vol. 3410: C.A. Coello Coello, A. Hernández Aguirre, E. Zitzler (Eds.), Evolutionary Multi-Criterion Optimization. XVI, 912 pages. 2005.

Vol. 3409: N. Guelfi, G. Reggio, A. Romanovsky (Eds.), Scientific Engineering of Distributed Java Applications. X, 127 pages. 2005.

Vol. 3408: D.E. Losada, J.M. Fernández-Luna (Eds.), Advances in Information Retrieval. XVII, 572 pages. 2005.

Vol. 3407: Z. Liu, K. Araki (Eds.), Theoretical Aspects of Computing - ICTAC 2004. XIV, 562 pages. 2005.

Vol. 3406: A. Gelbukh (Ed.), Computational Linguistics and Intelligent Text Processing. XVII, 829 pages. 2005.

Vol. 3404: V. Diekert, B. Durand (Eds.), STACS 2005. XVI, 706 pages. 2005.

Vol. 3403: B. Ganter, R. Godin (Eds.), Formal Concept Analysis. XI, 419 pages. 2005. (Subseries LNAI).

Vol. 3401: Z. Li, L.G. Vulkov, J. Waśniewski (Eds.), Numerical Analysis and Its Applications. XIII, 630 pages. 2005.

Vol. 3399: Y. Zhang, K. Tanaka, J.X. Yu, S. Wang, M. Li (Eds.), Web Technologies Research and Development - APWeb 2005. XXII, 1082 pages. 2005.

Vol. 3398: D.-K. Baik (Ed.), Systems Modeling and Simulation: Theory and Applications. XIV, 733 pages. 2005. (Subseries LNAI).

Vol. 3397: T.G. Kim (Ed.), Artificial Intelligence and Simulation. XV, 711 pages. 2005. (Subseries LNAI).

Vol. 3396: R.M. van Eijk, M.-P. Huget, F. Dignum (Eds.), Agent Communication. X, 261 pages. 2005. (Subseries LNAI).

Vol. 3395: J. Grabowski, B. Nielsen (Eds.), Formal Approaches to Software Testing. X, 225 pages. 2005.

Vol. 3394: D. Kudenko, D. Kazakov, E. Alonso (Eds.), Adaptive Agents and Multi-Agent Systems III. VIII, 313 pages. 2005. (Subseries LNAI).

Vol. 3393: H.-J. Kreowski, U. Montanari, F. Orejas, G. Rozenberg, G. Taentzer (Eds.), Formal Methods in Software and Systems Modeling. XXVII, 413 pages. 2005.

Vol. 3392: D. Seipel, M. Hanus, U. Geske, O. Bartenstein (Eds.), Applications of Declarative Programming and Knowledge Management. X, 309 pages. 2005. (Subseries LNAI).

Vol. 3391: C. Kim (Ed.), Information Networking. XVII, 936 pages. 2005.

Vol. 3390: R. Choren, A. Garcia, C. Lucena, A. Romanovsky (Eds.), Software Engineering for Multi-Agent Systems III. XII, 291 pages. 2005.

Vol. 3389: P. Van Roy (Ed.), Multiparadigm Programming in Mozart/OZ. XV, 329 pages. 2005.

Vol. 3388: J. Lagergren (Ed.), Comparative Genomics. VII, 133 pages. 2005. (Subseries LNBI).

Vol. 3387: J. Cardoso, A. Sheth (Eds.), Semantic Web Services and Web Process Composition. VIII, 147 pages. 2005.

Vol. 3386: S. Vaudenay (Ed.), Public Key Cryptography - PKC 2005. IX, 436 pages. 2005.

Vol. 3385: R. Cousot (Ed.), Verification, Model Checking, and Abstract Interpretation. XII, 483 pages. 2005.

Vol. 3383: J. Pach (Ed.), Graph Drawing. XII, 536 pages. 2005.

Vol. 3382: J. Odell, P. Giorgini, J.P. Müller (Eds.), Agent-Oriented Software Engineering V. X, 239 pages. 2005.

Vol. 3381: P. Vojtáš, M. Bieliková, B. Charron-Bost, O. Sýkora (Eds.), SOFSEM 2005: Theory and Practice of Computer Science. XV, 448 pages. 2005.

Vol. 3380: C. Priami, Transactions on Computational Systems Biology I. IX, 111 pages. 2005. (Subseries LNBI).

Vol. 3379: M. Hemmje, C. Niederee, T. Risse (Eds.), From Integrated Publication and Information Systems to Information and Knowledge Environments. XXIV, 321 pages. 2005.

Vol. 3378: J. Kilian (Ed.), Theory of Cryptography. XII, 621 pages. 2005.

Vol. 3377: B. Goethals, A. Siebes (Eds.), Knowledge Discovery in Inductive Databases. VII, 190 pages. 2005.

Vol. 3376: A. Menezes (Ed.), Topics in Cryptology – CT-RSA 2005. X, 385 pages. 2005.

Vol. 3375: M.A. Marsan, G. Bianchi, M. Listanti, M. Meo (Eds.), Quality of Service in Multiservice IP Networks. XIII, 656 pages. 2005.

Vol. 3374: D. Weyns, H.V.D. Parunak, F. Michel (Eds.), Environments for Multi-Agent Systems. X, 279 pages. 2005. (Subseries LNAI).

Vol. 3372: C. Bussler, V. Tannen, I. Fundulaki (Eds.), Semantic Web and Databases. X, 227 pages. 2005.

Vol. 3371: M.W. Barley, N. Kasabov (Eds.), Intelligent Agents and Multi-Agent Systems. X, 329 pages. 2005. (Subseries LNAI).

Vol. 3370: A. Konagaya, K. Satou (Eds.), Grid Computing in Life Science. X, 188 pages. 2005. (Subseries LNBI).

Vol. 3369: V.R. Benjamins, P. Casanovas, J. Breuker, A. Gangemi (Eds.), Law and the Semantic Web. XII, 249 pages. 2005. (Subseries LNAI).

Vol. 3368: L. Paletta, J.K. Tsotsos, E. Rome, G.W. Humphreys (Eds.), Attention and Performance in Computational Vision. VIII, 231 pages. 2005.

Vol. 3367: W.S. Ng, B.C. Ooi, A. Ouksel, C. Sartori (Eds.), Databases, Information Systems, and Peer-to-Peer Computing. X, 231 pages. 2005.

Vol. 3366: I. Rahwan, P. Moraitis, C. Reed (Eds.), Argumentation in Multi-Agent Systems. XII, 263 pages. 2005. (Subseries LNAI).

Vol. 3365: G. Mauri, G. Păun, M.J. Pérez-Jiménez, G. Rozenberg, A. Salomaa (Eds.), Membrane Computing. IX, 415 pages. 2005.

Vol. 3363: T. Eiter, L. Libkin (Eds.), Database Theory - ICDT 2005. XI, 413 pages. 2004.

Vol. 3362: G. Barthe, L. Burdy, M. Huisman, J.-L. Lanet, T. Muntean (Eds.), Construction and Analysis of Safe, Secure, and Interoperable Smart Devices. IX, 257 pages. 2005.

Vol. 3361: S. Bengio, H. Bourlard (Eds.), Machine Learning for Multimodal Interaction. XII, 362 pages. 2005.

Vol. 3360: S. Spaccapietra, E. Bertino, S. Jajodia, R. King, D. McLeod, M.E. Orlowska, L. Strous (Eds.), Journal on Data Semantics II. XI, 223 pages. 2005.

Vol. 3359: G. Grieser, Y. Tanaka (Eds.), Intuitive Human Interfaces for Organizing and Accessing Intellectual Assets. XIV, 257 pages. 2005. (Subseries LNAI).

Vol. 3358: J. Cao, L.T. Yang, M. Guo, F. Lau (Eds.), Parallel and Distributed Processing and Applications. XXIV, 1058 pages. 2004.

Vol. 3357: H. Handschuh, M.A. Hasan (Eds.), Selected Areas in Cryptography. XI, 354 pages. 2004.

Vol. 3356: G. Das, V.P. Gulati (Eds.), Intelligent Information Technology. XII, 428 pages. 2004.

Vol. 3355: R. Murray-Smith, R. Shorten (Eds.), Switching and Learning in Feedback Systems. X, 343 pages. 2005.

Vol. 3354: M. Margenstern (Ed.), Machines, Computations, and Universality. VIII, 329 pages. 2005.